Microhydrodynamics

Principles and Selected Applications

Microhydrodynamics

Principles and Selected Applications

Sangtae Kim
Donald W. Feddersen Distinguished Professor
of Mechanical Engineering and of Chemical Engineering
Purdue University

Seppo J. Karrila
Lilly Systems Biology - Singapore

Dover Publications, Inc.
Mineola, New York

Bibliographical Note

This Dover edition, first published in 2005, is an unabridged republication of the work originally published by Butterworth-Heinemann, Boston and London, in 1991 as part of the Butterworth-Heinemann Series in Chemical Engineering. The authors have written a new Preface specially for the Dover edition.

Library of Congress Cataloging-in-Publication Data

Microhydrodynamics : principles and selected applications / Sangtae Kim, Seppo J. Karrila.

p. cm.

Originally published: Boston : Butterworth-Heinemann, c1991.

Includes index.

ISBN 0-486-44219-5 (pbk.)

1. Hydrodynamics. I. Kim, Sangtae. II. Karrila, Seppo J.

QA911.K49 2005

532'.5—dc22

2005041157

Manufactured in the United States of America

Dover Publications, Inc., 31 East 2nd Street, Mineola, N.Y. 11501

To Julie and Moa;
without their patience and understanding,
this book would not have been possible.

Preface to the Dover Edition

The authors have had the gratifying opportunity to interact with a worldwide community of microhydrodynamicists in the decade and a half that have passed since the appearance of *Microhydrodynamics*. Here is a book that went out of print almost a decade ago and we continue to be surprised by the intensity of interest in our framework for the analysis of the motion of micron-scale particles (of arbitrary shape) in a viscous fluid. Indeed, there are sad anecdotes of missing copies in libraries around the world. So we share with the next generation of students the sense of excitement regarding this wonderful opportunity to release a reprint edition under the auspices of Dover Publications.

In light of the many other constraints on our time (and the authors having moved on to new challenges beyond microhydrodynamics) we have, with some reluctance, passed on the opportunity to create a second edition. Several important developments that would have gone in to a new edition merit mention here: discretization of the integral equation to pick out traction singularities at sharp corners and edges and validation of same by asymptotic analysis of the adjoint equation (the *Riesz-Rennerb* equation); elucidation of a "hydrodynamic steering" mechanism in self-assembly of micro objects (including biological macromolecules) in viscous fluids. During the past decade, the latter played an inspirational role in the development of the fluidic self assembly (FSA) process for the manufacture of ultra low-cost radio frequency identification (RFID) tags.

In the original edition, we mentioned (a now defunct) ftp server, flossie.che.wisc.edu, for updates to the list of *errata* and illustrative computer programs. Countless emails later, we take this opportunity to apologize for our lack of foresight. After some reflection, we should have anticipated this problem. Today, anonymous ftp and related cybertools have been supplanted by more powerful web-enabled distribution mechanisms. Ever the quick studies, this time we provide instructions that, hopefully, will stand the passage of time. Discounting the possibility of dramatic and disruptive technologies, we direct readers to search engines and the placement of key words "microhydrodynamics," author names, and current calendar year to retrieve news of updates to *Microhydrodynamics*.

The authors would like to thank John Grafton and his team at Dover Publications for considering the feasibility of a reprint edition. Our thanks also go to Prof. Tom Hughes (UT-Austin), for recommending this path. Beginning with a casual conversation with him at the USNC/TAM meeting at the National Academies in April, 2004, the "go" decision was received from Dover within three months! Sincere thanks also go to the National Science Foundation – two decades have passed since the first research awards in support of microhydrodynamics, and we are now just beginning to see the fruits of these efforts.

Lastly, we cannot emphasize enough our thanks and appreciation to the many builders of *shared cyberinfrastructure* – for enabling a small but dedicated community of microhydrodynamicists scattered around the globe to stay connected across space and time to resurrect what otherwise would have become an obscure, out-of-print monograph.

SANGTAE KIM
West Lafayette, Indiana
SEPPO J. KARRILA
Singapore

Contents

II Dynamics of a Single Particle

III Hydrodynamic Interactions

Preface

This four-part "methods-oriented" book covers analytical and numerical methods for describing the motion of small particles suspended in viscous fluids, at a level suitable for graduate students in engineering and applied mathematics. The first three parts cover the fundamental principles of low-Reynolds-number flow: the mathematical foundations, the dynamics of a single particle in a flow field, and hydrodynamic interactions between suspended particles. The fourth part of the book covers some recent advances in the mathematical and computational aspects of viscous particulate flows that point to new directions for large-scale simulations on parallel computers.

In the past 20 years, the literature on this subject (low Reynolds number hydrodynamics) and its applications have grown enormously. Instead of reviewing these advances in a complete and exhaustive fashion, our aim is to build a coherent framework that unifies the subject, thereby making the literature more accessible, especially for those wishing to build mathematical models of particulate systems.

Microhydrodynamics can be a very mathematical subject, for the fundamental equations are well established and have been studied in great depth. A student familiar with mathematical analysis will appreciate and understand the steps needed in establishing that a model is well posed and has a unique solution before attempting to find the solution, or that an iterative scheme like the method of reflections converges with respect to a norm in a suitably defined Hilbert space. On the other hand, the purely mathematical viewpoint of establishing existence, uniqueness, or convergence of solutions does not go far enough. For scientists and engineers the models provide the basis for quantitatively understanding the particulate phenomena — the numbers from microhydrodynamics are, in a sense, intermediate results. In this book, we attempt to blend these ideas. We hope that the more mathematical sections of this book will encourage and motivate students to learn the mathematical foundations from course material elsewhere, since even an introductory treatment of that material is beyond the scope of this book.

We thank the many people whose comments about earlier versions of the book have been so helpful, especially Osman Basaran (Oak Ridge) and Chris Lawrence (Illinois). Our student colleagues in the suspensions group at Wisconsin during the period 1983 – 1990 have contributed in many ways to this book: Doug Brune, Yuris Fuentes, John Geisz, Achim Gerstlauer, Gary Huber, Shih-Yuan Lu, Peyman Pakdel, Steve Strand, Zhengfang Xu, and B.J. Yoon. Special thanks go to Yuris Fuentes for helping us to incorporate his results on parallel computing, as well as lending his artistic talent with the illustrations. We also acknowledge the help received from Professors Mary Vernon and Mark Hill of UW-Madison's Computer Sciences Department for the work on the Sequent Symmetry.

The University of Illinois (Department of Chemical Engineering, Department of Theoretical and Applied Mechanics) and the University of Massachusetts (Department of Chemical Engineering) provided both financial support as well as a hospitable environment for book writing. We are especially grateful to the latter institution for giving one of us (SK) the opportunity to try the material in Part 4 in the classroom in the form of a special lecture series. Our Wisconsin colleagues have provided a fertile and constructive atmosphere, which is so necessary for bookwriting. We have especially benefitted in this respect from the traditions perpetuated by Bob Bird, Warren Stewart, Ed Lightfoot, and Arthur Lodge.

The term *microhydrodynamics* was suggested in 1974 by G.K. Batchelor as a way of defining a subject that retained its focus on fluid mechanics while keeping its spread into other fields within manageable proportions. Our perspective in this field has been greatly influenced, in chronological order, by Gary Leal (U.C. Santa Barbara), Howard Brenner (MIT), Bill Russel (Princeton), John Hinch (Cambridge), and David Jeffrey (Western Ontario). This book has greatly benefitted from our professional association with them.

We thank Howard Brenner, again, for encouragement and advice in his role as series editor. Also, it has been our pleasure to work with a very understanding staff at Butterworth–Heinemann Publishers. Finally, in this age of high-speed electronic communications, we feel that the communication between authors and readers does not stop at the printing press. We encourage readers to send (by e-mail, of course) corrections and suggestions for improvement. Readers may also log on to Flossie (as described on the following pages) for news on the latest corrections and updates to the program listings. With active participation from the readers, we hope to initiate a "microhydrodynamics program exchange."

Sangtae Kim
University of Wisconsin
Madison, Wisconsin
sang@flossie.che.wisc.edu

Seppo J. Karrila
Finnish Pulp and Paper Research Institute
Espoo, Finland
skarrila@aboy2.abo.fi

Organization Scheme

The book consists of 19 chapters, and all except the first are followed by exercises. These serve both to amplify the material already covered in each chapter and sometimes to introduce concepts not discussed elsewhere in the book. Some exercises are intended to help the student independently discover results that are otherwise well known in the literature. References cited in each chapter are collated (in alphabetical order by name of first author) at the end of each Part, instead of after each chapter. A number of illustrative (Fortran) programs are also available in connection with problems that require computational methods. In this electronic age, we have chosen to distribute this material via the national network; readers may log on to the *anonymous* user account on *flossie.che.wisc.edu* (IP number 128.104.170.10) and transfer source codes to their own machines. At the password prompt it is suggested that readers use their names (a password is not required) so that we may create a log record of the user community. The contents of the *anonymous* account have been subdivided along chapter lines. For example, the reader "John Doe" may access the boundary collocation program `mandr.for` of Chapter 13 by the following procedure:

```
% ftp flossie.che.wisc.edu
Name:  anonymous
Anonymous user OK. Enter real ident...  John Doe
*  cd chapter13
*  get mandr.for
*  bye
```

In Chapter 1, we introduce characteristic scales for a range of problems in microhydrodynamics and the dimensional analysis leading to the governing equations for Stokes flow. In Chapter 2, we discuss a number of general properties of the Stokes equations, including fundamental theorems dealing with uniqueness and minimum energy dissipation. We also discuss the Lorentz reciprocal theorem, which is perhaps the theorem most often used in our book. The fundamental solution of the Stokes equations (the Green's dyadic for the Stokes equation is called the *Oseen tensor*) is derived along with the integral representation for the velocity and pressure fields.

The rest of the book is divided into three parts: Part II on the dynamics of a single particle, Part III on hydrodynamic interactions, and Part IV on the computation of flow in complex geometries. Rather than attempt a comprehensive treatment of all aspects of viscous laminar flows, we shall focus our attention on a core set of themes for flow past a single particle that are readily generalized to encompass interactions between particles. We then show the connections between these analytical techniques and numerical techniques of Parts III and IV.

An understanding of the dynamics of a single particle is a prerequisite to everything else in the book, and many of the developments for multiparticle interactions are generalizations of results developed for a single particle. In Chapter 3, we derive the multipole expansion for a particle, starting from the integral representation for the velocity. For simple shapes, such as the sphere and ellipsoid, we show how the expansion can be truncated to obtain singularity solutions. The moments of the multipole expansion must be related to the given boundary conditions, and we show how this can be achieved by the use of the Faxén laws. We make a special effort to show the link between the singularity solutions and Faxén laws. The original derivation, by Faxén, required special properties of the sphere, thus obscuring the generality of this approach to particles of arbitrary shape.

Chapter 4 discusses classical solution techniques for flow past a single sphere, including Lamb's general solution and the Stokes streamfunction. The connection with the more general multipole theory for a particle of arbitrary shape is brought out, thereby reinforcing the message that there is a unified structure for particulate Stokes flow.

The resistance and mobility relations for a single particle in unbounded flow are introduced in Chapter 5, along with inversion formulae for obtaining one set from another. Their use is illustrated with examples from the suspensions literature, including the rheology of a dilute suspension of spheroids and electrophoresis of a charged particle in an electric field. The examples illustrate how the resistance and mobility functions appear beyond the obvious context of trajectory calculations.

Chapter 6 presents an introduction to time-dependent Stokes flows, including the concepts of added mass and Basset forces. It is shown that in the limit of slow temporal variations, time-dependent effects appear at $O(\mathrm{Re}^{1/2})$, and a general method of obtaining the $O(\mathrm{Re}^{1/2})$ coefficient for particles of arbitrary shape is derived.

Part III, on hydrodynamic interactions, starts in Chapter 7 with general concepts of resistance and mobility relations between interacting particles of arbitrary shape. The rest of Part III is divided according to the separation between the particles.

For particles that are widely separated (Chapter 8), we use the method of reflections. The disturbance fields for each particle are written as multipole expansions, with the moment coefficients determined by the Faxén laws of Chapter 3. The iterative nature of the method is brought out with detailed examples on the interaction between spheres. For many-body problems the inversion from

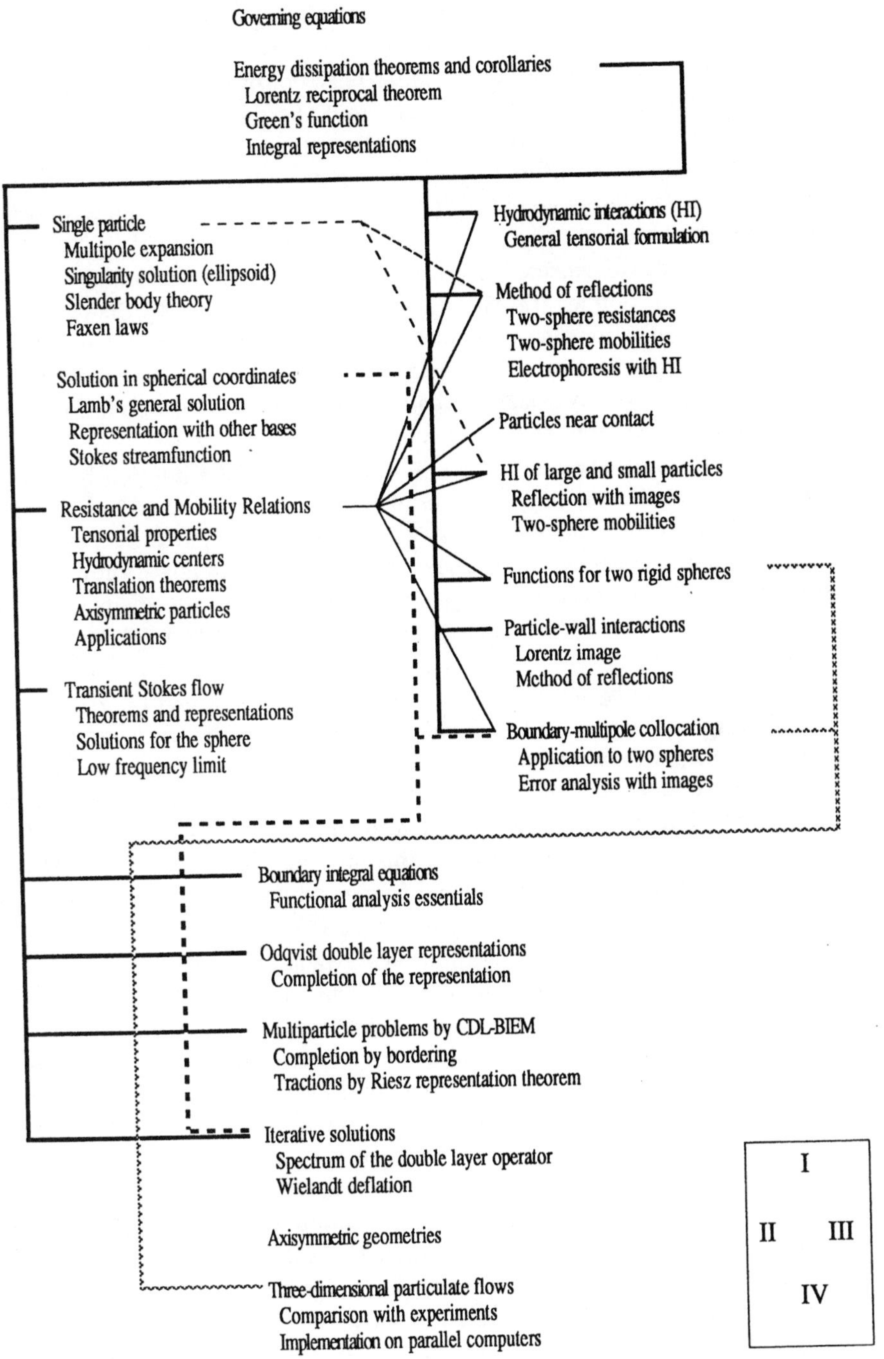
Governing equations
Energy dissipation theorems and corollaries
Lorentz reciprocal theorem
Green's function
Integral representations
Single particle
Multipole expansion
Singularity solution (ellipsoid)
Slender body theory
Faxen laws
Solution in spherical coordinates
Lamb's general solution
Representation with other bases
Stokes streamfunction
Resistance and Mobility Relations
Tensorial properties
Hydrodynamic centers
Translation theorems
Axisymmetric particles
Applications
Transient Stokes flow
Theorems and representations
Solutions for the sphere
Low frequency limit
Hydrodynamic interactions (HI)
General tensorial formulation
Method of reflections
Two-sphere resistances
Two-sphere mobilities
Electrophoresis with HI
Particles near contact
HI of large and small particles
Reflection with images
Two-sphere mobilities
Functions for two rigid spheres
Particle-wall interactions
Lorentz image
Method of reflections
Boundary-multipole collocation
Application to two spheres
Error analysis with images
Boundary integral equations
Functional analysis essentials
Odqvist double layer representations
Completion of the representation
Multiparticle problems by CDL-BIEM
Completion by bordering
Tractions by Riesz representation theorem
Iterative solutions
Spectrum of the double layer operator
Wielandt deflation
Axisymmetric geometries
Three-dimensional particulate flows
Comparison with experiments
Implementation on parallel computers
I
II
III
IV

resistance to mobility relations becomes cumbersome, so a direct computational procedure for both resistance and mobility problems is derived.

In Chapter 9, we review methods for calculating the interaction between two nearly touching particles. For surfaces in relative motion (squeezing flows and shearing flows) we discuss the singular behavior of the resistance functions and show that these leading terms come from the dominant contributions in the gap region. The higher order nonsingular terms in these problems, as well as the resistance functions for surfaces moving in tandem as a rigid body, have contributions from all regions of the flow; in general, these terms can be obtained only by numerical techniques.

In Chapter 10, we consider interactions between large and small particles, especially at separation lengths that fall between the two grossly different sizes. The method of reflections is still applicable, but we can no longer truncate the multipole expansion for the reflections at the larger particle. We show how the reflections can be done by image methods, and a number of examples involving interactions between large and small spheres are given.

The interactions between two spheres deserve special attention because of the great interest and the large number of applications in suspensions research. In Chapter 11 we present a summary of all known analytical results for the entire collection of resistance and mobility functions that relate the force and hydrodynamic dipole moment (torque and stresslet) for two unequal spheres in a linear flow field. Several examples are given to illustrate the use of the formulae and tables.

The treatment of hydrodynamic interactions between a small particle and a wall is derived in Chapter 12 by a limiting process of the interaction between a small and large particle. By this manner, we derive the image of a point force (Stokeslet) near a plane wall and obtain the *Lorentz reflection principle* for the plane wall in a natural way. The extension to a planar fluid-fluid interface, *Lee's reflection lemma*, is also synthesized in this fashion. Here again, our goal is to use methods that emphasize the unified framework for Stokes flow.

For interactions between particles of simple shape, we show that boundary collocation applied to a truncated multipole expansion for each particle yields an effective numerical method (Chapter 13). The basis must be chosen so that contributions from the higher basis elements decay rapidly, and this point is illustrated with examples for spheroids of extreme aspect ratios. The convergence of the method is illustrated by comparison with the exact solution for the point force near a sphere. This problem serves as an excellent model for particle-particle interactions because moments of all orders are excited in the sphere.

We call Part IV "Foundations of Parallel Computational Microhydrodynamics." For complex geometries, the only way to capture the details of the particle or container boundary shape is to bring the Stokes singularities to the particle surface. The resulting class of numerical techniques is known as the *boundary integral (equation) methods.* Their distinct advantage over spatial methods, such as finite elements or finite differences, is *the reduction of dimensionality.* That is, instead of a three-dimensional PDE, we solve a two-dimensional (boundary) integral equation, in which the unknowns are densities of the Stokes singularities distributed over the boundary of the fluid domain, and even infinite fluid domains without a container boundary cause no problems. Furthermore, although the mathematical theory places a smoothness requirement on the surface, this restriction can be loosened in the actual numerical procedure.

By going to a two-dimensional formulation, we reduce both the amount of work for discretization and the number of variables needed for getting some required numerical resolution. Yet at first glance it may appear that our emphasis on boundary methods goes against current trends in computational technology. For example, the emergence of fast and inexpensive processors suggests that the most interesting innovations will occur in parallel computational architectures. Indeed, fundamental physical barriers, such as the speed of light and the size of an atom, place a limitation on the speed and integration of a single processor. It is unlikely that in the future a single superprocessor would be orders of magnitude faster than those available today. Therefore the obvious trend to follow is parallel processing, which presents certain requirements on the algorithms to be implemented. It is easy to visualize how the equations from the finite element method or the finite difference method, which are just relations connecting only near neighbors, are implemented on parallel computers. Clearly these methods inherently possess a certain "graininess" that corresponds to a multiprocessor architecture. Boundary methods, on the other hand, lead to dense systems for the following reason. The fields emanating from the Stokes singularities on one boundary element propagate in all directions with a fairly persistent algebraic decay, so that each element interacts with all other elements to some extent. The mapping of such dense systems to parallel architectures is a nontrivial problem. Moreover, for the more popular boundary integral formulation for Stokes flow based on the single layer potential, we obtain the ill-posed Fredholm integral equation of the first kind and the associated ill-conditioning in the discretized equations. The onset of this ill-conditioning can be predicted using the techniques developed in Chapter 17 and represents a significant barrier to the application of this approach to complex geometries.

With this environment in mind, the emphasis of Part IV is on the construction of a new boundary integral method, which we call the Completed Double Layer Boundary Integral Equation Method (CDL-BIEM). The name is derived from the fact that Stokes double layer densities are employed, but since the double layers alone do not form a complete basis, a completion procedure is necessary. Using techniques from linear operator theory, we show that the solution of the general mobility problem of N particles in either a bounded or unbounded domain can be cast in terms of a fixed point problem, in which some finite power of the linear operator is a contraction mapping (the spectral radius of the operator is less than one). Furthermore, the three steps in the computational procedure (pre-processing or creation of the fixed point problem, iterative solution of the discretized problem, and post-processing to extract the physical variables of interest) can be formulated as parallel algorithms.

We will draw upon the developments of Parts II and III to show that in many-body suspension problems the interactions between elements on different particles are very small, in the sense that the eigenvalues determining the spectral radius are perturbed only slightly by hydrodynamic interactions between the particles. Thus we raise the proposition of attacking large-scale suspension simulations by a network of parallel processors, with each processor powerful enough to handle all boundary elements on a given particle. *In essence, we recover the nearest neighbor property of the spatial methods, only now the fundamental entity is not an element or a node, but the collection of elements on a particle.* The resulting iterative strategy will involve asynchronous iterations, accelerating the convergence relative to synchronous iterations similarly as the Gauss–Seidel iterations do in comparison with Gauss–Jacobi iterations. The many successes obtained to date with this new approach suggest that CDL-BIEM will be an important tool in our efforts to describe the motion of complex microstructures through a viscous fluid, although the full impact of the method may not be realized until it becomes popular with the research community.

Chapter 14 lists the mathematical background for comprehension of the material covered, to the extent that the reader should be able to apply the theory and algorithms presented for obtaining numerical results. We go over Fredholm integral equations of the first and the second kind, and some smoothness requirements for the kernels, which, for our application, imply smoothness requirements for the surface. The kernels encountered are *weakly singular*, so that an extension of the Fredholm theory, namely, the *Fredholm–Riesz–Schauder* theory for compact operators, is necessary. The fact that the integral operators are compact will be exploited in two ways. First, it implies that CDL-BIEM is a well-posed problem. Second, the spectrum of a compact operator is discrete, which in our case will be used to show that, even before discretization, the integral equations of CDL-BIEM can be solved by Neumann series expansion, in practice replaced by direct (Picard) iterations (also called *successive substitution*) corresponding to truncating the Neumann series.

The next two chapters put the integral representations for Stokes flow in the framework of linear operator theory. Chapter 15 presents the theory of Odqvist for a single particle, while Chapter 16 extends the theory to multiparticle problems in both bounded and unbounded domains. The range and the null space of the double layer integral operator are explored, and we show that in the multiparticle setting, no new null solutions are created by interactions between surfaces. The setting for multiparticle problems brings out another nontrivial result as well. In the analysis of the stress field generated by a double layer distribution on a Lyapunov-smooth container surface, we encounter improper integrals. Using just the techniques of classical functional analysis, the obvious way out is to increase the smoothness condition on the parametric representation of the surface, from Lyapunov-smooth to the class Bh (Hölder-continuous second derivatives), and this is the path followed by Odqvist. Thus, contrary to our physical intuition, Odqvist's theory requires a Bh-smooth container surface but only a Lyapunov-smooth particle surface. Our analysis in Section 16.2 of the more general multiparticle setting offers a new line of proof, which shows that the theory applies to Lyapunov-smooth containers and that the earlier asymmetric result is simply an artifact of the mathematical technique. The framework of linear operator theory is fully exploited in the construction of a direct procedure for calculating surface tractions for particles undergoing rigid-body motions. The proof requires the combined use of the Riesz representation theorem for linear functionals and the Lorentz reciprocal theorem.

Chapter 17 is perhaps the most important one in Part IV, from a conceptual point of view. The spectrum of the integral operator is analyzed and the procedure for shifting the outermost eigenvalues ($7N$ of them for an N-particle system plus seven more if the domain is bounded) inward, by the procedure known as *Wielandt deflation*, is explained. The final result is an iterative scheme, based on the fact that the Neumann series will converge and Picard iterations therefore are feasible. Physical results, such as the translational and rotational velocities of the particles, turn out to be exactly the projection of the solution onto the eigenspace of the deflated eigenvalues. In addition, the surface tractions for each particle are also obtained as an iterative solution of the adjoint problem. We show that multiple Wielandt deflations can be implemented as a parallel algorithm, provided that the basis for the eigenspace is orthonormal. Since this is not true in general, we show how the Gram matrix may be inverted analytically by projection methods, so that the Gram–Schmidt orthogonalization process is also readily implemented as a parallel operation.

In Chapter 18 (Fourier analysis for axisymmetric geometries) and Chapter 19 (three-dimensional geometries), we illustrate the method with numerical examples. Only relatively simple discretization schemes are considered. In part this is because of our belief that true optimization of codes for parallel processing must reflect the coupling between architecture and algorithm. Thus we focus our efforts on results that we feel will apply to a broad class of parallel computers. These examples include comparisons with other numerical methods and experimental data. An illustrative range of particle geometries, such as sharp corners and grooves is examined, and dense linear systems up to 10,000 $\times$ 10,000 in size are explored without difficulty. We stop at the frontiers of computational microhydrodyanmics research. For very large systems of equations, communication becomes a bottleneck, even for Jacobi iterations. A promising strategy for asynchronous iterations that minimizes communication bottlenecks is outlined.

Microhydrodynamics

Principles and Selected Applications

Part I

Governing Equations and Fundamental Theorems

Chapter 1

Microhydrodynamic Phenomena

1.1 Objective and Scope

Every student of science and engineering knows that a particle in a vacuum subject to an external force $\boldsymbol{F}^e$ undergoes an acceleration $\boldsymbol{a}$ given by Newton's famous law, $\boldsymbol{a} = \boldsymbol{F}^e/m$, where m is the mass of the particle. The particle trajectory is obtained immediately upon two integrations, and thus the fate of the particle can be determined. Every student of fluid dynamics learns that in a fluid medium, viscous resistance of the fluid decreases this acceleration so that the particle soon attains a terminal velocity. This resistive force, usually referred to as the hydrodynamic drag, $\boldsymbol{F}^H$ (*hydrodynamic* because water is the most prevalent example), is linear with respect to the instantaneous particle velocity relative to the fluid, so long as the particle is small, the fluid is sufficiently viscous, the motion is slow, or a combination of the above. Unfortunately, the hydrodynamic drag is shape-dependent, so the particle trajectories are no longer described by a lumped parameter like the mass. Instead the drag calculation involves the solution of a boundary value problem for the velocity, with the shape of the particle entering into the solution through the boundary conditions on the particle surface. For a small sphere of radius a moving with a relative velocity $\boldsymbol{U}$ through a fluid of viscosity μ, this procedure yields the Stokes law, $\boldsymbol{F}^H = -6\pi\mu a\boldsymbol{U}$.

Determining the motion of a particle or particles in bounded and unbounded flow is a central problem in the world of microhydrodynamics and the main objective of this book. The motion of the particles dictates the evolution of the suspension microstructure. The microstructure or multiparticle configuration in turn shapes the forces acting on the particles (which induce further motion). Thus the division of the subject matter (*i.e.*, colloid or particulate science) into hydrodynamics and statistical mechanics is somewhat arbitrary. Readers interested in a more complete treatment of the natural sciences that form the basis for understanding suspension phenomena are directed to [37, 42]. In some sense, this book provides an expanded view of the purely hydrodynamic aspects

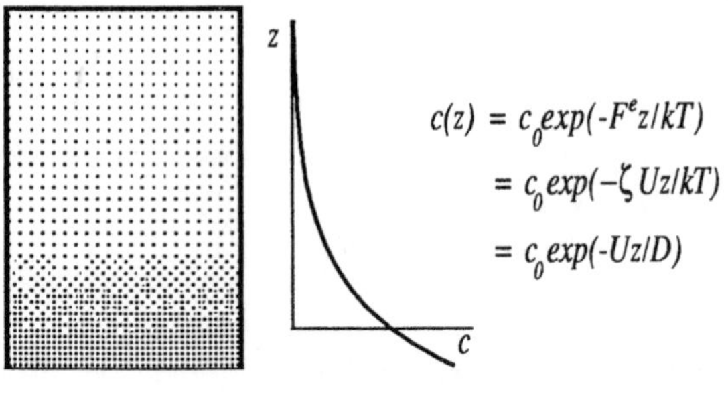

Figure 1.1: Brownian motion and an external field.

of the subject.

Since inertia plays a negligible role on the microscale, the hydrodynamic force balances the external forces, and the simplest paradigm may be stated as

$$\text{motion} \leftrightarrow \boldsymbol{F}^H = -\boldsymbol{F}^e .$$

In other words, if we know the relation (linear in our context) between particulate motion and the hydrodynamic force, we can determine the action of known external forces on the motion, or used in the opposite direction, infer the nature of unknown forces by observing the motion of the particle or particles. As in any introductory statements, there are qualifications. This simple paradigm applies only to noncolloidal particles, which are typically larger than a micron (10^{-6}m) but not so large that fluid inertia is important. This view is the fundamental underpinning of many important industrial operations, including flotation operations in ore processing [38], particle filtration operations common to many industries (an excellent review of particle capture from low-speed laminar flows is given by Spielman [41]), and aggregation and deposition of pulp fibers in paper manufacturing [40], to name a few.

In the colloidal size range (from the smallest continuum length scale such as the size of large globular protein molecules, to 0.1 microns) particle Brownian motion caused by the thermal agitation of the solvent molecules is important, and the particle motion is no longer deterministic. However, here also we have a role for the central theme of this book, the mobility relation between the particle motion and forces. The effect of Brownian motion in a liquid can be viewed as a diffusive process, an idea dating back to the early era of Einstein [13], with the diffusion coefficient given by kT/ζ, where k is the Boltzmann constant, T is the absolute temperature, and ζ^{-1} is the mobility of the particle. Thus the diffusion coefficient of a sphere of radius a is given by $D = kT/6\pi\mu a$.

To understand the source of this relation, we consider the situation depicted in Figure 1.1. A suspension of large particles will simply settle out and form a precipitated layer at the bottom of container. Colloidal particles, on the other hand, are subject to Brownian motion, and the formation of a concentration gradient is hindered by the action of Brownian motion; the flux of particles moving from the region of higher concentration will be greater than that from the region of lower concentration. In an equilibrium situation, the upwards flux due to Brownian motion, $-Ddc/dz$ must balance the flux due to gravity, $(\zeta^{-1}F^g c)$. (The velocity is given by $\zeta^{-1}F^g$, by definition of the mobility ζ^{-1}.) The resulting distribution $c(z)$ is the Boltzmann distribution only if $D = kT/\zeta$. The link between diffusivity and mobility is readily extended to rotary Brownian motion (a rotary Brownian diffusion coefficient related to the rotary mobility times kT) and diffusion for an assembly of particles [2, 8].

The stochastic nature of the trajectory of a single particle undergoing Brownian motion can also be modelled by using *stochastic differential equations* of the type known as the *Langevin equation.* In its simplest form, the Langevin equation incorporates the usual deterministic motion for the particle plus additional stochastic forces [15], and thus the viscous resistance of the fluid, again characterized by the mobility, is a key aspect of the computational procedure. In later chapters, we will see other subtle ways in which mobility relations and other concepts from the "standard" hydrodynamic calculations are incorporated as essential parts of the mathematical models of suspension behavior. Finally, readers interested in development of suspension simulations from a knowledge of the many-particle mobilities are directed to [7].

In Figure 1.2, we give the particle-based length scales for a number of natural and industrial particulate systems.

Going back to the central problem of relating particle motion and hydrodynamic force, we divide the task into two groups. For simple particle shapes in unbounded flow or those in flows bounded by simple container geometries, we describe a set of analytical solution techniques that enable one to relate the mobility to the key geometric parameters by elementary functions. Complex geometries for which the reader is convinced that the fine-scale details of the geometry are important can be handled by numerical methods that we describe, which are guaranteed to converge given sufficient computational power.

In the category of simple particle shapes, the most important examples are spheres and ellipsoids. The latter is particularly useful because it encompasses a wide range of shapes from disks to needles. Symmetry that is naturally inherent in certain growth processes often leads to particles of high symmetry; polystyrene latexes (perfect spheres) are one such example. The simplest boundary is the plane wall, but analytical and semi-analytical methods are available for spherical and cylindrical container geometries, as well as surfaces of mild curvature and those that fit a curvilinear coordinate system in which the equations separate. These categories of problems are described in Parts II and III.

In the category of *complex particle shapes*, we include flakes (clay particles, silver halide grains) and ribbons (pulp fibers), regular polyhedra (crystals), as well as particles without any symmetry whatsoever (globular proteins). Also, by combining simple shapes we end up with complex shapes, as in the steric layer formed by rigid rods (polymers) protruding from a perfect colloidal sphere.

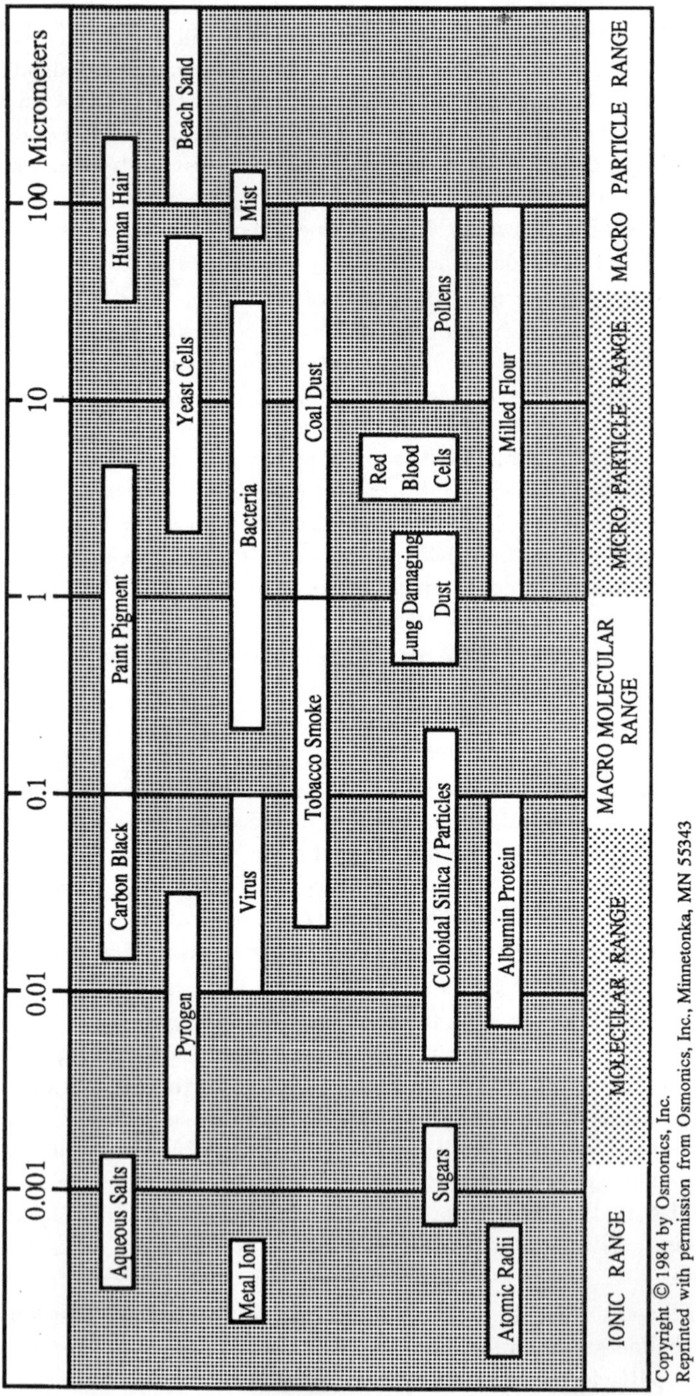

Figure 1.2: The particle size spectrum of microhydrodynamics.

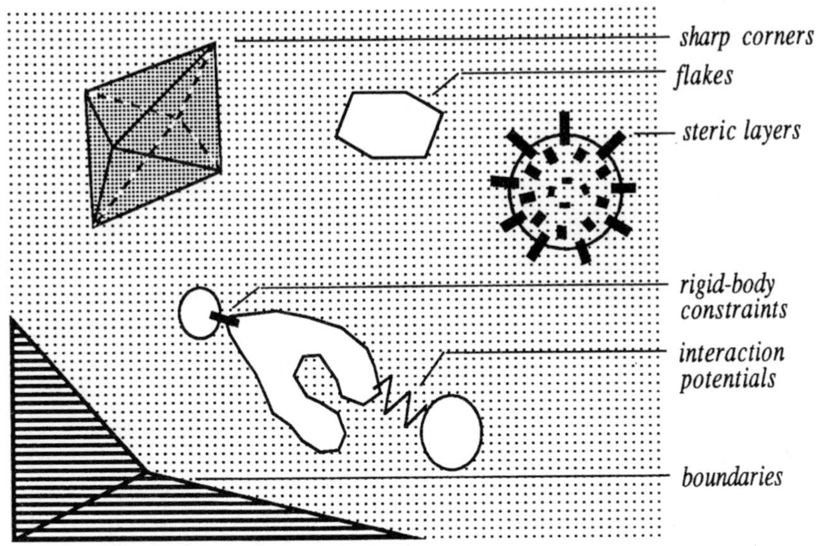

Figure 1.3: Particle shapes in microhydrodynamics.

These geometries are shown in Figure 1.3. To cite one example, the silver halide crystals or grains found in photographic emulsions are almost perfect hexagonal prisms with a thickness that is one tenth of the broadside dimensions. The analysis of the aggregation or deposition of these grains in a perfect pattern, as well as the stress distribution on a grain or aggregate, requires a local analysis with accurate hydrodynamics, especially near the sharp edges and corners. *Complex container geometries* are pervasive; examples include biological clefts and fiber mats and webs, and mathematical models that attempt to capture the geometry in a more complete manner will encounter them.

The dramatic growth in computational power over the past two decades has resulted in a similar growth in the range of problems solved in microhydrodynamics. Nevertheless, the complex geometries mentioned in the previous paragraph remain by and large outside the scope of current computational technology. These are after all three-dimensional flow problems with complex and evolving geometries (as the surfaces move relative to one another). Increased geometric resolution (a finer numerical mesh) scales into more unknowns at a quadratic to cubic growth rate and, given solution algorithms with solution times that also scale as quadratic to cubic powers of the number of unknowns, we see that even a dramatic increase (a factor of 500) corresponds only to a disappointingly small factor of two (or even smaller) increase in model complexity or resolution. Conventional supercomputers are unlikely to improve by such factors, and so the focus of Part IV is on the development of computational algorithms for parallel computers. We show how the central problem of microhydrodynamics can be split into independent concurrent streams, so that the projected growth rate in power of parallel computers can be exploited. Furthermore, the scaleup of time *vs.* number of unknowns is at the low end, approximately N^2, so that a factor of two increase in system size results in a factor of four increase in computational time.

1.2 The Governing Equations

The equations governing fluid flow are obtained by considering first conservation equations for mass, momentum, and energy. These equations contain some quantities, in particular the stress, that must be specified by supplying constitutive equations for the fluid.

1.2.1 The Equation of Continuity

The differential equation for conservation of mass, also known as the equation of continuity, is given by

$$\frac{\partial \rho}{\partial t} + \nabla \cdot (\rho \boldsymbol{v}) = 0 . \tag{1.1}$$

Here $\rho(\boldsymbol{x}, t)$ is the fluid density and $\boldsymbol{v}(\boldsymbol{x}, t)$, the Eulerian velocity. These field variables are evaluated instantaneously at time t at the point $\boldsymbol{x}$ fixed with respect to the laboratory.

Equation 1.1 can be derived by considering the mass balance over a differential volume element, as in [6], or from a balance over a macroscopic volume V fixed in space. The latter is given by

$$\frac{d}{dt} \int_V \rho \, dV = - \oint_S \rho \boldsymbol{v} \cdot \boldsymbol{n} \, dS .$$

The left side represents the time rate of change of the mass accumulated in V. Since mass is conserved, the change in this accumulation can occur only by mass entering or leaving through the bounding surface S, and that is what the right side of the equation represents. We apply the divergence theorem to obtain

$$\int_V \left(\frac{\partial \rho}{\partial t} + \nabla \cdot (\rho \boldsymbol{v}) \right) dV = 0 ,$$

and since this applies to all control volumes, the integrand must be identically zero and we obtain Equation 1.1. It is also possible to derive the equation of continuity by the use of a surface moving with the material points or *material surface* [31, 46].

The *substantial* or *material derivative* is defined by the operator equation

$$\frac{D(\cdot)}{Dt} = \frac{\partial(\cdot)}{\partial t} + \boldsymbol{v} \cdot \nabla(\cdot) \tag{1.2}$$

and is the result of applying the chain rule to differentiation with respect to time while following a fluid particle along its path. The equation of continuity when written in terms of the material derivative becomes

$$\frac{D\rho}{Dt} + \rho \nabla \cdot \boldsymbol{v} = 0 \ . \qquad (1.3)$$

If we exclude the possibility of change in the density as we follow the material so that $D\rho/Dt = 0$, then the equation of continuity reduces to the statement that the velocity field is solenoidal,

$$\nabla \cdot \boldsymbol{v} = 0 \ . \qquad (1.4)$$

In practice, we shall further restrict our consideration to fluids with uniform density, *i.e.*, ρ is a constant. Sometimes this more restrictive assumption is termed *incompressibility*. The physical validity of this approximation is now estimated. For a single-component fluid, the density may be taken as a known function of temperature and pressure given by the equation of state from thermodynamics. The systems of interest in this book are all under isothermal conditions, so we focus our attention on the dependence on pressure. Whereas gas densities are quite sensitive to changes in the pressure (the most familiar example is the perfect gas law, $\rho = p/RT$), those of liquids are not. For water at room temperatures, the engineer's steam table [21] charts a change in the fourth significant figure for a 50% increase from atmospheric pressure.

1.2.2 The Momentum Balance

The equation for conservation of momentum is

$$\rho \frac{D\boldsymbol{v}}{Dt} = \nabla \cdot \boldsymbol{\sigma} + \rho \boldsymbol{f} \ . \qquad (1.5)$$

Here, locally, $\boldsymbol{v}$ is the velocity, ρ is the density, $\boldsymbol{\sigma}$ is the stress tensor, and $\boldsymbol{f}$ is the external body force per unit mass. It is assumed that no external body couples act on the fluid. As we did earlier for the equation of continuity, this too is derived from a balance (in this case of momentum) on a macroscopic fixed control volume V,

$$\frac{d}{dt}\int_V \rho \boldsymbol{v}\, dV = -\oint_S \rho \boldsymbol{v}(\boldsymbol{v}\cdot\boldsymbol{n})\, dS + \oint_S \boldsymbol{\sigma}\cdot\boldsymbol{n}\, dS + \int_V \rho \boldsymbol{f}\, dV \ .$$

The left side represents the time rate of change of the momentum accumulated in V. Momentum can enter through the bounding surface S by virtue of material flow, and this is what the first term on the right side of the equation represents. The momentum accumulation can also change by virtue of surface and body forces acting on the fluid in V. Forces acting on the fluid in V over a differential element dS of the surface S are given by $\boldsymbol{\sigma}\cdot\boldsymbol{n}\, dS$, and the net effect is given by the surface integral over S. The action of body forces is given by the integral over V of the product of ρ and $\boldsymbol{f}$, the force per unit mass. In the case of gravity, we will denote the body force by the gravitational acceleration, $\boldsymbol{g}$.

We now apply the divergence theorem to obtain[1]

$$\frac{\partial(\rho \boldsymbol{v})}{\partial t} + \nabla \cdot (\rho \boldsymbol{v}\boldsymbol{v}) = \nabla \cdot \sigma + \rho \boldsymbol{f} \ .$$

To complete the equations for fluid flow, we must have a model for the relation between the stress in the fluid and the state of the fluid. Naturally, we would expect this relation to vary from one material to another. If the fluid is static, the surface force acts along the surface normal, $\boldsymbol{\sigma} \cdot \boldsymbol{n} = -p\boldsymbol{n}$, where p is simply the pressure acting on the surface, and we may write $\sigma = -p\boldsymbol{\delta}$.[2]

Now we consider deformation. The intuitive distinction between a solid and a fluid is that the stress in solids depends on the amount of deformation, while in fluids it depends on the *rate of deformation*. The instantaneous rate of deformation of a small material filament is characterized by the rate of strain tensor,

$$\boldsymbol{e} = \frac{1}{2}(\nabla \boldsymbol{v} + (\nabla \boldsymbol{v})^t) - \frac{1}{3}\boldsymbol{\delta}(\nabla \cdot \boldsymbol{v}) \ .$$

The simplest theory for a fluid is based on the hypothesis that the stress is linear with respect to $\boldsymbol{e}$, so that[3]

$$\boldsymbol{\sigma} = -p\boldsymbol{\delta} + 2\mu \boldsymbol{e} \ ,$$

and this equation is known as the *Newtonian constitutive equation.* Real fluids satisfying this relation to the limits of the purpose at hand are called *Newtonian fluids.* The material property μ is called the *dynamic viscosity,* and in most instances is simply referred to as the *viscosity.* The factor of two is in the constitutive relation only for historical reasons. Newton's experiments involved simple shear flow, $\boldsymbol{v} = \dot{\gamma} y \boldsymbol{e}_x$, and μ appears as the coefficient of proportionality between the shear stress component and the shear rate $\dot{\gamma}$, *i.e.*, $\sigma_{xy} = \mu\dot{\gamma}$. Non-Newtonian behavior is generally associated with long molecules (polymeric fluids) or fluids with microstructure (suspensions). Water and other fluids of compact molecular structure generally are Newtonian fluids [4, 5]. The methods described here can be used in the creation of mathematical models for non-Newtonian behavior of suspensions, but our main focus will be the microscale equations for the motion of a *Newtonian* solvent.

Combining the above constitutive equation with the momentum balance leads to the *Navier–Stokes equations* for an incompressible Newtonian fluid:

$$\rho\left(\frac{\partial \boldsymbol{v}}{\partial t} + (\boldsymbol{v} \cdot \nabla)\boldsymbol{v}\right) = -\nabla p + \mu \nabla^2 \boldsymbol{v} + \rho \boldsymbol{f} \qquad (1.6)$$

---

[1]To be precise, we should really use the transpose of the stress tensor in the first term on the right side, *i.e.*, use $\nabla \cdot \sigma^t$. This distinction will soon become irrelevant because the fluids under consideration here will have symmetric stress tensors, $\sigma = \sigma^t$. For fluids with nonsymmetric stress tensors, see [12].

[2]The identity tensor $\boldsymbol{\delta}$ has Cartesian components δ_{ij} that equal one or zero according as i equals j or not. δ_{ij} is often called the Kronecker delta.

[3]For a compressible fluid, we add an extra term, $\kappa(\nabla \cdot \boldsymbol{v})$, where κ is known as the bulk viscosity.

$$\nabla \cdot \boldsymbol{v} = 0 . \tag{1.7}$$

A more extensive discussion on the derivation of these equations may be found in many fluid mechanics texts [1, 16, 26, 27, 46, 48]. For a discussion on the kinematics of deformation and other continuum mechanical concepts, we refer the reader to Malvern [31].

1.2.3 The Stokes Equations

We note that the Navier–Stokes equations are nonlinear, since the substantial derivative contains a term that is homogeneous of order two in $\boldsymbol{v}$. Only in exceptional cases, such as steady laminar flow between parallel infinite plane walls or in an infinite circular pipe (both are called *Poiseuille flows*), will the nonlinear terms vanish from these equations.

The class of "solvable problems" is greatly expanded when we drop the nonlinear terms. The linearized Navier–Stokes equations for steady (*i.e.*, time-independent) motion are known as the *creeping motion* or (inhomogeneous) *Stokes* equations, and these are obtained by neglecting the substantial derivative of $\boldsymbol{v}$ in the Navier–Stokes equations:

$$\nabla \cdot \boldsymbol{\sigma} = -\nabla p + \mu \nabla^2 \boldsymbol{v} = -\rho \boldsymbol{f} \tag{1.8}$$

$$\nabla \cdot \boldsymbol{v} = 0 . \tag{1.9}$$

When the force field is conservative, such as gravity, it can be expressed as a gradient; then it may be lumped together with the pressure p, and this is the reason why the equations above are most often written without the force term. We shall call the equations thus obtained the *homogeneous Stokes equations* when we want to make this distinction explicitly. Actually the equations above are the *steady* Stokes equations, since sometimes the term $\rho\, \partial \boldsymbol{v}/\partial t$ is retained. To convince the reader that this linearization of a relatively complicated partial differential equations did not waste all the physical significance that the equations originally might have had, we now briefly discuss the relation between the solutions of the Stokes equations and those of the full Navier–Stokes equations.

The Stokes solutions provide a good approximation to the flow field near a particle when the *Reynolds number* defined by

$$\mathrm{Re} = \rho V L/\mu \tag{1.10}$$

is small. Here V and L are representative values of the speed (magnitude of the velocity) of the flow and the particle dimensions, respectively. As Proudman and Pearson [36] showed (for a sphere explicitly), matched asymptotic expansions may be used to correct the Stokes solution to that of the full equations, the correction being $O(\mathrm{Re})$ in the vicinity of the particle boundary, actually for distances up to $O(1/\mathrm{Re})$. Considering ρ and μ fixed, we see that the Stokes solutions require less correction, as either the particle size or the characteristic speed of the flow is reduced. The article by Brenner and Cox [11] demonstrates that these ideas apply to particles of arbitrary shape. Using the length scales from

Figure 1.2 and the properties of water (density of 1 gram per cubic centimeter and viscosity of 1 centipoise), and creeping motions ranging from microns to millimeters per second, we see that the Reynolds numbers range from 10^{-6} to 10^{-4}. Henceforth, the Stokes equations will be considered as the mathematical model to be used for particles in the realm of microhydrodynamics, or viewed in another way, microhydrodynamics is defined as such, but the tautology is physically relevant since many important systems fall into this category. In any case, an *a posteriori* check of the size of the neglected inertial terms could be made after a Stokes solution is obtained.

Even though a problem might be unsteady, in the sense that the geometric configuration of the boundaries will change with time, we shall continue to use the steady Stokes equations by making the assumption of quasi-steady flow, meaning that we assume the terms related to time dependence to be negligible. We will examine this in more detail in Chapter 6, the essential conclusion being that for the scales mentioned above, the quasi-steady approximation gives the correct behavior in the leading order with a small error proportional to $O(\mathrm{Re}^{1/2})$.

Modelling with Stokes equations has the following assets over the use of full Navier–Stokes equations. Firstly since the Stokes equations are linear, solutions may be superposed to get new solutions. Secondly, since inertial terms are neglected anyway, we need not restrict our coordinate frames to just the inertial ones; using, say, coordinates fixed to a rotating sphere will not give rise to any fictitious force field in the equations.

In spite of the simplifications indicated above, the number of analytical solutions known is still very small. A fairly comprehensive list can be obtained from Lamb's *Hydrodynamics* [25], Oseen's *Hydrodynamik* [35], Villat's *Leçons sur les Fluides Visquex* [44], and Happel and Brenner [16]. Chapter 1 of the last reference gives an excellent historical perspective to the low-Reynolds-number hydrodynamics literature from antiquity to the 1960s, including the first three of the preceding list. The 1930 paper by Odqvist [34] provides existence and uniqueness theorems for integral representation solutions of the Stokes equations. Although his work is in German, the main ideas and results are accessible in English in the monograph, *The Mathematical Theory of Viscous Incompressible Flow*, by Ladyzhenskaya [24].

1.2.4 Boundary Conditions for Fluid Flows

At the interface between a fluid and another inert medium (rigid solid or another fluid), the normal component of the velocity (in the frame moving with the boundary) vanishes identically; this is a kinematic condition that must be satisfied if the interface is to have an identity of its own. The conditions for the tangential component of the velocity are on less firm grounds. At a boundary with a solid, it is customary to use the *no-slip boundary condition*, based on empirical evidence. As the name implies, we assume that there is no slipping between the fluid and the boundary so that the tangential fluid velocities match those of the solid. Slip occurs for polymeric liquids and rarefied gases. In both cases the mechanism of interaction between the fluid molecules and those in the solid wall differs from that for compact molecules, such as water and the solid wall.

At the interface between two immiscible fluids, the tangential components of the velocity and tractions must match.[4] The normal component of the surface tractions jumps from one side to the other, with the jump balanced by the surface tension γ times the surface curvature. More explicitly, if the stresses on the two sides of the interface are $\boldsymbol{\sigma}'$ and $\boldsymbol{\sigma}''$, then

$$\boldsymbol{\sigma}' \cdot \boldsymbol{n} - \boldsymbol{\sigma}'' \cdot \boldsymbol{n} = -\gamma(\frac{1}{R_1} + \frac{1}{R_2})\boldsymbol{n} \ ,$$

with the surface normal pointing into the concave side where σ'' prevails. For charged interfaces, these equations must be modified by inclusion of surface charges and electric stresses into the balance [32].

1.2.5 The Energy Balance

We can derive the expression for the time rate of change of the total internal and kinetic energy in the control volume V by following the same procedure as that employed for the mass and the momentum balance. The end result is

$$\frac{\partial}{\partial t}\left(\frac{1}{2}\rho v^2 + \rho \hat{U}\right) = \rho \boldsymbol{v} \cdot \boldsymbol{f} + \nabla \cdot (\boldsymbol{\sigma} \cdot \boldsymbol{v}) + \nabla \cdot (k \nabla T) \ . \tag{1.11}$$

Here $\hat{U}$ is the internal energy per unit mass (the hat reminds us that it is per unit mass), k is the thermal conductivity of the fluid, and T is temperature. The heat flux $\boldsymbol{q}$ is given by Fourier's law as $\boldsymbol{q} = -k\nabla T$.

An expression for the time rate of change of the kinetic energy may be obtained from the momentum balance by taking the dot product of that equation with $\boldsymbol{v}$, and when this expression is subtracted from the energy balance we obtain the result

$$\rho\left(\frac{\partial}{\partial t} + \boldsymbol{v} \cdot \nabla\right)\hat{U} = 2\mu \boldsymbol{e} : \boldsymbol{e} + \nabla \cdot (k \nabla T) \ . \tag{1.12}$$

The first term on the right is always positive, unless $\boldsymbol{e}$ is identically zero; it represents the rate of production of internal energy by irreversible dissipation of kinetic energy. Under steady conditions, $\partial\hat{U}/\partial t = 0$, and the energy that is dissipated is either convected or conducted away from the material element. Thus the dissipation results in a raised temperature. For laminar viscous flows with velocity scale V, dimensional analysis reveals that this temperature increase scales as $\mu V^2/k$. In our problems in microhydrodynamics, the relative velocity is sufficiently small that we may neglect the temperature increase and assume that the system is under isothermal conditions.

[4] A tangential stress difference results if there are surface tension gradients on the interface, giving rise to the so-called Marangoni effect.

1.3 Colloidal Forces on Particles

Repulsion between small particles is governed by electrostatic forces, while attraction is governed primarily by London–van der Waals forces, also known as dispersion forces, and to a lesser extent, electrostatic forces. The relative importance of these to the hydrodynamic drag depends on two length scales: the size of the particles and the separation between the surfaces. The colloidal forces are quite short range in nature, and their range for a particle in the colloidal size range is tens to hundreds of angstroms ($100\,\text{Å} = 10^{-8}\,\text{m} = 0.01\,\mu\text{m}$). For problems involving aggregation and capture phenomena of larger particles, this suggests that colloidal forces can be neglected altogether for most of the particle trajectory, suggesting that the trajectory should be determined from hydrodynamic considerations, with colloidal forces appearing in the form of a sticking boundary condition below a critical particle-particle separation. On the other hand, the breakup of aggregates requires a careful analysis of the fairly difficult hydrodynamic calculation of stresses in flow past nearly touching particles, initially all moving in tandem as a rigid body. We show how to handle these situations in Part III (for the simple aggregate formed by two spheres) and Part IV (more general particle shapes).

The London–van der Waals attraction between a sphere of radius a and a plane wall may be quantified as [43]

$$F = \frac{A}{6}\frac{a}{h^2} ,$$

where A is Hamaker's constant. As the particle size a tends to 0, the attractive force decreases proportionally to a, while, for example, the gravitational forces decrease faster, as a^3. Hamaker's constant has units of energy and is commonly scaled with respect to the thermal energy scale kT. The values range from 0 to $20kT$, depending on the nature of the surface. A more rigorous foundation for the calculation of the dispersion force is given in the monograph by Mahanty and Ninham [30]. The review article by Kim and Lawrence [23] tabulates the relative importance of the attractive and hydrodynamic forces for a number of different sphere-plane interactions in air and water.

A charged surface in an aqueous environment attracts a layer (or cloud) of ions of the opposite charge (the counter-ions). Thus the electrostatic repulsion between two surfaces immersed in an aqueous medium is coulombic in nature, but is modified by the screening effect of the counter-ions. The calculation of this repulsive force in a flowing system thus requires the solution of a simultaneous system of partial differential equations governing the flow of the fluid (and the ions), as well as Maxwell's equations. The hydrodynamic and electrostatic forces are then obtained by surface integration of the hydrodynamic and Maxwell stresses, respectively. In Part II, we encounter the simplest example of this in the electrophoresis problem. A comprehensive treatment of both attractive and repulsive forces may be found in books on colloidal phenomena [19, 37, 39].

Chapter 2

General Properties and Fundamental Theorems

2.1 Introduction

On many occasions throughout this book, we will exploit a number of fundamental theorems satisfied by, and general properties of solutions of, the Stokes equations. These are collected and discussed at some length in this chapter. Theorems concerning energy dissipation are closely related to uniqueness proofs for solutions of the Stokes equations and are discussed in Section 2.2. Furthermore, they form the basis for *inclusion monotonicity theorems*, which place upper and lower bounds on energy dissipation rates, and thus the hydrodynamic drag and torque, for flows produced by the motion of rigid particles.

In Section 2.3, we discuss the *Lorentz reciprocal theorem*, a reciprocal relation between any two solutions of the Stokes equations. This theorem is essentially an analog of the more familiar Green's identity of potential theory, and the analogy carries as far as to the derivation of the integral representation for the velocity field. There are a host of other applications, but these will have to wait until subsequent chapters.

In Section 2.4, we present a derivation of the integral representation for Stokes flow. As in potential theory, these express the velocity fields as a weighted integral of the Green's function or fundamental solution. Instead of a formal derivation of the fundamental solution (which is possible with the use of Fourier transforms), we take the more elementary approach of guessing the form of the Green's function, and then verifying that it works. The integral representation then follows as a special case of the reciprocal theorem, *viz.*, the reciprocal relation between the Green's function and the velocity field of interest. The integral representation is used later on (especially in Part IV) as the basis for efficient solution strategies, especially in situations in which only particle motions are desired and the flow field in the background fluid is unimportant. Integral representations transform the governing three-dimensional partial differential equations into two-dimensional integral equations on the boundary of the fluid domain, and thus are particularly well suited for very large numerical computations. A special form of the integral representation for flows past particles in an unbounded domain will be derived for immediate use in the subsequent section on the multipole expansion.

The transition from the general theorems of this chapter to the specific theme of Chapter 3, flow past a single particle, occurs in the last section of Chapter 2, where we discuss the *multipole expansion* for the disturbance velocity field of a particle in an ambient flow field. Far away from the particle, the disturbance produced by the particle motion takes on a universal form (the multipole expansion), and as in electrostatics, follows from a Taylor series expansion of the Green's function in the integral representation, about a convenient reference point inside the particle. In place of moments of the surface charge distribution (net charge, dipole moment, quadrupole moment, *etc.*), we encounter moments of the surface traction (hydrodynamic drag on the particle, Stokes dipoles, Stokes quadrupoles, *etc.*).

2.2 Energy Dissipation Theorems

For a Newtonian fluid, the rate of energy dissipation due to viscosity, in a fluid region V is given by the expression [6]

$$\int_V \Phi_v \, dV = \int_V 2\mu \boldsymbol{e} : \nabla \boldsymbol{v} \, dV ,$$

where Φ_v is the rate of viscous energy dissipation per unit volume. For an incompressible Newtonian fluid, we may use the following identities:

$$\int_V 2\mu e_{ij} : \frac{\partial v_j}{\partial x_i} \, dV = \int_V 2\mu e_{ij} e_{ij} \, dV = \int_V \sigma_{ij} e_{ij} \, dV .$$

An analysis of energy dissipation yields two important theorems of Helmholtz [17] concerning the uniqueness of the Stokes solution and a minimum energy or variational principle.

2.2.1 Uniqueness

To prove uniqueness, assume that $\boldsymbol{v}$ and $\boldsymbol{v}'$ are two solutions of the Stokes equations that satisfy the same boundary conditions, so that $\boldsymbol{v} = \boldsymbol{v}'$ on S (the boundary of V). The rate of energy dissipated by the difference field, $\boldsymbol{v}' - \boldsymbol{v}$, may be manipulated as

$$\begin{aligned} 2\mu \int_V (e'_{ij} - e_{ij})(e'_{ij} - e_{ij}) \, dV &= 2\mu \int_V \left(\frac{\partial v'_i}{\partial x_j} - \frac{\partial v_i}{\partial x_j} \right) (e'_{ij} - e_{ij}) \, dV \\ &= 2\mu \oint_S (v'_i - v_i)(e'_{ij} - e_{ij}) n_j \, dS \\ &\quad - \mu \int_V (v'_i - v_i)(\nabla^2 v'_i - \nabla^2 v_i) \, dV \\ &= - \int_V (v'_i - v_i) \left(\frac{\partial p'}{\partial x_i} - \frac{\partial p}{\partial x_i} \right) \, dV \end{aligned}$$

$$= -\oint_S (v'_i - v_i)(p' - p) n_i \, dS$$
$$= 0 \, .$$

The preceding steps used the symmetry of $\boldsymbol{e}$ and $\boldsymbol{e}'$, integration by parts, the boundary conditions on S, and the divergence theorem (with the incompressibility condition $\nabla \cdot \boldsymbol{v}' = \nabla \cdot \boldsymbol{v} = 0$). Since the volume integral on the left-hand side is non-negative, the preceding equation can be satisfied if and only if $\boldsymbol{e}' = \boldsymbol{e}$ throughout V. Therefore, $\boldsymbol{v}'$ and $\boldsymbol{v}$ differ by at most a rigid-body motion (see Exercise 2.1). However, the boundary condition $\boldsymbol{v}' = \boldsymbol{v}$ precludes all nonzero rigid-body motions, so that $\boldsymbol{v}' = \boldsymbol{v}$ throughout V.

2.2.2 Minimum Energy Dissipation

Assume that $\boldsymbol{v}$ is a solution of the Stokes equations, and consider another field $\boldsymbol{v}^*$ that is solenoidal, *i.e.*, $\nabla \cdot \boldsymbol{v}^* = 0$. As before, we require $\boldsymbol{v} = \boldsymbol{v}^*$ on S. According to the *minimum energy dissipation theorem*, the Stokes solution dissipates less energy than any other solenoidal field with the same boundary velocities. We may prove this by starting with the relation

$$\int_V (e^*_{ij} - e_{ij}) e_{ij} \, dV = 0 \, , \tag{2.1}$$

which may be established using the same steps as in the uniqueness derivation. The volume integral on the LHS can be interpreted as an inner product, in the space of second order tensor fields, so that we have just shown an orthogonality relation. Now an application of the Pythagorean theorem immediately gives the minimum principle. (The length of the hypotenuse is always greater than or equal to the length of the base, as shown in Figure 2.1). Indeed, the following derivation is just a proof of the Pythagorean theorem in the notation relevant to the problem at hand. We write the rate of energy dissipated by $\boldsymbol{v}^*$ as

$$\begin{aligned}
2\mu \int_V e^*_{ij} e^*_{ij} \, dV &= 2\mu \int_V \{ e^*_{ij} e^*_{ij} - (e^*_{ij} - e_{ij}) e_{ij} \} \, dV \\
&= 2\mu \int_V \{ e_{ij} e_{ij} + (e^*_{ij} - e_{ij}) e^*_{ij} \} \, dV \\
&= 2\mu \int_V \{ e_{ij} e_{ij} + (e^*_{ij} - e_{ij}) e^*_{ij} - (e^*_{ij} - e_{ij}) e_{ij} \} \, dV \\
&= 2\mu \int_V \{ e_{ij} e_{ij} + (e^*_{ij} - e_{ij})(e^*_{ij} - e_{ij}) \} \, dV \, ,
\end{aligned}$$

which establishes that the dissipation rate is equal to that of $\boldsymbol{v}$ plus an integral of a non-negative integrand. This extra contribution is zero if and only if $\boldsymbol{e}^* = \boldsymbol{e}$, thus proving that the minimum dissipation occurs for $\boldsymbol{v}^* = \boldsymbol{v}$ by the same argument that ended the preceding section.

2.2.3 Lower Bounds on Energy Dissipation

By reversing the role of the velocity and stress fields, we may derive lower bounds for the rate of energy dissipation of Stokes velocity fields, as shown by

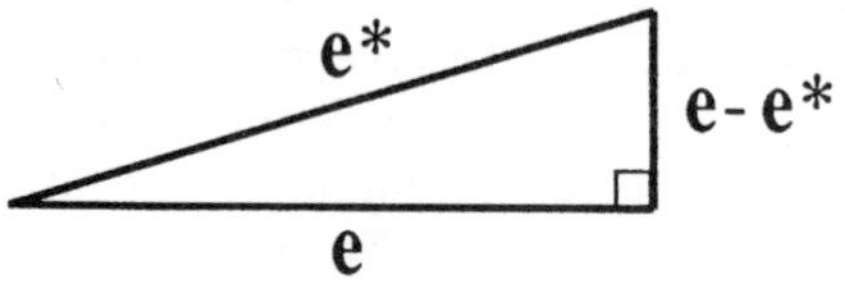

Figure 2.1: Minimum energy dissipation illustrated as a Pythagorean theorem.

Hill and Power [18]. Let $\boldsymbol{\sigma}'$ be an equilibrium stress field ($i.e.$, $\nabla \cdot \boldsymbol{\sigma}' = \mathbf{0}$). The field $\boldsymbol{e}' = (\boldsymbol{\sigma}' + p'\boldsymbol{\delta})/2\mu$ need not be associated with a velocity field.[1] We may obtain a lower bound for the energy dissipated by a solution of the Stokes equation in the same fluid domain, by considering a sequence of inequalities starting with $(\boldsymbol{e}' - \boldsymbol{e}) : (\boldsymbol{e}' - \boldsymbol{e}) \geq 0$:

$$\begin{aligned} e'_{ij}e'_{ij} - e_{ij}e_{ij} &\geq 2e_{ij}(e'_{ij} - e_{ij}) \\ \sigma'_{ij}e'_{ij} - \sigma_{ij}e_{ij} &\geq 2e_{ij}(\sigma'_{ij} - \sigma_{ij}) \\ \int_V (\sigma'_{ij}e'_{ij} - \sigma_{ij}e_{ij})\, dV &\geq 2\int_V \frac{\partial v_i}{\partial x_j}(\sigma'_{ij} - \sigma_{ij})\, dV \\ E' - E &\geq 2\oint_S v_i\sigma'_{ij}n_j\, dS - 2E \\ E &\geq 2\oint_S v_i\sigma'_{ij}n_j\, dS - E' . \end{aligned} \tag{2.2}$$

In the first part of this sequence of inequalities, we used the condition on the traces of $\boldsymbol{e}$ and $\boldsymbol{e}'$, $e.g.$, $e'_{ii} = \delta_{ij}e'_{ij} = 0$. In the case of $\boldsymbol{v}'$ equaling the Stokes solution $\boldsymbol{v}$, the last integral is exactly $2E' = 2E$ and the equality is obtained.

2.2.4 Energy Dissipation in Particulate Flows

For flow past suspended particles, the previous results on energy dissipation further simplify into bounds on the hydrodynamic drag and torque, as discussed by Hill and Power [18]. The rate of energy dissipation is identical to the rate at which work is done on the particle surface, as can be seen by the sequence of equations,[2]

$$\int_V \sigma_{ij}e_{ij}\, dV = \int_V \sigma_{ij}\frac{\partial v_i}{\partial x_j}\, dV$$

[1] Here p' is determined by requiring $\boldsymbol{e}'$ to be traceless.

[2] Reading this equation from right to left motivates the definition given earlier for the rate of energy dissipation in the fluid element, for Stokes flows.

$$= \oint_S v_i \sigma_{ij} n_j \, dS \, .$$

Here the surface normal $\boldsymbol{n}$ points out of the fluid domain and into the particle, so that $\boldsymbol{F}$ and $\boldsymbol{T}$ below are the force and torque imparted by the particle to the fluid. On the surface of a rigid particle moving with the rigid-body motion, $\boldsymbol{U} + \boldsymbol{\omega} \times \boldsymbol{x}$, where $\boldsymbol{U}$ is a constant translational velocity and $\boldsymbol{\omega}$ is a constant angular velocity, we get the equality

$$\begin{aligned} \oint_S v_i \sigma_{ij} n_j \, dS &= \oint_S \{U_i + (\boldsymbol{\omega} \times \boldsymbol{x})_i\} \, \sigma_{ij} n_j \, dS \\ &= U_i F_i + \omega_i T_i \, , \end{aligned}$$

where

$$\boldsymbol{F} = \oint \boldsymbol{\sigma} \cdot \boldsymbol{n} \, dS \qquad \text{and} \qquad \boldsymbol{T} = \oint \boldsymbol{x} \times (\boldsymbol{\sigma} \cdot \boldsymbol{n}) \, dS \, .$$

From this result, the inequalities concerning energy dissipation rates may be expressed as

$$2(\boldsymbol{F}' \cdot \boldsymbol{U} + \boldsymbol{T}' \cdot \boldsymbol{\omega}) - E' \leq E = \boldsymbol{F} \cdot \boldsymbol{U} + \boldsymbol{T} \cdot \boldsymbol{\omega} \leq E^* \, . \tag{2.3}$$

From these inequalities, Hill and Power deduce the following statements, which we shall collectively call *inclusion monotonicity*:

1. If a body with surface S_2 is completely contained within a body with surface S_1, and if F_1 and F_2 are the drag on S_1 and S_2, respectively, then $F_1 \geq F_2$, the translations and external boundary being the same in the two problems.

2. The drag F_1 on a particle translating inside a rigid stationary container C_1 is greater than or equal to the drag on the same particle translating in a container C_2 if C_2 completely encloses C_1.

3. The drag on a body S is increased by the presence of other bodies that are either held fixed or free to move.

The proofs for the three are quite similar. For statement 1, we consider the actual Stokes solution $\boldsymbol{v}_2$ generated by translation of S_2. For $\boldsymbol{v}_2^*$, we use $\boldsymbol{v}_1$, the solution for translation of S_1, but augmented by the uniform translational velocity $\boldsymbol{U}$ in the region between the bodies, $S_1 - S_2$. Then the minimum dissipation theorem implies the inequality

$$E_2 = \boldsymbol{F}_2 \cdot \boldsymbol{U} \leq E_2^* \, .$$

However the rate-of-strain in $S_1 - S_2$ is zero, so $E_2^* = E_1$ and

$$E_2 = \boldsymbol{F}_2 \cdot \boldsymbol{U} \leq E_1 = \boldsymbol{F}_1 \cdot \boldsymbol{U} \, .$$

The work of Hill and Power has been extended in a number of directions. Minimum and maximum principles with more general boundary conditions, such as traction boundary conditions and normal flows through surfaces, may be found in the work of Keller, Rubenfeld, and Molyneux [22]. Nir, Weinberger, and Acrivos, [33] consider particles in a shear flow. We will consider this again in Chapter 3, after the discussion of *resistance tensors* for particles of arbitrary shape. Finlayson [14] considers the full Navier–Stokes equations, and shows that no variational principles exist unless either $\boldsymbol{v} \times (\nabla \times \boldsymbol{v}) = \mathbf{0}$ or $\boldsymbol{v} \cdot \nabla \boldsymbol{v} = \mathbf{0}$. Variational theorems for non-Newtonian fluids are discussed in Johnson [20]. The inclusion monotonicity has also been established for sedimentation problems in Stokes flow by Weinberger [45]. We consider this case now in the framework of so-called mobility problems.

2.2.5 Energy Dissipation in Mobility Problems

In the preceding discussion we assumed boundary velocities and considered the resulting forces and torques in the usual mathematical format of boundary value problems, but actually, this is the inverse of most physical settings. Usually, the force and torque acting on the particle are given and the particle motion is to be determined; sedimentation of a particle of known mass illustrates this point. We shall call these physical problems *mobility* problems.

We start with inequality 2.2 encountered during the derivation of lower bounds to find

$$\begin{aligned}\int_V (\sigma'_{ij}e'_{ij} - \sigma_{ij}e_{ij})\, dV &\geq 2\int_V e_{ij}(\sigma'_{ij} - \sigma_{ij})\, dV \\ &= 2\oint_S v_i(\sigma'_{ij} - \sigma_{ij})n_j\, dS\end{aligned}$$

for the case $\nabla \cdot \boldsymbol{\sigma}' = \nabla \cdot \boldsymbol{\sigma} = \mathbf{0}$. Let V be the volume of fluid between a particle and a container, the particle in rigid-body motion, and $\boldsymbol{v} = 0$ on the container surface. Then the preceding inequality becomes

$$E'(V) - E(V) \geq 2\left\{(\boldsymbol{F}' - \boldsymbol{F}) \cdot \boldsymbol{U} + (\boldsymbol{T}' - \boldsymbol{T}) \cdot \boldsymbol{\omega}\right\} .$$

Setting $\boldsymbol{F}' = \boldsymbol{F}$ and $\boldsymbol{T}' = \boldsymbol{T}$, we get

$$E(V) \leq E'(V) \leq E'(V') \qquad \text{for } V \subseteq V' .$$

In words, *on keeping the total force and torque affecting each particle constant, expanding the particles or adding no-slip boundaries (like a container) will reduce the dissipation by at least the amount that occurred in the excluded fluid volume.*

With this result, we can, for example, handle the question of whether a screw-like groove in a particle slows its sedimentation velocity due to its rotation. It does not. By filling the grooves with neutrally buoyant filler, we get a smooth particle *of the same weight or buoyancy* ($\boldsymbol{F} = \boldsymbol{F}'$) and a smaller fluid volume ($V \subseteq V'$). Then the energy dissipation argument becomes

$$\boldsymbol{F} \cdot \boldsymbol{U}' = E'(V') \geq E(V) = \boldsymbol{F} \cdot \boldsymbol{U} ;$$

so the settling rate (projection of the velocity in the direction of gravity) is greater for the grooved particle.

2.3 Lorentz Reciprocal Theorem

Consider a closed region of fluid V bounded by a surface S. Suppose that the velocity fields $\boldsymbol{v}$ and $\boldsymbol{v}'$ both satisfy the Stokes equations. We denote their respective stress fields by $\boldsymbol{\sigma}$ and $\boldsymbol{\sigma}'$. The *Lorentz reciprocal theorem* [29] states that

$$\oint_S \boldsymbol{v}\cdot(\boldsymbol{\sigma}'\cdot\boldsymbol{n})\,dS = \oint_S \boldsymbol{v}'\cdot(\boldsymbol{\sigma}\cdot\boldsymbol{n})\,dS\ . \tag{2.4}$$

We choose to prove the relation, $\nabla\cdot(\boldsymbol{v}\cdot\boldsymbol{\sigma}') = \nabla\cdot(\boldsymbol{v}'\cdot\boldsymbol{\sigma})$, since the theorem statement follows directly by application of the divergence theorem. Consider the reduction for $\boldsymbol{\sigma}' : \boldsymbol{e}$,

$$\begin{aligned} \sigma'_{ij}e_{ij} &= (-p'\delta_{ij} + 2\mu e'_{ij})e_{ij} \\ &= 2\mu e'_{ij}e_{ij}\ . \end{aligned}$$

We have used the result $p'\delta_{ij}e_{ij} = p'e_{ii} = p'\,\nabla\cdot\boldsymbol{v} = 0$. Reversing the role of the primed and unprimed variables, we obtain also $\boldsymbol{\sigma} : \boldsymbol{e}' = 2\mu\boldsymbol{e} : \boldsymbol{e}'$, so that

$$\sigma'_{ij}e_{ij} = \sigma_{ij}e'_{ij}\ . \tag{2.5}$$

But consider the sequence of steps:

$$\begin{aligned} \sigma'_{ij}e_{ij} &= \frac{1}{2}\sigma'_{ij}\frac{\partial v_i}{\partial x_j} + \frac{1}{2}\sigma'_{ij}\frac{\partial v_j}{\partial x_i} \\ &= \sigma'_{ij}\frac{\partial v_i}{\partial x_j} \\ &= \frac{\partial}{\partial x_j}(\sigma'_{ij}v_i) - \left(\frac{\partial\sigma'_{ij}}{\partial x_j}\right)v_i \\ &= \frac{\partial}{\partial x_j}(\sigma'_{ij}v_i)\ . \end{aligned}$$

In the last step, we used the governing equation, $\nabla\cdot\boldsymbol{\sigma}' = \boldsymbol{0}$; for the penultimate step, we require $\sigma'_{ij} = \sigma'_{ji}$. If the same steps are applied to $\sigma_{ij}e'_{ij}$, it must reduce in an analogous fashion to $(\partial/\partial x_j)(\sigma_{ij}v'_i)$. So from Equation 2.5, we obtain the desired result, that for all points in V,

$$\nabla\cdot(\boldsymbol{v}\cdot\boldsymbol{\sigma}') = \nabla\cdot(\boldsymbol{v}'\cdot\boldsymbol{\sigma})\ ,$$

thus completing the proof.

Note that if the assumptions $\nabla\cdot\boldsymbol{\sigma}_1 = \boldsymbol{0}$ and $\nabla\cdot\boldsymbol{\sigma}_2 = \boldsymbol{0}$ are relaxed, then the corresponding result is

$$\begin{aligned} &\oint_S \boldsymbol{v}_1\cdot(\boldsymbol{\sigma}_2\cdot\boldsymbol{n})\,dS - \int_V \boldsymbol{v}_1\cdot(\nabla\cdot\boldsymbol{\sigma}_2)\,dV \\ &= \oint_S \boldsymbol{v}_2\cdot(\boldsymbol{\sigma}_1\cdot\boldsymbol{n})\,dS - \int_V \boldsymbol{v}_2\cdot(\nabla\cdot\boldsymbol{\sigma}_1)\,dV\ , \end{aligned} \tag{2.6}$$

which shall also prove to be quite useful.

In Exercise 2.2, we consider how the reciprocal theorem may be used to show that Stokes flow "transmits" unchanged the total force and torque from an inner closed surface to an outer enclosing surface.

2.4 Integral Representations

The integral representation for the velocity field can be viewed as a restatement of the governing equations from a three-dimensional partial differential equation to a two-dimensional integral equation for unknown densities over the boundary of the fluid domain. This reduction in dimensionality will be significant in the numerical solutions approach to the problems of microhydrodynamics. As a preliminary step, we derive the Green's function for Stokes flow.

2.4.1 The Green's Function for Stokes Flow

We seek the solution to the following problem:

$$\nabla \cdot \boldsymbol{\sigma} = -\nabla p + \mu \nabla^2 \boldsymbol{v} = -\boldsymbol{F}\delta(\boldsymbol{x}) \, , \qquad \nabla \cdot \boldsymbol{v} = 0 \, . \tag{2.7}$$

For a more detailed exposition on the Dirac delta function, $\delta(\boldsymbol{x})$, and the theory of distributions, we refer the reader to Lighthill [28]. For our present purposes, the meaning of the first equation is

1. For $\boldsymbol{x} \neq \mathbf{0}$, $\nabla \cdot \boldsymbol{\sigma} = 0$.

2. For any volume V that encloses the point $\boldsymbol{x} = \mathbf{0}$, $\int_V \nabla \cdot \boldsymbol{\sigma} \, dV = -\boldsymbol{F}$.

It is possible to derive the solution formally by the Fourier transform (see Exercise 2.9), or by using the solution for a translating sphere (Exercise 2.8). Here, we take the easier path of picking the solution out of thin air and proving that it works. We claim that the fundamental solution consists of the velocity and pressure pair,

$$\boldsymbol{v}(\boldsymbol{x}) = \boldsymbol{F} \cdot \frac{\mathcal{G}(\boldsymbol{x})}{8\pi\mu} \, , \qquad p(\boldsymbol{x}) = \boldsymbol{F} \cdot \frac{\mathcal{P}(\boldsymbol{x})}{8\pi\mu} \, , \tag{2.8}$$

where $\mathcal{G}(\boldsymbol{x})/8\pi\mu$ is a Green's dyadic. Many authors remove the factor of $8\pi\mu$ from the Green's dyadic (as we have done), so that the *Oseen tensor*, $\mathcal{G}(\boldsymbol{x})$, given by

$$\mathcal{G}_{ij}(\boldsymbol{x}) = \frac{1}{r}\delta_{ij} + \frac{1}{r^3}x_i x_j \, , \tag{2.9}$$

is purely a geometric quantity, *i.e.*, independent of fluid properties. In the literature we also encounter the terminology *Oseen-Burgers tensor* for the combination $\mathcal{G}(\boldsymbol{x})/8\pi\mu$ [5]. The pressure field of the Oseen tensor we denote by $\mathcal{P}(\boldsymbol{x})$, and it is given by

$$\mathcal{P}_j(\boldsymbol{x}) = 2\mu \frac{x_j}{r^3} + \mathcal{P}_j^{\infty} \, . \tag{2.10}$$

The proof consists of two steps. We first show that for $\boldsymbol{x} \neq \mathbf{0}$, Equations 2.9 and 2.10 satisfy the Stokes equations, and then show that if the fluid domain V encloses the origin, the divergence of the solution stress field behaves as a delta function.

By direct differentiation, noticing that $\partial r/\partial x_k = x_k/r$, we have for $\boldsymbol{x} \neq 0$,

$$\mathcal{G}_{ij,k} = -\frac{1}{r^3}\delta_{ij}x_k + \frac{1}{r^3}(\delta_{ik}x_j + \delta_{jk}x_i) - \frac{3}{r^5}x_i x_j x_k \ , \tag{2.11}$$

$$\nabla^2 \mathcal{G}_{ij} = \mathcal{G}_{ij,kk} = \frac{2}{r^3}\delta_{ij} - \frac{6}{r^5}x_i x_j \ . \tag{2.12}$$

Notice that from Equation 2.11 we may also verify that the equation of continuity is satisfied. Simply replace the index k with i, so that

$$8\pi\mu\nabla\cdot\boldsymbol{v} = F_j\mathcal{G}_{ij,i} = F_j\left[-\frac{1}{r^3}\delta_{ij}x_i + \frac{1}{r^3}(\delta_{ii}x_j + \delta_{ji}x_i) - \frac{3}{r^5}x_i x_j x_i\right] = 0 \ ,$$

since $\delta_{ij}x_i = x_j$, $\delta_{ii} = 3$ and $x_i x_i = r^2$. This shows that $\boldsymbol{v}$ satisfies continuity for $\boldsymbol{x} \neq \mathbf{0}$. The integral of $\mathcal{G}(\boldsymbol{x})\cdot\boldsymbol{n}$ over a small sphere of radius ϵ about the origin is

$$\oint_S \left(\frac{1}{\epsilon}\delta_{ij} + \frac{1}{\epsilon}n_i n_j\right) n_j \, dS \ ,$$

which, taking into account the smallness of dS, behaves as $O(\epsilon)$ for small ϵ. Thus the origin does not contain a source or sink and continuity is satisfied everywhere.

For $\boldsymbol{x} \neq \mathbf{0}$, the Laplacian term may be written as

$$\nabla^2\mathcal{G}_{ij} = \frac{2}{r^3}\delta_{ij} - \frac{6}{r^5}x_i x_j \ , = -2\frac{\partial^2}{\partial x_i \partial x_j}\left(\frac{1}{r}\right) \ ,$$

so the Oseen tensor has the pressure field

$$\mathcal{P}_j = -2\mu\frac{\partial}{\partial x_j}\left(\frac{1}{r}\right) + \mathcal{P}_j^\infty = 2\mu\frac{x_j}{r^3} + \mathcal{P}_j^\infty$$

and

$$p = \frac{\boldsymbol{F}\cdot\boldsymbol{x}}{4\pi r^3} + p^\infty \ .$$

This completes the demonstration that for $\boldsymbol{x} \neq \mathbf{0}$, Equation 2.8 satisfies the Stokes equations.

Now consider the stress field, denoted by $\boldsymbol{F}\cdot\boldsymbol{\Sigma}$, of the Green's function. The stress field of the Oseen tensor is then the triadic $8\pi\mu\boldsymbol{\Sigma}$, and its expression follows directly from the above expression for $\mathcal{P}$ and $\nabla\mathcal{G}$ as

$$8\pi\mu\Sigma_{ijk} = -\mathcal{P}_j\delta_{ik} + \mu(\mathcal{G}_{ij,k} + \mathcal{G}_{kj,i}) = -6\mu\frac{x_i x_j x_k}{r^5} \ .$$

Consider the volume integral

$$\int_V \nabla\cdot\boldsymbol{\Sigma}\, dV$$

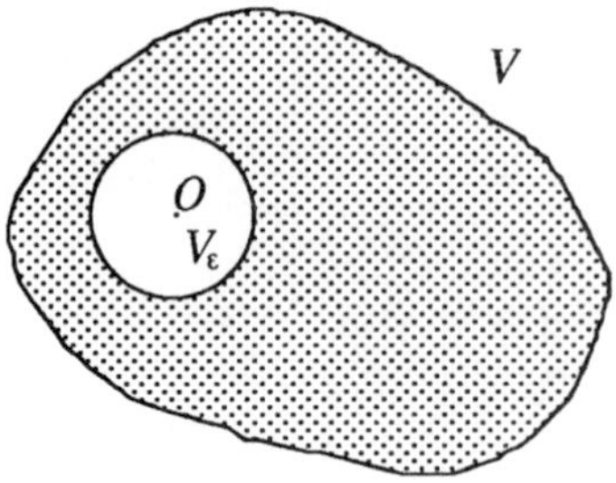

Figure 2.2: Breakdown of the fluid domain V into V_ϵ and $V - V_\epsilon$.

with V containing $\mathbf{0}$. Since $\nabla \cdot \boldsymbol{\Sigma} = \mathbf{0}$ everywhere except at the origin, the integral over V may be replaced with an integral over the volume of a sphere of radius ϵ, with ϵ arbitrarily small (see Figure 2.2). This volume integral can be converted to a surface integral using the divergence theorem and can be further reduced as follows:

$$F_j \oint_S \Sigma_{ijk} n_k \, dS(\boldsymbol{x}) = -\frac{3F_j}{4\pi} \oint_S \frac{n_i n_j n_k}{\epsilon^2} n_k \, dS = -F_i \; .$$

We have used $\boldsymbol{x} = \epsilon \boldsymbol{n}$ and the result (see Exercise 2.10) for the integral of $\boldsymbol{nn}$ over all orientations,

$$\int_0^{2\pi} \int_0^{\pi} n_i n_j \, \sin\theta \, d\theta \, d\phi = \frac{4\pi}{3} \delta_{ij} \; .$$

This completes the proof that $\boldsymbol{F} \cdot \mathcal{G}(\boldsymbol{x})/8\pi\mu$ is the Green's function for the Stokes equations.

Several comments concerning the Green's function are in order at this point. The term *Stokeslet* is often used for $\boldsymbol{F} \cdot \mathcal{G}(\boldsymbol{x})/8\pi\mu$. At other times, it is also called the *point-force solution*, the motivation for the terminology following naturally from the solution property, since the velocity field of the Green's function is force-free everywhere except at the origin, where it contains a force of strength $\boldsymbol{F}$. The analogy with the point charge solution in electrostatics is quite apparent. Finally, we may discard $\boldsymbol{F}$ from the problem statement, place the singular point at $\boldsymbol{\xi}$, and write the more general statements:

$$8\pi\mu \frac{\partial}{\partial x_k} \Sigma_{ijk}(\boldsymbol{x} - \boldsymbol{\xi}) = -\frac{\partial}{\partial x_i} \mathcal{P}_j(\boldsymbol{x} - \boldsymbol{\xi}) + \mu \nabla^2 \mathcal{G}_{ij} = -8\pi\mu \delta_{ij} \delta(\boldsymbol{x} - \boldsymbol{\xi}) \qquad (2.13)$$

$$\frac{\partial}{\partial x_i} \mathcal{G}_{ij}(\boldsymbol{x} - \boldsymbol{\xi}) = 0 \; . \qquad (2.14)$$

2.4.2 Integral Representation with Single and Double Layer Potentials

We now derive the integral representation for the velocity field. Consider the reciprocal theorem, Equation 2.6, with $\boldsymbol{v}_1$ replaced by $\mathcal{G}(\boldsymbol{\xi}-\boldsymbol{x})$ and $\boldsymbol{v}_2$ replaced by $\boldsymbol{v}$, the solution of the Stokes equations throughout a region V for which the integral representation is desired. Here, we fix $\boldsymbol{x}$ and use $\boldsymbol{\xi}$ as the integration variable in the reciprocal theorem. It is easy to show that the reciprocal theorem holds just as well for the dyadic component $\mathcal{G}_{i\ell}(\boldsymbol{\xi}-\boldsymbol{x})$, in place of the usual velocity component v_i. In fact, the index i undergoes the same set of steps, while the index ℓ is just carried along.

The reason for picking $\mathcal{G}(\boldsymbol{\xi}-\boldsymbol{x})$ as one of the two fields of the reciprocal relation becomes apparent when we recall that the divergence of its associated stress field is quite special: $\nabla_\xi \cdot \boldsymbol{\sigma}(\mathcal{G}) = -8\pi\mu\boldsymbol{\delta}\delta(\boldsymbol{\xi}-\boldsymbol{x})$. Equation 2.6 becomes

$$\begin{aligned}\oint_S \mathcal{G}_{i\ell}(\boldsymbol{\xi}-\boldsymbol{x})\,\sigma_{ik}(\boldsymbol{\xi})n_k\,dS(\boldsymbol{\xi}) &= 8\pi\mu\oint_S v_i(\boldsymbol{\xi})\,\Sigma_{i\ell k}(\boldsymbol{\xi}-\boldsymbol{x})n_k\,dS(\boldsymbol{\xi})\\ &+ 8\pi\mu\int_V v_i(\boldsymbol{\xi})\,\delta_{i\ell}\,\delta(\boldsymbol{\xi}-\boldsymbol{x})\,dV(\boldsymbol{\xi})\,,\end{aligned}$$

with the surface normal $\boldsymbol{n}$ *pointing out of the fluid region.* We may change the order of $\boldsymbol{\xi}$ and $\boldsymbol{x}$ in the arguments of $\mathcal{G}$ and $\boldsymbol{\Sigma}$ by recalling that $\mathcal{G}(\boldsymbol{\xi}-\boldsymbol{x}) = \mathcal{G}(\boldsymbol{x}-\boldsymbol{\xi})$ and $\boldsymbol{\Sigma}(\boldsymbol{\xi}-\boldsymbol{x}) = -\boldsymbol{\Sigma}(\boldsymbol{x}-\boldsymbol{\xi})$. The Kronecker delta leads to $v_i\delta_{i\ell} = v_\ell$ and finally, because of the properties of the Dirac delta function $\delta(\boldsymbol{\xi}-\boldsymbol{x})$, the volume integral simplifies to $\boldsymbol{v}(\boldsymbol{x})$ or zero, depending on whether $\boldsymbol{x}$ is inside or outside V. We simplify as just described and rearrange the preceding equation to obtain the following integral representation[3] for $\boldsymbol{v}$ (see the exercises for a corresponding representation of the associated pressure field):

$$\begin{aligned}\left.\begin{array}{ll}\text{if } \boldsymbol{x}\in V^o, & \boldsymbol{v}(\boldsymbol{x})\\ \text{if } \boldsymbol{x}\notin \bar{V}, & \boldsymbol{0}\end{array}\right\} &= -\oint_S [\boldsymbol{\sigma}(\boldsymbol{\xi})\cdot\hat{\boldsymbol{n}}(\boldsymbol{\xi})]\cdot\frac{\mathcal{G}(\boldsymbol{x}-\boldsymbol{\xi})}{8\pi\mu}\,dS(\boldsymbol{\xi})\\ &- \oint_S \boldsymbol{v}(\boldsymbol{\xi})\cdot\boldsymbol{\Sigma}(\boldsymbol{x}-\boldsymbol{\xi})\cdot\hat{\boldsymbol{n}}(\boldsymbol{\xi})\,dS(\boldsymbol{\xi})\,. \qquad (2.15)\end{aligned}$$

Here, the surface normal $\hat{\boldsymbol{n}}$ *points out of the particle and into the fluid region.*[4] *The integral representation is a significant statement — a Stokes velocity field may be reconstructed throughout a region V using only values of the velocity and traction fields on the boundary of V.*

The first term on the RHS is a velocity field generated by a distribution of surface forces of strength $\boldsymbol{\sigma}\cdot\hat{\boldsymbol{n}}\,dS$, since $\mathcal{G}(\boldsymbol{x}-\boldsymbol{\xi})$ is the velocity field generated by a point force at $\boldsymbol{\xi}$. By analogy with potential theory, this integral is called the *single layer potential* [24, 49] (the single layer of charges distributed over the surface of a electrical conductor is replaced by a single layer of forces).

---

[3]This integral representation says nothing (yet) about the boundary points $\boldsymbol{x}\in\bar{V}\setminus V = \partial V$, as V^o is the *interior* of V, explicitly excluding the boundary points, while $\bar{V}$ is the *closure* of V, explicitly including the boundary points.

[4]Throughout this book, we shall distinguish such surface normals from the generic brand by using a caret.

The second term is called the *double layer potential*, also by analogy with potential theory. In electrostatics, this term would correspond to a double layer of positive and negative charges separated by an infinitesimal gap, or equivalently, a surface distribution of electric dipoles. The structure of the hydrodynamic double layer is somewhat richer, and its physical interpretation is more involved. More explicitly, if we write

$$8\pi\mu v_j \Sigma_{jik}\hat{n}_k = \mathcal{P}_i \boldsymbol{v}\cdot\hat{\boldsymbol{n}} + \mu(\mathcal{G}_{ji,k} + \mathcal{G}_{ki,j})v_j\hat{n}_k \ ,$$

then the double layer distribution can be interpreted as a distribution of sources (or sinks) of strength $\boldsymbol{v}\cdot\hat{\boldsymbol{n}}$ and a true "double layer" of Stokeslets. This can also be deduced from the fact that $\boldsymbol{\Sigma}$ as a "velocity" field does not satisfy continuity, since $\nabla\cdot\boldsymbol{\Sigma}(\boldsymbol{x}-\boldsymbol{\xi}) = -\boldsymbol{\delta}\delta(\boldsymbol{x}-\boldsymbol{\xi})$, but $\nabla\mathcal{G}$ does, and so the $\mathcal{P}$-term must correspond to the sources and sinks. In fact, we see that

$$\oint_S \mathcal{P}_i\hat{n}_j\,dS = 2\mu\oint_S \frac{x_i\hat{n}_j}{r^3}\,dS = \frac{8}{3}\pi\mu\delta_{ij} \ .$$

The term $\mathcal{G}_{ji,k}v_j\hat{n}_k$ corresponds to opposing Stokeslets of dipole strength $\boldsymbol{v}$ displaced in the direction of $\hat{\boldsymbol{n}}$, while the term $\mathcal{G}_{ki,j}v_j\hat{n}_k$ corresponds to opposing Stokeslets of dipole strength $|\boldsymbol{v}|\hat{\boldsymbol{n}}$ displaced in the direction of $\boldsymbol{v}$. This symmetric placement of point forces implies that no net force or torque is exerted on the fluid. (We shall examine such structures in more detail in the next section.) Finally, despite the presence of sources and sinks over S, the total mass within S is conserved because the total source and sink strength is

$$\oint_S \boldsymbol{v}\cdot\hat{\boldsymbol{n}}\,dS = 0 \ .$$

The integral representation 2.15 gives the velocity of interest for a point $\boldsymbol{x}$ in the fluid region and zero outside. If we take the view that S is a virtual surface that encloses a subdomain V of a much larger region in which $\boldsymbol{v}$ is a meaningful Stokes velocity field, then the two surface integrals combine to give the correct result in the subdomain, but not outside. In fact, it fails catastrophically outside this subdomain; instead of the correct value for $\boldsymbol{v}$, we get zero. This jump discontinuity across the boundary must come from the double layer term, for we can show that the single layer potential is continuous. The kernel of the single layer potential, $\mathcal{G}(\boldsymbol{x}-\boldsymbol{\xi})$ behaves as $r_{\xi x}^{-1} = |\boldsymbol{x}-\boldsymbol{\xi}|^{-1}$ in the neighborhood of the point $\boldsymbol{\xi}$. In local polar coordinates about $\boldsymbol{\xi}$, $dS(\boldsymbol{\xi}) = r_{\xi x}dr_{\xi x}d\phi$, so the integrand of the single layer is nonsingular, and the velocity generated by the single layer distribution is continuous across the surface; letting $\boldsymbol{x}$ approach the surface from both sides, we simply get the limiting value that is exactly the value of the surface integral with $\boldsymbol{x}$ on S.

The double layer kernel, on the other hand, scales as $r_{\xi x}^{-2}$, so the integral exists as an improper integral for $\boldsymbol{x} = \boldsymbol{\eta}$ on S only if the dot product $\boldsymbol{n}(\boldsymbol{\xi})\cdot(\boldsymbol{\xi}-\boldsymbol{\eta})/r_{\xi\eta}$ vanishes as $\boldsymbol{\xi}$ approaches $\boldsymbol{\eta}$ along the surface. For "smooth" surfaces this is indeed the case; more explicitly, we say the surface is Lyapunov-smooth if $\boldsymbol{n}(\boldsymbol{\xi})\cdot(\boldsymbol{x}-\boldsymbol{\xi})/r_{\xi x} = O(r_{\xi x}^{\alpha})$ as $\boldsymbol{\xi}\to\boldsymbol{x}$, with $\alpha\in(0,1]$. Now as $\boldsymbol{x}$ approaches

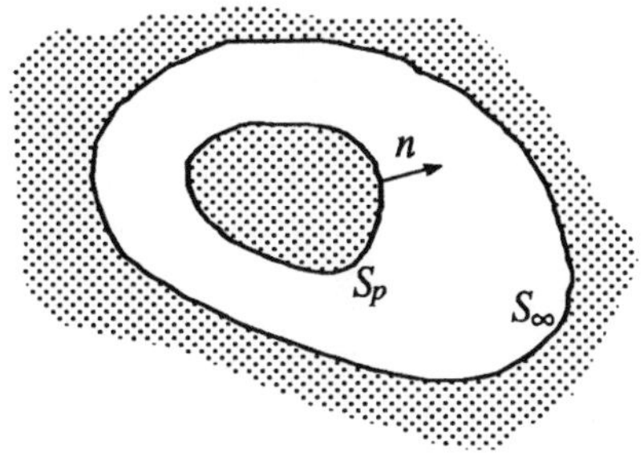

Figure 2.3: The region bounded by S_p and S_∞.

$\boldsymbol{\eta}$ from either side, the surface will locally look planar, and since the jump is due to the immediate neighborhood of $\boldsymbol{\eta}$, it can be claimed that by symmetry that the jump is equal in magnitude whether $\boldsymbol{x}$ comes to sit on the surface from the inside or the outside. This behavior is completely analogous to the jump in the potential as one passes through an electric double layer. Physically, we expect a jump in the *normal* component of the velocity field as we pass through a surface distribution of sources and sinks, since the velocity fields will be directed in opposite directions on the two sides. A surface distribution of symmetric force dipoles also causes a jump, but only in the *tangential* component of the velocity field. Indeed, since $\nabla \mathcal{G}$ is solenoidal, the standard "pillbox" argument shows that the normal component of the velocity produced by a true bilayer of Stokeslets is continuous across the surface. In Exercises 2.13 and 2.14 we ask the reader to evaluate the jump explicitly for some simple double layer distributions, where the integrals can be evaluated directly, and in Exercise 2.15 we examine jumps in the normal and tangential velocity components of the double layer potential.

2.4.3 Representation of Flows Outside a Rigid Particle

The integral representation takes on a special form involving only the single layer potential, for disturbance flows created by a particle undergoing rigid-body motion $\boldsymbol{U} + \boldsymbol{\omega} \times \boldsymbol{x}$ in an ambient flow field, $\boldsymbol{v}^\infty(\boldsymbol{x})$. The problem is meaningful only if $\boldsymbol{v}^\infty$ is a solution of the Stokes equation. For boundary conditions, we have

1. $\boldsymbol{v} = \boldsymbol{v}^{RBM} = \boldsymbol{U} + \boldsymbol{\omega} \times \boldsymbol{x}$ on the surface of the particle.

2. $\boldsymbol{v} = \boldsymbol{v}^\infty$ far away from the particle.

Define the disturbance field $\boldsymbol{v}^D = \boldsymbol{v} - \boldsymbol{v}^\infty$. Consider a fluid region, V_f, bounded by the particle surface and a large surface, S_∞, of dimension R, which we intend to expand by taking the limit of large R (see Figure 2.3). The integral representation for $\boldsymbol{v}^D$ is

$$\left.\begin{array}{ll}\text{if } \boldsymbol{x} \in V_f, & \boldsymbol{v}^D(\boldsymbol{x}) \\ \text{if } \boldsymbol{x} \in V_p, & \boldsymbol{0}\end{array}\right\} = -\frac{1}{8\pi\mu}\oint_{S_p}(\boldsymbol{\sigma}^D \cdot \hat{\boldsymbol{n}}) \cdot \mathcal{G}(\boldsymbol{x}-\boldsymbol{\xi})\, dS(\boldsymbol{\xi})$$
$$-\oint_{S_p}(\boldsymbol{v}^{RBM} - \boldsymbol{v}^{\infty}) \cdot \boldsymbol{\Sigma}(\boldsymbol{x}-\boldsymbol{\xi}) \cdot \hat{\boldsymbol{n}}\, dS(\boldsymbol{\xi})$$
$$-\frac{1}{8\pi\mu}\oint_{S_\infty}(\boldsymbol{\sigma}^D \cdot \hat{\boldsymbol{n}}) \cdot \mathcal{G}(\boldsymbol{x}-\boldsymbol{\xi})\, dS(\boldsymbol{\xi})$$
$$-\oint_{S_\infty}\boldsymbol{v}^D \cdot \boldsymbol{\Sigma}(\boldsymbol{x}-\boldsymbol{\xi}) \cdot \hat{\boldsymbol{n}}\, dS(\boldsymbol{\xi}) ,$$

where V_p denotes the region inside the particle, which of course is disjoint with V_f and $\hat{\boldsymbol{n}}$ points into V_f as before. We have also made use of the boundary condition on S_p. The integral over S_∞ will vanish in the limit of large R if $\boldsymbol{v}^D \to 0$ and $R\boldsymbol{\sigma}^D \to 0$, since $\mathcal{G}(\boldsymbol{x}-\boldsymbol{\xi}) \sim O(R^{-1})$, $\boldsymbol{\Sigma}(\boldsymbol{x}-\boldsymbol{\xi}) \sim O(R^{-2})$ and $dS \sim O(R^2)$. Now $\boldsymbol{v}^D \to 0$ is satisfied because of the problem statement. The other requirement also holds, since for large R we expect $\boldsymbol{\sigma}^D$ to scale as $\nabla\boldsymbol{v}^D$ and $O(\nabla\boldsymbol{v}^D) \sim O(\boldsymbol{v}^D)/R$.

The double layer potential in the integral over S_p may be eliminated as follows. Apply the velocity representation theorem to the field $\boldsymbol{U} + \boldsymbol{\omega} \times \boldsymbol{x} - \boldsymbol{v}^{\infty}$ *inside the particle*:

$$\left.\begin{array}{ll}\text{if } \boldsymbol{x} \in V, & \boldsymbol{0} \\ \text{if } \boldsymbol{x} \in V_p, & \boldsymbol{v}^{RBM} - \boldsymbol{v}^{\infty}(\boldsymbol{x})\end{array}\right\} = -\frac{1}{8\pi\mu}\oint_{S_p}[(\boldsymbol{\sigma}^{RBM} - \boldsymbol{\sigma}^{\infty}) \cdot \boldsymbol{n}] \cdot \mathcal{G}(\boldsymbol{x}-\boldsymbol{\xi})\, dS(\boldsymbol{\xi})$$
$$-\oint_{S_p}(\boldsymbol{v}^{RBM} - \boldsymbol{v}^{\infty}) \cdot \boldsymbol{\Sigma}(\boldsymbol{x}-\boldsymbol{\xi}) \cdot \boldsymbol{n}\, dS(\boldsymbol{\xi}) ,$$

with the surface normal $\boldsymbol{n}$ *now pointing into* V_p, *the region of interest*, and $\boldsymbol{\sigma}^{\infty}$ denoting the stress field of the ambient velocity field. The surface traction for rigid-body motion $\boldsymbol{\sigma}^{RBM} \cdot \boldsymbol{n}$ is just a constant pressure, $-p_0\boldsymbol{n}$, of no dynamic significance and contributes nothing to the integral. In fact, note that the integral of $\boldsymbol{n} \cdot \mathcal{G}$ over the surface of the particle is identically zero, since $\mathcal{G}$ is solenoidal.

We add the representations for $\boldsymbol{v}^D$ and $\boldsymbol{v}^{RBM} - \boldsymbol{v}^{\infty}$, keeping in mind $\hat{\boldsymbol{n}} = -\boldsymbol{n}$. The double layer terms cancel, and since $\boldsymbol{\sigma}^{\infty} + \boldsymbol{\sigma}^D = \boldsymbol{\sigma}$, and $\boldsymbol{v}^D = \boldsymbol{v} - \boldsymbol{v}^{\infty}$, we obtain the result

$$\boldsymbol{v}(\boldsymbol{x}) = \boldsymbol{v}^{\infty}(\boldsymbol{x}) - \frac{1}{8\pi\mu}\oint_{S_p}(\boldsymbol{\sigma}(\boldsymbol{\xi}) \cdot \boldsymbol{n}) \cdot \mathcal{G}(\boldsymbol{x}-\boldsymbol{\xi})\, dS(\boldsymbol{\xi}) , \qquad (2.16)$$

which shows that flows past a rigid particle can be represented with just single layer potentials. This is consistent with our intuition that the disturbance caused by the relative motion of a rigid body through a viscous fluid can be represented by a collection of point forces imparted to the fluid at the particle surface, and what initially appeared to be a collection of local sources, sinks, and force dipoles add up over the entire particle surface to simply rigid-body motions and a reconstruction of the ambient field. This simple result does not hold for the disturbance velocity field produced by the relative motion of a viscous drop. In the next section, we show how this special representation leads to the multipole expansion, a universal form for the disturbance flow field far away from the particle.

2.5 The Multipole Expansion

There is a one-to-one correspondence between the multipole expansion of hydrodynamics and the more familiar form from electrostatics. At great distances from the particle, *i.e.*, $|\boldsymbol{x}| \gg |\boldsymbol{\xi}|$, we cannot distinguish between the points $\boldsymbol{\xi}$ on the surface of the particle and a reference origin $\mathbf{0}$ located at a convenient point inside the particle, so that $\mathcal{G}(\boldsymbol{x}-\boldsymbol{\xi}) \sim \mathcal{G}(\boldsymbol{x})$ may be moved outside the integral in the single layer potential. The integral of what remains, $\boldsymbol{\sigma}\cdot\hat{\boldsymbol{n}}$, is simply the hydrodynamic drag on the particle, so that the single layer potential simplifies to

$$\boldsymbol{v}^D(\boldsymbol{x}) \sim -\boldsymbol{F}\cdot\frac{\mathcal{G}(\boldsymbol{x})}{8\pi\mu} ,$$

independent of the details of the particle shape. This is analogous to the single layer potential of electrostatics reducing to the field of a point charge far away from the conductor. The geometry and the analogy between hydrodynamics and electrostatics is illustrated in Figure 2.4. We may obtain higher order corrections for the disturbance field in which the coefficients are moments of the surface traction $\boldsymbol{\sigma}\cdot\hat{\boldsymbol{n}}$ and thus obtain the multipole expansion. Formally, we set $|\boldsymbol{x}| \gg |\xi|$ and take the Taylor series[5] of $\mathcal{G}(\boldsymbol{x}-\boldsymbol{\xi})$ in $\boldsymbol{\xi}$ about $\boldsymbol{\xi}=\mathbf{0}$,

$$\begin{aligned}
\mathcal{G}_{ij}(\boldsymbol{x}-\boldsymbol{\xi}) &= \sum_{n=0}^{\infty}\frac{1}{n!}(\boldsymbol{\xi}_c\cdot\nabla_{\xi})^n\mathcal{G}_{ij}(\boldsymbol{x}-\boldsymbol{\xi})|_{\xi=0} \\
&= \sum_{n=0}^{\infty}\frac{(-1)^n}{n!}(\boldsymbol{\xi}\cdot\nabla)^n\mathcal{G}_{ij}(\boldsymbol{x}) \\
&= \sum_{n=0}^{\infty}\frac{(-1)^n}{n!}\xi_{k_1}\cdots\xi_{k_n}\mathcal{G}_{ij,k_1\ldots k_n}(\boldsymbol{x}) ,
\end{aligned}$$

and insert the expansion into the velocity representation. The result is

$$\begin{aligned}
v_i(\boldsymbol{x}) - v_i^{\infty}(\boldsymbol{x}) &= -\frac{1}{8\pi\mu}\sum_{n=0}^{\infty}\frac{(-1)^n}{n!}\oint_{S_p}[(\boldsymbol{\sigma}\cdot\hat{\boldsymbol{n}})_j\xi_{k_1}\cdots\xi_{k_n}]\,dS\;\mathcal{G}_{ij,k_1\ldots k_n}(\boldsymbol{x}) \\
&= -\frac{F_j}{8\pi\mu}\mathcal{G}_{ij}(\boldsymbol{x}) + \frac{D_{jk}}{8\pi\mu}\mathcal{G}_{ij,k}(\boldsymbol{x}) + \cdots ,
\end{aligned}$$

with

$$\begin{aligned}
F_j &= \oint_{S_p}(\boldsymbol{\sigma}\cdot\hat{\boldsymbol{n}})_j\,dS , \\
D_{jk} &= \oint_{S_p}(\boldsymbol{\sigma}\cdot\hat{\boldsymbol{n}})_j\xi_k\,dS .
\end{aligned}$$

[5]The subscript c is used to indicate that the variable in question is considered a constant on carrying out differentiations, to make the first sum unambiguous.

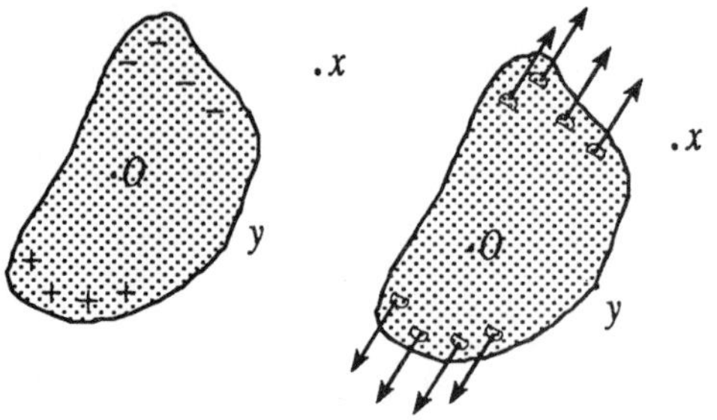

Figure 2.4: The multipole expansion for the electrostatic potential and the velocity field.

Thus in the far field, regardless of the details of the particle shape, all particle disturbance fields exhibit certain common features. The leading term, or monopole, will be a Stokeslet with coefficient $\boldsymbol{F}$ equal to the force exerted by the fluid on the particle. This field, decaying as $|\boldsymbol{x}|^{-1}$ away from the particle, will be present if and only if the particle and fluid exert a net force on each other. The next effect is a force dipole $\boldsymbol{D}$, a second-order tensor with a field that decays as $|\boldsymbol{x}|^{-2}$ away from the particle. Note that for Stokes flow the tensorial rank of the moment coefficients is one greater than the corresponding quantities in electrostatics, because the surface density $\boldsymbol{\sigma} \cdot \hat{\boldsymbol{n}}$ is a vector, whereas in electrostatics the surface charge density is a scalar.

The isotropic portion of $\boldsymbol{D}$, shown and subtracted in the equation below, is of no dynamic significance, since $\nabla \cdot \mathcal{G} = 0$. We split the rest into a symmetric part called the *stresslet* and an antisymmetric part,

$$D_{jk} - \frac{1}{3} D_{ii}\delta_{jk} = S_{jk} + T_{jk} \ ,$$

with

$$\begin{aligned} S_{jk} &= \frac{1}{2} \oint_S [(\boldsymbol{\sigma} \cdot \hat{\boldsymbol{n}})_j \xi_k + (\boldsymbol{\sigma} \cdot \hat{\boldsymbol{n}})_k \xi_j] \, dS - \frac{1}{3} \oint_S (\boldsymbol{\sigma} \cdot \hat{\boldsymbol{n}}) \cdot \boldsymbol{\xi} \, dS \ \delta_{jk} \ . \\ T_{jk} &= \frac{1}{2} \oint_S [(\boldsymbol{\sigma} \cdot \boldsymbol{n})_j \xi_k - (\boldsymbol{\sigma} \cdot \boldsymbol{n})_k \xi_j] \, dS \ . \end{aligned}$$

The antisymmetric portion may be identified with the hydrodynamic torque on the particle. Explicitly, we define a relation[6] between the components of an antisymmetric tensor and a pseudovector, $\boldsymbol{T}$,

$$-\epsilon_{ijk} T_{jk} = -\epsilon_{ijk} \oint_S (\boldsymbol{\sigma} \cdot \hat{\boldsymbol{n}})_j \xi_k \, dS = T_i \ ,$$

[6]The alternating tensor $\boldsymbol{\epsilon}$ can be defined by the relationship $\boldsymbol{a} \times \boldsymbol{b} = \epsilon_{ijk} a_j b_k$ being valid for arbitrary vectors $\boldsymbol{a}$ and $\boldsymbol{b}$, or by $\boldsymbol{\epsilon} = \boldsymbol{\delta} \times \boldsymbol{\delta}$.

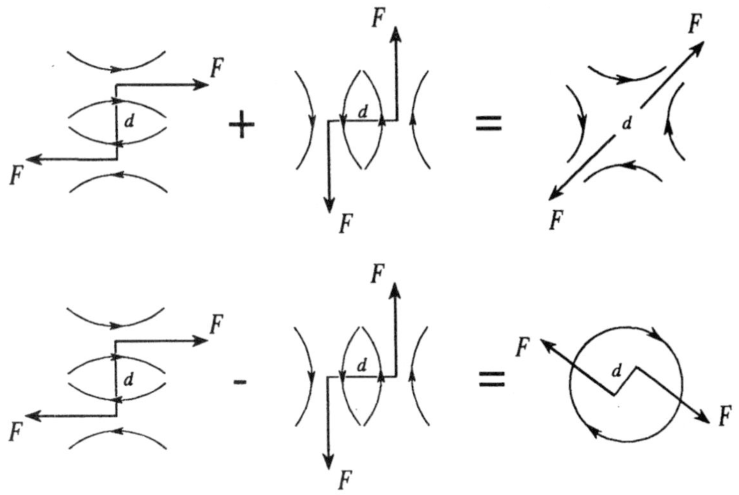

Figure 2.5: The stresslet and rotlet fields

which identifies $\boldsymbol{T}$ as the hydrodynamic torque exerted by the fluid on the particle. Equivalently, we may obtain the antisymmetric tensor components from the torque as

$$T_{jk} = -\frac{1}{2}\epsilon_{jk\ell}T_\ell \ .$$

The corresponding term in the multipole expansion may be written as

$$T_{jk}\mathcal{G}_{ij,k} = -\frac{1}{2}\epsilon_{jk\ell}T_\ell\mathcal{G}_{ij,k} = \frac{1}{2}(\boldsymbol{T}\times\nabla)_j\mathcal{G}_{ij} = \frac{1}{2}\boldsymbol{T}\cdot(\nabla\times\mathcal{G}) \ .$$

The antisymmetric dipole field can also be interpreted as the field of a point torque (also known as the *rotlet*) at the origin, since the torque on any surface enclosing the origin is precisely $\boldsymbol{T}$. This can also be seen by constructing a couple formed by two opposing Stokeslets of strength $\boldsymbol{F}$, separated by a displacement $\boldsymbol{d}$ with $\boldsymbol{d}\cdot\boldsymbol{F} = 0$ (see Figure 2.5). As we shrink the displacement, keeping $|\boldsymbol{d}||\boldsymbol{F}|$ constant, the antisymmetric portion approaches $\frac{1}{2}(F_jd_k - F_kd_j)\mathcal{G}_{ij,k}$.

The stresslet and symmetric dipole field are intimately connected with straining motions. Again referring to Figure 2.5, the field

$$\left[\frac{1}{2}(F_jd_k + F_kd_j) - \frac{1}{3}\boldsymbol{F}\cdot\boldsymbol{d}\,\delta_{jk}\right]\mathcal{G}_{ij,k}$$

produces a straining motion induced by Stokeslets oriented and displaced in the principal directions of the matrix $(F_jd_k + F_kd_j) - \frac{2}{3}\boldsymbol{F}\cdot\boldsymbol{d}\,\delta_{jk}$, as can be seen by a coordinate rotation, which transforms this matrix into diagonal form (in Figure 2.5, $\boldsymbol{F} \perp \boldsymbol{d}$ so the principal directions are at 45° between $\boldsymbol{F}$ and $\boldsymbol{d}$). We may draw the same conclusion starting with S_{jk}. (**Note:** Is the stresslet "transmitted" unchanged by Stokes flow? See Exercise 2.3).

A rigid particle in the ambient field $\boldsymbol{v}^{\infty}(\boldsymbol{x})$ may undergo a rigid body motion $\boldsymbol{U} + \boldsymbol{\omega} \times (\boldsymbol{x} - \boldsymbol{x}_0)$ that matches the ambient velocity and vorticity at $\boldsymbol{x}_0$, so that $\boldsymbol{F} = \boldsymbol{T} = \boldsymbol{0}$. Such particles produce a fairly weak disturbance of the ambient flow. However, rigid particles do not possess a mechanism for relieving the local straining motion in the fluid, hence in a general ambient field the stresslet cannot be nonzero. Thus in an arbitrary flow field, even a force-free and torque-free particle will produce a disturbance that decays as $|\boldsymbol{x}|^{-2}$ induced by the symmetric force dipole. This ultimately shows up as an increase in the rate of viscous dissipation of mechanical energy, and thus a suspension of rigid particles has an effective viscosity greater than that of the pure solvent. Indeed, the next example shows that the particle contribution to the effective stress in the suspension is connected intimately with the particle stresslets.

Example 2.1 Effective Stress in a Suspension of Rigid Particles [3, 9]

We start with the assumption that the macroscopic, observed stress tensor, *i.e.*, the effective stress in the suspension, is the ensemble average of the stress distribution in all realizations of the suspension. For homogeneous suspensions, this ensemble average is equivalent to a volume average over a volume that is large enough to contain a statistically significant number of particles, but is smaller than the scale of variations of interest in the macroscopic system. The formal expression for the effective stress, $\boldsymbol{\sigma}^{\text{eff}}$, can then be written as

$$\begin{aligned} \sigma_{ij}^{\text{eff}} &= \frac{1}{V}\int_V \sigma_{ij}\, dV \\ &= \frac{1}{V}\int_{V-\Sigma V_n} (-p\delta_{ij} + 2\mu e_{ij})\, dV + \sum_n \int_{V_n} \sigma_{ij}\, dV \ , \end{aligned}$$

where the volume integral has been decomposed into the portion over the fluid domain in which we may use the constitutive equation for the fluid stress and the portions over each rigid particle. The stresses in the rigid particle are indeterminate, but we may use the following relation,

$$\begin{aligned} \int_{V_n} \sigma_{ij}\, dV &= \int_{V_n} \frac{\partial}{\partial x_k}(\sigma_{ik}x_j)\, dV - \int_{V_n} \frac{\partial \sigma_{ik}}{\partial x_k} x_j\, dV \\ &= \oint_{S_n} \sigma_{ik}\hat{n}_k x_j\, dS - \int_{V_n} \frac{\partial \sigma_{ik}}{\partial x_k} x_j\, dV \ , \end{aligned}$$

to rewrite the stresses in terms of the surface tractions plus the first (volume) moment of $\nabla \cdot \boldsymbol{\sigma}$. If the particles are force-free, $\nabla \cdot \boldsymbol{\sigma} = \boldsymbol{0}$ and the last term is dropped.

The integral of $\boldsymbol{e}$ over the fluid region may be rewritten as follows:

$$\int_{V-V_n} 2e_{ij}\, dV = \int_V 2e_{ij}\, dV - \sum_n \int_{V_n} 2e_{ij}\, dV$$

$$= \int_V 2e_{ij}\,dV - \sum_n \oint_{S_n} (v_i\hat{n}_j + v_j\hat{n}_i)\,dS \ ,$$

so that

$$\frac{1}{V}\int_{V-V_n} 2e_{ij}\,dV = 2 < e_{ij} > \ - \frac{1}{V}\sum_n \int_{S_n} (v_i\hat{n}_j + v_j\hat{n}_i)\,dS \ .$$

Combining these results together, we find that

$$\sigma_{ij}^{\text{eff}} = -p^{\text{eff}}\delta_{ij} + 2\mu < e_{ij} > +\sigma_{ij}^{p} \ ,$$

with the isotropic part of the stress lumped into an effective pressure and in which the particle contribution to the stress is given by

$$\sigma_{ij}^{p} = \frac{1}{V}\sum_n \oint_{S_n} [\sigma_{ik}\hat{n}_k x_j - \mu(v_i\hat{n}_j + v_j\hat{n}_i)]\ dS \ .$$

For rigid particles, the velocity at the particle surface is a rigid-body motion so the velocity terms in the integral vanish identically. The rigid particles' contributions to the stress are given by the force dipoles from each particle in the volume.

We may examine the antisymmetric and symmetric parts separately to shed more light on the nature of the particle contribution to the suspension stress. The antisymmetric part of σ^p is seen to be directly related to the torque on each particle; thus the suspension stress contains an antisymmetric portion if and only if couples are generated within each particle. Otherwise, the suspension stress is symmetric, with the extra "particle" contributions coming directly from the sum of the stresslets of each particle in the volume. After this discussion of the role of the stresslet in suspension rheology, it should come as no surprise that much of the theoretical investigations in suspension rheology centers around computations of $\boldsymbol{S}$ as a function of system configuration. ◇

Exercises

Exercise 2.1 Zero Rate-of-Strain Fields.
If the rate-of-strain tensor $\boldsymbol{e}$ is zero throughout a volume of fluid V, show that the velocity field is at most a rigid-body motion throughout V.
Hint: Starting with $v_{i,j} = -v_{j,i}$, show that $v_{i,jk} = -v_{j,ki}$, which then implies that $v_{i,jk} = 0$ throughout V, so that $\boldsymbol{v}$ is linear in $\boldsymbol{x}$. Since $\boldsymbol{e} = \boldsymbol{0}$, show that this linear field can only be a uniform translation plus a solid-body rotation.

Exercise 2.2 Transmission of Force and Torque.
Using the reciprocal theorem, show that Stokes flow "transmits" unchanged the total force and torque from an inner closed surface to an outer enclosing surface.
Hint: Consider the fluid region between the surfaces and choose for $\boldsymbol{v}_2$ a rigid-body motion.

Exercise 2.3 Transmission of the Stresslet.
Does Stokes flow "transmit" unchanged the stresslet on an inner closed surface to an outer enclosing surface?
Hint: Consider the fluid region between the surfaces and choose $\boldsymbol{v}_2$ equal to $\boldsymbol{e} \cdot \boldsymbol{x}$, with a constant rate-of-strain, $\boldsymbol{e}$.

Exercise 2.4 Solenoidal Fields with Uniform Translations.
In this and two subsequent exercises, we illustrate the use of the minimum and maximum energy dissipation principles to obtain bounds as in Hill and Power [18]. Our first task is to find solenoidal trial solutions that satisfy the boundary condition, $\boldsymbol{v} = \boldsymbol{U}$. Consider the following trial solution, consisting of a Stokeslet and a field that decays as r^{-n}:

$$\boldsymbol{v} = \boldsymbol{U}\left(C\frac{a}{r} + (1-C)\frac{a^n}{r^n}\right) + C\frac{\boldsymbol{U}\cdot\boldsymbol{xx}}{a^2}\left(\frac{a^3}{r^3} - \frac{a^{n+2}}{r^{n+2}}\right) .$$

By inspection, we see that $\boldsymbol{v} = \boldsymbol{U}$ on the surface of a sphere of radius a. Since the Stokeslet is solenoidal, the trial solution will also be solenoidal if and only if the $O(r^{-n})$ field is solenoidal. Show that this requires $C = n/(2n-2)$. Only the case $n = 2$ is used in the example in Hill and Power.

Exercise 2.5 Upper Bounds.
Evaluate energy dissipation rates of the solenoidal fields, $\boldsymbol{v}^{(n)}$, of Exercise 2.4,

$$\boldsymbol{v}^{(n)} = \boldsymbol{U}\left(\frac{n}{2(n-1)}\frac{a}{r} + \frac{n-2}{2(n-1)}\frac{a^n}{r^n}\right) + \frac{\boldsymbol{U}\cdot\boldsymbol{xx}}{a^2}\frac{n}{2(n-1)}\left(\frac{a^3}{r^3} - \frac{a^{n+2}}{r^{n+2}}\right) .$$

With the notation, $E^{(n)} = E(\boldsymbol{v}^{(n)})$, your answers for $n = 2$ and $n = 3$ should be $E^{(2)} = \frac{56}{9}\pi\mu a U^2$ and $E^{(3)} = 6\pi\mu a U^2$. Show that $E^{(3)}$ is the least upper bound from this family of trial solutions. We will see in Chapter 3 that the case $n = 3$ is in fact the exact solution for the translating sphere.

Exercise 2.6 Lower Bounds.
Show that the traction field

$$\frac{a}{\mu}\sigma'_{ij} = \alpha\frac{U_k x_k x_i x_j}{a^3}\frac{a^5}{r^5} + \beta\left(2\frac{U_k x_k x_i x_j}{a^3}\frac{a^6}{r^6} - \frac{U_i x_j + U_j x_i}{2a}\frac{a^4}{r^4}\right)$$

satisfies $\nabla \cdot \boldsymbol{\sigma}' = \boldsymbol{0}$, and thus the condition of the dissipation theorem. Here, α and β are constants that we will use later on to tighten the bound from below. Show that

$$\begin{aligned}
\boldsymbol{F}' &= \oint_S \boldsymbol{\sigma}\cdot\boldsymbol{n}\, dS = \frac{4}{3}\pi\mu a\boldsymbol{U}\alpha \\
E' &= \frac{4}{9}\pi\mu a U^2\left(\alpha^2 + \alpha\beta + \frac{5}{6}\beta^2\right) \\
2\boldsymbol{F}'\cdot\boldsymbol{U}' - E' &= \frac{4}{9}\pi\mu a U^2\left(6\alpha - (\alpha^2 + \alpha\beta + \frac{5}{6}\beta^2)\right) \\
&= \frac{4}{9}\pi\mu a U^2\left(\frac{90}{7} - (\alpha + \frac{\beta}{2} - 3)^2 - \frac{7}{12}(\beta + \frac{18}{7})^2\right) .
\end{aligned}$$

The greatest lower bound $2\boldsymbol{F}' \cdot \boldsymbol{U}' - E' = (40/7)\pi\mu a U^2$ occurs for $\alpha = 30/7$, $\beta = -18/7$. In the original work of Hill and Power [18], the upper bound (Exercise 2.5) was set with $n = 2$ so that the bounds for the force on a translating sphere are given as

$$\frac{40}{7}\pi\mu a U \leq |F| \leq \frac{56}{9}\pi\mu a U .$$

In Chapter 3, we will derive the exact result, $F = 6\pi\mu a U$.

Exercise 2.7 Energy Dissipation Rates for the Translating Sphere.
In Chapter 3, we will show that the velocity field generated by a translating sphere is

$$\boldsymbol{u} = \boldsymbol{U}\left(\frac{3a}{4r} + \frac{a^3}{4r^3}\right) + \frac{\boldsymbol{U}\cdot\boldsymbol{x}\boldsymbol{x}}{r^2}\left(\frac{3a}{4r} - \frac{3a^3}{4r^3}\right) ,$$

where a is the sphere radius. Consider $\boldsymbol{v}^* = (1-C)\boldsymbol{u} + C\boldsymbol{v}^{(2)}$, with $\boldsymbol{v}^{(2)}$ as defined in Exercise 2.5. Evaluate $E(\boldsymbol{v}^*)$ and show that it is given by

$$E^* = E(\boldsymbol{u}) + \{E(\boldsymbol{v}) - E(\boldsymbol{u})\}\, C^2 = 6\pi\mu a U^2 + \frac{2}{9}\pi\mu a U^2 C^2 .$$

Note that $E(\boldsymbol{v}) > E(\boldsymbol{u})$, in accordance with the minimum energy dissipation principle for Stokes flow.

Exercise 2.8 Translating Sphere Shrunk to a Point.
Calculate the tractions on the translating sphere (see previous exercise), and integrate to get the total force. Keeping the force constant, let the sphere shrink to a point. What happens to the translation velocity and rate of energy dissipation? Do you recognize the resulting flow field?

Exercise 2.9 Derivation of the Oseen Tensor by Fourier Transform.
Define the Fourier transform pair,

$$\begin{aligned} \hat{\boldsymbol{v}}(\boldsymbol{k}) &= \left(\frac{1}{2\pi}\right)^{3/2} \int e^{i\boldsymbol{k}\cdot\boldsymbol{x}}\, \boldsymbol{v}(\boldsymbol{x})\, dV(\boldsymbol{x}) \\ \boldsymbol{v}(\boldsymbol{x}) &= \left(\frac{1}{2\pi}\right)^{3/2} \int e^{-i\boldsymbol{k}\cdot\boldsymbol{x}}\, \hat{\boldsymbol{v}}(\boldsymbol{k})\, dV(\boldsymbol{k}) , \end{aligned}$$

and apply the Fourier tranform to Equation 2.7. The result is (with F_j eliminated)

$$ik_i\frac{\hat{\mathcal{P}}_j}{8\pi\mu} - \frac{k^2}{8\pi}\hat{\mathcal{G}}_{ij} = -\delta_{ij}\left(\frac{1}{2\pi}\right)^{3/2} \qquad k_i\hat{\mathcal{G}}_{ij} = 0$$

Take the dot product with k_i in the transformed equation of motion to eliminate $\hat{\mathcal{G}}$ and show that this gives

$$\frac{\hat{\mathcal{P}}_j}{8\pi\mu} = ik_j\left(\frac{1}{2\pi}\right)^{3/2} k^{-2} .$$

Show that this leads to the correct result for the pressure,

$$\mathcal{P}_j(\boldsymbol{x}) = -2\mu\frac{\partial r^{-1}}{\partial x_j} \ .$$

Hint: First show by direct integration in $\boldsymbol{k}$-space, with $\boldsymbol{x} = (0,0,x_3)$, that

$$\mathcal{F}\left\{\frac{1}{4\pi r}\right\} = \left(\frac{1}{2\pi}\right)^{3/2} k^{-2} \ ,$$

and then apply the rule for transforming derivatives. Note that the preceding inverse transform constitutes the formal derivation of the Green's function for the Laplace equation.

Eliminate $\hat{\mathcal{P}}$ from the transformed equation of motion and show that

$$\frac{\hat{\mathcal{G}}_{ij}}{8\pi} = \delta_{ij}\left(\frac{1}{2\pi}\right)^{3/2} k^{-2} - \left(\frac{1}{2\pi}\right)^{3/2}\frac{k_i k_j}{k^4} \ .$$

Hint: Once again, the first term on the RHS is the transform of the Green's function for the Laplace equation, and after inversion, contributes $(4\pi r)^{-1}\delta_{ij}$. What remains is to show that

$$\mathcal{F}^{-1}\left\{\left(\frac{1}{2\pi}\right)^{3/2}\frac{k_i k_j}{k^4}\right\} = \frac{\delta_{ij}}{8\pi r} - \frac{x_i x_j}{8\pi r^3} \ .$$

Using symmetry arguments, deduce that the inverse tranform must be of the form

$$\mathcal{F}^{-1}\left\{\left(\frac{1}{2\pi}\right)^{3/2}\frac{k_i k_j}{k^4}\right\} = C_1\frac{\delta_{ij}}{8\pi r} + C_2\frac{x_i x_j}{8\pi r^3} \ ,$$

where C_1 and C_2 are numerical constants. Take the trace of both sides to get one condition, $3C_1 + C_2 = 2$. Show by direct integration in $\boldsymbol{k}$-space that for $i = j = 3$, and with $\boldsymbol{x} = (0,0,x_3)$, the equation yields another necessary condition, $C_1 + C_2 = 0$. The two equations for C_1 and C_2 imply that $C_1 = -C_2 = 1$ and this completes the formal derivation of the Oseen tensor.

The following definite integrals will be useful in this exercise:

$$\begin{aligned}
\int_0^\infty \frac{\sin k}{k}dk &= \frac{\pi}{2} \\
\int_0^\infty \frac{1-\cos k}{k^2}dk &= \frac{\pi}{2} \\
\int_0^\infty \frac{k-\sin k}{k^3}dk &= \frac{\pi}{4} \ .
\end{aligned}$$

The second and third integrals actually follow from the first (integration by parts).

Exercise 2.10 Angular Integration of the Dyadic $\boldsymbol{nn}$.
Show by direct integration in spherical polar coordinates that

$$\int_0^{2\pi}\int_0^{\pi} n_i n_j \sin\theta \, d\theta \, d\phi = \frac{4\pi}{3}\delta_{ij} \ .$$

Alternate derivations: Since the integral is isotropic, it must equal $C\delta_{ij}$. The constant C can be evaluated by considering the trace of the tensor. You could also use the divergence theorem on a unit sphere ($\boldsymbol{r} = \boldsymbol{n}$) in the form $\oint dS\, \boldsymbol{n}\boldsymbol{r} = \int dV\, \nabla\boldsymbol{r} = \ldots$

Exercise 2.11 Constant Single Layer Density on a Circular Disk. Consider a *constant* single layer density $\boldsymbol{\psi}$ distributed over a circular disk of radius b. Show by direct integration in cylindrical coordinates that the velocity field generated by this distribution is continuous passing through the disk. Are the tractions also continuous?

Exercise 2.12 A Special Single Layer Distribution on a Disk. Consider the single layer density

$$\boldsymbol{\psi}(\boldsymbol{\xi}) = \frac{\boldsymbol{F}}{2\pi b\sqrt{b^2 - (\xi_1^2 + \xi_2^2)}}$$

distributed over a circular disk of radius b. Show by direct integration in cylindrical coordinates that the velocity field generated by this distribution is continuous passing through the disk, and that in fact,

$$\begin{aligned} &\lim_{z\to 0+}\left\{-\iint \boldsymbol{\psi}(\boldsymbol{\xi})\cdot\frac{\mathcal{G}(\boldsymbol{x}-\boldsymbol{\xi})}{8\pi\mu}\,d\xi_1\,d\xi_2\right\} \\ &= \lim_{z\to 0-}\left\{-\iint \boldsymbol{\psi}(\boldsymbol{\xi})\cdot\frac{\mathcal{G}(\boldsymbol{x}-\boldsymbol{\xi})}{8\pi\mu}\,d\xi_1\,d\xi_2\right\} \\ &= \begin{cases} \frac{F}{16\mu b} & \text{if } \boldsymbol{F} \parallel \boldsymbol{e}_z, \\ \frac{3F}{32\mu b} & \text{if } \boldsymbol{F} \perp \boldsymbol{e}_z. \end{cases} \end{aligned}$$

This shows that our special choice for $\boldsymbol{\psi}(\boldsymbol{\xi})$ leads to the velocity field generated by a circular disk in uniform translation, and that F is the drag on the disk. Note that the tractions are singular at the rim, but the singularity is an integrable one and the hydrodynamic drag is finite.

Exercise 2.13 Jump in the Double Layer Potential with Constant Density on a Plane.
Consider a *constant* double layer density $\boldsymbol{\varphi}$ distributed over a coordinate plane (see Figure 2.6). Show by direct integration in cylindrical coordinates that at $\boldsymbol{x} = (x, y, z)$ the velocity field generated by the double layer distribution is

$$-\int_{-\infty}^{\infty}\int_{-\infty}^{\infty}(\mathrm{e}_3\cdot\boldsymbol{\Sigma}(\boldsymbol{x},\boldsymbol{\xi}))\cdot\boldsymbol{\varphi}\,d\xi_1 d\xi_2 = \begin{cases} \boldsymbol{\varphi}/2 & \text{if } z > 0 \\ \mathbf{0} & \text{if } z = 0 \\ -\boldsymbol{\varphi}/2 & \text{if } z < 0. \end{cases}$$

Note that the jump across the surface is equal to $|\boldsymbol{\varphi}|$ and that the jumps from the surface to either side are equal.
Hint: Without loss of generality, put $\boldsymbol{x}$ on the z-axis.

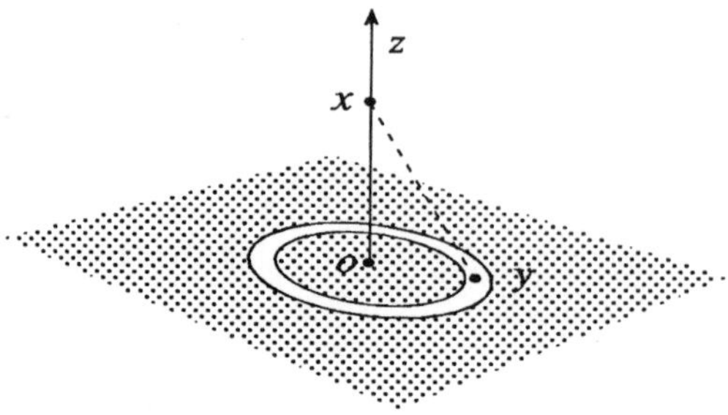

Figure 2.6: The geometry of the double layer distribution over an infinite plane.

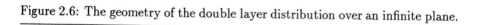

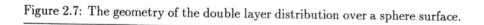

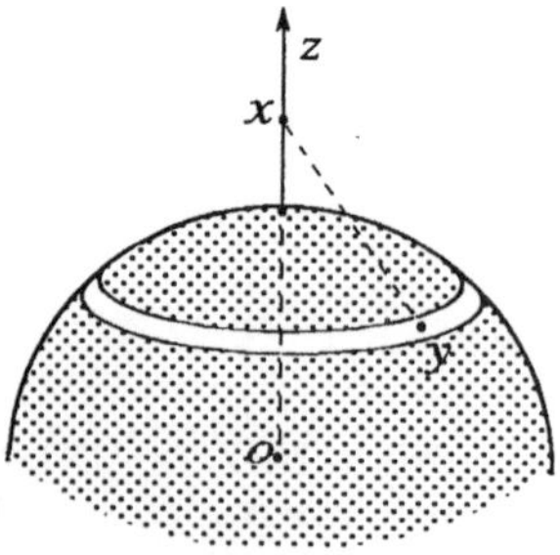

Figure 2.7: The geometry of the double layer distribution over a sphere surface.

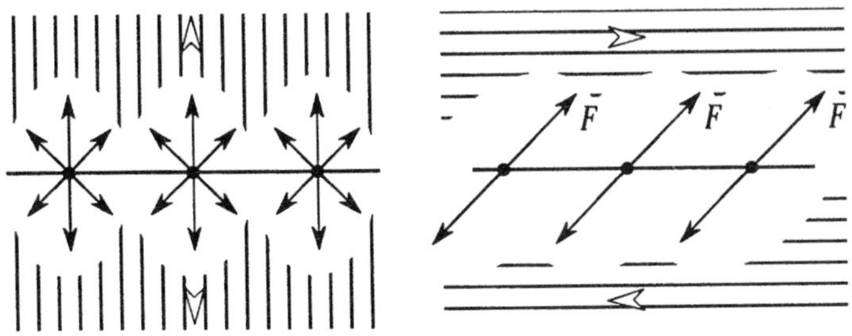

Figure 2.8: Distribution of sources/sinks and a true bilayer of Stokeslets.

Exercise 2.14 Jump in the Double Layer Potential with Constant Density on a Sphere.
Consider a *constant* double layer density $\boldsymbol{\varphi}$ distributed over the surface of a unit sphere. Show by direct integration in spherical polar coordinates that as the reference point $\boldsymbol{x}$ slides down the z-axis and passes through the "north pole" the velocity field generated by the double layer distribution jumps from zero to $-\boldsymbol{\varphi}/2$ to $-\boldsymbol{\varphi}$ for $\boldsymbol{x}$ just outside the sphere, on the sphere surface, and just inside the sphere surface, respectively.

Exercise 2.15 Normal and Tangential Components of the Double Layer Potential.
Consider the decomposition of the double layer kernel into the Oseen pressure field and the rate-of-strain field of the Oseen tensor,

$$8\pi\mu(\boldsymbol{\Sigma}(\boldsymbol{x}-\boldsymbol{\xi})\cdot\hat{\boldsymbol{n}})\cdot\boldsymbol{\varphi} = (\boldsymbol{\varphi}\cdot\hat{\boldsymbol{n}})\mathcal{P}(\boldsymbol{x}-\boldsymbol{\xi}) + \mu(\boldsymbol{\varphi}\hat{\boldsymbol{n}}+\hat{\boldsymbol{n}}\boldsymbol{\varphi}) : \nabla\mathcal{G}(\boldsymbol{x}-\boldsymbol{\xi}) \ .$$

The first term corresponds to sources and sinks, depending on the sign of the density, $\boldsymbol{\varphi}$, while the second group corresponds to a true bilayer of Stokeslets. From the figure, we can guess right away that the source/sink term gives a jump in the normal, but continuous tangential velocities, while for the bilayer the opposite occurs. Verify this by considering a local analysis using a constant density on a plane. In that situation, show also that only the tangential component of double layer density, $\boldsymbol{\varphi}^{\perp} = \boldsymbol{\varphi} - \boldsymbol{\varphi}\cdot\hat{\boldsymbol{n}}\hat{\boldsymbol{n}}$, contributes to the velocity field of the bilayer distribution.

Exercise 2.16 Self-Adjointness of the Oseen Tensor.
Use the Lorentz reciprocal theorem, Equation 2.6, and the fundamental property,

$$\frac{\partial}{\partial x_k}\Sigma_{ijk}(\boldsymbol{x}-\boldsymbol{\xi}) = -\delta_{ij}\delta(\boldsymbol{x}-\boldsymbol{\xi}) \ ,$$

to deduce that the Green's function for Stokes flow is *self-adjoint*:

$$\mathcal{G}_{ij}(\boldsymbol{x}-\boldsymbol{y}) = \mathcal{G}_{ji}(\boldsymbol{y}-\boldsymbol{x}) .$$

In words: the i-th component of the velocity field at $\boldsymbol{x}$ produced by a point force disturbance at $\boldsymbol{y}$, with the force directed along the j-th coordinate direction, is equal to the j-th component of the velocity field at $\boldsymbol{y}$ produced by a point force disturbance at $\boldsymbol{x}$, with the force directed along the i-th coordinate direction. This reciprocal property is closely tied to the concept of a self-adjoint operator, with $\mathcal{G}(\boldsymbol{x}-\boldsymbol{y})$ as the kernel of the integral operator.

Exercise 2.17 An Application of the Lorentz Reciprocal Theorem. Consider a rigid particle of arbitrary shape in an ambient flow field, $\boldsymbol{v}^\infty(\boldsymbol{x})$. We derive in this exercise Brenner's [10] result for the drag on the particle. Suppose that $\boldsymbol{v}'$ is the solution for a sphere translating with steady velocity $\boldsymbol{U}$ and $\boldsymbol{\sigma}'$ is the associated stress field. Denote the disturbance velocity and stress fields for the solution with the arbitrary ambient flow by $\boldsymbol{v}^D$ and $\boldsymbol{\sigma}^D$. An application of the Lorentz reciprocal theorem yields the following relation:

$$\oint_{S_p} \boldsymbol{v}' \cdot (\boldsymbol{\sigma}^D \cdot \hat{\boldsymbol{n}})\, dS = \oint_{S_p} \boldsymbol{v}^D \cdot (\boldsymbol{\sigma}' \cdot \hat{\boldsymbol{n}})\, dS .$$

Upon insertion of the boundary conditions, $\boldsymbol{v}' = \boldsymbol{U}$ and $\boldsymbol{v}^D = -\boldsymbol{v}^\infty$ on S_p, this simplifies to

$$\boldsymbol{U} \cdot \oint_{S_p} \boldsymbol{\sigma}^D \cdot \hat{\boldsymbol{n}}\, dS = -\oint_{S_p} \boldsymbol{v}^\infty \cdot (\boldsymbol{\sigma}' \cdot \hat{\boldsymbol{n}})\, dS .$$

Show that this implies that the drag on the particle in an arbitrary ambient field satisfying $\nabla \cdot \boldsymbol{\sigma}^\infty = \mathbf{0}$ may be related directly to that ambient field by the expression,

$$F_i(\boldsymbol{v}^\infty) = -\oint_{S_p} w_{ij} v_j^\infty\, dS ,$$

where $U_i w_{ij} = (\boldsymbol{\sigma}' \cdot \hat{\boldsymbol{n}})_j$ is the j-th component of the surface traction from the problem of the translating particle. The important conclusion here is: Given the surface tractions for the translation problem, one can compute directly the hydrodynamic drag for the same particle in an arbitrary field.

Using the development as a guide, derive a similar relation between the torque on the particle in an arbitrary ambient field and the solution for the rotating particle.

A final note: Using concepts from linear operator theory, we can also use this idea in reverse. In Part IV, we show how the previous equation may be interpreted as a linear functional mapping vectors from a Hilbert space to a scalar. According to the Riesz representation theorm, w_{ij} uniquely plays the role given above, so any algorithm for calculating F_i, when cast in the form above, identifies the surface traction for particles in rigid-body motion.

Exercise 2.18 Double Layers and Multipole Expansion.
The multipole expansion was derived starting from the single layer representation, assuming that there are no sources or sinks within the particle considered.

The general integral representation, on the other hand, also includes a double layer term. Inspect the effects of this on the multipole expansion, specifically for the first few terms. (An explicit formula for the contribution of a double layer to the stresslet will be given later, in Chapter 16.)

Exercise 2.19 Stokes Solutions from Stokes Solutions
Assume that the (velocity and pressure field) pair $(\boldsymbol{u}, p)$ satisfies the Stokes equations. Show that the pair $(\boldsymbol{\sigma}(;\boldsymbol{u}), \mu\nabla p)$ is also a Stokes solution, one order higher in tensorial rank. (The order can be reduced for example by taking dot products with some fixed vector.)
Hint: Take the Laplacian of $\boldsymbol{\sigma}(;\boldsymbol{u})$, writing out the stress field in terms of $\boldsymbol{u}$ and p and observing that the latter of these is harmonic (*i.e.*, its Laplacian vanishes identically). Use the fact that $\nabla^2\boldsymbol{u}$ can be expressed in terms of p according to Stokes equations.

Exercise 2.20 Integral Representation for the Pressure Field. The pressure field for the Oseen tensor has been given in the text. It is conventional to choose the indeterminate constant (base pressure) so that the pressure field will be decaying as infinity is approached, so $\mathcal{P}_j^\infty = 0$. Interpret the single layer as a superposition of Stokeslets and construct the corresponding pressure field also by superposition. Recall that $8\pi\mu\boldsymbol{\Sigma}$ is the stress field of the Oseen tensor, and use the previous exercise to express the related pressure field as a gradient of the pressure field used in the first superposition above. Show that this tensorial pressure field is a symmetric tensor.

References

[1] G. K. Batchelor. *An Introduction to Fluid Dynamics.* Cambridge University Press, Cambridge, 1967.

[2] G. K. Batchelor. Brownian diffusion of particles with hydrodynamic interaction. *J. Fluid Mech.*, 74:1–29, 1976.

[3] G. K. Batchelor. The stress system in a suspension of force-free particles. *J. Fluid Mech.*, 41:545–570, 1970.

[4] R. B. Bird, O. Hassager, and R. C. Armstrong. *Dynamics of Polymeric Liquids Volume 1: Fluid Mechanics*, 2nd edition. Wiley, New York, 1987.

[5] R. B. Bird, O. Hassager, R. C. Armstrong, and C. Curtiss. *Dynamics of Polymeric Liquids Volume 2: Kinetic Theory*, 2nd edition. Wiley, New York, 1987.

[6] R. B. Bird, W. E. Stewart, and E. N. Lightfoot. *Transport Phenomena.* Wiley, New York, 1960.

[7] J. F. Brady and G. Bossis. Stokesian dynamics. *Ann. Rev. Fluid Mech.*, 20:111–157, 1988.

[8] H. Brenner. Coupling between the translational and rotational Brownian motions of rigid particles of arbitrary shape. *J. Colloid Sci.*, 20:104–122, 1965.

[9] H. Brenner. Suspension rheology. *Prog. Heat and Mass Transfer*, 5:89–129, 1972.

[10] H. Brenner. The Stokes resistance of an arbitrary particle — IV. Arbitrary fields of flow. *Chem. Eng. Sci.*, 19:703–727, 1964.

[11] H. Brenner and R. G. Cox. The resistance to a particle of arbitrary shape in translational motion at small Reynolds numbers. *J. Fluid Mech.*, 17:561–595, 1963.

[12] J. S. Dahler and L. E. Scriven. Theory of structured continua. *Proc. R. Soc. Lond.*, A275:504–527, 1963.

[13] A. Einstein. Eine neue Bestimmung der Moleküldimensionen (A new determination of molecular dimensions). *Ann. Physik*, 19:289–306, 1911. See also *Ann. Physik* **34**, 591-592 (1911). Also in *Investigations on the Theory of the Brownian Movement*, Dover, New York, 1956.

[14] B. A. Finlayson. Existence of variational principles for the Navier-Stokes equation. *Phys. of Fluids*, 15((6)):963–967, 1972.

[15] C. W. Gardiner. *Handbook of Stochastic Methods: For Physics, Chemistry and the Natural Sciences*, 2nd edition. Springer-Verlag, Berlin, New York, 1985.

[16] J. Happel and H. Brenner. *Low Reynolds Number Hydrodynamics.* Martinus Nijhoff, The Hague, 1983.

[17] H. Helmholtz. Zur Theorie der stationären Ströme in reibenden Flüssigkeiten (A uniqueness theorem for viscous flow). *Verh. des naturh.-med. Vereins zu Heidelberg*, 5:1–7, 1868.

[18] R. Hill and G. Power. Extremum principles for slow viscous flow and the approximate calculation of drag. *Q. J. Mech. and Appl. Math*, 9:313–319, 1956.

[19] R. J. Hunter. *Zeta Potential in Colloid Science.* Academic Press, New York, 1981.

[20] Jr. Johnson, M. W. Some variational theorems for non-Newtonian flow. *Phys. of Fluids*, 3:871–878, 1960.

[21] J. H. Keenan, F. G. Keyes, P. G. Hill, and J. G. Moore. *Steam Tables.* Wiley, New York, 1969.

[22] J. B. Keller, L. A. Rubenfeld, and J. E. Molyneux. Extremum principles for slow viscous flows with applications to suspensions. *J. Fluid Mech.*, 30:97–125, 1967.

[23] S. Kim and C. J. Lawrence. Suspension mechanics for particle contamination control. *Chem. Eng. Sci.*, 43:991–1016, 1988.

[24] U. A. Ladyzhenskaya. *The Mathematical Theory of Viscous Incompressible Flow.* Gordon and Breach, New York, 1963.

[25] H. Lamb. *Hydrodynamics*, 6th edition. Dover, New York, 1932.

[26] L. D. Landau and E. M. Lifshitz. *Fluid Mechanics.* Pergamon Press, New York, 1959.

[27] W. E. Langlois. *Slow Viscous Flow.* MacMillan, New York, 1964.

[28] M. J. Lighthill. *An Introduction to Fourier Analysis and Generalized Functions.* Cambridge University Press, Cambridge, 1958.

[29] H. A. Lorentz. Ein allgemeiner Satz, die Bewegung einer reibenden Flüssigkeit betreffend, nebst einigen Anwendungen desselben (A general theorem concerning the motion of a viscous fluid and a few applications from it). *Versl. Kon. Akad. Wetensch. Amsterdam*, 5:168–174, 1896. Also in *Abhandlungen über Theoretische Physik*, 1:23–42 (1907) and *Collected Papers*, 4:7–14, Martinus Nijhoff, The Hague, 1937.

[30] J. Mahanty and B. W. Ninham. *Dispersion Forces.* Academic Press, New York, 1976.

[31] L. E. Malvern. *Introduction to the Mechanics of Continuous Medium.* Prentice-Hall, Englewood Cliffs, NJ, 1969.

[32] J. R. Melcher and G. I. Taylor. *Electrohydrodynamics: a review of the role of interfacial shear stresses. Ann. Rev. Fluid Mech.*, 1:111–146, 1969.

[33] A. Nir, H. F. Weinberger, and A. Acrivos. Variational inequalities for a body in a viscous shearing flow. *J. Fluid Mech.*, 68((4)):739–755, 1975.

[34] F. K. G. Odqvist. Über die Randwertaufgaben der Hydrodynamik zäher Flüssigkeiten (On the boundary value problems in hydrodynamics of viscous fluids). *Math. Z.*, 32:329–375, 1930.

[35] C. W. Oseen. *Hydrodynamik.* Akad. Verlagsgesellschaft, Leipzig, 1927.

[36] I. Proudman and J. R. A. Pearson. Expansions at small Reynolds numbers for the flow past a sphere and a circular cylinder. *J. Fluid Mech.*, 2:237–262, 1957.

[37] W. B. Russel, D. A. Saville, and W. R. Schowalter. *Colloidal Dispersions.* Cambridge University Press, Cambridge, 1989.

[38] H. J. Schulze. *Physico-Chemical Elementary Processes in Flotation.* Elsevier, Amsterdam, 1984.

[39] D. J. Shaw. *Introduction to Colloid and Surface Chemistry*, 3rd edition. Butterworths, London, 1980.

[40] G. A. Smook. *Handbook for Pulp and Paper Technologists.* Canadian Pulp and Paper Association, Montreal, 1982.

[41] L. A. Spielman. Particle capture from low-speed laminar flows. *Ann. Rev. Fluid Mech.*, 9:297–319, 1977.

[42] T. G. M. van de Ven. *Colloidal Hydrodynamics.* Academic Press, New York, 1989.

[43] E. J. W. Verwey and J. Th. G. Overbeek. *Theory of the Stability of Lyophobic Colloids.* Elsevier, Amsterdam, 1948.

[44] H. Villat. *Leçons sur les Fluides Visquex.* Gauthier-Villars, Paris, 1943.

[45] H. F. Weinberger. Variational properties of steady fall in Stokes flow. *J. Fluid Mech.*, 52:321–344, 1972.

[46] S. Whitaker. *Introduction to Fluid Mechanics.* Prentice-Hall, Englewood Cliffs, NJ, 1969.

[47] E. T. Whittaker and G. N. Watson. *A Course of Modern Analysis.* Cambridge University Press, New York, 1963.

[48] C. -S. Yih. *Fluid Mechanics.* West River Press, Ann Arbor, MI, 1979.

[49] G. K. Youngren and A. Acrivos. Stokes flow past a particle of arbitrary shape: a numerical method of solution. *J. Fluid Mech.*, 69:377–403, 1975.

Part II

Dynamics of a Single Particle

Chapter 3

The Disturbance Field of a Single Particle in a Steady Flow

3.1 Introduction

In this chapter we consider the slow motion of a single particle in a viscous fluid and the nature of the disturbance produced by such motions on the ambient flow field. The study of the motion of a single particle in Stokes flow is an old subject and the earliest solutions have been known for more than a century. Nevertheless, a thorough understanding of the subject is a prerequisite for the more difficult task of determining hydrodynamic interactions between particles, the subject of the following chapter. Furthermore, in Part IV we shall use ideas developed in this and other chapters of Part II to derive analytic solutions to furnish concrete examples to back up the theorems — the theorems alone being somewhat abstract in nature. The presentation here is therefore cast in a form that is somewhat different from existing books on fluid dynamics, since our motivations and ultimate goals are different from these references.

The topics covered in the sections of this chapter range in complexity from the simple introduction of our notation to illustration of some subtle properties of Stokes flow. The multipole expansion for the disturbance field of a rigid particle, which was mentioned in the previous chapter, provides a familiar starting point for the discussion. We consider the form taken by rigid particles and viscous drops, as the two are different because of the slip velocity at the mobile interface. In Section 3.3, we examine those special cases and shapes for which the multipole expansion terminates or reduces to an analytical form. We derive some exact results for ellipsoids and show that the multipole expansion has a finite radius of convergence. For needle-like particles (Section 3.4) we discuss an approximate solution method based on line distribution of singularities, also known as *slender body theory*. A direct relation (Faxén laws) between the ambient flow and the coefficients of the multipole expansion is introduced in Section 3.5, thus completing the circle for a given disturbance problem.

3.2 The Far Field Expansion: Rigid Particles and Drops

From Chapter 2 we have the multipole expansion for the disturbance velocity field of a rigid particle immersed in an ambient field $\boldsymbol{v}^{\infty}$,

$$\boldsymbol{v} = \sum_{n=0}^{\infty} \frac{(-1)^n}{n!} \boldsymbol{L}^{(n)} \cdot \frac{\mathcal{G}(\boldsymbol{x})}{8\pi\mu} , \tag{3.1}$$

with

$$\boldsymbol{L}^{(n)} = \oint_{S_p} [(\boldsymbol{\sigma} \cdot \hat{\boldsymbol{n}})\xi_{k_1} \ldots \xi_{k_n}] \, dS \frac{\partial}{\partial x_{k_1}} \cdots \frac{\partial}{\partial x_{k_n}} .$$

Equation 3.1 provides a completely general expression for $\boldsymbol{v}$. Each successive term introduces additional factors of a/r, where a is length scale based on the particle size; so the first few terms are sufficient to describe the effect of the particle far away from it. However, the computational procedure for the multipole coefficients (using the Faxén laws of Section 3.5) is quite tedious, so that in general the multipole expansion is not useful for velocity calculations, especially near the particle. The exception is for particles and flow fields possessing a high degree of symmetry, where the expansion terminates and further simplifications are possible. These so-called singularity solutions are discussed in the following section. We place the discussion of the Faxén relations after this because, for the general particle, the functional form of the Faxén relations is dictated by the singularity solution, as a consequence of the Lorentz reciprocal theorem. For the simplest geometry of all, the sphere, there are a number of other methods as well, but we defer to Chapter 4 a discussion of general solutions of the Stokes equations in spherical coordinates.

We now consider the extension of Equation 3.1 to the disturbance field of a viscous drop. There are subtle differences that can be traced to the slip velocity on the mobile interface and its effect on the integral representation. Of course, far away from the object the fundamental form of the solution cannot change; again we must have a multipole expansion in terms of Stokes monopoles, Stokes dipoles, Stokes quadrupoles, and so forth. Therefore, the differences between the disturbance field of a rigid particle and that produced by a viscous drop of the same dimension must manifest itself through the coefficients, *viz.*, the functional form taken by the force, stresslets, and other moments.

We repeat the steps of Section 2.4.3 for the integral representation for a disturbance field produced by a particle. For a stationary viscous drop, $\boldsymbol{v}\cdot\hat{\boldsymbol{n}} = 0$, but $\boldsymbol{v}$ itself is not zero – the interface is mobile. The integral representation is therefore (see Exercise 3.1)

$$\begin{aligned} \boldsymbol{v}(\boldsymbol{x}) - \boldsymbol{v}^{\infty}(\boldsymbol{x}) &= -\frac{1}{8\pi\mu} \oint_S (\boldsymbol{\sigma}(\boldsymbol{\xi}) \cdot \hat{\boldsymbol{n}}) \cdot \mathcal{G}(\boldsymbol{x} - \boldsymbol{\xi}) \, dS(\boldsymbol{\xi}) \\ &\quad - \oint_S \boldsymbol{v}(\boldsymbol{\xi}) \cdot (\boldsymbol{\Sigma} \cdot \hat{\boldsymbol{n}}) \, dS(\boldsymbol{\xi}) \end{aligned}$$

and has both the single layer and double layer terms.

The Taylor expansion of the kernel of the single layer potential leads to the same functional forms encountered with the rigid particle. For the double layer term, we recall that

$$8\pi\mu\boldsymbol{\Sigma} = \boldsymbol{\mathcal{P}}\boldsymbol{\delta} + \mu(\nabla\boldsymbol{\mathcal{G}} + (\nabla\boldsymbol{\mathcal{G}})^t) \ .$$

Since $\boldsymbol{v}\cdot\hat{\boldsymbol{n}} = 0$ at the interface, the fundamental pressure field makes no contribution to the velocity field generated by the double layer potential. In essence, the double layer potential is simply a distribution of symmetric Stokes dipoles over the interface, and the multipole expansion follows from the appropriate Taylor expansion. Consequently, the double layer potential contributes no force or torque, and to leading order in the expansion appears as a symmetric Stokes dipole located at the center of the drop.

We summarize the results for the viscous drop: The leading order terms are the Stokeslet and symmetric Stokes dipole, and the expressions for their coefficients (force and stresslet) are

$$\begin{aligned}
\boldsymbol{F} &= \oint_S (\boldsymbol{\sigma}\cdot\hat{\boldsymbol{n}})\, dS \\
\boldsymbol{S} &= \frac{1}{2}\oint_S [(\boldsymbol{\sigma}\cdot\hat{\boldsymbol{n}})\boldsymbol{\xi} + \boldsymbol{\xi}(\boldsymbol{\sigma}\cdot\hat{\boldsymbol{n}})]\, dS - \frac{1}{3}\oint_S (\boldsymbol{\sigma}\cdot\hat{\boldsymbol{n}})\cdot\boldsymbol{\xi}\, dS\ \boldsymbol{\delta} \\
&\quad - \ \mu\oint_S (\boldsymbol{v}\hat{\boldsymbol{n}} + \hat{\boldsymbol{n}}\boldsymbol{v})\, dS \ .
\end{aligned}$$

The identical result was obtained in Chapter 2 in the discussion on bulk stresses in a suspension. The "torque" on the drop is given formally by

$$\boldsymbol{T} = \oint_{S_p} \boldsymbol{\xi}\times(\boldsymbol{\sigma}\cdot\hat{\boldsymbol{n}})\, dS \ ,$$

but of course this is identically zero unless the drop fluid has an microstructure that endows it with an internal couple.

3.3 Singularity Solutions

For the general particle shape, the multipole expansion representation requires an infinite number of terms. Of course this is not a severe disadvantage in far field analyses, for in the far field the first few terms in the moment expansion capture the dominant behavior of the disturbance velocity field and at the same time provide information concerning the net hydrodynamic force, torque, *etc.* on the particle. In the near field, all terms in the expansion become comparable in magnitude, thus the multipole method becomes quite cumbersome unless some other simplifying factor enters into the picture. In this section, we shall show that for the sphere, the multipole expansion contains only a finite number of terms. For other simple shapes like the ellipsoid, a truncated expansion in just the lower order singularities is possible, provided that these singularities are distributed over a region. This collection of singularities and their region of

$$\begin{aligned}
\mathcal{G}_{ij} &= \frac{1}{r}\delta_{ij} + \frac{1}{r^3}x_i x_j \,, \qquad r = |\boldsymbol{x}| \\
\mathcal{G}_{ij,k} &= \frac{1}{r^3}(-\delta_{ij}x_k + \delta_{jk}x_i + \delta_{ik}x_j) - \frac{3}{r^5}x_i x_j x_k \\
\nabla^2\mathcal{G}_{ij} &= \frac{2}{r^3}\delta_{ij} - \frac{6}{r^5}x_i x_j \\
\nabla^2\mathcal{G}_{ij,k} &= -\frac{6}{r^5}(\delta_{ij}x_k + \delta_{jk}x_i + \delta_{ik}x_j) + \frac{30}{r^7}x_i x_j x_k
\end{aligned}$$

Table 3.1: The derivatives of the Oseen tensor and degenerate quadrupole.

distribution are collectively called the *image system*, and the resulting expressions are known as *singularity solutions.* A good starting point for a discussion of the singularity method for Stokes flow may be found in [20] and references therein, while [2] provides an excellent overview of the singularity method for a wide range of problems in mathematical physics.

Two final comments: The multipole expansion about the centroid of an ellipsoid and other particles for which singularity solutions are known can be expressed in closed form because the moments of the singularity distribution are easy to work out analytically. Also, we will restrict our attention in this section to the linear ambient flow, $\boldsymbol{U}^\infty + \boldsymbol{\Omega}^\infty \times \boldsymbol{x} + \boldsymbol{E}^\infty \cdot \boldsymbol{x}$.

3.3.1 Singularity System for Spheres

In electrostatics, we encounter the fact that the field of a point charge also provides the exact solution for a spherical perfect conductor. It is not surprising, then, that an analogous (but not identical) situation exists in Stokes flow. We shall consider the main flows of interest to microhydrodynamics (uniform stream, shear flows, extensional flows) and show that the multipole expansion does indeed contain a finite number of terms. At the beginning, we shall be content merely with demonstrations that the proposed solutions are correct (this is relatively straightforward). The more difficult task of explaining the solution philosophy and technique will be deferred until the end of the section.

Let us begin by examining the first few derivatives of the Oseen tensor $\mathcal{G}(\boldsymbol{x})$ and the degenerate quadrupole $\nabla^2\mathcal{G}(\boldsymbol{x})$. These expressions are used quite frequently and are listed in Table 3.1.

Consider the combination, $(1 + (a^2/6)\nabla^2)\mathcal{G}(\boldsymbol{x})$ evaluated on the surface of a sphere of radius a, $r = a$. From Table 3.1 we see that when $r = a$,

$$(1 + \frac{a^2}{6}\nabla^2)\mathcal{G}(\boldsymbol{x}) = \frac{4}{3a}\boldsymbol{\delta} \,.$$

From this, we immediately obtain the velocity field produced by a sphere translating with a steady velocity $\boldsymbol{U}$,

$$\boldsymbol{v}(\boldsymbol{x}) = 6\pi\mu a\boldsymbol{U} \cdot (1 + \frac{a^2}{6}\nabla^2)\frac{\mathcal{G}(\boldsymbol{x})}{8\pi\mu} .$$

It is a straightforward exercise to show that this expression is identical to the standard result expressed in spherical coordinates given in elementary books on fluid mechanics (see also Chapter 4). We have obtained two results that merit further discussion. Firstly, whereas the conducting sphere of electrostatics can be represented by just the point charge or monopole field, *the translating sphere in Stokes flow requires a degenerate quadrupole* $a^2\boldsymbol{F}\nabla^2\delta(\boldsymbol{x})$, *in addition to a monopole of strength* $6\pi\mu a\boldsymbol{U}$. This apparently is due to the more complicated structure of the Stokes equation as compared to the Laplace equation. Secondly, *we have derived Stokes law,* $\boldsymbol{F} = -6\pi\mu a\boldsymbol{U}$ *for the drag on a sphere undergoing steady translation, without explicit computation of surface tractions.* This illustrates the point that was raised earlier that solutions expressed as a multipole expansion yield quantities of interest, such as the hydrodynamic force, in a straightforward fashion.

Example 3.1 The Rotating Sphere.
From Table 3.1 we see that when $r = a$, the antisymmetric part of the Stokes dipole satisfies

$$\frac{1}{2}(\mathcal{G}_{ij,k} - \mathcal{G}_{ik,j}) = \frac{1}{a^3}(\delta_{ik}x_j - \delta_{ij}x_k) .$$

The velocity field of a rotating sphere in the ambient flow $\boldsymbol{\Omega}^\infty \times \boldsymbol{x}$ may be written as

$$\begin{aligned} \boldsymbol{v}(\boldsymbol{x}) - \boldsymbol{\Omega}^\infty \times \boldsymbol{x} &= -4\pi\mu a^3[\boldsymbol{\epsilon} \cdot (\boldsymbol{\Omega}^\infty - \boldsymbol{\omega}) \cdot \nabla] \cdot \frac{\mathcal{G}(\boldsymbol{x})}{8\pi\mu} \\ &= a^3(\boldsymbol{\Omega}^\infty - \boldsymbol{\omega}) \times \nabla\frac{1}{r} = (\boldsymbol{\omega} - \boldsymbol{\Omega}^\infty) \times \boldsymbol{x}\frac{a^3}{r^3} , \end{aligned}$$

in agreement with Lamb's general solution. In conclusion, the rotating sphere may be represented with just a *rotlet* and $\boldsymbol{T} = 8\pi\mu a^3(\boldsymbol{\Omega}^\infty - \boldsymbol{\omega})$ is the torque exerted by the fluid on the sphere.◇

Example 3.2 The Sphere in a Rate-of-Strain Field.
From Table 3.1 we see that when $r = a$,

$$(1 + \frac{a^2}{10}\nabla^2)(\mathcal{G}_{ij,k} + \mathcal{G}_{ik,j}) = -\frac{6}{5a^3}(\delta_{ik}x_j + \delta_{ij}x_k - \frac{2}{3}\delta_{jk}x_i) .$$

From this, we get immediately the velocity field produced by a fixed sphere in the straining field $\boldsymbol{E}^\infty \cdot \boldsymbol{x}$,

$$\boldsymbol{v}(\boldsymbol{x}) - \boldsymbol{E}^\infty \cdot \boldsymbol{x} = \frac{20}{3}\pi\mu a^3(\boldsymbol{E}^\infty \cdot \nabla) \cdot (1 + \frac{a^2}{10}\nabla^2)\frac{\mathcal{G}(\boldsymbol{x})}{8\pi\mu} .$$

The fixed sphere in a straining field may be represented by a stresslet of strength $(20/3)\pi\mu a^3\boldsymbol{E}^\infty$ and a degenerate octupole $a^2(\boldsymbol{S} \cdot \nabla)\nabla^2\delta(\boldsymbol{x})$.◇

For large λ, we obtain the singularity solution for the rigid sphere, while for $\lambda = 0$ (bubble), the degenerate quadrupole vanishes. The Stokeslet alone provides the exact solution for a translating bubble.

Other interior solutions appear in the context of a viscous drop in the n-th order field. The gradient of the Stokeson is a third-order tensor that is linear in r (because the Stokeson is quadratic in r). The symmetric and antisymmetric derivatives are known as the *roton* and *stresson* [20] in analogy with the rotlet and stresslet. These are in fact rarely used as such, since the roton and stresson simply correspond to a rigid-body rotation and a constant rate-of-strain field. They (*i.e.*, the rotational and rate-of-strain fields), together with a field that is cubic in r. are used in the construction of the interior flow field for a spherical drop in the linear field $\boldsymbol{\Omega} \times \boldsymbol{x} + \boldsymbol{E} \cdot \boldsymbol{x}$. The cubic field is in fact the less obvious portion of the solution, but it can be obtained either from Lamb's general solution or by the appropriate linear combination of $\boldsymbol{E} : \boldsymbol{xxx}$ and $\boldsymbol{E} \cdot \boldsymbol{x} r^2$.

3.3.3 Singularity System for Ellipsoids

The ellipsoid is encountered frequently in mathematical physics because it is a relatively simple nonaxisymmetric shape. Our analysis for the ellipsoid also provides a unified approach to a number of hydrodynamic problems, for the general ellipsoid contains a broad class of shapes ranging from disks to needles. Admittedly, a discussion on spheroids (ellipsoids of revolution) would introduce a smaller jump and would provide a more gradual rise in the demands placed on the reader. However, our major goal here is to show that a surprisingly simple and general approach to the flow past an ellipsoid exists.

Consider a general triaxial ellipsoid with coordinates chosen so that the surface of the ellipsoid is given by the expression

$$\frac{x^2}{a^2} + \frac{y^2}{b^2} + \frac{z^2}{c^2} = 1 \; , \tag{3.2}$$

with $a \geq b \geq c$. Initially, we expect a great jump in complexity going from the singularity solution for the sphere to that for the general ellipsoid. The three fundamental problems — *translation*, *rotation*, and *linear ambient field* — are solved in the papers of Oberbeck [63], Edwardes [26], and Jeffery [46]. These original solutions in ellipsoidal coordinates, collected in the first part of Table 3.2, are certainly intricate and apparently defy further analysis. (Readers interested in further study of ellipsoidal harmonics are directed to Hobson's treatise [42].) Our goal here is to formulate a solution method based on the multipole expansion and certain theorems from potential theory that provides a uniform framework for derivation of these as well as other flow fields involving the ellipsoid. We shall first present the final result and then provide the derivation. Various degenerate cases, such as the ellipsoid of revolution, are of great interest and take up the remainder of the discussion.

The solutions in Table 3.2 are special cases of the solutions for the n-th order ambient velocity field,

$$v_i^\infty = H_{ik_1k_2\ldots k_n} x_{k_1} x_{k_2} \ldots x_{k_n} \; . \tag{3.3}$$

Bilevel indices are used to avoid the introduction of n different subscript variables.

The solution to this problem will be presented, first in the singularity form, and then in terms of ellipsoidal harmonics, which includes the classical solutions as special cases. The transformation from the singularity solution to the multipole expansion is shown at the end of this subsection.

The disturbance velocity for an $(n-1)$-th order ambient field is explicitly

$$v_i = \sum_{m=0}^{[(n-1)/2]} L_{(n-2m)\ j} \int_E f_{(n-2m)}(\boldsymbol{x}')\{1 + \frac{c^2 q^2 \nabla^2}{4(n-2m)-2}\} \frac{\mathcal{G}_{ij}(\boldsymbol{x}-\boldsymbol{x}')}{8\pi\mu} dA(\boldsymbol{x}') \,, \tag{3.4}$$

with

$$\begin{aligned} f_{(n)}(\boldsymbol{x}) &= \frac{(2n-1)q^{(2n-3)}}{2\pi a_E b_E} \,, \qquad q(\boldsymbol{x}) = \sqrt{1 - x^2/a_E^2 - y^2/b_E^2} \,, \\ a_E &= \sqrt{a^2 - c^2} \,, \quad b_E = \sqrt{b^2 - c^2} \,, \\ L_{(n)j} &= \frac{(-1)^n}{(n-1)!} P_{jk_1 k_2 \ldots k_{n-1}} \frac{\partial}{\partial x_{k_1}} \frac{\partial}{\partial x_{k_2}} \cdots \frac{\partial}{\partial x_{k_{n-1}}} \,. \end{aligned}$$

Thus, an n-th order ambient field with n even (odd), induces a distribution of all even (odd) multipole moments up to and including the n-th moment. The pressure, p, is obtained by an analogous distribution of the fundamental pressure field. The constant tensors $\boldsymbol{P}$ give the strengths of the distributed multipole moments. The first two, P_j and P_{jk}, are the force and stress dipole. The precise relation between the $\boldsymbol{P}$'s and the multipole moments taken about the particle center is examined later. For now, we simply note that the solution method includes a procedure for their evaluation. Finally, the focal ellipse $E(\boldsymbol{x}')$,

$$\frac{x'^2}{a_E^2} + \frac{y'^2}{b_E^2} \le 1 \,, \quad z = 0 \,, \tag{3.5}$$

is the degenerate elliptical disk in the family of ellipsoids that are confocal to the particle ellipsoid. Their role as the image system in potential flow is discussed in Miloh [61].

We will also derive an alternate form of Equation 3.4 expressed in terms of the ellipsoidal harmonic G_n of Table 3.2. This alternate solution is

$$\begin{aligned} v_i &= \sum_{m=0}^{[(n-1)/2]} \frac{(-1)^{n-2m-1}(n-2m)(2n-4m)!}{2^{2n-4m}\,(n-2m)!\,(n-2m)!} \frac{L_{(n-2m)\ j}}{8\pi\mu} \\ &\quad \times \{(\delta_{ij} - x_j \frac{\partial}{\partial x_i}) G_{n-2m-1} + \frac{a_j^2}{2(n-2m)} \frac{\partial^2}{\partial x_i \partial x_j} G_{n-2m}\} \,. \end{aligned} \tag{3.6}$$

There is a summation over j and the notation a_j for $j = 1, 2, 3$ has been introduced for a, b, c. For $n = 1$ and 2, this form is precisely the appropriate classical solution in Table 3.2.

1. Solution in Ellipsoidal Coordinates [26, 46, 63]

$$v_i - U_i^\infty = -\frac{F_j}{16\pi\mu}\left[\delta_{ij}G_0 - x_j\frac{\partial G_0}{\partial x_i} + \frac{a_j^2}{2}\frac{\partial^2 G_1}{\partial x_i \partial x_j}\right]$$

$$v_i - (\boldsymbol{\Omega}^\infty \times \boldsymbol{x} + \boldsymbol{E}^\infty \cdot \boldsymbol{x})_i = \frac{-3}{32\pi\mu}(\boldsymbol{S}\cdot\nabla + \frac{1}{2}\boldsymbol{T}\times\nabla)_j\left[\delta_{ij}G_1 - x_j\frac{\partial G_1}{\partial x_i} + \frac{a_j^2}{4}\frac{\partial^2 G_2}{\partial x_i \partial x_j}\right]$$

$$G_n = \int_\lambda^\infty \left(\frac{x^2}{a^2+t} + \frac{y^2}{b^2+t} + \frac{z^2}{c^2+t} - 1\right)^n \frac{dt}{\Delta(t)},$$

with $\Delta(t) = \sqrt{(a^2+t)(b^2+t)(c^2+t)}$ and the ellipsoidal coordinate $\lambda(x,y,z)$ defined as the postive root of

$$\frac{x^2}{a^2+t} + \frac{y^2}{b^2+t} + \frac{z^2}{c^2+t} = 1 .$$

The relation between $\boldsymbol{F}$, $\boldsymbol{T}$, and $\boldsymbol{S}$ and the $\boldsymbol{U}^\infty - \boldsymbol{U}$, $\boldsymbol{\Omega}^\infty - \boldsymbol{\omega}$, and $\boldsymbol{E}^\infty$ are given in Table 3.3.

2. Singularity Solutions

$$\boldsymbol{v} - \boldsymbol{U}^\infty = -\boldsymbol{F}\cdot\int_E f_{(1)}(\boldsymbol{x}')\{1 + \frac{c^2q^2\nabla^2}{2}\}\frac{\mathcal{G}(\boldsymbol{x}-\boldsymbol{x}')}{8\pi\mu}dA(\boldsymbol{x}')$$

$$\boldsymbol{v} - \boldsymbol{\Omega}^\infty \times \boldsymbol{x} - \boldsymbol{E}^\infty \cdot \boldsymbol{x} = (\boldsymbol{S}\cdot\nabla + \frac{1}{2}\boldsymbol{T}\times\nabla)\cdot\int_E f_{(2)}(\boldsymbol{x}')\{1 + \frac{c^2q^2\nabla^2}{6}\}\frac{\mathcal{G}(\boldsymbol{x}-\boldsymbol{x}')}{8\pi\mu}dA(\boldsymbol{x}')$$

$$f_{(n)}(\boldsymbol{x}) = \frac{(2n-1)q^{2n-3}}{2\pi a_E b_E}, \qquad q(\boldsymbol{x}) = \sqrt{1 - x^2/a_E^2 - y^2/b_E^2},$$

$$a_E = \sqrt{a^2 - c^2}, \quad b_E = \sqrt{b^2 - c^2} .$$

3. Multipole Expansions

$$\boldsymbol{v} - \boldsymbol{U}_i^\infty = -\boldsymbol{F}\cdot\left(\frac{\sinh D}{D}\right)\frac{\mathcal{G}(\boldsymbol{x})}{8\pi\mu}$$

$$\boldsymbol{v} - \boldsymbol{\Omega}^\infty \times \boldsymbol{x} - \boldsymbol{E}^\infty \cdot \boldsymbol{x} = (\boldsymbol{S}\cdot\nabla + \frac{1}{2}\boldsymbol{T}\times\nabla)\cdot\left(\frac{3}{D}\frac{\partial}{\partial D}\right)\left(\frac{\sinh D}{D}\right)\frac{\mathcal{G}(\boldsymbol{x})}{8\pi\mu}$$

$$D^2 = a^2\frac{\partial^2}{\partial x^2} + b^2\frac{\partial^2}{\partial y^2} + c^2\frac{\partial^2}{\partial z^2} .$$

Table 3.2: The solutions for an ellipsoid in rigid-body motion $\boldsymbol{U} + \boldsymbol{\omega}\times\boldsymbol{x}$ in the ambient fields $\boldsymbol{U}^\infty$ and $\boldsymbol{\Omega}^\infty \times \boldsymbol{x} + \boldsymbol{E}^\infty \cdot \boldsymbol{x}$.

$$\begin{aligned}
F_x &= \frac{16\pi\mu abc}{\chi_0 + \alpha_0 a^2}(U_x^\infty - U_x) \\
T_x &= \frac{16\pi\mu abc}{3(b^2\beta_0 + c^2\gamma_0)}\left[(b^2+c^2)(\Omega_x^\infty - \omega_x) + (b^2-c^2)E_{yz}^\infty\right] \\
S_{xx} &= \frac{16\pi\mu abc}{9(\beta_0''\gamma_0''\gamma_0''\alpha_0'' + \alpha_0''\beta_0'')}\left[2\alpha_0''E_{xx}^\infty - \beta_0''E_{yy}^\infty - \gamma_0''E_{zz}^\infty\right] \\
S_{xy} = S_{yx} &= \frac{8\pi\mu abc}{3(a^2\alpha_0 + b^2\beta_0)}\left[(a^2-b^2)(\Omega_z^\infty - \omega_z) + \frac{(\alpha_0+\beta_0)}{\gamma_0'}E_{xy}^\infty\right]
\end{aligned}$$

The χ_0, α_0, β_0, γ_0, α_0', β_0', γ_0', α_0'', β_0'', and γ_0'' are constant (elliptic integrals) that are obtained by evaluating the ellipsoidal harmonics at the surface $\lambda = 0$:

$$\chi_0 = abcG_0(0)\ , \qquad \alpha_0 = abc\int_0^\infty \frac{dt}{(a^2+t)\Delta(t)}\ .$$

The constants β_0 and γ_0 are obtained by successive cycling of the dependence on a, b, and c. The primed and double-primed constants are defined by

$$\alpha' = \frac{\gamma - \beta}{b^2 - c^2}\ , \qquad \alpha'' = \frac{b^2\beta - c^2\gamma}{b^2 - c^2}\ ,$$

with the rest defined by cycling a, b, and c and α, β, and γ. The notation shows the common pattern in these solutions; consequently our notation differs from the original works. There is a summation over j in the velocity expressions and the notation a_j for $j = 1, 2, 3$ is used for a, b, c.

Table 3.3: The *resistance relations* between the force, torque, and stresslet and the ambient flow in the solutions of Oberbeck, Edwardes, and Jeffery.

The harmonics $\chi(\lambda)$ and Ω (the Dirichlet potential) that appear in the solution of Oberbeck (see also [35]) are proportional to G_0 and G_1:

$$\chi = abcG_0 \ , \qquad \Omega = \pi abcG_1 \ .$$

We now show that the two solutions, Equations 3.4 and 3.6, are equivalent and that they satisfy the appropriate boundary conditions.

We first establish the singularity representations for the ellipsoidal harmonic G_n. We define

$$H_n = \int_E \frac{q^{2n-1}}{2\pi a_E b_E} \frac{dA(\boldsymbol{x}')}{|\boldsymbol{x} - \boldsymbol{x}'|} \tag{3.7}$$

and show that

$$H_n = \frac{(-1)^n \ (2n)!}{2^{2n+1} \ n! \ n!} \ G_n \ . \tag{3.8}$$

This also establishes that G_n is a harmonic.

For $n = 0$, Equation 3.8 is equivalent to the singularity representation for $\chi(\lambda)$ and follows from an application of Gauss' law to the potential field of an elliptic disk at constant potential [61]. Therefore, we assume that Equation 3.8 holds for n and show that this, in turn, implies that it holds for $n + 1$. To do this, we first establish the following relations:

$$G_{n+1}(\boldsymbol{x}; 1) = -2(n+1) \int_0^1 u^{2n+2} G_n(\boldsymbol{x}; u) \, du \ , \tag{3.9}$$

$$H_{n+1}(\boldsymbol{x}; 1) = (2n+1) \int_0^1 u^{2n+2} H_n(\boldsymbol{x}; u) \, du \ . \tag{3.10}$$

The parameter u in $G_n(\boldsymbol{x}; u)$ and $H_n(\boldsymbol{x}; u)$ denotes that the particle size has been scaled with u, *i.e.*, a, b, and c are replaced by ua, ub, and uc. The original ellipsoid corresponds to $u = 1$. We shall refer to the integral operations on the RHS of the preceding equations as *geometric scaling transformations.*

These relations are verified by direct integration. For example, Equation 3.9 is obtained in the following manner: Start with the definition of $G_n(\boldsymbol{x}, u)$ and integrate both sides with respect to u, changing variables from t to $\tau = u^2 t$. This yields

$$\int_0^1 u^{2n+2} G_n(\boldsymbol{x}; u) du = \int_0^1 u \int_{v(u)}^{\infty} (\frac{x^2}{a^2 + \tau} + \frac{y^2}{b^2 + \tau} + \frac{z^2}{c^2 + \tau} - u^2)^n \frac{d\tau}{\Delta(\tau)} du,$$

where $v(u) = \lambda / u^2$. Now change the order of integration so that the double integral is performed over the region $\lambda \le \tau \le \infty$ and $v^{-1}(\tau) \le u \le 1$, with

$$v^{-1}(\tau) = [\frac{x^2}{a^2 + \tau} + \frac{y^2}{b^2 + \tau} + \frac{z^2}{c^2 + \tau}]^{1/2} \ .$$

Consequently,

$$\int_0^1 u^{2n+2} G_n(\boldsymbol{x}; u) du = \int_\lambda^\infty \left[\int_{v^{-1}(\tau)}^1 (v^{-1}(\tau)^2 - u^2)^n u \ du \right] \frac{d\tau}{\Delta(\tau)} \ ,$$

and the integration with respect to u leads to Equation 3.9. Analogous steps may be used to derive Equation 3.10.

The induction to the case $n+1$ from the case n is accomplished by u-integration of both sides of Equation 3.8 with weights chosen as in Equations 3.9 and 3.10. The result is

$$H_{n+1} = \frac{(-1)^{n+1}\,(2n+2)!}{2^{2n+3}\,(n+1)!\,(n+1)!} G_{n+1} ,$$

which completes the proof.

The singularity solution may be rearranged into the harmonic representation by using the following identities involving the Oseen tensor:

$$\mathcal{G}_{ij}(\boldsymbol{x}-\boldsymbol{x}') = \delta_{ij}\frac{1}{|\boldsymbol{x}-\boldsymbol{x}'|} - (x-x')_j\frac{\partial}{\partial x_i}[\frac{1}{|\boldsymbol{x}-\boldsymbol{x}'|}] \quad (3.11)$$

$$\nabla^2\mathcal{G}(\boldsymbol{x}-\boldsymbol{x}') = -2\nabla\nabla\frac{1}{|\boldsymbol{x}-\boldsymbol{x}'|} \quad (3.12)$$

$$\int_E x'_j q^n \frac{dA(\boldsymbol{x}')}{|\boldsymbol{x}-\boldsymbol{x}'|} = -\frac{(a_j^2-c^2)}{(n+2)}\frac{\partial}{\partial x_j}\int_E \frac{q^{n+2}dA(\boldsymbol{x}')}{|\boldsymbol{x}-\boldsymbol{x}'|} . \quad (3.13)$$

The last equation follows from an integration by parts.

From Equation 3.6 it is clear that the evaluation of $\boldsymbol{v}$ on the ellipsoid surface requires knowledge of the values taken by $\partial^n G_n$ and $\partial^{n+1}G_n$ on the surface. (When the tensorial subscripts are obvious, the notation will be simplified by using ∂^n as the n-th derivative.) These derivatives are obtained by successive application of the Leibniz rule, with the additional simplification that the integrand of G_n evaluated at λ vanishes (by definition of λ). Thus

$$\partial G_n = 2nx_{k_1}\int_\lambda^\infty \frac{F^{n-1}}{(a_{k_1}^2+t)}\frac{dt}{\Delta(t)}$$

$$\partial^2 G_n = 4n(n-1)x_{k_1}x_{k_2}\int_\lambda^\infty \frac{F^{n-2}}{(a_{k_1}^2+t)(a_{k_2}^2+t)}\frac{dt}{\Delta(t)} + 2n\delta_{k_1k_2}\int_\lambda^\infty \frac{F^{n-1}}{(a_{k_1}^2+t)}\frac{dt}{\Delta(t)} ,$$

with

$$F = \frac{x^2}{a^2+t} + \frac{y^2}{b^2+t} + \frac{z^2}{c^2+t} - 1 ,$$

until

$$\partial^n G_n = \sum_{m=0}^{[n/2]} 2^{n-m}\frac{n!}{m!}\{\delta_{k_1k_2}\ldots\delta_{k_{2m-1}k_{2m}}x_{k_{2m+1}}\ldots x_{k_n} \times \int_\lambda^\infty \frac{F^m}{[\Pi_{\alpha=1}^n(a_{k_\alpha}^2+t)]}\frac{dt}{\Delta(t)} + \cdots\ (sym)\ \}$$

($\alpha \neq k_2, k_4, \ldots, k_{2m}$ for term shown). The "(sym)" in the preceding expression denotes that at each m we must take all permutations of the indices k_i and that only one representative term has been shown explicitly.

At the summation index m, there are $n!/(2^m m!(n-2m)!)$ terms corresponding to all possible permutations of $\{k_1 \ldots k_n\}$ in the representative term. The key idea is that on the ellipsoid surface, $\lambda = 0$, the definite integrals become numerical constants that depend only on the shape of the ellipsoid. *Thus $\partial^n G_n$ reduces to a polynomial in $\boldsymbol{x}$ of degree n.*

The velocity expression also involves terms of the form $\partial^{n+1} G_n$, which we now proceed to evaluate at the ellipsoid surface:

$$\begin{aligned} \partial^{n+1} G_n &= -2^n n! x_{k_1} \ldots x_{k_n} \frac{1}{[\Pi_{\alpha=1}^n a_{k_\alpha}^2] \Delta(\lambda)} \frac{\partial \lambda}{\partial x_{k_{n+1}}} \\ &+ \text{ polynomial of degree } n-1. \end{aligned} \tag{3.14}$$

At the surface $\lambda = 0$,

$$\frac{\partial \lambda}{\partial x_k} = \frac{2x_k}{a_k^2} \left[\sum_{l=1}^{3} \frac{x_l^2}{a_l^4} \right]^{-1} .$$

Thus the leading order term in Equation 3.14 is not of the form specified in the boundary condition for the $(n-1)$-th order ambient field. In other words, the proposed velocity representation will have the proper behavior at the ellipsoid surface if and only if this $(n+1)$-th order field is canceled. There are only two such terms in the velocity representation of Equation 3.6. They occur in the second and third terms when $m = 0$, *i.e.*,

$$-x_j \partial^n G_n = 2^n (n-1)! x_i x_j x_{k_1} \ldots x_{k_{n-1}} ([\Pi_{\alpha=1}^n a_{k_\alpha}^2] \Delta(0) [\sum_{l=1}^{3} \frac{x_l^2}{a_l^4}])^{-1} + \cdots$$

$$\frac{a_j^2}{2n} \partial^{n+1} G_n = -2^n (n-1)! x_i x_j x_{k_1} \ldots x_{k_{n-1}} a_j^2 ([\Pi_{\alpha=1}^{n+1} a_{k_\alpha}^2] \Delta(0) [\sum_{l=1}^{3} \frac{x_l^2}{a_l^4}])^{-1} + \cdots .$$

In the second equation, a_j^2 may be eliminated so the two unwanted terms cancel each other. Thus, the velocity field evaluated at the surface of the ellipsoid is in fact a polynomial in $\boldsymbol{x}$ of degree $(n-1)$. The lower order fields (with the order successively decreasing by two) may be eliminated by repeated use of the preceding argument, *i.e.*, by mathematical induction. Thus the velocity may be expressed as in Equations 3.4 or 3.6 as claimed.

The last step in the solution procedure requires the determination of the $\boldsymbol{P}$ in terms of $\boldsymbol{H}$, the known gradients of the ambient field. This final step is accomplished by inverting the set of linear equations for the unknown tensorial coefficients. The symmetry in the ellipsoid geometry decouples the system so that the lower order cases $n = 1, 2, 3$ may be inverted analytically.

The singularity and ellipsoidal-harmonic forms of the velocity field are needed when exact computations of the velocity are required. We now derive the form most useful for far field analyses, the *multipole expansion*. The central identity in this discussion is the following:

$$\begin{aligned} \int_E f_{(1)}(\boldsymbol{x}') \{1 + \frac{c^2 q^2 \nabla^2}{2}\} \mathcal{G}(\boldsymbol{x} - \boldsymbol{x}') \, dA(\boldsymbol{x}') &= \left(\frac{\sinh D}{D} \right) \mathcal{G}(\boldsymbol{x}) \\ &= \left(1 + \frac{D^2}{3!} + \frac{D^4}{5!} + \cdots \right) \mathcal{G}(\boldsymbol{x}) , \end{aligned} \tag{3.15}$$

where, as shown, the operator on the right-hand side is defined formally by the power series in D^2, with $D^2 = a^2\partial^2/\partial x^2 + b^2\partial^2/\partial y^2 + c^2\partial^2/\partial z^2$. We note that D^2 can also be written as $D^2 = \tilde{D}^2 + c^2\nabla^2$ with $\tilde{D}^2 = a_E^2\partial^2/\partial x^2 + b_E^2\partial^2/\partial y^2$.

Since the operands are *biharmonic* functions, the higher powers of D^2 may be expanded as $D^{2k} = \tilde{D}^{2k} + c^2 k\tilde{D}^{2k-2}\nabla^2$, and, therefore, the symbolic operator on the RHS of Equation 3.15 becomes

$$\begin{aligned}\frac{\sinh D}{D} &= \frac{\sinh\tilde{D}}{\tilde{D}} + c^2\sum_{k=1}^{\infty}\frac{k\tilde{D}^{2k-2}}{(2k+1)!}\nabla^2 \\ &= \frac{\sinh\tilde{D}}{\tilde{D}} + \frac{c^2}{2}\left(\frac{1}{\tilde{D}}\frac{\partial}{\partial\tilde{D}}\right)\left(\frac{\sinh\tilde{D}}{\tilde{D}}\right)\nabla^2 .\end{aligned}$$

Equation 3.15 now follows by matching the terms in $\mathcal{G}$ and $\nabla^2\mathcal{G}$. We show that

$$\begin{aligned}\int_E f_{(1)}(\boldsymbol{x}')\mathcal{G}(\boldsymbol{x}-\boldsymbol{x}')dA(\boldsymbol{x}') &= \left(\frac{\sinh\tilde{D}}{\tilde{D}}\right)\mathcal{G}(\boldsymbol{x}) \\ \int_E f_{(1)}(\boldsymbol{x}')q^2\nabla^2\mathcal{G}(\boldsymbol{x}-\boldsymbol{x}')dA(\boldsymbol{x}') &= \left(\frac{1}{\tilde{D}}\frac{\partial}{\partial\tilde{D}}\right)\left(\frac{\sinh\tilde{D}}{\tilde{D}}\right)\nabla^2\mathcal{G}(\boldsymbol{x}) .\end{aligned}$$

These two equations follow from identities concerning the surface tractions on an ellipsoid established by Brenner [12].[1]

For the singularity solution of arbitrary order we must establish the identity,

$$\begin{aligned}&\int_E f_{(n)}(\boldsymbol{x}')\{1+\frac{c^2q^2\nabla^2}{4n-2}\}(\nabla)^{n-1}\mathcal{G}(\boldsymbol{x}-\boldsymbol{x}')dA(\boldsymbol{x}') \\ &= \frac{(2n)!}{2^n n!}\{\left(\frac{1}{D}\frac{\partial}{\partial D}\right)^{n-1}\left(\frac{\sinh D}{D}\right)\}(\nabla)^{n-1}\mathcal{G}(\boldsymbol{x}) \\ &= \frac{(2n)!}{n!}\sum_{k=1}^{\infty}\frac{(k+n-1)!\;D^{2k-2}}{(k-1)!\;(2k+2n-2)!}(\nabla)^{n-1}\mathcal{G}(\boldsymbol{x}) .\end{aligned} \tag{3.16}$$

The left-hand side of Equation 3.16 can be generated by successive applications of the geometric similarity transformation. If we define

$$\boldsymbol{J}_n(\boldsymbol{x}) = \int_E f_{(n)}(\boldsymbol{x}')[1+\frac{c^2q^2}{4n-2}\nabla^2]\frac{\mathcal{G}(\boldsymbol{x}-\boldsymbol{x}')}{8\pi\mu}dA(\boldsymbol{x}') , \tag{3.17}$$

then

$$\boldsymbol{J}_{n+1}(\boldsymbol{x};1) = (2n+1)\int_0^1 u^{2n}\boldsymbol{J}_n(\boldsymbol{x};u)du \quad n=1,2,3,\ldots , \tag{3.18}$$

where again the parameter u denotes that the ellipsoid dimensions have been rescaled by u. The derivation is analogous to the discussion subsequent to Equation 3.8.

[1]See his equations (26) and (27). For $c = 0$, the ellipsoid degenerates into an elliptical disk, so $f_1(\boldsymbol{x}')$ is identically the sum of the surface tractions on the two faces of this disk and his identities apply.

The same procedure applied to the right-hand side of Equation 3.16 gives, since $D^2(u) = u^2 D^2(1)$,

$$\begin{aligned}
\int_0^1 u^{2n} \sum_{k=1}^{\infty} \frac{2^n (k+n-1)!\, D^{2k-2}(u)}{(k-1)!\,(2k+2n-2)!} du
&= \sum_{k=1}^{\infty} \frac{2^n (k+n-1)!\, D^{2k-2}}{(k-1)!\,(2k+2n-2)!} \int_0^1 u^{2n+2k-2} du \\
&= \sum_{k=1}^{\infty} \frac{2^{n+1} (k+n)!\, D^{2k-2}}{(k-1)!\,(2k+2n)!} \\
&= \left(\frac{1}{D}\frac{\partial}{\partial D}\right)^n \left(\frac{\sinh D}{D}\right),
\end{aligned}$$

as required. The final result for the multipole expansion is given in Table 3.2.

3.3.4 Singularity System for Prolate Spheroids

For ellipsoids of revolution, most of the expressions in the preceding discussions reduce to much simpler forms. The elliptic integrals reduce to more elementary functions, and the axisymmetry may be exploited to simplify the relations between the multipole moments and particle motions. We defer the general discussion on the role of axisymmetry to a later section, but this discussion will furnish examples for later use. We will consider the prolate spheroid $(a > b = c)$ here. Oblate spheroids, with $a = b > c$, will be considered in the following subsection.

The ellipsoidal harmonics in the solutions of Oberbeck, Edwardes, and Jeffery are now *spheroidal harmonics*. The relation between our G_n and the standard notation for spheroidal harmonics may be established without great difficulty (Exercise 3.3), but this will not be of further use to us. Instead, we shall focus our attention on the rather interesting simplifications that occur in the singularity and multipole solutions. Prolate spheroids become slender bodies in the limit $c/a \ll 1$, and these exact solutions shall prove quite useful in our discussion of *slender body theory* (Section 3.4).

For a prolate spheroid, the focal ellipse $E(\boldsymbol{x}')$ is degenerate; it is the line segment between the foci of the generating ellipse. The reduction of the image system of the ellipsoid can be achieved by letting $b_E \to 0$ and by taking the Taylor series of $\mathcal{G}(\boldsymbol{x} - \boldsymbol{x}')$ with respect to y' about $y' = 0$ (see Exercise 3.4). Then the integrals in the singularity solution simplify to a line distribution of the form

$$\begin{aligned}
\boldsymbol{J}_n(\boldsymbol{x}) &= \frac{(2n-1)!}{[2^{n-1}(n-1)!]^2} \int_{-ae}^{ae} \left(1 - \frac{x^2}{a^2e^2}\right)^{n-1} \left[1 + \frac{c^2\nabla^2}{4n}\right] \frac{\mathcal{G}(\boldsymbol{x} - \boldsymbol{\xi})}{8\pi\mu} \frac{d\xi}{2ae}, \\
& \quad n = 1, 2, \ldots,
\end{aligned}$$

where $e = (a^2 - c^2)^{1/2}/a$ is the eccentricity of the generating ellipse.

The multipole expansions in Table 3.2 may be simplified by using the relation

$$\begin{aligned} D^2 &= a^2\frac{\partial}{\partial x^2} + c^2\left(\frac{\partial}{\partial y^2} + \frac{\partial}{\partial z^2}\right) \\ &= (a^2 - c^2)\frac{\partial}{\partial x^2} + c^2\nabla^2 \\ &= D_x^2 + c^2\nabla^2 , \end{aligned}$$

with the last equality defining D_x^2. The differential operators that appear in the multipole solution can then be combined into two groups, as shown in the following equation:

$$\begin{aligned} \left(\frac{1}{D}\frac{\partial}{\partial D}\right)^n\left(\frac{\sinh D}{D}\right) &= \left(\frac{1}{D_x}\frac{\partial}{\partial D_x}\right)^n\left(\frac{\sinh D_x}{D_x}\right) \\ &+ \left(\frac{1}{D_x}\frac{\partial}{\partial D_x}\right)^{n+1}\left(\frac{\sinh D_x}{D_x}\right)\frac{c^2}{2}\nabla^2 . \end{aligned}$$

Here as before, the operator identity is restricted to operations on biharmonic functions. Consequently, the multipole solutions may be written solely in terms of $\mathcal{G}$ and $\nabla^2\mathcal{G}$ and their derivatives along the axial direction. Thus for the three solutions of greatest interest, we have

$$\begin{aligned} \boldsymbol{v} - \boldsymbol{U}_i^\infty &= -\boldsymbol{F}\cdot\left(\frac{\sinh D_x}{D_x}\right)\frac{\mathcal{G}(\boldsymbol{x})}{8\pi\mu} \\ &- \boldsymbol{F}\cdot\left(\frac{1}{D_x}\frac{\partial}{\partial D_x}\right)\left(\frac{\sinh D_x}{D_x}\right)\frac{c^2}{2}\nabla^2\frac{\mathcal{G}(\boldsymbol{x})}{8\pi\mu} \end{aligned}$$

$$\begin{aligned} \boldsymbol{v} &= \boldsymbol{\Omega}^\infty \times \boldsymbol{x} + \boldsymbol{E}^\infty\cdot\boldsymbol{x} \\ &+ (\boldsymbol{S}\cdot\nabla + \frac{1}{2}\boldsymbol{T}\times\nabla)\cdot\left[\left(\frac{3}{D_x}\frac{\partial}{\partial D_x}\right)\left(\frac{\sinh D_x}{D_x}\right)\frac{\mathcal{G}(\boldsymbol{x})}{8\pi\mu}\right. \\ &\left. + \left(\frac{1}{D_x}\frac{\partial}{\partial D_x}\right)^2\left(\frac{\sinh D_x}{D_x}\right)\frac{3c^2}{2}\frac{\nabla^2\mathcal{G}(\boldsymbol{x})}{8\pi\mu}\right] . \end{aligned}$$

The final task is to relate the force, torque, and stresslet on the spheroid to $\boldsymbol{U}^\infty - \boldsymbol{U}$, $\boldsymbol{\Omega}^\infty - \boldsymbol{\omega}$, and $\boldsymbol{E}$. The expressions for the force, torque, and stresslet may be written in a way that highlights the orientation of the various vectors and tensors with respect to the symmetry axis. Denoting the unit directional vector along the axis by $\boldsymbol{d}$, we have

$$\begin{aligned} F_i &= 6\pi\mu a\left[X^A d_i d_j + Y^A(\delta_{ij} - d_i d_j)\right](U_j^\infty - U_j) \\ T_i &= 8\pi\mu a^3\left[X^C d_i d_j + Y^C(\delta_{ij} - d_i d_j)\right](\Omega_j^\infty - \omega_j) \\ &- 8\pi\mu a^3 Y^H \epsilon_{ijl} d_l d_k E_{jk}^\infty \\ S_{ij} &= \frac{20}{3}\pi\mu a^3\left[X^M d_{ijkl}^{(0)} + Y^M d_{ijkl}^{(1)} + Z^M d_{ijkl}^{(2)}\right]E_{kl}^\infty \\ &+ \frac{20}{3}\pi\mu a^3\frac{3}{5}Y^H(\epsilon_{ikl}d_j + \epsilon_{jkl}d_i)d_l(\Omega_k^\infty - \omega_k) , \end{aligned}$$

where

$$\begin{aligned}
d^{(0)} &= \frac{3}{2}(d_i d_j - \frac{1}{3}\delta_{ij})(d_k d_l - \frac{1}{3}\delta_{kl}) \\
d^{(1)} &= \frac{1}{2}(d_i\delta_{jl}d_k + d_j\delta_{il}d_k + d_i\delta_{jk}d_l + d_j\delta_{ik}d_l - 4d_i d_j d_k d_l) \\
d^{(2)} &= \frac{1}{2}(\delta_{ik}\delta_{jl} + \delta_{jk}\delta_{il} - \delta_{ij}\delta_{kl} + d_i d_j \delta_{kl} + \delta_{ij} d_k d_l \\
&\quad - d_i\delta_{jl}d_k - d_j\delta_{il}d_k - d_i\delta_{jk}d_l - d_j\delta_{ik}d_l + d_i d_j d_k d_l) ,
\end{aligned}$$

and X, Y, and Z are scalar resistance functions. The X functions are associated with axisymmetric problems, *e.g.*, X^A is used for the force due to translation along the axis of symmetry. Further details and explanations are furnished in Section 3.5.4. The expressions for the scalar functions of the prolate spheroid are given in Table 3.4.

Example 3.3 A Torque-Free Prolate Spheroid in a Linear Field. We use the results for the resistance functions to obtain an important result concerning the angular velocity of a torque-free spheroid in the linear field, $\boldsymbol{\Omega}^\infty \times \boldsymbol{x} + \boldsymbol{E}^\infty \cdot \boldsymbol{x}$. In the expression for the torque, we simply set $\boldsymbol{T} = \boldsymbol{0}$, and dot multiply through with the dyad,

$$(8\pi\mu a^3)^{-1}\left[(X^C)^{-1}\boldsymbol{dd} + (Y^C)^{-1}(\boldsymbol{\delta} - \boldsymbol{dd})\right]$$

to obtain

$$\begin{aligned}
\omega_m - \Omega_m^\infty &= -\left[(X^C)^{-1}d_m d_i + (Y^C)^{-1}(\delta_{mi} - d_m d_i)\right] Y^H \epsilon_{ijl} d_l d_k E_{jk}^\infty \\
&= -(Y^H/Y^C)\epsilon_{mjl} d_l d_k E_{jk}^\infty .
\end{aligned}$$

Using the relation,

$$\frac{Y^H}{Y^C} = \frac{a^2 - b^2}{a^2 + b^2} ,$$

we find that the rotational motion follows as

$$\boldsymbol{\omega} \times \boldsymbol{x} = \boldsymbol{\Omega}^\infty \times \boldsymbol{x} + \left(\frac{a^2 - b^2}{a^2 + b^2}\right)\left(\boldsymbol{E}^\infty \cdot \boldsymbol{d}\,(\boldsymbol{d} \cdot \boldsymbol{x}) - \boldsymbol{E}^\infty : \boldsymbol{dxd}\right) .$$

In words, a prolate spheroid rotates with the angular velocity of the ambient fluid, plus a contribution from the rate-of-strain field, which acts to align the axis along a principal direction of $\boldsymbol{E}$. The relative importance of the alignment effect depends on the aspect ratio: It is most important for slender spheroids, less important for nearly spherical spheroids, and nonexistent for spheres. In Chapter 5, we will study this problem in more detail and show that as the spheroid rotates in the linear field, its axis traces a periodic trajectory known as a *Jeffery orbit.*◇

$$\begin{aligned}
F_i &= 6\pi\mu a \left[X^A d_i d_j + Y^A(\delta_{ij} - d_i d_j)\right](U_j^\infty - U_j) \\
T_i &= 8\pi\mu a^3 \left[X^C d_i d_j + Y^C(\delta_{ij} - d_i d_j)\right](\Omega_j^\infty - \omega_j) \\
&\quad - 8\pi\mu a^3 Y^H \epsilon_{ijl} d_l d_k E_{jk}^\infty \\
S_{ij} &= \frac{20}{3}\pi\mu a^3 \left[X^M d_{ijkl}^{(0)} + Y^M d_{ijkl}^{(1)} + Z^M d_{ijkl}^{(2)}\right] E_{kl}^\infty \\
&\quad + \frac{20}{3}\pi\mu a^3 \frac{3}{5} Y^H (\epsilon_{ikl} d_j + \epsilon_{jkl} d_i) d_l (\Omega_k^\infty - \omega_k)
\end{aligned}$$

$$\begin{aligned}
d^{(0)} &= \frac{3}{2}(d_i d_j - \frac{1}{3}\delta_{ij})(d_k d_l - \frac{1}{3}\delta_{kl}) \\
d^{(1)} &= \frac{1}{2}(d_i \delta_{jl} d_k + d_j \delta_{il} d_k + d_i \delta_{jk} d_l + d_j \delta_{ik} d_l - 4 d_i d_j d_k d_l) \\
d^{(2)} &= \frac{1}{2}(\delta_{ik}\delta_{jl} + \delta_{jk}\delta_{il} - \delta_{ij}\delta_{kl} + d_i d_j \delta_{kl} + \delta_{ij} d_k d_l \\
&\quad - d_i \delta_{jl} d_k - d_j \delta_{il} d_k - d_i \delta_{jk} d_l - d_j \delta_{ik} d_l + d_i d_j d_k d_l)
\end{aligned}$$

$$\begin{aligned}
X^A &= \frac{8}{3} e^3 \left[-2e + (1+e^2)L\right]^{-1} \\
Y^A &= \frac{16}{3} e^3 \left[2e + (3e^2 - 1)L\right]^{-1} \\
X^C &= \frac{4}{3} e^3 (1-e^2) \left[2e - (1-e^2)L\right]^{-1} \\
Y^C &= \frac{4}{3} e^3 (2-e^2) \left[-2e + (1+e^2)L\right]^{-1} \\
Y^H &= \frac{4}{3} e^5 \left[-2e + (1+e^2)L\right]^{-1} \\
X^M &= \frac{8}{15} e^5 \left[(3-e^2)L - 6e\right]^{-1} \\
Y^M &= \frac{4}{5} e^5 \left[2e(1-2e^2) - (1-e^2)L\right] \\
&\quad \times \left[\left(2e(2e^2-3) + 3(1-e^2)L\right)\left(-2e + (1+e^2)L\right)\right]^{-1} \\
Z^M &= \frac{16}{5} e^5 (1-e^2) \left[3(1-e^2)^2 L - 2e(3-5e^2)\right]^{-1} \\
L(e) &= \ln\left(\frac{1+e}{1-e}\right)
\end{aligned}$$

Table 3.4: The *resistance functions* for the prolate spheroid.

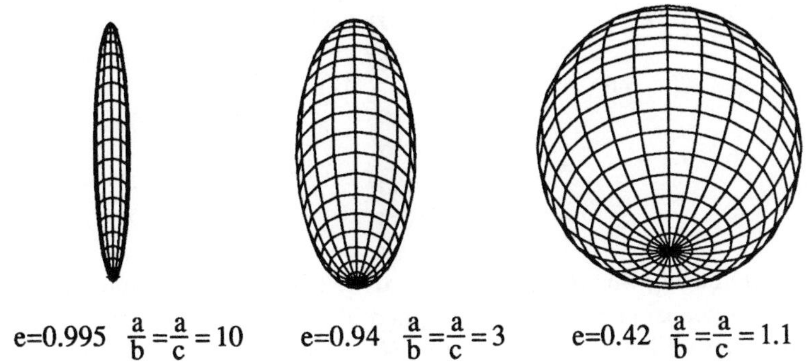

Figure 3.1: Prolate spheroids of various aspect ratios.

	$e \ll 1$	Needles: $\epsilon = \sqrt{1-e^2} \ll 1$, $E = (\ln(2/\epsilon))^{-1}$
X^A	$1 - \frac{2e^2}{5} - \frac{17e^4}{175} + \cdots$	$\frac{4E}{6-3E} - \frac{(8-6E)E\epsilon^2}{12-12E+3E^2} + \cdots$
Y^A	$1 - \frac{3e^2}{10} - \frac{57e^4}{700} + \cdots$	$\frac{8E}{6+3E} - \frac{4E^2\epsilon^2}{12+12E+3E^2} + \cdots$
X^C	$1 - \frac{6e^2}{5} + \frac{27e^4}{175} + \cdots$	$\frac{2\epsilon^2}{3} + \frac{(2-2E)\epsilon^4}{3E} + \cdots$
Y^C	$1 - \frac{9e^2}{10} + \frac{18e^4}{175} + \cdots$	$\frac{2E}{6-3E} + \frac{E^2\epsilon^2}{12-12E+3E^2} + \cdots$
Y^H	$\frac{e^2}{2} - \frac{e^4}{5} + \cdots$	$\frac{2E}{6-3E} - \frac{(8-5E)E\epsilon^2}{12-12E+3E^2} + \cdots$
X^M	$1 - \frac{6e^2}{7} - \frac{e^4}{49} + \cdots$	$\frac{4E}{30-45E} - \frac{(24-26E)E\epsilon^2}{60-180E+135E^2} + \cdots$
Y^M	$1 - \frac{13e^2}{14} + \frac{44e^4}{735} + \cdots$	$\frac{2E}{10-5E} - \frac{(16-32E+13E^2)\epsilon^2}{20-20E+5E^2} + \cdots$
Z^M	$1 - \frac{8e^2}{7} + \frac{17e^4}{147} + \cdots$	$\frac{4\epsilon^2}{5} + \frac{2\epsilon^4}{5} + \cdots$

Table 3.5: Asymptotic behavior of the resistance functions for prolate spheroids: near-spheres ($e \ll 1$) and needles ($e \to 1$).

3.3.5 Singularity System for Oblate Spheroids

For an oblate spheroid ($a = b > c$), the focal ellipse $E(\boldsymbol{x}')$ becomes a circular disk of radius $a_E = b_E = ae$. Now consider the relation

$$\begin{aligned} D^2 &= a^2\left(\frac{\partial}{\partial x^2}+\frac{\partial}{\partial y^2}\right)+c^2\frac{\partial}{\partial z^2} \\ &= -(a^2-c^2)\frac{\partial}{\partial z^2}+a^2\nabla^2 \\ &= -D_z^2+a^2\nabla^2 \, . \end{aligned}$$

We see, by comparing this result with the earlier one for prolate spheroids, that the multipole expansion for the oblate spheroid can be obtained from that for the prolate spheroid by replacing D_x with iD_z and $c^2\nabla^2$ with $a^2\nabla^2$. Therefore, the multipole expansion for the oblate spheroid simplifies as follows:

$$\begin{aligned} \left(\frac{1}{D}\frac{\partial}{\partial D}\right)^n\left(\frac{\sinh D}{D}\right) &= \left(\frac{-1}{D_z}\frac{\partial}{\partial D_z}\right)^n\left(\frac{\sin D_z}{D_z}\right) \\ &+ \left(\frac{-1}{D_z}\frac{\partial}{\partial D_z}\right)^{n+1}\left(\frac{\sin D_z}{D_z}\right)\frac{a^2}{2}\nabla^2 \, , \end{aligned}$$

where, as before, the operator identity is restricted to operations on biharmonic functions. Again, the solutions are expressed in terms of $\mathcal{G}$ and $\nabla^2\mathcal{G}$ and their derivatives along the axial direction. Thus for the three solutions of greatest interest, we have

$$\begin{aligned} \boldsymbol{v}-\boldsymbol{U}_i^\infty &= -\boldsymbol{F}\cdot\left(\frac{\sin D_z}{D_z}\right)\frac{\mathcal{G}(\boldsymbol{x})}{8\pi\mu} \\ &+ \boldsymbol{F}\cdot\left(\frac{1}{D_z}\frac{\partial}{\partial D_z}\right)\left(\frac{\sin D_z}{D_z}\right)\frac{a^2}{2}\nabla^2\frac{\mathcal{G}(\boldsymbol{x})}{8\pi\mu} \end{aligned}$$

$$\begin{aligned} \boldsymbol{v} &= \boldsymbol{\Omega}^\infty\times\boldsymbol{x}+\boldsymbol{E}^\infty\cdot\boldsymbol{x} \\ &+ (\boldsymbol{S}\cdot\nabla+\frac{1}{2}\boldsymbol{T}\times\nabla)\cdot\left[\left(\frac{-3}{D_z}\frac{\partial}{\partial D_z}\right)\left(\frac{\sin D_z}{D_z}\right)\frac{\mathcal{G}(\boldsymbol{x})}{8\pi\mu}\right. \\ &+ \left.\left(\frac{1}{D_z}\frac{\partial}{\partial D_z}\right)^2\left(\frac{\sin D_z}{D_z}\right)\frac{3a^2}{2}\frac{\nabla^2\mathcal{G}(\boldsymbol{x})}{8\pi\mu}\right] \, . \end{aligned}$$

The force, torque, and stresslet on the oblate spheroid are also written in terms of $\boldsymbol{d}$ and the scalar functions X, Y, and Z (see Table 3.6). The expressions for the oblate and prolate ellipsoids of revolution are closely related. In fact, each resistance function for one may be obtained from the counterpart of the other by replacing e with $ie/\sqrt{1-e^2}$. This is equivalent to switching the roles played by a and c (see Exercise 3.5).

The multipole expansion is the appropriate form when $|\boldsymbol{x}| \gg a$, but does it converge for $|\boldsymbol{x}| \sim c$? To answer this question, let us consider the disturbance field of an oblate spheroid translating in the direction of its axis and examine $\boldsymbol{v}(\boldsymbol{x}_1)$ for points $\boldsymbol{x}_1$ on the axis of symmetry. We use the following identities derived with the help of formulae linking Stokes multipoles to the spherical harmonics (see Section 4.2.1):

$$
\begin{aligned}
\frac{(\boldsymbol{d}\cdot\nabla)^n}{n!}\boldsymbol{F}^{\|}\cdot\mathcal{G}(\boldsymbol{x})|_{x=x_1} &= 2\boldsymbol{F}^{\|}r^{-(n+1)}\\
\frac{(\boldsymbol{d}\cdot\nabla)^n}{n!}\boldsymbol{F}^{\|}\cdot\nabla^2\mathcal{G}(\boldsymbol{x})|_{x=x_1} &= -2(n+1)(n+2)\boldsymbol{F}^{\|}r^{-(n+3)}\ ,
\end{aligned}
$$

where $r=|\boldsymbol{x}_1|$, $\boldsymbol{d}=\boldsymbol{x}_1/r$ and $\boldsymbol{F}^{\|}$ is used to denote that $\boldsymbol{F}$ is parallel to $\boldsymbol{d}$. If we insert these results into the multipole expansion, we obtain

$$
\boldsymbol{v}(\boldsymbol{x}_1) = -\frac{\boldsymbol{F}^{\|}}{8\pi\mu}\frac{3}{2}\frac{a}{r}\sum_{k=0}^{\infty}(-1)^k\left[\frac{1}{2k+1}-\frac{a^2}{r^2}\frac{k+1}{2k+3}\right]\left(\frac{a_E}{r}\right)^{2k}\ .
$$

The infinite series in r^{-1} converges only if $r^{-1}\leq a_E=ae$. We may visualize this constraint as a "hemispheric dome" of radius ae over the focal circle of the spheroid. As we slide down the axis of symmetry toward the spheroid, the infinite series will converge only if $\boldsymbol{x}_1$ lies outside the hemisphere. This example thus illustrates that, in general, the multipole expansion has a finite radius of convergence in r^{-1}, and that the exact value of this radius of convergence depends on the details of the particle geometry. In particular, it need not correspond to the particle surface nor the surface of an "effective sphere" that circumscribes the particle.

For translations transverse to the axis, we use

$$
\begin{aligned}
\frac{(\boldsymbol{d}\cdot\nabla)^n}{n!}\boldsymbol{F}^{\perp}\cdot\mathcal{G}(\boldsymbol{x})|_{x=x_1} &= \boldsymbol{F}^{\perp}r^{-(n+1)}\\
\frac{(\boldsymbol{d}\cdot\nabla)^n}{n!}\boldsymbol{F}^{\perp}\cdot\nabla^2\mathcal{G}(\boldsymbol{x})|_{x=x_1} &= (n+1)(n+2)\boldsymbol{F}^{\perp}r^{-(n+3)}
\end{aligned}
$$

to obtain the corresponding result,

$$
\boldsymbol{v}(\boldsymbol{x}_1) = \frac{\boldsymbol{F}^{\perp}}{8\pi\mu}\frac{3}{4}\frac{a}{r}\sum_{k=0}^{\infty}(-1)^k\left[\frac{1}{2k+1}+\frac{a^2}{r^2}\frac{k+1}{2k+3}\right]\left(\frac{a_E}{r}\right)^{2k}\ .
$$

Again, the radius of convergence is at $r^{-1}=ae$. Similar results hold for the prolate spheroid.

3.4 Slender Body Theory

Consider a needle-like rigid body whose length $2a$ is much greater than its width $2b$. Such particles are encountered in many settings, such as pulp fibers, rigid biopolymers like the tobacco mosaic virus, and various metal oxides used in the manufacture of magnetic storage devices. Understanding the motion of such particles in a viscous fluid is therefore of some importance.

$$
\begin{aligned}
F_i &= 6\pi\mu a \left[X^A d_i d_j + Y^A(\delta_{ij} - d_i d_j)\right](U_j^\infty - U_j) \\
T_i &= 8\pi\mu a^3 \left[X^C d_i d_j + Y^C(\delta_{ij} - d_i d_j)\right](\Omega_j^\infty - \omega_j) \\
&\quad - 8\pi\mu a^3 Y^H \epsilon_{ijl} d_l d_k E_{jk}^\infty \\
S_{ij} &= \frac{20}{3}\pi\mu a^3 \left[X^M d_{ijkl}^{(0)} + Y^M d_{ijkl}^{(1)} + Z^M d_{ijkl}^{(2)}\right] E_{kl}^\infty \\
&\quad + \frac{20}{3}\pi\mu a^3 \frac{3}{5} Y^H(\epsilon_{ikl} d_j + \epsilon_{jkl} d_i) d_l (\Omega_k^\infty - \omega_k)
\end{aligned}
$$

$$
\begin{aligned}
d^{(0)} &= \frac{3}{2}(d_i d_j - \frac{1}{3}\delta_{ij})(d_k d_l - \frac{1}{3}\delta_{kl}) \\
d^{(1)} &= \frac{1}{2}(d_i \delta_{jl} d_k + d_j \delta_{il} d_k + d_i \delta_{jk} d_l + d_j \delta_{ik} d_l - 4 d_i d_j d_k d_l) \\
d^{(2)} &= \frac{1}{2}(\delta_{ik}\delta_{jl} + \delta_{jk}\delta_{il} - \delta_{ij}\delta_{kl} + d_i d_j \delta_{kl} + \delta_{ij} d_k d_l \\
&\quad - d_i \delta_{jl} d_k - d_j \delta_{il} d_k - d_i \delta_{jk} d_l - d_j \delta_{ik} d_l + d_i d_j d_k d_l)\ ,
\end{aligned}
$$

$$
\begin{aligned}
X^A &= \frac{4}{3} e^3 \left[(2e^2 - 1)C + e\sqrt{1 - e^2}\right]^{-1} \\
Y^A &= \frac{8}{3} e^3 \left[(2e^2 + 1)C - e\sqrt{1 - e^2}\right]^{-1} \\
X^C &= \frac{2}{3} e^3 \left[C - e\sqrt{1 - e^2}\right]^{-1} \\
Y^C &= \frac{2}{3} e^3 (2 - e^2) \left[e\sqrt{1 - e^2} - (1 - 2e^2)C\right]^{-1} \\
Y^H &= -\frac{2}{3} e^5 \left[e\sqrt{1 - e^2} - (1 - 2e^2)C\right]^{-1} \\
X^M &= \frac{4}{15} e^5 \left[(3 - 2e^2)C - 3e\sqrt{1 - e^2}\right]^{-1} \\
Y^M &= \frac{2}{5} e^5 \left[e(1 + e^2) - \sqrt{1 - e^2}C\right] \\
&\quad \times \left[\left(3e - e^3 - 3\sqrt{1 - e^2}C\right)\left(e\sqrt{1 - e^2} - (1 - 2e^2)C\right)\right]^{-1} \\
Z^M &= \frac{8}{5} e^5 \left[3C - (2e^3 + 3e)\sqrt{1 - e^2}\right]^{-1} \\
C(e) &= \cot^{-1}\left(\frac{\sqrt{1 - e^2}}{e}\right)
\end{aligned}
$$

Table 3.6: The *resistance functions* for the oblate spheroid.

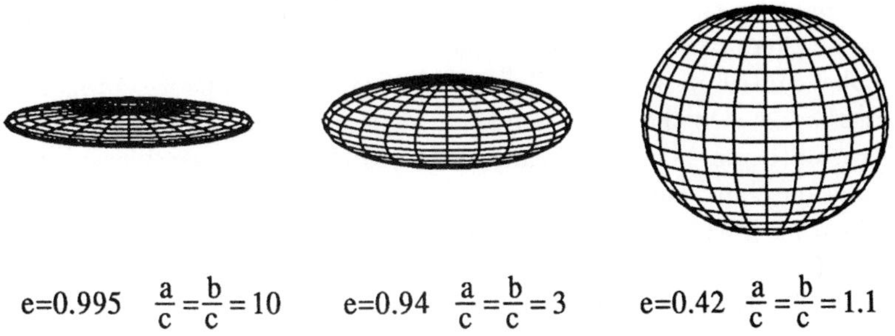

Figure 3.2: Oblate spheroids of various aspect ratios.

	$e \ll 1$	Disks: $\epsilon = \sqrt{1-e^2} \ll 1$
X^A	$1 - \frac{e^2}{10} - \frac{31e^4}{1400} + \cdots$	$\frac{8}{3\pi}(1 + \frac{1}{2}\epsilon^2 + \cdots)$
Y^A	$1 - \frac{e^2}{5} - \frac{79e^4}{1400} + \cdots$	$\frac{16}{9\pi}(1 + \frac{8}{3\pi}\epsilon - \frac{15\pi^2-128}{18\pi^2}\epsilon^2 + \cdots)$
X^C	$1 - \frac{3e^2}{10} - \frac{99e^4}{1400} + \cdots$	$\frac{4}{3\pi}(1 + \frac{4}{\pi}\epsilon + \frac{32-3\pi^2}{2\pi^2}\epsilon^2 + \cdots)$
Y^C	$1 - \frac{3e^2}{5} + \frac{39e^4}{1400} + \cdots$	$\frac{4}{3\pi}(1 + \frac{3}{2}\epsilon^2 + \cdots)$
Y^H	$-\frac{1}{2}e^2(1 - \frac{e^2}{10} - \frac{31e^4}{1400} + \cdots)$	$-\frac{4}{3\pi}(1 - \frac{1}{2}\epsilon^2 + \cdots)$
X^M	$1 - \frac{9e^2}{14} - \frac{13e^4}{392} + \cdots$	$\frac{8}{15\pi}(1 + \frac{8}{\pi}\epsilon + \frac{128-9\pi^2}{2\pi^2}\epsilon^2 + \cdots)$
Y^M	$1 - \frac{4e^2}{7} - \frac{173e^4}{5880} + \cdots$	$\frac{4}{5\pi}(1 + \frac{\pi}{2}\epsilon + \frac{3\pi^2-20}{8}\epsilon^2 + \cdots)$
Z^M	$1 - \frac{5e^2}{14} - \frac{95e^4}{1176} + \cdots$	$\frac{16}{15\pi}(1 + \frac{16}{3\pi}\epsilon + \frac{512-45\pi^2}{18\pi^2}\epsilon^2 + \cdots)$

Table 3.7: Asymptotic behavior of the resistance functions for oblate spheroids: near-spheres ($e \ll 1$) and disks ($e \to 1$).

Let $\boldsymbol{x}$ represent a point in the fluid, the origin taken at the particle center. For $|\boldsymbol{x}| \gg a$, we obtain the usual result: The exact shape and overall dimensions of the particle are not important and the disturbance field is described accurately by a multipole expansion about the origin. The Stokeslet and rotlet fields account for the force and torque on the particle and the stresslet field describes the effect of the particle on the rate-of-strain in the fluid. Higher order multipoles contribute ever smaller corrections, their influence diminishing by factors of $a/|\boldsymbol{x}|$. On the other hand, for $|\boldsymbol{x}|$ less than a, the details concerning the shape of the particle, specifically the overall length, is clearly important and an expansion about the particle center no longer makes any sense; in fact, we learned earlier in this chapter that these expansions can be nonconvergent near the particle.

Recall that the multipole expansion originates from the integral representation for the disturbance field. The idea was that at great distances from the particle, the surface distribution of Stokeslets is expanded in a Taylor series about the particle center. For a slender body, we instead expand the Stokeslets about points on the particle axis, thus taking advantage of the "smallness" of b. The result is a line distribution of the lower order singularities, much like those encountered for prolate spheroids.

Burgers [19] gave an elementary version of this idea and correctly deduced that, to leading order, the disturbance field of a slender body in uniform translation is given by a constant line distribution of Stokeslets over the particle length. The field was revived in the 1960s [17, 18, 84, 87] and placed on a more rigorous foundation using the theory of matched asymptotic expansions [86]. The theory has been generalized in a number of directions, for example, noncircular cross sections [5], more general ambient fields [5, 23, 47], and center-line curvature [47]. In this introduction, we consider only the simplest case of a slender body of circular cross section and straight backbone (*i.e.*, an axisymmetric particle). Nevertheless, the solution will be general in that explicit formulae will be obtained for particles of arbitrary shape, save for the slenderness condition.

The development of the theory for inviscid potential flow preceded, and thus influenced, the development of the theory for Stokes flow. This has led to some unfortunate consequences, such as the use of singularities from potential flow, with the result that some of the unifying structures are obscured. For example, the solutions for translations parallel to and transverse to the axis are actually quite similar when represented in terms of Stokes singularities, but appear quite different when written in terms of singularities from potential flow. For this reason, we have rewritten the solutions here in terms of $\mathcal{G}$ and its derivatives, but they may be derived by a formal procedure analogous to that in Tillett [86].

The geometry of this problem is given in the Figure 3.3 We place the particle along the x-axis, with the particle shape described by the cylindrical radial coordinate, ρ:

$$\rho(x) = \sqrt{y^2 + z^2} = \epsilon R(x) \qquad -a \leq x \leq a\ .$$

Since $\epsilon = b/a \ll 1$ has been scaled from the width, $max\{R(x)\} = a$. The ends

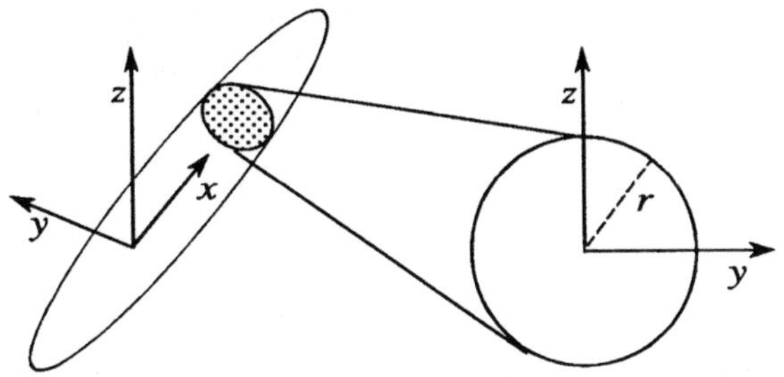

Figure 3.3: The geometry of the slender body.

are assumed to be round, with $R^2 \sim A^+(a-x)$ near $x = a$ and $R^2 \sim A^-(a+x)$ near $x = -a$. For the prolate spheroid, $A^+ = A^- = 2a$.

Consider a slender body in steady translation through a quiescent fluid. We obtain the leading order terms of slender body theory by following the exposition in [5]. Consider the velocity field produced by a constant line distribution of Stokeslets,

$$v_i = f_j \int_{-a}^{a} \frac{\mathcal{G}_{ij}(\boldsymbol{x}-\boldsymbol{\xi})}{8\pi\mu}\, d\xi \ .$$

With $\boldsymbol{\xi}$ on the x-axis, the nine components of the dyadic integral are

$$\begin{pmatrix} I+\mathcal{I}_2 & \mathcal{I}_1\cos\phi & \mathcal{I}_1\sin\phi \\ \mathcal{I}_1\cos\phi & I+\mathcal{I}_0\cos^2\phi & \mathcal{I}_0\sin\phi\cos\phi \\ \mathcal{I}_1\sin\phi & \mathcal{I}_0\sin\phi\cos\phi & I+\mathcal{I}_0\sin^2\phi \end{pmatrix} ,$$

where I and $\mathcal{I}_n$ are elementary integrals defined by

$$I = \int_{-a}^{a} \frac{d\xi}{\{(x-\xi)^2+\rho^2\}^{1/2}} , \qquad \mathcal{I}_n = \int_{-a}^{a} \frac{\rho^{2-n}(x-\xi)^n\, d\xi}{\{(x-\xi)^2+\rho^2\}^{3/2}} .$$

For $\rho \sim b \ll a$, these integrals take the asymptotic values,

$$I \sim 2\ln(2/\epsilon) + O(\epsilon^2) = 2E^{-1} + O(\epsilon^2) ,$$

$$\mathcal{I}_0 \sim 2 + O(\epsilon^2) , \quad \mathcal{I}_1 \sim O(\epsilon^2) , \quad \mathcal{I}_2 \sim 2E^{-1} - 2 + O(\epsilon^2) ,$$

and the dyadic integral becomes, to $O(\epsilon^2)$,

$$\begin{pmatrix} 4E^{-1}-2 & 0 & 0 \\ 0 & 2E^{-1}+2\cos^2\phi & 2\sin\phi\cos\phi \\ 0 & 2\sin\phi\cos\phi & 2E^{-1}+2\sin^2\phi \end{pmatrix} .$$

For convenience, we have defined the small parameter $E = 1/\ln(2/\epsilon)$. Thus for $\rho \ll a$ (and provided also that we are away from the ends) the velocity field is a constant of the form

$$\boldsymbol{v} \sim \frac{1}{4\pi\mu E}\boldsymbol{f}\cdot[2\boldsymbol{dd} + (\boldsymbol{\delta} - \boldsymbol{dd}) + O(E)] \ ,$$

and thus describes a rigid translation of the slender body. This suggests that for a slender body translating with a velocity $\boldsymbol{U}$, the disturbance field is approximately that produced by a line distribution of Stokeslets, with a constant force density given by

$$\begin{aligned} f^{\|} &= 2\pi\mu E U^{\|} + O(E^2) \\ f^{\perp} &= 4\pi\mu E U^{\perp} + O(E^2) . \end{aligned}$$

The drag on the particle is simply the net Stokeslet strength obtained by integration over the axis,

$$\begin{aligned} F^{\|} &= 4\pi\mu a E U + O(E^2) \\ F^{\perp} &= 8\pi\mu a E U + O(E^2) , \end{aligned}$$

and thus we recover Burger's deduction that the drag for axial translation is one-half that for transverse translation.

This simple version of slender body theory may be refined by including higher order singularities and by computing the line integrals to higher order [5, 86]. The singular behavior at the ends is handled by a separate expansion in those regions, following the standard procedure of matched asymptotic analysis [34, 88]. The end result for the disturbance field produced by translation of a slender body may be written as

$$\begin{aligned} \boldsymbol{v} &= -\boldsymbol{U}\cdot\int_{\alpha}^{\beta}\left[f^{\|}(\xi)\boldsymbol{dd} + f^{\perp}(\xi)(\boldsymbol{\delta}-\boldsymbol{dd})\right]\cdot\left\{1+\frac{1}{4}(\epsilon R(\xi))^2\nabla^2\right\}\mathcal{G}(\boldsymbol{x}-\boldsymbol{\xi})\,d\xi \\ &+ \int_{\alpha}^{\beta} h(\xi)(\boldsymbol{U}\cdot\nabla)\boldsymbol{d}\cdot\mathcal{G}(\boldsymbol{x}-\boldsymbol{\xi})\,d\xi , \end{aligned}$$

with

$$\begin{aligned} f^{\|}(\xi) &= \frac{E}{4}\left\{\frac{1-E\ln[(a^2-\xi^2)^{1/2}/R(\xi)]}{1-\frac{1}{2}E}\right\} + O(1/\ln\epsilon)^3 \\ f^{\perp}(\xi) &= \frac{E}{2}\left\{\frac{1-E\ln[(a^2-\xi^2)^{1/2}/R(\xi)]}{1+\frac{1}{2}E}\right\} + O(1/\ln\epsilon)^3 \\ h(\xi) &= \frac{1}{4}\epsilon^2\frac{d}{d\xi}\left(R(\xi)^2 f^{\perp}(\xi)\right) \\ &\quad + \frac{1}{2}(\epsilon R(\xi))^2\left\{\frac{d}{d\xi}\left[\left(\ln\left(\frac{(a^2-\xi^2)^{1/2}}{R(\xi)}\right)+\frac{1}{E}-1\right)f^{\perp}(\xi)\right]\right. \\ &\qquad + \frac{R'(\xi)}{R(\xi)}f^{\perp}(\xi) \\ &\quad \left. + \frac{1}{2}\int_{\alpha}^{\beta}\frac{f^{\perp}(\xi)-f^{\perp}(x)-(\xi-x)f^{\perp\prime}(x)}{(\xi-x)^2}\operatorname{sgn}(\xi-x)\,d\xi\right\} . \end{aligned}$$

The disturbance field is represented by a line distribution of Stokeslets, degenerate Stokes quadrupoles, and Stokes dipoles (stresslets and rotlets). In the limit of small b/a, the singularity solution for the prolate spheroid is consistent with that obtained by slender body theory.

By analyzing the asymptotic behavior at the ends, Tillet has shown that the distribution provides a uniform flow near a and $-a$ if the distribution ends at the foci of the osculating spheroid, *i.e.*,

$$\alpha = -a + \frac{1}{4}\epsilon^2 A^- \ , \quad \beta = a - \frac{1}{4}\epsilon^2 A^+ \ .$$

The ratio of the Stokeslet densities for axisymmetric and transverse motions are

$$\frac{f^{\parallel}}{f^{\perp}} = \frac{1}{2}\left(\frac{1+\frac{1}{2}E}{1-\frac{1}{2}E}\right) = \frac{1}{2}(1 + E + \cdots) \ ,$$

so that the drag for axisymmetric motion is one-half that for transverse motions, with a small $(\ln \epsilon^{-1})^{-1}$ correction that is independent of the particle shape.

3.5 Faxén Laws

So far in this chapter, we have discussed various aspects of the multipole expansion and have presented several examples and applications. However, with the exception of a few simple examples (where the solution can be obtained by other means), we have not addressed a key point – a method for determining the multipole moments. Thus, the question is: Given only the ambient flow $\boldsymbol{v}^\infty(\boldsymbol{x})$ and the motion of the particle, how do we determine the moments?

The answer to the question takes the form of direct relations, known as Faxén laws, that express the moments in terms of $\boldsymbol{v}^\infty$ and its derivatives. Actually, Hilding Faxén's original work [28, 29] concerned only the force and torque on a sphere; the original Faxén laws are essentially

$$\boldsymbol{F} = 6\pi\mu a\left(1 + \frac{a^2}{6}\nabla^2\right)\boldsymbol{v}^\infty(\boldsymbol{x})|_{x=0} - 6\pi\mu a\boldsymbol{U} \quad (3.19)$$

$$\boldsymbol{T} = 8\pi\mu a^3(\boldsymbol{\Omega}^\infty(\boldsymbol{x}) - \boldsymbol{\omega})|_{x=0} \ , \quad (3.20)$$

and were derived by calculating the force and torque from Lamb's general solution. Thus Faxén's original approach is not readily extended to the general particle.

The law for the force contains two interesting features. The first is that the correction to Stokes' law required by the general ambient flow $\boldsymbol{v}^\infty$ is relatively simple, being proportional to the ambient pressure gradient $\nabla^2\boldsymbol{v}^\infty = \nabla p^\infty/\mu$. The second, and perhaps the more remarkable, aspect is that we have encountered the grouping $(1 + (a^2/6)\nabla^2)$ before, in our discussion of the singularity solution for the translating sphere. We shall see that this reoccurrence is not accidental, but follows as a corollary to the Lorentz reciprocal theorem. This observation leads to a method for construction of moments for the general particle.

We recall from Chapter 2 the following more general statement of the Lorentz reciprocal theorem:

$$\oint_S \boldsymbol{v}_1 \cdot (\boldsymbol{\sigma}_2 \cdot \boldsymbol{n})\, dS - \int_V \boldsymbol{v}_1 \cdot (\nabla \cdot \boldsymbol{\sigma}_2)\, dV$$

This procedure is readily generalized to higher order fields. The reader is encouraged to examine the linear combination

$$\mathcal{G}_{ij,k_1\ldots k_n} + \frac{a^2}{4n+6}\nabla^2 \mathcal{G}_{ij,k_1\ldots k_n}$$

to investigate its suitability as a solution for a sphere in an n-th order ambient field.

3.3.2 The Spherical Drop and Interior Flows

For interior flows the *Stokeson* is the analog of the Stokeslet [20], since it too is linear with respect to a constant vector $\boldsymbol{U}$. The velocity field of the Stokeson may be written as

$$\begin{aligned} v_i = \mathcal{H}_{ij}U_j &= (2r^2\delta_{ij} - x_i x_j)U_j \\ p &= 10\mu\boldsymbol{x}\cdot\boldsymbol{U} \ . \end{aligned}$$

It should come as no surprise that the Stokeson enters into the solution for the translating spherical drop, since the interior solution must be linear with respect to $\boldsymbol{U}$, and $\boldsymbol{U}\cdot\mathcal{H}$ meets this criterion. Indeed, the solution for the translating drop can be constructed using a Stokeslet and a degenerate quadrupole outside and a Stokeson and a uniform field $\boldsymbol{U}$ inside the drop. The velocity fields inside and outside the drop are written as

$$\begin{aligned} \boldsymbol{v}^{(i)} &= D_0\hat{\boldsymbol{U}} + D_2 a^{-2}\boldsymbol{U}\cdot\mathcal{H}(\boldsymbol{x}) \\ \boldsymbol{v}^{(o)} &= \frac{3a}{4}\boldsymbol{U}\cdot(C_0 + C_2 a^2\nabla^2)\mathcal{G}(\boldsymbol{x}) \end{aligned}$$

to satisfy the boundary condition for large r and at $r = 0$. The boundary conditions at the drop interface provide the equations to determine the four unknown constants.

At $r = a$, the boundary conditions on the radial and tangential velocities and the tangential component of the surface traction $\boldsymbol{\sigma}\cdot\boldsymbol{n}$ are

1. $\hat{\boldsymbol{n}}\cdot\boldsymbol{v}^{(o)} = \hat{\boldsymbol{n}}\cdot\boldsymbol{U}$

2. $\hat{\boldsymbol{n}}\cdot\boldsymbol{v}^{(i)} = \hat{\boldsymbol{n}}\cdot\boldsymbol{U}$

3. $\boldsymbol{v}^{(o)} - \hat{\boldsymbol{n}}\hat{\boldsymbol{n}}\cdot\boldsymbol{v}^{(o)} = \boldsymbol{v}^{(i)} - \hat{\boldsymbol{n}}\hat{\boldsymbol{n}}\cdot\boldsymbol{v}^{(i)}$

4. $(\boldsymbol{e}^{(o)}\cdot\hat{\boldsymbol{n}})\cdot(\boldsymbol{\delta} - \hat{\boldsymbol{n}}\hat{\boldsymbol{n}}) = \lambda(\boldsymbol{e}^{(i)}\cdot\hat{\boldsymbol{n}})\cdot(\boldsymbol{\delta} - \hat{\boldsymbol{n}}\hat{\boldsymbol{n}})$

where $\lambda = \mu^{(i)}/\mu^{(o)}$, the ratio of the drop and solvent viscosities. Here, C_0 and C_2, are the quantities of greatest interest, and the final results are

$$\begin{aligned} C_0 &= \frac{2+3\lambda}{3(1+\lambda)} \\ C_2 &= \frac{\lambda}{6(1+\lambda)} \ . \end{aligned}$$

$$= \oint_S \boldsymbol{v}_2 \cdot (\boldsymbol{\sigma}_1 \cdot \boldsymbol{n})\, dS - \int_V \boldsymbol{v}_2 \cdot (\nabla \cdot \boldsymbol{\sigma}_1)\, dV \;, \tag{3.21}$$

where V is the fluid volume bounded by the particle surface and a large sphere "at infinity." The surface integrals are over these boundaries of V. Recall that the simpler and more frequently used form of the reciprocal theorem follows from the assumption $\nabla \cdot \boldsymbol{\sigma} = 0$. We relax this condition here for reasons that will soon become apparent.

In all applications of the reciprocal theorem in this section, the fields $\boldsymbol{v}_1$, $\boldsymbol{\sigma}_1$, $\boldsymbol{v}_2$, and $\boldsymbol{\sigma}_2$ *decay far away from the particle*, to the extent that the surface contributions from S_∞ vanish in the limit of large S_∞. The integral over the particle surface S_p is written in terms of the outward normal (which points into the fluid), so that Equation 3.21 may be rewritten as

$$\oint_{S_p} \boldsymbol{v}_1 \cdot (\boldsymbol{\sigma}_2 \cdot \boldsymbol{n})\, dS + \int_V \boldsymbol{v}_1 \cdot (\nabla \cdot \boldsymbol{\sigma}_2)\, dV$$
$$= \oint_{S_p} \boldsymbol{v}_2 \cdot (\boldsymbol{\sigma}_1 \cdot \boldsymbol{n})\, dS + \int_V \boldsymbol{v}_2 \cdot (\nabla \cdot \boldsymbol{\sigma}_1)\, dV \;. \tag{3.22}$$

Let us now take for $\boldsymbol{v}_1$ the solution for a particle translating with velocity $\boldsymbol{U}$ in a quiescent fluid. We note that $\nabla \cdot \boldsymbol{\sigma}_1 = 0$ in V. For $\boldsymbol{v}_2$, we take the velocity field generated by a point force $\boldsymbol{F} \cdot \mathcal{G}(\boldsymbol{x} - \boldsymbol{y})/8\pi\mu$, where $\boldsymbol{y}$ lies outside the particle, and the particle is stationary. Then $\boldsymbol{v}_2 = 0$ on S_p, and $\nabla \cdot \boldsymbol{\sigma}_2 = -\boldsymbol{F}\delta(\boldsymbol{x} - \boldsymbol{y})$ in V. When these boundary conditions and identities are inserted into Equation 3.22, the result is

$$\boldsymbol{U} \cdot \boldsymbol{F}_2 - \boldsymbol{v}_1(\boldsymbol{y}) \cdot \boldsymbol{F} = 0 \;, \tag{3.23}$$

where $\boldsymbol{F}_2$ is the force on the particle generated by the surface traction $\boldsymbol{\sigma}_2 \cdot \boldsymbol{n}$. Due to linearity of the Stokes equation, we may factor $\boldsymbol{U}$ from $\boldsymbol{v}_1$. Furthermore, suppose that $\boldsymbol{v}_1$ is written as a singularity solution, so that

$$\boldsymbol{v}_1(\boldsymbol{x}) = \boldsymbol{U} \cdot \mathcal{F}\{\mathcal{G}(\boldsymbol{x} - \boldsymbol{\xi})/8\pi\mu\} \;, \tag{3.24}$$

where $\mathcal{F}$ is a linear functional and $\boldsymbol{\xi}$ represents the region inside the translating particle over which the singularities are distributed. Then Equation 3.23 becomes

$$(\boldsymbol{F}_2)_i = \mathcal{F}\{F_j \mathcal{G}_{ji}(\boldsymbol{y} - \boldsymbol{\xi})/8\pi\mu\} \;. \tag{3.25}$$

However, $F_j \mathcal{G}_{ji}(\boldsymbol{y} - \boldsymbol{\xi})/8\pi\mu = F_j \mathcal{G}_{ij}(\boldsymbol{\xi} - \boldsymbol{y})/8\pi\mu$ is[2] the ambient field of the $\boldsymbol{v}_2$ problem, evaluated over the image region, $\boldsymbol{x} = \boldsymbol{\xi}$. So for this case we have shown that

$$\boldsymbol{F}_2 = \mathcal{F}\{\boldsymbol{v}^\infty(\boldsymbol{\xi})\} \;. \tag{3.26}$$

But all ambient fields $\boldsymbol{v}^\infty$ that satisfy the Stokes equation can be constructed from an appropriate set of images, so Equation 3.26 holds for the general ambient field. For spheres, recall $\boldsymbol{\xi}$ is the sphere center and that

$$\mathcal{F} = 6\pi\mu a\{1 + \frac{a^2}{6}\nabla^2\} \;, \tag{3.27}$$

[2]This equality can be verified directly, but it also follows from the self-adjoint property of the Green's dyadic (see Exercise 2.16).

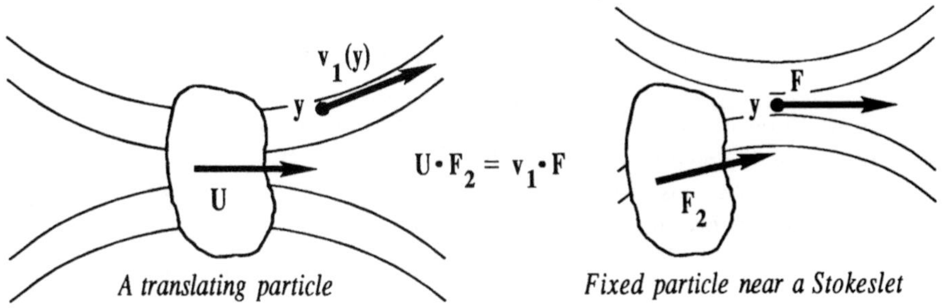

Figure 3.4: A reciprocal relation: the translational disturbance field and the drag on a sphere near a Stokeslet.

so we recover Faxén's original result for the force on the sphere.[3] The analogous development of the Faxén law for the torque and stresslet on a stationary particle yields (see Exercise 3.6)

$$\boldsymbol{T} = \mathcal{T}\{\boldsymbol{v}^\infty(\boldsymbol{\xi})\} \tag{3.28}$$

$$\boldsymbol{S} = \mathcal{S}\{\boldsymbol{v}^\infty(\boldsymbol{\xi})\} \tag{3.29}$$

where $\mathcal{T}$ and $\mathcal{S}$ are the linear functionals of $\mathcal{G}(\boldsymbol{x}-\boldsymbol{\xi})/8\pi\mu$ associated with the singularity solutions with boundary conditions $\boldsymbol{\omega}\times\boldsymbol{x}$ and $\boldsymbol{E}\cdot\boldsymbol{x}$. The Faxén laws for a moving particle are obtained by adding the contributions for the particle moving through quiescent fluid to the results for the stationary particle.

Example 3.4 The Faxén Laws for the Torque and Stresslet on a Sphere.

We first derive the results for a stationary sphere. The singularity solution for the rotating sphere is

$$v_i = -4\pi\mu a^3 \epsilon_{jk\ell}\omega_\ell \frac{\partial}{\partial\xi_k}\frac{\mathcal{G}_{ij}(\boldsymbol{x}-\boldsymbol{\xi})}{8\pi\mu} ,$$

so that

$$T_\ell = -4\pi\mu a^3 \epsilon_{jk\ell}\frac{\partial v_j^\infty}{\partial\xi_k}|_{x=0} = 8\pi\mu a^3 \Omega_\ell^\infty(\boldsymbol{x})|_{x=0} .$$

The result for a rotating sphere, Equation 3.20, follows by linear superposition.

The singularity solution for the straining field is

$$v_i = E_{ij}x_j - \frac{20}{3}\pi\mu a^3 E_{jk}\{1+\frac{a^2}{10}\nabla^2\} \times\frac{1}{2}\left(\frac{\partial}{\partial\xi_k}\frac{\mathcal{G}_{ij}(\boldsymbol{x}-\boldsymbol{\xi})}{8\pi\mu}+\frac{\partial}{\partial\xi_j}\frac{\mathcal{G}_{ik}(\boldsymbol{x}-\boldsymbol{\xi})}{8\pi\mu}\right) .$$

[3]This simple derivation was pointed out to us in a private communication from Dr. E.J. Hinch, of Cambridge University.

It is now clear which terms generate $\boldsymbol{E}\cdot\boldsymbol{x}$ on the sphere surface, so that

$$S_{jk} = \frac{20}{3}\pi\mu a^3\{1 + \frac{a^2}{10}\nabla^2\}e_{jk}(\boldsymbol{x})|_{x=0} \ .$$

The stresslet on a sphere in an arbitrary field is the sum of two contributions: an ambient rate-of-strain contribution (with the rate-of-strain evaluated at the sphere center), which is of the same form as when the sphere is in a linear field, plus a correction proportional to the Laplacian of the rate-of-strain (a third derivative of the velocity field).◇

3.5.1 Ellipsoids and Spheroids

From the singularity solutions and multipole expansions for the ellipsoid, we obtain two forms of the Faxén laws for the ellipsoid:

$$\begin{aligned}
\boldsymbol{F} &= \mu\boldsymbol{A}\cdot\int_E f_{(1)}(\boldsymbol{x}')\{1+\frac{c^2q^2\nabla^2}{2}\}\boldsymbol{v}^\infty(\boldsymbol{x}')dA(\boldsymbol{x}') - \mu\boldsymbol{A}\cdot\boldsymbol{U} \\
&= \mu\boldsymbol{A}\cdot\left(\frac{\sinh D}{D}\right)\boldsymbol{v}^\infty(\boldsymbol{x})|_{x=0} - \mu\boldsymbol{A}\cdot\boldsymbol{U}
\end{aligned}$$

$$\begin{aligned}
\boldsymbol{T} &= \mu\boldsymbol{C}\cdot\int_E f_{(2)}(\boldsymbol{x}')\{\frac{1}{2}\nabla\times\boldsymbol{v}^\infty - \boldsymbol{\omega}\}dA(\boldsymbol{x}') \\
&+ \mu\widetilde{\boldsymbol{H}} : \int_E f_{(2)}(\boldsymbol{x}')\{1+\frac{c^2q^2\nabla^2}{6}\}\boldsymbol{e}^\infty(\boldsymbol{x}')dA(\boldsymbol{x}') \\
&= \mu\boldsymbol{C}\cdot\left(\frac{3}{D}\frac{\partial}{\partial D}\right)\left(\frac{\sinh D}{D}\right)\frac{1}{2}\nabla\times\boldsymbol{v}^\infty(\boldsymbol{x})|_{x=0} - \mu\boldsymbol{C}\cdot\boldsymbol{\omega} \\
&+ \mu\widetilde{\boldsymbol{H}} : \left(\frac{3}{D}\frac{\partial}{\partial D}\right)\left(\frac{\sinh D}{D}\right)\boldsymbol{e}^\infty(\boldsymbol{x})|_{x=0} \ .
\end{aligned}$$

$$\begin{aligned}
\boldsymbol{S} &= \mu\boldsymbol{M} : \int_E f_{(2)}(\boldsymbol{x}')\{1+\frac{c^2q^2\nabla^2}{6}\}\boldsymbol{e}^\infty(\boldsymbol{x}')dA(\boldsymbol{x}') \\
&+ \mu\boldsymbol{H}\cdot\int_E f_{(2)}(\boldsymbol{x}')\{\frac{1}{2}\nabla\times\boldsymbol{v}^\infty - \boldsymbol{\omega}\}dA(\boldsymbol{x}') \\
&= \mu\boldsymbol{M} : \left(\frac{3}{D}\frac{\partial}{\partial D}\right)\left(\frac{\sinh D}{D}\right)\boldsymbol{e}^\infty(\boldsymbol{x})|_{x=0} \\
&+ \mu\boldsymbol{H}\cdot\left(\frac{3}{D}\frac{\partial}{\partial D}\right)\left(\frac{\sinh D}{D}\right)\frac{1}{2}\nabla\times\boldsymbol{v}^\infty(\boldsymbol{x})|_{x=0} - \mu\boldsymbol{H}\cdot\boldsymbol{\omega} \ .
\end{aligned}$$

The symbolic operator (power series in D^2) form is efficient if the higher order terms in $(D^2)^k$ vanish rapidly with increasing k. However, the symbolic operator sum will not always converge. For example, consider the ambient field generated by the disturbance produced by a small particle translating near the ellipsoid. The series in $(D^2)^k\boldsymbol{v}^\infty = (D^2)^k\boldsymbol{F}\cdot\mathcal{G}(\boldsymbol{x}-\boldsymbol{y})/8\pi\mu$ will diverge if $\boldsymbol{y}$ is closer than a critical distance, as shown in Chapter 13. The integral form of the Faxén law is always well defined for all $\boldsymbol{y}$ outside the ellipsoid and thus is the analytical continuation of the series.

The Faxén laws for the spheroid may be obtained by reduction of the results for the ellipsoid or by starting with the singularity solutions and applying the reciprocal theorem. The results for *prolate* spheroids are the most interesting, since the focal disk becomes the line segment between the two foci at $\pm k$ on the axis of symmetry, with $k = ae$. Consequently, the Faxén laws may be written in terms of line integrals of the ambient velocity field and its derivatives [50]:

$$\begin{aligned}
F_i &= 6\pi\mu a\left[X^A d_i d_j + Y^A(\delta_{ij} - d_i d_j)\right] \\
&\quad \times \frac{1}{2k}\int_{-k}^{k}\left\{1 + (k^2-\xi^2)\frac{(1-e^2)}{4e^2}\nabla^2\right\}(v_j^\infty - U_j)\,d\xi \\
T_i &= 8\pi\mu a^3\left[X^C d_i d_j + Y^C(\delta_{ij} - d_i d_j)\right] \\
&\quad \times \frac{3}{8k^3}\int_{-k}^{k}(k^2-\xi^2)((\nabla\times\boldsymbol{v}^\infty)_j - 2\omega_j)\,d\xi \\
&\quad - 8\pi\mu a^3 Y^H \epsilon_{ijl} d_l d_k \\
&\quad \times \frac{3}{4k^3}\int_{-k}^{k}\left\{(k^2-\xi^2)(1 + (k^2-\xi^2)\frac{(1-e^2)}{8e^2}\nabla^2\right\} e_{jk}^\infty\,d\xi \\
S_{ij} &= \frac{20}{3}\pi\mu a^3\left[X^M d_{ijkl}^{(0)} + Y^M d_{ijkl}^{(1)} + Z^M d_{ijkl}^{(2)}\right] \\
&\quad \times \frac{3}{4k^3}\int_{-k}^{k}\left\{(k^2-\xi^2)(1 + (k^2-\xi^2)\frac{(1-e^2)}{8e^2}\nabla^2\right\} e_{kl}^\infty\,d\xi \\
&\quad + \frac{20}{3}\pi\mu a^3\frac{3}{5}Y^H(\epsilon_{ikl}d_j + \epsilon_{jkl}d_i)d_l \\
&\quad \times \frac{3}{8k^3}\int_{-k}^{k}(k^2-\xi^2)((\nabla\times\boldsymbol{v}^\infty)_k - 2\omega_k)\,d\xi\ .
\end{aligned}$$

For a force-free and torque-free spheroid, we set $\boldsymbol{F} = \boldsymbol{T} = \boldsymbol{0}$ and obtain the following expressions for the translational and angular velocities:

$$\boldsymbol{U} = \frac{1}{2k}\int_{-k}^{k}\left\{1 + (k^2-\xi^2)\frac{(1-e^2)}{4e^2}\nabla^2\right\}\boldsymbol{v}^\infty\,d\xi$$

$$\begin{aligned}
\boldsymbol{\omega} &= \frac{3}{8k^3}\int_{-k}^{k}(k^2-\xi^2)\nabla\times\boldsymbol{v}^\infty\,d\xi \\
&+ \frac{e^2}{2-e^2}\frac{3}{4k^3}\int_{-k}^{k}\left\{(k^2-\xi^2)(1 + (k^2-\xi^2)\frac{(1-e^2)}{8e^2}\nabla^2\right\}\boldsymbol{d}\times(\boldsymbol{e}^\infty\cdot\boldsymbol{d})\,d\xi\ .
\end{aligned}$$

The orientation of the spheroid changes with time as

$$\begin{aligned}
\dot{\boldsymbol{d}} = \boldsymbol{\omega}\times\boldsymbol{d} &= \frac{3}{8k^3}\int_{-k}^{k}(k^2-\xi^2)(\nabla\times\boldsymbol{v}^\infty)\times\boldsymbol{d}\,d\xi \\
&+ \frac{e^2}{2-e^2}\frac{3}{4k^3}\int_{-k}^{k}\left\{(k^2-\xi^2)(1 + (k^2-\xi^2)\frac{(1-e^2)}{8e^2}\nabla^2\right\} \\
&\times\ (\boldsymbol{e}^\infty\cdot\boldsymbol{d} - \boldsymbol{e}^\infty : \boldsymbol{ddd})\,d\xi\ .
\end{aligned}$$

3.5.2 The Spherical Drop

The Faxén laws for viscous drops are also of the same functional form as the singularity solutions for the drop. The proof requires some extensions of the steps used for the rigid particle. We start with the reciprocal theorem and for $\boldsymbol{v}_1$ use the velocity field produced by a translating drop, while $\boldsymbol{v}_2$ is taken to be the field of a *stationary* drop near a point force at $\boldsymbol{y}$. The proof proceeds along identical lines to the one used for the rigid particle, up to the point where the boundary conditions are inserted into the reciprocal relation. We must have

$$\oint_S \boldsymbol{v}_1 \cdot (\boldsymbol{\sigma}_2 \cdot \boldsymbol{n})\, dS + \boldsymbol{v}_1(\boldsymbol{y}) \cdot \boldsymbol{F} = \oint_S \boldsymbol{v}_2 \cdot (\boldsymbol{\sigma}_1 \cdot \boldsymbol{n})\, dS\ . \tag{3.30}$$

At this point, it proves convenient to extract $\boldsymbol{U}$ from $\boldsymbol{v}_1$ so that the preceding equation becomes

$$\boldsymbol{U} \cdot \boldsymbol{F}_2 + \boldsymbol{v}_1(\boldsymbol{y}) \cdot \boldsymbol{F} = -\oint_S (\boldsymbol{v}_1 - \boldsymbol{U}) \cdot (\boldsymbol{\sigma}_2 \cdot \boldsymbol{n})\, dS + \oint_S \boldsymbol{v}_2 \cdot (\boldsymbol{\sigma}_1 \cdot \boldsymbol{n})\, dS\ . \tag{3.31}$$

The boundary conditions at the surface of the drop require that $\boldsymbol{n} \cdot \boldsymbol{v}_2 = 0$ and $\boldsymbol{n} \cdot (\boldsymbol{v}_1 - \boldsymbol{U}) = 0$, so only the *tangential* components of the surface traction are retained in the dot products. An application of the reciprocal theorem to the inner fields associated with $\boldsymbol{v}_1 - \boldsymbol{U}$ and $\boldsymbol{v}_2$ yields the relation,

$$\oint_S (\boldsymbol{v}_1^{(i)} - \boldsymbol{U}) \cdot (\boldsymbol{\sigma}_2^{(i)} \cdot \boldsymbol{n})\, dS = \oint_S \boldsymbol{v}_2^{(i)} \cdot (\boldsymbol{\sigma}_1^{(i)} \cdot \boldsymbol{n})\, dS\ . \tag{3.32}$$

Here again, it is understood that only the tangential component of the surface traction is retained. But in both problem 1 and 2, the tangential component of the traction is continuous across the interface. The preceding equation then implies that the two surface integrals in Equation 3.31 cancel. Again, we suppose that $\boldsymbol{v}_1$, the field produced by the translating drop, is available as a singularity solution, so that

$$\boldsymbol{v}_1(\boldsymbol{x}) = \boldsymbol{U} \cdot \mathcal{F}\{\mathcal{G}(\boldsymbol{x} - \boldsymbol{\xi})/8\pi\mu\}\ ,$$

where $\mathcal{F}$ is a linear functional and $\boldsymbol{\xi}$ represents the region over which the singularities are distributed. Using the same arguments as before, Equation 3.31 becomes

$$\boldsymbol{F}_2 = \mathcal{F}\{\boldsymbol{v}^\infty(\boldsymbol{\xi})\}\ ,$$

and we conclude that the Faxén law is of the same functional form as the singularity solution. For a spherical drop, we know the singularity solution and the Faxén law for the force follows immediately as

$$\boldsymbol{F} = 2\pi\mu_o a \left(\frac{2+3\lambda}{1+\lambda}\right) \left\{1 + \frac{\lambda a^2 \nabla^2}{2(2+3\lambda)}\right\} \boldsymbol{v}^\infty(\boldsymbol{x})|_{x=0} - 2\pi\mu_o a \left(\frac{2+3\lambda}{1+\lambda}\right) \boldsymbol{U}\ .$$

This was first shown by Hetsroni and Haber [38], who followed Faxén's approach.

The Faxén law for the stresslet on a spherical drop was first derived by Rallison [69]. For the sake of consistency we employ the standard proof, but Rallison's original proof contains essentially the same ideas.

In the reciprocal theorem, use for v_1 the disturbance field of a drop in a rate-of-strain field, $E \cdot x$, and for v_2, use the field of a *stationary* drop near a point force at y. Because of the boundary condition $v_1 = -E \cdot x$ on S_p, we obtain

$$\begin{aligned} -E : \frac{1}{2} \oint_{S_p} (x(\sigma \cdot n) + x(\sigma \cdot n)) \, dS + v_1(y) \cdot F \\ = -\oint_{S_p} (v_1 + E \cdot x) \cdot (\sigma_2 \cdot n) \, dS + \oint_{S_p} v_2 \cdot (\sigma_1 \cdot n) \, dS \, . \end{aligned} \tag{3.33}$$

The boundary conditions at the surface of the drop require that $n \cdot v_2 = 0$ and $n \cdot (v_1 + E \cdot x) = 0$, so again, only the *tangential* components of the surface traction are retained in the dot products. An application of the reciprocal theorem to the inner fields associated with $v_1 - U$ and v_2 yields the relation,

$$\oint_S (v_1^{(i)} + E \cdot x) \cdot (\sigma_2^{(i)} \cdot n) \, dS = \oint_S v_2^{(i)} \cdot (\sigma_1^{(i)} \cdot n + 2\mu_i E \cdot n) \, dS \, ,$$

where again it is understood that only the tangential components of the surface tractions are retained. Again, the tangential components of the traction are continuous across the interface, so that the preceding equation becomes

$$\oint_S (v_1 + E \cdot x) \cdot (\sigma_2 \cdot n) \, dS = \oint_S v_2 \cdot (\sigma_1 \cdot n + 2\mu_o E \cdot n) \, dS \, ,$$

and Equation 3.33 simplifies to

$$E : \left[\frac{1}{2} \oint_S (x(\sigma_2 \cdot n) + x(\sigma_2 \cdot n)) \, dS - \mu_o \oint_S (v_2 n + n v_2) \, dS \right] = v_1(y) \cdot F \, .$$

We recognize the quantity inside the brackets on the left hand side of this equation as the stresslet, S_2.

Again, suppose that v_1, the disturbance field of the drop in a rate-of-strain field is in the form of a singularity solution,

$$v_1(x) = E : \mathcal{S}\{\mathcal{G}(x - \xi)/8\pi\mu\} \, ,$$

where $\mathcal{S}$ is a linear functional and ξ represents the region over which the singularities are distributed. Using the same arguments as before, we conclude that

$$S_2 = \mathcal{F}\{v^\infty(\xi)\} \, .$$

This singularity solution is derived for a spherical drop in Exercise 3.7, and based on that result, the Faxén law for the stresslet follows as

$$S = \frac{4}{3}\pi\mu_o a \left(\frac{2 + 5\lambda}{1 + \lambda} \right) \left\{ 1 + \frac{\lambda a^2 \nabla^2}{2(2 + 5\lambda)} \right\} e^\infty(x)|_{x=0} \, .$$

Exercises

Exercise 3.1 The Integral Representation for the Disturbance Field of a Viscous Drop.
Derive the integral representation for the disturbance field of a viscous drop:

$$\begin{aligned} \boldsymbol{v}(\boldsymbol{x}) - \boldsymbol{v}^{\infty}(\boldsymbol{x}) &= -\frac{1}{8\pi\mu}\oint_S (\boldsymbol{\sigma}(\boldsymbol{\xi})\cdot\hat{\boldsymbol{n}})\cdot\mathcal{G}(\boldsymbol{x}-\boldsymbol{\xi})\, dS(\boldsymbol{\xi}) \\ &\quad - \oint_S \boldsymbol{v}(\boldsymbol{\xi})\cdot(\boldsymbol{\Sigma}\cdot\hat{\boldsymbol{n}})\, dS(\boldsymbol{\xi}) \,. \end{aligned}$$

Use the analogous representation for the flow inside the drop to derive the following representation in terms of a traction jump across the interface and a double layer potential [68, 70]

$$\begin{aligned} \frac{1}{2}\boldsymbol{v}(\boldsymbol{\xi}) &= -\frac{1}{8\pi\mu}\frac{1}{1+\lambda}\oint_S([\boldsymbol{\sigma}^{(o)} - \boldsymbol{\sigma}^{(i)}](\boldsymbol{\xi})\cdot\hat{\boldsymbol{n}})\cdot\mathcal{G}(\boldsymbol{x}-\boldsymbol{\xi})\, dS(\boldsymbol{\xi}) \\ &\quad - \left(\frac{1-\lambda}{1+\lambda}\right)\oint_S \boldsymbol{v}(\boldsymbol{\xi})\cdot(\boldsymbol{\Sigma}(\boldsymbol{x}-\boldsymbol{\xi})\cdot\hat{\boldsymbol{n}})\, dS(\boldsymbol{\xi}) \,. \end{aligned}$$

Exercise 3.2 The Stresslet for a Viscous Drop.
Expand the integral representation of the previous exercise and show that the stresslet on a viscous drop is given by

$$\boldsymbol{S} = \frac{1}{2}\oint_S [(\boldsymbol{\sigma}\cdot\hat{\boldsymbol{n}})\boldsymbol{\xi} + \boldsymbol{\xi}(\boldsymbol{\sigma}\cdot\hat{\boldsymbol{n}})]\, dS - \frac{1}{3}\oint_S (\boldsymbol{\sigma}\cdot\hat{\boldsymbol{n}})\cdot\boldsymbol{\xi}\, dS\, \boldsymbol{\delta} - \mu\oint_{S_p}(\boldsymbol{v}\hat{\boldsymbol{n}} + \hat{\boldsymbol{n}}\boldsymbol{v})\, dS \,.$$

Exercise 3.3 The Spheroidal Harmonics.
When the Laplace equation $\nabla^2\Phi = 0$ is separated in prolate spheroidal coordinates (η, θ, ϕ),

$$\begin{aligned} x &= c \sinh\eta \sin\theta \cos\phi \\ y &= c \sinh\eta \sin\theta \sin\phi \\ z &= c \cosh\eta \cos\theta \,, \end{aligned}$$

the solution may be written as [42]

$$\Phi(\eta, \theta, \phi) = P_n(\cosh\eta) P_n(\cos\theta) e^{im\phi} \,.$$

Find the relation between these solutions and G_n.

Exercise 3.4 Degenerate Limits for the Ellipsoid.
Show that the singularity solution and image system of the general triaxial ellipsoid reduce to that for the prolate spheroid in the limit of vanishing $b_E = \sqrt{b^2 - c^2}$.
Hint: In the limit of small b_E/a_E, replace $\mathcal{G}(\boldsymbol{x}, \boldsymbol{\xi})$ with its Taylor series in ξ_2 about $\xi_2 = 0$; then show that the higher order terms vanish and that the leading term, after integration with respect to ξ_2, gives the singularity density of the prolate spheroid solution.

Exercise 3.5 Relation Between the Solutions for Prolate and Oblate Spheroids.
Show that the resistance functions for the prolate and oblate spheroids can be obtained from one another by switching the roles played by a and c, or equivalently, by replacing the eccentricity e everywhere by $ie/\sqrt{1-e^2}$, using the properties of the complex trigonometric and logarithmic functions.

Exercise 3.6 The Faxén Law for the Torque and Stresslet.
Relate the Faxén law for the torque and stresslet on a general particle and the particle's singularity solutions for rotation and rate-of-strain fields.

Exercise 3.7 The Singularity Solution for a Drop in a Rate-of-Strain Field.
Show that the singularity solution for a spherical drop in a rate of strain field is

$$\boldsymbol{v} = \boldsymbol{E} \cdot \boldsymbol{x} + 2\pi\mu_o a(\boldsymbol{E} \cdot \nabla) \cdot \left(\frac{2+5\lambda}{1+\lambda}\right) \left\{1 + \frac{\lambda a^2 \nabla^2}{2(2+5\lambda)}\right\} \frac{\mathcal{G}(\boldsymbol{x})}{8\pi\mu} .$$

Note that as $\lambda \to \infty$, we recover the solution for the rigid sphere, while $\lambda = 0$ (bubble) requires just the stresslet field. The singularity solution pertains to the exterior flow field. Inside the drop, the velocity should be represented as a linear combination to $\boldsymbol{E} \cdot \boldsymbol{x}$ and a cubic field.

In this solution for the viscous drop, the normal forces *are not* balanced. When surface tension forces dominate over viscous forces, the shape of the surface may be determined in an iterative fashion, with the above solution providing the leading term [22]. For the general case, a numerical approach is required, *e.g.*, [70].

Chapter 4

Solutions in Spherical Coordinates

4.1 Introduction

In this chapter we consider the solution of the Stokes equations in spherical polar coordinates. The slow motion of a single sphere in various flow fields is an old subject, and the earliest solutions have been known for more than a century. However, our aim here is to present the subject in a unified framework and to build an inventory of analytical examples that will serve us well throughout the rest of the book. Indeed, recent work in the numerical analysis for the Stokes equations provides the motivation for new viewpoints for an old subject.

In Section 4.2, we describe Lamb's general solution for the Stokes equation. The connection with the multipole expansion is brought out, and the solutions for the sphere in translation, rotation, and linear fields are derived to illustrate the use of Lamb's solution. The elements of Lamb's solution do not form an orthogonal basis on the sphere surface. Hence, the matching of boundary conditions is far more involved than, for example, the corresponding problems in potential theory. In Sections 4.3 and 4.4, we discuss two variations of Lamb's solution, the so-called *adjoint method* involving the use of a dual basis, and an *orthonormal basis* for the surface vector fields of solutions to the Stokes equations. We conclude this chapter with a brief survey of the Stokes streamfunction; again the main emphasis is the connection of this special technique for axisymmetric flows to more general methods.

4.2 Lamb's General Solution

In spherical polar coordinates (r, θ, ϕ) (see Figure 4.1), the Laplace equation $\nabla^2\Phi = 0$, separates to yield

$$\Phi_{nm}(r, \theta, \phi) = r^n P_n^m(\cos\theta)e^{im\phi} ,$$

where P_n^m, $n = 0, 1, 2, \ldots$, $-n \leq m \leq n$ are the associated Legendre functions [1, 42]. Lamb [53] has derived the analogous solutions for the Stokes equations.

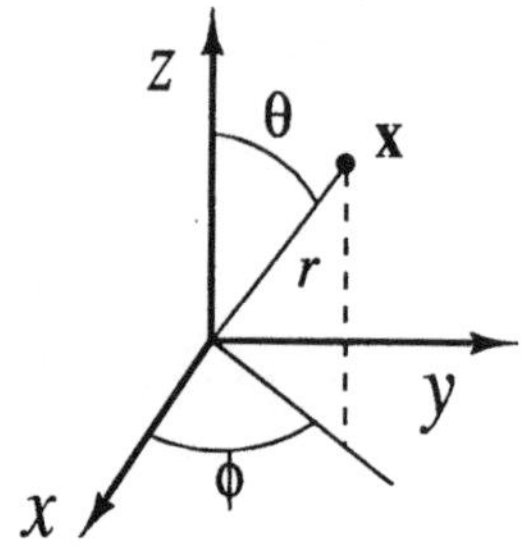

Figure 4.1: Spherical polar coordinates.

Since the pressure satisfies the Laplace equation, we may expand it as above,

$$p = \sum_{n=-\infty}^{\infty} p_n \ ,$$

where p_n is a solid spherical harmonic of order n,

$$p_n = r^n \sum_{m=0}^{n} P_n^m(\cos\theta)\,(a_{mn}\cos m\phi + \tilde{a}_{mn}\sin m\phi) \ .$$

The ϕ-dependence is often written in terms of the trigonometric functions as above [35, 53], to avoid the use of complex variables.

We write the momentum balance as

$$\nabla^2 \boldsymbol{v} = \frac{1}{\mu}\nabla p$$

and construct homogeneous and particular solutions of this partial differential equation. The result is *Lamb's general solution* (see also [35]):

$$\begin{aligned} \boldsymbol{v} &= \sum_{\substack{n=-\infty \\ n\neq 1}}^{\infty} \left[\frac{(n+3)r^2\nabla p_n}{2\mu(n+1)(2n+3)} - \frac{n\boldsymbol{x}p_n}{\mu(n+1)(2n+3)} \right] \\ &\quad + \sum_{n=-\infty}^{\infty} \left[\nabla\Phi_n + \nabla\times(\boldsymbol{x}\chi_n)\right] \ , \end{aligned} \tag{4.1}$$

where Φ_n and χ_n are also solid spherical harmonics of order n. The homogeneous solution is constructed from a potential (the $\nabla\Phi$ term) and a toroidal field (the $\nabla\times(\boldsymbol{x}\chi)$ term). In Exercise 4.1, we verify that the p_n terms form a particular solution.

In essence, Equation 4.1 reduces the task to that of solving the much easier problems, $\nabla^2 p_n = \nabla^2\Phi_n = \nabla^2\chi_n = 0$. The general solution for the harmonics

Φ_n and χ_n are also of the standard form,

$$\Phi_n = r^n \sum_{m=0}^{n} P_n^m(\cos\theta)\left(b_{mn}\cos m\phi + \tilde{b}_{mn}\sin m\phi\right) \tag{4.2}$$

$$\chi_n = r^n \sum_{m=0}^{n} P_n^m(\cos\theta)\left(c_{mn}\cos m\phi + \tilde{c}_{mn}\sin m\phi\right) . \tag{4.3}$$

Axisymmetric flow problems correspond to the case $m = 0$ and the azimuthal angle ϕ does not appear in the solution. The three scalar functions are then written as

$$p_n = a_n r^n P_n(\cos\theta) \tag{4.4}$$

$$\Phi_n = b_n r^n P_n(\cos\theta) \tag{4.5}$$

$$\chi_n = c_n r^n P_n(\cos\theta) , \tag{4.6}$$

where $P_n = P_n^0$ are the Legendre polynomials. Only p_n and Φ_n are needed to describe axisymmetric flows of the type $\boldsymbol{v} = v_r(r,\theta)\boldsymbol{e}_r + v_\theta(r,\theta)\boldsymbol{e}_\theta$, while only χ_n appears for swirling flows, $\boldsymbol{v} = v_\phi(r,\theta)\boldsymbol{e}_\phi$.

For *interior* flows such as that inside a spherical drop, we discard harmonics with $n < 0$ in Equation 4.1, since these fields are singular at the origin, and the expansion becomes

$$\boldsymbol{v} = \sum_{n=1}^{\infty}\left[\frac{(n+3)r^2\nabla p_n}{2\mu(n+1)(2n+3)} - \frac{n\boldsymbol{x}p_n}{\mu(n+1)(2n+3)}\right] + \sum_{n=1}^{\infty}\left[\nabla\Phi_n + \nabla\times(\boldsymbol{x}\chi_n)\right] . \tag{4.7}$$

(The case $n = 0$ gives $\boldsymbol{v}$ identically zero.)

On the other hand, for *exterior* flows such as a disturbance field outside a sphere, we discard the positive harmonics since we expect the disturbance fields to decay as $r \to \infty$. We also find it more convenient to work with positive integers, so we replace n in Equation 4.1 with $-n-1$ and obtain

$$\boldsymbol{v} = \sum_{n=1}^{\infty}\left[-\frac{(n-2)r^2\nabla p_{-n-1}}{2\mu n(2n-1)} + \frac{(n+1)\boldsymbol{x}p_{-n-1}}{\mu n(2n-1)}\right] + \sum_{n=1}^{\infty}\left[\nabla\Phi_{-n-1} + \nabla\times(\boldsymbol{x}\chi_{-n-1})\right] , \tag{4.8}$$

with

$$p_{-n-1} = r^{-n-1}\sum_{m=0}^{n} P_n^m(\cos\theta)\left(A_{mn}\cos m\phi + \tilde{A}_{mn}\sin m\phi\right) \tag{4.9}$$

$$\Phi_{-n-1} = r^{-n-1}\sum_{m=0}^{n} P_n^m(\cos\theta)\left(B_{mn}\cos m\phi + \tilde{B}_{mn}\sin m\phi\right) \tag{4.10}$$

$$\chi_{-n-1} = r^{-n-1}\sum_{m=0}^{n} P_n^m(\cos\theta)\left(C_{mn}\cos m\phi + \tilde{C}_{mn}\sin m\phi\right) . \tag{4.11}$$

The case $n = 0$, *e.g.*, Φ_{-1}, is associated with sources and sinks, and thus is not encountered in the physical description of disturbance flows produced by the motion of rigid particles. However, we will encounter these in the mathematical setting in Part IV.

As a representation in spherical polar coordinates, Lamb's solution is clearly ideal for flow problems involving a single sphere, but the utility of Lamb's solution extends beyond this obvious application. The velocity representation forms a *complete set* and thus provides the basis for numerical solution of multisphere problems in Parts III and IV.

4.2.1 The Connection with the Multipole Expansion

Lamb's original work simply pulls this solution out of thin air; given the expression, it is a simple matter to verify that it satisfies the Stokes equations. However, we may derive Lamb's solution in a systematic fashion starting with the multipole expansion, and work our way towards an expression involving simple combination of the scalar harmonics. In this manner, we show that Lamb's solution is simply a convenient parametrization of the more general multipole expansion in spherical polar coordinates.

We consider axisymmetric flows outside a sphere; the multipole expansion for $\boldsymbol{v}$ takes the form (see Exercise 4.2):

$$v_i = d_j \sum_{n=0}^{\infty} \left\{ a_n \frac{(\boldsymbol{d}\cdot\nabla)^n}{n!} \mathcal{G}_{ij}(\boldsymbol{x}-\boldsymbol{x}_1) + b_n \frac{(\boldsymbol{d}\cdot\nabla)^n}{n!} \nabla^2 \mathcal{G}_{ij}(\boldsymbol{x}-\boldsymbol{x}_1) \right\}, \quad (4.12)$$

with $\boldsymbol{d} = -\boldsymbol{e}_z$.

First, compare the two pressure representations,

$$p = -2\mu \sum_{n=0}^{\infty} a_n \frac{(\boldsymbol{d}\cdot\nabla)^{n+1}}{n!} \frac{1}{r} = -2\mu \sum_{n=1}^{\infty} n a_{n-1} \frac{(\boldsymbol{d}\cdot\nabla)^n}{n!} \frac{1}{r},$$

for the multipole expansion and

$$p = \sum_{n=1}^{\infty} A_{0n} r^{-n-1} P_n^0(\cos\theta)$$

for Lamb's general solution. Many useful properties of the Legendre polynomials and associated Legendre functions are found in [1] and [42]. One of the more important result is

$$r^{-n-1} P_n^0(\cos\theta) = \frac{(\boldsymbol{d}\cdot\nabla)^n}{n!} \frac{1}{r}$$

so that the pressure in Lamb's general solution may be rewritten as

$$p = \sum_{n=1}^{\infty} A_{0n} \frac{(\boldsymbol{d}\cdot\nabla)^n}{n!} \frac{1}{r}.$$

The two representations for the pressure match if we set $A_{0n} = -2\mu n a_{n-1}$ for $n \geq 1$.

We now consider the full velocity field. We start with

$$\boldsymbol{d}\cdot\mathcal{G} = -r^2\nabla(\boldsymbol{d}\cdot\nabla)\frac{1}{r} - 4\boldsymbol{x}(\boldsymbol{d}\cdot\nabla)\frac{1}{r} .$$

Apply $(\boldsymbol{d}\cdot\nabla)^n$ to both sides and use the Leibniz rule for the n-th derivative of a product. The result is

$$(\boldsymbol{d}\cdot\nabla)^n\boldsymbol{d}\cdot\mathcal{G} = -r^2\nabla\pi_{n+1} - 2n(\boldsymbol{d}\cdot\boldsymbol{x})\nabla\pi_n - n(n-1)\nabla\pi_{n-1} - 4\boldsymbol{x}\pi_{n+1} - 4n\boldsymbol{d}\pi_n .$$

We have introduced the notation $\pi_n = (\boldsymbol{d}\cdot\nabla)^n(1/r)$.

To recover Lamb's solution, we must eliminate terms in $(\boldsymbol{d}\cdot\boldsymbol{x})\nabla\pi_n$ and $\boldsymbol{d}\pi_n$ in the preceding equation. The second is readily accomplished with the aid of the following recursion formula for the Legendre polynomials:

$$\boldsymbol{d}\pi_n = \frac{-r^2\nabla\pi_{n+1}}{(n+1)(2n+1)} + \frac{n}{2n+1}\nabla\pi_{n-1} - \frac{(2n+3)\boldsymbol{x}\pi_{n+1}}{(n+1)(2n+1)} .$$

The other formula can also be derived from this one by applying the inner product with $\boldsymbol{d}$ to both sides, which gives

$$\pi_{n-1} = \frac{-r^2}{n(2n-1)}\pi_{n+1} + \frac{n-1}{2n-1}\pi_{n-1} - \frac{(2n+1)(\boldsymbol{d}\cdot\boldsymbol{x})}{n(2n-1)}\pi_n ,$$

(the index has been lowered by 1) and then taking the gradient, which gives

$$\begin{aligned}\nabla\pi_{n-1} &= \frac{-2\boldsymbol{x}}{n(2n-1)}\pi_{n+1} - \frac{r^2}{n(2n-1)}\nabla\pi_{n+1} + \frac{n-1}{2n-1}\nabla\pi_{n-1} \\ &\quad - \frac{(2n+1)(\boldsymbol{d}\cdot\boldsymbol{x})}{n(2n-1)}\nabla\pi_n - \frac{(2n+1)}{n(2n-1)}\boldsymbol{d}\pi_n .\end{aligned}$$

This may be solved for $(\boldsymbol{d}\cdot\boldsymbol{x})\nabla\pi_n$, with the result,

$$(\boldsymbol{d}\cdot\boldsymbol{x})\nabla\pi_n = \frac{-2\boldsymbol{x}}{2n+1}\pi_{n+1} - \frac{r^2}{2n+1}\nabla\pi_{n+1} - \frac{n^2}{2n+1}\nabla\pi_{n-1} - \boldsymbol{d}\pi_n .$$

Thus, the a_n term in the multipole expansion can be rewritten as

$$\begin{aligned}a_n(\boldsymbol{d}\cdot\nabla)^n\boldsymbol{d}\cdot\mathcal{G} &= a_n\left[\frac{(n-1)r^2\nabla\pi_{n+1}}{(n+1)(2n+1)} - \frac{2(n+2)\,\boldsymbol{x}\pi_{n+1}}{(n+1)(2n+1)}\right] \\ &\quad - a_n\frac{n(n-1)}{(2n+1)}\nabla\pi_{n-1} .\end{aligned}$$

The b_n terms in the multipole expansion are transformed with less effort since $\nabla^2\mathcal{G} = -2\nabla\nabla(1/r)$. Simply re-index the summation, collect terms in π_n to obtain the result:

$$\begin{aligned}\boldsymbol{v} &= \sum_{n=1}^{\infty} a_{n-1}\left[\frac{(n-2)r^2}{(2n-1)}\nabla\frac{(\boldsymbol{d}\cdot\nabla)^n}{n!}\frac{1}{r} - \frac{2(n+1)\boldsymbol{x}}{(2n-1)}\frac{(\boldsymbol{d}\cdot\nabla)^n}{n!}\frac{1}{r}\right] \\ &\quad - \nabla\sum_{n=1}^{\infty}\left[\frac{n}{2n+3}a_{n+1}\frac{(\boldsymbol{d}\cdot\nabla)^n}{n!}\frac{1}{r} + 2nb_{n-1}\right] ,\end{aligned}$$

which is Lamb's solution with

$$A_{0n} = -2\mu n a_{n-1} \qquad B_{0n} = -\frac{n}{2n+3}a_{n+1} - 2nb_{n-1} .$$

The relation between the χ_n term and multipoles of the form

$$\epsilon_{jkl}T_j \frac{(\boldsymbol{d}\cdot\nabla)^n}{n!}\frac{\partial}{\partial x_k}\mathcal{G}_{il} = 2\epsilon_{ijk}T_j \frac{(\boldsymbol{d}\cdot\nabla)^n}{n!}\frac{\partial}{\partial x_k}\frac{1}{r}$$

may be established in a similar manner (see Exercise 4.3).

We have successfully derived the coefficients in the p_n terms, although our synthesis applies only for the special case of axisymmetric and exterior flows. It is clear, however, that more general solutions, for both interior as well as exterior flows, are obtained by the use of harmonics of the form $r^n P_n^m(\cos\theta)\sin m\phi$ and $r^n P_n^m(\cos\theta)\cos m\phi$ for all integral n (see also Exercise 4.1).

4.2.2 Force, Torque, and Stresslet

Since the force and torque on any particle are given by the coefficients of the Stokeslet and rotlet in the multipole expansion, we examine the fields that decay as r^{-1} and r^{-2} in the exterior part of Lamb's general solution. We find (see Exercise 4.4) the simple formulae,

$$F_x = -4\pi A_{11} \qquad F_y = -4\pi \tilde{A}_{11} \qquad F_z = -4\pi A_{01}$$

$$T_x = -8\pi C_{11} \qquad T_y = -8\pi \tilde{C}_{11} \qquad T_z = -8\pi C_{01} ,$$

for the force and torque on a spherical surface enclosing the origin. A more succinct form is given in [35]

$$\boldsymbol{F} = -4\pi\nabla(r^3 p_{-2}) \tag{4.13}$$

$$\boldsymbol{T} = -8\pi\mu\nabla(r^3\chi_{-2}) . \tag{4.14}$$

An analogous result for the stresslet,

$$\boldsymbol{S} = -\frac{2\pi}{3}\nabla\nabla(r^5 p_{-3}) ,$$

is discussed in Exercise 4.4.

4.2.3 Matching of Boundary Conditions

The demonstration that an arbitrary velocity can be matched at the sphere boundary completes the proof that Lamb's solution is "general." Given boundary conditions for $\boldsymbol{v}$, the r, θ, and ϕ components of $\boldsymbol{v}$ provide three equations for the three unknown coefficients associated with the mode (m,n) (one coefficient per p, Φ, and χ). In the following discussion, we show three different implementations of this idea.

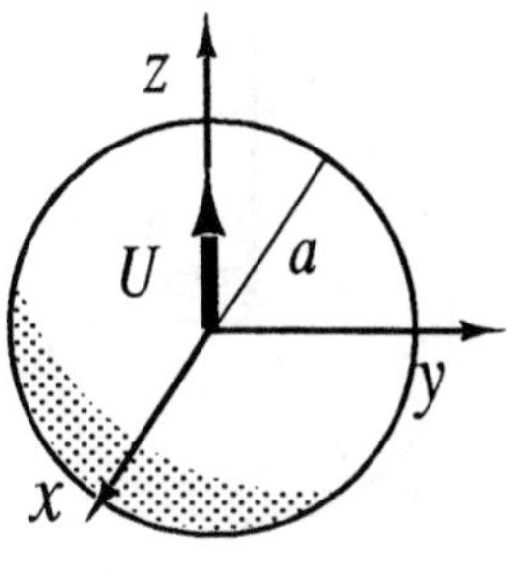

Figure 4.2: The translating sphere.

Matching of Radial Velocity, Surface Divergence, and Surface Curl

Given a boundary condition of the form $\boldsymbol{v} = \boldsymbol{V}_s(\theta, \phi)$ at $r = a$, we match the radial velocity, and in place of V_θ and V_ϕ, the surface divergence and radial component of the surface curl. Lamb's coefficients then follow as [11, 35]

$$V_r = \sum_{n=-\infty}^{\infty} \left\{ \frac{na\ p_n|_{r=a}}{2\mu(2n+3)} + \frac{n}{a}\Phi_n|_{r=a} \right\} \tag{4.15}$$

$$-r\nabla_s \cdot \boldsymbol{V}|_{r=a} = \sum_{n=-\infty}^{\infty} \left\{ \frac{n(n+1)a\ p_n|_{r=a}}{2\mu(2n+3)} + \frac{n(n-1)}{a}\Phi_n|_{r=a} \right\} \tag{4.16}$$

$$\boldsymbol{x} \cdot \nabla_s \times \boldsymbol{V} = \sum_{n=-\infty}^{\infty} \left\{ n(n+1)\chi_n|_{r=a} \right\} . \tag{4.17}$$

A complete discussion on the *surface divergence* and *surface curl* operators and their applications in matching interfacial boundary conditions is available in [74]. In spherical polar coordinates, the explicit expressions are

$$-r\nabla_s \cdot \boldsymbol{V} = -2V_r - \frac{1}{\sin\theta}\frac{\partial}{\partial\theta}(V_\theta \sin\theta) - \frac{1}{\sin\theta}\frac{\partial V_\phi}{\partial\phi} \tag{4.18}$$

$$\boldsymbol{x} \cdot \nabla_s \times \boldsymbol{V} = \frac{1}{\sin\theta}\frac{\partial}{\partial\theta}(V_\phi \sin\theta) - \frac{1}{\sin\theta}\frac{\partial V_\theta}{\partial\phi} . \tag{4.19}$$

In most applications, the matching procedure is further facilitated because modes not present in the surface velocity, $\boldsymbol{V}$, need not be carried along in the analysis, as shown in the examples.

Example 4.1 The Translating Sphere.
We denote the sphere speed and radius by U and a, and take the translation to be in the z-direction, so that the surface velocity field is given by

$$\boldsymbol{V}_s(\theta, \phi) = U\boldsymbol{e}_z = U\cos\theta\ \boldsymbol{e}_r - U\sin\theta\ \boldsymbol{e}_\theta .$$

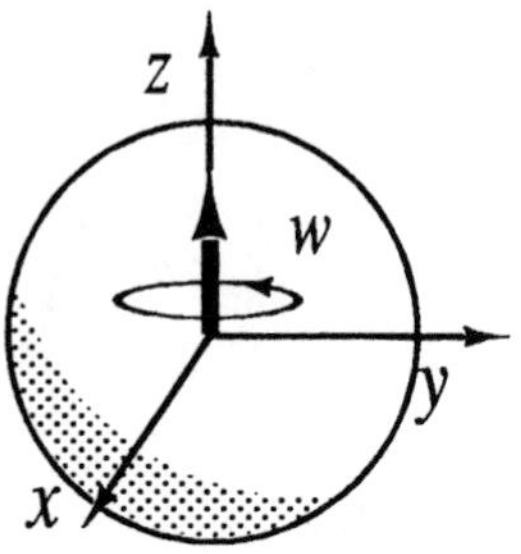

Figure 4.3: The rotating sphere.

From Equations 4.16 to 4.17, the relevant expressions for the boundary conditions are

$$\begin{aligned} V_r &= U\cos\theta \\ -r\nabla_s \cdot \boldsymbol{V}|_{r=a} &= 0 \\ \boldsymbol{x}\cdot\nabla\times\boldsymbol{V} &= 0\,. \end{aligned}$$

These expressions reveal that only $P_1(\cos\theta) = \cos\theta$ is required, the relevant harmonics being p_{-2} and Φ_{-2}, and that χ is not needed. With $n = -2$ in Equations 4.16 to 4.17, we find that $(\mu a U)^{-1}A_1 = 3/2$ and $(Ua^3)^{-1}B_1 = 1/4$, so that

$$p_{-2} = \frac{3}{2}\mu U a r^{-2}\cos\theta\,, \qquad \Phi_{-2} = \frac{1}{4}Ua^3r^{-2}\cos\theta\,,$$

and

$$\begin{aligned} v_r &= U\cos\theta\left[\frac{3}{2}\frac{a}{r} - \frac{1}{2}\left(\frac{a}{r}\right)^3\right] \\ v_\theta &= -U\sin\theta\left[\frac{3}{4}\frac{a}{r} + \frac{1}{4}\left(\frac{a}{r}\right)^3\right]\,. \end{aligned}$$

Using Equation 4.13 we obtain *Stokes law* for the drag exerted by the fluid on the translating sphere,

$$F_z = -4\pi A_1 U = -6\pi\mu a U\,.$$

The negative sign is appropriate since the fluid is resisting the motion. This result can also be obtained by integrating the surface traction, $\boldsymbol{\sigma}\cdot\boldsymbol{n}$, over the sphere surface.◇

Example 4.2 The Rotating Sphere.
Consider a rigid sphere of radius a rotating about the z-axis with angular

velocity, ω, in an otherwise quiescent fluid. At the sphere surface, we have $\boldsymbol{V}(\boldsymbol{x}) = \omega \boldsymbol{e}_z \times \boldsymbol{x}$, or simply

$$\boldsymbol{V} = \omega a \sin\theta\, \boldsymbol{e}_\phi \ .$$

The radial velocity and surface divergence vanish identically. The radial component of the surface vorticity yields the desired solution,

$$\begin{aligned} \boldsymbol{x}\cdot\nabla\times\boldsymbol{V} \ = \ 2\chi_{-2} &= 2C_1 a^{-2}\cos\theta \\ \frac{1}{\sin\theta}\frac{\partial}{\partial\theta}(V_\phi\sin\theta) - \frac{1}{\sin\theta}\frac{\partial v_\theta}{\partial\phi} &= 2\omega a\cos\theta \ , \end{aligned}$$

so that $C_1 = \omega a^3$ and

$$\boldsymbol{v} = \nabla\times\left(\boldsymbol{x}\left\{C_1 r^{-2}\cos\theta\right\}\right) = \omega\boldsymbol{e}_z\times\boldsymbol{x}\frac{a^3}{r^3} \ .$$

From Equation 4.14 we deduce that the fluid resists this rotation by exerting a hydrodynamic torque, $T_z = -8\pi\mu a^3\omega$.◇

Example 4.3 Motion of a Spherical Drop Revisited (The Hadamard–Rybczynski Problem).
Consider a spherical drop of viscosity μ_i translating in another fluid of viscosity μ_o. Again, we denote the sphere speed and radius by U and a, and take the translation to be in the z-direction. Surface tension acts to retain the spherical shape while fluid motion distorts it. For large surface tensions, we may attempt a perturbation solution, assuming a spherical shape. The kinematic conditions require the velocities normal to the fluid-fluid interface to match the motion of the interface. In addition, we assume a no-slip condition so that the tangential components of the velocities and tractions are continuous across the interface. This leaves just the normal component of the tractions unsatisfied.

By assuming a drop shape, we have arbitrarily set a degree of freedom and, as a consequence, cannot satisfy this condition exactly. For large surface tensions, this small error may be corrected by perturbing the shape of the drop, in effect using this last condition to determine the shape of the drop. For the translation problem, rather fortuitously, the normal component of the tractions is matched properly. However, the more general situation is encountered when the drop is in an extensional flow.

From our experience with the boundary conditions for the rigid sphere of the previous example, we expect only $n = -2$ to appear in the exterior disturbance field (or $n = 1$ in the exterior representation, Equation 4.8). This in turn suggests that only $n = 1$ is used in the interior field. The exterior solution is then of the form,

$$v_r^{(o)} = \left\{A_1\left(\frac{a}{r}\right) - 2B_1\left(\frac{a}{r}\right)^3\right\}U\cos\theta \qquad (4.20)$$

$$v_\theta^{(o)} = -\left\{\frac{A_1}{2}\left(\frac{a}{r}\right) + B_1\left(\frac{a}{r}\right)^3\right\}U\sin\theta \ . \qquad (4.21)$$

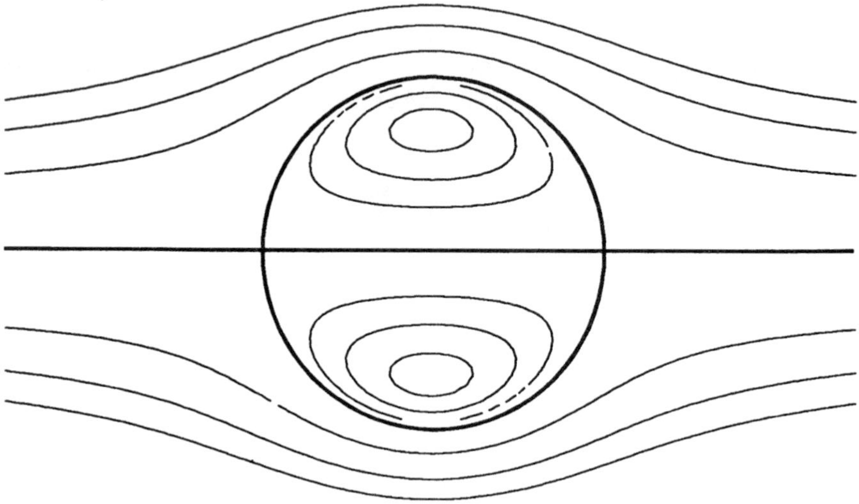

Figure 4.4: The translating drop, for $\mu_i = \mu_o$.

The velocity representation inside the drop is obtained from the p_1 and Φ_1 terms. We express these two harmonics as

$$p_1 = a_1 \frac{\mu U}{a} \frac{r}{a} \cos\theta \ , \qquad \Phi_1 = b_1 U r \cos\theta \ .$$

The interior solution simplifies to

$$v_r^{(i)} = \left\{ \frac{a_1}{10} \left(\frac{r}{a} \right)^2 + b_1 \right\} U \cos\theta \qquad (4.22)$$

$$v_\theta^{(i)} = -\left\{ \frac{a_1}{5} \left(\frac{r}{a} \right)^2 + b_1 \right\} U \sin\theta \ . \qquad (4.23)$$

The boundary conditions at $r = a$ are

1. $v_r^{(o)} = U \cos\theta$ (a kinematic condition)
2. $v_r^{(i)} = U \cos\theta$ (a kinematic condition)
3. v_θ continuous (the no-slip condition)
4. $d_{\theta r}$ continuous (matching of the tangential component of the surface tractions)

This leads to the following system of equations:

$$A_1 - 2B_1 = 1$$

$$\begin{aligned}\frac{1}{10}a_1 + b_1 &= 1\\ \frac{1}{2}A_1 + B_1 - \frac{1}{5}a_1 - b_1 &= 0\\ 6B_1 + \frac{3}{10}\lambda a_1 &= 0\,.\end{aligned}$$

Here, $\lambda = \mu_i/\mu_o$ is the ratio of inner and outer fluid viscosities. To obtain the fourth equation, we have used the following expression for the stress component $d_{\theta r}$:

$$d_{\theta r} = \mu\left[r\frac{\partial}{\partial r}\left(\frac{v_\theta}{r}\right) + \frac{1}{r}\frac{\partial v_r}{\partial \theta}\right]\,.$$

The solution of the preceding system of equations is

$$A_1 = \frac{2+3\lambda}{2(1+\lambda)}\,,\qquad B_1 = \frac{\lambda}{4(1+\lambda)}\,,\qquad a_1 = -\frac{5}{1+\lambda}\,,\qquad b_1 = \frac{3+2\lambda}{2(1+\lambda)}\,.$$

The force exerted by the fluid on the drop is obtained from the exterior field as $-4\pi\mu_o a A_1 U$, or

$$F_z = -2\pi\mu_o a U\,\frac{2+3\lambda}{1+\lambda}\,.$$

The negative sign indicates that the fluid is resisting the motion. Note the limiting cases:

1. Rigid sphere; as $\lambda \to \infty$, $F_z \to -6\pi\mu_o a U$ and we recover Stokes law.
2. Bubble; for $\lambda = 0$, we obtain $F_z = -4\pi\mu_o a U$.

As an exercise, we ask the reader to examine the jump, $\boldsymbol{\sigma}^{(o)} : \hat{\boldsymbol{n}}\hat{\boldsymbol{n}} - \boldsymbol{\sigma}^{(i)} : \hat{\boldsymbol{n}}\hat{\boldsymbol{n}}$ to check whether the assumed spherical drop shape is consistent with the force balance.◇

4.3 The Adjoint Method

There is another solution method for spherical coordinates that has been developed by Schmitz and Felderhof [73]. It is closely related to Lamb's solution, and we may call it the *adjoint method.* Define the velocity and pressure fields of Lamb's interior solution as

$$\boldsymbol{v}^{(0)}_{nm}(\boldsymbol{x}) = \nabla\left\{r^n \widetilde{Y}^m_n(\theta,\phi)\right\} = r^{n-1}\boldsymbol{A}_{nm}(\boldsymbol{x}) \tag{4.24}$$

$$\boldsymbol{v}^{(1)}_{nm}(\boldsymbol{x}) = i\nabla\times\left\{\boldsymbol{x}r^n \widetilde{Y}^m_n(\theta,\phi)\right\} = ir^n\boldsymbol{C}_{nm}(\boldsymbol{x}) \tag{4.25}$$

$$\begin{aligned}\boldsymbol{v}^{(2)}_{nm}(\boldsymbol{x}) &= \left\{\frac{(2n+1)(n+3)}{2n}r^2\nabla - (2n+1)\boldsymbol{x}\right\}\left\{r^n\widetilde{Y}^m_n(\theta,\phi)\right\}\\ &= r^{n+1}\left[\frac{(n+1)(2n+3)}{2n}\boldsymbol{A}_{nm}(\boldsymbol{x}) + \boldsymbol{B}_{nm}(\boldsymbol{x})\right]\end{aligned} \tag{4.26}$$

$$p_{nm}(\boldsymbol{x}) = p^{(2)}_{nm}(\boldsymbol{x}) = \mu(n+1)(2n+1)(2n+3)r^n\widetilde{Y}^m_n(\theta,\phi)\,. \tag{4.27}$$

The explicit expressions for the surface vector fields $\boldsymbol{A}_{nm}$, $\boldsymbol{B}_{nm}$ and $\boldsymbol{C}_{nm}$ are obtained readily from the vector operators and are

$$
\begin{aligned}
\boldsymbol{A}_{nm} &= n\widetilde{Y}_n^m \boldsymbol{e}_r + \frac{\partial \widetilde{Y}_n^m}{\partial \theta}\boldsymbol{e}_\theta + \frac{1}{\sin\theta}\frac{\partial \widetilde{Y}_n^m}{\partial \phi}\boldsymbol{e}_\phi \\
\boldsymbol{B}_{nm} &= -(n+1)\widetilde{Y}_n^m \boldsymbol{e}_r + \frac{\partial \widetilde{Y}_n^m}{\partial \theta}\boldsymbol{e}_\theta + \frac{1}{\sin\theta}\frac{\partial \widetilde{Y}_n^m}{\partial \phi}\boldsymbol{e}_\phi \\
\boldsymbol{C}_{nm} &= \frac{1}{\sin\theta}\frac{\partial \widetilde{Y}_n^m}{\partial \phi}\boldsymbol{e}_\theta - \frac{\partial \widetilde{Y}_n^m}{\partial \theta}\boldsymbol{e}_\phi \ ,
\end{aligned}
$$

where the functions $\widetilde{Y}_n^m(\theta, \phi)$ are related to the normalized spherical harmonics $Y_n^m(\theta, \phi)$ by

$$
\widetilde{Y}_n^m(\theta,\phi) = \eta_{nm} Y_n^m(\theta,\phi) = (-1)^m P_n^m(\cos\theta)e^{im\phi}
$$

and η_{nm} is the normalization constant,

$$
\eta_{nm} = \left[\frac{4\pi}{2n+1}\frac{(n+m)!}{(n-m)!}\right]^{1/2} .
$$

The surface integral of products of $\boldsymbol{A}_{nm}$, $\boldsymbol{B}_{nm}$, and $\boldsymbol{C}_{nm}$ are derived in Exercise 4.5 and are summarized below:

$$
\begin{aligned}
\oint_S \boldsymbol{A}_{lk}^* \cdot \boldsymbol{A}_{nm}\, dS &= n(2n+1)\eta_{nm}^2 a^2 \delta_{ln}\delta_{km} \\
\oint_S \boldsymbol{B}_{lk}^* \cdot \boldsymbol{B}_{nm}\, dS &= (n+1)(2n+1)\eta_{nm}^2 a^2 \delta_{ln}\delta_{km} \\
\oint_S \boldsymbol{C}_{lk}^* \cdot \boldsymbol{C}_{nm}\, dS &= n(n+1)\eta_{nm}^2 a^2 \delta_{ln}\delta_{km} \\
\oint_S \boldsymbol{A}_{lk}^* \cdot \boldsymbol{B}_{nm}\, dS &= \oint_S \boldsymbol{B}_{lk}^* \cdot \boldsymbol{A}_{nm}\, dS = 0 \\
\oint_S \boldsymbol{B}_{lk}^* \cdot \boldsymbol{C}_{nm}\, dS &= \oint_S \boldsymbol{C}_{lk}^* \cdot \boldsymbol{B}_{nm}\, dS = 0 \\
\oint_S \boldsymbol{C}_{lk}^* \cdot \boldsymbol{A}_{nm}\, dS &= \oint_S \boldsymbol{A}_{lk}^* \cdot \boldsymbol{C}_{nm}\, dS = 0 .
\end{aligned}
$$

The essential idea behind the method is the existence of an adjoint set (or dual basis) $\{\boldsymbol{w}_{nm}^{(0)}, \boldsymbol{w}_{nm}^{(1)}, \boldsymbol{w}_{nm}^{(2)}\}$,

$$
\boldsymbol{w}_{nm}^{(0)} = \frac{r^{-n}}{n(2n+1)\eta_{nm}^2}\left[\boldsymbol{A}_{nm} - \frac{2n+3}{2}\boldsymbol{B}_{nm}\right] \tag{4.28}
$$

$$
\boldsymbol{w}_{nm}^{(1)} = \frac{ir^{-n-1}}{n(n+1)\eta_{nm}^2}\boldsymbol{C}_{nm} \tag{4.29}
$$

$$
\boldsymbol{w}_{nm}^{(2)} = \frac{r^{-n-2}}{(n+1)(2n+1)\eta_{nm}^2}\boldsymbol{B}_{nm} \ , \tag{4.30}
$$

that satisfies the orthonormality relations (see Exercise 4.5),

$$
\frac{1}{a}\oint_S (\boldsymbol{w}_{lk}^{(i)})^* \cdot \boldsymbol{v}_{nm}^{(j)}\, dS = \delta_{ij}\delta_{ln}\delta_{km} .
$$

Here, a is the sphere radius and the asterisk ($*$) denotes complex conjugation. The surface integration of the vector dot product is in fact a well-known inner product (denoted $\langle \bullet, \bullet \rangle$), which will be encountered again in Part IV.

Suppose that we wish to match the boundary condition, $\boldsymbol{v} = \boldsymbol{V}(\theta, \phi)$, on the sphere surface. We write Lamb's general solution as

$$\boldsymbol{v} = \sum_n \sum_m \left\{ a_{mn}^{(0)} \boldsymbol{v}_{nm}^{(0)} + a_{mn}^{(1)} \boldsymbol{v}_{nm}^{(1)} + a_{mn}^{(2)} \boldsymbol{v}_{nm}^{(2)} \right\}$$

and take the inner product of both sides with fields from the adjoint set. The orthonormal property yields immediately the explicit solution for the coefficients,

$$a_{mn}^{(i)} = \frac{1}{a} \oint_S \boldsymbol{V}^* \cdot \boldsymbol{w}_{nm}^{(i)} \, dS \ .$$

For *exterior* fields, we write Lamb's solution as

$$p_{nm}(\boldsymbol{x}) = p_{nm}^{(0)}(\boldsymbol{x}) = \frac{\mu}{2n+1} r^{-n-1} \tilde{Y}_n^m(\theta, \phi) \tag{4.31}$$

$$\boldsymbol{v}_{nm}^{(0)}(\boldsymbol{x}) = \frac{r^{-n}}{(2n+1)^2} \left[\frac{n+1}{n(2n-1)} \boldsymbol{A}_{nm}(\boldsymbol{x}) - \frac{1}{2} \boldsymbol{B}_{nm}(\boldsymbol{x}) \right] \tag{4.32}$$

$$\boldsymbol{v}_{nm}^{(1)}(\boldsymbol{x}) = \frac{i r^{-n-1}}{n(n+1)(2n+1)} \boldsymbol{C}_{nm}(\boldsymbol{x}) \tag{4.33}$$

$$\boldsymbol{v}_{nm}^{(2)}(\boldsymbol{x}) = \frac{n r^{-n-2}}{(n+1)(2n+1)^2(2n+3)} \boldsymbol{B}_{nm}(\boldsymbol{x}) \ , \tag{4.34}$$

and the adjoint fields are given by the expressions,

$$\boldsymbol{w}_{nm}^{(0)} = \frac{(2n-1)(2n+1)}{(n+1)\eta_{nm}^2} r^{n-1} \boldsymbol{A}_{nm} \tag{4.35}$$

$$\boldsymbol{w}_{nm}^{(1)} = \frac{i(2n+1)}{\eta_{nm}^2} r^n \boldsymbol{C}_{nm} \tag{4.36}$$

$$\boldsymbol{w}_{nm}^{(2)} = \frac{(2n+1)(2n+3)}{n\eta_{nm}^2} r^{n+1} \left[\frac{2n-1}{2} \boldsymbol{A}_{nm} + \boldsymbol{B}_{nm} \right] . \tag{4.37}$$

The problem of the translating sphere can also be solved by the adjoint method. We set $m = 0$, since the problem is axisymmetric and consider the representation for exterior flows. In Exercise 4.7 we ask the reader to show that the inner product, $\langle \boldsymbol{w}_{n0}^{(i)}, U\boldsymbol{e}_z \rangle$, vanishes except for $n = 1$. This then leads to the same solution as before.

4.4 An Orthonormal Basis for Stokes Flow

The solid spherical harmonics, in addition to being solutions of the Laplace equation, form an orthonormal basis for *scalar* functions on the sphere surface. In other words, the solid spherical harmonics $\{\psi_n\}$ form a convenient basis for the expansion of a surface function, $f(\theta,\phi) = \sum_n a_n \psi_n(\theta,\phi)$, because of the orthonormal property satisfied by the ψ_n,

$$\int_0^{2\pi}\int_0^{\pi} \psi_m \psi_n \sin\theta \, d\theta d\phi = \delta_{mn} \; .$$

We have already seen that the elements of Lamb's solution do not form an orthonormal basis for *vector* functions on the sphere surface. However, the solution set is almost decoupled, since the different modes are mutually orthogonal, and since for a given mode the toroidal field is orthogonal to the other two. By taking the appropriate linear combination of the particular and homogeneous solutions, we induce a Gram-Schmidt orthogonalization and obtain an orthonormal basis:

$$\begin{aligned} \boldsymbol{\varphi}_{nm}^{(1)} &= a^{n-1}\sqrt{n(2n+1)}\frac{(2n-1)}{\eta_{nm}(n+1)} \\ &\quad \times \left\{ -\frac{(n-2)r^2\nabla}{2n(2n-1)} + \frac{(n+1)\boldsymbol{x}}{n(2n-1)} + \frac{a^2\nabla}{2(2n+1)} \right\} \left\{ r^{-n-1}\widetilde{Y}_n^m \right\} \\ &= \frac{\sqrt{n(2n+1)}}{a\eta_{nm}} \left[\frac{(a/r)^n \boldsymbol{A}_{nm}}{n(2n+1)} + \frac{(2n-1)}{(2n+1)} \frac{\left((a/r)^2-1\right)(a/r)^n \boldsymbol{B}_{nm}}{2(n+1)} \right] \end{aligned} \tag{4.38}$$

$$\boldsymbol{\varphi}_{nm}^{(2)} = \frac{a^{n+1}\nabla\left\{ r^{-n-1}\widetilde{Y}_n^m \right\}}{\eta_{nm}\sqrt{(n+1)(2n+1)}} = \frac{(a/r)^{n+2}\boldsymbol{B}_{nm}}{a\eta_{nm}\sqrt{(n+1)(2n+1)}} \tag{4.39}$$

$$\boldsymbol{\varphi}_{nm}^{(3)} = \frac{ia^n}{\eta_{nm}\sqrt{n(n+1)}} \nabla \times \left\{ \boldsymbol{x} r^{-n-1}\widetilde{Y}_n^m \right\} = \frac{i(a/r)^{n+1}\boldsymbol{C}_{nm}}{a\eta_{nm}\sqrt{n(n+1)}} \tag{4.40}$$

$$n \geq 1, \qquad -n \leq m \leq n \; .$$

The reader should verify that on the sphere surface $r = a$, the orthonormal relations,

$$\oint_S (\boldsymbol{\varphi}_{lk}^{(\alpha)})^* \cdot \boldsymbol{\varphi}_{nm}^{(\beta)} \, dS = \delta_{\alpha\beta}\delta_{ln}\delta_{km} \; ,$$

are satisfied. This basis also has the nice property that $\boldsymbol{\varphi}_{1m}^{(1)}$ and $\boldsymbol{\varphi}_{1m}^{(3)}$, with $(m = -1, 0, 1)$, correspond to disturbance fields produced by the six independent rigid-body motions of a sphere.

Suppose that we wish to match a general boundary velocity, $\boldsymbol{V}(\theta,\phi)$, on the sphere surface. We may now write the general solution as

$$\boldsymbol{v} = \sum_n \sum_m \left\{ a_{mn}^{(1)}\boldsymbol{\varphi}_{nm}^{(1)} + a_{mn}^{(2)}\boldsymbol{\varphi}_{nm}^{(2)} + a_{mn}^{(3)}\boldsymbol{\varphi}_{nm}^{(3)} \right\}$$

and take the inner product of both sides with a particular basis element. The orthonormal property yields immediately the explicit solution for the coefficients,

$$a_{mn}^{(\alpha)} = \oint_S \boldsymbol{V}^* \cdot \boldsymbol{\varphi}_{nm}^{(\alpha)} \, dS \; .$$

In Part IV, we will see that the elements of this basis are in fact eigenfunctions of a self-adjoint operator, and thus obtain a deeper understanding for the orthonormal property.

We illustrate this method by considering again the problem of the translating sphere. The element $\varphi_{10}^{(1)}$ is the disturbance velocity field produced by a constant translational velocity in the z-direction, so on the sphere surface it becomes a constant vector in the z-direction. The explicit results for the surface vector fields and the basis element are

$$\begin{aligned}
\boldsymbol{A}_{10} &= \cos\theta \boldsymbol{e}_r - \sin\theta \boldsymbol{e}_\theta = \boldsymbol{e}_z \\
\boldsymbol{B}_{10} &= -2\cos\theta \boldsymbol{e}_r - \sin\theta \boldsymbol{e}_\theta \\
\varphi_{10}^{(1)} &= \frac{(a/r)}{a\sqrt{4\pi}}\left[\boldsymbol{A}_{10} + \frac{1}{4}\left(\frac{a^2}{r^2} - 1\right)\boldsymbol{B}_{10}\right] \\
&= \frac{1}{a\sqrt{4\pi}}\left\{\boldsymbol{e}_r \cos\theta\left[\frac{3}{2}\frac{a}{r} - \frac{1}{2}\left(\frac{a}{r}\right)^3\right] - \boldsymbol{e}_\theta\left[\frac{3}{4}\frac{a}{r} + \frac{1}{4}\left(\frac{a}{r}\right)^3\right]\right\} .
\end{aligned}$$

Since $a_{01}^{(1)} = \langle U\boldsymbol{e}_z, \varphi_{10}^{(1)}\rangle = \sqrt{4\pi}aU$, we recover the same solution as before.

4.5 The Stokes Streamfunction

Although the emphasis in this book is on general solution methods needed to tackle the problems of microhydrodynamics, we present here a brief overview of the Stokes' streamfunction method for axisymmetric flows. The student will encounter it frequently in the literature, and we shall call upon it occasionally from our repertoire of solution methods to check more general numerical and analytic methods. For a more complete and self-contained treatment of the method, especially for axisymmetric curvilinear coordinates, we direct the reader to Chapter 4 of Happel and Brenner [35].

4.5.1 Relation to the Vector Potential

Consider an axisymmetric Stokes flow field given by $\boldsymbol{v} = v_r(r,\theta) + v_\theta(r,\theta)$. A solenoidal velocity field must be the curl of a vector potential, *i.e.*, $\boldsymbol{v} = \nabla \times \boldsymbol{A}$. Only one component of the vector potential appears in two-dimensional flows, and for our axisymmetric flow it is the ϕ-component, so that $\boldsymbol{A} = A_\phi(r,\theta)\boldsymbol{e}_\phi$. As we shall show presently, it is convenient to express this one component of the vector potential as $A_\phi(r,\theta) = -\psi(r,\theta)/r\sin\theta$, where ψ is the Stokes streamfunction. In terms of this streamfunction, the two velocity components are

$$v_r = \frac{-1}{r^2\sin\theta}\frac{\partial\psi}{\partial\theta} , \qquad v_\theta = \frac{1}{r\sin\theta}\frac{\partial\psi}{\partial r} . \tag{4.41}$$

By the use of the Stokes (curl) theorem, we can show that the difference $\psi(r_2,\theta_2) - \psi(r_1,\theta_1)$ is related to the flow through an axisymmetric surface generated by rotating (about the axis of symmetry) any simple curve connecting the points (see Figure 4.5). The flow rate through this surface of revolution is given by

$$Q = \int_S \boldsymbol{v}\cdot\boldsymbol{n}\, dS = \int_S (\nabla\times\boldsymbol{A})\cdot\boldsymbol{n}\, dS$$

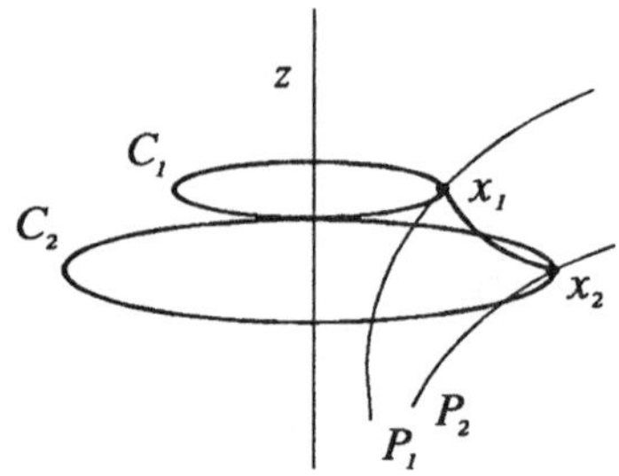

Figure 4.5: The volumetric flow and the streamfunction.

$$\begin{aligned} &= \int_{C1} \boldsymbol{A} \cdot \boldsymbol{e}_\phi\, r \sin\theta \, d\phi - \int_{C2} \boldsymbol{A} \cdot \boldsymbol{e}_\phi\, r \sin\theta \, d\phi \\ &= 2\pi\psi_2 - 2\pi\psi_1 \, . \end{aligned}$$

The Stokes theorem was used to convert the surface integrals into line integrals along the edges of the surface, with the directed differential arc lengths parametrized as $d\boldsymbol{s} = \boldsymbol{e}_\phi r \sin\theta d\phi$. We have thus derived an important property of the Stokes streamfunction: *Lines with ψ constant are the streamlines.*

Another interesting property of the streamfunction is uncovered upon two successive applications of the curl operator:

$$\begin{aligned} \nabla \times \frac{\boldsymbol{e}_\phi \psi}{r\sin\theta} &= \frac{\boldsymbol{e}_r}{r^2 \sin\theta}\frac{\partial\psi}{\partial\theta} - \frac{\boldsymbol{e}_\theta}{r\sin\theta}\frac{\partial\psi}{\partial r} \\ \nabla \times \nabla \times \frac{\boldsymbol{e}_\phi \psi}{r\sin\theta} &= \frac{\boldsymbol{e}_\phi}{r}\left[\frac{\partial}{\partial r}\left(\frac{-1}{\sin\theta}\frac{\partial\psi}{\partial r}\right) - \frac{\partial}{\partial\theta}\left(\frac{1}{r^2\sin\theta}\frac{\partial\psi}{\partial\theta}\right)\right] \\ &= \frac{-\boldsymbol{e}_\phi E^2\psi}{r\sin\theta} \, , \end{aligned}$$

where the streamfunction operator E^2 is defined by

$$E^2\psi = \frac{\partial^2\psi}{\partial r^2} + \frac{\sin\theta}{r^2}\frac{\partial}{\partial\theta}\left(\frac{1}{\sin\theta}\frac{\partial\psi}{\partial\theta}\right) \, . \tag{4.42}$$

A simple mnemonic for this result is: Commute $\nabla \times \nabla\times$ with $\boldsymbol{e}_\phi / r\sin\theta$, and convert the differential operator to $-E^2$. Note that we have derived an expression for the vorticity in terms of the streamfunction,

$$\nabla \times \boldsymbol{v} = -\nabla \times \nabla \times \frac{\boldsymbol{e}_\phi \psi}{r\sin\theta} = \frac{\boldsymbol{e}_\phi E^2\psi}{r\sin\theta} \, .$$

4.5.2 The Stokes Equations and the Streamfunction

We may write the governing Stokes equations in terms of the streamfunction. The vector Laplacian may be rewritten as

$$\nabla^2 \boldsymbol{v} = \nabla(\nabla \cdot \boldsymbol{v}) - \nabla \times \nabla \times \boldsymbol{v} = -\nabla \times \nabla \times \boldsymbol{v} \, ,$$

since $\nabla \cdot \boldsymbol{v} = 0$. Taking the curl of the Stokes equation, we eliminate the pressure gradient and arrive at

$$-\mu \nabla \times (\nabla \times \nabla \times \boldsymbol{v}) = -\mu \nabla \times \nabla \times \frac{\boldsymbol{e}_\phi E^2 \psi}{r \sin\theta} = \mu \frac{\boldsymbol{e}_\phi E^2(E^2\psi)}{r \sin\theta} = 0 ,$$

so that the governing equations reduce to $E^4\psi = 0$.

4.5.3 Boundary Conditions for the Streamfunction

The most common axisymmetric flow problems is that of a fixed axisymmetric body in a uniform stream flowing in the axial direction, or the closely related problem of a rigid body translating along the direction of its axis of symmetry. For the first problem, we set ψ to a constant along the particle profile, to satisfy the kinematic condition of no flow into the body. If the surface of the body intersects the axis of symmetry (this condition would exclude the torus, for example), it is convenient to set this constant to zero, so that with $\psi = 0$ also along the axis of symmetry, we obtain the desired condition of no flow through the particle surface. The no-slip boundary condition requires the vanishing of the normal derivative of ψ, *i.e.*, $\partial\psi/\partial n = 0$. Far away from the particle, the disturbance field decays and we should have just the uniform stream, $\boldsymbol{v}^\infty = U^\infty \boldsymbol{e}_z$. The streamfunction for this flow is readily derived as

$$\psi^\infty = -\frac{1}{2} U^\infty r^2 \sin^2\theta .$$

For a particle in uniform translation, we must have $r^{-2}\psi \to 0$ as $r \to \infty$, to get vanishing velocity fields, while on the surface of the particle, the boundary conditions to match the rigid body translation become

$$\psi + \frac{1}{2} U r^2 \sin^2\theta = C , \qquad \frac{\partial}{\partial n}(\psi + \frac{1}{2} U r^2 \sin^2\theta) = 0 . \tag{4.43}$$

Here again, the constant is set at $C = 0$, the axial value if the body intersects the axis of symmetry.

4.5.4 The Axisymmetric Stokeslet and the Drag on a Body

The velocity field produced by a Stokeslet of strength F directed along the z-axis, $F\boldsymbol{e}_z \cdot \mathcal{G}(\boldsymbol{x})/8\pi\mu$, generates an axisymmetric flow field. It is easy to show that the streamfunction for this flow field is (see Exercise 4.10)

$$\psi = \frac{-F}{8\pi\mu} r \sin^2\theta . \tag{4.44}$$

Note that the corresponding velocity field decays as r^{-1}.

The hydrodynamic drag exerted by the fluid on a particle follows from this result as

$$\lim_{r\to\infty} \frac{8\pi\mu\psi}{r\sin^2\theta} ,$$

where ψ is the streamfunction for the disturbance field, since far away from the particle the velocity field must be dominated by the Stokeslet field. More explicitly, all other terms in the multipole expansion (for the streamfunction) decay to zero after multiplication with r. An alternate derivation of this result is given in Payne and Pell [65].

A different expression for the drag on the body was developed by Stimson and Jeffery [80] starting from the surface tractions on the particle surface. The expressions for the relevant components of $\sigma \cdot \hat{\boldsymbol{n}}$ in terms of the streamfunction and the key steps in the derivation are given in Happel and Brenner [35]. The final result for the drag is

$$\pi\mu \int (r\sin\theta)^3 \frac{\partial}{\partial n}\left(\frac{E^2\psi}{(r\sin\theta)^2}\right) ds ,$$

where the line integral is along a meridian of the body.

4.5.5 Separation in Spherical Coordinates

The separability of the operator E^2 is closely related to the separability of the Laplacian operator ∇^2 and the solutions of $E^2\psi = 0$ in a number of coordinate systems are given in [35, 53, 60]. The problem of finding solutions to $E^4\psi = 0$ is usually split into the homogeneous solution ψ^h and the particular solution ψ^p, $\psi = \psi^h + \psi^p$, with

$$\begin{aligned} E^2\psi^h &= 0 , \\ E^2\psi^p &\neq 0 , \qquad \text{but} \quad E^2(E^2\psi^p) = 0 . \end{aligned}$$

The first general solution of this problem (in spherical coordinates) appears in Sampson [71]. We write

$$E^2 = \frac{\partial^2}{\partial r^2} + \frac{1-\xi^2}{r^2}\frac{\partial^2}{\partial \xi^2}$$

with $\xi = \cos\theta$, and look for a solution to $E^2\psi^h = 0$ of the form $\psi^h = R(r)X(\xi)$. The equation separates to yield

$$\frac{r^2}{R}\frac{d^2R}{dr^2} = -\frac{1-\xi^2}{X}\frac{d^2X}{d\xi^2} = n(n-1) .$$

As with the Laplace equation, the separation constant must be of a special form, in this case $n(n-1)$ with integral n, to keep ψ bounded at both poles. The solutions for $R(r)$ are simply r^n and r^{-n+1}, while for $X(\xi)$, we have Gegenbauer's equation, whose solutions are the Gegenbauer functions of degree $-1/2$, which, following [35], we denote by $\mathcal{G}_n$ and $\mathcal{H}_n$ (see also [1]). These functions are related to the Legendre functions of the first and second kind by

$$\mathcal{G}_n(\xi) = \frac{P_{n-2} - P_n}{2n-1}, \qquad n \geq 2$$
$$\mathcal{H}_n(\xi) = \frac{Q_{n-2} - Q_n}{2n-1}, \qquad n \geq 2$$

The first two functions may be *defined* as $\mathcal{G}_0(\xi) = 1$, $\mathcal{G}_1(\xi) = -\xi$, $\mathcal{H}_0(\xi) = -\xi$, and $\mathcal{H}_1(\xi) = -1$.

The functions of the first kind are polynomials in ξ, since they are simple combinations of the Legendre polynomials, and are therefore regular everywhere in $-1 \leq \xi \leq 1$. The explicit formula of the Rodriguez type is

$$\mathcal{G}_n(\xi) = \frac{-1}{(n-1)!} \frac{d^{n-2}}{d\xi^{n-2}} \left(\frac{\xi^2 - 1}{2} \right)^{n-1},$$

and the first few are

$$\mathcal{G}_2(\xi) = \frac{1}{2}(1 - \xi^2), \qquad \mathcal{G}_3(\xi) = \frac{1}{2}\xi(1 - \xi^2),$$

$$\mathcal{G}_4(\xi) = \frac{1}{8}(1 - \xi^2)(5\xi^2 - 1), \qquad \mathcal{G}_5(\xi) = \frac{1}{8}\xi(1 - \xi^2)(7\xi^2 - 3).$$

The functions of the second kind follow the form usually seen in "second" solutions to second-order ordinary differential equations obtained by the method of Fröbenius,

$$\mathcal{H}_n(\xi) = \frac{1}{2}\mathcal{G}_n(\xi) \ln \frac{1+\xi}{1-\xi} + \mathcal{K}_n(\xi),$$

where

$$\mathcal{K}_n(\xi) = -\sum_{k}^{[(n+1)/2]} \frac{(2n-4k+1)}{(2k-1)(n-k)} \left[1 - \frac{(2k-1)(n-k)}{n(n-1)} \right] \mathcal{G}_{n+1-2k}(\xi).$$

Since $\mathcal{G}_n$ is regular, $\mathcal{H}_n$ is unbounded at both poles, as required by the theory concerning singular points of linear second-order ordinary differential equations. Thus, in most applications of interest, we disregard the second solutions, and write the solution to $E^2\psi^h = 0$ as

$$\psi^h(r, \xi) = \sum_{n=0}^{\infty} (A_n r^n + B_n r^{-n+1}) \mathcal{G}_n(\xi).$$

For ψ_n^p, we try solutions of the form $r^\alpha \mathcal{G}_n(\xi)$, such that $E^2\psi_n^h$ is of the form given above. This immediately leads to the result that α can be either $n+2$ or $-n+3$. Thus the general solution for the streamfunction is

$$\psi = \sum_{n=2}^{\infty} (A_n r^n + B_n r^{-n+1} + C_n r^{n+2} + D_n r^{-n+3}) \mathcal{G}_n(\xi).$$

Geometry	Reference
Conical diffuser	Happel and Brenner [35]
Lens, spherical cap	Payne and Pell [65], Collins [21]
Multiple spheres/spheroids	Gluckman *et al.* [32]
Sphere near a plane wall	Brenner [9]
Spheroid (disk)	Lamb [53], Payne and Pell [65], Sampson [71]
Torus	Pell and Payne [66]
Two spheres (translation)	Stimson and Jeffery [80]
Two spheres (linear field)	Green (Ph.D. Dissertation, Cambridge)
Two spherical drops (transl.)	Haber *et al.* [33]
Two spheroids	Liao and Krueger [57]
Venturi tube	Happel and Brenner [35]

Table 4.1: Streamfunction solutions for various axisymmetric geometries.

For the disturbance flow produced by a translating sphere, we need only $n = 2$ for which $\mathcal{G}_2(\xi) = (1-\xi^2)/2 = \frac{1}{2}\sin^2\theta$, so that

$$\psi = \frac{1}{2}(A_2 r^2 + B_2 r^{-1} + C_2 r^4 + D_2 r)\mathcal{G}_2(\xi) \; .$$

We must set $A_2 = C_2 = 0$, since the disturbance velocity vanishes at infinity. The boundary conditions at the sphere surface, Equation 4.43, yield two equations for the remaining two unknowns, and the final result for the streamfunction for the sphere (translating in the positive z-direction) is

$$\psi = \frac{1}{4}Ur^2\sin^2\theta\left[-3\left(\frac{a}{r}\right) + \left(\frac{a}{r}\right)^3\right] \; .$$

This concludes our brief overview of the streamfunction. Table 4.1 lists solutions available in the literature for various geometries. The solution for the torus [66] is particularly noteworthy, since $\psi \neq 0$ on the surface of the torus. In fact, this (initially unknown) constant corresponds to the net flow through the hole of the torus. The constant must be determined by solution of an implicit transcendental equation for the one separating streamline coming from infinity that hits the surface of the torus. The works on flow past multiple spheres and spheroids [32, 57] use the streamfunction, but the equations are solved numerically by the boundary collocation technique of Chapter 13.

Exercises

Exercise 4.1 Derivation of Lamb's General Solution.
Verify that if Φ and χ are harmonics, then $\nabla\Phi$ and $\nabla \times (\boldsymbol{x}\chi)$ satisfy $\nabla^2\boldsymbol{v} = \mathbf{0}$ and $\nabla \cdot \boldsymbol{v} = 0$.

Also show that if $\mu\boldsymbol{v} = \alpha_n r^2 \nabla p_n + \beta_n \boldsymbol{x} p_n$, then

1. $\nabla \cdot \boldsymbol{v} = 0$ only if $2n\alpha_n + (n+3)\beta_n = 0$.
2. $\mu\nabla^2\boldsymbol{v} = \nabla p$ only if $2(2n+1)\alpha_n + 2\beta_n = 1$.

Solve these two equations for α_n and β_n and show that this gives Lamb's general solution.
Hint: $\boldsymbol{x} \cdot \nabla p_n = np_n$ and $\boldsymbol{x} \cdot \nabla\nabla p_n = (n-1)\nabla p_n$.

Exercise 4.2 The Multipole Expansion for Axisymmetric Flows.
Show that for axisymmetric flows, with $\boldsymbol{v} = v_r(r,\theta)\boldsymbol{e}_r + v_\theta(r,\theta)\boldsymbol{e}_\theta$, the multipole expansion for the velocity field can be reduced to the form

$$\boldsymbol{v} = \boldsymbol{d}\sum_{n=0}^{\infty}\left\{a_n\frac{(\boldsymbol{d}\cdot\nabla)^n}{n!}\mathcal{G}_{ij}(\boldsymbol{x}-\boldsymbol{x}_c) \;+\; b_n\frac{(\boldsymbol{d}\cdot\nabla)^n}{n!}\nabla^2\mathcal{G}_{ij}(\boldsymbol{x}-\boldsymbol{x}_c)\right\} .$$

Here, $\boldsymbol{d}$ is a unit vector parallel to the symmetry axis and reference point $\boldsymbol{x}_c$ lies inside the body on the axis of symmetry.
Hint: Start with the nonvanishing components of $\boldsymbol{\sigma}\cdot\hat{\boldsymbol{n}}$ on the particle surface, and expand $\boldsymbol{\sigma}\cdot\hat{\boldsymbol{n}}\cdot\mathcal{G}(\boldsymbol{x}-\boldsymbol{\xi})$ in a Taylor series in $\boldsymbol{\xi}$ about $\boldsymbol{\xi}=0$. Consider the surface integral as a series of rings around the z-axis (the axis of symmetry). Symmetry arguments should convince you that the differential operator $\nabla_\xi\nabla_\xi$ has only terms of the type

$$\frac{\partial^m}{\partial\xi_3^m}\left(\frac{\partial^2}{\partial\xi_1^2}+\frac{\partial^2}{\partial\xi_2^2}-C\frac{\partial^n}{\partial\xi_3^n}\right) = \frac{\partial^m}{\partial\xi_3^m}\left(\nabla_\xi^2-\frac{\partial^2}{\partial\xi_3^2}-C\frac{\partial^n}{\partial\xi_3^n}\right)$$

(non-negative integers m, n with $0 \le m+n \le 2$) left over after integration around any ring; now successive applications of this differential operator gives the desired velocity representation.

Exercise 4.3 The Toroidal Fields.
Derive the relation between the toroidal fields $\nabla\times(\boldsymbol{x}\chi_{-n-1})$ and the derivatives of the rotlet field.

Exercise 4.4 Force, Torque, and Stresslet from Lamb's Solution.
Examine Lamb's general solution for $n=-2$, rewrite these fields as Stokes singularities, and then derive the following expressions for the force, torque, and stresslet:

$$\boldsymbol{F} = -4\pi\nabla(r^3p_{-2})\ , \quad \boldsymbol{T} = -8\pi\mu\nabla(r^3\chi_{-2})\ , \quad \boldsymbol{S} = -\frac{2}{3}\pi\nabla\nabla(r^5p_{-3})\ .$$

Exercise 4.5 Inner Products for the Adjoint Method.
Derive expressions for the inner products between the surface vector fields $\boldsymbol{A}$, $\boldsymbol{B}$, and $\boldsymbol{C}$ of the adjoint method, and verify that the elements in the dual basis are orthogonal to the basis elements in Lamb's general solution.
Hint: First verify that

$$\int_0^\pi \frac{P_n^mP_n^k}{\sin^2\theta}\sin\theta d\theta = \frac{(n+m)!}{m(n-m)!}\delta_{km}$$

$$\int_0^\pi\left[\frac{m^2P_n^mP_l^m}{\sin^2\theta}+\frac{dP_n^m}{d\theta}\frac{dP_l^m}{d\theta}\right]\sin\theta d\theta = \frac{2n(n+1)}{2n+1}\frac{(n+m)!}{m(n-m)!}\delta_{ln}\ .$$

Exercise 4.6 Surface Tractions from Lamb's Solution.
Show that the traction fields for Lamb's general solution are

$$\begin{aligned}\boldsymbol{\sigma}\cdot\hat{\boldsymbol{n}} &= \frac{\mu}{r}\sum_{n=-\infty}^{\infty}\left[(n-1)\nabla\times(\boldsymbol{x}\chi_n)+2(n-1)\nabla\Phi_n\right]\\ &\quad+\frac{1}{r}\sum_{n=-\infty}^{\infty}\left[\frac{n(n+2)r^2\nabla p_n}{(n+1)(2n+3)}-\frac{(2n^2+4n+3)\boldsymbol{x}p_n}{(n+1)(2n+3)}\right].\end{aligned}$$

Use this result to derive the expressions for the force, torque, and stresslet on a sphere.

Exercise 4.7 The Translating Sphere by the Adjoint Method.
Show that $\langle \boldsymbol{w}_n^{(i)}, \boldsymbol{e}_z\rangle = 0$ for all $n \neq 1$, by writing $\boldsymbol{e}_z$ as a linear combination of $\boldsymbol{A}_{10}(\theta,\phi)$ and $\boldsymbol{B}_{10}(\theta,\phi)$.

Exercise 4.8 The Sphere in a Rate-of-Strain Field.
Use Lamb's general solution to derive the solution for the disturbance field for a fixed sphere of radius a in a rate-of-strain field $\boldsymbol{E}\cdot\boldsymbol{x}$. Note that there are five cases:

$$\boldsymbol{E}^{(1)} = \begin{pmatrix} -1 & 0 & 0\\ 0 & -1 & 0\\ 0 & 0 & 2\end{pmatrix} \qquad \boldsymbol{E}^{(2)} = \begin{pmatrix} 1 & 0 & 0\\ 0 & -1 & 0\\ 0 & 0 & 0\end{pmatrix}$$

$$\boldsymbol{E}^{(3)} = \begin{pmatrix} 0 & 1 & 0\\ 1 & 0 & 0\\ 0 & 0 & 0\end{pmatrix} \qquad \boldsymbol{E}^{(4)} = \begin{pmatrix} 0 & 0 & 1\\ 0 & 0 & 0\\ 1 & 0 & 0\end{pmatrix}$$

$$\boldsymbol{E}^{(5)} = \begin{pmatrix} 0 & 0 & 0\\ 0 & 0 & 1\\ 0 & 1 & 0\end{pmatrix}.$$

The general field may be reconstructed as

$$\boldsymbol{E} = \frac{1}{2}E_{zz}\boldsymbol{E}^{(1)} + \frac{1}{2}(E_{xx}-E_{yy})\boldsymbol{E}^{(2)} + E_{xy}\boldsymbol{E}^{(3)} + E_{xz}\boldsymbol{E}^{(4)} + E_{yz}\boldsymbol{E}^{(5)}.$$

Write the surface velocity $\boldsymbol{V}(\theta,\phi) = -\boldsymbol{E}\cdot\boldsymbol{x}$ and identify $m=0$ (degeneracy 1), $m=1$ (degeneracy 2), and $m=2$ (degeneracy 2). The multiple solutions at $m \neq 0$ correspond to symmetry about the z-axis. Show that, in all cases, the stresslet is given by $\boldsymbol{S} = (20/3)\pi\mu a^3\boldsymbol{E}$.

Exercise 4.9 The Translating Spherical Drop: Jump in the Normal Stresses.
Evaluate $\boldsymbol{\sigma}^{(i)} : \hat{\boldsymbol{n}}\hat{\boldsymbol{n}}$ and $\boldsymbol{\sigma}^{(o)} : \hat{\boldsymbol{n}}\hat{\boldsymbol{n}}$ for the translating spherical drop. Then show that the force balance at the drop surface is satisfied exactly.

Exercise 4.10 Streamfunctions for Stokes Singularities.
Derive the streamfunction ψ for the following Stokes singularities:

1. *The Stokeslet,* $F\boldsymbol{e}_z \cdot \mathcal{G}(\boldsymbol{x})$. Plot the streamfunction and show that one streamline profiles a sphere (the radius of this sphere depends on the strength of the Stokeslet). So the Stokeslet alone is an exact solution for a flow past a spherical object. Clearly, it is not the rigid sphere, so what is this spherical object?

2. *The Stresslet,* $[S(\boldsymbol{e}_z\boldsymbol{e}_z - \frac{1}{3}\boldsymbol{\delta})\cdot\nabla]\cdot\mathcal{G}(\boldsymbol{x})$. Derive an expression for the stresslet in terms of ψ, analogous to the result of Payne and Pell for the drag.

3. *The Axisymmetric Stokeson.* Combine the solution elements for the rigid sphere with the Stokeson and the uniform stream to construct the solution for a viscous drop. Plot the streamlines inside and outside the drop.

Exercise 4.11 The Streamfunction and Lamb's General Solution.
By rewriting the Gegenbauer functions in terms of the Legendre polynomials, and by comparing expressions for the radial and tangential velocity components, show that the Sampson's expansion for the streamfunction in spherical coordinates and the axisymmetric version of Lamb's general solution are equivalent. Find the relation between the two sets of coefficients.

Chapter 5

Resistance and Mobility Relations

5.1 Introduction

In this chapter, we address a specific issue, the linear relation between the moments of the surface tractions on a rigid particle and the parameters that govern the relative motion of the particle through a flowing viscous fluid, and follow that with examples to illustrate the use of these relations in suspension theory. As before, we assume that the length scale for the macroscopic problem and that for the microstructure (particle size) are well separated. So on an intermediate length scale that is still much greater than the particle size, we may take the ambient fluid flow field as a combination of a uniform streaming velocity plus a linear field (constant velocity gradient). In other words, the particle interacts with a fluid environment described by the ambient flow field, $\boldsymbol{U}^{\infty} + \boldsymbol{\Omega}^{\infty} \times \boldsymbol{x} + \boldsymbol{E}^{\infty} \cdot \boldsymbol{x}$, with the usual decomposition of the linear term into rotational and rate-of-strain fields. Steady external forces and torques acting on the particles (*e.g.*, gravity, electromagnetic fields) will produce a rigid-body motion of the particle, $\boldsymbol{U} + \boldsymbol{\omega} \times \boldsymbol{x}$, of just the right proportion to balance the resulting hydrodynamic forces and torques with these external agents, as required by Newton's law.

In the next four sections of this chapter we focus our attention on three quantities (the hydrodynamic force $\boldsymbol{F}$, torque $\boldsymbol{T}$, and stresslet $\boldsymbol{S}$ exerted by the fluid on the particle). These are followed by two sections on applications. We examine the theory for the rheological behavior of a dilute suspension of nonspherical particles and show how the resistance functions are used in the development of the theory. In Section 5.7, we discuss *electrophoresis*, or the motion of a charged particle through a viscous electrolyte under the influence of an applied electric field. The role of the resistance functions in the balance between hydrodynamic and electrical forces is examined. A derivation of the electrophoretic mobility for the important limiting cases of thin (Smoluchowski limit) and thick (Hückel limit) are included. We show how the Lorentz reciprocal theorem gives directly the famous result of Smoluchowski that the electrophoretic mobility is independent of the particle geometry in the limit of thin double layers.

From these two applications, it will become clear that microhydrodynamic calculations alone do not provide a complete description of the physics. For example, the rheological behavior of a suspension is also dependent on the orientation distribution function and Brownian motion plays an important role therein, while the analysis of electrophoretic motion requires an additional set of physical laws to describe the effect of the ion (charge) distribution. However, these examples illustrate the ways in which the resistance and mobility function play an integral role in the mathematical analysis; for as long as viscous forces are important, microhydrodynamics will be called upon to convert information concerning forces acting on a particle into a prescription for particle motion.

5.2 The Resistance Tensor

The setting for this and three subsequent sections is a particle undergoing rigid-body motion in a linear ambient field. We focus our attention on three quantities — the hydrodynamic force $\boldsymbol{F}$, torque $\boldsymbol{T}$ and stresslet $\boldsymbol{S}$ — exerted by the fluid on the particle. From the viewpoint of a mathematical boundary value problem, it is quite natural to set the particle and fluid velocities (the boundary conditions) and then calculate $\boldsymbol{F}$, $\boldsymbol{T}$ and $\boldsymbol{S}$. This so-called resistance problem is the subject of this section, and the discussion follows closely the exposition of Brenner [10]. In *mobility* problems, $\boldsymbol{F}$ and $\boldsymbol{T}$ are given and the relative motion of the particle through the fluid is to be determined. These problems are discussed in Section 5.3 and the connection between the two are given in Section 5.4.

In the problem at hand, the disturbance velocity field has the boundary condition at the particle surface

$$\boldsymbol{v}^D(\boldsymbol{x}) = \boldsymbol{U} - \boldsymbol{U}^\infty + (\boldsymbol{\omega} - \boldsymbol{\Omega}^\infty) \times \boldsymbol{x} - \boldsymbol{E}^\infty \cdot \boldsymbol{x} , \qquad \boldsymbol{x} \in S .$$

Since the governing equations are linear, we may decompose the resistance problem into three simpler problems:

1. Translation of the particle with a steady velocity $\boldsymbol{U} - \boldsymbol{U}^\infty$ through a quiescent fluid.

2. Rotation of the particle with a steady angular motion, $(\boldsymbol{\omega} - \boldsymbol{\Omega}^\infty) \times \boldsymbol{x}$, in a quiescent fluid. We defer until Section 5.2.2 a discussion on the choice of the center of rotation.

3. A fixed particle in a rate-of-strain field $\boldsymbol{E}^\infty \cdot \boldsymbol{x}$.

For the first problem, we may again exploit linearity plus dimensional analysis to assert that the relation between the $\boldsymbol{F}$ and $\boldsymbol{U} - \boldsymbol{U}^\infty$ must be of the form

$$\boldsymbol{F} = \mu \boldsymbol{A} \cdot (\boldsymbol{U}^\infty - \boldsymbol{U}) ,$$

where $\boldsymbol{A}$ is a second order-tensor with dimension of length that depends only on the shape of the particle. While the derivation along these lines is straightforward, Brenner [10] gives the following more elegant proof based on the Lorentz reciprocal theorem. Consider two separate fluid motions produced by translational velocities $\boldsymbol{U}'$ and $\boldsymbol{U}''$. Denote the resulting forces by $\boldsymbol{F}'$ and $\boldsymbol{F}''$. Then from the reciprocal theorem (Chapter 2) we have

$$\boldsymbol{U}' \cdot \boldsymbol{F}'' = \boldsymbol{U}'' \cdot \boldsymbol{F}' .$$

We write the relation between the force and translational velocity as $\boldsymbol{F} = \boldsymbol{f}(\boldsymbol{U})$, and replace $\boldsymbol{U}'' \rightarrow \boldsymbol{U}_1 + \boldsymbol{U}_2$ and $\boldsymbol{U}' \rightarrow \boldsymbol{U}_1$. We do this again, only now using $\boldsymbol{U}' \rightarrow \boldsymbol{U}_2$. The resulting equations are

$$\begin{aligned} (\boldsymbol{U}_1 + \boldsymbol{U}_2) \cdot \boldsymbol{f}(\boldsymbol{U}_1) &= \boldsymbol{U}_1 \cdot \boldsymbol{f}(\boldsymbol{U}_1 + \boldsymbol{U}_2) \\ (\boldsymbol{U}_1 + \boldsymbol{U}_2) \cdot \boldsymbol{f}(\boldsymbol{U}_2) &= \boldsymbol{U}_2 \cdot \boldsymbol{f}(\boldsymbol{U}_1 + \boldsymbol{U}_2) , \end{aligned}$$

and by combining the two, we find the relation,

$$(\boldsymbol{U}_1 + \boldsymbol{U}_2) \cdot (\boldsymbol{f}(\boldsymbol{U}_1 + \boldsymbol{U}_2) - \boldsymbol{f}(\boldsymbol{U}_1) - \boldsymbol{f}(\boldsymbol{U}_2)) = 0 .$$

Since $\boldsymbol{U}_1$ and $\boldsymbol{U}_2$ are arbitrary, we conclude that $\boldsymbol{f}(\boldsymbol{U}_1 + \boldsymbol{U}_2) = \boldsymbol{f}(\boldsymbol{U}_1) + \boldsymbol{f}(\boldsymbol{U}_2)$, *i.e.*, the relation is linear.[1]

Generalizing these ideas, the linearity of the Stokes equations leads to the set of linear relations between the moments and the flow parameters [15]:

$$\begin{pmatrix} \boldsymbol{F} \\ \boldsymbol{T} \\ \boldsymbol{S} \end{pmatrix} = \mu \begin{pmatrix} \boldsymbol{A} & \widetilde{\boldsymbol{B}} & \widetilde{\boldsymbol{G}} \\ \boldsymbol{B} & \boldsymbol{C} & \widetilde{\boldsymbol{H}} \\ \boldsymbol{G} & \boldsymbol{H} & \boldsymbol{M} \end{pmatrix} \begin{pmatrix} \boldsymbol{U}^\infty - \boldsymbol{U} \\ \boldsymbol{\Omega}^\infty - \boldsymbol{\omega} \\ \boldsymbol{E}^\infty \end{pmatrix} . \tag{5.1}$$

The square matrix is the *resistance matrix*, sometimes called the grand resistance matrix to distinguish from the subcases (sub-block matrices); it contains second-rank tensors $\boldsymbol{A}$, $\boldsymbol{B}$, and $\boldsymbol{C}$, third-rank tensors $\boldsymbol{G}$ and $\boldsymbol{H}$, and a fourth-rank tensor $\boldsymbol{M}$. The fluid viscosity has been scaled from these quantities so that the matrix elements have dimensions of length to the first ($\boldsymbol{A}$), second ($\boldsymbol{B}$, $\boldsymbol{G}$), or third ($\boldsymbol{C}$, $\boldsymbol{H}$, and $\boldsymbol{M}$) powers. The multiplication rules for the matrix elements in the above equation are the tensorial inner products, as in

$$\begin{aligned} F_i &= \mu A_{ij}(U_j^\infty - U_j) + \mu \widetilde{B}_{ij}(\Omega_j^\infty - \omega_j) + \mu \widetilde{G}_{ijk} E_{jk}^\infty , \\ S_{ij} &= \mu G_{ijk}(U_k^\infty - U_k) + \mu H_{ijk}(\Omega_k^\infty - \omega_k) + \mu M_{ijkl} E_{kl}^\infty . \end{aligned}$$

Tildes are employed for the upper triangular elements of the resistance matrix because the tensors are not independent, as we now show.

[1]Strictly speaking, this only shows that projections onto the sum vector satisfy linearity. To finish the proof, use $f(U) = f(U) \cdot e_i e_i = f(e_i) \cdot U e_i = (e_i f(e_i)) \cdot U$.

5.2.1 Symmetry and Positive Definiteness

We will show that the grand resistance matrix is symmetric, *i.e.*,

$$\begin{aligned} A_{ij} = A_{ji} , \qquad C_{ij} &= C_{ji} , \qquad M_{ijkl} = M_{klij} , \\ B_{ij} = \tilde{B}_{ji} , \qquad G_{ijk} &= \tilde{G}_{kij} , \qquad H_{ijk} = \tilde{H}_{kij} , \end{aligned}$$

as a consequence of the Lorentz reciprocal theorem. The proofs for the six symmetry relations are similar. We shall prove the relation for $\boldsymbol{A}$ and $\boldsymbol{H}$, and leave the rest for Exercise 5.1.

To show that $\boldsymbol{A}$ is symmetric, start once more from $\boldsymbol{U}' \cdot \boldsymbol{F}'' = \boldsymbol{U}'' \cdot \boldsymbol{F}'$, and let $\boldsymbol{U}'$ and $\boldsymbol{U}''$ correspond to motions in the i-th and j-th coordinate directions. For the $\boldsymbol{H}$ relation, the same idea applies. We start with the following statement of the Lorentz reciprocal theorem:

$$\oint_S \boldsymbol{v}_1 \cdot \boldsymbol{\sigma}_2 \cdot \boldsymbol{n} \, dS = \oint_S \boldsymbol{v}_2 \cdot \boldsymbol{\sigma}_1 \cdot \boldsymbol{n} \, dS . \tag{5.2}$$

For the first velocity field, we take the disturbance field of the particle fixed in a straining motion $\boldsymbol{E}^\infty \cdot \boldsymbol{x}$, while for the second we choose the disturbance field generated by the particle rotating with velocity $\boldsymbol{\omega}$ in the ambient field $\boldsymbol{\Omega}^\infty \times \boldsymbol{x}$. These two velocity fields vanish at infinity, while on the particle surface the boundary conditions require

$$\boldsymbol{v}_1(\boldsymbol{x}) = -\boldsymbol{E}^\infty \cdot \boldsymbol{x} , \qquad \boldsymbol{v}_2(\boldsymbol{x}) = \boldsymbol{\omega} \times \boldsymbol{x} - \boldsymbol{\Omega}^\infty \times \boldsymbol{x} . \tag{5.3}$$

The particle torque and stresslet in these two flows follow from the definitions of the resistance functions as

$$T_k = \oint_{S_p} [\boldsymbol{x} \times (\boldsymbol{\sigma}_2 \cdot \boldsymbol{n})]_k \, dS = \tilde{H}_{kij} E^\infty_{ij} \tag{5.4}$$

and

$$S_{ij} = \frac{1}{2} \oint_{S_p} [x_i(\boldsymbol{\sigma}_1 \cdot \boldsymbol{n})_j + (\boldsymbol{\sigma}_1 \cdot \boldsymbol{n})_i x_j] \, dS = H_{ijk}(\Omega^\infty_k - \omega_k) . \tag{5.5}$$

The torque evaluated with the disturbance stress field $\boldsymbol{\sigma}_1$ equals that of the complete solution because the undisturbed straining motion, $\boldsymbol{E}^\infty \cdot \boldsymbol{x}$, does not generate a torque.

For the surface in the reciprocal theorem, we take the boundaries of the fluid region between the particle and a large spherical surface "at infinity." The contributions from the latter vanish because the fields decay far away from the particle at least as fast as $\boldsymbol{v}_1 \sim O(r^{-1})$, $\boldsymbol{v}_2 \sim O(r^{-1})$, $\boldsymbol{\sigma}_1 \sim O(r^{-2})$, $\boldsymbol{\sigma}_2 \sim O(r^{-2})$.[2] The contributions on the particle surface are simplified using the boundary conditions, Equation 5.3.

[2]A straining field about, or rotational motions of, the general particle may contain a monopole field.

Since $\boldsymbol{v}_1$ is the disturbance in the straining field, we have

$$\begin{aligned}
-\oint_S (\boldsymbol{v}_1)_i(\boldsymbol{\sigma}_2\cdot\boldsymbol{n})_i\,dS &= \oint_S E^\infty_{ij}x_j(\boldsymbol{\sigma}_2\cdot\boldsymbol{n})_i\,dS \\
&= E^\infty_{ij}\oint_S x_j(\boldsymbol{\sigma}_2\cdot\boldsymbol{n})_i\,dS \\
&= E^\infty_{ij}\frac{1}{2}\oint_S (x_i(\boldsymbol{\sigma}_2\cdot\boldsymbol{n})_j + x_j(\boldsymbol{\sigma}_2\cdot\boldsymbol{n})_i)\,dS \\
&= E^\infty_{ij}H_{ijk}(\Omega^\infty_k - \omega_k)\ . \qquad (5.6)
\end{aligned}$$

To go from the second to the third line, we used $E^\infty_{ij} = E^\infty_{ji}$ to symmetrize the force dipole into the stresslet, and the last step invoked Equation 5.5.

Similarly, with $\boldsymbol{v}_2$ as the disturbance field of the rotational problem, we have

$$\begin{aligned}
-\oint_S (\boldsymbol{v}_2)_i(\boldsymbol{\sigma}_1\cdot\boldsymbol{n})_i\,dS &= \oint_S \epsilon_{ikj}(\boldsymbol{\Omega}^\infty_k - \boldsymbol{\omega}_k)x_j(\boldsymbol{\sigma}_1\cdot\boldsymbol{n})_i\,dS \\
&= (\boldsymbol{\Omega}^\infty_k - \boldsymbol{\omega}_k)\oint_S \epsilon_{kji}x_j(\boldsymbol{\sigma}_1\cdot\boldsymbol{n})_i\,dS \\
&= (\boldsymbol{\Omega}^\infty_k - \boldsymbol{\omega}_k)\tilde{H}_{kij}E^\infty_{ij}\ . \qquad (5.7)
\end{aligned}$$

To go from the second to the third line, we used Equation 5.4 for the torque on the particle in a straining field.

Combining Equations 5.6 and 5.7 gives

$$E^\infty_{ij}H_{ijk}(\Omega^\infty_k - \omega_k) = (\boldsymbol{\Omega}^\infty_k - \boldsymbol{\omega}_k)\tilde{H}_{kij}E^\infty_{ij}\ . \qquad (5.8)$$

But $\boldsymbol{\Omega}^\infty - \boldsymbol{\omega}$ and $\boldsymbol{E}^\infty$ are arbitrary constants, so we have

$$H_{ijk} = \tilde{H}_{kij}\ , \qquad (5.9)$$

as claimed. Having shown that the grand matrix is symmetric, we now show that it is positive-definite.

The rate of energy dissipation (Chapter 2) of the disturbance field is given by

$$\int_V e^D_{ij}e^D_{ij}\,dV = -\oint_{S_p} v^D_i\sigma^D_{ij}\hat{n}_j\,dS\ .$$

The disturbance stress field is given everywhere by $\boldsymbol{\sigma}^D = \boldsymbol{\sigma} - 2\mu\boldsymbol{E}^\infty$, and on the particle surface we have $\boldsymbol{v}^D = \boldsymbol{U} - \boldsymbol{U}^\infty + (\boldsymbol{\omega} - \boldsymbol{\Omega}^\infty)\times\boldsymbol{x} - \boldsymbol{E}^\infty\cdot\boldsymbol{x}$. Inserting these results in above gives the inequality

$$(\boldsymbol{U}^\infty - \boldsymbol{U})\cdot\boldsymbol{F} + (\boldsymbol{\Omega}^\infty - \boldsymbol{\omega})\cdot\boldsymbol{T} + \boldsymbol{E}^\infty : \boldsymbol{S} \geq \boldsymbol{E}^\infty : \boldsymbol{S}^\infty = 2\mu V_p\boldsymbol{E}^\infty : \boldsymbol{E}^\infty\ ,$$

where $\boldsymbol{S}^\infty = 2\mu V_p\boldsymbol{E}^\infty$ is the stresslet of the undisturbed rate-of-strain field,

$$\boldsymbol{S}^\infty = 2\mu V_p\boldsymbol{E}^\infty = \frac{1}{2}\oint_{S_p}[(\boldsymbol{\sigma}^\infty\cdot\hat{\boldsymbol{n}})\boldsymbol{x} + \boldsymbol{x}(\boldsymbol{\sigma}^\infty\cdot\hat{\boldsymbol{n}})]\,dS\ .$$

We replace $\boldsymbol{F}$, $\boldsymbol{T}$, and $\boldsymbol{S}$ using the resistance relation, and the resulting inequality establishes that the grand resistance matrix is positive-definite. We may take special cases, *e.g.*, translational motion alone, or just rotational motion, or pure straining motion about a fixed particle, to show that the diagonal elements $\boldsymbol{A}$, $\boldsymbol{C}$, and $\boldsymbol{M}$ are positive-definite tensors.

Now that we know that $\boldsymbol{A}$ is a symmetric, positive-definite, second order tensor, we arrive at the following conclusions:

1. There are three mutually orthogonal axes, given by the principal directions of $\boldsymbol{A}$. Translations through a quiescent fluid along one of these give forces directed only along that direction. This result follows directly from the orthogonality properties of eigenvectors of symmetric matrices [83]. *The important conclusion is that, even for a particle of arbitrary shape, there is a natural choice for the coordinate axes.*

2. The angle θ_{UF} between $\boldsymbol{U}$ and $\boldsymbol{F}$ is obtuse. This follows from the energy dissipation argument, $-\boldsymbol{U}\cdot\boldsymbol{F} = -|\boldsymbol{U}|\,|\boldsymbol{F}|\cos\theta_{UF} > 0$.

3. Given $\boldsymbol{A}\cdot\boldsymbol{v}_i = A_i\boldsymbol{v}_i$, *i.e.*, the drag is colinear along three directions $\boldsymbol{v}_i$, and no pair of which are mutually orthogonal, except perhaps one pair. Then $\boldsymbol{A}$ is isotropic. Proof: Take $\boldsymbol{v}_1$ and $\boldsymbol{v}_2$ with $\boldsymbol{v}_1\cdot\boldsymbol{v}_2 \neq 0$. We have $A_1\boldsymbol{v}_1\cdot\boldsymbol{v}_2 = \boldsymbol{v}_2\cdot\boldsymbol{A}\cdot\boldsymbol{v}_1 = \boldsymbol{v}_1\cdot\boldsymbol{A}\cdot\boldsymbol{v}_2 = A_2\boldsymbol{v}_1\cdot\boldsymbol{v}_2$, which forces $A_1 = A_2$ if $\boldsymbol{v}_1\cdot\boldsymbol{v}_2 \neq 0$. This is a direct way of establishing that regular polyhedra (the Platonic solids) have isotropic resistance tensor $\boldsymbol{A}$.

Similar conclusions may be derived for the rotational resistance tensor $\boldsymbol{C}$.

5.2.2 The Hydrodynamic Center of Resistance

Consider the axisymmetric pear-shaped object of Figure 5.1. For transverse translations, there will be a torque exerted in the direction $\boldsymbol{U}\times\boldsymbol{d}$, the exact magnitude (and sign) of the torque depending on the choice of the center. This is readily apparent from the figure, as we move the center along the axis from point P to Q. At some point between P and Q, (the hydrodynamic center) we expect the torque to vanish. The skew-shaped propeller in the same figure also experiences a torque when it is moved (without rotation) along the axial direction. Here, however, the hydrodynamic torque is in the axial direction, so a shift in the center along the axis has no effect on the torque, so there is no point for which the coupling tensor $\boldsymbol{B}$ vanishes. On the other hand, the centroid is still a special point, because there the coupling tensor is symmetric. This is most readily seen by observing that translations along the axial direction and any transverse direction give torques in the same direction as the motion. Thus the tensor has three orthogonal principal directions. We will see below that the hydrodynamic center exists for particles of arbitrary shape and that it is unique. We note that if $\boldsymbol{B}$ vanishes at some point, it can do so only at that point, for this is simply a special case of the coupling tensor being symmetric.[3]

[3]In the original work [10] vanishing of $\boldsymbol{B}$ is given as the condition for the hydrodynamic center. The more complete result and proof given here follow those given in the later work [35], which first appeared in 1965.

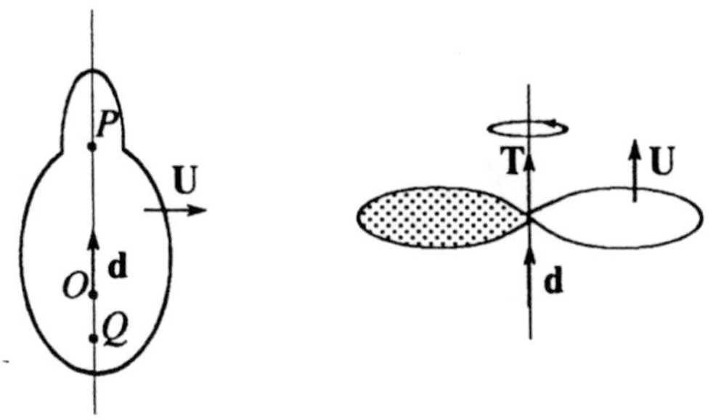

Figure 5.1: Hydrodynamic centers of resistance: an axisymmetric body and a propeller.

This discussion then motivates a search for the "natural origin" for a particle of arbitrary shape to complement the natural coordinate directions found above.

Consider a particle undergoing just translational motion and the resulting torques $\boldsymbol{T}(\boldsymbol{x}_1)$ and $\boldsymbol{T}(\boldsymbol{x}_2)$ with respect to two centers $\boldsymbol{x}_1$ and $\boldsymbol{x}_2$. The difference is calculated as

$$\begin{aligned}\boldsymbol{T}(\boldsymbol{x}_2) - \boldsymbol{T}(\boldsymbol{x}_1) &= \oint_{S_p} (\boldsymbol{x} - \boldsymbol{x}_2) \times (\boldsymbol{\sigma} \cdot \hat{\boldsymbol{n}})\, dS - \oint_{S_p} (\boldsymbol{x} - \boldsymbol{x}_1) \times (\boldsymbol{\sigma} \cdot \hat{\boldsymbol{n}})\, dS \\ &= \oint_{S_p} (\boldsymbol{x}_1 - \boldsymbol{x}_2) \times (\boldsymbol{\sigma} \cdot \hat{\boldsymbol{n}})\, dS \\ &= (\boldsymbol{x}_1 - \boldsymbol{x}_2) \times \boldsymbol{F} \ ,\end{aligned}$$

or in terms of the coupling tensor,

$$B_{ij}^{(2)} = B_{ij}^{(1)} - \epsilon_{ikl}(\boldsymbol{x}_2 - \boldsymbol{x}_1)_k A_{lj} \ .$$

Here, the superscripts on the coupling tensor denote the respective centers. Suppose that at $\boldsymbol{x}_2 = \boldsymbol{x}_{cr}$ (this will be the hydrodynamic center of resistance) $\boldsymbol{B}^{(cr)}$ is symmetric, so that the antisymmetric part of the previous equation becomes

$$B_{ij}^{(1)} - B_{ji}^{(1)} = \epsilon_{ikl}(\boldsymbol{x}_{cr} - \boldsymbol{x}_1)_k A_{lj} - \epsilon_{jkl}(\boldsymbol{x}_{cr} - \boldsymbol{x}_1)_k A_{li} \ . \qquad (5.10)$$

To obtain an explicit expression for the hydrodynamic center of resistance, we apply ϵ_{mij} to both sides and use

$$\epsilon_{mij}\epsilon_{ikl} = \delta_{jk}\delta_{lm} - \delta_{jl}\delta_{km} \ , \qquad \epsilon_{mij}\epsilon_{jkl} = \delta_{km}\delta_{il} - \delta_{lm}\delta_{ik} \ ,$$

to arrive at

$$\left[(\boldsymbol{A} - (\mathrm{tr}\boldsymbol{A})\boldsymbol{\delta}) \cdot (\boldsymbol{x}_{cr} - \boldsymbol{x}_1)\right]_m = \frac{1}{2}\epsilon_{mij}\left[B_{ij}^{(1)} - B_{ji}^{(1)}\right] \ .$$

The matrix associated with $\boldsymbol{A} - (\mathrm{tr}\boldsymbol{A})\boldsymbol{\delta}$ is always nonsingular. In fact, with the principal directions $\boldsymbol{e}_i$, $i = 1, 2, 3$ as basis, $\boldsymbol{A}$ is diagonal and the matrix in

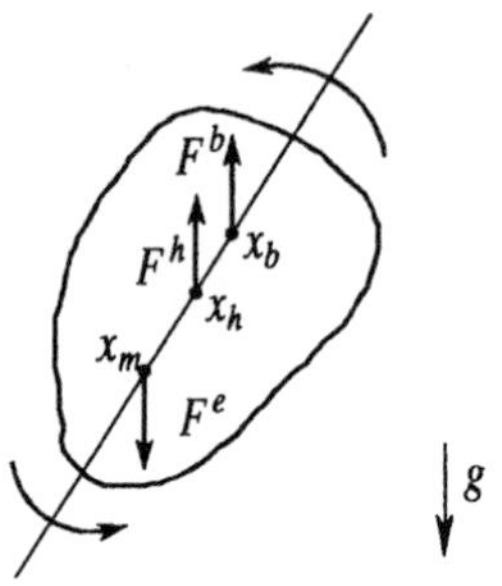

Figure 5.2: The couples acting on a settling body.

question has the representation

$$\boldsymbol{A} - (\mathrm{tr}\boldsymbol{A})\boldsymbol{\delta} = -\begin{pmatrix} A_2 + A_3 & 0 & 0 \\ 0 & A_3 + A_1 & 0 \\ 0 & 0 & A_1 + A_2 \end{pmatrix},$$

where $A_i > 0$, $i = 1, 2, 3$ are the principal values. The determinant, $|\boldsymbol{A} - (\mathrm{tr}\boldsymbol{A})\boldsymbol{\delta}|$, cannot vanish, and the explicit expression for the hydrodynamic center follows as

$$\boldsymbol{x}_{cr} - \boldsymbol{x}_1 = \frac{1}{2}\left((\mathrm{tr}\boldsymbol{A})\boldsymbol{\delta} - \boldsymbol{A}\right)^{-1} \cdot \left[\boldsymbol{\epsilon} : \left(\boldsymbol{B}^{(1)} - \boldsymbol{B}^{t(1)}\right)\right] .$$

In terms of the principal axes of $\boldsymbol{A}$, this relation may be written as

$$\begin{aligned} \boldsymbol{x}_{cr} - \boldsymbol{x}_1 &= \frac{B_{32}^{(1)} - B_{23}^{(1)}}{(A_2 + A_3)} \boldsymbol{e}_1 + \frac{B_{13}^{(1)} - B_{31}^{(1)}}{(A_3 + A_1)} \boldsymbol{e}_2 \\ &+ \frac{B_{21}^{(1)} - B_{12}^{(1)}}{(A_1 + A_2)} \boldsymbol{e}_3 . \end{aligned} \tag{5.11}$$

Note that from the method of construction, the converse is readily demonstrated: At the hydrodynamic center of resistance, Equation 5.10 holds so that the coupling tensor is symmetric.

In summary, to locate the hydrodynamic center of resistance of a particle of arbitrary shape, we first determine the tensor $\boldsymbol{A}$, and then find its eigenvalues $\{A_1, A_2, A_3\}$ and principal directions $\{\boldsymbol{e}_1, \boldsymbol{e}_2, \boldsymbol{e}_3\}$. We then pick *any* convenient point $\boldsymbol{x}_1$ and calculate the coupling tensor $\boldsymbol{B}(\boldsymbol{x}_1)$ between the translation velocity and the torque with respect to that point. Equation 5.11 gives the location of the hydrodynamic center relative to $\boldsymbol{x}_1$. Although the reference point appears in the formulae, it is easily verified that the hydrodynamic center is well defined and independent of the choice for $\boldsymbol{x}_1$ (see Exercises 5.2 and 5.3).

For an illustration of the use of the hydrodynamic center, consider an axisymmetric solid object of nonuniform density distribution (see Figure 5.2). (Another illustration is given later in the chapter.) In general, the hydrodynamic center, $\boldsymbol{x}_{cr}$, the center of buoyancy (center of mass of the displaced fluid),

$\boldsymbol{x}_b$, and center of mass, $\boldsymbol{x}_m$, are different, as shown in Figure 5.2. Also, while the first two are purely geometric entities determined by the surface geometry, the third center depends on the details of the mass distribution within the body. The rigid-body rotation of the particle as it settles under the influence of gravity can be determined by balancing all three couples on the body, with the forces placed at the appropriate centers.

5.2.3 Translation Theorems for Resistance Tensors

In the preceding discussion, we saw that the coupling tensor depended on the choice of the center in the calculation of the hydrodynamic torque. Consider two different parametrizations of a rigid-body motion, $\boldsymbol{U}_1 + \boldsymbol{\omega} \times (\boldsymbol{x} - \boldsymbol{x}_1) = \boldsymbol{U}_2 + \boldsymbol{\omega} \times (\boldsymbol{x} - \boldsymbol{x}_2)$, with the torques in each case measured relative to the center of rotation. The so-called translation theorems [35] furnish the relation between tensors defined using the different origins:

$$\begin{aligned}
\boldsymbol{A}^{(2)} &= \boldsymbol{A}^{(1)} = \boldsymbol{A} \\
\boldsymbol{B}^{(2)} &= \boldsymbol{B}^{(1)} - (\boldsymbol{x}_2 - \boldsymbol{x}_1) \times \boldsymbol{A} \\
\boldsymbol{C}^{(2)} &= \boldsymbol{C}^{(1)} - (\boldsymbol{x}_2 - \boldsymbol{x}_1) \times \boldsymbol{A} \times (\boldsymbol{x}_2 - \boldsymbol{x}_1) \\
&+ \boldsymbol{B}^{(1)} \times (\boldsymbol{x}_2 - \boldsymbol{x}_1) - (\boldsymbol{x}_2 - \boldsymbol{x}_1) \times \boldsymbol{B}^{t(1)} .
\end{aligned}$$

These identities are readily established using arguments encountered in the derivation of the hydrodynamic center. The translation theorems are quite useful, as we will see later. In the following section, we consider the resistance tensors of a "composite" particle formed by joining simpler units. The tensor for the composite particle can be constructed from those of the units by applying the translation theorems. In Part IV, we will encounter hydrodynamic calculations with respect to centers dictated by the mathematical analysis, while the interpretation of the results and subsequent physical applications are most natural at the hydrodynamic center. The translation theorems provide the needed linkage.

5.3 The Mobility Tensor

Many physical problems in microhydrodynamics require the solution of the motion of the particle in response to prescribed forces and torques in a known ambient flow. We call such situations *mobility* problems. We can invoke linearity again to write

$$\begin{pmatrix} \boldsymbol{U}^\infty - \boldsymbol{U} \\ \boldsymbol{\Omega}^\infty - \boldsymbol{\omega} \\ \mu^{-1}\boldsymbol{S} \end{pmatrix} = \begin{pmatrix} \boldsymbol{a} & \tilde{\boldsymbol{b}} & \tilde{\boldsymbol{g}} \\ \boldsymbol{b} & \boldsymbol{c} & \tilde{\boldsymbol{h}} \\ \boldsymbol{g} & \boldsymbol{h} & \boldsymbol{m} \end{pmatrix} \begin{pmatrix} \mu^{-1}\boldsymbol{F} \\ \mu^{-1}\boldsymbol{T} \\ \boldsymbol{E}^\infty \end{pmatrix} . \tag{5.12}$$

The square matrix is the *mobility matrix*; it contains second-rank tensors $\boldsymbol{a}$, $\boldsymbol{b}$, and $\boldsymbol{c}$, third-rank tensors $\boldsymbol{g}$ and $\boldsymbol{h}$, and a fourth-rank tensor $\boldsymbol{m}$. The fluid viscosity has been scaled from these quantities so that the matrix elements have dimensions of length to the -1 ($\boldsymbol{a}$), -2 ($\boldsymbol{b}$), -3 ($\boldsymbol{c}$), zero ($\boldsymbol{h}$), first ($\boldsymbol{g}$), and third ($\boldsymbol{m}$) powers.

The elements of the mobility matrix also obey symmetry relations as a consequence of the Lorentz reciprocal theorem:

$$
\begin{aligned}
a_{ij} = a_{ji} \,, \qquad c_{ij} &= c_{ji} \,, \qquad m_{ijkl} = m_{klij} \,, \\
b_{ij} = \widetilde{b}_{ji} \,, \qquad g_{ijk} &= -\widetilde{g}_{kij} \,, \qquad h_{ijk} = -\widetilde{h}_{kij} \,.
\end{aligned}
$$

Note the occurrence of minus signs in the relations for $\boldsymbol{g}$ and $\boldsymbol{h}$. The proofs for symmetry are similar to the ones for the resistance matrix (see Exercise 5.1). For example, the relation for $\boldsymbol{h}$ can be derived as follows. We start with Equation 5.2 and take for $\boldsymbol{v}_1$ the disturbance field of a *force-free, torque-free particle fixed in a straining motion* $\boldsymbol{E}^\infty \cdot \boldsymbol{x}$. For $\boldsymbol{v}_2$, we choose the disturbance field generated by a force-free particle, with external torque $-\boldsymbol{T}$ in a quiescent fluid. Of course, our choices for $\boldsymbol{v}_1$ and $\boldsymbol{v}_2$ are dictated by the definitions of $\boldsymbol{h}$ and $\widetilde{\boldsymbol{h}}$. These two velocity fields vanish at infinity, while on the particle surface the boundary conditions are

$$
\boldsymbol{v}_1(\boldsymbol{x}) = -\boldsymbol{E}^\infty \cdot \boldsymbol{x} + \boldsymbol{U}_1 + \boldsymbol{\omega}_1 \times \boldsymbol{x} \,, \qquad \boldsymbol{v}_2(\boldsymbol{x}) = \boldsymbol{U}_2 + \boldsymbol{\omega}_2 \times \boldsymbol{x} \,.
$$

The rotational velocity $\boldsymbol{\omega}_1$ and stresslet $\boldsymbol{S}_2$ in these two flows follow from the definitions of the mobility functions as

$$
(\boldsymbol{\omega}_1)_k = -\widetilde{h}_{kij} E^\infty_{ij}
$$

and

$$
(\boldsymbol{S}_2)_{ij} = h_{ijk} (\boldsymbol{T}_2)_k \,.
$$

We apply the reciprocal theorem over the same fluid region as before. One side of the equation simplifies as

$$
\begin{aligned}
-\oint_S (\boldsymbol{v}_1)_i (\boldsymbol{\sigma}_2 \cdot \boldsymbol{n})_i \, dS &= \oint_S \left[E^\infty_{ij} x_j - (\boldsymbol{U}_1)_i - (\boldsymbol{\omega}_1 \times \boldsymbol{x})_i \right] (\boldsymbol{\sigma}_2 \cdot \boldsymbol{n})_i \, dS \\
&= E^\infty_{ij} (\boldsymbol{S}_2)_{ij} - (\boldsymbol{U}_1)_i (\boldsymbol{F}_2)_i - (\boldsymbol{\omega}_1)_i (\boldsymbol{T}_2)_i \\
&= E^\infty_{ij} h_{ijk} (\boldsymbol{T}_2)_k + \widetilde{h}_{kij} E^\infty_{ij} (\boldsymbol{T}_2)_k \,. \qquad (5.13)
\end{aligned}
$$

In the last step, we replaced $\boldsymbol{S}_2$ and $-\boldsymbol{\omega}_1$ with $\boldsymbol{h} \cdot \boldsymbol{T}_2$ and $\widetilde{\boldsymbol{h}} : \boldsymbol{E}^\infty$, respectively, and we also used $\boldsymbol{F}_2 = \boldsymbol{0}$. The other side of the reciprocal relation simplifies as

$$
\begin{aligned}
-\oint_S (\boldsymbol{v}_2)_i (\boldsymbol{\sigma}_1 \cdot \boldsymbol{n})_i \, dS &= -\oint_S \left[(\boldsymbol{U}_2)_i + (\boldsymbol{\omega}_2 \times \boldsymbol{x})_i \right] (\boldsymbol{\sigma}_1 \cdot \boldsymbol{n})_i \, dS \\
&= -(\boldsymbol{U}_2)_i (\boldsymbol{F}_1)_i - (\boldsymbol{\omega}_2)_i (\boldsymbol{T}_1)_i \\
&= \boldsymbol{0} \,. \qquad (5.14)
\end{aligned}
$$

Combining Equations 5.13 and 5.14 gives, since $\boldsymbol{T}_2$ and $\boldsymbol{E}$ are arbitrary constants, $h_{ijk} = -\widetilde{h}_{kij}$, as claimed.

5.3.1 Translation Theorems for Mobility Tensors

The translation theorems for the mobility tensor are as follows:

$$\begin{aligned} \boldsymbol{a}^{(2)} &= \boldsymbol{a}^{(1)} - (\boldsymbol{x}_2 - \boldsymbol{x}_1) \times \boldsymbol{c} \times (\boldsymbol{x}_2 - \boldsymbol{x}_1) - (\boldsymbol{x}_2 - \boldsymbol{x}_1) \times \boldsymbol{b}^{(1)} + \boldsymbol{b}^{t(1)} \times (\boldsymbol{x}_2 - \boldsymbol{x}_1) \\ \boldsymbol{b}^{(2)} &= \boldsymbol{b}^{(1)} + \boldsymbol{c} \times (\boldsymbol{x}_2 - \boldsymbol{x}_1) \\ \boldsymbol{c}^{(2)} &= \boldsymbol{c}^{(1)} = \boldsymbol{c} \,. \end{aligned}$$

If we compare this with the translation theorems for the resistance tensors, we see that the roles played by $\{\boldsymbol{a}, \boldsymbol{b}, \boldsymbol{b}^t, \boldsymbol{c}\}$ and $\{\boldsymbol{A}, \boldsymbol{B}, \boldsymbol{B}^t, \boldsymbol{C}\}$ are reversed.

5.3.2 The Hydrodynamic Center of Mobility

Here, we correct a widely-held misconception and show that the hydrodynamic center for the mobility tensors in general differs from the hydrodynamic center of resistance. The hydrodynamic center for the mobility, $\boldsymbol{x}_{cm}$, is given by a relation analogous to the one obtained earlier for the resistance tensor:

$$\boldsymbol{x}_{cm} - \boldsymbol{x}_1 = \frac{b^{(1)}_{23} - b^{(1)}_{32}}{(c_2 + c_3)} e_1 + \frac{b^{(1)}_{31} - b^{(1)}_{13}}{(c_3 + c_1)} e_2 + \frac{b^{(1)}_{12} - b^{(1)}_{21}}{(c_1 + c_2)} e_3 \,.$$

This result can also be deduced from the corresponding result for the resistance formulation, by reversing the roles played by $\{\boldsymbol{a}, \boldsymbol{b}, \boldsymbol{b}^t, \boldsymbol{c}\}$ and $\{\boldsymbol{A}, \boldsymbol{B}, \boldsymbol{B}^t, \boldsymbol{C}\}$.

The distance between the two hydrodynamic centers is obtained by solving the following relation[4]:

$$\boldsymbol{c}^{-1} \times (\boldsymbol{x}_{cm} - \boldsymbol{x}_{cr}) + (\boldsymbol{x}_{cm} - \boldsymbol{x}_{cr}) \times \boldsymbol{c}^{-1} = -\boldsymbol{\beta}^{(cr)} + \boldsymbol{\beta}^{t(cr)} \,,$$

where $\boldsymbol{\beta}^{(cr)} = [\boldsymbol{C}\boldsymbol{A}^{-1}\boldsymbol{B}]^{(cr)}$ is a product of resistance tensors evaluated at the hydrodynamic center of resistance. Therefore, *the hydrodynamic centers of resistance and mobility coincide if and only if* $\boldsymbol{\beta}^{(cr)} = \boldsymbol{\beta}^{t(cr)}$. In terms of the principal directions and values of the mobility tensor, $\boldsymbol{c}$, the result is

$$\boldsymbol{x}_{cm} - \boldsymbol{x}_{cr} = \frac{\beta^{(cr)}_{23} - \beta^{(cr)}_{32}}{(c_2^{-1} + c_3^{-1})} e_1 + \frac{\beta^{(cr)}_{31} - \beta^{(cr)}_{13}}{(c_3^{-1} + c_1^{-1})} e_2 + \frac{\beta^{(cr)}_{12} - \beta^{(cr)}_{21}}{(c_1^{-1} + c_2^{-1})} e_3 \,.$$

We interject a historical note here. Brenner's "proof" [13] that the hydrodynamic centers of resistance and mobility are coincident contains an error, equivalent to the erroneous conclusion that $\boldsymbol{\beta}$ is symmetric for all particle shapes. Bernal and de la Torre [6] and Wegener [90] appear to have been the first to (independently) notice this error. These studies of the diffusivity of proteins required mobility calculations of bent and hinged rods, and for such objects $\boldsymbol{\beta}$ is not symmetric. Later in this chapter, we will examine other shapes for which this is also the case.

---

[4]We thank Douglas Brune for bringing this relation to our attention.

5.4 Relations Between the Resistance and Mobility Tensors

Clearly, the resistance and mobility formulations are related and but for $\boldsymbol{S}$ appearing as a dependent variable in both problems, the two formulations would be formal inverses of each other. The force and torque portion of the resistance equation may be written as

$$\begin{aligned}\begin{pmatrix} \boldsymbol{F} \\ \boldsymbol{T} \end{pmatrix} &= \mu \begin{pmatrix} \boldsymbol{A} & \tilde{\boldsymbol{B}} \\ \boldsymbol{B} & \boldsymbol{C} \end{pmatrix} \begin{pmatrix} \boldsymbol{U}^\infty - \boldsymbol{U} \\ \boldsymbol{\Omega}^\infty - \boldsymbol{\omega} \end{pmatrix} + \mu \begin{pmatrix} \tilde{\boldsymbol{G}} \\ \tilde{\boldsymbol{H}} \end{pmatrix} \begin{pmatrix} \boldsymbol{E}^\infty \end{pmatrix} \\ &= \mu \begin{pmatrix} \boldsymbol{A} & \tilde{\boldsymbol{B}} \\ \boldsymbol{B} & \boldsymbol{C} \end{pmatrix} \left[\begin{pmatrix} \boldsymbol{a} & \tilde{\boldsymbol{b}} \\ \boldsymbol{b} & \boldsymbol{c} \end{pmatrix} \begin{pmatrix} \mu^{-1}\boldsymbol{F} \\ \mu^{-1}\boldsymbol{T} \end{pmatrix} + \begin{pmatrix} \tilde{\boldsymbol{g}} \\ \tilde{\boldsymbol{h}} \end{pmatrix} \begin{pmatrix} \boldsymbol{E}^\infty \end{pmatrix} \right] \\ &\quad + \mu \begin{pmatrix} \tilde{\boldsymbol{G}} \\ \tilde{\boldsymbol{H}} \end{pmatrix} \begin{pmatrix} \boldsymbol{E}^\infty \end{pmatrix}, \end{aligned} \tag{5.15}$$

which implies that

$$\begin{pmatrix} \boldsymbol{A} & \tilde{\boldsymbol{B}} \\ \boldsymbol{B} & \boldsymbol{C} \end{pmatrix} \begin{pmatrix} \boldsymbol{a} & \tilde{\boldsymbol{b}} \\ \boldsymbol{b} & \boldsymbol{c} \end{pmatrix} = \begin{pmatrix} \boldsymbol{\delta} & \boldsymbol{0} \\ \boldsymbol{0} & \boldsymbol{\delta} \end{pmatrix} \tag{5.16}$$

and

$$\begin{pmatrix} \boldsymbol{A} & \tilde{\boldsymbol{B}} \\ \boldsymbol{B} & \boldsymbol{C} \end{pmatrix} \begin{pmatrix} \tilde{\boldsymbol{g}} \\ \tilde{\boldsymbol{h}} \end{pmatrix} + \begin{pmatrix} \tilde{\boldsymbol{G}} \\ \tilde{\boldsymbol{H}} \end{pmatrix} = \begin{pmatrix} \boldsymbol{0} \\ \boldsymbol{0} \end{pmatrix}. \tag{5.17}$$

In a similar manner, the resistance expression for the stresslet may be combined with the mobility expressions to obtain relations for $\boldsymbol{g}$, $\boldsymbol{h}$, and $\boldsymbol{m}$ in terms of the resistance tensors. The complete set of relations follows as

$$\begin{pmatrix} \boldsymbol{a} & \tilde{\boldsymbol{b}} \\ \boldsymbol{b} & \boldsymbol{c} \end{pmatrix} = \begin{pmatrix} \boldsymbol{A} & \tilde{\boldsymbol{B}} \\ \boldsymbol{B} & \boldsymbol{C} \end{pmatrix}^{-1} \tag{5.18}$$

$$\begin{pmatrix} \tilde{\boldsymbol{g}} \\ \tilde{\boldsymbol{h}} \end{pmatrix} = -\begin{pmatrix} \boldsymbol{A} & \tilde{\boldsymbol{B}} \\ \boldsymbol{B} & \boldsymbol{C} \end{pmatrix}^{-1} \begin{pmatrix} \tilde{\boldsymbol{G}} \\ \tilde{\boldsymbol{H}} \end{pmatrix} \tag{5.19}$$

$$\begin{pmatrix} \boldsymbol{g} & \boldsymbol{h} \end{pmatrix} = \begin{pmatrix} \boldsymbol{G} & \boldsymbol{H} \end{pmatrix} \begin{pmatrix} \boldsymbol{A} & \tilde{\boldsymbol{B}} \\ \boldsymbol{B} & \boldsymbol{C} \end{pmatrix}^{-1} \tag{5.20}$$

$$\boldsymbol{m} = \boldsymbol{M} - \begin{pmatrix} \boldsymbol{G} & \boldsymbol{H} \end{pmatrix} \begin{pmatrix} \boldsymbol{A} & \tilde{\boldsymbol{B}} \\ \boldsymbol{B} & \boldsymbol{C} \end{pmatrix}^{-1} \begin{pmatrix} \tilde{\boldsymbol{G}} \\ \tilde{\boldsymbol{H}} \end{pmatrix}. \tag{5.21}$$

Note that the stresslet on a particle when the particle is force-free and torque-free is in general different from that when it is stationary. In fact, using energy dissipation arguments, we may state $\boldsymbol{m} \leq \boldsymbol{M}$.

Some further reduction is possible, using inversion relations for block matrices. The explicit results for the mobility tensors $\boldsymbol{a}$, $\boldsymbol{b}$, and $\boldsymbol{c}$ are

$$\boldsymbol{a} = (\boldsymbol{A} - \boldsymbol{B}^t\boldsymbol{C}^{-1}\boldsymbol{B})^{-1} \tag{5.22}$$

$$\boldsymbol{b} = -\boldsymbol{C}^{-1}\boldsymbol{B}(\boldsymbol{A} - \boldsymbol{B}^t\boldsymbol{C}^{-1}\boldsymbol{B})^{-1} \tag{5.23}$$

$$\boldsymbol{c} = (\boldsymbol{C} - \boldsymbol{B}\boldsymbol{A}^{-1}\boldsymbol{B}^t)^{-1}. \tag{5.24}$$

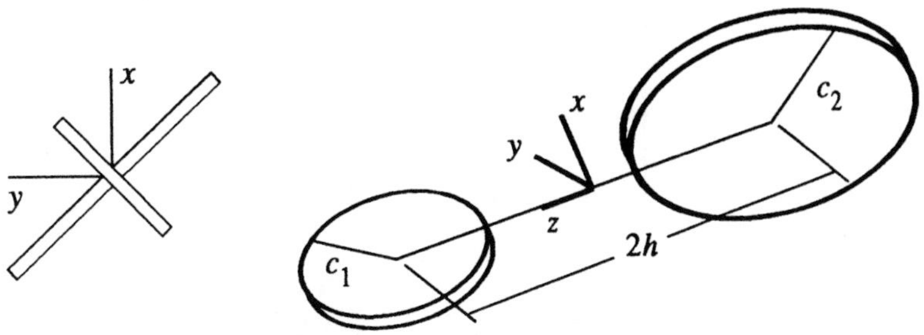

Figure 5.3: A propeller with unequal blades.

In principle, when faced with a mobility problem, we may proceed by first solving an appropriate set of resistance problems for which the particle motions are prescribed (boundary conditions are specified) and then use the relations above to obtain the mobility functions. For many-body problems, this approach is feasible in principle, but infeasible in practice — the computational costs are prohibitive. Furthermore, with respect to the gap width, the resistance functions diverge in a singular fashion as contact is approached (see Chapter 9), so the resistance matrix becomes ill-conditioned, *i.e.*, hard to invert numerically. For this reason, we will devote some effort in later chapters to the issue of direct solution of mobility problems.

We conclude this section with an example calculation of the resistance and mobility tensors of a "propeller" formed by attaching two thin circular disks to a frictionless rod, in the skew-symmetric fashion, pitched at an angle of 90° to each other, as shown in Figure 5.3. The center-to-center distance along the rod will be denoted by $2h$. We generalize the original treatment in Happel and Brenner [35] (pp. 179–180) by taking disks of different sizes, which introduces a key asymmetry in the problem that is necessary to demonstrate that the hydrodynamic centers of resistance and mobility are distinct. We will neglect hydrodynamic interactions between the disks, a good assumption if h is great compared to the disk radii.

In the first step, we write the resistance tensor with respect to an origin taken at the midpoint between the disk centers. The resistance tensor for the composite object is most easily computed by applying the translation theorems to the resistances of each disk, *i.e.*, shifting the reference point for each resistance tensor from the disk centers O_1 and O_2 to the midpoint, O. If the disk radii are $c_1 = c$ and $c_2 = 3c$ (this choice leads to nice numbers in the subsequent illustration; a general treatment is of little interest), and lengths are scaled with c, the resistance tensors become:

$$\begin{pmatrix} \boldsymbol{A} & \boldsymbol{B}^t \\ \boldsymbol{B} & \boldsymbol{C} \end{pmatrix} = \frac{16}{3}\left(\begin{array}{ccc|ccc} 10 & 1 & 0 & 2h & -5h & 0 \\ 1 & 10 & 0 & 5h & -2h & 0 \\ 0 & 0 & 8 & 0 & 0 & 0 \\ \hline 2h & 5h & 0 & 56+10h^2 & -h^2 & 0 \\ -5h & -2h & 0 & -h^2 & 56+10h^2 & 0 \\ 0 & 0 & 0 & 0 & 0 & 56 \end{array}\right).$$

The symmetry of the overall matrix and the submatrices on the diagonal blocks is in accordance with the general theory.

We find that the eigenvalues and eigenvectors of the matrix representing the resistance tensor $\boldsymbol{A}$ are $8, 9, 11$, and

$$\begin{pmatrix} 1/\sqrt{2} \\ -1/\sqrt{2} \\ 0 \end{pmatrix}, \quad \begin{pmatrix} 1/\sqrt{2} \\ 1/\sqrt{2} \\ 0 \end{pmatrix}, \quad \begin{pmatrix} 0 \\ 0 \\ 1 \end{pmatrix}.$$

We may diagonalize $\boldsymbol{A}$ by using the orthogonal matrix Q with columns consisting of the eigenvectors above. In the new coordinate system, the matrices are given by $Q^t\boldsymbol{A}Q$, $Q^t\boldsymbol{B}Q$, *etc.*, and the grand resistance matrix becomes

$$\frac{16}{3}\left(\begin{array}{ccc|ccc} 9 & 0 & 0 & 0 & 0 & 0 \\ 0 & 11 & 0 & 7h & -3h & 0 \\ 0 & 0 & 8 & 0 & 0 & 0 \\ \hline 0 & 7h & 0 & 56+11h^2 & 0 & 0 \\ 0 & -3h & 0 & 0 & 56+11h^2 & 0 \\ 0 & 0 & 0 & 0 & 0 & 56 \end{array}\right).$$

In this example, the principal directions of $\boldsymbol{A}$ and $\boldsymbol{C}$ just happen to coincide, so that $\boldsymbol{C}$ is diagonalized as well.

We continue the example by locating the hydrodynamic center (for disks of equal size, the center would be at the midpoint). Using Equation 5.11, we find $\boldsymbol{x}_{cr} = -(7/20)\mathbf{e}_3$, and after applying the translation theorems for the resistance tensors, we arrive at

$$\frac{16}{3}\left(\begin{array}{ccc|ccc} 9 & 0 & 0 & 0 & \frac{63}{20}h & 0 \\ 0 & 11 & 0 & \frac{63}{20}h & -3h & 0 \\ 0 & 0 & 8 & 0 & 0 & 0 \\ \hline 0 & \frac{63}{20}h & 0 & 56+\frac{2979}{400}h^2 & \frac{21}{20}h^2 & 0 \\ \frac{63}{20}h & -3h & 0 & \frac{21}{20}h^2 & 56+\frac{4041}{400}h^2 & 0 \\ 0 & 0 & 0 & 0 & 0 & 56 \end{array}\right)$$

for rotations and torques about the hydrodynamic center of resistance. The coupling tensor $\boldsymbol{B}^{(cr)}$ is now symmetric, as required by the general theory. The following result for the coupling tensor $\boldsymbol{b}$ with respect to $\boldsymbol{x}_{cr}$ can be derived by using Equation 5.24,

$$\boldsymbol{b} = \frac{1}{D(h)} \begin{pmatrix} 147h^3 & -18h(35h^2+196) & 0 \\ -56h(9h^2+77) & 3h(171h^2+1120) & 0 \\ 0 & 0 & 0 \end{pmatrix} ,$$

where $D(h) = 20(549h^4 + 9072h^2 + 34496)$ and $\boldsymbol{b}$ *is not symmetric*, since $\boldsymbol{x}_{cm} \neq \boldsymbol{x}_{cr}$ for this object. Both hydrodynamic centers are located on the side of the rod closer to the bigger disk. For the limiting case of $h \gg 1$, we have

$$\begin{aligned} \boldsymbol{x}_{cr} &= -0.175h\boldsymbol{e}_3 \\ \boldsymbol{x}_{cm} &= -0.408h\boldsymbol{e}_3 , \end{aligned}$$

and the distance between the two hydrodynamic centers is approximately $0.233h$.

5.5 Axisymmetric Particles

5.5.1 General Resistance Formulation in a Linear Field

For axisymmetric shapes, we obtain further reductions in the algebraic structure of the resistance and mobility problems. Perhaps the simplest manifestation of this statement concerns the drag coefficient on a translating particle. This coefficient will take on the same values for translations in the two directions transverse to the axis of symmetry. If the coordinate system is chosen so that the z-axis is along this axis of symmetry, then the resistance relation becomes

$$\begin{aligned} F_1 &= \mu A^{\perp}(U_1^{\infty} - U_1) \\ F_2 &= \mu A^{\perp}(U_2^{\infty} - U_2) \\ F_3 &= \mu A^{\|}(U_3^{\infty} - U_3) , \end{aligned}$$

where $A_{33} = A^{\|}$ and $A_{11} = A_{22} = A^{\perp}$. Net translation along one of the coordinate directions gives a force component in only that direction, due to symmetry. The equality of the resistances in the x- and y-directions is also a consequence of the symmetry in the particle shape.

Before proceeding further, we digress here to introduce a slight generalization in the notation. We do this because in later chapters we will want to align coordinate axes to suit other considerations such as the direction of external fields, *etc.* If we denote the particle symmetry axis with the unit vector $\boldsymbol{d}$, then in the preceding equations $\boldsymbol{e}_z = \boldsymbol{d}$ and every occurrence of $\boldsymbol{e}_x\boldsymbol{e}_x + \boldsymbol{e}_y\boldsymbol{e}_y$ may be replaced with $\boldsymbol{\delta} - \boldsymbol{dd}$. Thus, the resistance relation may be written as

$$\boldsymbol{F} = \mu \left[A^{\|}\boldsymbol{dd} + A^{\perp}(\boldsymbol{\delta} - \boldsymbol{dd})\right] \cdot (\boldsymbol{U}^{\infty} - \boldsymbol{U}) .$$

We may apply analogous operations on the set of resistance tensors to obtain

$$A_{ij} = X^A d_i d_j + Y^A(\delta_{ij} - d_i d_j) \tag{5.25}$$

$$\begin{aligned} B_{ij} &= Y^B \epsilon_{ijk} d_k & (5.26)\\ C_{ij} &= X^C d_i d_j + Y^C(\delta_{ij} - d_i d_j) & (5.27)\\ G_{ijk} &= X^G(d_i d_j - \frac{1}{3}\delta_{ij})d_k + Y^G(d_i\delta_{jk} + d_j\delta_{ik} - 2d_i d_j d_k) & (5.28)\\ H_{ijk} &= \widetilde{H}_{kij} = Y^H(\epsilon_{ikl}d_j + \epsilon_{jkl}d_i)d_l & (5.29)\\ M_{ijkl} &= X^M d^{(0)}_{ijkl} + Y^M d^{(1)}_{ijkl} + Z^M d^{(2)}_{ijkl} \ , & (5.30) \end{aligned}$$

where

$$\begin{aligned} d^{(0)} &= \frac{3}{2}(d_i d_j - \frac{1}{3}\delta_{ij})(d_k d_l - \frac{1}{3}\delta_{kl})\\ d^{(1)} &= \frac{1}{2}(d_i\delta_{jl}d_k + d_j\delta_{il}d_k + d_i\delta_{jk}d_l + d_j\delta_{ik}d_l - 4d_i d_j d_k d_l)\\ d^{(2)} &= \frac{1}{2}\delta_{ik}\delta_{jl} + \delta_{jk}\delta_{il} - \delta_{ij}\delta_{kl} + d_i d_j\delta_{kl} + \delta_{ij}d_k d_l\\ &\quad - d_i\delta_{jl}d_k - d_j\delta_{il}d_k - d_i\delta_{jk}d_l - d_j\delta_{ik}d_l + d_i d_j d_k d_l) \ , \end{aligned}$$

and X, Y, and Z are scalar resistance functions. The X functions are associated with axisymmetric problems, *e.g.*, $X^A = A^{\|}$, and the Y functions with transverse cases for the A and C tensors. More explicitly, if we use spherical coordinates about the particle axis $\boldsymbol{d}$, then the ambient flows in the resistance problems all have ϕ-dependences of the form $\exp(im\phi)$, as shown in Chapter 4. We assign X, Y, and Z to $m = 0$, ± 1, and ± 2, respectively.

The forms given in Equations 5.25–5.30 may be verified by considering $\boldsymbol{d} = \boldsymbol{e}_z$ and allowing for symmetry. For example, the coupling between translation and torque on an axisymmetric body is of the form $\boldsymbol{T} = -\mu Y^B \boldsymbol{U} \times \boldsymbol{d}$, and of course this coupling occurs only if we are interested in the torque about a point other than the hydrodynamic center. The analysis for $\boldsymbol{M}$ is more involved, and since it is perhaps less familiar to most readers, we consider it in some greater detail.

Example 5.1 The Resistance Tensor $\boldsymbol{M}$ for Axisymmetric Particles. We know that the rate-of-strain field $\boldsymbol{E}$ has five independent components because $E_{ij} = E_{ji}$ and $E_{jj} = 0$. For an axisymmetric particle with the axis of symmetry in the z-direction, we may select the following five rate-of-strain fields as the "basis":

$$\boldsymbol{E}^{(1)} = \begin{pmatrix} -1 & 0 & 0 \\ 0 & -1 & 0 \\ 0 & 0 & 2 \end{pmatrix} \qquad \boldsymbol{E}^{(2)} = \begin{pmatrix} 1 & 0 & 0 \\ 0 & -1 & 0 \\ 0 & 0 & 0 \end{pmatrix}$$

$$\boldsymbol{E}^{(3)} = \begin{pmatrix} 0 & 1 & 0 \\ 1 & 0 & 0 \\ 0 & 0 & 0 \end{pmatrix} \qquad \boldsymbol{E}^{(4)} = \begin{pmatrix} 0 & 0 & 1 \\ 0 & 0 & 0 \\ 1 & 0 & 0 \end{pmatrix}$$

$$\boldsymbol{E}^{(5)} = \begin{pmatrix} 0 & 0 & 0 \\ 0 & 0 & 1 \\ 0 & 1 & 0 \end{pmatrix} .$$

This may be shown explicitly by the following decomposition:

$$\boldsymbol{E} = \frac{1}{2}E_{zz}\boldsymbol{E}^{(1)} + \frac{1}{2}(E_{xx} - E_{yy})\boldsymbol{E}^{(2)} + E_{xy}\boldsymbol{E}^{(3)} + E_{xz}\boldsymbol{E}^{(4)} + E_{yz}\boldsymbol{E}^{(5)}.$$

Since $\boldsymbol{d} = \boldsymbol{e}_z$, the previous decomposition can be written as

$$E_{ij} = d^{(0)}_{ijkl}E_{kl} + d^{(1)}_{ijkl}E_{kl} + d^{(2)}_{ijkl}E_{kl} \; .$$

The $\boldsymbol{d}^{(0)}$ term corresponds to $(E_{zz}/2)\boldsymbol{E}^{(1)}$, the $\boldsymbol{d}^{(1)}$ term corresponds to $E_{xz}\boldsymbol{E}^{(4)} + E_{yz}\boldsymbol{E}^{(5)}$, and the last term corresponds to $\frac{1}{2}(E_{xx} - E_{yy})\boldsymbol{E}^{(2)} + E_{xy}\boldsymbol{E}^{(3)}$. Note that this decomposition actually holds for any tensor, because all terms canel except $\boldsymbol{E}$, but in general its usefulness is restricted to axisymmetric geometries.

The stresslet can also be decomposed in the same way, *i.e.*,

$$S_{ij} = d^{(0)}_{ijkl}S_{kl} + d^{(1)}_{ijkl}S_{kl} + d^{(2)}_{ijkl}S_{kl} \; . \tag{5.31}$$

One can identify coordinate transformations that leave only one group invariant. Thus for each of the three groupings in the rate-of-strain field, assume initially that all three groupings of the stresslet expression are required and then show that the coefficients of the two "unwanted" groupings are zero.

For example, consider the rate-of-strain field

$$E_{ij} = d^{0}_{ijkl}E_{kl} \; ,$$

and first assume that the stresslet is of the general form in Equation 5.31. An arbitrary rotation by an angle ϕ about the z-axis has no effect on $\boldsymbol{d}^{(0)} : \boldsymbol{E}$ and $\boldsymbol{d}^{(0)} : \boldsymbol{S}$. However, $\boldsymbol{d}^{(m)} : \boldsymbol{S}$, $m = 1, 2$ contains factors of $\cos m\phi$ and $\sin m\phi$, thus the equality is possible only if these terms are absent. Analogous arguments apply for the other rate-of-strain fields and thus the final result is obtained.◇

5.5.2 A Torque-Free Axisymmetric Particle in a Linear Field: The Jeffery Orbits

We generalize the result derived earlier in this chapter, concerning the rotation of a prolate spheroid in a linear field, to the more general case of any axisymmetric particle in a linear field. Again, we set $\boldsymbol{T} = \boldsymbol{0}$ in the expression for the torque and dot multiply through with the dyad:

$$\left[(X^C)^{-1}\boldsymbol{dd} + (Y^C)^{-1}(\boldsymbol{\delta} - \boldsymbol{dd})\right] \; .$$

Again, we find that the rotational motion follows as

$$\boldsymbol{\omega} \times \boldsymbol{x} = \boldsymbol{\Omega}^\infty \times \boldsymbol{x} + \left(\frac{Y^H}{Y^C}\right)\left[\boldsymbol{E}^\infty \cdot \boldsymbol{d}\,(\boldsymbol{d} \cdot \boldsymbol{x}) - \boldsymbol{E}^\infty : \boldsymbol{x}\boldsymbol{d}\,\boldsymbol{d}\right] \; . \tag{5.32}$$

Thus the general axisymmetric particle rotates with the angular velocity of the ambient fluid, plus a contribution from the rate-of-strain field that acts to align the axis along a principal direction of $\boldsymbol{E}$. The relative importance of the

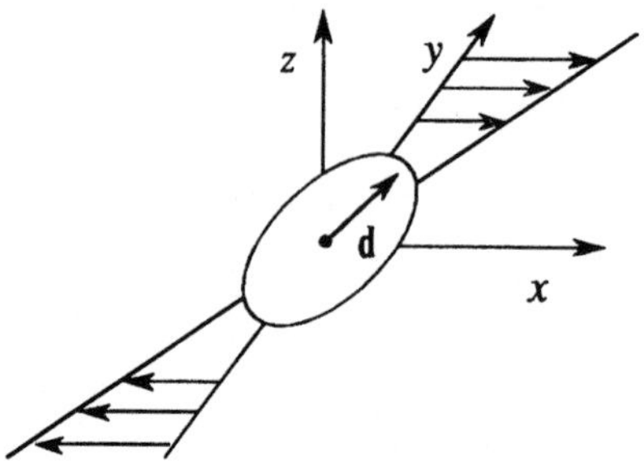

Figure 5.4: An axisymmetric particle in a shear field.

alignment effect depends on the aspect ratio: It becomes less important for nearly spherical shapes. In fact, the Bretherton constant, $B = Y^H/Y^C$ may be viewed as a hydrodynamic measure of nonsphericity. For ellipsoids of revolution with aspect ratio r, we recall that $B = (r^2-1)/(r^2+1)$, and thus $B > 0$ for prolate objects, $B = 0$ for a sphere,[5] and $B < 0$ for an oblate object. For what axisymmetric shapes would we expect $|B| \leq 1$?

We now consider the special case of a neutrally buoyant torque-free axisymmetric particle in the shear field $\boldsymbol{v}^\infty = \dot{\gamma} y \boldsymbol{e}_1$. The orientation of the particle axis $\boldsymbol{d}$ is parametrized using spherical polar coordinates so that

$$\begin{aligned} d_1 &= \sin\theta\cos\phi \\ d_2 &= \sin\theta\sin\phi \\ d_3 &= \cos\theta \;. \end{aligned}$$

The geometry is depicted in Figure 5.4. For this flow field, the ambient rotation and rate-of-strain are given by

$$\boldsymbol{\Omega}^\infty = -\frac{1}{2}\dot{\gamma}\boldsymbol{e}_3\;, \qquad \boldsymbol{E}^\infty = \frac{\dot{\gamma}}{2}\begin{pmatrix} 0 & 1 & 0 \\ 1 & 0 & 0 \\ 0 & 0 & 0 \end{pmatrix} .$$

The time evolution of the particle orientation, $\dot{\boldsymbol{d}}$, is obtained from the general mobility relation for the rotation of an axisymmetric particle in a linear field, Equation 5.32, by simply replacing $\boldsymbol{x} \rightarrow \boldsymbol{d}$:

$$\dot{\boldsymbol{d}} = \boldsymbol{\omega} \times \boldsymbol{d} = \boldsymbol{\Omega}^\infty \times \boldsymbol{d} + B(\boldsymbol{E}^\infty \cdot \boldsymbol{d} - \boldsymbol{E}^\infty : \boldsymbol{d}\boldsymbol{d}\,\boldsymbol{d})\;. \tag{5.33}$$

For the shear flow, the various terms simplify as

$$\boldsymbol{\Omega}^\infty \times \boldsymbol{d} \;=\; \frac{\dot{\gamma}}{2}(d_2\boldsymbol{e}_1 - d_1\boldsymbol{e}_2)$$

[5]And any other particle that experiences no torque when kept fixed in a rate-of-strain field.

$$\begin{aligned} \boldsymbol{E}^{\infty} \cdot \boldsymbol{d} &= \frac{\dot{\gamma}}{2}(d_2 \boldsymbol{e}_1 + d_1 \boldsymbol{e}_2) \\ \boldsymbol{E}^{\infty} : \boldsymbol{d}\boldsymbol{d}\boldsymbol{d} &= \dot{\gamma} d_1 d_2 (d_1 \boldsymbol{e}_1 + d_2 \boldsymbol{e}_2 + d_3 \boldsymbol{e}_3) \ . \end{aligned}$$

To go from the equation for $\dot{\boldsymbol{d}}$ to equations for $\dot{\theta}$ and $\dot{\phi}$, we examine $\dot{d}_3$ and $d_1 \dot{d}_2 - d_2 \dot{d}_1$. From above, we have

$$\begin{aligned} \dot{d}_3 &= -\sin\theta\,\dot{\theta} = -B\dot{\gamma}\sin^2\theta\cos\theta\sin\phi\cos\phi \\ d_1\dot{d}_2 - d_2\dot{d}_1 &= \sin^2\theta\,\dot{\phi} = -\frac{\dot{\gamma}}{2}\sin^2\theta + B\frac{\dot{\gamma}}{2}\sin^2\theta(\sin^2\phi - \cos^2\phi) \ , \end{aligned}$$

or

$$\begin{aligned} \dot{\theta} &= -\left(\frac{r^2-1}{r^2+1}\right)\frac{\dot{\gamma}}{4}\sin 2\theta\,\sin 2\phi \\ \dot{\phi} &= \frac{-\dot{\gamma}}{r^2+1}(r^2\cos^2\phi + \sin^2\phi) \ . \end{aligned}$$

We have replaced the Bretherton constant by *defining* a hydrodynamic aspect ratio with $B = (r^2-1)/(r^2+1)$.

The differential equations may be solved exactly and yield periodic trajectories known as the *Jeffery orbits*,

$$\begin{aligned} \tan\theta &= \frac{Cr}{\left[r^2\cos^2\phi + \sin^2\phi\right]^{1/2}} \\ \tan\phi &= -r\tan\left(\frac{\dot{\gamma}t}{r+r^{-1}}\right) \ . \end{aligned}$$

The period $T = 2\pi(r + r^{-1})/\dot{\gamma}$ scales with the inverse shear rate and becomes longer with increasing nonsphericity. The orbits in this orientation space (θ, ϕ) may be visualized by plotting the trajectories on the unit sphere, as shown in Figure 5.4. The constant of integration, C, is known as the *orbital constant* and values in $0 \le C < \infty$ determine orbits that range from the equator to a "degenerate orbit" consisting of a point at the "north pole." The exact shape of the orbit depends on the particle aspect ratio *via* the Bretherton constant B, as shown in Figure 5.5. For the sphere, we label a point on the surface to create artificially an axis of symmetry. For this situation, clearly the Jeffrey orbits correspond to lines of latitude; as the sphere spins about the z-axis with a constant angular velocity, the point label traces a cone. On the other hand, for slender bodies, $B \to 1$, the orbits approach great circles linking the positive and negative x-axes. The velocity along a given orbit now varies greatly. For example, at the orientation given by the point P_1, the slender body is swept out of alignment and reaches a maximum velocity at the point P_2. At point P_3, the orbital velocity is quite small, and the particle will spend most of the time in this orientation. However, as soon as the particle axis crosses the xz-plane, the cycle is repeated.

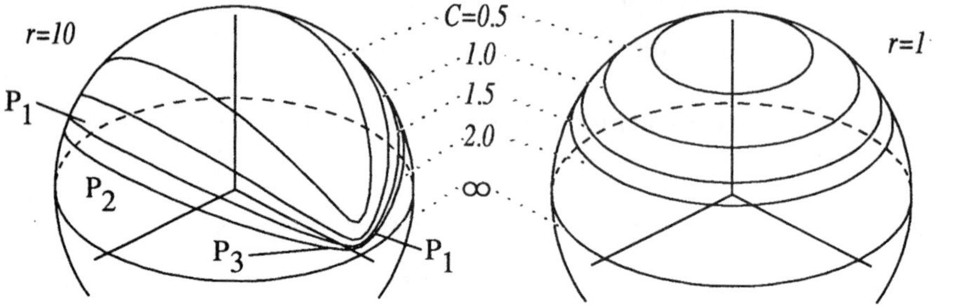

Figure 5.5: Jeffery orbits for different aspect ratios.

5.6 Rheology of a Dilute Suspension of Spheroids

The theoretical description of the rheological behavior of a dilute suspension of axisymmetric particles is one of the great "success stories" of microhydrodynamics. This endeavor has a long and distinguished history and the list of participants encompasses virtually all branches of the physical sciences, for this problem has served quite well as a model of the behavior (orientation *vs.* flow) of polymers in solutions. Our modest goal here is to illustrate how the resistance and mobility functions are used in the theoretical analysis; readers interested in a more complete physical picture are directed to the excellent reviews in in [7, 14]. A fairly complete review of the literature *circa* 1989 is available in Strand's doctoral dissertation [81].[6]

Jeffery [46] was the first to study the motion of a spheroid in a shear field, as a key step in an effort to extend Einstein's viscosity calculation to nonspherical (orientable) particles. After discovering this periodicity in the microhydrodynamics, he realized that a direct application of Einstein's idea would lead to a time periodic rheology. For instance, if a suspension of aligned prolate spheroids is sheared, the viscosity would undergo periodic oscillations, since the stresslet is smaller for a spheroid aligned in the direction of shear than for one transverse to the flow, $X^M < Y^M$. Real suspensions, on the other hand, exhibit fading memory, so that after several cycles these oscillations are damped out.

For a dilute suspension of submicron particles, rotary Brownian diffusion (Chapter 1) is the most plausible mechanism for the fading memory. With the inclusion of Brownian motion, the equations are no longer deterministic; rather, we describe the system as an ensemble with an orientation distribution function, $\psi(\boldsymbol{d}, t)$. The quantity $\psi(\boldsymbol{d}, t) \sin\theta \, d\theta \, d\phi$ gives the fraction of systems in the ensemble with particle orientations in the differential region about (θ, ϕ). The implications for the resulting rheological behavior of the suspension are now

[6]We acknowledge the assistance of Dr. Steven R. Strand in the preparation of this section.

clear. The ambient flow may orient the particle in certain preferred directions, but a nonuniform distribution of particle orientations leads to a "diffusive flux" from regions of high "concentration" to low "concentration."The resulting orientation distribution function determines the relative weighting of the stresslet functions in the particle contributions to the suspension stress. The balance between the flow-induced orientation and the randomizing influence of Brownian motion is affected by the flow strength, and thus the rheological property exhibits non-Newtonian behavior, such as shear thinning.

We start with the equation of continuity for ψ,

$$\frac{\partial \psi}{\partial t} + \nabla_d \cdot (\dot{\boldsymbol{d}}\psi) = 0 \ ,$$

and assign the following expression for $\dot{\boldsymbol{d}} = \omega \times \boldsymbol{d}$:

$$\dot{\boldsymbol{d}} = \boldsymbol{\omega} \times \boldsymbol{d} = \boldsymbol{\Omega}^\infty \times \boldsymbol{d} + \left(\frac{Y^H}{Y^C}\right)(\boldsymbol{E}^\infty \cdot \boldsymbol{d} - \boldsymbol{E}^\infty : \boldsymbol{ddd}) - (Y^C)^{-1} kT \frac{\partial}{\partial \boldsymbol{d}} \ln \psi \ .$$

Note that the effect of rotary Brownian motion comes in as a diffusive velocity. This equation may be obtained formally by starting with an angular momentum balance,

$$\boldsymbol{T}^{\mathrm{Hyd}} + \boldsymbol{T}^{\mathrm{Br}} = \boldsymbol{0} \ ,$$

with

$$\boldsymbol{T}^{\mathrm{Br}} = -\boldsymbol{d} \times \frac{\partial}{\partial \boldsymbol{d}}(kT \ln \psi) \ ,$$

and solving for $\boldsymbol{\omega}$.

The governing equation for ψ is thus a Fokker-Planck equation balancing Brownian diffusion and convection,

$$\frac{\partial \psi}{\partial t} + \left[\boldsymbol{\Omega}^\infty \times \boldsymbol{d} + \left(\frac{Y^H}{Y^C}\right) \boldsymbol{E}^\infty \cdot \boldsymbol{d}\right] \cdot \nabla_d \psi - 3 \left(\frac{Y^H}{Y^C}\right) \boldsymbol{E}^\infty : \boldsymbol{dd}\,\psi = D_r \nabla_d^2 \psi \ , \quad (5.34)$$

with the rotary diffusion coefficent given by $D_r = kT/Y^C$.

For the shear flow problem and the parametrization of the particle axis in spherical polar coordinates, the diffusion equation becomes

$$\begin{aligned} \frac{1}{D_r}\frac{\partial \psi}{\partial t} &= -\frac{\dot{\gamma}}{D_r}\left[\frac{\sin\phi\cos\phi}{\sin\theta}\frac{\partial}{\partial\theta}(\psi\sin^2\theta\cos\theta) - \frac{\partial}{\partial\phi}(\psi\sin^2\phi)\right] \\ &+ \left[\frac{1}{\sin\theta}\frac{\partial}{\partial\theta}\left(\sin\theta\frac{\partial\psi}{\partial\theta}\right) + \frac{1}{\sin^2\theta}\frac{\partial^2\psi}{\partial\phi^2}\right] , \qquad (5.35) \end{aligned}$$

where the second line is simply the angular parts of the Laplacian operator in spherical coordinates. The dimensionless group $P = \dot{\gamma}/D_r$ is often designated as a *Péclet* number, since it is a ratio of diffusive and convective time scales. Small P corresponds to fast diffusion over a weak flow, while large P corresponds to strong flows over weak diffusion.

We summarize a number of techniques for solving the orientation diffusion equation:

1. **Low Péclet Numbers.** When diffusion dominates over convection, we may solve for ψ using a regular perturbation in P. Without flow, the problem is quite simple and can be solved exactly. The next few terms in the small-P expansion for ψ are also obtained without too much difficulty [7, 31]. A procedure for generating these series on a computer is described in [49].

2. **High Péclet Numbers.** When strong convection dominates over weak diffusion, the problem may also be solved by a perturbation method, but a singular perturbation problem results, as expected by analogy with related problems in heat and mass transfer. Sharp gradients in ψ result near the slow regions of the Jeffrey orbits (the regions near the x-axis) and so, as shown in the work of Leal and Hinch [56], there are small "boundary layer" regions where diffusion is always important.

3. **Nearly Spherical Particles.** As mentioned above, the Jeffrey orbits take a particularly simple form for spheres. A perturbation solution in small deviations from sphericity is possible [67].

4. **Numerical Solutions (Galerkin Method).** The distribution function is expanded with the spatial dependence in terms of the spherical harmonics. The expansion is truncated at a finite number of terms, and the unknown coefficients are obtained by orthogonality conditions on the residual (the Galerkin method). Since the basis functions originate from the diffusion problem, although only a few terms are needed at small P, the number of terms in the expansion increases with P. However, with modest computational resources (minicomputers), this approach provides reliable solutions up to $P \sim 10^2$, thus bridging the gap between low and high Péclet number perturbation solutions. The Galerkin method was first applied to the orientation diffusion equation by Stewart and Sørensen [77] and the extension to the time-dependent equation is given in Strand *et al.* [82].

Given the orientation distribution, the bulk rheology is obtained by the ensemble average of the stresslet,

$$\begin{aligned} \boldsymbol{S} &= \boldsymbol{m} : \boldsymbol{E}^\infty - \boldsymbol{h} : \boldsymbol{T}^{\mathrm{Br}} \\ \sigma^{\mathrm{P}} &= < \boldsymbol{m} : \boldsymbol{E}^\infty > - < \boldsymbol{h} : \boldsymbol{T}^{\mathrm{Br}} > . \end{aligned}$$

The second term in the expression for the stresslet leads to the *direct Brownian contribution* to the bulk stress. (The *indirect* contribution is *via* the influence of Brownian motion on the orientation distribution and the resulting orientation averages). We may insert the resistance functions for the torque and stresslet on an axisymmetric particle to obtain the following expression relating the stress tensor with averages of moments of the particle orientation vector:

$$\sigma^{\mathrm{P}} = 2\mu c\,\{2A_H \boldsymbol{E}^\infty :< \boldsymbol{dddd} >$$

Hinch & Leal	Scheraga	Brenner
A	$J+K-L$	$(5/4)(3Q_2+4Q_3^0)$
B	$(L-M)/2$	$(5/2)Q_3^0$
C	M	$5Q_1$
F	N	$15N$

Hinch & Leal	Resistance functions
$8A$	$15X_1^M - 20Y_1^M + 5Z_1^M - 12B(r)Y_1^H$
$4B$	$5Y_1^M - 5Z_1^M + 3B(r)Y_1^H$
$2C$	$5Z_1^M$
F	$-9Y_1^H$

Table 5.1: The stress coefficients for steady shear flow.

$$+ \; 2B_H(\boldsymbol{E}^{\infty}\cdot < \boldsymbol{dd} > + < \boldsymbol{dd} > \cdot \boldsymbol{E}^{\infty} - \frac{2}{3}\delta \boldsymbol{E}^{\infty} :< \boldsymbol{dd} >$$

$$+ \; C_H \boldsymbol{E}^{\infty} \; + \; F_H D_r(< \boldsymbol{dd} > - \frac{1}{3}\delta)\Big\} \; .$$

The notation follows Hinch and Leal [40] (which differs slightly from their later paper [41]), and the functions (A_H, B_H, C_H, F_H) are identical to their functions (A, B, C, F). The function D_r is the same rotary diffusivity that appears in the differential equation for the orientation distribution function. The relations to the standard resistance functions, as well as other notations found in the literature for the bulk stress, are given in Table 5.1.[7]

Figure 5.6 shows the behavior of the stress coefficients A_H, B_H, C_H, and F_H over a range of aspect ratios. The magnitude of all the functions becomes unbounded as the aspect ratio approaches zero. Functions A_H and F_H are also unbounded as r becomes large. Due to the complicated manner in which the stress coefficients A_H,B_H,C_H, and F_H depend on the spheroid aspect ratio, it is of interest to determine their limiting behavior for various important geometries: disk-like particles ($r \to 0$), nearly spherical particles ($r \to 1$), and rod-like particles ($r \to \infty$). The asymptotic forms are tabulated here and are in agreement with those of [40] except for the forms for C_H in the limits $r \to 1$ and $r \to 0$, which are in error in [40].

Hinch and Leal use the limiting values for the stress coefficients in developing expressions for the bulk stress for various limiting cases of particle shape

[7]Brenner [16], Hinch and Leal [40], and Kim [52] present expressions for the full stress tensor accounting for hydrodynamic and Brownian effects. Scheraga [72] presents an expression for the intrinsic viscosity; his notation is closely related to Jeffery's and has been widely cited.

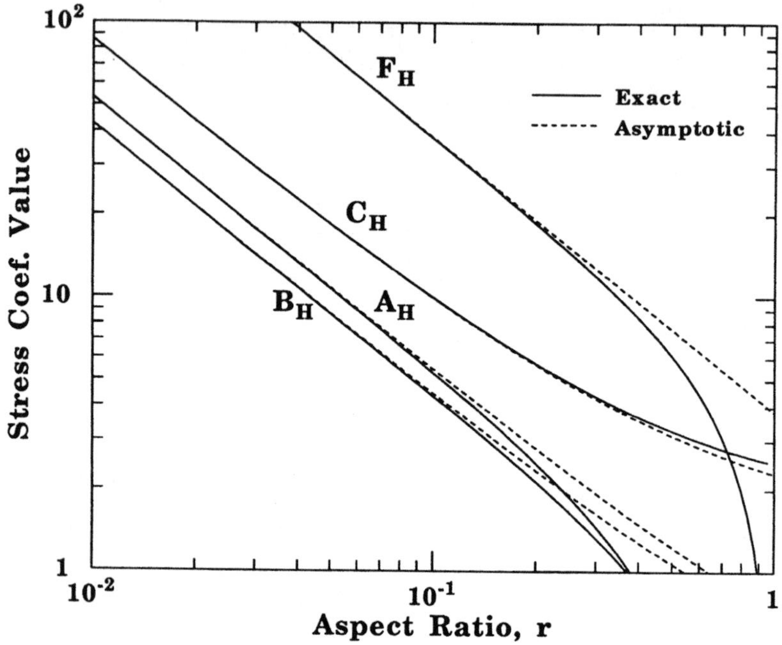

Figure 5.6: Comparison of asymptotic expressions with exact values of stress coefficients for oblate spheroids.

Coef.	$r \to \infty$	$r=1+\epsilon$, $\epsilon \to 0$	$r \to 0$
A_H	$\dfrac{r^2}{4(\ln(2r)-3/2)}$	$\dfrac{395}{294}\epsilon^2$	$\dfrac{5}{3\pi r}+\left(\dfrac{104}{9\pi^2}-1\right)$
B_H	$\dfrac{3\ln(2r)-11/2}{r^2}$	$\dfrac{15}{28}\epsilon-\dfrac{895}{1176}\epsilon^2$	$\dfrac{-4}{3\pi r}+\left(\dfrac{1}{2}-\dfrac{64}{9\pi^2}\right)$
C_H	2	$\dfrac{5}{2}-\dfrac{5}{7}\epsilon+\dfrac{235}{294}\epsilon^2$	$\dfrac{8}{3\pi r}+\dfrac{128}{9\pi^2}+O(r^2)$
F_H	$\dfrac{3r^2}{\ln(2r)-1/2}$	9ϵ	$-\dfrac{12}{\pi r}+O(r)$

Table 5.2: The limiting behavior of the stress coefficients.

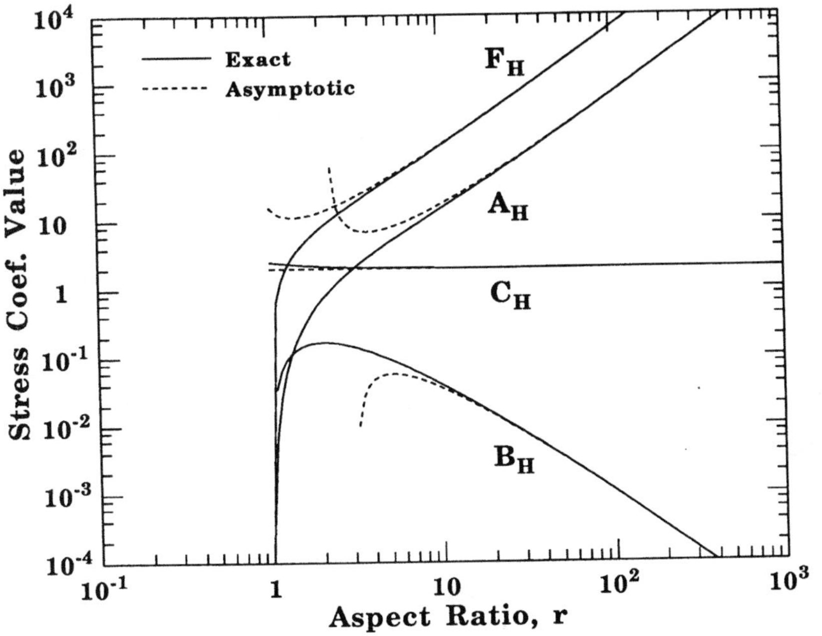

Figure 5.7: Comparison of asymptotic expressions with exact values of stress coefficients for prolate spheroids.

and shear strength. The range of validity of these asymptotic expressions is illustrated in the figures. The asymptotic expressions are quite close to the exact values over much of their individual ranges of interest, although it is apparent that the exact formulae are still needed for certain intermediate aspect ratios.

The intrinsic viscosity is an important material function for simple shear flow and is defined by

$$[\eta] = \lim_{c \to 0} \left(\frac{\sigma_{12} - \mu\dot{\gamma}}{c\mu\dot{\gamma}} \right) .$$

The results are plotted in the figure. As mentioned earlier, the balance between flow-induced orientation and Brownian motion results in non-Newtonian behavior. The viscosity exhibits shear-thinning; with increasing shear rate, a greater fraction of particles align with the flow and the particle contribution to the stress *via* the stresslet diminishes. The suspension also exhibits nonzero normal stress differences. For a review of these properties, as well as time-dependent rheological properties, we direct the reader to [7, 81] and references therein.

5.7 Electrophoresis

Electrophoresis, the motion of a charged particle under an applied electric field in an viscous electrolyte, is one of the most widely applied experimental methods

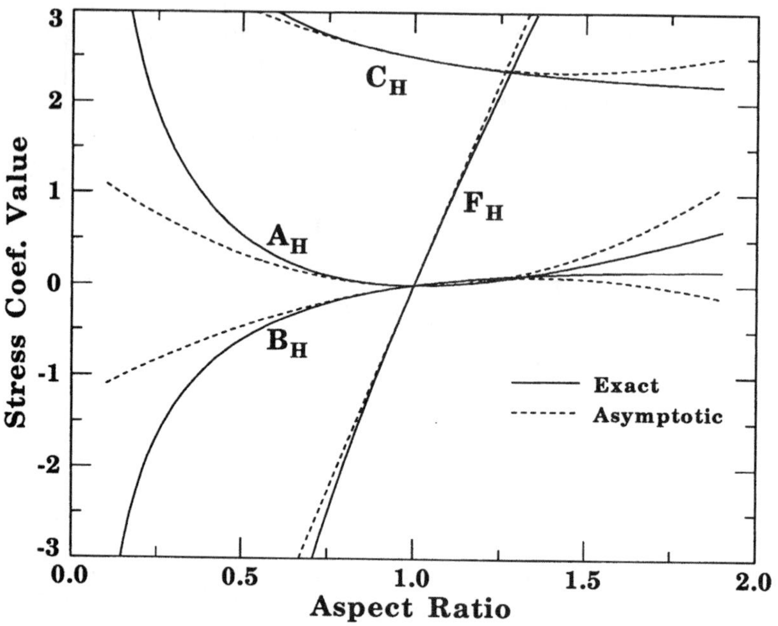

Figure 5.8: Comparison of asymptotic expressions with exact values of stress coefficients for spheroids of aspect ratio near unity.

in colloid research.[8] Separation and characterization of biochemical materials such as proteins by means of their electrophoretic properties are important operations in biochemistry and biophysics. In addition to such applications, electrophoresis is still used as a basic tool in studying the surface properties of colloidal substances.

The analysis of electrophoresis is essentially a study of the balance between the applied electrical force and the viscous resistance of the fluid. Indeed, for the so-called Hückel limit, the analysis is quite similar to that for sedimentation, in that the particle velocity follows from a balance of the external force (net charge on the particle $\times$ strength of the electric field) and the Stokes drag. If the counter-ions were not present (this is the limiting case of a very diffuse double layer), the electrophoretic velocity would be determined by equating the hydrodynamic drag force and the electrostatic force on the particle in an undisturbed applied field, as first proposed by Debye and Hückel [24, 43]. Thus for a spherical particle the electrophoretic mobility is given by

$$\frac{U}{E} = \frac{Q}{6\pi\mu a} = \frac{4\pi\epsilon\psi_0}{6\pi\mu a} = \frac{2\epsilon\psi_0}{3\mu} . \qquad (5.36)$$

Here ψ_0 is the surface potential of the particle and ϵ is the permittivity of the

[8]We acknowledge the assistance of Dr. Byung J. Yoon in the preparation of this section.

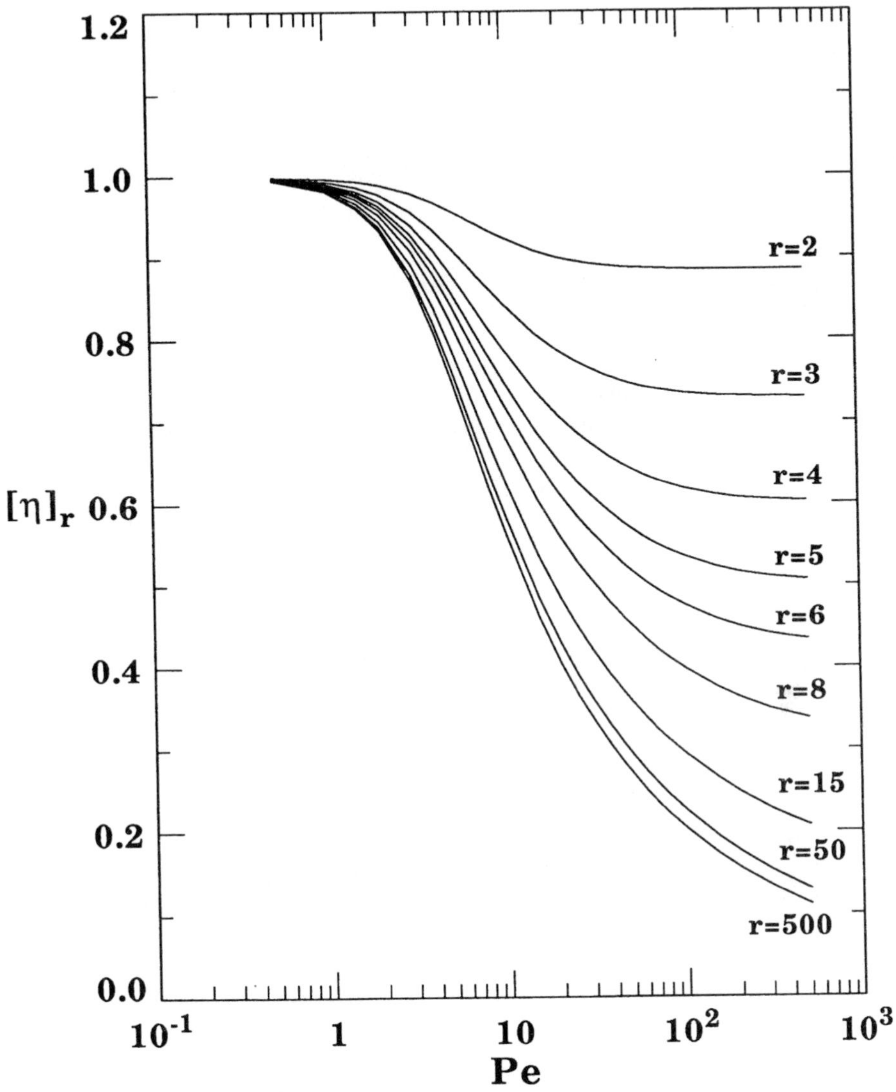

Figure 5.9: Intrinsic viscosity of a suspension of prolate spheroids in steady shear flow.

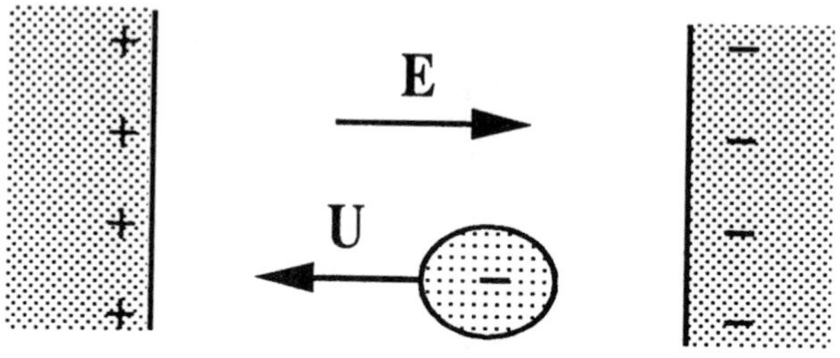

Figure 5.10: Electrophoresis of a charged particle in an electric field.

fluid.[9] The corresponding result for a spheroid using the appropriate resistance functions has also been noted [36].

For very thin double layers, or more precisely, when the radius of curvature of the particle surface is much larger than the double layer thickness, there is another simple limit known as the Helmholtz–Smoluchowski theory [76]. In this limit the electrophoretic velocity $\boldsymbol{U}$ of a rigid, nonconducting particle in an applied electric field $\boldsymbol{E}$ is given by

$$\frac{U}{E} = \frac{\epsilon \psi_0}{\mu} \; . \tag{5.37}$$

It is rather remarkable that Equation 5.37 holds for any particle shape if the above conditions plus a few more listed below are met. The proof is available in Morrison [62] and Teubner [85]. Essentially, the shape dependence of the viscous drag is canceled exactly by the same shape dependence of the electric force — this coming from the sheet of counter-ions fitting the particle form like a tight bodysuit. Later in this section we show that this is connected with the Lorentz reciprocal theorem and other machinery developed in this chapter. For situations of interest outside the two limiting regimes of very thin and very diffuse double layers, a more detailed analysis of the electroviscous forces is necessary.

The first critical analysis that takes into account the finite thickness of the double layer was reported by Henry [37]. Henry considered the electrophoretic motion of spherical particles and infinite cylinders aligned parallel or perpendicular to the electric field. He obtained the expressions for the electrophoretic

[9]The permittivity of a medium is defined by its appearance in Coulomb's law, $F = (Q_1 Q_2)/4\pi\epsilon r^2$, where F is the force between point charges Q_1 and Q_2 separated by a distance r. The permittivity of a vacuum, denoted by ϵ_0, is $8.854 \times 10^{-12}\mathrm{kg}^{-1}\mathrm{m}^{-3}\mathrm{s}^4\mathrm{A}^2$. The dielectric constant of a material is the ratio between its permittivity and ϵ_0, and thus is a dimensionless quantity. Care should be exercised, since many authors (see, for example, the original papers by Debye and Hückel) prefer to use Gaussian units in which the unit of charge is defined so that $\epsilon_0 = 1$. For electrophoresis problems, the Gaussian result is readily obtained from ours by the formal operation: $\epsilon \to \epsilon/4\pi$. Jackson [45] contains a lucid description on units and dimensions of electrodynamics.

mobility as functions of the ratio between the particle radius and the thickness of the double layer, and showed that the Smoluchowski and the Hückel results were correct limits for spherical particles with a thin double layer and a thick double layer, respectively. In his analysis he assumed additivity of the imposed electric field in the presence of the particle and the field due to the equilibrium electric double layer. For the latter he used the solution of the linearized Poisson–Boltzmann equation (which we discuss shortly), which is valid only for small double layer potentials.

There are two major limitations in Henry's theory. Firstly, the deformation of the electric double layer around the particle during the electrophoretic motion is neglected. Since then many authors have treated this relaxation effect for spherical particles [8, 64, 91] and infinite cylinders [78, 79]. The relaxation effect gives a significant contribution when the thickness of the double layer is comparable to the particle radius and when the potentials are high. Secondly, only simple particle shapes, *i.e.*, spherical or cylindrical, are considered. Obviously, when the double layer is very thin, one can use the Smoluchowski equation, 5.37, irrespective of the particle shape. The thick double layer limit also can be easily determined for any particle geometry for which the resistance tensor is known. However, most colloidal particles are highly nonspherical, and the thickness of the double layer is comparable to the size of the particle in many cases.

5.7.1 Particles with Finite Electric Double Layers

In this subsection, we derive an expression for the electrophoretic mobility of nonspherical particles with double layers of finite thickness, following the exposition in Yoon [92]. We neglect the relaxation effect and assume small potentials around particles. We then discuss the special results obtained for sphere and spheroids, the latter being particularly important because many colloidal particles can be modelled as prolate spheroids (needle-like) or oblate spheroids.

A charged particle in an electrolyte (a solution of positive and negative ions), attracts a "cloud" of ions of opposite sign, also known as the counter-ions. In most colloidal systems, the particles are negatively charged, so the counter-ion cloud is composed of positive ions or cations. The charge distribution on the particle surface and in the counter-ion cloud together comprise the *diffuse double layer*, as shown in Figure 5.11. For a more complete discussion on the theory of colloidal phenomena, the diffuse electric double layer, and electrophoresis, we direct the reader to [25, 44, 75, 89].

Under equilibrium conditions, the ion concentrations follow the Boltzmann distribution,

$$n_\nu = n_\nu^\infty \exp(-z_\nu e\psi/kT) \ , \tag{5.38}$$

where z_ν is the valence of the ν-th ion species, n_ν^∞ is the ambient concentration far away from the particle, e is the fundamental charge, ψ is the electrosatic potential, k is the Boltzmann constant, and T is the absolute temperature. In other words, for each ion species, the distribution is influenced by the electric

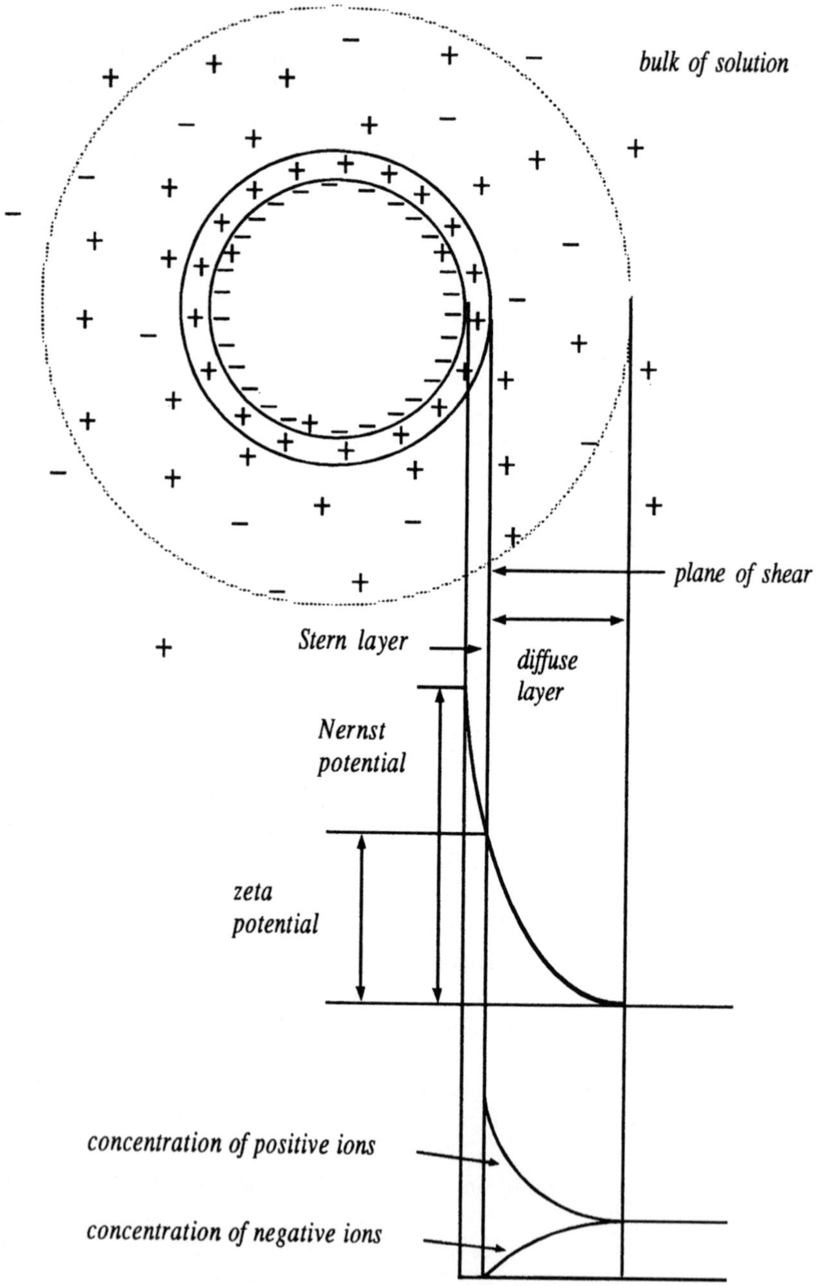

Figure 5.11: The diffuse double layer.

potential energy $z^{\nu}e\psi$, which pushes ions in the direction of lower potential energy, but Brownian diffusion of the ion, in the form of the thermal energy scale kT, imparts a diffusive flux that pushes the ions against the potential.

The charge distribution in the fluid, ρ_e, is obtained from these ionic concentrations as

$$\rho_e = e\sum_{\nu=1}^{n} n_\nu z_\nu \, . \tag{5.39}$$

Implicit in Equation 5.39 is the assumption that, on the length scale of interest, the ions in a differential volume element appear as a continuous charge distribution. Equations 5.38 and 5.39 dictate how the electrostatic potential influences the charge distribution, ρ_e. However, the charge distribution also determines the potential by Poisson's equation,

$$\nabla^2\psi = -\frac{\rho_e}{\epsilon} \, . \tag{5.40}$$

Combining Equations 5.39 and 5.40, we obtain the Poisson–Boltzmann equation,

$$\nabla^2\psi = -\frac{\rho_e(\psi)}{\epsilon} \, ,$$

which under condition of low potentials, $e\psi \ll kT$, reduces to the *linearized Poisson–Boltzmann equation*,

$$\nabla^2\psi - \kappa^2\psi = 0 \, , \tag{5.41}$$

with

$$\kappa^2 = \frac{e^2}{\epsilon kT}\sum_{\nu=1}^{n} n_\nu^\infty z_\nu^2 \, .$$

Here κ^{-1} has units of length and is known as the *Debye length.* It is the length scale associated with the exponential decay in the potential from the surface value to that of the ambient fluid. At distances much greater than κ^{-1}, the counter-ion cloud screens the charge on the particle and the potential decays exponentially to zero, in contrast to the inverse-separation decay of the Coulombic potential of a charged particle in a vacuum.

We now apply a uniform external electric field $\boldsymbol{E}$. When the applied field is small compared with the field due to the equilibrium double layer, we can assume additivity of the potentials. Neglecting the deformation of the double layer due to the applied field, we decompose the potential around the particle into two independent contributions: ψ from the equilibrium double layer and ϕ from the ambient field. The potential fields satisfy the equations

$$\begin{aligned} \nabla^2\psi &= \kappa^2\psi \, , \\ \nabla^2\phi &= 0 \, , \end{aligned}$$

with the boundary conditions

$$\begin{aligned} &\psi\,|_{r\to\infty} = 0 \, , &&\psi\,|_S = \psi_0 \, , \\ &\phi\,|_{r\to\infty} = -\boldsymbol{E}\cdot\boldsymbol{x} \, , &&\hat{\boldsymbol{n}}\cdot\nabla\phi\,|_S = 0 \, . \end{aligned}$$

We assume that the surface potential is constant and the dielectric constant of the particle is much smaller than that of the surrounding medium. But our results are not restricted to the situation with these boundary conditions, because the electrophoretic mobility is independent of the dielectric constant of the particle and the electrostatic boundary conditions on the particle surface [64].

The electric force on the particle has two distinct contributions. There is an electrostatic contribution given by

$$\int_{V_f} \rho_e \nabla\phi \, dV$$

and a hydrodynamic contribution due to the distribution of body forces $-\rho_e\nabla\phi$ throughout the fluid. Because of the Lorentz reciprocal theorem (see the discussion in Chapter 3 on the derivation of the Faxén law), this contribution is given exactly by

$$-\left[\int_{V_f} \tilde{u}_{ij}\rho_e \frac{\partial\phi}{\partial x_j}\, dV\right] ,$$

as first derived by Teubner [85]. The velocity field $\tilde{u}_{ij}$ represents the disturbance velocity field (j-th component) due to the translational motion of an uncharged particle (with a unit velocity along the i-axis). The hydrodynamic force is caused by the motion of the mobile ions in the diffuse double layer — in effect, the particle is moving not in a quiescent fluid, but against a local current of counter-ions flowing in the opposite direction under the action of the electric field. Since this hydrodynamic force acts against the motion of the particle, it is called the *electrophoretic retardation force*.

The force balance for the particle may be written as

$$\mu A_{ij}U_j = -\epsilon\left[\int_{V_f} \nabla^2\psi\frac{\partial\phi}{\partial x_i}\, dV\right] + \epsilon\left[\int_{V_f} \tilde{u}_{ij}\nabla^2\psi\frac{\partial\phi}{\partial x_j}\, dV\right] . \tag{5.42}$$

The left-hand side is the hydrodynamic force due to the motion of the particle and $\boldsymbol{A}$ is the resistance matrix for translation, discussed earlier in the chapter.

Using Green's second identity and the boundary condition on ϕ and ψ, we can rewrite the first volume integral into the surface integral over the particle surface S_p:

$$-\epsilon\left[\int_{V_f} \nabla^2\psi\frac{\partial\tilde{\phi}_k}{\partial x_i}\, dV\right] E_k = \epsilon\left[\oint_{S_p} \frac{\partial\psi}{\partial n}\frac{\partial\tilde{\phi}_k}{\partial x_i}\, dS\right] E_k = -\left[\oint_{S_p} \sigma\frac{\partial\tilde{\phi}_k}{\partial x_i}\, dS\right] E_k .$$

We now separate $\tilde{\phi}_k$ into two parts, one due to the ambient field, $-x_k$, and the second due to the disturbance field, denoted by $\tilde{\chi}_k$. The final result is

$$\begin{aligned}\mu A_{ij}U_j \;=\; & \overbrace{[\int_{S_p} \sigma\, dS]E_i}^{\text{I}} + \overbrace{[\int_{S_p} -\sigma\frac{\partial\tilde{\chi}_k}{\partial x_i} dS]E_k}^{\text{II}} + \overbrace{\epsilon[\int_{V_f} -\tilde{u}_{ik}\nabla^2\psi\, dV]E_k}^{\text{III}} \\ & + \overbrace{\epsilon[\int_{V_f} \tilde{u}_{ij}\nabla\psi\frac{\partial\tilde{\chi}_k}{\partial x_j}\, dV]E_k}^{\text{IV}} .\end{aligned} \tag{5.43}$$

Here I and II are the electrostatic forces on the surface charges due to the undisturbed applied electric field and the disturbance dipole electric field, respectively, while III and IV are the hydrodynamic drag forces on the particle due to the motion of the mobile ions. In III the ions move along the undisturbed applied field **E**, whereas in IV the ions move along the disturbance dipole field.

To solve the electrophoresis problem for an arbitrary particle under the assumptions of this section, we must first solve three boundary value problems: the linearized Poisson–Boltzmann equation to obtain ψ, the Laplace equation to get ϕ, and the Stokes equation for the translation of an uncharged particle through a quiescent fluid to get the velocity field $\tilde{u}_{ij}$. For the sphere, this is easily accomplished, and each integral in Equation 5.43 reduces to

$$\mathrm{I} = 4\pi\epsilon a\psi_0(1+\kappa a)\boldsymbol{E} \tag{5.44}$$

$$\mathrm{II} = \mathbf{0} \tag{5.45}$$

$$\mathrm{III} = -4\pi\epsilon a\psi_0\kappa a\boldsymbol{E} \tag{5.46}$$

$$\mathrm{IV} = 2\pi\epsilon a\psi_0\boldsymbol{E} + 12\pi\epsilon\int_a^\infty \psi\left[\left(\frac{a}{r}\right)^4 - \frac{5}{2}\left(\frac{a}{r}\right)^6\right] dr\,\boldsymbol{E}\ , \tag{5.47}$$

and summing these equations we recover[10] the result first obtained by Henry [37]:

$$\frac{U}{E} = \frac{\epsilon\psi_0}{\mu}\left[\frac{1}{6}(1+\kappa a) + \frac{1}{6}\left(12-(\kappa a)^2\right)\int_1^\infty \frac{e^{\kappa a(1-t)}}{t^5}\,dt\right]\ .$$

The ambient field contributions, I and III, are both linear to κa and have the same proportionality constants. Thus $O(\kappa a)$ terms cancel each other exactly and only the constant term $4\pi\epsilon a\psi_0\boldsymbol{E}$ survives. Obviously, this constant term corresponds to the Hückel limit.

The electrostatic force due to the disturbance field vanishes identically because of the isotropy of spherical particles. Since there is no effect of the double layer thickness on I + II + III, the hydrodynamic force in the disturbance dipole field is the only force that causes the transition from the Hückel limit to the Smoluchowski limit. This hydrodynamic force has the following asymptotic expressions:

$$\text{When } \kappa a \ll 1,\quad \mathrm{IV} = 4\pi\epsilon a\psi_0(\frac{1}{16})(\kappa a)^2\boldsymbol{E}$$

$$\text{When } \kappa a \gg 1,\quad \mathrm{IV} = 2\pi\epsilon a\psi_0\boldsymbol{E} - 18\pi\epsilon a\psi_0(\frac{1}{\kappa a})\boldsymbol{E}\ . \tag{5.48}$$

It can be shown that this force is a monotonically increasing function of κa, and so the electrophoretic velocity of spheres increases with decreasing double layer thickness.

We note the following general points regarding Equation 5.43:

[10] The precise expression for the equivalent result in the original paper by Henry differs from ours, since his work is also in CGS units. As noted earlier, we may obtain his expressions by replacing our ϵ with $\epsilon/4\pi$.

1. A mobile ion moving perpendicular to the particle surface exerts more hydrodynamic force on the particle than one moving parallel to the particle surface. (This is the same reason why transverse flow around a prolate spheroid exerts more drag on the particle than an axisymmetric flow.) Thus the sign and magnitude of the hydrodynamic force (III or IV) is mainly determined by the amount of mobile ions moving perpendicular to the particle surface.

2. For nearly spherical particles the electrostatic force in a dipole field is almost zero because of its isotropic shape. But the hydrodynamic force in a dipole field has a finite value in consequence of (1). Since the dipole is oriented opposite to the applied field in the region where ions move perpendicular to the particle surface, this force is an enhancing one.

3. When the longest axis of spheroidal particles is aligned perpendicular to the applied field (Figure 5.12), one can easily see that the electrostatic force in a dipole field is a retarding one but the hydrodynamic force is an enhancing one. Since the magnitude of the hydrodynamic force is greater than that of the electrostatic force, the net force due to the dipole field is an enhancing one. When the longest axis of spheroidal particles is aligned parallel to the applied field (Figure 5.12), the signs of two forces are opposite to the perpendicular case and the electrostatic force is bigger than the hydrodynamic force. Thus the net force in a dipole field is again an enhancing one.

4. For particles with a thick double layer, the disturbance field contribution (II+IV) is negligible because the disturbance dipole field decays as r^{-3}. For particles with a thin double layer the domain that gives a significant contribution to the volume integrals is very small so that the contribution from the ambient field and the contribution from the disturbance field are of the same orders of magnitude. (Compare Equations 5.44 and 5.48, for example.)

5. When the double layer is very thick the electrophoretic motion of the particle is mainly determined by the balancing between the electrostatic force I (enhancing one) and the hydrodynamic force III (retarding one). The electrostatic force increases monotonically as κa increases and is independent of the particle orientation. The hydrodynamic force also increases monotonically, but does depend on the particle orientation. Thus the electrophoretic mobility of nonspherical particles may not be a monotonic function of κa in this region. This effect has been shown to occur for spheroids [92].

Although these arguments are based on Equation 5.43, where the deformation of the double layer is neglected, we can make the following qualitative observation on the electrophoretic motion of spherical particles with a deformed double layer. During the electrophoretic motion of the particle the total surface

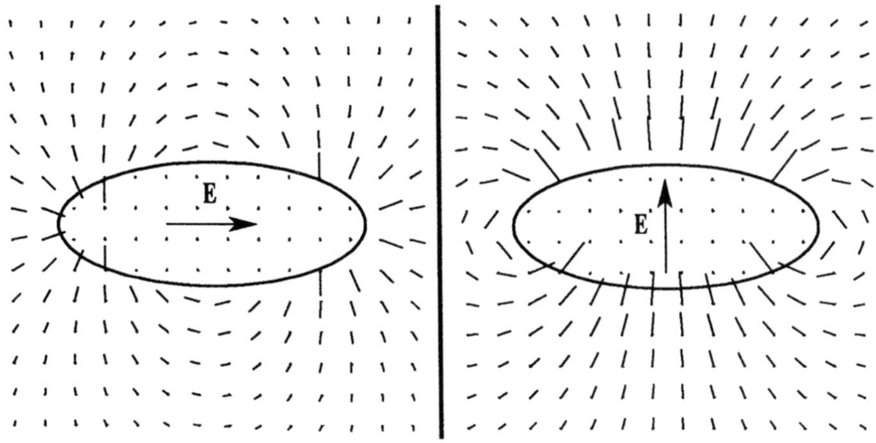

Figure 5.12: Alignment of spheroids in an applied electric field.

charge changes only slightly from the equilibrium (undeformed double layer) value, but the distribution of mobile ions in the diffuse double layer changes significantly. More counter-ions swarm behind the particle and move transversely. Thus, the hydrodynamic retardation force increases faster than the electrostatic force. Therefore, the electrophoretic velocity decreases for small κa. As κa increases the additional hydrodynamic force due to the disturbance field (enhancing one) becomes significant and the electrophoretic velocity increases accordingly.

The results for the electrophoretic mobility for oblate spheroids using Equation 5.43 are shown in Figure 5.13. For details of the calculation, we direct the reader to [92]. As noted above, the transition from the Hückel to the Smoluchowski limit is not monotonic.

5.7.2 Nonuniform Surface Potentials

We derive an expression for the electrophoretic velocity for a charged particle with nonuniform surface potential (also known as the zeta potential), assuming a thin electric double layer. The analysis follows the work of Fair and Anderson [27]. The fluid flow is driven by electrical stesses and the governing equations for the velocity and pressure are

$$\begin{aligned} -\nabla p + \mu \nabla^2 \boldsymbol{v} + \nabla \cdot \boldsymbol{m} &= \boldsymbol{0}\,, \qquad \nabla \cdot \boldsymbol{m} = \rho_e \boldsymbol{E}\,, \\ \nabla \cdot \boldsymbol{v} &= 0 \\ \text{on } S_p \quad \boldsymbol{v} &= \boldsymbol{U} + \boldsymbol{\omega} \times \boldsymbol{x}\,. \end{aligned}$$

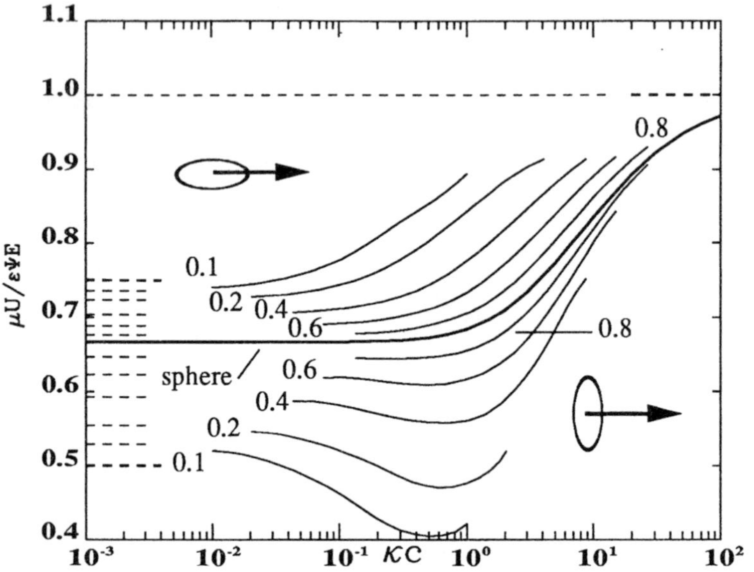

Figure 5.13: Electrophoretic mobility of oblate spheroids.

The electrical stress $\boldsymbol{m}$ is also known as the Maxwell stress. We assume a thin double layer everywhere so that $\kappa\bar{R} \gg 1$, *i.e.*, everywhere the local radius of curvature $\bar{R}$ is much greater than the double layer thickness. We may use the method of matched asymptotic expansions. To leading order, in the inner region $y \sim O(\kappa^{-1})$, where y is the normal coordinate for the local plane geometry near the surface. The governing equations reduce to

$$\begin{aligned} \mu\frac{\partial^2 \boldsymbol{v}}{\partial y^2} &= -\rho_e \boldsymbol{E}_s \\ \rho_e &= -\epsilon\frac{\partial^2}{\partial y^2}\frac{(\phi - \phi_s)}{} \\ \text{on } S_p \quad \boldsymbol{v} &= \boldsymbol{U} + \boldsymbol{\omega} \times \boldsymbol{x} \ , \end{aligned}$$

where $\phi_s = \lim_{\kappa y \to \infty} \phi$ and $\boldsymbol{E}_s = -\nabla\phi_s$. The velocity profile may be integrated over the double layer region to give the matching condition,

$$\text{on } S_p^+ \quad \boldsymbol{v} = \boldsymbol{U} + \boldsymbol{\omega} \times \boldsymbol{x} + \boldsymbol{v}^{(S)} \ , \qquad \boldsymbol{v}^{(S)} = -\frac{\epsilon}{\mu}\psi_0 \boldsymbol{E}_s \ . \tag{5.49}$$

Outside the double layer ρ_e is negligible so the electrical stresses vanish. The governing equations in the outer region are thus the Stokes equations, with the modified boundary condition, Equation 5.49. Since $\nabla \cdot (\boldsymbol{\sigma} + \boldsymbol{m}) = \boldsymbol{0}$ in the outer fluid, we find that the net force and torque is zero on S_p^+.

Even though we apply the usual no-slip boundary condition for $\boldsymbol{v}$, the outer problem sees an apparent *Smoluchowski slip velocity* $\boldsymbol{v}^{(S)}$, but not the sharp gradient inside the double layer. Using the linearity of the Stokes equations, we may write the force and torque on S_p^+ as the sum of contributions due to the rigid-body motion and that due to the slip velocity

$$\boldsymbol{F} = \oint_{S_p^+} \sigma \cdot \hat{\boldsymbol{n}} \;=\; -\mu \boldsymbol{A} \cdot \boldsymbol{U} - \tilde{\boldsymbol{B}} \cdot \boldsymbol{\omega} + \boldsymbol{F}^{(S)} = \boldsymbol{0} \;, \tag{5.50}$$

$$\boldsymbol{T} = \oint_{S_p^+} \boldsymbol{x} \times \sigma \cdot \hat{\boldsymbol{n}} \;=\; -\mu \boldsymbol{B} \cdot \boldsymbol{U} - \boldsymbol{C} \cdot \boldsymbol{\omega} + \boldsymbol{T}^{(S)} = \boldsymbol{0} \;. \tag{5.51}$$

The contributions from the slip velocity may be cast in an alternate form using the Lorentz reciprocal theorem, *e.g.*,

$$\boldsymbol{F}^{(S)} = \oint_{S_p^+} \boldsymbol{\delta} \cdot \boldsymbol{\sigma}^{(S)} \cdot \hat{\boldsymbol{n}}\, dS = \oint_{S_p^+} \boldsymbol{v}^{(S)} \cdot \boldsymbol{\sigma}(;\boldsymbol{\delta}) \cdot \hat{\boldsymbol{n}}\, dS \;,$$

where $\boldsymbol{U} \cdot \boldsymbol{\sigma}(;\boldsymbol{\delta}) \cdot \hat{\boldsymbol{n}}$ is the surface traction generated by the particle translating with velocity $\boldsymbol{U}$. After substitution of the expresson for the slip velocity, we obtain the result,

$$\mu \boldsymbol{A} \cdot \boldsymbol{U} + \tilde{\boldsymbol{B}} \cdot \boldsymbol{\omega} \;=\; -\frac{\epsilon}{\mu} \oint_{S_p^+} \psi_0 \boldsymbol{\sigma}(;\boldsymbol{\delta}) \cdot \hat{\boldsymbol{n}}\, dS \cdot \boldsymbol{E}_s \tag{5.52}$$

$$\mu \boldsymbol{B} \cdot \boldsymbol{U} + \boldsymbol{C} \cdot \boldsymbol{\omega} \;=\; -\frac{\epsilon}{\mu} \oint_{S_p^+} \psi_0 \boldsymbol{\sigma}(;\boldsymbol{\delta} \times \boldsymbol{x}) \cdot \hat{\boldsymbol{n}}\, dS \cdot \boldsymbol{E}_s \;, \tag{5.53}$$

which may be inverted using the mobility functions $\boldsymbol{a}$, $\boldsymbol{b}$, and $\boldsymbol{c}$ to yield explicit expressions for the particle motion. Equations 5.52 and 5.53 are especially applicable in situations in which the potential varies along the surface of the particle.

If the particle is at a constant potential, we may move ψ_0 out of the integral. The surface integrations of the RBM-tractions then yield the resistance functions $\boldsymbol{A}$ and $\boldsymbol{C}$. *For a particle with* $\boldsymbol{B} = \boldsymbol{0}$, *the shape dependence disappears from the electrophoretic mobility* and we obtain the vector form of Smoluchowski's celebrated result (Equation 5.37). This idea is readily generalized to show that an assembly of particles at constant zeta potential and thin double layer all move with the same Smoluchowski velocity [48]. This can lead to quite dramatic effects. Since hydrodynamic interactions are essentially zero for this situation, theories developed for dilute suspensions still apply at very high concentrations as shown in [93].

Exercises

Exercise 5.1 Symmetry Relations for the Resistance and Mobility Functions.
Derive all of the symmetry relations for the resistance and mobility functions [39].
Hint: The selection of $\boldsymbol{v}_1$ and $\boldsymbol{v}_2$ should be motivated by the resistance and mobility tensors of interest.

Exercise 5.2 The Hydrodynamic Center is Well Defined.
Show that result for the hydrodynamic center, Equation 5.11, is independent of the choice of the reference point, $\boldsymbol{x}_1$.

Exercise 5.3 The Hydrodynamic Center of a Dumbbell.
Consider a dumbbell formed by two unequal spheres joined by a rigid rod of negligible friction. Use Equation 5.11 to show that, neglecting hydrodynamic interactions between the spheres, the hydrodynamic center is at

$$\boldsymbol{x}_{cr} = \frac{a_1\, \boldsymbol{x}_1}{a_1 + a_2} + \frac{a_2\, \boldsymbol{x}_2}{a_1 + a_2} \,.$$

In other words, the hydrodynamic center lies on the dumbbell axis with the distance from the sphere centers weighted by sphere radii.

Exercise 5.4 The Hydrodynamic Center of a Propeller.
Consider the propeller-shaped body of Figure 5.1. Use symmetry to show that there exists a point on the axis where the three principal directions satisfy $\boldsymbol{B}(\mathbf{0}) \cdot \boldsymbol{e}_1 = \mathbf{0}$, $\boldsymbol{B}(\mathbf{0}) \cdot \boldsymbol{e}_2 \parallel \boldsymbol{e}_2$, and $\boldsymbol{B}(\mathbf{0}) \cdot \boldsymbol{e}_3 \parallel \boldsymbol{e}_3$, where $\boldsymbol{e}_1$ is directed along the axis of one of the blades.

From Equation 5.11 for the hydrodynamic center, we conclude that this point is the hydrodynamic center.

Exercise 5.5 Bounds for an Axisymmetric Body in a Linear Field.
Consider a force-free axisymmetric particle in a two-dimensional linear flow field confined to the xy-plane, *i.e.*, $\boldsymbol{v}^\infty = \boldsymbol{\kappa} \cdot \boldsymbol{x}$ with

$$\boldsymbol{\kappa} = \begin{pmatrix} \kappa_{11} & \kappa_{12} & 0 \\ \kappa_{21} & \kappa_{22} & 0 \\ 0 & 0 & 0 \end{pmatrix} .$$

Show that at the instant that the particle is oriented along the y-axis, the torque and stresslet in this ambient field require only the resistance functions Y^C, Y^H, and Y^M. Then use energy dissipation arguments to derive an inequality of the form

$$(Y^H)^2 < \alpha Y^C Y^M \,, \qquad \text{or} \qquad \left(\frac{Y^H}{Y^C}\right)^2 < \alpha \frac{Y^M}{Y^C} \,.$$

Find the value of α. Note that the last expression gives an upper bound for the Bretherton constant.

Exercise 5.6 Inclusion Monotonicity for the Resistance Tensor.
In this exercise, we derive the inclusion monotonicity for the resistance tensor. Let $\mathcal{R}$ and $\mathcal{R}'$ denote the resistance matrices of two particles, with the primed particle completely inside the other. Denote the "vector" formed by concatenation of $\boldsymbol{U}^\infty - \boldsymbol{U}$, $\boldsymbol{\Omega}^\infty - \boldsymbol{\omega}$, and $\boldsymbol{E}^\infty$ by $\mathcal{U}$. Show that the inclusion monotonicity of Chapter 2 may be expressed as a statement concerning the resistance matrices, that for all $\mathcal{U}$,

$$0 < \mathcal{U} \cdot \mathcal{R}' \cdot \mathcal{U} \le \mathcal{U} \cdot \mathcal{R} \cdot \mathcal{U} \,.$$

Exercise 5.7 Inclusion Monotonicity for the Mobility Tensor.
In this exercise, we derive the inclusion monotonicity for the mobility tensor. Let $\mathcal{M}$ and $\mathcal{M}'$ denote the mobility matrices of two particles, with the primed particle completely inside the other. Denote the "vector" formed by concatenation of $\boldsymbol{F}$ and $\boldsymbol{T}$ by $\mathcal{F}$. Show that the inclusion monotonicity of Chapter 2 may be expressed as a statement concerning the mobility matrices, that for all $\mathcal{F}$,

$$0 < \mathcal{F} \cdot \mathcal{M} \cdot \mathcal{F} \leq \mathcal{F} \cdot \mathcal{M}' \cdot \mathcal{F} \ .$$

Does this result follow directly from Exercise 3.7? If A and B are two positive-definite matrices, does $A < B$ imply $A^{-1} > B^{-1}$? Here, the ordering $A < B$ is defined by the difference $B - A$ being positive-definite.
Hint: Use the properties of positive-definite matrices, in particular the existence of $A^{-1/2}$ (see [30, 83]).

Exercise 5.8 Electrophoresis: Thick Electric Double Layers.
Show that in the limit of a thick electric double layer, $\kappa a \to 0$, Equation 5.43 reduces to the Hückel limit for the electrophoretic velocity,

$$\boldsymbol{U} = \frac{Q}{\mu} \boldsymbol{A}^{-1} \cdot \boldsymbol{E} \ .$$

Exercise 5.9 Electrophoresis: Thin Electric Double Layers.
Show that in the limit of a thin electric double layer, $\kappa a \to \infty$, Equation 5.43 reduces to the Smoluchowski limit for the electrophoretic mobility,

$$\boldsymbol{U} = \frac{\epsilon}{\mu} \psi_0 \boldsymbol{E} \ ,$$

so that the electrophoretic mobility is independent of particle size and shape.

Exercise 5.10 Electrophoresis of Spheres at Nonuniform Surface Potentials.
Show that for a sphere, Equation 5.52 simplifies to

$$\boldsymbol{U} = \frac{\epsilon}{\mu} \bar{\psi}_0 \boldsymbol{E}_s \ ,$$

where $\bar{\psi}_0$ is the area-averaged surface potential on the sphere.

Exercise 5.11 Electrophoresis of a Prolate Spheroid at Nonuniform Surface Potentials.
Consider a surface potential of the form

$$\psi_0 = \beta_0 + \beta_2 z^2 \ .$$

Show that by interpreting $\psi_0 \boldsymbol{E}_s$ as an ambient velocity field for the effective Stokes flow problem, the tedious algebra involved with the direct use of Equation 5.52 may be bypassed by the use of the Faxén law for the prolate spheroid. Show that it is possible for a prolate spheroid with zero net charge to undergo electrophoresis.

Chapter 6

Transient Stokes Flows

6.1 Time Scales

Throughout Part II, up to this point, we have restricted our attention to steady flow. Strictly speaking, we do not have a steady problem since the moving particle naturally changes the problem geomery. This is especially true for the multiparticle geometries of Part III. However, our analysis made the implicit assumption that these unsteady problems could be tackled as a sequence of steady-state problems consisting of a series of "snap shots" of the flow. Each frame yields a solution for the particle velocity, the geometry is updated by moving the particle by an incremental amount over the small time step Δt, and then the whole process is repeated. (We are focusing, for the moment, on deterministic problems, such as trajectory analysis.) In this chapter we test the validity of this approximate procedure more closely by analyzing the leading corrections due to time-dependent flow.

In Chapter 1 we showed that the importance of the Eulerian acceleration term, $\rho\partial \boldsymbol{v}/\partial t$, depended on the time scale of the phenomena of interest. When this time scale is of order U/ℓ, the acceleration term is $O(\mathrm{Re})$ smaller than the viscous terms. On the other hand, when the phenomena of interest occur over the much faster time scale, ℓ^2/ν, that it takes vorticity to diffuse over a length ℓ, the acceleration term is of the same order as the viscous terms. The time-dependent flows encountered in this chapter will illustrate this, as well as other interesting concepts of transient low Reynolds number flow.

The formal derivation *via* the dimensional analysis proceeds as follows. We pick velocity, length, and time scales V, ℓ, and τ so that the Navier–Stokes equations may be rendered into a dimensionless form:

$$\mathrm{Re}\,\mathrm{Sl}\,\frac{\partial \boldsymbol{v}}{\partial t} + \mathrm{Re}\,\boldsymbol{v}\cdot\nabla\boldsymbol{v} = -\nabla p + \nabla^2\boldsymbol{v} \ .$$

Now if the Reynolds number and the Strouhal number satisfy

$$\mathrm{Re} = \frac{U\ell}{\nu} \to 0$$

$$\mathrm{Sl} = \frac{\ell}{U\tau} \to \infty \ ,$$

but with $\text{Re}\,\text{Sl} = \ell^2/\nu\tau = O(1)$, then all terms except the nonlinear term will be $O(1)$. The nonlinear term will be $O(\text{Re})$ smaller than the rest. If we retain only those $O(1)$ dominant terms, we obtain, in dimensional form, the time-dependent Stokes equations:

$$\rho\frac{\partial \boldsymbol{v}}{\partial t} = -\nabla p + \mu\nabla^2\boldsymbol{v} , \qquad \nabla\cdot\boldsymbol{v} = 0 .$$

When $\boldsymbol{v} = \widehat{\boldsymbol{v}}\exp(-i\omega t)$ and $p = \widehat{p}\exp(-i\omega t)$ are inserted in these equations, we obtain the Fourier transformed equations,

$$-\nabla\widehat{p} + \mu\nabla^2\widehat{\boldsymbol{v}} - \mu\alpha^2\widehat{\boldsymbol{v}} = \mathbf{0} , \qquad \nabla\cdot\widehat{\boldsymbol{v}} = 0 , \tag{6.1}$$

where $\alpha^2 = -i\omega/\nu$. Here $|\alpha|^{-1}$ has dimensions of length, in fact, the distance over which the vorticity diffuses through this fluid (of kinematic viscosity ν) during a time scale $\tau \sim \omega^{-1}$.

Suppose that the phenomena of interest has a characteristic velocity U, lengths are scaled with $|\alpha|^{-1}$, and pressure with $\mu|\alpha|U$. The governing equations are then rendered into the following dimensionless form:

$$-\nabla\widehat{p} + \nabla^2\widehat{\boldsymbol{v}} - \widehat{\boldsymbol{v}} = \mathbf{0} , \qquad \nabla\cdot\widehat{\boldsymbol{v}} = 0 . \tag{6.2}$$

The boundary conditions (at particle surfaces) will introduce in a natural way the characteristic length scale, $\ell \to a$, where a is a characteristic dimension of the particle. The dimensionless group $\lambda = |\alpha|a$ is a ratio of the length scales, $|\alpha|^{-1}$ and a, and is the key parameter of this chapter. Small λ corresponds to slow temporal variations, and for steady flows λ is identically zero.

If we render the governing equations dimensionless using a for the characteristic length scale, then the scaled equations become

$$-\nabla\widehat{p} + \nabla^2\widehat{\boldsymbol{v}} - \lambda^2\widehat{\boldsymbol{v}} = \mathbf{0} , \qquad \nabla\cdot\widehat{\boldsymbol{v}} = 0 , \tag{6.3}$$

and since λ appears explicitly, this form is useful in studying the limits for small λ. In fact, we will find uses for the dimensional and both nondimensional forms, hence the scaling will be declared explicitly unless the situation is clear from the context.

Our task in this chapter is to determine the leading $O(\lambda)$ transient corrections to the steady solutions of the preceding chapters. These calculations are most readily accomplished in the transformed variables, but their meaning in the time domain will be examined as well. In the next section, we derive the fundamental solution, which forms the basis for much of the subsequent discussion. Borrowing a concept from steady flow, we solve transient flows for the sphere by the singularity method. Among these are Stokes' classical solution for the oscillating sphere (translation motions) and Lamb's solution for a sphere in oscillatory rotation. The $O(\lambda)$ and $O(\lambda^2)$ terms introduce the concept of Basset forces and added mass.

In Section 6.3, we show that the transient Stokes equations also possesses a reciprocal theorem, and from this we derive the integral representation and

Faxén laws. The latter furnishes a good *entré* to the literature on particle tracers of Laser-Doppler Velocimetry (LDV). In the last section, we introduce a general expression for the $O(\lambda)$ correction for a particle of arbitrary shape. The correction is proportional to the square of the steady Stokes resistance tensor, and *thus can be determined from an analysis of steady flows only.*

6.2 The Fundamental Solution

We seek a solution to the fundamental problem,

$$-\nabla\widehat{p}+\nabla^2\widehat{\boldsymbol{v}}-\widehat{\boldsymbol{v}}=-\boldsymbol{F}\delta(\boldsymbol{x}) , \qquad \nabla\cdot\widehat{\boldsymbol{v}}=0 . \tag{6.4}$$

We write this fundamental solution as $\widehat{\boldsymbol{v}}=\boldsymbol{F}\cdot\mathcal{G}(\boldsymbol{x})/8\pi$, where $\mathcal{G}$ is (the Fourier transform of) the transient Oseen tensor and $\widehat{p}=\boldsymbol{F}\cdot\mathcal{P}(\boldsymbol{x})/8\pi$. The transient Oseen tensor is deduced using the same line of reasoning as in the steady problem, the main difference being that the fundamental solution is expressed in terms of a scalar function that satisfies the modified Helmholtz equation,

$$\nabla^2\psi-\psi=0 ,$$

instead of the Laplace equation. The final (nondimensional) result is[1]

$$\begin{aligned}\mathcal{G}(\boldsymbol{x}) &= (\nabla\nabla-\boldsymbol{\delta}\nabla^2)\frac{2}{r}(1-e^{-r}) \\ &= \frac{4}{r^3}[1-(1+r)e^{-r}]\frac{\boldsymbol{x}\boldsymbol{x}}{r^2} \\ &\quad +\frac{2}{r^3}[(1+r+r^2)e^{-r}-1](\boldsymbol{\delta}-\frac{\boldsymbol{x}\boldsymbol{x}}{r^2}) ,\end{aligned} \tag{6.5}$$

$$\mathcal{P}(\boldsymbol{x}) = -2\nabla\frac{1}{r} = 2\frac{\boldsymbol{x}}{r^3} . \tag{6.6}$$

For small r we obtain the expansion,

$$\mathcal{G}(\boldsymbol{x})=\mathcal{G}_0(\boldsymbol{x})-\frac{4}{3}\boldsymbol{\delta}+\frac{r}{4}(3\boldsymbol{\delta}-\frac{\boldsymbol{x}\boldsymbol{x}}{r^2})+O(r^2) ,$$

where $\mathcal{G}_0(\boldsymbol{x})$ is the Oseen tensor for steady flow. The dimensional forms are given by

$$\begin{aligned}\mathcal{G}(\boldsymbol{x};\alpha) &= \frac{4\alpha}{(\alpha r)^3}[1-(1+\alpha r)e^{-\alpha r}]\frac{\boldsymbol{x}\boldsymbol{x}}{r^2} \\ &\quad +\frac{2\alpha}{(\alpha r)^3}[(1+\alpha r+(\alpha r)^2)e^{-\alpha r}-1](\boldsymbol{\delta}-\frac{\boldsymbol{x}\boldsymbol{x}}{r^2}) , \\ \mathcal{P}(\boldsymbol{x};\alpha) &= 2\mu\frac{\boldsymbol{x}}{r^3} ,\end{aligned}$$

and are obtained from the nondimensional forms by reintroducing the proper scales. The notation $(;\alpha)$ reminds us that this is the dimensional form. In the small αr expansion, the leading term, $\mathcal{G}_0$, does not depend on α as required.

[1]We have used a notational convention to simplify the expressions. Lengths have been "scaled" with α^{-1} instead of $|\alpha|^{-1}$. The distinction is important since α is a complex number.

6.2.1 The Oscillating Sphere

Stokes' celebrated work concerning a sphere oscillating with velocity $\widehat{\boldsymbol{U}}e^{-i\omega t}$ provides an illustrative example of time-dependent Stokes flow. Here it is natural to set $\tau = \omega^{-1}$, where ω is the frequency of oscillation.

We seek a solution of the unsteady Stokes equation, with the boundary conditions,

$$\begin{aligned}
\widehat{\boldsymbol{v}} &= \widehat{\boldsymbol{U}} \quad \text{on the sphere surface, } i.e., r = a, \\
\widehat{\boldsymbol{v}} &\to \boldsymbol{0} \quad \text{for } r \to \infty \\
\widehat{p} &\to p_0 \quad \text{for } r \to \infty .
\end{aligned}$$

Following the strategy that proved so successful in the steady problem, we conjecture that a singularity solution of the form

$$\widehat{\boldsymbol{v}} = 6\pi\mu a\widehat{\boldsymbol{U}} \cdot (B_0 + B_2 a^2\nabla^2)\alpha\frac{\mathcal{G}(\boldsymbol{x})}{8\pi\mu}$$

exists, where B_0 and B_2 are dimensionless functions of $\lambda = \alpha a$. For steady flows, recall that $B_0 = 1$ and $B_2 = 1/6$. Now at $r = a$, we have

$$\begin{aligned}
\widehat{\boldsymbol{v}} &= \frac{3}{4}\widehat{\boldsymbol{U}} \cdot \left\{ \frac{\boldsymbol{x}\boldsymbol{x}}{r^2}\left[\frac{6B_0}{\lambda^2} - \left(\frac{B_0}{\lambda^2} + B_2\right)e^{-\lambda}[6 + 6\lambda + 2\lambda^2]\right] \right. \\
&\quad \left. + \boldsymbol{\delta}\left[-\frac{2B_0}{\lambda^2} + 2\left(\frac{B_0}{\lambda^2} + B_2\right)e^{-\lambda}[1 + \lambda + \lambda^2]\right]\right\} ,
\end{aligned}$$

so it follows from these boundary conditions that

$$\begin{aligned}
6B_0 - [6 + 6\lambda + 2\lambda^2]e^{-\lambda}(B_0 + \lambda^2 B_2) &= 0 \\
-B_0 + [1 + \lambda + \lambda^2]e^{-\lambda}(B_0 + \lambda^2 B_2) &= \frac{2}{3}\lambda^2 ,
\end{aligned}$$

or

$$B_0 = 1 + \lambda + \frac{1}{3}\lambda^2 , \qquad B_2 = \lambda^{-2}(e^{\lambda} - B_0) .$$

As in the steady Stokes problem, the force $\widehat{\boldsymbol{F}}e^{-i\omega t}$ exerted by the fluid on the sphere is readily extracted from the singularity solution using the properties of the Dirac delta function:

$$\begin{aligned}
\widehat{\boldsymbol{F}} &= \oint_{S_p} \widehat{\boldsymbol{\sigma}} \cdot \boldsymbol{n}\, dS = \int_{V_p} \nabla \cdot \widehat{\boldsymbol{\sigma}}\, dV \\
&= \int_{V_p} (-6\pi\mu a B_0 \widehat{\boldsymbol{U}}\delta(\boldsymbol{x}) + \mu\alpha^2\widehat{\boldsymbol{U}})\, dV \\
&= -6\pi\mu a[B_0 - \frac{2}{9}(\alpha a)^2]\widehat{\boldsymbol{U}} \\
&= -6\pi\mu a[1 + \lambda + \frac{1}{9}\lambda^2]\widehat{\boldsymbol{U}} .
\end{aligned}$$

Thus the force on an oscillating sphere consists of three parts: a term in phase with the particle motion, a term proportional to $\omega^{1/2}$ and out of phase by $\pi/4$, and a term out of phase by $\pi/2$ or in phase with the acceleration. Note also that the Stokes law for steady translation is recovered in the limit $\lambda \to 0$.

Later on, in the derivation of the Faxén law for unsteady Stokes flow, we will need the expression for the surface traction from this problem. We may obtain this by first determining the rate-of-strain $\hat{e}$ and pressure $\hat{p}$ from the velocity field. The final expression for the surface tracton is

$$\hat{\boldsymbol{\sigma}} \cdot \boldsymbol{n}\big|_{r=a} = -\frac{3\mu}{2a}\hat{\boldsymbol{U}} \cdot \left[(1+\lambda)\boldsymbol{\delta} + \frac{\lambda^2}{3}\boldsymbol{nn}\right] . \tag{6.7}$$

We recover the same expression for $\hat{\boldsymbol{F}}$ as above by integrating this result for the surface tractions. It is interesting to note that for unsteady Stokes flow, the traction is no longer directed along $\hat{\boldsymbol{U}}$; in fact, the $O(\lambda^2)$ term is directed normal to the sphere surface (this is precisely the inviscid pressure).

Basset [4] extended Stokes' solution to arbitrary motion, $\boldsymbol{U}(t)$, and his solution, given below, may be obtained by Fourier inversion[2] of the preceding result (see Landau and Lifshitz [54] for the derivation):

$$\boldsymbol{F}(t) = -6\pi\mu a\boldsymbol{U}(t) - 6\mu a^2\sqrt{\pi/\nu}\int_{-\infty}^{t} \dot{\boldsymbol{U}}(t')\frac{dt'}{\sqrt{t-t'}} - \frac{2}{3}\pi\rho a^3\frac{d\boldsymbol{U}}{dt} .$$

The origins of the first and third terms are readily identified. The first term is the pseudo-steady Stokes drag, and it originates from the O(1), in-phase term in $\hat{\boldsymbol{F}}$. The third term, a force contribution proportional to the particle acceleration, originates from the $\lambda^2 = -i\omega a^2/\nu$ term in $\hat{\boldsymbol{F}}$. This term is also known as the *added mass* term, because the extra force expended to overcome the inertia of the neighboring fluid is equivalent to an apparent increase in the mass of the sphere equal to one half of the mass of the displaced fluid. The second term, a convolution integral involving the sphere's history of motion, is known as the *Basset force* or *Basset memory integral*, and comes from the $O(\lambda)$ term in $\hat{\boldsymbol{F}}$. Note that we may write $\mu a^2\nu^{-1/2}$ as $(\mu a)^{1/2}(\rho a^3)^{1/2}$ so that the intermediate term is in some sense a geometric mean of the other two. The convolution integral is in fact the definition of the "half-derivative."

6.2.2 Sphere Released from Rest

For the case where an external force is applied with $F^e(t) = 0$ for $t < 0$ and arbitrary for $t > 0$ (this includes release from rest as a special case), Basset's formula may be inverted [3] so that in nondimensional form the sphere velocity is given by

$$\boldsymbol{U}(t) = \boldsymbol{F}^e(0)g(t) + \int_0^t \boldsymbol{G}(t')g(t-t')dt' , \tag{6.8}$$

where

$$\boldsymbol{G}(t) \;=\; \boldsymbol{F}^e + \frac{d\boldsymbol{F}^e}{dt} - \sqrt{\frac{\beta}{\pi}}\int_0^t \dot{\boldsymbol{F}}^e(t')\frac{dt'}{\sqrt{t-t'}} - \sqrt{\frac{\beta}{\pi t}}\boldsymbol{F}^e(0) ,$$

[2]Basset's work predates the widespread application of integral transforms, hence his solution is not by the Fourier integral.

$$g(t) = \frac{1}{\sqrt{\beta(\beta-4)}}[\exp(m_+t) - \exp(m_-t)] ,$$
$$\beta = \frac{9}{2}\Big[\frac{\rho_p}{\rho} + \frac{1}{2}\Big]^{-1}, \qquad m_\pm = \frac{\beta}{2} - 1 \pm \frac{1}{2}\sqrt{\beta(\beta-4)} .$$

Here, $\boldsymbol{F}^e$, $\boldsymbol{U}$, and t have been scaled with a characteristic force F_0, velocity $F_0/6\pi\mu a$, and time $a^2/\nu\beta$, respectively. We now give an outline of the derivation.

Since we have only $t \geq 0$, the derivation of the preceding result is most natural in terms of the Laplace transform. We start with the force balance,

$$\boldsymbol{F}^e(t) - 6\pi\mu a\boldsymbol{U}(t) - 6\mu a^2\sqrt{\pi/\nu}\int_0^t \dot{\boldsymbol{U}}(t')\frac{dt'}{\sqrt{t-t'}} - \frac{2}{3}\pi\rho a^3\frac{d\boldsymbol{U}}{dt} = \frac{4}{3}\pi\rho_p a^3\frac{d\boldsymbol{U}}{dt} .$$

The transformed equation is

$$\widehat{\boldsymbol{F}}^e(s) = 6\pi\mu a\widehat{\boldsymbol{U}}(s)\left[1 + \frac{as^{1/2}}{\sqrt{\nu}} + \frac{a^2 s}{\nu\beta}\right] .$$

If we now scale this equation as stated earlier and solve for $\widehat{\boldsymbol{U}}$, the result is

$$\widehat{\boldsymbol{U}}(s) = \frac{\widehat{\boldsymbol{F}}^e(s)}{1 + \sqrt{\beta}s^{1/2} + s} .$$

At this point, the inversion may be performed by partial fraction expansions, leading to an expression in the time domain in terms of the error function of complex argument. The solution given here is obtained by clearing the denominator of $s^{1/2}$; we rewrite $\widehat{\boldsymbol{U}}$ as

$$\widehat{\boldsymbol{U}}(s) = \frac{1}{(1+s)^2 - \beta s} \times \widehat{\boldsymbol{F}}^e(s)(1 + s - \sqrt{\beta}s^{1/2}) = \widehat{g}(s)\widehat{\boldsymbol{h}}(s) .$$

From the convolution theorem, we have

$$\boldsymbol{U}(t) = \int_0^t g(t-t')\boldsymbol{h}(t')dt' .$$

The first factor, $\widehat{g}(s)$, is inverted by partial fractions by noting that

$$(1+s)^2 - \beta s = (s - m_+)(s - m_-) .$$

Thus we have

$$\widehat{g}(s) = \frac{1}{\sqrt{\beta(\beta-4)}}\left[\frac{1}{s-m_+} - \frac{1}{s-m_-}\right] , \qquad g(t) = \frac{e^{m_+t} - e^{m_-t}}{\sqrt{\beta(\beta-4)}} .$$

The second factor, $\widehat{\boldsymbol{h}}(s)$, requires another application of the convolution theorem to handle the $s^{1/2}\widehat{\boldsymbol{F}}^e(s)$ term. We define $\widehat{\boldsymbol{h}}(s) = \widehat{\boldsymbol{G}}(s) + \boldsymbol{F}^e(0)$, so that

$$\widehat{\boldsymbol{G}}(s) = \widehat{\boldsymbol{F}}^e(s)(1 + s - \sqrt{\beta}s^{1/2}) - \boldsymbol{F}^e(0)$$
$$= \widehat{\boldsymbol{F}}^e(s) + s\widehat{\boldsymbol{F}}^e(s) - \boldsymbol{F}^e(0) - \frac{\sqrt{\beta}}{s^{1/2}}(s\widehat{\boldsymbol{F}}^e(s) - \boldsymbol{F}^e(0)) - \frac{\sqrt{\beta}}{s^{1/2}}\boldsymbol{F}^e(0)$$

and $\boldsymbol{G}(t)$ is as claimed. In the final step, we insert $\boldsymbol{h}(t) = \boldsymbol{G}(t) + \boldsymbol{F}^e(0)\delta(t)$ into the convolution integral to obtain Equation 6.8.

At very short times $t \ll a^2/\nu$, all manifestations of the viscous resistance of the fluid are negligible. For example, a sphere released from rest experiences a constant acceleration and we expect $\boldsymbol{U}(t) \sim \boldsymbol{F}^e(\rho_p + \frac{1}{2}\rho)V_p)^{-1}t$.

The exact solution has the small time behavior given by

$$\boldsymbol{U}(t) \sim \boldsymbol{F}^e t\left[1 - \frac{4}{3}\sqrt{\frac{\beta t}{\pi}}\right] ,$$

or in dimensional form,

$$\boldsymbol{U}(t) \sim \boldsymbol{F}^e\left[(\rho_p + \frac{1}{2}\rho)V_p\right]^{-1} t\left[1 - \frac{4\beta}{3}\sqrt{\frac{\nu t}{\pi a^2}}\right] ,$$

which is consistent with our expectations. The correction term scales with time as $t^{3/2}$ and comes from the Basset integral. In some elementary treatments of transient settling, the Basset effect is neglected. The pseudo-steady Stokes drag introduces a simple damping effect, which leads to a first order differential equation for $\boldsymbol{U}(t)$ and the solution,

$$\boldsymbol{U}(t) = \frac{\boldsymbol{F}^e}{6\pi\mu a}\left[1 - \exp\left(\frac{-6\pi\mu a t}{(\rho_p + \frac{1}{2}\rho)V_p}\right)\right] \qquad \text{(Basset effect neglected).}$$

At small times, this gives the *erroneous* result

$$\boldsymbol{U}(t) \sim \boldsymbol{F}^e\left[(\rho_p + \frac{1}{2}\rho)V_p\right]^{-1} t\left[1 - \frac{\beta\nu t}{2a^2}\right] .$$

On the vorticity diffusion time scale a^2/ν (at low Reynolds numbers, this still corresponds to the brief instant just after the release of the particle), the contributions from the pseudo-steady Stokes drag, the Basset force, and the added mass are comparable and all three effects must be included. Much later, on the time scale of a/U, the pseudo-steady Stokes drag gives the dominant contribution, with the Basset and added mass effects decaying in relative importance as $\text{Re}^{1/2}$ and Re. For a particle flow problem with Re $= 0.01$, the Basset correction is 10% and the cumulative effect can be significant.

6.2.3 Oscillatory Rotation of a Sphere

The velocity field produced by a sphere undergoing oscillatory rotation $\hat{\boldsymbol{\omega}}e^{-i\omega t}\times\boldsymbol{x}$ in a quiescent fluid can be represented by the transient rotlet,

$$\hat{\boldsymbol{v}} = (\hat{\boldsymbol{T}}\cdot\nabla)\cdot\mathcal{G}(\boldsymbol{x})/(8\pi\mu) .$$

On the sphere surface, the rotlet reduces to the desired rotational motion, provided that

$$T_{ij} = \epsilon_{ijk}\hat{\omega}_k\frac{4\pi\mu a^3 e^{\lambda}}{1+\lambda} .$$

The torque on the sphere follow as

$$\begin{aligned}\widehat{T}_i &= \oint_{S_p} \epsilon_{ijk} x_j (\widehat{\boldsymbol{\sigma}} \cdot \boldsymbol{n})_k \, dS \\ &= -\epsilon_{ijk} B_0(\lambda) e^{-\lambda} T_{jk} \\ &= -8\pi\mu a^3 \frac{1+\lambda+\lambda^2/3}{1+\lambda} \widehat{\omega}_i \,. \end{aligned} \tag{6.9}$$

The corresponding result in the time domain is

$$\begin{aligned}\boldsymbol{T}(t) &= -8\pi\mu a^3 \boldsymbol{\omega}(t) \\ &- \frac{8}{3}\int_{-\infty}^{t} \dot{\boldsymbol{\omega}}(t') \left[\frac{\mu a^4 \pi^{1/2}}{\sqrt{\nu(t-t')}} - \pi a^5 \rho e^{t-t'} \operatorname{erfc}\sqrt{t-t'} \right] dt' \,. \end{aligned} \tag{6.10}$$

For large λ, $\widehat{\boldsymbol{T}}$ is linear in λ (unlike $\widehat{\boldsymbol{F}}$, which was quadratic) and, correspondingly, in the time domain there is no term in phase with the angular acceleration, $\dot{\boldsymbol{\omega}}(t)$, *i.e.*, no "added moment of inertia."

6.3 Reciprocal Theorem and Applications

The derivation of the reciprocal theorem for unsteady Stokes flow is quite similar to that given in Chapter 2 for steady flow. For our purposes here, we express the theorem as

$$\begin{aligned}&\oint_S \widehat{\boldsymbol{v}}_1 \cdot (\widehat{\boldsymbol{\sigma}}_2 \cdot \boldsymbol{n}) \, dS + \int_V \widehat{\boldsymbol{v}}_1 \cdot (\nabla \cdot \widehat{\boldsymbol{\sigma}}_2) \, dV \\ &= \oint_S \widehat{\boldsymbol{v}}_2 \cdot (\widehat{\boldsymbol{\sigma}}_1 \cdot \boldsymbol{n}) \, dS + \int_V \widehat{\boldsymbol{v}}_2 \cdot (\nabla \cdot \widehat{\boldsymbol{\sigma}}_1) \, dV \,, \end{aligned} \tag{6.11}$$

where the fields $\widehat{\boldsymbol{v}}_1$, $\widehat{\boldsymbol{\sigma}}_1$, $\widehat{\boldsymbol{v}}_2$, and $\widehat{\boldsymbol{\sigma}}_2$ decay far away from the particle, to the extent that the surface contributions are taken from only the particle surface. Note that we may choose to subtract a volume integral of $\widehat{\boldsymbol{v}}_1 \cdot \widehat{\boldsymbol{v}}_2$ from both sides, to obtain a form closer to the governing equation.

As in steady Stokes flow, the reciprocal theorem may be used to derive the integral representation, Faxén laws, and symmetry relations for the resistance and mobility functions. We shall expand on these ideas in this section.

6.3.1 Integral Representations

We obtain the integral representation for disturbance fields by setting $\widehat{\boldsymbol{v}}_1$ as the fundamental solution and letting $\widehat{\boldsymbol{v}}_2$ be the solution of interest. The procedure is analogous to that used for steady flow and we obtain single and double layer potentials:

$$\begin{aligned}\widehat{\boldsymbol{v}}(\boldsymbol{x}) &= -\frac{1}{8\pi} \oint_{S_p} (\widehat{\boldsymbol{\sigma}} \cdot \boldsymbol{n}) \cdot \mathcal{G}(\boldsymbol{x}-\boldsymbol{\xi}) \, dS(\boldsymbol{\xi}) \\ &- \oint_{S_p} \widehat{\boldsymbol{v}} \cdot \boldsymbol{\Sigma}(\boldsymbol{x}-\boldsymbol{\xi}) \cdot \boldsymbol{n} \, dS(\boldsymbol{\xi}) \,. \end{aligned}$$

For a translating particle, the double layer potential may be replaced by the formula obtained by applying the representation to the uniform field $\widehat{\boldsymbol{U}}$ inside the particle. Here, however, the uniform field carries a pressure field equal to $-\widehat{\boldsymbol{U}} \cdot \boldsymbol{x}$, and so the final result reads

$$\begin{aligned} \widehat{\boldsymbol{v}}(\boldsymbol{x}) &= -\frac{1}{8\pi} \oint_{S_p} (\widehat{\boldsymbol{\sigma}} \cdot \boldsymbol{n} - \widehat{\boldsymbol{U}} \cdot \boldsymbol{x}\boldsymbol{n}) \cdot \mathcal{G}(\boldsymbol{x} - \boldsymbol{\xi})\, dS(\boldsymbol{\xi}) \\ &= -\frac{1}{8\pi} \oint_{S_p} (\widehat{\boldsymbol{\sigma}} \cdot \boldsymbol{n}) \cdot \mathcal{G}(\boldsymbol{x} - \boldsymbol{\xi})\, dS(\boldsymbol{\xi}) + \frac{\widehat{\boldsymbol{U}} \cdot}{8\pi} \int_{V_p} \mathcal{G}(\boldsymbol{x} - \boldsymbol{\xi})\, dV(\boldsymbol{\xi}) \,. \end{aligned}$$

6.3.2 The Faxén Law: Particles of Arbitrary Shape

Faxén laws for unsteady Stokes flow can be derived using the same approach as that employed in the steady problem. Starting from the reciprocal theorem, we show that the Faxén law must be of the same functional form as the associated transient singularity solution. This is illustrated here for the force law and the translational singularity solution.

We set $\widehat{\boldsymbol{v}}_1$ to be the solution for a particle oscillating with velocity $\widehat{\boldsymbol{U}}_1$ in a quiescent fluid. We note that $\nabla \cdot \widehat{\boldsymbol{\sigma}}_1 - \widehat{\boldsymbol{v}}_1 = 0$ in V. For $\widehat{\boldsymbol{v}}_2$, we take the velocity field generated by a point force $\widehat{\boldsymbol{F}} \cdot \mathcal{G}(\boldsymbol{x} - \boldsymbol{y})/8\pi$, where $\boldsymbol{y}$ lies outside the particle, and the particle stationary. Then $\widehat{\boldsymbol{v}}_2 = 0$ on S, and in V, $\nabla \cdot \widehat{\boldsymbol{\sigma}}_2 = -\widehat{\boldsymbol{F}}\delta(\boldsymbol{x} - \boldsymbol{y})$. When these boundary conditions and identities are inserted into Equation 6.11, the result is

$$\widehat{\boldsymbol{U}}_1 \cdot \widehat{\boldsymbol{F}}_2 - \widehat{\boldsymbol{v}}_1(\boldsymbol{y}) \cdot \widehat{\boldsymbol{F}} = 0 \,, \tag{6.12}$$

where $\widehat{\boldsymbol{F}}_2$ is the force on the particle generated by the surface traction $\widehat{\boldsymbol{\sigma}}_2 \cdot \boldsymbol{n}$. Now, due to linearity of the Stokes equation, we may factor $\widehat{\boldsymbol{U}}_1$ from $\widehat{\boldsymbol{v}}_1$. Furthermore, suppose that $\widehat{\boldsymbol{v}}_1$ is written as a singularity solution, then

$$\widehat{\boldsymbol{v}}_1(\boldsymbol{x}) = \widehat{\boldsymbol{U}}_1 \cdot \mathcal{F}\{\mathcal{G}(\boldsymbol{x} - \boldsymbol{\xi})/8\pi\} \,, \tag{6.13}$$

where $\mathcal{F}$ is a linear functional and $\boldsymbol{\xi}$ represents the region over which the singularities are distributed. Then Equation 6.12 becomes

$$(\widehat{\boldsymbol{F}}_2)_i = \mathcal{F}\{\widehat{F}_j \mathcal{G}_{ji}(\boldsymbol{y} - \boldsymbol{\xi})/8\pi\} \,, \tag{6.14}$$

but since $\widehat{F}_j \mathcal{G}_{ji}(\boldsymbol{y} - \boldsymbol{\xi})/8\pi = F_j \mathcal{G}_{ij}(\boldsymbol{\xi} - \boldsymbol{y})/8\pi$ is the ambient field evaluated over the image region, we have shown that

$$\widehat{\boldsymbol{F}}_2 = \mathcal{F}\{\widehat{\boldsymbol{v}}^\infty(\boldsymbol{\xi})\} \,. \tag{6.15}$$

But all ambient fields $\widehat{\boldsymbol{v}}^\infty$ that satisfy the unsteady Stokes equation can be constructed from an appropriate set of images, so we have derived the general result as well.

The Faxén law for a moving particle is obtained by adding the contributions for the particle moving through a quiescent fluid to the results for the stationary particle.

6.3.3 The Faxén Law: Force on a Rigid Sphere

We shall derive two different but equivalent forms of the Faxén law. The first form is obtained by starting with the integral representation for the oscillating sphere (the surface tractions was given earlier in this chapter):

$$\begin{aligned}\widehat{\boldsymbol{v}}(\boldsymbol{x}) &= 6\pi\mu a\widehat{\boldsymbol{U}}\cdot\left[\frac{1+\lambda}{4\pi a^2}\oint_{S_p}\frac{\mathcal{G}(\boldsymbol{x}-\boldsymbol{\xi};\alpha)}{8\pi\mu}\,dS(\boldsymbol{\xi})\right.\\ &\quad\left.+\frac{\lambda^2}{12\pi a^2}\oint_{S_p}\boldsymbol{nn}\cdot\frac{\mathcal{G}(\boldsymbol{x}-\boldsymbol{\xi};\alpha)}{8\pi\mu}\,dS(\boldsymbol{\xi})\right]\\ &\quad+\frac{\lambda^2}{8\pi a^2}\int_{V_p}\widehat{\boldsymbol{U}}\cdot\mathcal{G}(\boldsymbol{x}-\boldsymbol{\xi};\alpha)\,dV(\boldsymbol{\xi})\ .\end{aligned}$$

The Faxén relation for the stationary sphere follows as

$$\begin{aligned}\widehat{\boldsymbol{F}} &= 6\pi\mu a\left[\frac{1+\lambda}{4\pi a^2}\oint_{S_p}\widehat{\boldsymbol{v}}^{\infty}(\boldsymbol{\xi})dS(\boldsymbol{\xi})+\frac{\lambda^2}{12\pi a^2}\oint_{S_p}\boldsymbol{nn}\cdot\widehat{\boldsymbol{v}}^{\infty}(\boldsymbol{\xi})dS(\boldsymbol{\xi})\right.\\ &\quad\left.+\frac{\lambda^2}{6\pi a^3}\int_{V_p}\widehat{\boldsymbol{v}}^{\infty}(\boldsymbol{\xi})dV(\boldsymbol{\xi})\right]\ .\end{aligned}$$

We wish to collect all terms of $O(\lambda^2)$ as a volume integral. This may be accomplished along the lines

$$\begin{aligned}\oint_{S_p} n_i n_j \widehat{v}_j^{\infty}(\boldsymbol{\xi})\,dS(\boldsymbol{\xi}) &= a^{-1}\oint_{S_p} x_i n_j \widehat{v}_j^{\infty}(\boldsymbol{\xi})\,dS(\boldsymbol{\xi})\\ &= a^{-1}\int_{V_p}\frac{\partial}{\partial x_j}(x_i\widehat{v}_j^{\infty}(\boldsymbol{\xi})\,dS(\boldsymbol{\xi})\\ &= a^{-1}\int_{V_p}[\delta_{ij}\widehat{v}_j^{\infty}+x_i(\nabla\cdot\widehat{\boldsymbol{v}}^{\infty})]\,dS(\boldsymbol{\xi})\\ &= a^{-1}\int_{V_p}\widehat{v}_i^{\infty}(\boldsymbol{\xi})\,dS(\boldsymbol{\xi})\ .\end{aligned}$$

Thus we obtain the Faxén law, derived by Mazur and Bedeaux [59], for the force on an oscillating sphere in an arbitrary Stokes ambient field:

$$\widehat{\boldsymbol{F}}=6\pi\mu a\left[\frac{1+\lambda}{4\pi a^2}\oint_{S_p}\widehat{\boldsymbol{v}}^{\infty}(\boldsymbol{\xi})\,dS(\boldsymbol{\xi})+\frac{\lambda^2}{4\pi a^3}\int_{V_p}\widehat{\boldsymbol{v}}^{\infty}(\boldsymbol{\xi})\,dV(\boldsymbol{\xi})-\widehat{\boldsymbol{U}}\left(1+\lambda+\frac{\lambda^2}{9}\right)\right]$$

where for the general case we have simply added the additional contribution for a sphere moving through a quiescent fluid. In the time domain, these results become

$$\begin{aligned}\boldsymbol{F}(t) &= 6\pi\mu a[\boldsymbol{v}_S^{\infty}(t)-\boldsymbol{U}(t)]+6\mu a^2\sqrt{\pi/\nu}\int_{-\infty}^{t}(\dot{\boldsymbol{v}}_S^{\infty}(t')-\dot{\boldsymbol{U}}(t'))\frac{dt'}{\sqrt{t-t'}}\\ &\quad+\frac{2}{3}\pi\rho a^3\left(\frac{\partial}{\partial t}\boldsymbol{v}_V^{\infty}-\frac{d\boldsymbol{U}}{dt}\right)+\frac{4}{3}\pi\rho a^3\frac{\partial}{\partial t}\boldsymbol{v}_V^{\infty}\ ,\end{aligned}\qquad(6.16)$$

a form also given by Mazur and Bedeaux, where the ambient velocity appears in the form of averages over the sphere surface and volume,

$$\begin{aligned} \boldsymbol{v}_S^\infty(t) &= \frac{1}{4\pi a^2}\oint_{S_p} \widehat{\boldsymbol{v}}^\infty(\boldsymbol{\xi},t)\,dS(\boldsymbol{\xi})\ , \\ \boldsymbol{v}_V^\infty(t) &= \frac{3}{4\pi a^3}\int_{V_p} \widehat{\boldsymbol{v}}^\infty(\boldsymbol{\xi},t)\,dV(\boldsymbol{\xi})\ . \end{aligned}$$

We summarize the result for unsteady, spatially inhomogeous flows: The drag depends on the *relative* velocity and acceleration between the sphere and ambient fluid, plus a "buoyancy" contribution equal to the force required to accelerate the mass of fluid displaced by the sphere. The Faxén correction for curvature in the ambient velocity field is implicit in the averaged fields $\boldsymbol{v}_S^\infty$ and $\boldsymbol{v}_V^\infty$. While our development is aimed at transient effects in Stokes flow, Equation 6.16 also plays a prominent role in the analysis of tracer particle motions in Laser-Doppler Velocimetry (LDV) at finite Reynolds numbers (see, for example, [58] and references therein). The key idea is that a tracer particle essentially moves with the fluid element, and thus in the frame of reference moving with the particle the flow is at low Reynolds number (based on particle size and relative velocity). The force calculations involve Equation 6.16, with appropriate corrections for the fictitious forces in this noninertial reference frame.

A second form of the Faxén law follows from the singularity solution as

$$\widehat{\boldsymbol{F}} = 6\pi\mu a[B_0(\lambda) + B_2(\lambda)a^2\nabla^2]\widehat{\boldsymbol{v}}^\infty(\boldsymbol{\xi})|_{x=0} - 6\pi\mu a\widehat{\boldsymbol{U}}\left(1+\lambda+\frac{\lambda^2}{9}\right)\ .$$

This form can also be derived directly from the first, by noting that with $(\nabla^2)^n\widehat{\boldsymbol{v}}^\infty = \alpha^{2n-2}\nabla^2\widehat{\boldsymbol{v}}^\infty$,

$$\begin{aligned} \frac{1}{4\pi a^2}\oint_{S_p} \widehat{\boldsymbol{v}}^\infty(\boldsymbol{\xi})\,dS(\boldsymbol{\xi}) &= \widehat{\boldsymbol{v}}^\infty(0) + a^2\sum_{n=1}^{\infty}\frac{\lambda^{2n-2}\nabla^2\widehat{\boldsymbol{v}}^\infty(0)}{(2n+1)!} \\ \frac{3}{4\pi a^3}\int_{V_p} \widehat{\boldsymbol{v}}^\infty(\boldsymbol{\xi})\,dV(\boldsymbol{\xi}) &= \widehat{\boldsymbol{v}}^\infty(0) + a^2\sum_{n=1}^{\infty}\frac{3\lambda^{2n-2}\nabla^2\widehat{\boldsymbol{v}}^\infty(0)}{(2n+3)(2n+1)!}\ . \end{aligned}$$

The two infinite series together yield precisely the B_2 term in the second form of the Faxén law. This second form is useful at small λ, since all field variables are evaluated at a single point.

6.3.4 The Faxén Law: Viscous Drop

The Faxén law for the force on a viscous drop must also be of the same functional form as the singularity solution for the oscillating drop. The proof is similar to that used for the rigid particle. We start with the reciprocal theorem and insert the velocity field produced by an oscillating drop for $\widehat{\boldsymbol{v}}_1$, while $\widehat{\boldsymbol{v}}_2$ is taken to be the field of a *stationary* drop near a point force at $\boldsymbol{y}$. The proof proceeds along identical lines to the one used for the rigid particle, up to the point where the boundary conditions are inserted into Equation 6.11. Now we must have

$$\oint_{S_p} \widehat{\boldsymbol{v}}_1\cdot(\widehat{\boldsymbol{\sigma}}_2\cdot\boldsymbol{n})\,dS + \widehat{\boldsymbol{v}}_1(\boldsymbol{y})\cdot\widehat{\boldsymbol{F}} = \oint_{S_p} \widehat{\boldsymbol{v}}_2\cdot(\widehat{\boldsymbol{\sigma}}_1\cdot\boldsymbol{n})\,dS\ . \qquad (6.17)$$

At this point, it proves convenient to extract $\widehat{\boldsymbol{U}}$ from $\widehat{\boldsymbol{v}}_1$ so that the preceding equation becomes

$$\widehat{\boldsymbol{U}} \cdot \widehat{\boldsymbol{F}}_2 + \widehat{\boldsymbol{v}}_1(\boldsymbol{y}) \cdot \widehat{\boldsymbol{F}} = -\oint_{S_p} (\widehat{\boldsymbol{v}}_1 - \widehat{\boldsymbol{U}}) \cdot (\widehat{\boldsymbol{\sigma}}_2 \cdot \boldsymbol{n})\, dS + \oint_{S_p} \widehat{\boldsymbol{v}}_2 \cdot (\widehat{\boldsymbol{\sigma}}_1 \cdot \boldsymbol{n})\, dS \ . \tag{6.18}$$

The boundary conditions at the surface of the drop require that $\boldsymbol{n} \cdot \widehat{\boldsymbol{v}}_2 = 0$ and $\boldsymbol{n} \cdot (\widehat{\boldsymbol{v}}_1 - \widehat{\boldsymbol{U}}) = 0$, so only the *tangential* components of the surface traction are retained in the dot products. But an application of the reciprocal theorem to the inner fields associated with $\widehat{\boldsymbol{v}}_1 - \widehat{\boldsymbol{U}}$ and $\widehat{\boldsymbol{v}}_2$ yields the relation,

$$\oint_{S_p} (\widehat{\boldsymbol{v}}_1^{(i)} - \widehat{\boldsymbol{U}}) \cdot (\widehat{\boldsymbol{\sigma}}_2^{(i)} \cdot \boldsymbol{n})\, dS = \oint_{S_p} \widehat{\boldsymbol{v}}_2^{(i)} \cdot (\widehat{\boldsymbol{\sigma}}_1^{(i)} \cdot \boldsymbol{n})\, dS \ , \tag{6.19}$$

where again it is understood that only the tangential component of the surface traction is retained. But in both problem 1 and 2, the tangential component of the traction is continuous across the interface. The preceding equation then implies that the two surface integrals in Equation 6.18 cancel.

Again, we suppose that $\widehat{\boldsymbol{v}}_1$, the field produced by the oscillating drop, is available as a singularity solution, so that

$$\widehat{\boldsymbol{v}}_1(\boldsymbol{x}) = \widehat{\boldsymbol{U}}_1 \cdot \mathcal{F}\{\mathcal{G}(\boldsymbol{x} - \boldsymbol{\xi})/8\pi\} \ , \tag{6.20}$$

where $\mathcal{F}$ is a linear functional and $\boldsymbol{\xi}$ represents the region over which the singularities are distributed. We now may use the same arguments as used before for the rigid particle to arrive at the conclusion that the Faxén law is of the same functional form as the singularity solution.

6.3.5 The Oscillating Spherical Drop

We consider a viscous drop oscillating with velocity $\widehat{\boldsymbol{U}} e^{-i\omega t}$ in a quiescent fluid. We assume that surface tension forces dominate over viscous forces so that the drop retains its spherical shape. Once again, for convenience, we shall adopt the notational convention of scaling lengths with α^{-1}, and the associated scales for the pressure, *etc.*

The following solutions of the transient Stokes equations are bounded in the interior of a sphere of radius a:

$$\widehat{\boldsymbol{v}} = \widehat{\boldsymbol{U}} \ , \qquad \widehat{\boldsymbol{v}} = (\nabla \widehat{\boldsymbol{U}} \cdot \nabla - \boldsymbol{U} \nabla^2) \frac{\sinh r}{r} \ ;$$

in fact, the linear combination

$$\widehat{\boldsymbol{v}} = \frac{1}{\lambda^2} \widehat{\boldsymbol{U}} + \frac{3}{2} (\nabla \widehat{\boldsymbol{U}} \cdot \nabla - \boldsymbol{U} \nabla^2) \frac{\sinh r}{r}$$

reduces to the Stokeson of Chapter 3 in the limit of small λ.[3] Therefore, for the velocity field inside the drop, we assume the form

$$\widehat{\boldsymbol{v}} = D_0 \widehat{\boldsymbol{U}} + D_2 (\nabla \widehat{\boldsymbol{U}} \cdot \nabla - \boldsymbol{U} \nabla^2) \frac{\sinh r}{r} \ , \tag{6.21}$$

[3] More explicitly, in terms of the dimensional variables, we rewrite αr as $\lambda r / a$ and take the limit of small λ while keeping r/a fixed.

while for outside the drop, we assume the familiar form,

$$\widehat{\boldsymbol{v}} = \frac{3\lambda}{4}\widehat{\boldsymbol{U}} \cdot (C_0 + C_2\lambda^2\nabla^2)\mathcal{G}(\boldsymbol{x}) \,, \tag{6.22}$$

which already satisfies the boundary condition for large r. At $r = a$, the boundary conditions on the radial and tangential velocities and the tangential component of the surface traction $\widehat{\boldsymbol{\sigma}} \cdot \boldsymbol{n}$ are

1. $\boldsymbol{n} \cdot \widehat{\boldsymbol{v}}^{(o)} = \boldsymbol{n} \cdot \widehat{\boldsymbol{U}}$
2. $\boldsymbol{n} \cdot \widehat{\boldsymbol{v}}^{(i)} = \boldsymbol{n} \cdot \widehat{\boldsymbol{U}}$
3. $\widehat{\boldsymbol{v}}^{(o)} - \boldsymbol{nn} \cdot \widehat{\boldsymbol{v}}^{(o)} = \widehat{\boldsymbol{v}}^{(i)} - \boldsymbol{nn} \cdot \widehat{\boldsymbol{v}}^{(i)}$
4. $(\widehat{\boldsymbol{e}}^{(o)} \cdot \boldsymbol{n}) \cdot (\boldsymbol{\delta} - \boldsymbol{nn}) = \kappa(\widehat{\boldsymbol{e}}^{(i)} \cdot \boldsymbol{n}) \cdot (\boldsymbol{\delta} - \boldsymbol{nn})$,

where $\kappa = \mu^{(i)}/\mu^{(o)}$, the ratio of the drop and solvent viscosities.[4]

These conditions yield, respectively, the four equations,

$$\begin{aligned}
W - (1+\lambda)X &= \frac{\lambda^2}{3} \\
Y + (2\tanh\lambda - 2\lambda)Z &= 1 \\
-W + (1+\lambda+\lambda^2)X - \frac{2}{3}\lambda^2 Y + \frac{2}{3}\lambda^2(\lambda^2\tanh\lambda - \lambda + \tanh\lambda)Z &= 0 \\
9W - \frac{3}{2}(6 + 6\lambda + 3\lambda^2 + \lambda^3) - \kappa\lambda^2\left[(6+3\lambda^2)\tanh\lambda - 6\lambda - \lambda^3\right] Z &= 0 \,,
\end{aligned}$$

where $W = C_0$, $X = e^{-\lambda}(C_0 + \lambda^2 C_2)$, $Y = D_0$, and $Z = D_2(\cosh\lambda)/\lambda^3$. The solution is

$$\begin{aligned}
C_0 &= B_0(\lambda) - \frac{(1+\lambda)^2 f(\lambda)}{D(\lambda,\kappa)} \\
C_2 &= B_2(\lambda) - \frac{e^{\lambda} - (1+\lambda)}{\lambda^2}\frac{(1+\lambda)f(\lambda)}{D(\lambda,\kappa)} \\
D_0 &= 1 - \frac{(1+\lambda)(3\tanh\lambda - 3\lambda)}{D(\lambda,\kappa)} \\
D_2 &= \frac{3\lambda^3 \mathrm{sech}\lambda(1+\lambda)}{2D(\lambda,\kappa)} \,,
\end{aligned}$$

where

$$f(\lambda) = \lambda^2\tanh\lambda - 3\lambda + 3\tanh\lambda$$

and

$$D(\lambda,\kappa) = \kappa[\lambda^3 - \lambda^2\tanh\lambda - 2f(\lambda)] + (\lambda+3)f(\lambda) \,.$$

[4]In much of the literature, the viscosity ratio is denoted by λ. Unfortunately, λ is also the accepted notation for the frequency parameter αa. To avoid confusion, throughout this chapter we will use κ for the viscosity ratio.

In the limit of large κ (a very viscous drop) with λ fixed, we recover the solution for the rigid sphere, with C_0 and C_2 becoming B_0 and B_2. On the other hand, with κ fixed, we obtain at low frequencies (small λ) the solution

$$\begin{aligned} C_0 &= \frac{2+3\kappa}{3(1+\kappa)}\left[1+\frac{2+3\kappa}{3(1+\kappa)}\lambda\right]+O(\lambda^2) \\ C_2 &= \frac{\kappa}{6(1+\kappa)}\left[1+\frac{2+3\kappa}{3(1+\kappa)}\lambda\right]+O(\lambda^2)\,. \end{aligned}$$

The $O(1)$ terms give the Hadamard–Rybczynski solution for steady translation. Note also that for C_0, the coefficient of the $O(\lambda)$ term, is the square of the $O(1)$ coefficient. We will show at the end of this chapter that this is the general form taken by the low-frequency correction.

In the high-frequency limit, the drag coefficient has the asymptotic behavior

$$C_0 - \frac{2\lambda^2}{9} = \frac{\lambda^2}{9} + \frac{\kappa\lambda}{\kappa+1}\,,$$

which shows that the added mass is independent of κ, as expected, since the potential solution outside the sphere applies for all values of κ. Following Lawrence and Weinbaum [55], we may write the force on the drop as

$$\widehat{\boldsymbol{F}} = 6\pi\mu a\widehat{U}\left[\frac{2+3\kappa}{3(1+\kappa)} + B^{\infty}\lambda + \frac{\lambda^2}{9} + (B^0 - B^{\infty})\lambda L(\lambda,\kappa)\right]\,,$$

where

$$B^0 = \frac{(2+3\kappa)^2}{9(1+\kappa)^2} \qquad \text{and} \qquad B^{\infty} = \frac{\kappa}{\kappa+1}$$

are the $O(\lambda)$ (or Basset) coefficients for small and large λ, and $L(\lambda,\kappa)$ is a dimensionless function of λ, which varies from 1 for small λ to 0 for large λ. For the rigid sphere, the L-term is not present because $B^0 = B^{\infty}$.

6.3.6 The Faxén Law: Force on a Spherical Drop

The Faxén law for the drag on a viscous drop in an arbitrary unsteady Stokes flow is

$$\widehat{\boldsymbol{F}} = 6\pi\mu a[C_0(\lambda) + C_2(\lambda)a^2\nabla^2]\widehat{\boldsymbol{v}}^{\infty}(\boldsymbol{\xi})|_{x=0} - 6\pi\mu a\widehat{U}[C_0 - \frac{2}{9}\lambda^2]\,,$$

with C_0 and C_2 from the preceding discussion. The form analogous to the one derived by Mazur and Bedeaux for the rigid sphere follows immediately as

$$\begin{aligned} \widehat{\boldsymbol{F}} &= 6\pi\mu a\left[\frac{\gamma_0(\lambda)}{4\pi a^2}\oint_{S_p}\widehat{\boldsymbol{v}}^{\infty}(\boldsymbol{\xi})dS(\boldsymbol{\xi}) + \frac{3\gamma_2(\lambda)}{4\pi a^3}\int_{V_p}\widehat{\boldsymbol{v}}^{\infty}(\boldsymbol{\xi})dV(\boldsymbol{\xi})\right] \\ &\quad - 6\pi\mu a\widehat{U}\left[C_0 - \frac{2}{9}\lambda^2\right]\,, \end{aligned}$$

with

$$\gamma_0(\lambda) = 1 + \lambda - \frac{\lambda^2}{3}\left(1 + \frac{\lambda^2 \sinh\lambda}{3(\lambda e^\lambda - B_0 \sinh\lambda)}\right)\frac{(1+\lambda)f}{D(\lambda,\kappa)} ,$$

$$\gamma_2(\lambda) = \frac{\lambda^2}{3} + \frac{\lambda^2}{3}\left(1 + \frac{\lambda^2 \sinh\lambda}{3(\lambda e^\lambda - B_0 \sinh\lambda)}\right)\frac{(1+\lambda)f}{D(\lambda,\kappa)} ,$$

and f and D defined as in solution for the oscillating drop.

6.4 The Low-Frequency Limit

From the preceding discussion, it appears that in the limit of small frequencies, the unsteady Stokes results reduce to the steady Stokes solution plus a correction of $O(\lambda)$, with the $O(\lambda)$ coefficient expressed as a square of the steady result. Here, we investigate this more closely to establish a general description of the low-frequency limit.

A convenient place to start is the integral representation, with the fundamental solution expanded in small λ. The integral equation (dimensionless form) for the traction becomes

$$\hat{\boldsymbol{U}} = -\frac{1}{8\pi}\oint_{S_p} (\hat{\boldsymbol{\sigma}} \cdot \boldsymbol{n}) \cdot [\mathcal{G}_0(\boldsymbol{x} - \boldsymbol{\xi}) - \frac{4}{3}\lambda\boldsymbol{\delta}]\, dS + O(\lambda)^2 ,$$

or, if we expand the stress as $\hat{\boldsymbol{\sigma}} = \hat{\boldsymbol{\sigma}}_0 + \lambda\hat{\boldsymbol{\sigma}}_1 + \cdots$,

$$\hat{\boldsymbol{U}} - \frac{\lambda \boldsymbol{F}_0}{6\pi} = -\frac{1}{8\pi}\oint_{S_p} [(\hat{\boldsymbol{\sigma}}_0 + \lambda\hat{\boldsymbol{\sigma}}_1) \cdot \boldsymbol{n}] \cdot \mathcal{G}_0(\boldsymbol{x} - \boldsymbol{\xi})\, dS .$$

Thus we obtain the integral equation for *steady* Stokes flow, but with an effective uniform velocity[5] $\hat{\boldsymbol{U}} - \lambda\boldsymbol{F}_0/6\pi$.

If the surface traction and force in the steady Stokes problem have the form

$$\boldsymbol{\sigma} \cdot \boldsymbol{n} = -\frac{3}{2}\mathcal{S} \cdot \boldsymbol{U} , \qquad \boldsymbol{F} = -6\pi \boldsymbol{A} \cdot \boldsymbol{U} ,$$

then the low-frequency limit must be of the form

$$\hat{\boldsymbol{\sigma}} \cdot \boldsymbol{n} = -\frac{3}{2}\mathcal{S} \cdot (\boldsymbol{U} + \lambda \boldsymbol{A} \cdot \boldsymbol{U})$$
$$\boldsymbol{F} = -6\pi \boldsymbol{A} \cdot (\boldsymbol{U} + \lambda \boldsymbol{A} \cdot \boldsymbol{U}) .$$

The result derived earlier for the sphere is consistent with these general expressions, with $\boldsymbol{A} = \mathcal{S} = \boldsymbol{\delta}$.

The Faxén law also follows directly from the surface tractions for the translating particle. Thus, at low frequencies the Faxén law (dimensional form) for the general particle, correct to $O(\lambda)$, is

$$\hat{\boldsymbol{F}} = 6\pi\mu a(\boldsymbol{\delta} + \lambda\boldsymbol{A}) \cdot \frac{1}{4\pi a^2}\oint_{S_p} \mathcal{S} \cdot \hat{\boldsymbol{v}}^\infty dS + \cdots .$$

[5]The steady force F_0 here has been scaled with $\mu a U$.

We may apply the low-frequency expansion to the resistance expressions for an arbitrary particle undergoing translation and rotation. Suppose that the steady forces and torques are described by the resistance relation

$$\begin{aligned} \boldsymbol{F} &= -6\pi \boldsymbol{A}\cdot\boldsymbol{U} - 6\pi\widetilde{\boldsymbol{B}}\cdot\boldsymbol{\omega} \\ \boldsymbol{T} &= -6\pi\boldsymbol{B}\cdot\boldsymbol{U} - 8\pi\boldsymbol{C}\cdot\boldsymbol{\omega} \ . \end{aligned}$$

Once again, the $O(\lambda)$ correction appears *via* an effective uniform velocity given by $-\lambda \boldsymbol{F}_0/(6\pi)$, so the resistance relations follow as

$$\begin{aligned} \widehat{\boldsymbol{F}} &= -6\pi\boldsymbol{A}\cdot\left[\widehat{\boldsymbol{U}} + \lambda\boldsymbol{A}\cdot\widehat{\boldsymbol{U}} + \lambda\widetilde{\boldsymbol{B}}\cdot\widehat{\boldsymbol{\omega}}\right] - 6\pi\widetilde{\boldsymbol{B}}\cdot\widehat{\boldsymbol{\omega}} \\ \widehat{\boldsymbol{T}} &= -6\pi\boldsymbol{B}\cdot\left[\widehat{\boldsymbol{U}} + \lambda\boldsymbol{A}\cdot\widehat{\boldsymbol{U}} + \lambda\widetilde{\boldsymbol{B}}\cdot\widehat{\boldsymbol{\omega}}\right] - 8\pi\boldsymbol{C}\cdot\widehat{\boldsymbol{\omega}} \ . \end{aligned}$$

We draw the following observations from the preceding example. The tensors that couple $\widehat{\boldsymbol{F}}$ to $\widehat{\boldsymbol{\omega}}$ and $\widehat{\boldsymbol{T}}$ to $\widehat{\boldsymbol{U}}$ are

$$-6\pi(\boldsymbol{\delta} + \lambda\boldsymbol{A})\cdot\widetilde{\boldsymbol{B}} \qquad \text{and} \qquad -6\pi\boldsymbol{B}\cdot(\boldsymbol{\delta}+\lambda\boldsymbol{A}) \ ,$$

respectively, so the resistance tensor is symmetric, as required by the reciprocal theorem. Finally, we see that the coupling of $\widehat{\boldsymbol{F}}$ to $\widehat{\boldsymbol{\omega}}$ and $\widehat{\boldsymbol{T}}$ to $\widehat{\boldsymbol{U}}$ occurs at $O(\lambda)$ if and only if it also exists for steady Stokes flow.

Exercises

Exercise 6.1 Oscillatory Rotation of a Sphere.
Consider the velocity field produced by a sphere undergoing oscillatory rotation $\widehat{\boldsymbol{\omega}}e^{-i\omega t}\times\boldsymbol{x}$ in a quiescent fluid. Starting with the transient rotlet,

$$\widehat{\boldsymbol{v}} = (\widehat{\boldsymbol{T}}\cdot\nabla)\cdot\mathcal{G}(\boldsymbol{x})/(8\pi\mu) \ .$$

Show that on the sphere surface, the rotlet reduces to the desired rotational motion if

$$T_{ij} = \epsilon_{ijk}\widehat{\omega}_k \frac{4\pi\mu a^3 e^{\lambda}}{1+\lambda} \ .$$

Calculate the surface tractions and verify that the torque on the sphere is given by Equation 6.9. Invert this result and show that in the time domain

$$\begin{aligned} \boldsymbol{T}(t) &= -8\pi\mu a^3\boldsymbol{\omega}(t) \\ &- \frac{8}{3}\int_{-\infty}^{t}\dot{\boldsymbol{\omega}}(t')\left[\frac{\mu a^4\pi^{1/2}}{\sqrt{\nu(t-t')}} - \pi a^5 \rho e^{t-t'}\operatorname{erfc}\sqrt{t-t'}\right]dt' \ . \end{aligned}$$

Exercise 6.2 The Reciprocal Theorem for Unsteady Flow.
Derive the reciprocal theorem for unsteady flow, Equation 6.11.

Exercise 6.3 The Integral Representation for Unsteady Flow.
Derive the integral representation (in terms of single layer and double layer potentials) for unsteady Stokes flow. Show that the disturbance field can be expressed using just the single layer potential.

Now consider the vector field defined by $\hat{\boldsymbol{\Omega}}e^{-i\omega t} \times \boldsymbol{x}$ over some region V (with $\boldsymbol{\Omega}$ a constant pseudo-vector). Does this vector field satisfy the time-dependent Stokes equations?

Hint: Consider the following experiment. A closed container completely filled with a viscous fluid is subjected to oscillatory rigid-body rotation. Will the fluid inside undergo oscillatory rigid-body rotation in step with the boundary of the container? What are the implications for the existence of a single layer representation for unsteady particulate Stokes flows?

Exercise 6.4 Viscous Drop at the Low-Frequency Limit.
Show that the $O(\lambda)$ term for the drag on the viscous drop is given by the square of the resistance tensor for the drop in steady flow.

Exercise 6.5 Rigid Sphere in a Linear Field.
Consider a fixed rigid sphere centered in the rate-of-strain field, $\hat{\boldsymbol{v}}^{\infty} = \boldsymbol{E}^{\infty} \cdot \boldsymbol{x}$. Construct the disturbance solution using the transient stresslet and degenerate octupole.

Exercise 6.6 Impulse on a Rigid Sphere.
Consider an impulsive external force, $\boldsymbol{F}^e(t) = \boldsymbol{P}\delta(t)$, and find the formal expression for the motion of the sphere, $\boldsymbol{U}(t)$.

Exercise 6.7 Sphere Released from Rest.
For a sphere released from rest, we have $\boldsymbol{F}^e(t) = 0$, for $t < 0$, and $\boldsymbol{F}^e(t) = \boldsymbol{F}_0$ constant, for $t \geq 0$. Show that the expression for $\boldsymbol{U}(t)$ may be written in the more simple form,

$$\boldsymbol{U}(t) = \boldsymbol{F}_0 g(t) + \boldsymbol{F}_0 \int_0^t g(t-t')\,dt' - \boldsymbol{F}_0 \left[\frac{e^{m_+ t}\operatorname{erf}(\sqrt{m_+ t})}{\sqrt{m_+(\beta-4)}} - \frac{e^{m_- t}\operatorname{erf}(\sqrt{m_- t})}{\sqrt{m_-(\beta-4)}} \right] ,$$

$$g(t) = \frac{e^{m_+ t} - e^{m_- t}}{m_+ - m_-} = 2\exp\left\{ \left(\frac{\beta}{2} - 1\right) t \right\} \frac{\sin\left(\frac{1}{2}\sqrt{\beta(4-\beta)}t\right)}{\sqrt{\beta(4-\beta)}} .$$

The force, velocity, and time have been scaled with F_0, $F_0/6\pi\mu a$, and $a^2/\nu\beta$. Use this result to derive asymptotic expansions for small t and for large t. For $\rho_p/\rho = 1 + \epsilon$, $0 < \epsilon \ll 1$, (for example, a particle tracer) what is the nature of the decay to the terminal Stokes velocity?

Although m_+ and m_- can become complex, the above can always be rewritten as a real expression, using integral representations for the error function (see [1]). In any event, the complex form is more convenient for working out the asymptotic expansions.

References

[1] M. Abramowitz and I. A. Stegun. *Handbook of Mathematical Functions.* Dover, New York, 1972.

[2] A. D. Alawneh and R. P. Kanwal. Singularity methods in mathematical physics. *SIAM Rev.*, 19:437–470, 1977.

[3] L. Arminski and S. Weinbaum. Effect of waveform and duration of impulse on the solution to the Basset–Langevin equation. *Phys. Fluids*, 22:404–411, 1979.

[4] A. B. Basset. *A Treatise on Hydrodynamics*, Volume 2. Deighton, Bell and Co., Cambridge, 1888.

[5] G. K. Batchelor. Slender-body theory for particles of arbitrary cross-section in Stokes flow. *J. Fluid Mech.*, 44:419–440, 1970.

[6] J. M. Bernal and J. de la Torre. Transport properties and hydrodynamic centers of rigid macromolecules with arbitrary shape. *Biopolymers*, 19:751–766, 1980.

[7] R. B. Bird, O. Hassager, R. C. Armstrong, and C. Curtiss. *Dynamics of Polymeric Liquids*, Volume 2, 2nd edition. Wiley, New York, 1987.

[8] F. Booth. The cataphoresis of spherical particles in strong fields. *J. Chem. Phys.*, 18:1361–1364, 1950.

[9] H. Brenner. The slow motion of a sphere through a viscous fluid towards a plane surface. *Chem. Eng. Sci.*, 16:242–251, 1961.

[10] H. Brenner. The Stokes resistance of an arbitrary particle. *Chem. Eng. Sci.*, 18:1–25, 1963.

[11] H. Brenner. The Stokes resistance of a slightly deformed sphere. *Chem. Eng. Sci.*, 19:519–539, 1964.

[12] H. Brenner. The Stokes resistance of an arbitrary particle — V. Symbolic operator representation of intrinsic resistance. *Chem. Eng. Sci.*, 21:97–109, 1966.

[13] H. Brenner. Coupling between the translational and rotational Brownian motions of rigid particles of arbitrary shape. *J. Colloid Interface Sci.*, 23:407–436, 1967.

[14] H. Brenner. Suspension rheology. *Prog. Heat and Mass Transfer*, 5:89–129, 1972.

[15] H. Brenner and M. E. O'Neill. On the Stokes resistance of multiparticle systems in a linear shear field. *Chem. Eng. Sci.*, 27:1421–1439, 1972.

[16] H. Brenner. Rheology of a dilute suspension of axisymmetric Brownian particles. *Int. J. Multiphase Flow*, 1:195–341, 1974.

[17] S. Broersma. Rotational diffusion coefficient of a cylindrical particle. *J. Chem. Phys.*, 32:1626–1631, 1960.

[18] S. Broersma. Viscous force constant for a closed cylinder. *J. Chem. Phys.*, 32:1632–1635, 1960.

[19] J. M. Burgers. On the motion of small particles of elongated form suspended in a viscous liuid. Second report on viscosity and plasticity, Chap. III. *Kon. Ned. Akad. Wet.*, 16:113–184, 1938.

[20] A. T. Chwang and T. Y. Wu. Hydromechanics of low-Reynolds-number flow. Part 2. Singularity method for Stokes flows. *J. Fluid Mech.*, 67:787–815, 1975.

[21] W. D. Collins. A note on the axisymmetric Stokes flow of viscous fluid past a spherical cap. *Mathematika*, 10:72–78, 1963.

[22] R. G. Cox. The deformation of a drop in a general time-dependent fluid flow. *J. Fluid Mech.*, 37:601–623, 1969.

[23] R. G. Cox. The motion of long slender bodies in a viscous fluid. Part 1. General theory. *J. Fluid Mech.*, 44:791–810, 1970.

[24] P. Debye and E. Hückel. Bemerkungen zu einem Satze über die kataphoretische Wanderungsgeschwindigkeit suspendierter Teilchen (Remarks on a theorem concerning the cataphoretic velocity of suspended particles). *Physik. Z.*, 25:49–52, 1924. Also in *Collected Papers*, Interscience, New York, 1954.

[25] S. S. Dukhin and B. V. Derjaguin. *Electrokinetic Phenomena in Surface and Colloid Science*, Vol. 7, E. Matijević, Editor. John Wiley and Sons, New York, 1974.

[26] D. Edwardes. Steady motion of a viscous liquid in which an ellipsoid is constrained to rotate about a principal axis. *Q. J. Math.*, 26:70–78, 1892.

[27] M. Fair and J. L. Anderson. Electrophoresis of nonuniformly charged ellipsoidal particles. *J. Colloid Interface Sci.*, 127:388–400, 1989.

[28] H. Faxén. Der Widerstand gegen die Bewegung einer starren Kugel in einer zähen Flüssigkeit, die zwischen zwei parallelen Ebenen Wänden eingeschlossen ist (The resistance against the movement of a rigid sphere in a viscous fluid enclosed between two parallel planes). *Annalen der Physik*, 4(68):89–119, 1922.

[29] H. Faxén. Der Widerstand gegen die Bewegung einer starren Kugel in einer zähen Flüssigkeit, die zwischen zwei parallelen Ebenen Wänden eingeschlossen ist (The resistance against the movement of a rigid sphere in a viscous fluid enclosed between two parallel planes). *Arkiv fur Matematik, Astronomi och Fysik*, 18(29):1–52, 1924.

[30] F. R Gantmacher. *The Theory of Matrices.* Chelsea, New York, 1960.

[31] H. Giesekus. Elektro-viskose Flüssigkeiten, für die in stationären Schichtströmungen sämtliche Normalspannungskomponenten verschieden groß sind (Electro-viscous fluids with different normal stress components in steady-state laminar flow). *Rheol. Acta*, 2:50–62, 1962.

[32] M. J. Gluckman, R. Pfeffer, and S. Weinbaum. A new technique for treating multiparticle slow viscous flow: axisymmetric flow past spheres and spheroids. *J. Fluid Mech.*, 50:705–740, 1971.

[33] S. Haber, G. Hetsroni, and A. Solan. On the low Reynolds number motion of two droplets. *Intl. J. Multiphase Flow*, 1:57–71, 1973.

[34] R. A. Handelsman and J. B. Keller. Axially symmetric potential flow around a slender body. *J. Fluid Mech.*, 28:131–147, 1967.

[35] J. Happel and H. Brenner. *Low Reynolds Number Hydrodynamics.* Martinus Nijhoff, The Hague, 1983.

[36] L. B. Harris. Simplified calculation of electrophoretic mobility of nonspherical particles when the electric double layer is very extended. *J. Colloid Interface Sci.*, 34:322–325, 1970.

[37] D. C. Henry. The cataphoresis of suspended particles. Part I. The equation of cataphoresis. *Proc. R. Soc.*, A133:106–129, 1931.

[38] G. Hetsroni and S. Haber. Flow in and around a droplet or bubble submerged in an unbounded arbitrary velocity field. *Rheol. Acta*, 9:488–496, 1970.

[39] E. J. Hinch. Note on the symmetries of certain material tensors for a particle in Stokes flow. *J. Fluid Mech.*, 54:423–425, 1972.

[40] E. J. Hinch and L. G. Leal. The effect of Brownian motion on the rheological properties of a suspension of non-spherical particles. *J. Fluid Mech.*, 52:683–712, 1972.

[41] E. J. Hinch and L. G. Leal. Time-dependent shear flows of a suspension of particles with weak Brownian rotations. *J. Fluid Mech.*, 57:753–767, 1973.

[42] E. W. Hobson. *The Theory of Spherical and Ellipsoidal Harmonics.* Chelsea, New York, 1965.

[43] E. Hückel. Die Kataphorese der Kugel (The cataphoresis of the sphere). *Physik. Z.*, 25:204–210, 1924.

[44] R. J. Hunter. *Zeta Potential in Colloid Science.* Academic Press, New York, 1981.

[45] J. D. Jackson. *Classical Electrodynamics*, 2nd edition. Wiley, New York, 1975.

[46] G. B. Jeffery. The motion of ellipsoidal particles immersed in a viscous fluid. *Proc. R. Soc.*, A102:161–179, 1922.

[47] R. E. Johnson. An improved slender-body theory for Stokes flow. *J. Fluid Mech.*, 99:411–431, 1980.

[48] H. J. Keh and J. L. Anderson. Boundary effects on electrophoretic motion of colloidal spheres. *J. Fluid Mech.*, 153:417–439, 1985.

[49] S. Kim and X. Fan. A perturbation solution for rigid dumbbell suspensions in steady shear flow. *J. Rheol.*, 28(2):117–122, 1984.

[50] S. Kim. A note on Faxén laws for nonspherical particles. *Intl. J. Multiphase Flow*, 11(5):713–719, 1985.

[51] S. Kim and R. T. Mifflin. The resistance and mobility functions of two equal spheres in low-Reynolds-number flow. *Phys. Fluids*, 28:2033–2045, 1985.

[52] S. Kim. Singularity solutions for ellipsoids in low-Reynolds-number flows: with applications to the calculation of hydrodynamic interactions in suspensions of ellipsoids. *Intl. J. Multiphase Flow*, 12:469–491, 1986.

[53] H. Lamb. *Hydrodynamics*, 6th edition. Dover, New York, 1932.

[54] L. D. Landau and E. M. Lifshitz. *Fluid Mechanics.* Pergamon Press, New York, 1959.

[55] C. J. Lawrence and S. Weinbaum. The force on an axisymmetric body in linearized, time-dependent motion: a new memory term. *J. Fluid Mech.*, 171:209–218, 1987.

[56] L. G. Leal and E. J. Hinch. The effect of weak Brownian rotations on particles in shear flow. *J. Fluid Mech.*, 46:685–703, 1971.

[57] W. H. Liao and D. Krueger. Multipole expansion calculation of slow viscous flow about spheroids of different sizes. *J. Fluid Mech.*, 96:223–241, 1980.

[58] M. R. Maxey and J. J. Riley. Equation of motion for a small rigid sphere in a nonuniform flow. *Phys. Fluids*, 26:883–889, 1983.

[59] P. Mazur and D. Bedeaux. A generalization of Faxén's theorem to non-steady motion of a sphere through an incompressible fluid in arbitrary flow. *Physica*, 76:235–246, 1974.

[60] L. M. Milne-Thomson. *Theoretical Hydrodynamics*, 4th edition. Macmillan, New York, 1960.

[61] T. Miloh. The ultimate image singularities for external ellipsoidal harmonics. *SIAM J. Appl. Math.*, 26(2):334–344, 1974.

[62] Jr. Morrison, F. A. Electrophoresis of a particle of arbitrary shape. *J. Colloid Interface Sci.*, 34:210–214, 1970.

[63] A. Oberbeck. Über stationäre Flüssigkeitsbewegungen mit Berücksichtigung der inneren Reibung (On steady-state flow under consideration of inner friction). *J. Reine. Angew. Math.*, 81:62–80, 1876.

[64] R. W. O'Brien and L. R. White. Electrophoretic mobility of a spherical colloidal particle. *J. Chem. Soc., Faraday Trans.*, 74(2):1607–1626, 1978.

[65] L. E. Payne and W. H. Pell. The Stokes flow problem for a class of axially symmetric bodies. *J. Fluid Mech.*, 7:529–549, 1960.

[66] W. H. Pell and L. E. Payne. On Stokes flow about a torus. *Mathematika*, 7:78–92, 1960.

[67] A. Peterlin. Über die Viskosität von verdünnten Lösungen und Suspensionen in Abhängigkeit von der Teilchenform (On the viscosity of dilute solutions and suspensions governed by particle shape). *Z. Phys.*, 111:232–263, 1938.

[68] C. Pozrikidis. The instability of a moving viscous drop. *J. Fluid Mech.*, 210:1–21, 1990.

[69] J. M. Rallison. Note on the Faxén relations for a particle in Stokes flow. *J. Fluid Mech.*, 88:529–533, 1978.

[70] J. M. Rallison and A. Acrivos. A numerical study of the deformation and burst of a viscous drop in an extensional flow. *J. Fluid Mech.*, 89((1)):191–200, 1978.

[71] R. A. Sampson. On Stokes' current function. *Phil. Trans. R. Soc. Lond.*, A182:449–518, 1891.

[72] H. A. Scheraga. Non-Newtonian viscosity of solutions of ellipsoidal particles. *J. Chem. Phys.*, 23:1526–1532, 1955.

[73] R. Schmitz and B. U. Felderhof. Creeping flow about a spherical particle. *Physica*, 113A:90–102, 1982.

[74] L. E. Scriven. Dynamics of a fluid interface. *Chem. Eng. Sci.*, 12:98–108, 1960.

[75] D. J. Shaw. *Introduction to Colloid and Surface Chemistry*, 3rd edition. Butterworths, London, 1980.

[76] M. Smoluchowski. Elektrische Endosmose und Strömungsströme (Electoosmosis and current flow), in *Handbuch der Elektrizität und des Magnetismus Volume 2*, L. Graetz, Editor. Barth, Leipzig, 1921.

[77] W. E. Stewart and J. P. Sϕrensen. Hydrodynamic interaction effects in rigid dumbbell suspensions. II. Computations for steady shear flow. *Trans. Soc. Rheol.*, 16:1–13, 1972.

[78] D. Stigter. Electrophoresis of highly-charged colloidal cylinders in univalent salt solutions. 1. Mobility in transverse field. *J. Phys. Chem.*, 82:1417–1423, 1978.

[79] D. Stigter. Electrophoresis of highly-charged colloidal cylinders in univalent salt solutions. 2. Random orientation in external field and application to polyelectrolytes. *J. Phys. Chem.*, 82:1424–1429, 1978.

[80] M. Stimson and G. B. Jeffery. The motion of two spheres in a viscous fluid. *Proc. R. Soc.*, A111:110–116, 1926.

[81] S. R. Strand. *Rheological and Rheo-optical Properties of Dilute Suspensions of Dipolar Brownian Particles.* Ph.D. Dissertation, University of Wisconsin, Madison, WI, 1989.

[82] S. R. Strand, S. Kim, and S. J. Karrila. Computation of rheological properties of suspensions of rigid rods: stress growth after inception of steady shear flow. *J. Non-Newtonian Fluid Mech.*, 24:311–329, 1987.

[83] G. Strang. *Linear Algebra and Its Applications.* Academic Press, New York, 1976.

[84] G. I. Taylor. Motion of axisymmetric bodies in viscous fluids, in *Problems of Hydrodynamics and Continuum Mechanics*, SIAM, Philadelphia, 1969.

[85] M. Teubner. The motion of charged colloidal particles in electric fields. *J. Chem. Phys.*, 76:5564–5573, 1982.

[86] J. P. K. Tillett. Axial and transverse Stokes flow past slender axisymmetric bodies. *J. Fluid Mech.*, 44:401–417, 1970.

[87] E. O. Tuck. Some methods for flows past blunt slender bodies. *J. Fluid Mech.*, 18:619–635, 1964.

[88] M. van Dyke. *Perturbation Methods in Fluid Mechanics.* Parabolic Press, Stanford, CA, 1975.

[89] E. J. W. Verwey and J. Th. G. Overbeek. *Theory of The Stability of Lyophobic Colloids.* Elsevier, Amsterdam, 1948.

[90] W. A. Wegener. Hydrodynamic resistance and diffusion coefficients of a freely-hinged rod. *Biopolymers*, 19:1899–1908, 1980.

[91] P. H. Wiersema, A. L. Loeb, and J. Th. G. Overbeek. Calculation of the electrophoretic mobility of a spherical colloid particle. *J. Colloid Interface Sci.*, 22:78–99, 1966.

[92] B. J. Yoon and S. Kim. Electrophoresis of spheroidal particles. *J. Colloid and Interface Sci.*, 128:275–288, 1988.

[93] C. F. Zukoski and D. A. Saville. Electrokinetic properties of particles in concentrated suspensions. *J. Colloid Interface Sci.*, 115:422–436, 1987.

Part III

Hydrodynamic Interactions

Chapter 7

General Formulation of Resistance and Mobility Relations

7.1 Introduction

In our effort to understand the behavior of suspensions, we must take into account the effect of particle-particle interactions, for such interactions exist in all but very dilute suspensions. In this part of the book, we consider several methods for calculating *hydrodynamic interactions* between particles. Our task will be guided in part by the knowledge gained in Part II of the flow that induces or is induced by the motion of a particle.

What do we mean by hydrodynamic interaction? When two particles suspended in a viscous fluid approach each other, the motion of each particle is influenced by the other, even in the absence of interparticle interactions, such as van der Waals and electrostatic forces. The velocity field generated by the motion of one particle is transmitted through the fluid medium and influences the motion of, as well as the hydrodynamic force, torque, and stresslet on the other particle. Thus when two particles move towards each other, for example, as a result of an attractive colloidal force, the hydrodynamic interaction between the two retards the motion; using just the single particle mobility in the attractive force law overestimates the rate of aggregation.

For the sake of organization, we divide the discussion on interactions according to particle-particle and particle-wall interactions (see Figure 7.1). For each, the method of attack depends on the separation between the surfaces. For widely separated particles (the distance between closest points on the surfaces is much greater than particle size), a general asymptotic method known as the method of reflection is available. The solution can be expressed analytically as a series in terms of (the small parameter) particle size over separation.

Surfaces near contact present a far more challenging problem, both from an analytical and computational viewpoint. For rigid surfaces in relative motion, the flow in the gap region dominates and *lubrication theory* provides the leading

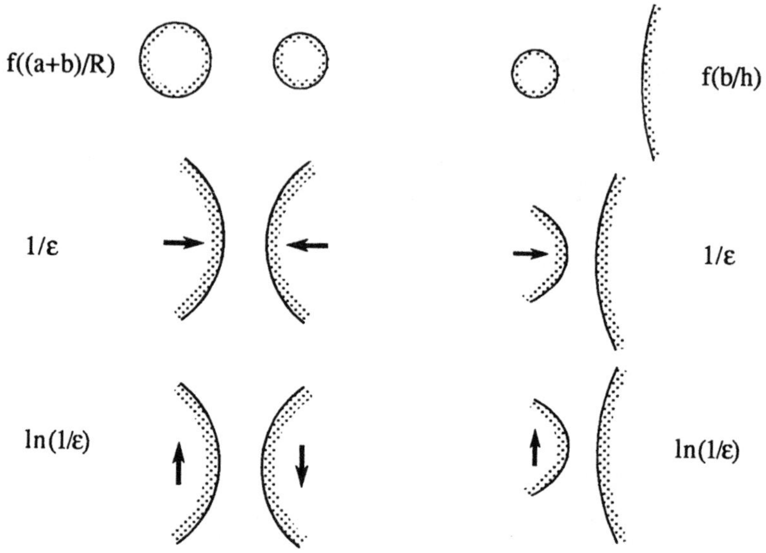

Figure 7.1: The range of hydrodynamic interaction geometries.

terms in an asymptotic expansion. Here, the small parameter ϵ is the gap distance divided by the characteristic size of the particle (usually proportional to the local radius of curvature). For squeezing motions of two rigid surfaces, this leading term is $O(\epsilon^{-1})$ and truly dominates over higher order corrections. On the other hand, for shearing motions (curved surfaces sliding past each other) the leading term is only $O(\ln \epsilon^{-1})$, and thus dominates the solution only in the mathematical sense. Usually, the next term is $O(1)$, and even if we take ϵ based on the ratio of molecular to macroscopic dimensions, $\ln \epsilon^{-1}$ is not a very large number. Unfortunately, the next term cannot be obtained by an asymptotic analysis of the region near the gap; all regions of the particle surface contribute to the $O(1)$ term, and thus a numerical solution is unavoidable.

Relative motion of two viscous drops near contact presents a new wrinkle. Consider the simplest problem of two spherical drops (high surface tension). Squeezing flows now produce a weaker singularity in the leading term of $O(\epsilon^{-1})$ since the surfaces are mobile. For the bubble, the singularity is an even weaker $O(\ln \epsilon^{-1})$. Shearing motions of one drop past another is not singular at all. The drops simply slip past each other with a finite mobility. Since all surface regions contribute to this $O(1)$ result, a matched asymptotic expansion approach is not practical.

Lubrication methods do not apply at all for rigid surfaces moving in tandem as a single rigid body. In fact, the velocity gradient is small in the gap region. In the frame moving with the particles, the gap region is stagnant. Here again, all regions near the bodies contribute to the $O(1)$ result for the resistance and mobility functions and a numerical computation is necessary. Although the flow in the gap is not singular, the gap region provides a challenge to the computational methods. These computations are of additional interest because these $O(1)$ results are closely related to the $O(1)$ corrections to the results of lubrication theory.

In the next section, we discuss the general framework for resistance and mobility problems in the multiparticle setting. As in Part II, we encounter resistance and mobility matrices and rules for transforming one to the other.

In Chapter 8, we discuss the method of reflections for interactions between widely separated particles. We emphasize the importance of direct solutions of mobility problems. For two spheres, we discuss how the method of refletions can be greatly streamlined by the use of addition theorems for spherical harmonics. In essence, the higher order terms in the expansion (which become more important at smaller separations) can be generated on a computer.

In Chapter 9, we discuss the asymptotic methods for two almost-touching particles. As alluded to earlier, we draw a distinction between lubrication flows, such as those produced by shearing or squeezing motions, and non-lubrication flows, for example, the flow produced by two almost-touching particles moving in tandem as a rigid body.

In Chapter 10, we consider interactions between a small "satellite" particle in the vicinity of a much larger particle. These can be described quite efficiently by solutions obtained by image methods. The characteristic feature is the specific relation between three length scales. The separation between the particles is much larger than the dimension of the smaller particle, but is comparable to the dimension of the larger particle. Thus the larger particle is in the "far field" of the smaller particle, so the disturbance field produced by the satellite may be described quite accurately by the first few terms of the multipole expansion. In other words, the smaller particle may be represented by a small collection of Stokes singularities. On the other hand, the smaller particle is not in the far field of the larger particle, and the details of the geometry of the larger particle will be quite important. For each Stokes singularity used in the representation of the smaller particle, we find the exact image field, to satisfy the boundary condition on the surface of the larger particle.

The interaction between two spheres is one of the more frequently used result from microhydrodynamics. Many colloidal systems contain particles that are nearly spherical, as a consequence of the domination of Brownian forces in the aggregation phase of the synthesis. The relatively simple geometry permits the application of a wealth of analytical and numerical techniques. In fact, numerical methods designed for more general problems are quite often tested against the wealth of information available for the two-sphere problem. For these reasons, we gather these results, as well as the numerical results of the later chapters, into a concise summary in Chapter 11.

In Chapter 12, we discuss particle-wall interactions, a degenerate but important case of interactions between two particles. Our discussion also includes a formal derivation of the image system for a planar wall, the so-called Lorentz reflection theorem. We conclude with a brief extension of these ideas to interactions between a drop and a fluid-fluid interface (*Lee reflection lemma*).

In Chapter 13, we present a powerful numerical method for the interaction between a small number of particles, the *boundary-multipole collocation method.* The velocity is expanded using a basis set of solutions to the Stokes equations. The coefficients of the expansion are determined by setting the boundary conditions at the collocation points. The key to the method is the availability of a good basis set, so in practice this restricts the method to interactions between particles where each particle alone has a shape that fits a separable coordinate system. An exposition on the convergence of this numerical method is provided.

7.2 Resistance and Mobility Relations

In Chapter 5 we used the linearity of the Stokes equations to affect a decomposition of the resistance and mobility relations for a single particle into smaller subproblems (force on a translating particle, torque on a rotating particle, *etc.*). Here, we apply the same approach to multiparticle hydrodynamic interactions. In the following discussion, it should be clear that the general construction for an N-particle system is, with only minor changes in the notation, essentially the same as the framework for the pair interaction problem. Accordingly, we limit our exposition to pair interactions.

As before, we are primarily interested in the force $\boldsymbol{F}_\alpha$, the torque $\boldsymbol{T}_\alpha$, and the stresslet $\boldsymbol{S}_\alpha$ *exerted by the fluid on the particles.* The particles (labeled by $\alpha = 1, 2$) are in rigid-body motion, $\boldsymbol{U}_\alpha + \boldsymbol{\omega} \times (\boldsymbol{x} - \boldsymbol{x}_\alpha)$, where $\boldsymbol{x}_\alpha$ is a reference point inside particle α, in the ambient field $\boldsymbol{v}^\infty = \boldsymbol{U}^\infty + \boldsymbol{\Omega}^\infty \times \boldsymbol{x} + \boldsymbol{E}^\infty \cdot \boldsymbol{x}$.

7.2.1 The Resistance Matrix

For the resistance problem, the specified quantities are the particle velocities and the ambient field. The unknown moments of the stress distribution over each particle surface may be expressed, as was done earlier for the single particle, as [5, 45]

$$\begin{pmatrix} \boldsymbol{F}_1 \\ \boldsymbol{F}_2 \\ \boldsymbol{T}_1 \\ \boldsymbol{T}_2 \\ \boldsymbol{S}_1 \\ \boldsymbol{S}_2 \end{pmatrix} = \mu \begin{pmatrix} \boldsymbol{A}_{11} & \boldsymbol{A}_{12} & \tilde{\boldsymbol{B}}_{11} & \tilde{\boldsymbol{B}}_{12} & \tilde{\boldsymbol{G}}_{11} & \tilde{\boldsymbol{G}}_{12} \\ \boldsymbol{A}_{21} & \boldsymbol{A}_{22} & \tilde{\boldsymbol{B}}_{21} & \tilde{\boldsymbol{B}}_{22} & \tilde{\boldsymbol{G}}_{21} & \tilde{\boldsymbol{G}}_{22} \\ \boldsymbol{B}_{11} & \boldsymbol{B}_{12} & \boldsymbol{C}_{11} & \boldsymbol{C}_{12} & \tilde{\boldsymbol{H}}_{11} & \tilde{\boldsymbol{H}}_{12} \\ \boldsymbol{B}_{21} & \boldsymbol{B}_{22} & \boldsymbol{C}_{21} & \boldsymbol{C}_{22} & \tilde{\boldsymbol{H}}_{21} & \tilde{\boldsymbol{H}}_{22} \\ \boldsymbol{G}_{11} & \boldsymbol{G}_{12} & \boldsymbol{H}_{11} & \boldsymbol{H}_{12} & \boldsymbol{M}_{11} & \boldsymbol{M}_{12} \\ \boldsymbol{G}_{21} & \boldsymbol{G}_{22} & \boldsymbol{H}_{21} & \boldsymbol{H}_{22} & \boldsymbol{M}_{21} & \boldsymbol{M}_{22} \end{pmatrix} \begin{pmatrix} \boldsymbol{v}^\infty(\boldsymbol{x}_1) - \boldsymbol{U}_1 \\ \boldsymbol{v}^\infty(\boldsymbol{x}_2) - \boldsymbol{U}_2 \\ \boldsymbol{\Omega}^\infty - \boldsymbol{\omega}_1 \\ \boldsymbol{\Omega}^\infty - \boldsymbol{\omega}_2 \\ \boldsymbol{E}^\infty \\ \boldsymbol{E}^\infty \end{pmatrix} .$$

As before, the resistance matrix contains second-rank tensors $\boldsymbol{A}$, $\boldsymbol{B}$, and $\boldsymbol{C}$; third-rank tensors $\boldsymbol{G}$ and $\boldsymbol{H}$; and a fourth-rank tensors $\boldsymbol{M}$. The subscripts on a typical tensor element in the resistance matrix, as in $\boldsymbol{P}_{\alpha\beta}$, denote that the relation is between the appropriate stress moment on particle α and the appropriate velocity quantity for particle β. As in Chapter 5, tildes are employed to highlight symmetry relations between the tensors.

The proofs for the symmetry relations for multiparticle systems follow from the Lorentz reciprocal theorem in a manner that is completely analogous to the single-particle proofs (see Exercise 7.1). The end results, with the components of $\boldsymbol{P}_{\alpha\beta}$ written as $P_{ij}^{(\alpha\beta)}$, are

$$\begin{aligned} A_{ij}^{(\alpha\beta)} &= A_{ji}^{(\beta\alpha)} && (7.1)\\ C_{ij}^{(\alpha\beta)} &= C_{ji}^{(\beta\alpha)} && (7.2)\\ M_{ijkl}^{(\alpha\beta)} &= M_{klij}^{(\beta\alpha)} && (7.3)\\ B_{ij}^{(\alpha\beta)} &= \widetilde{B}_{ji}^{(\beta\alpha)} && (7.4)\\ G_{ijk}^{(\alpha\beta)} &= \widetilde{G}_{kij}^{(\beta\alpha)} && (7.5)\\ H_{ijk}^{(\alpha\beta)} &= \widetilde{H}_{kij}^{(\beta\alpha)} . && (7.6) \end{aligned}$$

7.2.2 The Mobility Matrix

The mobility problem, in which the particle motion and stresslets are the unknowns that are to be related to the given quantities, $\boldsymbol{F}_\alpha$, $\boldsymbol{T}_\alpha$, and the ambient field arises frequently in many physical problems. We can invoke linearity again to write

$$\begin{pmatrix} \boldsymbol{v}^\infty(\boldsymbol{x}_1) - \boldsymbol{U}_1 \\ \boldsymbol{v}^\infty(\boldsymbol{x}_2) - \boldsymbol{U}_2 \\ \boldsymbol{\Omega}^\infty - \boldsymbol{\omega}_1 \\ \boldsymbol{\Omega}^\infty - \boldsymbol{\omega}_2 \\ \mu^{-1}\boldsymbol{S}_1 \\ \mu^{-1}\boldsymbol{S}_2 \end{pmatrix} = \begin{pmatrix} \boldsymbol{a}_{11} & \boldsymbol{a}_{12} & \widetilde{\boldsymbol{b}}_{11} & \widetilde{\boldsymbol{b}}_{12} & \widetilde{\boldsymbol{g}}_1 \\ \boldsymbol{a}_{21} & \boldsymbol{a}_{22} & \widetilde{\boldsymbol{b}}_{21} & \widetilde{\boldsymbol{b}}_{22} & \widetilde{\boldsymbol{g}}_2 \\ \boldsymbol{b}_{11} & \boldsymbol{b}_{12} & \boldsymbol{c}_{11} & \boldsymbol{c}_{12} & \widetilde{\boldsymbol{h}}_1 \\ \boldsymbol{b}_{21} & \boldsymbol{b}_{22} & \boldsymbol{c}_{21} & \boldsymbol{c}_{22} & \widetilde{\boldsymbol{h}}_2 \\ \boldsymbol{g}_{11} & \boldsymbol{g}_{12} & \boldsymbol{h}_{11} & \boldsymbol{h}_{12} & \boldsymbol{m}_1 \\ \boldsymbol{g}_{21} & \boldsymbol{g}_{22} & \boldsymbol{h}_{21} & \boldsymbol{h}_{22} & \boldsymbol{m}_2 \end{pmatrix} \begin{pmatrix} \mu^{-1}\boldsymbol{F}_1 \\ \mu^{-1}\boldsymbol{F}_2 \\ \mu^{-1}\boldsymbol{T}_1 \\ \mu^{-1}\boldsymbol{T}_2 \\ \boldsymbol{E}^\infty \end{pmatrix} .$$

The mobility matrix contains second-rank tensors $\boldsymbol{a}$, $\boldsymbol{b}$, and $\boldsymbol{c}$; third-rank tensors $\boldsymbol{g}$ and $\boldsymbol{h}$; and a fourth-rank tensor $\boldsymbol{m}$, as before. The same convention that was employed for the resistance matrix is used again, so that in $\boldsymbol{p}_{\alpha\beta}$, α and β denote the connection between a velocity on particle α and forces or higher order moments on particle β.

The elements of the mobility matrix also obey symmetry relations (see Exercise 7.1). These are

$$\begin{aligned} a_{ij}^{(\alpha\beta)} &= a_{ji}^{(\beta\alpha)} && (7.7)\\ c_{ij}^{(\alpha\beta)} &= c_{ji}^{(\beta\alpha)} && (7.8)\\ m_{ijkl}^{(1)} + m_{ijkl}^{(2)} &= m_{klij}^{(1)} + m_{klij}^{(2)} && (7.9)\\ b_{ij}^{(\alpha\beta)} &= \widetilde{b}_{ji}^{(\beta\alpha)} && (7.10)\\ g_{ijk}^{(1\alpha)} + g_{ijk}^{(2\alpha)} &= -\widetilde{g}_{kij}^{(\alpha)} && (7.11)\\ h_{ijk}^{(1\alpha)} + h_{ijk}^{(2\alpha)} &= -\widetilde{h}_{kij}^{(\alpha)} , && (7.12) \end{aligned}$$

with the components of $\boldsymbol{p}_{\alpha\beta}$ written as $p_{ij}^{(\alpha\beta)}$. The minus signs appear once more in the relations for $\boldsymbol{g}$ and $\boldsymbol{h}$.

7.2.3 Relations Between the Resistance and Mobility Tensors

The relation between the resistance and mobility matrices may be established by the same procedure as that used in the single-particle problem. The formal expressions have been arranged in the following set of equations so that the mobility tensors may be obtained from the resistance tensors when complete information on the latter is available.

$$\begin{pmatrix} \boldsymbol{a}_{11} & \boldsymbol{a}_{12} & \widetilde{\boldsymbol{b}}_{11} & \widetilde{\boldsymbol{b}}_{12} \\ \boldsymbol{a}_{21} & \boldsymbol{a}_{22} & \widetilde{\boldsymbol{b}}_{21} & \widetilde{\boldsymbol{b}}_{22} \\ \boldsymbol{b}_{11} & \boldsymbol{b}_{12} & \boldsymbol{c}_{11} & \boldsymbol{c}_{12} \\ \boldsymbol{b}_{21} & \boldsymbol{b}_{22} & \boldsymbol{c}_{21} & \boldsymbol{c}_{22} \end{pmatrix} = \begin{pmatrix} \boldsymbol{A}_{11} & \boldsymbol{A}_{12} & \widetilde{\boldsymbol{B}}_{11} & \widetilde{\boldsymbol{B}}_{12} \\ \boldsymbol{A}_{21} & \boldsymbol{A}_{22} & \widetilde{\boldsymbol{B}}_{21} & \widetilde{\boldsymbol{B}}_{22} \\ \boldsymbol{B}_{11} & \boldsymbol{B}_{12} & \boldsymbol{C}_{11} & \boldsymbol{C}_{12} \\ \boldsymbol{B}_{21} & \boldsymbol{B}_{22} & \boldsymbol{C}_{21} & \boldsymbol{C}_{22} \end{pmatrix}^{-1} \tag{7.13}$$

$$\begin{pmatrix} \boldsymbol{g}_{11} & \boldsymbol{g}_{12} & \boldsymbol{h}_{11} & \boldsymbol{h}_{12} \\ \boldsymbol{g}_{21} & \boldsymbol{g}_{22} & \boldsymbol{h}_{21} & \boldsymbol{h}_{22} \end{pmatrix} = \begin{pmatrix} \boldsymbol{G}_{11} & \boldsymbol{G}_{12} & \boldsymbol{H}_{11} & \boldsymbol{H}_{12} \\ \boldsymbol{G}_{21} & \boldsymbol{G}_{22} & \boldsymbol{H}_{21} & \boldsymbol{H}_{22} \end{pmatrix} \times \begin{pmatrix} \boldsymbol{a}_{11} & \boldsymbol{a}_{12} & \widetilde{\boldsymbol{b}}_{11} & \widetilde{\boldsymbol{b}}_{12} \\ \boldsymbol{a}_{21} & \boldsymbol{a}_{22} & \widetilde{\boldsymbol{b}}_{21} & \widetilde{\boldsymbol{b}}_{22} \\ \boldsymbol{b}_{11} & \boldsymbol{b}_{12} & \boldsymbol{c}_{11} & \boldsymbol{c}_{12} \\ \boldsymbol{b}_{21} & \boldsymbol{b}_{22} & \boldsymbol{c}_{21} & \boldsymbol{c}_{22} \end{pmatrix} \tag{7.14}$$

$$\begin{pmatrix} \boldsymbol{m}_1 \\ \boldsymbol{m}_2 \end{pmatrix} = \begin{pmatrix} \boldsymbol{M}_{11} + \boldsymbol{M}_{12} \\ \boldsymbol{M}_{21} + \boldsymbol{M}_{22} \end{pmatrix} + \begin{pmatrix} \boldsymbol{G}_{11} & \boldsymbol{G}_{12} & \boldsymbol{H}_{11} & \boldsymbol{H}_{12} \\ \boldsymbol{G}_{21} & \boldsymbol{G}_{22} & \boldsymbol{H}_{21} & \boldsymbol{H}_{22} \end{pmatrix} \begin{pmatrix} \widetilde{\boldsymbol{g}}_1 \\ \widetilde{\boldsymbol{g}}_2 \\ \widetilde{\boldsymbol{h}}_1 \\ \widetilde{\boldsymbol{h}}_2 \end{pmatrix} . \tag{7.15}$$

7.2.4 Axisymmetric Geometries

We have already discussed the special forms taken by the resistance and mobility tensors when the particle shape is axisymmetric. An analogous development is possible for multiparticle resistance and mobility tensors in axisymmetric geometries. An important example of such axisymmetry is encountered in the geometry of two spheres.

The forms taken by the tensors are identical to those encountered earlier in the single-particle problem. These are reproduced in Table 7.1. Labels α and β are employed to keep track of the various particle-particle interactions. Our task then is to find the relation between the *scalar* functions that describe the resistance and mobility properties. Note that since we will be dealing with matrices with scalar elements, the inversion operations that appear in the following discussion are the usual matrix inverse operations.

It is intuitively obvious, and also readily shown by algebraic arguments, that the mobility functions x depend only on the X resistance functions, that the y functions depend only on the Y functions, and that the z functions depend only on the Z functions. For example, translational motions along the particle axis induce forces that are also directed *only* along the axis. Conversely, forces directed along the axis produce translational motions *only* in the axial direction. These two problems are inverses of each other, and thus the functions $x^a_{\alpha\beta}$ are related only to $X^A_{\alpha\beta}$.

For axisymmetric geometries, problems involving translation along the axis and rotation about the axis are decoupled as far as the forces and torques are concerned (this is why the scalar functions of the form X^B do not appear in the table). Thus the reduced version of Equation 7.13 for the scalar function is block diagonal, so that we obtain

$$\begin{pmatrix} x^a_{11} & x^a_{12} \\ x^a_{12} & x^a_{22} \end{pmatrix} = \begin{pmatrix} X^A_{11} & X^A_{12} \\ X^A_{12} & X^A_{22} \end{pmatrix}^{-1} \tag{7.16}$$

and

$$\begin{pmatrix} x^c_{11} & x^c_{12} \\ x^c_{12} & x^c_{22} \end{pmatrix} = \begin{pmatrix} X^C_{11} & X^C_{12} \\ X^C_{12} & X^C_{22} \end{pmatrix}^{-1} . \tag{7.17}$$

Note that the symmetry relations have been used. The x^g functions may now be obtained from the following:

$$\begin{pmatrix} x^g_{11} & x^g_{12} \\ x^g_{21} & x^g_{22} \end{pmatrix} = \begin{pmatrix} X^G_{11} & X^G_{12} \\ X^G_{21} & X^G_{22} \end{pmatrix} \begin{pmatrix} x^a_{11} & x^a_{12} \\ x^a_{12} & x^a_{22} \end{pmatrix} . \tag{7.18}$$

The relations for x^m are given below along with the expressions for y^m and z^m.

Relations between the y and Y functions involve simultaneous inversion of the full complement of the functions associated with the A, B, and C tensors, because calculations of the force-translation and torque-rotation relations are coupled for transverse motions. Explicitly, we have

$$\begin{pmatrix} y^a_{11} & y^a_{12} & y^b_{11} & y^b_{21} \\ y^a_{12} & y^a_{22} & y^b_{12} & y^b_{22} \\ y^b_{11} & y^b_{12} & y^c_{11} & y^c_{12} \\ y^b_{21} & y^b_{22} & y^c_{12} & y^c_{22} \end{pmatrix} = \begin{pmatrix} Y^A_{11} & Y^A_{12} & Y^B_{11} & Y^B_{21} \\ Y^A_{12} & Y^A_{22} & Y^B_{12} & Y^B_{22} \\ Y^B_{11} & Y^B_{12} & Y^C_{11} & Y^C_{12} \\ Y^B_{21} & Y^B_{22} & Y^C_{12} & Y^C_{22} \end{pmatrix}^{-1} \tag{7.19}$$

$$\begin{pmatrix} y^g_{11} & y^g_{12} \\ y^g_{21} & y^g_{22} \end{pmatrix} = \begin{pmatrix} Y^G_{11} & Y^G_{12} & -Y^H_{11} & -Y^H_{12} \\ Y^G_{21} & Y^G_{22} & -Y^H_{21} & -Y^H_{22} \end{pmatrix} \begin{pmatrix} y^a_{11} & y^a_{12} \\ y^a_{12} & y^a_{22} \\ y^b_{11} & y^b_{12} \\ y^b_{21} & y^b_{22} \end{pmatrix} \tag{7.20}$$

$$\begin{pmatrix} -y^h_{11} & -y^h_{12} \\ -y^h_{21} & -y^h_{22} \end{pmatrix} = \begin{pmatrix} Y^G_{11} & Y^G_{12} & -Y^H_{11} & -Y^H_{12} \\ Y^G_{21} & Y^G_{22} & -Y^H_{21} & -Y^H_{22} \end{pmatrix} \begin{pmatrix} y^b_{11} & y^b_{21} \\ y^b_{12} & y^a_{22} \\ y^c_{11} & y^c_{12} \\ y^c_{21} & y^c_{22} \end{pmatrix} \tag{7.21}$$

$$
\begin{aligned}
A_{ij}^{(\alpha\beta)} &= X_{\alpha\beta}^{A} d_i d_j + Y_{\alpha\beta}^{A}(\delta_{ij} - d_i d_j) \\
B_{ij}^{(\alpha\beta)} &= Y_{\alpha\beta}^{B} \epsilon_{ijk} d_k \\
C_{ij}^{(\alpha\beta)} &= X_{\alpha\beta}^{C} d_i d_j + Y_{\alpha\beta}^{C}(\delta_{ij} - d_i d_j) \\
G_{ijk}^{(\alpha\beta)} &= X_{\alpha\beta}^{G}(d_i d_j - \frac{1}{3}\delta_{ij}) d_k + Y_{\alpha\beta}^{G}(d_i \delta_{jk} + d_j \delta_{ik} - 2 d_i d_j d_k) \\
H_{ijk}^{(\alpha\beta)} &= Y_{\alpha\beta}^{H}(\epsilon_{ikl} d_l d_j + \epsilon_{jkl} d_l d_i) \\
M_{ijkl}^{(\alpha\beta)} &= X_{\alpha\beta}^{M} d_{ijkl}^{(0)} + Y_{\alpha\beta}^{M} d_{ijkl}^{(1)} + Z_{\alpha\beta}^{M} d_{ijkl}^{(2)}
\end{aligned}
$$

$$
\begin{aligned}
a_{ij}^{(\alpha\beta)} &= x_{\alpha\beta}^{a} d_i d_j + y_{\alpha\beta}^{a}(\delta_{ij} - d_i d_j) \\
b_{ij}^{(\alpha\beta)} &= y_{\alpha\beta}^{b} \epsilon_{ijk} d_k \\
c_{ij}^{(\alpha\beta)} &= x_{\alpha\beta}^{c} d_i d_j + y_{\alpha\beta}^{c}(\delta_{ij} - d_i d_j) \\
g_{ijk}^{(\alpha\beta)} &= x_{\alpha\beta}^{g}(d_i d_j - \frac{1}{3}\delta_{ij}) d_k + y_{\alpha\beta}^{g}(d_i \delta_{jk} + d_j \delta_{ik} - 2 d_i d_j d_k) \\
h_{ijk}^{(\alpha\beta)} &= y_{\alpha\beta}^{h}(\epsilon_{ikl} d_l d_j + \epsilon_{jkl} d_l d_i) \\
m_{ijkl}^{(\alpha)} &= x_{\alpha\beta}^{m} d_{ijkl}^{(0)} + y_{\alpha\beta}^{m} d_{ijkl}^{(1)} + z_{\alpha\beta}^{m} d_{ijkl}^{(2)}
\end{aligned}
$$

where

$$
\begin{aligned}
d^{(0)} &= \frac{3}{2}(d_i d_j - \frac{1}{3}\delta_{ij})(d_k d_l - \frac{1}{3}\delta_{kl}) \\
d^{(1)} &= \frac{1}{2}(d_i \delta_{jl} d_k + d_j \delta_{il} d_k + d_i \delta_{jk} d_l + d_j \delta_{ik} d_l - 4 d_i d_j d_k d_l) \\
d^{(2)} &= \frac{1}{2}(\delta_{ik}\delta_{jl} + \delta_{jk}\delta_{il} - \delta_{ij}\delta_{kl} + d_i d_j \delta_{kl} + \delta_{ij} d_k d_l \\
&\quad - d_i \delta_{jl} d_k - d_j \delta_{il} d_k - d_i \delta_{jk} d_l - d_j \delta_{ik} d_l + d_i d_j d_k d_l)
\end{aligned}
$$

Table 7.1: Resistance and mobility tensors for axisymmetric geometries.

$$
\begin{aligned}
x_\alpha^m &= X_{\alpha 1}^M + X_{\alpha 2}^M - \frac{2}{3}X_{\alpha 1}^G(x_{11}^g + x_{21}^g) - \frac{2}{3}X_{\alpha 2}^G(x_{12}^g + x_{22}^g) && (7.22)\\
y_\alpha^m &= Y_{\alpha 1}^M + Y_{\alpha 2}^M - 2Y_{\alpha 1}^G(y_{11}^g + y_{21}^g) - 2Y_{\alpha 2}^G(y_{12}^g + y_{22}^g)\\
&\quad - 2Y_{\alpha 1}^H(y_{11}^h + y_{21}^h) - 2Y_{\alpha 2}^H(y_{12}^h + y_{22}^h) && (7.23)\\
z_\alpha^m &= Z_{\alpha 1}^M + Z_{\alpha 2}^M\,. && (7.24)
\end{aligned}
$$

This completes the derivation of the relation between the resistance and mobility functions for axisymmetric gemetries. In a given problem, if we know the complete set of resistance functions, we may then construct any mobility function of interest. Since it is in the resistance formulation that the boundary conditions are given, this solution strategy is quite popular. However, its usefulness is limited to problems involving only a small number of particles. In simulations involving many particles, inversion of the resistance formulation is impractical, and we shall pay particular attention to the question of direct solution of mobility problems.

Exercises

Exercise 7.1 Symmetry Relations for the Resistance and Mobility Tensors.
Use the Lorentz reciprocal theorem to derive the following symmetry relations for the resistance and mobility tensors.

Resistance tensors:

$$
\begin{aligned}
A_{ij}^{(\alpha\beta)} &= A_{ji}^{(\beta\alpha)}\\
C_{ij}^{(\alpha\beta)} &= C_{ji}^{(\beta\alpha)}\\
M_{ijkl}^{(\alpha\beta)} &= M_{klij}^{(\beta\alpha)}\\
B_{ij}^{(\alpha\beta)} &= \widetilde{B}_{ji}^{(\beta\alpha)}\\
G_{ijk}^{(\alpha\beta)} &= \widetilde{G}_{kij}^{(\beta\alpha)}\\
H_{ijk}^{(\alpha\beta)} &= \widetilde{H}_{kij}^{(\beta\alpha)}
\end{aligned}
$$

Mobility tensors:

$$
\begin{aligned}
a_{ij}^{(\alpha\beta)} &= a_{ji}^{(\beta\alpha)}\\
c_{ij}^{(\alpha\beta)} &= c_{ji}^{(\beta\alpha)}\\
m_{ijkl}^{(1)} + m_{ijkl}^{(2)} &= m_{klij}^{(1)} + m_{klij}^{(2)}\\
b_{ij}^{(\alpha\beta)} &= \widetilde{b}_{ji}^{(\beta\alpha)}\\
g_{ijk}^{(1\alpha)} + g_{ijk}^{(2\alpha)} &= -\widetilde{g}_{kij}^{(\alpha)}\\
h_{ijk}^{(1\alpha)} + h_{ijk}^{(2\alpha)} &= -\widetilde{h}_{kij}^{(\alpha)}
\end{aligned}
$$

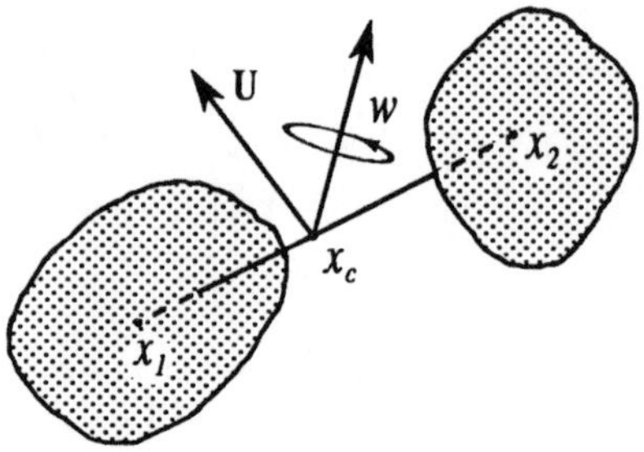

Figure 7.2: The generalized dumbbell.

Exercise 7.2 Another Definition of the Resistance Relations.
Consider once again particles (labelled by $\alpha = 1, 2$) in rigid-body motion, $\boldsymbol{U}_\alpha + \boldsymbol{\omega} \times (\boldsymbol{x} - \boldsymbol{x}_\alpha)$, where $\boldsymbol{x}_\alpha$ is a reference point inside particle α, in the ambient field $\boldsymbol{v}^\infty(\boldsymbol{x}) = \boldsymbol{U}^\infty + \boldsymbol{\Omega}^\infty \times \boldsymbol{x} + \boldsymbol{E}^\infty \cdot \boldsymbol{x}$. In the resistance formulation, we may choose to use the uniform velocity relative to the origin, $\boldsymbol{U}^\infty - \boldsymbol{U}_1$, instead of the relative velocity, $\boldsymbol{v}^\infty(\boldsymbol{x}_\alpha) - \boldsymbol{U}_\alpha$, resulting in

$$\begin{pmatrix} \boldsymbol{F}_1 \\ \boldsymbol{F}_2 \\ \boldsymbol{T}_1 \\ \boldsymbol{T}_2 \\ \boldsymbol{S}_1 \\ \boldsymbol{S}_2 \end{pmatrix} = \mu \begin{pmatrix} \mathcal{A}_{11} & \mathcal{A}_{12} & \tilde{\mathcal{B}}_{11} & \tilde{\mathcal{B}}_{12} & \tilde{\mathcal{G}}_{11} & \tilde{\mathcal{G}}_{12} \\ \mathcal{A}_{21} & \mathcal{A}_{22} & \tilde{\mathcal{B}}_{21} & \tilde{\mathcal{B}}_{22} & \tilde{\mathcal{G}}_{21} & \tilde{\mathcal{G}}_{22} \\ \mathcal{B}_{11} & \mathcal{B}_{12} & \mathcal{C}_{11} & \mathcal{C}_{12} & \tilde{\mathcal{H}}_{11} & \tilde{\mathcal{H}}_{12} \\ \mathcal{B}_{21} & \mathcal{B}_{22} & \mathcal{C}_{21} & \mathcal{C}_{22} & \tilde{\mathcal{H}}_{21} & \tilde{\mathcal{H}}_{22} \\ \mathcal{G}_{11} & \mathcal{G}_{12} & \mathcal{H}_{11} & \mathcal{H}_{12} & \mathcal{M}_{11} & \mathcal{M}_{12} \\ \mathcal{G}_{21} & \mathcal{G}_{22} & \mathcal{H}_{21} & \mathcal{H}_{22} & \mathcal{M}_{21} & \mathcal{M}_{22} \end{pmatrix} \begin{pmatrix} \boldsymbol{U}^\infty - \boldsymbol{U}_1 \\ \boldsymbol{U}^\infty - \boldsymbol{U}_2 \\ \boldsymbol{\Omega}^\infty - \boldsymbol{\omega}_1 \\ \boldsymbol{\Omega}^\infty - \boldsymbol{\omega}_2 \\ \boldsymbol{E}^\infty \\ \boldsymbol{E}^\infty \end{pmatrix}.$$

Find the relations between these tensors and the standard ones of Section 7.2.

Exercise 7.3 The Generalized Dumbbell.
Consider a generalized dumbbell formed by two particles of arbitrary shape connected by a rigid rod with negligible drag coefficient. Consider the rigid-body motion, $\boldsymbol{U} + \boldsymbol{\omega} \times (\boldsymbol{x} - \boldsymbol{x}_c)$, in a quiescent fluid. Here $\boldsymbol{x}_c$ is a point on the rod (see Figure 7.2). Relate the single particle resistance tensors $\boldsymbol{A}$ and $\boldsymbol{B}$ to the two-particle tensors of Section 7.2.

Exercise 7.4 An Important Mobility Tensor.
Consider the two particle interaction problem with $\boldsymbol{F}_2 = -\boldsymbol{F}_1$. Most pair interaction forces are of this form. In phenomena ranging from particle aggregation to Brownian diffusion, the key quanity is the relative velocity, $\boldsymbol{U}_2 - \boldsymbol{U}_1$, and it is common to relate this to the interparticle force using the mobility tensor, $\boldsymbol{\zeta}^{-1}$, as in $\boldsymbol{U}_2 - \boldsymbol{U}_1 = \boldsymbol{\zeta}^{-1} \cdot \boldsymbol{F}_2$. Relate $\boldsymbol{\zeta}^{-1}$ to the two-particle mobility tensors, $\boldsymbol{a}_{\alpha\beta}$. Derive a special form for axisymmetric geometries (*e.g.*, two spheres) in terms of the mobility functions $x^a_{\alpha\beta}$ and $y^a_{\alpha\beta}$.

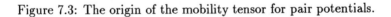

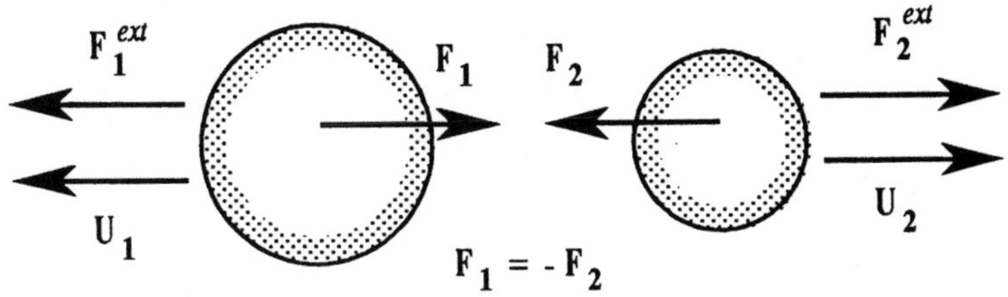

Figure 7.3: The origin of the mobility tensor for pair potentials.

Chapter 8

Particles Widely Separated: The Method of Reflections

8.1 The Far Field

A quite general asymptotic method valid for particles of arbitrary shape is at our disposal for two widely separated particles (large R/a, where R is the particle-particle separation and a is the characteristic particle size). This perturbation scheme in small a/R was developed by Smoluchowski [64] and has come to be known as the *method of reflections* [27]. A recent work by Luke [54] furnishes a proof based on energy dissipation arguments that the scheme converges for particles of quite general shape. (See also the discussion at the end of Chapter 15.)

In the zeroeth order approximation, the solution for two widely separated particles is formed by superposition of the fields produced by the isolated particle solutions. In other words, we neglect hydrodynamic interactions between particles. We have seen earlier in Chapter 3 that the disturbance field of an isolated particle may be written as a multipole expansion and that such an expansion is particularly useful in the analysis of the far field. The method of reflections is based on the idea that the ambient field about each particle consists of the original ambient field plus the disturbance field produced by the other particle(s). The method is iterative, since a correction of the ambient field about a given particle generates a new disturbance solution for that particle, which in turn modifies the ambient field about another particle.

The process of incorporating the effect of an ambient field with a new disturbance field is called a *reflection*, hence the name, *method of reflections.* The ambient field used in the reflection is denoted as the *incident* field, and the new disturbance solution is denoted as the *reflected* field. The solution procedure may be envisioned as illustrated in Figure 8.1.

For the two-particle problem, we take $\boldsymbol{x}_1$ and $\boldsymbol{x}_2$ as the reference positions for each particle and denote the isolated particle solutions as $\boldsymbol{v}_1$ and $\boldsymbol{v}_2$. Now on the surfaces S_1 and S_2 of each particle, we have the boundary conditions,

$$\boldsymbol{v}_1 = \boldsymbol{U}_1 + \boldsymbol{\omega}_1 \times (\boldsymbol{x} - \boldsymbol{x}_1) - \boldsymbol{v}^\infty \quad \text{on } S_1$$

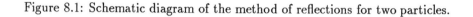

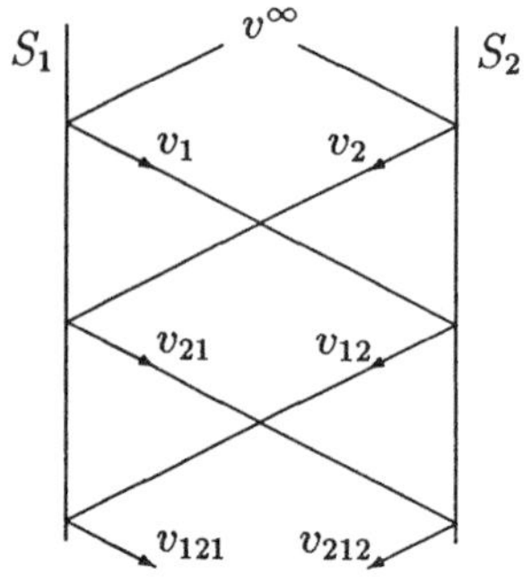

Figure 8.1: Schematic diagram of the method of reflections for two particles.

$$\boldsymbol{v}_2 = \boldsymbol{U}_2 + \boldsymbol{\omega}_2 \times (\boldsymbol{x} - \boldsymbol{x}_2) - \boldsymbol{v}^\infty \quad \text{on } S_2 ,$$

and so if we set $\boldsymbol{v} = \boldsymbol{v}^\infty + \boldsymbol{v}_1 + \boldsymbol{v}_2$, we see that the errors in the boundary condition follow as $\boldsymbol{v}_2(x)$ for a point x on the surface of particle 1 and $\boldsymbol{v}_1(x)$ for a point on particle 2. This error will be at least as small as a/R, since the decay in $\boldsymbol{v}_1$ and $\boldsymbol{v}_2$ is at most that of a Stokes monopole. The next reflection will reduce this error.

The first reflection fields $\boldsymbol{v}_{12}$ and $\boldsymbol{v}_{21}$ are defined formally as the solutions to the Stokes equations vanishing at infinity with the additional boundary conditions

$$\begin{aligned} \boldsymbol{v}_{12} &= -\boldsymbol{v}_1 \quad \text{on } S_2 \\ \boldsymbol{v}_{21} &= -\boldsymbol{v}_2 \quad \text{on } S_1 . \end{aligned}$$

Now the errors in the boundary condition scale as $\boldsymbol{v}_{12}(\boldsymbol{x})$ for a point $\boldsymbol{x}$ on the surface of particle 1 and $\boldsymbol{v}_{21}(\boldsymbol{x})$ for a point on particle 2. The error has been reduced, since the the far field values of $\boldsymbol{v}_{12}(\boldsymbol{x})$ and $\boldsymbol{v}_{21}(\boldsymbol{x})$ are smaller than the near field values of the same, and these in turn are of the same order of magnitude as the far field values of $\boldsymbol{v}_2(\boldsymbol{x})$ and $\boldsymbol{v}_1(\boldsymbol{x})$.

The higher order reflections are obtained in exactly the same manner. For example, the next reflections are obtained with $\boldsymbol{v}_{12}$ and $\boldsymbol{v}_{21}$ playing the roles previously played by $\boldsymbol{v}_2$ and $\boldsymbol{v}_1$. The reflected fields will be denoted by $\boldsymbol{v}_{121}$ and $\boldsymbol{v}_{212}$. In general, we shall keep track of all fields by assigning unique subscripts formed by augmenting the subscripts of the incident field by the subscript corresponding to the particle on which the reflection is taking place. The error in the boundary condition is given by the values taken by the highest order reflected fields evaluated at the surface of the other particle (which is also a far field value).

Finally, the most natural form for the reflected fields is the multipole expansion, with the moments determined by application of the Faxén law (on the incident fields). For simple shapes such as spheres and ellipsoids, analytical forms of the Faxén laws are available (Chapter 3), and so the method of reflections will yield analytical solutions. For the general particle shape, the Faxén relations are obtained by applying the reciprocal theorem to the dual boundary value problem. If the latter are obtained from a numerical solution, then the method of reflections will yield semi-analytical solutions, since the coefficients in the a/R expansion originate from the numerical solutions.

We now examine the formal structure of the method of reflections, in particular the differences in the method as applied to resistance problems and mobility problems. While the method works equally well for both problems, for two reasons its main application in microhydrodynamics lies in mobility problems. First, mobility problems arise more frequently in microhydrodynamics and the method produces the desired solutions directly, *i.e.*, without an inversion of the resistance problem. The second reason is that the far field forms of the mobility functions are quite accurate, even when the particles are fairly close together. The reason, as we shall see shortly, is that the higher order reflections in mobility problems consist of dipole-dipole interactions, which are much weaker than the monopole-monopole interactions usually encountered in resistance problems.

8.2 Resistance Problems

In resistance problems, particle motions and the ambient field are prescribed, and we must determine the force, torque, and higher order moments of the surface traction. Also, the zeroeth order fields $\boldsymbol{v}_1$ and $\boldsymbol{v}_2$ produce exactly, the prescribed particle motions. In subsequent reflections, the reflected fields satisfy the no-slip condition. Thus at each reflection, the multipole expansion for the reflected field always leads off with a Stokes monopole. The strength of the Stokes monopole, *i.e.*, the hydrodynamic force on the particle, always scales as the difference between the ambient velocity and the particle velocity. Since the latter is zero at the higher order reflections, the strength of the monopole will be equal to the magnitude of the incident field, which in turn is simply the far field limit of the previous reflection. Thus in resistance problems we typically get the following behavior:

$$\boldsymbol{F}_\alpha^{(n+1)} \sim O\left(\frac{a}{R}\right) \boldsymbol{F}_\beta^{(n)} ;$$

the $(n+1)$-th reflection's contribution to the hydrodynamic force on particle α will be $O(a/R)$ smaller than n-th reflection's contribution to particle β. In particular, for flow past two stationary bodies, we have

$$\boldsymbol{F}_\alpha^{(N)} \sim O\left(\frac{a}{R}\right)^N .$$

The scaling for other moments, such as the torque and stresslet, may be obtained by reference to the Faxén relations; the end result is that for a given reflection, the n-th moment of the surface traction is $O(a/R)$ smaller than the $(n-1)$-th order moment. In actual applications we must first set the desired order of accuracy in a/R. This then will determine how many multipole moments are required, with the number of required moments going down as the order of the reflection is raised. These ideas are illustrated in the following examples on resistance problems for spheres.

Example 8.1 Resistance Tensors for Two Translating Spheres. We take two nonrotating spheres centered at $\boldsymbol{x}_1$ and $\boldsymbol{x}_2$, with radii a and b, and translational velocities $\boldsymbol{U}_1$ and $\boldsymbol{U}_2$, respectively. We define the parameter $\beta = b/a$ for the ratio of sphere radii. Our goal is to calculate the force, torque, and stresslet on the spheres, accurate to $O(R^{-4})$.

The zeroeth order solution is simply the Stokes solution for the disturbance caused by an isolated, translating sphere in a uniform stream. Thus we have

$$
\begin{aligned}
\boldsymbol{v}_1 &= -\boldsymbol{F}_1^{(0)} \cdot \left\{1 + \frac{a^2}{6}\nabla^2\right\} \frac{\mathcal{G}(\boldsymbol{x}-\boldsymbol{x}_1)}{(8\pi\mu)} \\
\boldsymbol{v}_2 &= -\boldsymbol{F}_2^{(0)} \cdot \left\{1 + \frac{b^2}{6}\nabla^2\right\} \frac{\mathcal{G}(\boldsymbol{x}-\boldsymbol{x}_2)}{(8\pi\mu)} ,
\end{aligned}
$$

with

$$
\begin{aligned}
\boldsymbol{F}_1^{(0)} &= 6\pi\mu a(\boldsymbol{U}^\infty - \boldsymbol{U}_1) \\
\boldsymbol{T}_1^{(0)} &= 0 \\
\boldsymbol{S}_1^{(0)} &= 0 ,
\end{aligned}
$$

and a similar set of results for sphere 2.

The first reflection fields, $\boldsymbol{v}_{21}$ and $\boldsymbol{v}_{12}$, are expanded as

$$
\begin{aligned}
\boldsymbol{v}_{21} &= -\boldsymbol{F}_1^{(1)} \cdot \left\{1 + \frac{a^2}{6}\nabla^2\right\} \frac{\mathcal{G}(\boldsymbol{x}-\boldsymbol{x}_1)}{(8\pi\mu)} \\
&\quad + \left[\boldsymbol{S}_1^{(1)} \cdot \nabla + \frac{1}{2}\boldsymbol{T}_1^{(1)} \times \nabla\right] \cdot \frac{\mathcal{G}(\boldsymbol{x}-\boldsymbol{x}_1)}{(8\pi\mu)} + \cdots \\
\boldsymbol{v}_{12} &= -\boldsymbol{F}_2^{(1)} \cdot \left\{1 + \frac{b^2}{6}\nabla^2\right\} \frac{\mathcal{G}(\boldsymbol{x}-\boldsymbol{x}_2)}{(8\pi\mu)} \\
&\quad + \left[\boldsymbol{S}_2^{(1)} \cdot \nabla + \frac{1}{2}\boldsymbol{T}_2^{(1)} \times \nabla\right] \cdot \frac{\mathcal{G}(\boldsymbol{x}-\boldsymbol{x}_2)}{(8\pi\mu)} + \cdots ,
\end{aligned}
$$

with multipole moments obtained from the Faxén laws as

$$
\begin{aligned}
\boldsymbol{F}_1^{(1)} &= 6\pi\mu a(1 + \frac{a^2}{6}\nabla^2)\boldsymbol{v}_2|_{x=x_1} \\
&= \boldsymbol{F}_2^{(0)} \cdot \left[\left(-\frac{3}{2}\frac{a}{R} + \frac{1}{2}(1+\beta^2)\left(\frac{a}{R}\right)^3\right)\boldsymbol{dd}\right. \\
&\quad \left. - \left(\frac{3}{4}\frac{a}{R} + \frac{1}{4}(1+\beta^2)\left(\frac{a}{R}\right)^3\right)(\boldsymbol{\delta} - \boldsymbol{dd})\right]
\end{aligned}
$$

$$
\begin{aligned}
\boldsymbol{T}_1^{(1)} &= 4\pi\mu a^3 \nabla \times \boldsymbol{v}_2|_{x=x_1} \\
&= \frac{a^3}{R^2}\boldsymbol{F}_2^{(0)} \times \boldsymbol{d} \\
\boldsymbol{S}_1^{(1)} &= \frac{20}{3}\pi\mu a^3 (1 + \frac{a^2}{10}\nabla^2) e_2|_{x=x_1} \\
&= \left(-\frac{5}{2}\frac{a^3}{R^2} + \frac{3}{2}\frac{a^5}{R^4}(1+\frac{5}{3}\beta^2)\right)(\boldsymbol{dd} - \frac{1}{3}\boldsymbol{\delta})\boldsymbol{d}\cdot\boldsymbol{F}_2^{(0)} \\
&- \frac{1}{2}\frac{a^5}{R^4}(1+\frac{5}{3}\beta^2)(\boldsymbol{F}_2^{(0)}\boldsymbol{d} + \boldsymbol{d}\boldsymbol{F}_2^{(0)} - 2\boldsymbol{ddd}\cdot\boldsymbol{F}_2^{(0)}) \; .
\end{aligned}
$$

Here, $\boldsymbol{d}$ denotes the unit vector, $(\boldsymbol{x}_2 - \boldsymbol{x}_1)/|\boldsymbol{x}_2 - \boldsymbol{x}_1|$. The algebraic reductions were obtained by inserting the appropriate expressions for $\mathcal{G}$ and its derivatives. Note that $\boldsymbol{T}^{(1)}$ does not have a term of $O(R^{-4})$ because $\nabla^2\mathcal{G}$ is irrotational. Expressions for $\boldsymbol{F}_2^{(1)}$, $\boldsymbol{T}_2^{(1)}$, and $\boldsymbol{S}_2^{(1)}$ may be obtained from the preceding by switching a and b and the indices 1 and 2.

At the next (second) reflection the fields $\boldsymbol{v}_{121}$ and $\boldsymbol{v}_{212}$ are expanded as

$$
\begin{aligned}
\boldsymbol{v}_{121} &= -\boldsymbol{F}_1^{(2)}\cdot\left\{1 + \frac{a^2}{6}\nabla^2\right\}\frac{\mathcal{G}(\boldsymbol{x}-\boldsymbol{x}_1)}{(8\pi\mu)} \\
&+ \left[\boldsymbol{S}_1^{(2)}\cdot\nabla + \frac{1}{2}\boldsymbol{T}_1^{(2)}\times\nabla\right]\cdot\frac{\mathcal{G}(\boldsymbol{x}-\boldsymbol{x}_1)}{(8\pi\mu)} + \cdots \\
\boldsymbol{v}_{212} &= -\boldsymbol{F}_2^{(2)}\cdot\left\{1 + \frac{b^2}{6}\nabla^2\right\}\frac{\mathcal{G}(\boldsymbol{x}-\boldsymbol{x}_2)}{(8\pi\mu)} \\
&+ \left[\boldsymbol{S}_2^{(2)}\cdot\nabla + \frac{1}{2}\boldsymbol{T}_2^{(2)}\times\nabla\right]\cdot\frac{\mathcal{G}(\boldsymbol{x}-\boldsymbol{x}_2)}{(8\pi\mu)} + \cdots \; ,
\end{aligned}
$$

with multipole moments obtained from the Faxén laws as

$$
\begin{aligned}
\boldsymbol{F}_1^{(2)} &= 6\pi\mu a(1 + \frac{a^2}{6}\nabla^2)\boldsymbol{v}_{12}|_{x=x_1} \\
&= \boldsymbol{F}_2^{(1)}\cdot\left[\left(-\frac{3}{2}\frac{a}{R} + \frac{1}{2}(1+\beta^2)\left(\frac{a}{R}\right)^3\right)\boldsymbol{dd}\right. \\
&\quad \left. - \left(\frac{3}{4}\frac{a}{R} + \frac{1}{4}(1+\beta^2)\left(\frac{a}{R}\right)^3\right)(\boldsymbol{\delta} - \boldsymbol{dd})\right] \\
&= \boldsymbol{F}_1^{(0)}\cdot\left[\left(\frac{9}{4}\beta\left(\frac{a}{R}\right)^2 - (\frac{3\beta}{2} + \frac{3\beta^3}{4})\left(\frac{a}{R}\right)^4\right)\boldsymbol{dd}\right. \\
&\quad \left. + \left(\frac{9}{16}\beta\left(\frac{a}{R}\right)^2 + (\frac{3\beta}{8} + \frac{3\beta^3}{16})\left(\frac{a}{R}\right)^4\right)(\boldsymbol{\delta} - \boldsymbol{dd})\right] \\
\boldsymbol{T}_1^{(2)} &= 4\pi\mu a^3\nabla\times\boldsymbol{v}_{12}|_{x=x_1} \\
&= \frac{a^3}{R^2}\boldsymbol{F}_2^{(1)}\times\boldsymbol{d} \\
&= -\frac{3\beta}{4}\frac{a^4}{R^3}\boldsymbol{F}_1^{(0)}\times\boldsymbol{d} + O(R^{-5})
\end{aligned}
$$

$$
\begin{aligned}
\boldsymbol{S}_1^{(2)} &= \frac{20}{3}\pi\mu a^3(1+\frac{a^2}{10}\nabla^2)\boldsymbol{e}_{12}|_{x=x_1} \\
&= -\frac{5}{2}\frac{a^3}{R^2}(\boldsymbol{dd}-\frac{1}{3}\boldsymbol{\delta})\boldsymbol{d}\cdot\boldsymbol{F}_2^{(1)} + O(R^{-5}) \\
&= \frac{15\beta}{4}\frac{a^4}{R^3}(\boldsymbol{dd}-\frac{1}{3}\boldsymbol{\delta})\boldsymbol{d}\cdot\boldsymbol{F}_1^{(0)} + O(R^{-5})\ .
\end{aligned}
$$

As before, expressions for $\boldsymbol{F}_2^{(2)}$, $\boldsymbol{T}_2^{(2)}$, and $\boldsymbol{S}_2^{(2)}$ may be obtained from the analogous quantity for sphere 1 by switching a and b and the indices 1 and 2.

Since the final results are desired accurate to $O(R^{-4})$, the contributions from the third and fourth reflections may be obtained by retaining only the leading order (Stokes monopole) terms at the earlier reflections. The results are

$$
\begin{aligned}
\boldsymbol{F}_1^{(3)} &= \left(-\frac{3a}{4}\mathcal{G}(\boldsymbol{x}_1-\boldsymbol{x}_2)\right)\cdot\left(-\frac{3b}{4}\mathcal{G}(\boldsymbol{x}_2-\boldsymbol{x}_1)\right)\cdot\left(-\frac{3a}{4}\mathcal{G}(\boldsymbol{x}_1-\boldsymbol{x}_2)\right)\cdot\boldsymbol{F}_2^0 \\
&= -\boldsymbol{F}_2^{(0)}\cdot\left[\frac{27\beta}{8}\left(\frac{a}{R}\right)^3\boldsymbol{dd}+\frac{27\beta}{64}\left(\frac{a}{R}\right)^3(\boldsymbol{\delta}-\boldsymbol{dd})\right] \\
\boldsymbol{T}_1^{(3)} &= \frac{9\beta}{16}\frac{a^5}{R^4}\boldsymbol{F}_2^{(0)}\times\boldsymbol{d}+O(R^{-6}) \\
\boldsymbol{S}_1^{(3)} &= -\frac{45\beta}{32}\frac{a^5}{R^4}(\boldsymbol{dd}-\frac{1}{3}\boldsymbol{\delta})\boldsymbol{d}\cdot\boldsymbol{F}_2^{(0)}+O(R^{-6})
\end{aligned}
$$

and

$$
\begin{aligned}
\boldsymbol{F}_1^{(4)} &= \left(-\frac{3a}{4}\mathcal{G}(\boldsymbol{x}_1-\boldsymbol{x}_2)\right)\cdot\left(-\frac{3b}{4}\mathcal{G}(\boldsymbol{x}_2-\boldsymbol{x}_1)\right)\cdot\left(-\frac{3a}{4}\mathcal{G}(\boldsymbol{x}_1-\boldsymbol{x}_2)\right) \\
&\qquad \cdot\left(-\frac{3b}{4}\mathcal{G}(\boldsymbol{x}_2-\boldsymbol{x}_1)\right)\cdot\boldsymbol{F}_1^0 \\
&= \boldsymbol{F}_1^{(0)}\cdot\left[\frac{81\beta^2}{16}\left(\frac{a}{R}\right)^4\boldsymbol{dd}+\frac{81\beta^2}{256}\left(\frac{a}{R}\right)^4(\boldsymbol{\delta}-\boldsymbol{dd})\right] \\
\boldsymbol{T}_1^{(4)} &\sim O(R^{-5}) \\
\boldsymbol{S}_1^{(4)} &\sim O(R^{-5})\ .
\end{aligned}
$$

The combined result for $\boldsymbol{F}_1$, $\boldsymbol{T}_1$, and $\boldsymbol{S}_1$ obtained by adding the contributions from the zeroeth through the fourth reflection is

$$
\begin{aligned}
\boldsymbol{F}_1 &= \boldsymbol{F}_1^{(0)}\cdot\left[\left(1+\frac{9}{4}\beta\left(\frac{a}{R}\right)^2-(\frac{3\beta}{2}+\frac{81\beta^2}{16}+\frac{3\beta^3}{4})\left(\frac{a}{R}\right)^4\right)\boldsymbol{dd}\right. \\
&\qquad \left.+\left(1+\frac{9}{16}\beta\left(\frac{a}{R}\right)^2+(\frac{3\beta}{8}-\frac{81\beta^2}{256}+\frac{3\beta^3}{16})\left(\frac{a}{R}\right)^4\right)(\boldsymbol{\delta}-\boldsymbol{dd})\right] \\
&+\ \boldsymbol{F}_2^{(0)}\cdot\left[\left(-\frac{3}{2}\frac{a}{R}+\frac{1}{2}(1-\frac{27\beta}{4}+\beta^2)\left(\frac{a}{R}\right)^3\right)\boldsymbol{dd}\right. \\
&\qquad \left.-\left(\frac{3}{4}\frac{a}{R}+\frac{1}{4}(1+\frac{27\beta}{16}+\beta^2)\left(\frac{a}{R}\right)^3\right)(\boldsymbol{\delta}-\boldsymbol{dd})\right]
\end{aligned}
$$

$$
\begin{aligned}
\boldsymbol{T}_1 &= -\frac{3\beta}{4}\frac{a^4}{R^3}\boldsymbol{F}_1^{(0)} \times \boldsymbol{d} + \left(\frac{a^3}{R^2} + \frac{9\beta}{16}\frac{a^5}{R^4}\right)\boldsymbol{F}_2^{(0)} \times \boldsymbol{d} \\
\boldsymbol{S}_1 &= \frac{15\beta}{4}\frac{a^4}{R^3}(\boldsymbol{dd} - \frac{1}{3}\boldsymbol{\delta})\boldsymbol{d}\cdot\boldsymbol{F}_1^{(0)} \\
&+ \left(-\frac{5}{2}\frac{a^3}{R^2} + \frac{3}{2}(1 - \frac{15\beta}{16} + \frac{5\beta^2}{3})\frac{a^5}{R^4}\right)(\boldsymbol{dd} - \frac{1}{3}\boldsymbol{\delta})\boldsymbol{d}\cdot\boldsymbol{F}_2^{(0)} \\
&- \frac{1}{2}(1 + \frac{5}{3}\beta^2)\frac{a^5}{R^4}(\boldsymbol{F}_2^{(0)}\boldsymbol{d} + \boldsymbol{d}\boldsymbol{F}_2^{(0)} - 2\boldsymbol{ddd}\cdot\boldsymbol{F}_2^{(0)}) ,
\end{aligned}
$$

which is consistent with the forms obtained for axisymmetric geometries in the previous chapter ◊

The preceding example illustrates the general statements made earlier about the method of reflections. Later on we shall consider a much more efficient implementation of the method of reflection for spheres that is based on the addition theorem for spherical harmonics.

We know quite a bit about pair interactions. It would be very convenient if multiparticle interactions could be described in a pairwise-additive fashion. In the next example, we solve a three-body problem by the method of reflections, and find that three-body effects come in at $O(a/R)^2$. This is a fairly strong interaction, which implies that pairwise addition of the resistance functions is not as accurate as the corresponding treatments in molecular simulations.

Example 8.2 Three-Body Effects.
Consider three identical, fixed, nonrotating spheres with centers at $\boldsymbol{x}_1$, $\boldsymbol{x}_2$, and $\boldsymbol{x}_3$, in a uniform stream $\boldsymbol{U}^\infty$. We can easily generalize the analysis to handle unequal spheres. As in the two-sphere problem, we label reflection fields by appending the sphere-number to the label of the incident field.

The zeroeth order solution is simply the linear superposition of the three Stokes solutions:

$$
\begin{aligned}
\boldsymbol{v}_1 &= -\boldsymbol{F}^{(0)}\cdot\left\{1 + \frac{a^2}{6}\nabla^2\right\}\frac{\mathcal{G}(\boldsymbol{x} - \boldsymbol{x}_1)}{(8\pi\mu)} \\
\boldsymbol{v}_2 &= -\boldsymbol{F}^{(0)}\cdot\left\{1 + \frac{a^2}{6}\nabla^2\right\}\frac{\mathcal{G}(\boldsymbol{x} - \boldsymbol{x}_2)}{(8\pi\mu)} \\
\boldsymbol{v}_3 &= -\boldsymbol{F}^{(0)}\cdot\left\{1 + \frac{a^2}{6}\nabla^2\right\}\frac{\mathcal{G}(\boldsymbol{x} - \boldsymbol{x}_3)}{(8\pi\mu)} ,
\end{aligned}
$$

with

$$
\boldsymbol{F}^{(0)} = 6\pi\mu a\boldsymbol{U}^\infty .
$$

There are six reflection fields — $\boldsymbol{v}_{21}$, $\boldsymbol{v}_{31}$, $\boldsymbol{v}_{12}$, $\boldsymbol{v}_{32}$, $\boldsymbol{v}_{13}$, and $\boldsymbol{v}_{23}$ — at the next reflection, since each sphere sees incident fields from two sources. The expression for $\boldsymbol{v}_{21}$ is identical to that encountered in the two-sphere problem:

$$
\begin{aligned}
\boldsymbol{v}_{21} &= -\boldsymbol{F}_1^{(2)}\cdot\left\{1 + \frac{a^2}{6}\nabla^2\right\}\frac{\mathcal{G}(\boldsymbol{x} - \boldsymbol{x}_1)}{(8\pi\mu)} \\
&+ \left[\boldsymbol{S}_1^{(2)}\cdot\nabla + \frac{1}{2}\boldsymbol{T}_1^{(2)} \times \nabla\right]\cdot\frac{\mathcal{G}(\boldsymbol{x} - \boldsymbol{x}_1)}{(8\pi\mu)} + \cdots ,
\end{aligned}
$$

with multipole moments[1] obtained from the Faxén laws as

$$\begin{aligned}
\boldsymbol{F}_1^{(2)} &= 6\pi\mu a(1+\frac{a^2}{6}\nabla^2)\boldsymbol{v}_2|_{x=x_1} \\
&= \boldsymbol{F}^{(0)}\cdot\left[\left(-\frac{3}{2}\frac{a}{R_{12}}+\left(\frac{a}{R_{12}}\right)^3\right)\boldsymbol{d}_{12}\boldsymbol{d}_{12}\right. \\
&\quad \left. -\left(\frac{3}{4}\frac{a}{R_{12}}+\frac{1}{2}\left(\frac{a}{R_{12}}\right)^3\right)(\boldsymbol{\delta}-\boldsymbol{d}_{12}\boldsymbol{d}_{12})\right] \\
\boldsymbol{T}_1^{(2)} &= 4\pi\mu a^3\nabla\times\boldsymbol{v}_2|_{x=x_1} \\
&= \frac{a^3}{R_{12}^2}\boldsymbol{F}^{(0)}\times\boldsymbol{d}_{12} \\
\boldsymbol{S}_1^{(2)} &= \frac{20}{3}\pi\mu a^3(1+\frac{a^2}{10}\nabla^2)\boldsymbol{e}_2|_{x=x_1} \\
&= \left(-\frac{5}{2}\frac{a^3}{R_{12}^2}+4\frac{a^5}{R_{12}^4}\right)(\boldsymbol{d}_{12}\boldsymbol{d}_{12}-\frac{1}{3}\boldsymbol{\delta})\boldsymbol{d}_{12}\cdot\boldsymbol{F}^{(0)} \\
&\quad -\frac{4}{3}\frac{a^5}{R_{12}^4}(\boldsymbol{F}^{(0)}\boldsymbol{d}_{12}+\boldsymbol{d}_{12}\boldsymbol{F}^{(0)}-2\boldsymbol{d}_{12}\boldsymbol{d}_{12}\boldsymbol{d}_{12}\cdot\boldsymbol{F}^{(0)}) ,
\end{aligned}$$

with $R_{12} = |\boldsymbol{x}_1 - \boldsymbol{x}_2|$ and $\boldsymbol{d}_{12} = (\boldsymbol{x}_1 - \boldsymbol{x}_2)/R_{12}$. The corresponding expressions for the other five reflection fields may be obtained from these expressions by the appropriate permutation of sphere labels. Note that three-body effects have not entered at this order (each quantity has at most two of the three sphere labels).

At the next (second) reflection, let us examine only reflections at sphere 1, with the understanding that the same events occur at spheres 2 and 3. Of the six first-reflection fields, only the four that emanate from either sphere 2 or 3 may act as incident fields on sphere 1. For each incident field, we have a reflection field, namely, $\boldsymbol{v}_{121}$, $\boldsymbol{v}_{131}$, $\boldsymbol{v}_{231}$, and $\boldsymbol{v}_{321}$. The last two are three-body effects and we see that if $R_{12} \sim R_{13} \sim R_{23} \sim R$, then three-body effects come in at $O(R^{-2})$.

For each reflection field, we use the multipole expansion, with moments determined from Faxén laws. The expressions for $\boldsymbol{v}_{121}$ and $\boldsymbol{v}_{231}$ are given below:

$$\begin{aligned}
\boldsymbol{v}_{121} &= -\boldsymbol{F}_1^{(12)}\cdot\left\{1+\frac{a^2}{6}\nabla^2\right\}\frac{\mathcal{G}(\boldsymbol{x}-\boldsymbol{x}_1)}{(8\pi\mu)} \\
&\quad +\left[\boldsymbol{S}_1^{(12)}\cdot\nabla+\frac{1}{2}\boldsymbol{T}_1^{(12)}\times\nabla\right]\cdot\frac{\mathcal{G}(\boldsymbol{x}-\boldsymbol{x}_1)}{(8\pi\mu)}+\cdots ,
\end{aligned}$$

with

$$\begin{aligned}
\boldsymbol{F}_1^{(12)} &= 6\pi\mu a(1+\frac{a^2}{6}\nabla^2)\boldsymbol{v}_{12}|_{x=x_1} \\
\boldsymbol{T}_1^{(12)} &= 4\pi\mu a^3\nabla\times\boldsymbol{v}_{12}|_{x=x_1} \\
\boldsymbol{S}_1^{(12)} &= \frac{20}{3}\pi\mu a^3(1+\frac{a^2}{10}\nabla^2)\boldsymbol{e}_{12}|_{x=x_1} ,
\end{aligned}$$

[1]In the three-sphere problem, we depart from the convention used earlier and indicate the reflection order by using the same label as the incident field.

$$
\begin{aligned}
\boldsymbol{v}_{231} &= -\boldsymbol{F}_1^{(23)} \cdot \left\{1 + \frac{a^2}{6}\nabla^2\right\} \frac{\mathcal{G}(\boldsymbol{x} - \boldsymbol{x}_1)}{(8\pi\mu)} \\
&\quad + \left[\boldsymbol{S}_1^{(23)} \cdot \nabla + \frac{1}{2}\boldsymbol{T}_1^{(23)} \times \nabla\right] \cdot \frac{\mathcal{G}(\boldsymbol{x} - \boldsymbol{x}_1)}{(8\pi\mu)} + \cdots ,
\end{aligned}
$$

with

$$
\begin{aligned}
\boldsymbol{F}_1^{(23)} &= 6\pi\mu a(1 + \frac{a^2}{6}\nabla^2)\boldsymbol{v}_{23}|_{x=x_1} \\
\boldsymbol{T}_1^{(23)} &= 4\pi\mu a^3 \nabla \times \boldsymbol{v}_{23}|_{x=x_1} \\
\boldsymbol{S}_1^{(23)} &= \frac{20}{3}\pi\mu a^3(1 + \frac{a^2}{10}\nabla^2)\boldsymbol{e}_{23}|_{x=x_1} .
\end{aligned}
$$

The expressions for $\boldsymbol{v}_{131}$ and $\boldsymbol{v}_{321}$ are obtained in a similar manner.

The force on sphere 1 may be written as

$$
\begin{aligned}
\boldsymbol{F}_1 &= 6\pi\mu a\boldsymbol{U}^\infty + \boldsymbol{F}_1^{(2)} + \boldsymbol{F}_1^{(3)} \\
&\quad + \boldsymbol{F}_1^{(12)} + \boldsymbol{F}_1^{(13)} + \boldsymbol{F}_1^{(23)} + \boldsymbol{F}_1^{(32)} \\
&\quad + \boldsymbol{F}_1^{(123)} + \boldsymbol{F}_1^{(132)} + \ldots + \boldsymbol{F}_1^{(323)} \\
&\quad + \boldsymbol{F}_1^{(1212)} + \boldsymbol{F}_1^{(1232)} + \ldots + \boldsymbol{F}_1^{(3213)} \\
&\quad + \boldsymbol{F}_1^{(12123)} + \boldsymbol{F}_1^{(12132)} + \ldots + \boldsymbol{F}_1^{(32323)} ,
\end{aligned}
$$

where the contributions from the zeroeth through the fifth reflection are as indicated. The tracking of the reflection fields (there are $3 \times 2^{n+1}$ fields at the n-th reflection) is a straightforward exercise in bookkeeping skills; the essential aspects of the method differ very little from that encountered earlier in the two-sphere problem.

For spheres centered at the vertices of an equilateral triangle (so that $R_{12} = R_{13} = R_{23} = R$, see Figure 8.2), the final result is quite simple, since the single parameter R/a describes the problem geometry, and the force on sphere 1 may be written as [46]

$$
\boldsymbol{F}_1 = 6\pi\mu a\boldsymbol{U}^\infty \cdot [f_1\boldsymbol{e}_1\boldsymbol{e}_1 + f_2\boldsymbol{e}_2\boldsymbol{e}_2 + f_3\boldsymbol{e}_3\boldsymbol{e}_3] ,
$$

with

$$
\boldsymbol{e}_1 = \frac{1}{\sqrt{3}}(\boldsymbol{d}_{12} + \boldsymbol{d}_{13}) , \qquad \boldsymbol{e}_2 = \boldsymbol{d}_{23} , \qquad \boldsymbol{e}_3 = \boldsymbol{e}_1 \times \boldsymbol{e}_2 ,
$$

and

$$
\begin{aligned}
f_1 &= 1 - \frac{21}{8}\left(\frac{a}{R}\right) + \frac{45}{8}\left(\frac{a}{R}\right)^2 - \frac{6191}{512}\left(\frac{a}{R}\right)^3 + \frac{135327}{4096}\left(\frac{a}{R}\right)^4 \\
&\quad - \frac{689823}{8192}\left(\frac{a}{R}\right)^5 + \cdots \\
f_2 &= 1 - \frac{15}{8}\left(\frac{a}{R}\right) + \frac{153}{32}\left(\frac{a}{R}\right)^2 - \frac{5447}{512}\left(\frac{a}{R}\right)^3 + \frac{102885}{4096}\left(\frac{a}{R}\right)^4 \\
&\quad - \frac{1085373}{16384}\left(\frac{a}{R}\right)^5 + \cdots \\
f_3 &= 1 - \frac{3}{2}\left(\frac{a}{R}\right) + \frac{9}{4}\left(\frac{a}{R}\right)^2 - \frac{35}{8}\left(\frac{a}{R}\right)^3 + \frac{165}{16}\left(\frac{a}{R}\right)^4 - \frac{675}{32}\left(\frac{a}{R}\right)^5 + \cdots
\end{aligned}
$$

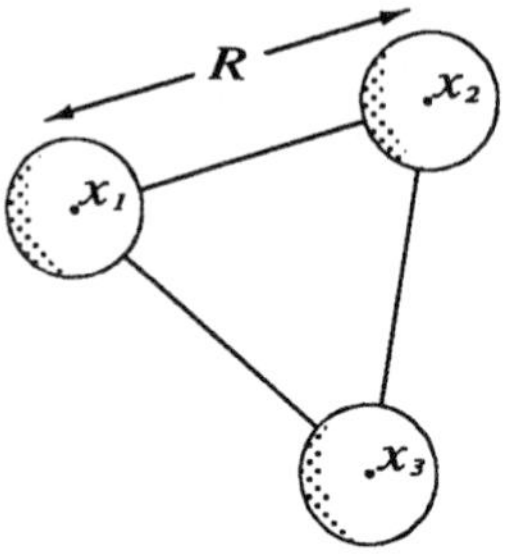

Figure 8.2: The three-sphere geometry.

The effects of three-body interactions are shown in Figure 8.2, where these results are plotted along with the pairwise additive approximations. In general, resistance functions are not pairwise additive in the far field. In Exercise 8.3, the corresponding mobility problem is considered, and the reader will see that the analysis is considerably easier. ◇

8.3 Mobility Problems

In mobility problems, particle motions in a specified ambient field are to be determined. The motions arise from prescribed forces and torques on each particle. We may also want to calculate higher order moments of the traction such as the stresslet. As in the resistance problem, we start with the single-particle solutions $\boldsymbol{v}_1$ and $\boldsymbol{v}_2$, but in the reflection procedure there are essential differences.

In mobility problems, the zeroeth order fields $\boldsymbol{v}_1$ and $\boldsymbol{v}_2$ produce exactly the *prescribed forces and torques.* Therefore, in subsequent reflections the particle motions must be such that the reflected fields are force-free and torque-free. These translational and rotational velocities will scale as the ambient velocity and velocity gradient, respectively. Furthermore, the reflected field's multipole expansion will lead off with a *stresslet* of strength of the same order as the ambient velocity gradient. Thus in mobility problems we typically get the following behavior:

$$\boldsymbol{U}_\alpha^{(N+1)} \sim O\left(\frac{a}{R}\right)^3 \boldsymbol{U}_\beta^{(N)} ,$$

that is, the $(N+1)$-th reflection's contribution to the translational velocity of particle α will be $O(a/R)^3$ smaller than the N-th reflection's contribution to the velocity of particle β. Two powers of a/R are due to the far field decay of the stresslet field, while the additional factor of a/R is due to the relative magnitudes of $\boldsymbol{U}_\beta^{(N)}$ and $\boldsymbol{S}_\beta^{(N)}$. The scaling for other higher order moments of

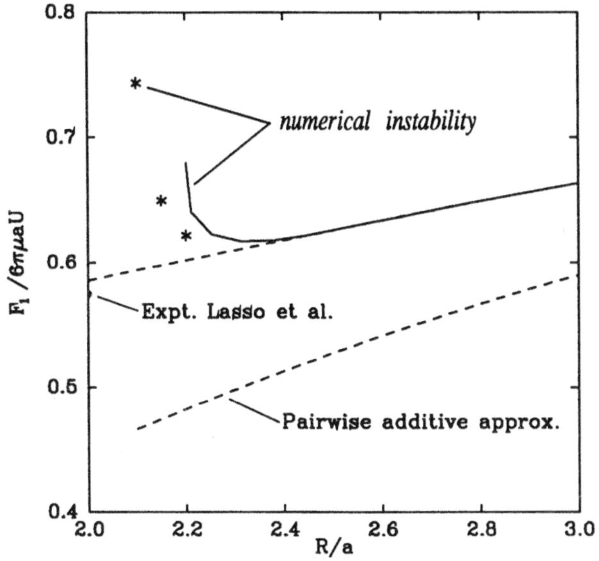

Figure 8.3: Three-sphere interactions and pairwise additive approximations.

the traction is as before and the n-th moment of the surface traction will be $O(a/R)$ smaller than the $(n-1)$-th order moment.

The end result is that, *given the same amount of information concerning the Faxén relations for the moments, and after the same number of reflections, the results for the mobility functions will be accurate to higher order in a/R than the result for the resistance functions.* Thus, contrary to one's initial perception and the approach in the older literature, when applying the method of reflections, one should first solve the entire collection of mobility problems and then invert these if the resistance solutions are also required. These ideas will be illustrated with the mobility calculation for spheres. We will perform the same number of reflections as in the resistance problem, but will obtain solutions accurate to $O(a/R)^7$. These may then be inverted to obtain the resistance relations accurate to $O(a/R)^7$.

Example 8.3 Two Torque-Free Spheres Acted on by External Forces in a Quiescent Fluid.
We shall derive, accurate to $O(R^{-5})$, and errors of $O(R^{-7})$, the translational and rotational velocities of two torque-free spheres acted on by external forces in a quiescent fluid (this includes the sedimentation problem).

We take two spheres centered at $\boldsymbol{x}_1$ and $\boldsymbol{x}_2$, with radii a and b, and denote the external forces as $\boldsymbol{F}_1^e$ and $\boldsymbol{F}_2^e$, respectively. As in the resistance problem, we define $\beta = b/a$. The zeroeth order solution is the Stokes solution for an isolated, translating sphere, *i.e.*,

$$\begin{aligned} \boldsymbol{v}_1 &= \boldsymbol{F}_1^e \cdot \left\{1 + \frac{a^2}{6}\nabla^2\right\} \frac{\mathcal{G}(\boldsymbol{x}-\boldsymbol{x}_1)}{(8\pi\mu)} \\ \boldsymbol{v}_2 &= \boldsymbol{F}_2^e \cdot \left\{1 + \frac{b^2}{6}\nabla^2\right\} \frac{\mathcal{G}(\boldsymbol{x}-\boldsymbol{x}_2)}{(8\pi\mu)} , \end{aligned}$$

with

$$\begin{aligned} 6\pi\mu a \boldsymbol{U}_1^{(0)} &= \boldsymbol{F}_1^e = -\boldsymbol{F}_1 \\ \boldsymbol{T}_1^{(0)} &= 0 \\ 6\pi\mu b \boldsymbol{U}_2^{(0)} &= \boldsymbol{F}_2^e = -\boldsymbol{F}_2 \\ \boldsymbol{T}_2^{(0)} &= 0 . \end{aligned}$$

The first reflection fields, $\boldsymbol{v}_{21}$ and $\boldsymbol{v}_{12}$, are expanded as

$$\begin{aligned} \boldsymbol{v}_{21} &= (\boldsymbol{S}_1^{(1)} \cdot \nabla) \cdot \frac{\mathcal{G}(\boldsymbol{x}-\boldsymbol{x}_1)}{(8\pi\mu)} + \cdots \\ \boldsymbol{v}_{12} &= (\boldsymbol{S}_2^{(1)} \cdot \nabla) \cdot \frac{\mathcal{G}(\boldsymbol{x}-\boldsymbol{x}_2)}{(8\pi\mu)} + \cdots , \end{aligned}$$

with particle motions and stresslets obtained from the Faxén laws as

$$\begin{aligned} \boldsymbol{U}_1^{(1)} &= (1 + \frac{a^2}{6}\nabla^2)\boldsymbol{v}_2|_{x=x_1} \\ &= \boldsymbol{U}_2^{(0)} \cdot \left[\left(\frac{3\beta}{2}\frac{a}{R} - \frac{1}{2}\beta(1+\beta^2)\left(\frac{a}{R}\right)^3 \right) \boldsymbol{dd} \right. \\ &\quad \left. + \left(\frac{3\beta}{4}\frac{a}{R} + \frac{1}{4}\beta(1+\beta^2)\left(\frac{a}{R}\right)^3 \right)(\boldsymbol{\delta} - \boldsymbol{dd}) \right] \\ \boldsymbol{\omega}_1^{(1)} &= \frac{1}{2}\nabla \times \boldsymbol{v}_2|_{x=x_1} \\ &= -\frac{3}{4}\frac{b}{R^2}\boldsymbol{U}_2^{(0)} \times \boldsymbol{d} \\ \boldsymbol{S}_1^{(1)} &= \frac{20}{3}\pi\mu a^3(1 + \frac{a^2}{10}\nabla^2)\boldsymbol{e}_2|_{x=x_1} \\ &= \left(-\frac{5}{2}\frac{a^3}{R^2} + \frac{3}{2}\frac{a^5}{R^4}(1 + \frac{5}{3}\beta^2) \right)(\boldsymbol{dd} - \frac{1}{3}\boldsymbol{\delta})\boldsymbol{d} \cdot \boldsymbol{F}_2 \\ &\quad - \frac{1}{2}\frac{a^5}{R^4}(1 + \frac{5}{3}\beta^2)(\boldsymbol{F}_2\boldsymbol{d} + \boldsymbol{d}\boldsymbol{F}_2 - 2\boldsymbol{ddd} \cdot \boldsymbol{F}_2) . \end{aligned}$$

The algebraic reductions were obtained by inserting the appropriate expressions for $\mathcal{G}$ and its derivatives. Expressions for $\boldsymbol{U}_2^{(1)}$, $\boldsymbol{\omega}_2^{(1)}$, and $\boldsymbol{S}_2^{(1)}$ may be obtained from the preceding by switching a and b and the indices 1 and 2.

At the next (second) reflection, the fields $\boldsymbol{v}_{121}$ and $\boldsymbol{v}_{212}$ are expanded as

$$\begin{aligned} \boldsymbol{v}_{121} &= (\boldsymbol{S}_1^{(2)} \cdot \nabla) \cdot \frac{\mathcal{G}(\boldsymbol{x}-\boldsymbol{x}_1)}{(8\pi\mu)} + \cdots \\ \boldsymbol{v}_{212} &= (\boldsymbol{S}_2^{(2)} \cdot \nabla) \cdot \frac{\mathcal{G}(\boldsymbol{x}-\boldsymbol{x}_2)}{(8\pi\mu)} + \cdots , \end{aligned}$$

with motions and stresslets given by

$$\begin{aligned}
\boldsymbol{U}_1^{(2)} &= (1+\frac{a^2}{6}\nabla^2)\boldsymbol{v}_{12}|_{x=x_1} \\
&= \boldsymbol{U}_1^{(0)}\cdot\left[\left(-\frac{15}{4}\beta^3\left(\frac{a}{R}\right)^4\right)\boldsymbol{dd}+\left(O(R^{-6})(\boldsymbol{\delta}-\boldsymbol{dd})\right)\right] \\
\boldsymbol{\omega}_1^{(2)} &= \frac{1}{2}\nabla\times\boldsymbol{v}_{12}|_{x=x_1} = O(R^{-7}) \\
\boldsymbol{S}_1^{(2)} &= \frac{20}{3}\pi\mu a^3(1+\frac{a^2}{10}\nabla^2)\boldsymbol{e}_{12}|_{x=x_1} \\
&= \frac{25}{2}\beta^3\frac{a^6}{R^5}(\boldsymbol{dd}-\frac{1}{3}\boldsymbol{\delta})\boldsymbol{d}\cdot\boldsymbol{F}_1\ .
\end{aligned}$$

The translational velocity contribution from the second reflection is $O(R^{-4})$ if the force on sphere 1 is directed along the axis, but is only $O(R^{-6})$ if the force is directed orthogonal to the axis. The second situation results in a weaker contribution, because the leading order term in $\boldsymbol{v}_{12}$ is a quadrupole field for that case. The angular velocity contribution from the second reflection is $O(R^{-7})$ instead of $O(R^{-5})$, because the leading order term in $\boldsymbol{v}_{12}$, a dipole field of cumulative strength of $O(R^{-4})$, is irrotational. As before, expressions for $\boldsymbol{U}_2^{(2)}$, $\boldsymbol{\omega}_2^{(2)}$, and $\boldsymbol{S}_2^{(2)}$ may be obtained from the analogous quantity for sphere 1 by switching a and b and the indices 1 and 2.

Since the final results are desired accurate to $O(R^{-5})$, the contributions from the third and fourth reflections are not needed. These will be $O(R^{-7})$ and $O(R^{-10})$, respectively. The results for $\boldsymbol{U}_1$ and $\boldsymbol{\omega}_1$ are obtained by adding the contributions from the reflections and are

$$\begin{aligned}
6\pi\mu a\boldsymbol{U}_1 &= -\boldsymbol{F}_1\cdot\left[\left(1-\frac{15}{4}\beta^3\left(\frac{a}{R}\right)^4\right)\boldsymbol{dd}+\left(1+O(R^{-6})\right)(\boldsymbol{\delta}-\boldsymbol{dd})\right] \\
&- \boldsymbol{F}_2\cdot\left[\left(\frac{3}{2}\frac{a}{R}-\frac{1}{2}(1+\beta^2)\left(\frac{a}{R}\right)^3\right)\boldsymbol{dd}\right. \\
&\qquad\left.+\left(\frac{3}{4}\frac{a}{R}+\frac{1}{4}(1+\beta^2)\left(\frac{a}{R}\right)^3\right)(\boldsymbol{\delta}-\boldsymbol{dd})\right] \\
6\pi\mu a^2\boldsymbol{\omega}_1 &= \boldsymbol{F}_1\times\boldsymbol{d}\left[O(R^{-7})\right]+\boldsymbol{F}_2\times\boldsymbol{d}\left[\frac{3}{4}\left(\frac{a}{R}\right)^2\right] \\
\boldsymbol{S}_1 &= \frac{25}{2}\beta^3\frac{a^6}{R^5}(\boldsymbol{dd}-\frac{1}{3}\boldsymbol{\delta})\boldsymbol{d}\cdot\boldsymbol{F}_1 \\
&+ \left(-\frac{5}{2}\frac{a^3}{R^2}+\frac{3}{2}\frac{a^5}{R^4}(1+\frac{5}{3}\beta^2)\right)(\boldsymbol{dd}-\frac{1}{3}\boldsymbol{\delta})\boldsymbol{d}\cdot\boldsymbol{F}_2 \\
&- \frac{1}{2}\frac{a^5}{R^4}(1+\frac{5}{3}\beta^2)(\boldsymbol{F}_2\boldsymbol{d}+\boldsymbol{d}\boldsymbol{F}_2-2\boldsymbol{ddd}\cdot\boldsymbol{F}_2)
\end{aligned}$$

and are consistent with the forms obtained for axisymmetric geometries, as discussed in the previous section. ◇

In the preceding example, the result for the case with forces along the line of centers may be inverted to obtain the solution of the corresponding resistance problem, to show that the results are consistent with the $O(R^{-4})$ calculation of Example 8.1 (see Exercise 8.5). For motions and forces perpendicular to the line of centers, we will also need the translational and rotational velocities of two force-free spheres subject to external torques (see Exercise 8.4). The two mobility problems can be combined to recover the resistance solution of Example 8.1 to $O(a/R)^5$. The important conclusion from these two examples is that the mobility solution can be obtained to higher order (R^{-5} *vs.* R^{-4}) with fewer reflections (two instead of four) than the corresponding resistance solution.

Example 8.4 Mobility Functions for Two Spherical Drops.
Consider two spherical *drops* centered at $\boldsymbol{x}_1$ and $\boldsymbol{x}_2$, and immersed in a fluid of viscosity μ. The sphere radii a and b, external forces $\boldsymbol{F}_1^e$ and $\boldsymbol{F}_2^e$, and $\beta = b/a$ are as in the preceding example. We keep things fairly general by allowing different drop viscosities, μ_1 and μ_2. Define viscosity ratios $\lambda_i = \mu_i/\mu$, $i = 1, 2$ and let $\Lambda_i = \lambda_i/(1+\lambda_i)$.

We generalize the preceding example to viscous drops by using the appropriate Faxén relations from Chapter 3,

$$\begin{aligned}\boldsymbol{U}_1 &= \left(1 + \frac{\Lambda_1 a^2\nabla^2}{2(2+\Lambda_1)}\right)\boldsymbol{v}^{\infty}(\boldsymbol{x})|_{x=x_1} \\ \boldsymbol{S}_1 &= \frac{4}{3}(2+3\Lambda_1)\pi\mu a^3\left(1 + \frac{\Lambda_1 a^2\nabla^2}{2(2+3\Lambda_1)}\right)\boldsymbol{e}^{\infty}(\boldsymbol{x})|_{x=x_1} ,\end{aligned}$$

in place of the ones for the rigid sphere. The other parts of the analysis are identical to those used in the preceding example, and the final result for the translational velocity on drop 1 may be written as

$$\begin{aligned}6\pi\mu a\boldsymbol{U}_1 &= -\boldsymbol{F}_1\cdot\left[\left(\frac{3}{2+\Lambda_1} - \frac{9\Lambda_2+6}{4}\beta^3\left(\frac{a}{R}\right)^4\right)\boldsymbol{dd}\right. \\ &\quad \left. + \left(\frac{3}{2+\Lambda_1} + O(R^{-6})\right)(\boldsymbol{\delta} - \boldsymbol{dd})\right] \\ &\quad - \boldsymbol{F}_2\cdot\left[\left(\frac{3}{2}\frac{a}{R} - \frac{3}{2}\left(\frac{\Lambda_1}{2+\Lambda_1} + \frac{\Lambda_2\beta^2}{2+\Lambda_2}\right)\left(\frac{a}{R}\right)^3\right)\boldsymbol{dd}\right. \\ &\quad \left. + \left(\frac{3}{4}\frac{a}{R} + \frac{3}{4}\left(\frac{\Lambda_1}{2+\Lambda_1} + \frac{\Lambda_2\beta^2}{2+\Lambda_2}\right)\left(\frac{a}{R}\right)^3\right)(\boldsymbol{\delta} - \boldsymbol{dd})\right] .\end{aligned}$$

◇

Example 8.5 Sedimentation of Two Spheroids in a Quiescent Fluid.
The hydrodynamic interaction between two prolate spheroids is of considerable interest as a model problem for suspensions consisting of elongated particles. Here, we generalize Example 8.3 (spheres) to prolate spheroids, following the treatment in Kim [44].

We take two spheroids centered at $\boldsymbol{x}_1$ and $\boldsymbol{x}_2$, and acted on by external forces $\boldsymbol{F}_1^e$ and $\boldsymbol{F}_2^e$. The zeroeth order solution for an isolated, translating prolate spheroid is expressed in the singularity form from Chapter 3:

$$\begin{aligned} \boldsymbol{v}_1 &= \boldsymbol{F}_1^e \cdot \frac{1}{2k_1} \int_{-k_1}^{k_1} \left\{ 1 + (k_1^2 - \xi_1^2) \frac{(1-e_1^2)^2}{4e_1^2} \nabla^2 \right\} \frac{\mathcal{G}(\boldsymbol{x} - \boldsymbol{\xi}_1)}{(8\pi\mu)} \, d\xi_1 \\ \boldsymbol{v}_2 &= \boldsymbol{F}_2^e \cdot \frac{1}{2k_2} \int_{-k_2}^{k_2} \left\{ 1 + (k_2^2 - \xi_2^2) \frac{(1-e_2^2)^2}{4e_2^2} \nabla^2 \right\} \frac{\mathcal{G}(\boldsymbol{x} - \boldsymbol{\xi}_2)}{(8\pi\mu)} \, d\xi_2 \ , \end{aligned}$$

with

$$6\pi\mu a \boldsymbol{U}_1^{(0)} = \boldsymbol{F}_1^e \cdot \left[(X^A)^{-1} \boldsymbol{d}_1 \boldsymbol{d}_1 + (Y^A)^{-1} (\boldsymbol{\delta} - \boldsymbol{d}_1 \boldsymbol{d}_1) \right] \ .$$

The axis of spheroid 1 is denoted by $\boldsymbol{d}_1$; $k_1 = ae_1$ is the focal parameter. The expression for $\boldsymbol{U}_2^{(0)}$ is quite similar and is obtained simply by switching labels 1 and 2.

The first reflection fields, $\boldsymbol{v}_{21}$ and $\boldsymbol{v}_{12}$, are represented to leading order by stresslets distributed over the spheroid axis. The explicit expression for the reflection at spheroid 1, $\boldsymbol{v}_{21}$, is

$$\boldsymbol{v}_{21} = (\boldsymbol{S}_1^{(1)} \cdot \nabla) \cdot \frac{3}{4k_1^3} \int_{-k_1}^{k_1} (k_1^2 - \xi_1^2) \left\{ 1 + (k_1^2 - \xi_1^2) \frac{(1-e_1^2)^2}{8e_1^2} \nabla^2 \right\} \frac{\mathcal{G}(\boldsymbol{x} - \boldsymbol{\xi}_1)}{(8\pi\mu)} \, d\xi_1 + \cdots$$

with particle motions and stresslets obtained from the Faxén laws applied to the incident field, $\boldsymbol{v}_1$, *i.e.*,

$$\begin{aligned} \boldsymbol{U}_1^{(1)} &= \frac{1}{2k_1} \int_{-k_1}^{k_1} \left\{ 1 + (k_1^2 - \xi_1^2) \frac{(1-e_1^2)^2}{4e_1^2} \nabla^2 \right\} \boldsymbol{v}_2(\boldsymbol{\xi}_1) \, d\xi_1 \\ &= \boldsymbol{F}_2^e \cdot \int_{-k_1}^{k_1} \frac{d\xi_1}{2k_1} \int_{-k_2}^{k_2} \frac{d\xi_2}{2k_2} \left\{ 1 + (k_1^2 - \xi_1^2) \frac{(1-e_1^2)^2}{4e_1^2} \nabla_1^2 \right. \\ &\qquad \left. + (k_2^2 - \xi_2^2) \frac{(1-e_2^2)^2}{4e_2^2} \nabla_1^2 \right\} \frac{\mathcal{G}(\boldsymbol{\xi}_1 - \boldsymbol{\xi}_2)}{(8\pi\mu)} \\ \boldsymbol{\omega}_1^{(1)} &= \frac{3}{8k_1^3} \int_{-k_1}^{k_1} (k_1^2 - \xi_1^2) \nabla \times \boldsymbol{v}_2(\boldsymbol{\xi}_1) \, d\xi_1 \\ &+ \frac{e_1^2}{(2-e_1^2)} \frac{3}{4k_1^3} \int_{-k_1}^{k_1} (k_1^2 - \xi_1^2) \left\{ 1 + (k_1^2 - \xi_1^2) \frac{(1-e_1^2)^2}{8e_1^2} \nabla^2 \right\} \\ &\qquad \times \boldsymbol{d}_1 \times (\boldsymbol{e}_2(\boldsymbol{x}_1) \cdot \boldsymbol{d}_1) \, d\xi_1 \\ (\boldsymbol{S}_1^{(1)})_{ij} &= \frac{20}{3} \pi\mu a_1^3 \left[X^M d_{ijkl}^{(0)} + Y^M d_{ijkl}^{(1)} + Z^M d_{ijkl}^{(2)} \right] \\ &\qquad \times \frac{3}{4k_1^3} \int_{-k_1}^{k_1} (k_1^2 - \xi_1^2) \left\{ 1 + (k_1^2 - \xi_1^2) \frac{(1-e_1^2)^2}{8e_1^2} \nabla^2 \right\} \boldsymbol{e}_2(\boldsymbol{\xi}_1) \, d\xi_1 \\ &+ 4\pi\mu a_1^3 Y^H (d_i \epsilon_{jkl} + d_j \epsilon_{ikl}) d_l \frac{3}{8k_1^3} \int_{-k_1}^{k_1} (k_1^2 - \xi_1^2) (\nabla \times \boldsymbol{v}_2(\boldsymbol{\xi}_1))_k \, d\xi_1 \ . \end{aligned}$$

Expressions for $\boldsymbol{U}_2^{(1)}$, $\boldsymbol{\omega}_2^{(1)}$, and $\boldsymbol{S}_2^{(1)}$ may be obtained from these preceding expressions by switching the spheroid labels 1 and 2.

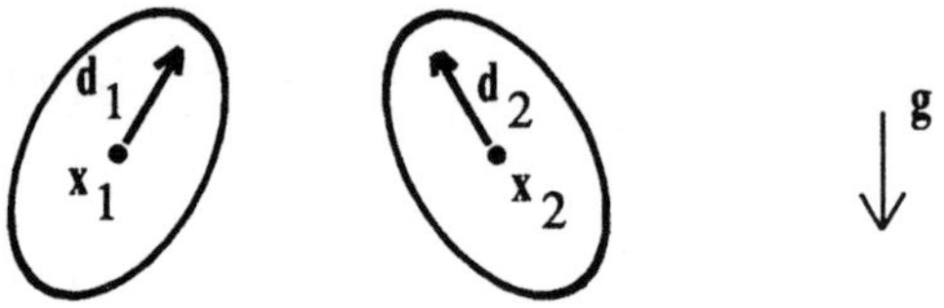

Figure 8.4: Two inclined prolate spheroids with axes in a common vertical plane.

The contributions from the next (second) reflection are obtained in exactly the same manner, with the incident field $\boldsymbol{v}_{12}$ in place of $\boldsymbol{v}_2$. While the final results are not as readily expressed in a form as compact as that encountered earlier for spheres, we may write a simple computer program to evaluate these integrals numerically, *e.g.*, by Gaussian quadratures, to obtain results for the mobility functions.

For example, consider two identical, inclined spheroids placed with their axes in a common vertical plane, as shown in Figure 8.4. Without hydrodynamic interaction, each spheroid would settle vertically, with a slight sideward drift induced by the anisotropy in the mobility tensor. However, each spheroid, being torque-free, must rotate in the common plane, with the angular velocity scaling with the vorticity of the Stokeslet field produced by the other (more precisely, an integrated weight of that vorticity field, as prescribed by the Faxén law). The change in orientation modifes the drift velocity so that the trajectory of the centroid is as shown in Figure 8.5. Note that both "single encounter" and periodic, meandering trajectories are possible. ◇

8.4 Renormalization Theory

In Part II, we saw how bulk suspension properties, such as the effective viscosity, can be obtained from an analysis of the flow past a single particle. Naturally, those results were limited to dilute systems in which hydrodynamic interactions are negligible. We may now use our knowledge of pair interactions to obtain the first corrections, and it is reasonable to expect that the solution takes the form of a virial expansion in the volume fraction c. For example, the effective viscosity of a suspension of rigid spheres should be of the form

$$\frac{\mu^{\text{eff}}}{\mu} = 1 + \frac{5}{2}c + Bc^2 + \dots ,$$

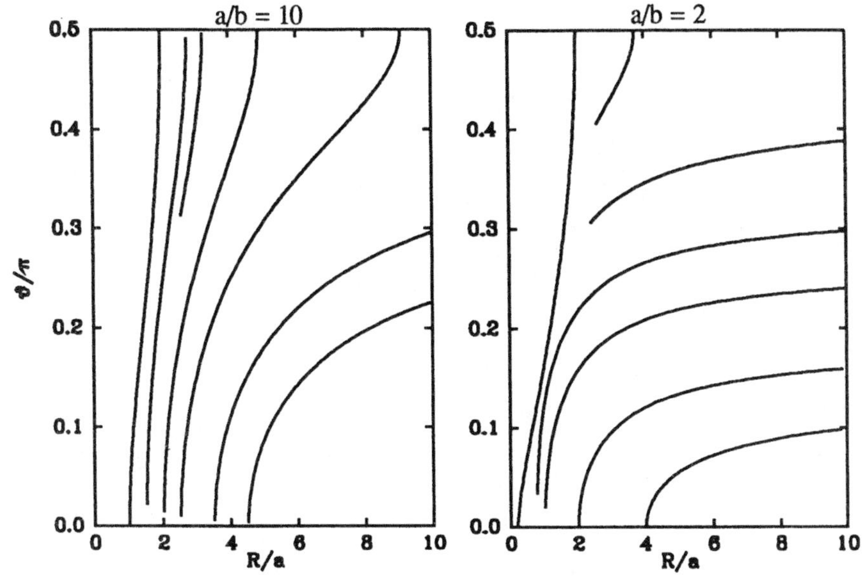

Figure 8.5: Spheroid (centroid) trajectories.

with the precise value of B depending on the details of the two-particle problem. Indeed, going through the steps described in Chapter 2, we have

$$\boldsymbol{\sigma}^{\text{eff}} = \boldsymbol{\sigma}^{\text{fluid}} + \boldsymbol{\sigma}^{\text{particle}} ,$$

with the particle contribution to the bulk stress given by an ensemble average of the stresslet,

$$\boldsymbol{\sigma}^{\text{particle}} = n < \boldsymbol{S} > .$$

Following the usual procedure of statistical physics, we would expect a result in terms of the configurational integral,

$$< \boldsymbol{S} >= \boldsymbol{S}_0 + \int_{|x_2 - x_1 \geq 2a|} P(\boldsymbol{x}_2|\boldsymbol{x}_1)(\boldsymbol{S}_{12} - \boldsymbol{S}_0)\, dV(\boldsymbol{x}_2) + O(c^2) .$$

Here, $\boldsymbol{S}_0 = \frac{20}{3}\pi\mu a^3 \boldsymbol{E}^\infty$ is the single-sphere result for the stresslet; $\boldsymbol{S}_{12} - \boldsymbol{S}_0$ is the "excess" stresslet on sphere 1 due to hydrodynamic interactions with another sphere at $\boldsymbol{x}_2$. This pair contribution to the bulk stress is obtained by integration over all allowed positions of sphere 2, with contributions weighted according to the conditional pair probability $P(\boldsymbol{x}_2|\boldsymbol{x}_1)$. But this simple approach encounters difficulties in the problems of microhydrodynamics, because the interactions between the particles are too strong.

For a dilute suspension, we expect $P(\boldsymbol{x}_2|\boldsymbol{x}_1) \sim n$, the number density, as $|\boldsymbol{x}_2 - \boldsymbol{x}_1| \rightarrow \infty$. While this leads to the expected scaling of $O(c^2)$, the excess

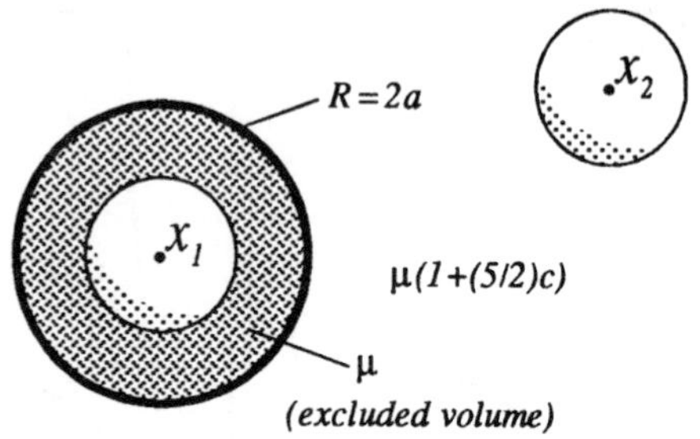

Figure 8.6: The effective medium in the renormalized formulation.

stresslet, $S_{12} - S_0$ decays as R^{-3} so that the configurational integral is of the form

$$\int_{2a}^{R^\infty} R^{-3} R^2 \, dR \; ,$$

which is nonconvergent as $R^\infty \to \infty$.

The resolution of this paradox is given in a number of important papers in the development of suspension rheology [2, 29, 36, 57]. The successive limits of large R^∞ and small c do not commute. Indeed, fixing c first at a small value we see that there is a new length scale $a/c^{1/3}$, and in a volume much larger than these dimensions there will be many spheres. Thus for R^∞ much greater than this length, the pair interactions S_{12} occur in an effective medium quite different from the pure solvent. In fact, for the purposes of the $O(c^2)$ problem, we may approximate the effective medium as a Newtonian fluid with the Einstein viscosity. In general, nonconvergent interactions indicate that the limiting processes mentioned above do not commute and that screening in an effective medium should be taken into consideration [29].

To pursue this in a more concrete setting, we average the governing balance equation over all realizations containing a sphere centered at x_1, leaving the conditionally averaged equation,

$$\begin{aligned} & -\nabla < p > (x|x_1) + \mu \nabla^2 < v > (x|x_1) \\ & = \int_{R \geq 2a} P(x_2|x_1) \oint_{|\xi - x_2| = a} < \sigma > (\xi|x_1, x_2) \cdot n(\xi) \delta(x - \xi) \, dS(\xi) \; . \quad (8.1) \end{aligned}$$

The physical meaning of this equation is that in all realizations with a sphere at x_1, the fluid satisfies the inhomogeneous Stokes equations, with the body forces given exactly by the surface tractions on the other spheres. The latter effect is of course weighted by the conditional probability of a second sphere configured at x_2. If we make what appears to be a reasonable approximation, replace $< \sigma > (\xi|x_1, x_2)$ with that of two spheres in the pure Newtonian solvent and a small relative error of $O(c)$, then the error is made and the noncovergent integrals arise. As mentioned above, this approximation is not uniformly valid in $V(\boldsymbol{x}_2)$. However, if we approximate $<\boldsymbol{\sigma}>(\boldsymbol{\xi}|\boldsymbol{x}_1,\boldsymbol{x}_2)$ with that of two spheres in the Einstein medium with viscosity $\mu(1+\frac{5}{2}c)$, the error is still of $O(c)$, but is now uniformly valid throughout $V(\boldsymbol{x}_2)$. Indeed, borrowing from cell and self-consistent field theories, we may use the Einstein viscosity only beyond a certain value of R. The exact location of this viscosity jump does not alter the final result for the virial expansion, but does shift favorably the relative contributions of the analytical over the numerical portions in the final expression for the viscosity [43].

Hinch, in his original work [29], places the viscosity jump at $R = 2a$, or the excluded volume radius, and the renormalized formulation is

$$\begin{aligned}
&-\nabla <p>(\boldsymbol{x}|\boldsymbol{x}_1) + \mu\nabla^2 <\boldsymbol{v}>(\boldsymbol{x}|\boldsymbol{x}_1) \\
&+ \int_{R\geq 2a} \frac{5}{2}\mu c \nabla_2^2 <\boldsymbol{v}>(\boldsymbol{x}_2|\boldsymbol{x}_1)\delta(\boldsymbol{x}-\boldsymbol{x}_2)\, dV(\boldsymbol{x}_2) \\
&+ \left(\oint_{R=2a} - \oint_{\infty}\right) 5\mu c <\boldsymbol{e}>(\boldsymbol{x}_2|\boldsymbol{x}_1)\cdot \boldsymbol{n}_2\delta(\boldsymbol{x}-\boldsymbol{x}_2)\, dS(\boldsymbol{x}_2) \\
&= \int_{R\geq 2a} \left\{ P(\boldsymbol{x}_2|\boldsymbol{x}_1) \oint_{|\xi - x_2| = a} <\boldsymbol{\sigma}>(\boldsymbol{\xi}|\boldsymbol{x}_1,\boldsymbol{x}_2)\cdot \boldsymbol{n}(\boldsymbol{\xi})\delta(\boldsymbol{x}-\boldsymbol{\xi})\, dS(\boldsymbol{\xi}) \right. \\
&\qquad \left. -5\mu c <\boldsymbol{e}>(\boldsymbol{x}_2|\boldsymbol{x}_1)\cdot \nabla_2\delta(\boldsymbol{x}-\boldsymbol{x}_2)\right\} dV(\boldsymbol{x}_2)\,. \qquad (8.2)
\end{aligned}$$

Using the properties of the Dirac delta function, we see that the left-hand side of this equation has a viscosity jump at $R = 2a$, from the pure solvent value to the Einstein viscosity. Comparing Equations 8.1 and 8.2, as long as the exact conditionally averaged field variables are retained, the two equations are identical. The new terms appearing in the renormalized equation cancel exactly, by an application of the Green's identity and the divergence theorem. However, the renormalized equation is the one in which the effect of the second test sphere (the inhomogeneous terms on the right-hand side) can be uniformly truncated with respect to small c.

The stresslet on sphere 1 for Equation 8.2 can be written directly using the Faxén relation, and ultimately leads to the following expression for the bulk stress:

$$\begin{aligned}
\boldsymbol{\sigma}(\boldsymbol{x}) &= -p(\boldsymbol{x})\boldsymbol{\delta} + 2\mu(1+\frac{5}{2}c)\boldsymbol{e}(\boldsymbol{x}) + \frac{515}{64}\mu c^2 \boldsymbol{e}(\boldsymbol{x}) \\
&+ P(\boldsymbol{x})\int_{R\geq 2a} \{P(\boldsymbol{x}_2|\boldsymbol{x}_1)(\boldsymbol{S}_{12}(\boldsymbol{x}|\boldsymbol{x}_2) - \boldsymbol{S}_0) \\
&\qquad \left. -5\mu c \boldsymbol{E}^D\left((\boldsymbol{x},\boldsymbol{x}_2,\boldsymbol{e}+\boldsymbol{E}^\infty)\right)\right\} dV(\boldsymbol{x}_2)\,. \qquad (8.3)
\end{aligned}$$

The term $\frac{515}{64}\mu c^2 \boldsymbol{e}(\boldsymbol{x})$ comes from the viscosity jump, *i.e.*, the stresslet on an isolated sphere in a pure rate-of-strain field in a fluid medium with a viscosity jump at $R = 2a$.

The distribution of dipoles subtracted on the RHS of Equation 8.2 originated as part of the renormalization procedure. Thus the original nonconvergent integrand, $(\boldsymbol{S}_{12}(\boldsymbol{x}|\boldsymbol{x}_2) - \boldsymbol{S}_0)$, has been modified by a disturbance rate-of-strain field $\boldsymbol{E}^D(\boldsymbol{x},\boldsymbol{x}_2,\boldsymbol{e}+\boldsymbol{E}^D)$. The arguments denote that this is the extra rate-of-strain due to the presence of a second sphere at $\boldsymbol{x}_2$ and that the sphere at $\boldsymbol{x}_2$

is immersed in an ambient flow with the rate of strain $e + E^D$. (To get the usual notation in this book, map $e \to E^\infty$ and $E^D \to e_1$, the rate-of-strain field of the first reflection from sphere 1.) The reader should verify that the definition for $E^D((x, x_2, e + E^D))$ is precisely that required for the zeroeth, first and second reflection contributions to the stresslet on sphere 1, and thus the modified integrand now decays as $O(R^{-8})$.

A numerical solution of the two-sphere problem is necessary to obtain the excess stresslet to complete the solution [2, 45, 69]. It should be noted that the pair probability, $P(x_2|x_1)$, for the flowing suspension cannot be set arbitrarily, but must be obtained by solving the pair conservation equation:

$$\frac{\partial}{\partial t} P(x_2|x_1) + \nabla_2 \cdot ((U_2 - U_1)P(x_2|x_1)) = 0 . \tag{8.4}$$

For pure rate-of-strain flow, $U_2 - U_1$ can be written in terms of two mobility functions A and B (see Chapter 11) as

$$U_2 - U_1 = E^\infty \cdot (x_2 - x_1) - [A(R)dd + B(R)(\delta - dd)] \cdot E^\infty \cdot (x_2 - x_1) .$$

With the condition that $P(x_2|x_1) \to n$ as $R \to \infty$, Equation 8.4 has the exact solution,

$$P(x_2|x_1) = \frac{n}{1 - A} \exp\left\{ \int_R^\infty \frac{3(B(\rho) - A(\rho))}{\rho(1 - A(\rho))} \, d\rho \right\} .$$

Numerical integration of the last term in Equation 8.3 is all that remains in the calculation of the $O(C^2)$ coefficient of the viscosity.

Equation 8.4 cannot be solved for simple shear flows, because the spheres form doublets (see Chapter 11) whose history cannot be traced to conditions at infinity. Higher order effects, such as Brownian diffusion, must be included [14, 15]. For pure rate-of-strain fields, the conservation equation can be solved and $P(x_1|x_2)$ can be expressed in terms of the two-sphere mobility functions x^g and y^g, and the final result for the effective viscosity is[2]

$$\frac{\mu^{\text{eff}}}{\mu} = 1 + \frac{5}{2}c + 6.95c^2 + \dots .$$

8.5 Multipole Expansions for Two Spheres

The method of reflections as outlined in the previous section is a general solution strategy that generates a series solution suitable for determining interactions between particles of arbitrary shape, as long as they are far apart. The method is viable under these conditions because the algebraic manipulations for the first few terms in the series are quite simple. As the particles approach each other, the number of terms required in this series increases. The effort required to

[2]In 1972, accurate results for the stresslets of the two-sphere problem were not available. The often-quoted value 7.6 for the $O(c^2)$ coefficient given in the original work [2] was obtained from a rough interpolation that overestimated the contributions from the sphere-sphere interactions.

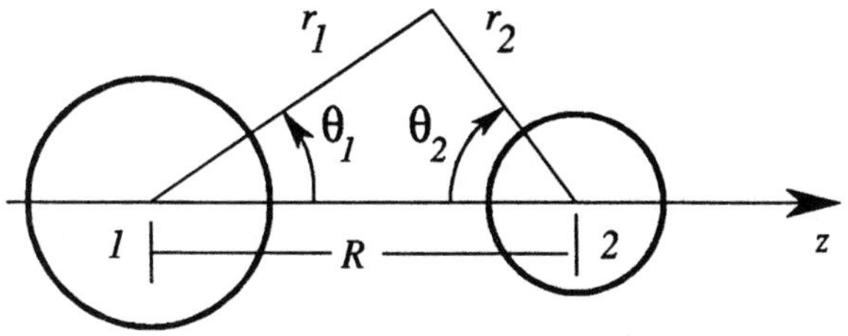

Figure 8.7: The two-sphere geometry

obtain the higher-order terms is prohibitive and the method becomes less attractive. However, for the special case of two spheres, we can still obtain series solutions by the judicious use of transformation formulas relating Lamb's solution expanded with respect to two different origins. The method was developed by Jeffrey and Onishi [38], and the discussion in this section borrows heavily from their work.

In a nutshell, we borrow the form of the velocity representation from the method of reflections and expand the disturbance field about the two sphere centers, but now the boundary conditions at each sphere are set simultaneously. Of course, the disturbance field from the other sphere presents some problems since those fields are cast in terms of a coordinate system based on a different origin (located at the other sphere's center). However, spherical harmonics based at one origin can be recast in terms of those expanded about another origin, by the so-called addition theorems [30], and since our velocity fields are closely related to, and in fact derived from spherical harmonics, we find that the disturbance field about one sphere may be written in terms of velocity fields expanded about the other.

The two-sphere geometry is shown in Figure 8.7 The spheres are centered at $\boldsymbol{x}_1$ and $\boldsymbol{x}_2$, and their radii are denoted by $a_1 = a$ and $a_2 = b$. Spherical polar coordinates (r_1, θ_1, ϕ) about the origin $\boldsymbol{x}_1$ are defined in the usual fashion, with $\boldsymbol{d} = (\boldsymbol{x}_2 - \boldsymbol{x}_1)/R$ as the polar $(z-)$ axis. For (r_2, θ_2, ϕ) we use a *left-handed* system, with $\boldsymbol{e}_r^{(2)} \times \boldsymbol{e}_\theta^{(2)} = -\boldsymbol{e}_\phi$, to simplify the form taken by addition theorem for the spherical harmonics. Based on Hobson [30], we write the addition theorem as

$$\left(\frac{a_\alpha}{r_\alpha}\right)^{n+1} Y_{mn}(\theta_\alpha, \phi) = \left(\frac{a_\alpha}{R}\right)^{n+1} \sum_{s=m}^{\infty} \left(\begin{array}{c} n+s \\ s+m \end{array}\right) \left(\frac{r_{3-\alpha}}{R}\right)^s Y_{ms}(\theta_{3-\alpha}, \phi) \ , \quad (8.5)$$

with $\alpha = 1, 2$, depending on whether we wish to expand harmonics based on $\boldsymbol{x}_1$ in terms of those at $\boldsymbol{x}_2$ or *vice-versa.* A right-handed system can be defined at $\boldsymbol{x}_2$, *i.e.*, by letting $\theta_2 \to \pi - \theta_2$, but this will introduce a factor of -1 in a number of places in the following discussion.

The pressure and velocity fields are written as a linear combination of two sets of Lamb's solution (one for each sphere center):

$$p = p^{(1)} + p^{(2)} , \qquad \boldsymbol{v} = \boldsymbol{v}^{(1)} + \boldsymbol{v}^{(2)} ,$$

with

$$p^{(\alpha)} = \frac{\mu}{a_\alpha} \sum_{m=0}^{\infty} \sum_{n=m}^{\infty} A_{mn}^{(\alpha)} \left(\frac{a_\alpha}{r_\alpha}\right)^{n+1} Y_{mn}(\theta_\alpha, \phi)$$

$$\begin{aligned}
\boldsymbol{v}^{(\alpha)} &= \sum_{m=0}^{\infty} \sum_{n=m}^{\infty} \left\{ \frac{-(n-2)r_\alpha^2}{2n(2n-1)a_\alpha} \nabla \left[A_{mn}^{(\alpha)} \left(\frac{a_\alpha}{r_\alpha}\right)^{n+1} Y_{mn}(\theta_\alpha, \phi) \right] \right. \\
&\quad + \frac{(n+1)\boldsymbol{r}_\alpha}{n(2n-1)a_\alpha} A_{mn}^{(\alpha)} \left(\frac{a_\alpha}{r_\alpha}\right)^{n+1} Y_{mn}(\theta_\alpha, \phi) \\
&\quad + a_\alpha \nabla \left[B_{mn}^{(\alpha)} \left(\frac{a_\alpha}{r_\alpha}\right)^{n+1} Y_{mn}(\theta_\alpha, \phi) \right] \\
&\quad \left. + \nabla \times \left[C_{mn}^{(\alpha)} \left(\frac{a_\alpha}{r_\alpha}\right)^{n+1} Y_{mn}(\theta_\alpha, \phi) \right] \right\} .
\end{aligned}$$

The existence of such representations can be argued by appealing to the form taken by the usual Taylor series operation with the surface variable of the integral representation for the two-sphere velocity field. Note that the equations have been scaled so that the coefficients $A_{mn}^{(\alpha)}$, $B_{mn}^{(\alpha)}$, and $C_{mn}^{(\alpha)}$ have dimensions of velocities.

On the sphere surfaces, we take the known boundary velocity $\boldsymbol{V}^{(\alpha)}(\theta_\alpha, \phi)$ and set the radial velocity, surface divergence, and surface vorticity as explained in Chapter 4. For the sake of argument, we assume these known quantities can be written as

$$\begin{aligned}
\boldsymbol{v} \cdot \boldsymbol{e}_r^{(\alpha)} &= \boldsymbol{V}^{(\alpha)} \cdot \boldsymbol{e}_r^{(\alpha)} &= \sum_{m=0}^{\infty} \sum_{n=m}^{\infty} R_{mn}^{(\alpha)} Y_{mn}(\theta_\alpha, \phi) \\
\boldsymbol{r}_\alpha \cdot \nabla(\boldsymbol{v} \cdot \boldsymbol{e}_r^{(\alpha)}) &= -r_\alpha \nabla \cdot \boldsymbol{V}^{(\alpha)} \cdot \boldsymbol{e}_r^{(\alpha)} &= \sum_{m=0}^{\infty} \sum_{n=m}^{\infty} D_{mn}^{(\alpha)} Y_{mn}(\theta_\alpha, \phi) \\
\boldsymbol{r}_\alpha \cdot \nabla \times \boldsymbol{v} &= \boldsymbol{r}_\alpha \nabla \times \boldsymbol{V}^{(\alpha)} &= \sum_{m=0}^{\infty} \sum_{n=m}^{\infty} \omega_{mn}^{(\alpha)} Y_{mn}(\theta_\alpha, \phi) .
\end{aligned}$$

As in Chapter 4, the boundary conditions yield three sets of equations, after collection of terms corresponding to each spherical harmonic. The new wrinkle comes from terms that originate from the other sphere, which must be recast in the more usable form by applying the addition theorem, Equation 8.5. Because we are dealing with vector quantities, we will also need the following geometric relations:

$$\begin{aligned}
\boldsymbol{e}_r^{(\alpha)} &= (r_{3-\alpha} - R\cos\theta_{3-\alpha})\boldsymbol{e}_r^{(3-\alpha)} + R\sin\theta_{3-\alpha}\boldsymbol{e}_\theta^{(3-\alpha)} \\
r_\alpha^2 &= R^2 + r_{3-\alpha}^2 - 2Rr_{3-\alpha}\cos\theta_{3-\alpha} .
\end{aligned}$$

The following relations are obtained after some considerable algebra:

$$(n+1)(2n+1)B_{mn}^{(\alpha)} - \frac{1}{2}(n+1)A_{mn}^{(\alpha)} + \frac{n}{2n+3}\sum_{s=m}^{\infty}\begin{pmatrix} n+s \\ n+m \end{pmatrix} A_{ms}^{(3-\alpha)} t_\alpha^{n+1} t_{3-\alpha}^s$$
$$= D_{mn}^{(\alpha)} - (n-1)R_{mn}^{(\alpha)} \tag{8.6}$$

$$\frac{n+1}{2n-1}A_{mn}^{(\alpha)} + \sum_{s=m}^{\infty}\begin{pmatrix} n+s \\ n+m \end{pmatrix} t_\alpha^{n-1} t_{3-\alpha}^s \Big[(-1)^\alpha m(2n+1) i C_{ms}^{(3-\alpha)} t_{3-\alpha}$$
$$+ n(2n+1)B_{ms}^{(3-\alpha)} t_{3-\alpha}^2$$
$$+ \left(\frac{n+1}{2n-1}\right)\frac{ns(n+s-2ns-2) - m^2(2ns-4s-4n+2)}{2s(2s-1)(n+s)} A_{ms}^{(3-\alpha)}$$
$$+ \frac{n}{2}A_{ms}^{(3-\alpha)} t_\alpha^2\Big]$$
$$= D_{mn}^{(\alpha)} + (n+2)R_{mn}^{(\alpha)} \tag{8.7}$$

$$n(n+1)C_{mn}^{(\alpha)} + \sum_{s=m}^{\infty}\begin{pmatrix} n+s \\ n+m \end{pmatrix} t_\alpha^n t_{3-\alpha}^s \left[-nsC_{ms}^{(3-\alpha)} t_{3-\alpha} + (-1)^\alpha \frac{m}{s} i A_{ms}^{(3-\alpha)}\right]$$
$$= \omega_{mn}^{(\alpha)}, \tag{8.8}$$

where $t_\alpha = a_\alpha / R$. These equations may be solved by a number of different methods, but we shall focus our attention on getting the final solution in the form of a series in R^{-1}.

8.5.1 Translations Along the Axis

The problem of motions along the sphere-sphere axis are decomposed into two subproblems; the first involving two spheres approaching each other with equal velocities and the second involving two spheres translating together at the same velocity. We may then exploit certain symmetries in these two problems to reduce the work involved.

Resistance Functions for Two Approaching Spheres

In the first problem, we set $\boldsymbol{V}_1 = \boldsymbol{U}_1 = U\boldsymbol{d}$, $\boldsymbol{V}_2 = \boldsymbol{U}_2 = -U\boldsymbol{d}$, so that

$$R_{mn}^{(\alpha)} = U\delta_{m0}\delta_{n1}, \qquad D_{mn}^{(\alpha)} = \omega_{mn}^{(\alpha)} = 0 .$$

Since the problem is axisymmetric, only terms corresponding to $m = 0$ are relevant, and since we are not dealing with a "swirling" flow, we do not need $C_{0n}^{(\alpha)}$. We expand $A_{0n}^{(\alpha)}$ and $B_{0n}^{(\alpha)}$ as a power series in R^{-1}, *i.e.*,

$$A_{0n}^{(\alpha)} = \frac{3}{2}U\sum_{p=0}^{\infty}\sum_{q=0}^{\infty} A_{npq}^{(\alpha)} t_\alpha^p t_{3-\alpha}^q$$
$$B_{0n}^{(\alpha)} = \frac{3}{4}U\sum_{p=0}^{\infty}\sum_{q=0}^{\infty} \frac{1}{2n+1} B_{npq}^{(\alpha)} t_\alpha^p t_{3-\alpha}^q .$$

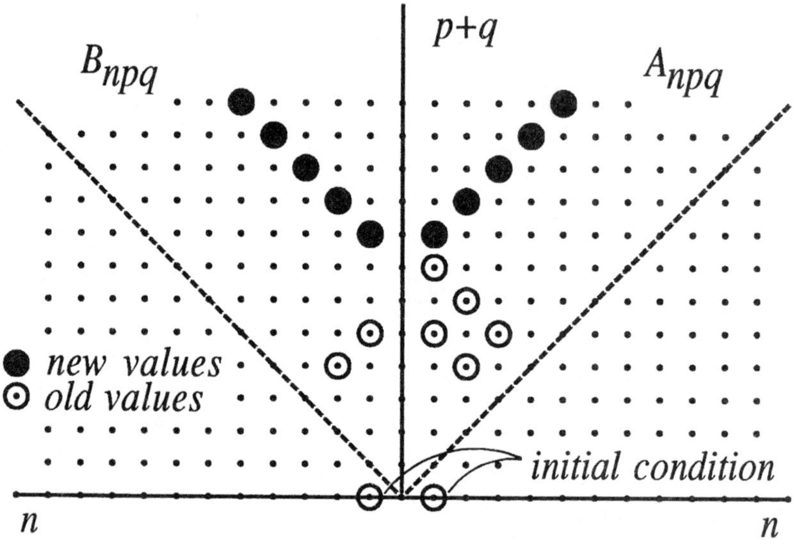

Figure 8.8: The recursion relations for A_{npq} and B_{npq}.

(We have made use of the final result to redefine the expansion parameters so as to simplify the initial conditions and recursion relations.)

We insert these expansions into Equations 8.6 and 8.7 to obtain the recursion relations,

$$A_{n00} = B_{n00} = \delta_{1n}$$

$$\begin{aligned} A_{npq} &= \sum_{s=1}^{q} \binom{n+s}{n} \left[\frac{n(2n+1)(2ns-n-s+2)}{2(n+1)(2s-1)(n+s)} A_{s(q-s)(p-n+1)} \right. \\ &\quad \left. - \frac{n(2n-1)}{2(n+1)} A_{s(q-s)(p-n-1)} - \frac{n(2n-1)(2n+1)}{2(n+1)(2s+1)} B_{s(q-s-2)(p-n+1)} \right] \end{aligned}$$

$$B_{npq} = A_{npq} - \frac{2n}{(n+1)(2n+3)} \sum_{s=1}^{q} \binom{n+s}{n} A_{s(q-s)(p-n-1)} \; .$$

The label α has been dropped since the coefficients are identical for the two spheres. (This can be shown explicitly from the recursion relations and is ultimately due to the symmetry in the problem.) The operation of the recursion relation is shown schematically in Figure 8.8. The essential advantage of this approach over the method of reflections is that the recursion relations can be easily programmed into a computer, and many terms can be generated.

The force on sphere α is given by

$$\boldsymbol{F}_\alpha = 4\pi\mu a_\alpha (-1)^\alpha A_{01}^{(\alpha)} \boldsymbol{d} \; .$$

With both spheres 1 and 2 moving, the force on sphere 1 is given by

$$X_{11}^A - X_{12}^A = 6\pi a \sum_{p=0}^{\infty}\sum_{q=0}^{\infty} A_{1pq} t_1^p t_2^q .$$

Resistance Functions for Two Spheres Moving in Tandem

In the second problem, we set $\boldsymbol{V}_1 = \boldsymbol{V}_2 = U\boldsymbol{d}$, and here we find

$$R_{mn}^{(\alpha)} = (-1)^{3-\alpha} U \delta_{m0}\delta_{n1} , \qquad D_{mn}^{(\alpha)} = \omega_{mn}^{(\alpha)} = 0 .$$

We repeat the procedure just used for spheres approaching each other and obtain a similar set of recursion relations. In fact, an examination of the new recursion relations reveals that it may be transformed to the old recursion relations if A_{npq} and B_{npq} are replaced everywhere by $(-1)^{n+p+q+\alpha}A_{npq}$ and $(-1)^{n+p+q+\alpha}B_{npq}$. This, of course, was the reason why the translation problem was decomposed into two subproblems in the first place. Since the spheres are moving in tandem, the force on sphere 1 is now given by

$$X_{11}^A + X_{12}^A = 6\pi a \sum_{p=0}^{\infty}\sum_{q=0}^{\infty} (-1)^{p+q} A_{1pq} t_1^p t_2^q ,$$

where A_{1pq} is that obtained in the previous problem.

Since we know the sum and difference of X_{11}^A and X_{12}^A, we can derive explicit expressions for both. We find that X_{11}^A is a series in even powers of R^{-1}, whereas X_{12}^A is a series in odd powers of R^{-1}. The first few terms of the final result are tabulated in Chapter 11.

Mobility Functions for Two Approaching Spheres

In the mobility problems, $\boldsymbol{F}_1$ and $\boldsymbol{F}_2$ are known and $\boldsymbol{U}_1$ and $\boldsymbol{U}_2$ are to be determined. Of course, having obtained the resistance functions, we could simply invert the resistance matrix. However, it is also possible to obtain the mobility functions directly, with a minor modification of the recursion scheme. First, assume that the forces are in opposite directions such that the scaled forces satisfy

$$(6\pi\mu a_1)^{-1}\boldsymbol{F}_1 = -(6\pi\mu a_2)^{-1}\boldsymbol{F}_2 = U\boldsymbol{d} .$$

The physical significance of U is that it is the approach velocity that each sphere, when isolated, would experience when subject to these forces, as a consequence of Stokes' law. Our task is to determine the actual velocities, $\boldsymbol{U}_1$ and $\boldsymbol{U}_2$, which differ from the widely separated value because of hydrodynamic interactions. We shall write these unknown velocities as

$$\boldsymbol{U}_1 = U^{(1)}\boldsymbol{d} , \qquad \boldsymbol{U}_2 = -U^{(2)}\boldsymbol{d} .$$

Since $\boldsymbol{\omega}_1 = \boldsymbol{\omega}_2 = \boldsymbol{0}$ we may set

$$R_{mn}^{(\alpha)} = U_\alpha \delta_{m0}\delta_{n1} , \qquad D_{mn}^{(\alpha)} = \omega_{mn}^{(\alpha)} = 0 ,$$

and drop the $C_{0n}^{(\alpha)}$ terms. We expand $A_{0n}^{(\alpha)}$, $B_{0n}^{(\alpha)}$, and $U^{(\alpha)}$ as

$$\begin{aligned} A_{0n}^{(\alpha)} &= \frac{3}{2}U\sum_{p=0}^{\infty}\sum_{q=0}^{\infty} A_{npq}t_\alpha^p t_{3-\alpha}^q \\ B_{0n}^{(\alpha)} &= \frac{3}{4}U\sum_{p=0}^{\infty}\sum_{q=0}^{\infty} \frac{1}{2n+1}B_{npq}t_\alpha^p t_{3-\alpha}^q \\ U^{(\alpha)} &= U\sum_{p=0}^{\infty}\sum_{q=0}^{\infty} U_{pq}t_\alpha^p t_{3-\alpha}^q \, , \end{aligned}$$

and, as before, the two spheres have the same coefficients and the label α is not needed.

Since the force is known, the initial conditions of the recursion relations now read

$$A_{1pq} = \delta_{0p}\delta_{0q} \, .$$

For $n = 1$, Equation 8.6 provides an equation for U_{pq},

$$U_{pq} = -\sum_{s=1}^{q}(s+1)\left[\frac{3A_{s(q-s)p}}{4(2s-1)} - \frac{1}{4}A_{s(q-s)(p-2)} - \frac{3B_{s(q-s-2)p}}{4(2s+1)}\right] ,$$

while for $n > 1$, it is a recursion relation for A_{npq},

$$\begin{aligned} A_{npq} &= \sum_{s=1}^{q}\binom{n+s}{n}\left[\frac{n(2n+1)(2ns-n-s+2)}{2(n+1)(2s-1)(n+s)}A_{s(q-s)(p-n+1)}\right. \\ &\quad \left. -\frac{n(2n-1)}{2(n+1)}A_{s(q-s)(p-n-1)} - \frac{n(2n-1)(2n+1)}{2(n+1)(2s+1)}B_{s(q-s-2)(p-n+1)}\right] . \end{aligned}$$

Equation 8.7 remains valid for all $n \geq 1$ and provides a recursion relation for B_{npq}.

$$B_{npq} = A_{npq} - \frac{2n}{(n+1)(2n+3)}\sum_{s=1}^{q}\binom{n+s}{n}A_{s(q-s)(p-n-1)} \, .$$

With the assumed hydrodynamic forces acting on spheres 1 and 2, the translation velocity of sphere 1 provides the relation,

$$6\pi a x_{11}^a - 6\pi b x_{12}^a = \sum_{p=0}^{\infty}\sum_{q=0}^{\infty} U_{pq}t_1^p t_2^q \, .$$

Mobility Functions for Forces in the Same Direction

In the second mobility problem, we set

$$(6\pi\mu a_1)^{-1}\boldsymbol{F}_1 = (6\pi\mu a_2)^{-1}\boldsymbol{F}_2 = U\boldsymbol{d} \, ,$$

and ultimately we find that the coefficients are given by $(-1)^{p+q}U_{pq}$, where U_{pq} is that obtained in the first problem:

$$6\pi a x_{11}^a + 6\pi b x_{12}^a = \sum_{p=0}^{\infty}\sum_{q=0}^{\infty}(-1)^{p+q} U_{pq}t_1^p t_2^q \, .$$

As was the case in the resistance problem, the $1-1$ function, x_{11}^a, is a series in even powers of R^{-1} while the $1-2$ function, x_{12}^a, is a series in odd powers of R^{-1}. The first few terms of the final result are tabulated in Chapter 11.

8.6 Electrophoresis of Particles with Thin Double Layers

8.6.1 Hydrodynamic Interaction Between Spheres

In Chapter 5, we considered electrophoresis of particle with thin electrical double layers and showed that the mobility was independent of particle shape. We now consider the role of hydrodynamic interactions for two widely separated particles. The method of reflection analysis shown here follows closely that of Chen and Keh [6], who considered the interaction between two spherical colloidal particles.

Consider two spheres with surface (zeta) potentials ζ_1 and ζ_2 undergoing electrophoresis in response to an applied electric field $\boldsymbol{E}^\infty$. In all other respects, the two-sphere geometry is as in earlier sections of this chapter. The governing equations for the electric field, electrostatic potential, the hydrodynamic variables are

$$\begin{aligned}
\nabla \cdot \boldsymbol{E} &= -\nabla^2 \psi = 0 \\
\hat{\boldsymbol{n}} \cdot \boldsymbol{E} &= 0 \quad \text{on } S_\alpha \\
\boldsymbol{E} &\to \boldsymbol{E}^\infty \quad \text{as } |\boldsymbol{x}| \to \infty \\
-\nabla p + \mu \nabla^2 \boldsymbol{v} &= 0 \\
-\nabla \cdot \boldsymbol{v} &= 0 \\
\boldsymbol{v} &= \boldsymbol{U}_\alpha - \frac{\epsilon \zeta_\alpha}{\mu} \boldsymbol{E} + \boldsymbol{\omega} \times (\boldsymbol{x} - b x_\alpha) \quad \text{on } S_\alpha \\
\boldsymbol{v} &\to 0 \quad \text{as } |\boldsymbol{x}| \to \infty .
\end{aligned}$$

The boundary condition on $\boldsymbol{v}$ consists of the rigid-body motion plus the Smoluchowski slip velocity seen by an observer in the outer region beyond the thin double layer (see Section 5.7).

As in the hydrodynamic problems discussed earlier in this chapter, this problem can be solved by the method of reflections, with each reflection contribution calculated by an application of the Faxén law to the disturbance field produced by distant particle(s). The Faxén law for the electrophoretic velocity of a particle in an ambient velocity field $\boldsymbol{v}^{Amb}$ and electric field $\boldsymbol{E}^{Amb}$ has been derived by Keh and Anderson [42] as

$$\boldsymbol{U} = \frac{\epsilon \zeta}{\mu} \boldsymbol{E}^{Amb}|_{x=0} + \left(1 + \frac{a^2}{6} \nabla^2\right) \boldsymbol{v}^{Amb}|_{x=0} \, . \tag{8.9}$$

Here, we temporarily use "*Amb*" instead of ∞ to denote the ambient field to avoid confusion with the widely accepted notation $\boldsymbol{E}^\infty$ for the applied electric field. Clearly, Equation 8.9 is simply the usual Faxén relation for Stokes flow, but with an effective Stokes ambient velocity field consisting of the true physical ambient field plus the Smoluchowski slip velocity. (The Faxén curvature term vanishes identically for the electric field.)

We expand the velocity and electric fields as

$$\begin{aligned} \boldsymbol{v} &= \boldsymbol{v}_1 + \boldsymbol{v}_2 + \boldsymbol{v}_{21} + \boldsymbol{v}_{12} + \cdots \\ \boldsymbol{E} &= \boldsymbol{E}^\infty + \boldsymbol{E}_1 + \boldsymbol{E}_2 + \boldsymbol{E}_{21} + \boldsymbol{E}_{12} + \cdots , \end{aligned}$$

with the subscripts tracing the reflection pattern as before. The zeroeth reflection at sphere 1 is the single particle result:

$$\begin{aligned} \boldsymbol{U}_1^{(0)} &= \frac{\epsilon\zeta_1}{\mu}\boldsymbol{E}^\infty \\ \boldsymbol{E}_1 &= -\frac{1}{2}\left(\frac{a}{r_1}\right)^3 (3\boldsymbol{e}_{r1}\boldsymbol{e}_{r1} - \boldsymbol{\delta}) \cdot \boldsymbol{E}^\infty . \end{aligned}$$

On S_1, the disturbance velocity field $\boldsymbol{v}_1$ satisfies the boundary condition,

$$\boldsymbol{v}_1 = \boldsymbol{U}_1^{(0)} - \frac{\epsilon\zeta_1}{\mu}\boldsymbol{E} = \frac{\epsilon\zeta_1}{\mu}(\boldsymbol{E}^\infty - \boldsymbol{E}) = -\frac{\epsilon\zeta_1}{\mu}\boldsymbol{E}_1 .$$

Since $\boldsymbol{E}_1$ is a vector harmonic, *an extension of this condition throughout the fluid region trivially gives the velocity field that satisfies the Stokes equations (with zero pressure gradient) and the above boundary condition*, so we have the desired result:

$$\boldsymbol{v}_1 = \frac{\epsilon\zeta_1}{\mu}\left[\frac{1}{2}\left(\frac{a}{r_1}\right)^3 (3\boldsymbol{e}_{r1}\boldsymbol{e}_{r1} - \boldsymbol{\delta}) \cdot \boldsymbol{E}^\infty\right] .$$

The disturbance field is a degenerate Stokes quadrupole and decays far away from the sphere as r^{-3}.

The first reflection at sphere 2 gives the leading order effect of hydrodynamic interactions, and this contribution is evaluated by appling the Faxén law of Keh and Anderson, Equation 8.9, to the incident electric and velocity fields:

$$\begin{aligned} \boldsymbol{U}_2^{(1)} &= \frac{\epsilon\zeta_2}{\mu}\boldsymbol{E}_1|_{x=x_2} + \left(1 + \frac{b^2}{6}\nabla^2\right)\boldsymbol{v}_1|_{x=x_2} \\ &= \frac{\epsilon\zeta_2}{\mu}\boldsymbol{E}_1|_{x=x_2} - \frac{\epsilon\zeta_1}{\mu}\boldsymbol{E}_1|_{x=x_2} \\ &= \frac{\epsilon}{\mu}(\zeta_2 - \zeta_1)\boldsymbol{E}_1|_{x=x_2} . \end{aligned}$$

The electrophoretic velocity to these leading order terms is thus given by

$$\boldsymbol{U}_2 = \frac{\epsilon\zeta_2}{\mu}\boldsymbol{E}^\infty - \frac{\epsilon}{\mu}(\zeta_2 - \zeta_1)\left[\frac{1}{2}\left(\frac{a}{R}\right)^3 (3\boldsymbol{dd} - \boldsymbol{\delta}) \cdot \boldsymbol{E}^\infty\right] .$$

We can readily generalize this result to all reflection orders and state that spheres with thin double layers do not interact hydrodynamically, unless they differ in surface potential, in which case they interact rather weakly as R^{-3}, *via* their degenerate Stokes quadrupole fields.

8.6.2 Multiple Ellipsoids

We may extend the analysis of Chen and Keh [6] to ellipsoids, including the important case of oblate (disk-like) spheroids, by combining the singularity solutions of Chapter 3 with the preceding line of reasoning. The Faxén law can be written immediately as

$$\boldsymbol{U} = \int_E f_{(1)} \left\{ 1 + \frac{1}{2} c^2 q^2 \nabla^2 \right\} \left\{ \boldsymbol{v}^{Amb} + \frac{\epsilon\zeta}{\mu} \boldsymbol{E}^{Amb} \right\} dA \ ,$$

with the ambient velocity and electric fields evaluated at the focal ellipse. If $\nabla^2 \boldsymbol{v} = 0$, as is the case with the incident fields, then the Faxén relation simplifies to

$$\boldsymbol{U} = \int_E f_{(1)} \left\{ \boldsymbol{v}^{Amb} + \frac{\epsilon\zeta}{\mu} \boldsymbol{E}^{Amb} \right\} dA \ .$$

The disturbance electric field $\boldsymbol{E}_1$ produced by ellipsoid 1 can be expressed as a singularity solution [53],

$$\boldsymbol{E}_1 = (\boldsymbol{M} \cdot \boldsymbol{E}^\infty) \cdot \nabla\nabla \int_E \frac{f_2(\boldsymbol{\xi})\, dA(\boldsymbol{\xi})}{|\boldsymbol{x} - \boldsymbol{\xi}|} \ ,$$

where $\boldsymbol{M}$ is a diagonal tensor expressible in terms of elliptic integrals, *viz.*,

$$M_{ii} = \left[\frac{3}{abc} - \frac{3}{2} \int_0^\infty \frac{dt}{(a_i^2 + t)\Delta t} \right]^{-1} \qquad \text{(no sum on } i\text{)}.$$

By following the same path as in the discussion for spheres, we find the following expression for the electrophoretic velocity of ellipsoid 2:

$$\boldsymbol{U}_2 = \frac{\epsilon\zeta_2}{\mu} \boldsymbol{E}^\infty + \boldsymbol{U}_2^{(1)} \ ,$$

with

$$\begin{aligned} \boldsymbol{U}_2^{(1)} &= -\frac{\epsilon}{\mu} (\zeta_2 - \zeta_1)(\boldsymbol{M} \cdot \boldsymbol{E}^\infty) \cdot \\ & \int_{E2} \int_{E1} \frac{f_{(1)}(\boldsymbol{\xi}_2) f_{(2)}(\boldsymbol{\xi}_1)}{R^3} (3\boldsymbol{dd} - \boldsymbol{\delta})\, dA_1(\boldsymbol{\xi}_1) dA_2(\boldsymbol{\xi}_2) \ . \qquad (8.10) \end{aligned}$$

The separation $R(\boldsymbol{\xi}_1, \boldsymbol{\xi}_2)$ and unit vector $\boldsymbol{d}$ now denote, respectively, the distance and direction between two points $\boldsymbol{\xi}_1$ and $\boldsymbol{\xi}_2$ on the focal ellipses. The analysis of the interaction between two ellipsoids and degenerate cases (spheroids) in various configurations of interest in colloidal hydrodynamics now follows in a straightforward fashion from this general result.

Exercises

Exercise 8.1 The Force and Torque on Two Rotating Spheres.
Consider two stationary spheres centered at $\boldsymbol{x}_1$ and $\boldsymbol{x}_2$ undergoing independent rotational motions, $\boldsymbol{\omega}_1 \times (\boldsymbol{x} - \boldsymbol{x}_1)$ and $\boldsymbol{\omega}_2 \times (\boldsymbol{x} - \boldsymbol{x}_2)$, in a quiescent fluid. Use the method of reflections to calculate the force and torque on the spheres accurate to $O(R^{-6})$.

Exercise 8.2 The Force on Two Translating Spheres.
Extend the results of Example 8.1 to $O(R^{-6})$ by including contributions from the fifth and sixth reflections. Note that the quadrupole terms in $\boldsymbol{v}_{12}$ and $\boldsymbol{v}_{21}$ must be retained so that the Faxén relation for the quadrupole moment,

$$\begin{aligned}\oint_S (\boldsymbol{\sigma}\cdot\hat{\boldsymbol{n}})_i \xi_j \xi_k \, dS(\boldsymbol{\xi}) &= C_1 \pi \mu a^5 \left\{1 + \frac{a^2\nabla^2}{14}\right\} \boldsymbol{v}^{\infty}(\boldsymbol{x})_{i,jk}|_{x=0} \\ &+ C_2 a^2 F_i \delta_{jk} + C_3 a^2 (F_j \delta_{ik} + F_k \delta_{ij}) \ ,\end{aligned}$$

must be derived. The method devised in Chapter 3 may be used [43], or, in this case, some simple arguments to fix the constants C_1, C_2 and C_3. Compare your answer for the force with the result given in Chapter 11.

Exercise 8.3 Sedimentation of Three Spheres.
Consider the mobility version of the three-sphere example: three identical, torque-free spheres with centers at $\boldsymbol{x}_1$, $\boldsymbol{x}_2$, and $\boldsymbol{x}_3$, settling under the action of gravity. To reduce the number of parameters, consider centers at the vertices of an equilateral triangle, so that $|\boldsymbol{x}_\alpha - \boldsymbol{x}_\beta| = R$ for all α, β. Find the sedimentation velocities accurate to $O(R^{-5})$ using just two reflections.

Exercise 8.4 External Torques on Two Spheres.
Consider two spheres with external torques acting about $\boldsymbol{x}_1$ and $\boldsymbol{x}_2$ in a quiescent fluid. Use the method of reflections to calculate the resulting rigid-body motions accurate to $O(R^{-5})$.

Exercise 8.5 Resistance Tensors by Inversion of Mobility Tensors.
The results from the previous exercise and the mobility example given earlier in the chapter provide to $O(R^{-5})$ the complete description of the mobility tensor for two spheres. Show that the resistance tensors obtained by inversion of these results are consistent with those obtained by direct calculation.

Exercise 8.6 Two Force-Free and Torque-Free Spheres in a Linear Field.
Consider two force-free and torque-free spheres in the linear field,

$$\boldsymbol{v}^{\infty} = \boldsymbol{\Omega}^{\infty} \times \boldsymbol{x} + \boldsymbol{E}^{\infty} \cdot \boldsymbol{x} \ .$$

Use the method of reflections to calculate the rigid body motion and stresslet for each sphere. Compare your solution to the result given in Chapter 11. Results for the relative velocity, $\boldsymbol{U}_2 - \boldsymbol{U}_1$, and the rotational velocities are also available from [1]. Their expression for the relative velocity,

$$\boldsymbol{U}_2 - \boldsymbol{U}_1 = \boldsymbol{\Omega}^{\infty} \times (\boldsymbol{x}_2 - \boldsymbol{x}_1) + \boldsymbol{E}^{\infty} \cdot (\boldsymbol{x}_2 - \boldsymbol{x}_1) - [A\boldsymbol{dd} + B(\boldsymbol{\delta} - \boldsymbol{dd})] \cdot \boldsymbol{E}^{\infty} \cdot (\boldsymbol{x}_2 - \boldsymbol{x}_1) \ ,$$

with

$$\begin{aligned} A &= \frac{5}{2}\frac{a^3 + b^3}{R^3} - \frac{3(a^5 + b^5) + 5a^2b^2(a+b)}{2R^5} + 25\frac{a^3b^3}{R^6} + o\left(\left(\frac{a+b}{R}\right)^6\right) \ , \\ B &= \frac{3(a^5 + b^5) + 5a^2b^2(a+b)}{3R^5} + o\left(\left(\frac{a+b}{R}\right)^6\right) \ , \end{aligned}$$

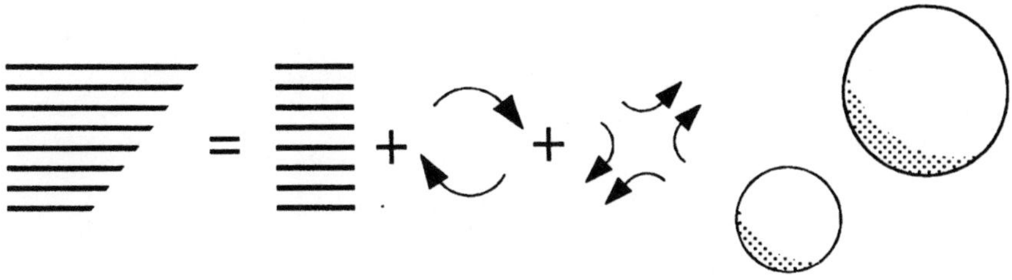

Figure 8.9: Two spheres in a linear field.

has gained wide usage. The functions A and B correspond, respectively, to motions along and transverse to the line of centers.

Exercise 8.7 Sedimentation of Two Oblate Spheroids.
Use the Faxén relations for the oblate spheroid to derive sedimentation velocities for two oblate spheroids. Compare the strength of hydrodynamic interactions with that in the corresponding problem for prolate spheroids.

Exercise 8.8 Electrophoresis of Two Oblate Spheroids.
Simplify Equation 8.10 for the case of two equal oblate spheroids. Devise a numerical scheme for the integrations over the focal ellipse and examine the electrophoretic mobility of two spheroids in edge-edge, edge-face, and face-face configurations at various orientations to the applied electric field.

9.1 Overview

Knowledge of the interactions between particles in close proximity is essential in understanding the suspension phenomena. For a certain class of problems, lubrication theory provides the leading order and singular terms in the resistance functions. If ℓ denotes the characteristic length of the particles, then the gap, defined as the separation between the closest points on the two particle surfaces, will be written as $\epsilon\ell$, with $\epsilon \ll 1$, and ϵ will be the key small paramater throughout this chapter.

For two rigid particles in relative motion, the dominant contributions to the force and torque may be determined by the methods of lubrication theory. In particular, we shall show *via* the lubrication equations that these leading order terms in the force and torque are singular as ϵ^{-1} and $\ln \epsilon^{-1}$ for small ϵ. Two representative problems (next two sections) involving the relative motion of spheres illustrate the essential ideas of lubrication theory. A sphere moving towards a second stationary sphere generates a "squeezing" flow, and we will see that the hydrodynamic resistance scales as ϵ^{-1}. On the other hand, a sphere moving past a second stationary sphere, in a direction transverse to the sphere-sphere axis, generates a "shearing" flow, and we will see that the hydrodynamic force and torque scales as $\ln \epsilon^{-1}$.

In Sections 9.4 and 9.5, we consider relative motion of two viscous drops near contact. We shall see that squeezing flows (Section 9.4) may involve a weaker singularity in the leading term, *viz.* $O(\epsilon^{-1/2})$, while for a bubble the singularity is an even weaker $O(\ln \epsilon^{-1})$. Shearing motions of one drop past another are not singular at all — the drops simply slip past each other with a finite mobility. Since all surface regions contribute to this $O(1)$ result, a matched asymptotic expansions approach is not practical. The numerical procedure for this problem is presented in Chapter 13, but the results are discussed in this chapter to provide a measure of continuity.

As mentioned at the beginning of Part III, lubrication methods do not apply at all for rigid surfaces moving in tandem as a single rigid body, because all regions near the bodies contribute to the $O(1)$ result for the resistance functions.

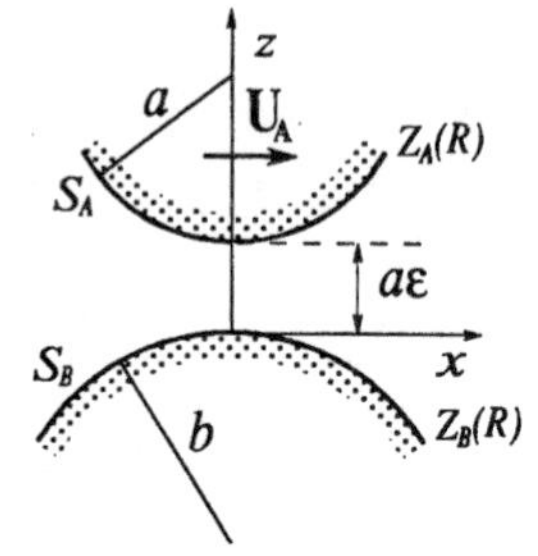

Figure 9.1: Flow produced by shearing motion of a sphere past another.

The numerical computations for these problems are also presented in Chapter 13, but these results will be used here to generate the higher order corrections to the terms obtained from lubrication theory.

9.2 Shearing Motions of Rigid Surfaces

Consider a moving sphere (denoted by S_A) of radius a and a stationary sphere (S_B) of radius $b = \beta a$. As shown in Figure 9.1, we choose a Cartesian coordinate system in which sphere S_A moves in the positive x-direction with velocity U and in which the sphere-sphere axis is coincident with the z-axis. The stationary sphere lies in the lower half space with the xy-plane as a tangent plane at the origin. The gap or clearance between the two spheres is denoted by $a\epsilon$.

The symmetries in the flow field near the origin suggest[1] the use of cylindrical coordinates (r, θ, z), with the corresponding velocity components (v_r, v_θ, v_z) and p written as

$$\begin{aligned} v_r &= U_A U(r,z)\cos\theta \\ v_\theta &= U_A V(r,z)\sin\theta \\ v_z &= U_A W(r,z)\cos\theta \\ p &= \frac{\mu U_A}{a} P(r,z)\cos\theta \,. \end{aligned}$$

The governing equations then become:

$$\frac{1}{a}\frac{\partial P}{\partial r} = L_0^2 U - \frac{2}{r^2}(U+V) \tag{9.1}$$

$$-\frac{1}{a}\frac{P}{r} = L_0^2 V - \frac{2}{r^2}(U+V) \tag{9.2}$$

$$\frac{1}{a}\frac{\partial P}{\partial z} = L_1^2 W \tag{9.3}$$

[1]See Chapter 18 for a more general discussion of the Fourier components.

$$\frac{\partial U}{\partial r} + \frac{U+V}{r} + \frac{\partial W}{\partial z} = 0 \; . \tag{9.4}$$

The operator L_m^2 is defined by

$$L_m^2 = \frac{\partial^2}{r^2} + \frac{1}{r}\frac{\partial}{\partial r} - \frac{m^2}{r^2} + \frac{\partial^2}{z^2} \; .$$

Near the origin, the surfaces of the spheres are given by

$$z_A = a(1+\epsilon) - (a^2 - r^2)^{1/2} \;\sim\; a\epsilon + \frac{r^2}{2a}$$

and

$$z_B = -\beta a + (\beta^2 a^2 - r^2)^{1/2} \;\sim\; -\frac{1}{2\beta a} r^2$$

for $r \ll a$. These results suggest a new, dimensionless "stretched" r coordinate,

$$R = \epsilon^{-1/2}\frac{r}{a} \; ,$$

while for the z-direction we scale with the gap $a\epsilon$ to obtain

$$Z = \epsilon^{-1}\frac{z}{a} \; .$$

In these new coordinates, the sphere surfaces are given by

$$Z_A = 1 + \frac{1}{2}R^2 + \frac{1}{8}\epsilon R^4 + O(\epsilon^2) \tag{9.5}$$

$$Z_B = -\frac{1}{2\beta}R^2 - \frac{1}{(2\beta)^3}\epsilon R^4 + O(\epsilon^2) \; . \tag{9.6}$$

Now the boundary conditions imply that both U and V are O(1), so from the equation of continuity, Equation 9.4, we obtain the scaling for W as $\epsilon^{1/2}$. The scaling for the pressure then follows immediately from the governing equations as $\epsilon^{3/2}$. Our first (and fortunately correct) inclination is to assume the simplest form, a regular expansion in ϵ of the form

$$P(R,Z) = \epsilon^{-3/2}P_0(R,Z) + \epsilon^{-1/2}P_1(R,Z) + \ldots \tag{9.7}$$

$$U(R,Z) = U_0(R,Z) + \epsilon U_1(R,Z) + \ldots \tag{9.8}$$

$$V(R,Z) = V_0(R,Z) + \epsilon V_1(R,Z) + \ldots \tag{9.9}$$

$$W(R,Z) = \epsilon^{1/2}W_0(R,Z) + \epsilon^{3/2}W_1(R,Z) + \ldots \; . \tag{9.10}$$

Substitution of the asymptotic expansions into the governing equations yields a hierarchy of equations, of which the lowest order is

$$\frac{\partial P_0}{\partial R} = \frac{\partial^2 U_0}{\partial Z^2} \; , \tag{9.11}$$

$$-\frac{P_0}{R} = \frac{\partial^2 V_0}{\partial Z^2} \; , \tag{9.12}$$

$$\frac{\partial P_0}{\partial Z} = 0 \; , \tag{9.13}$$

and

$$\frac{\partial U_0}{\partial R} + \frac{U_0 + V_0}{R} + \frac{\partial W_0}{\partial Z} = 0 \,. \tag{9.14}$$

Not surprisingly, these equations are identical to those encountered in lubrication theory [63]. The boundary conditions on the surfaces of the upper and lower sphere are

$$U_0(R, Z_A) = -V_0(R, Z_A) = 1, \qquad W_0(R, Z_A) = 0,$$
$$U_0(R, Z_B) = V_0(R, Z_B) = W_0(R, Z_B) = 0.$$

At the next order, we obtain the following equations for U_1, V_1, W_1, and P_1:

$$\frac{\partial P_1}{\partial R} = \frac{\partial^2 U_1}{\partial Z^2} + \frac{\partial^2 U_0}{\partial R^2} + \frac{1}{R}\frac{\partial U_0}{\partial R} - \frac{2(U_0 + V_0)}{R^2}\,, \tag{9.15}$$

$$-\frac{P_1}{R} = \frac{\partial^2 V_1}{\partial Z^2} + \frac{\partial^2 V_0}{\partial R^2} + \frac{1}{R}\frac{\partial V_0}{\partial R} - \frac{2(U_0 + V_0)}{R^2}\,, \tag{9.16}$$

$$\frac{\partial P_1}{\partial Z} = \frac{\partial^2 W_0}{\partial Z^2}\,, \tag{9.17}$$

and

$$\frac{\partial U_1}{\partial R} + \frac{U_1 + V_1}{R} + \frac{\partial W_1}{\partial Z} = 0 \,, \tag{9.18}$$

with boundary conditions

$$U_1(R, Z_A) = -\frac{1}{8}R^4\frac{\partial U_0}{\partial Z}\,,$$
$$V_1(R, Z_A) = -\frac{1}{8}R^4\frac{\partial V_0}{\partial Z}\,,$$
$$W_1(R, Z_A) = -\frac{1}{8}R^4\frac{\partial W_0}{\partial Z}\,,$$

and

$$U_1(R, Z_B) = \frac{1}{(2\beta)^3}R^4\frac{\partial U_0}{\partial Z}\,,$$
$$V_1(R, Z_B) = \frac{1}{(2\beta)^3}R^4\frac{\partial V_0}{\partial Z}\,,$$
$$W_1(R, Z_B) = \frac{1}{(2\beta)^3}R^4\frac{\partial W_0}{\partial Z}\,,$$

accounting for the $O(\epsilon)$ discrepancy in the placement of the sphere surface in the lower order equations.

We now turn our attention to the solution of these equations and the associated results for the force and torque on the spheres. From Equation 9.13, we see that P_0 is a function only of R whence we may integrate Equations 9.11 and 9.12 to obtain the following parabolic profiles for U_0 and V_0:

$$U_0 = -\frac{1}{2}P_0'(Z - Z_B)(Z_A - Z) + \frac{Z - Z_B}{H} \tag{9.19}$$

$$V_0 = \frac{1}{2}\frac{P_0}{R}(Z - Z_B)(Z_A - Z) - \frac{Z - Z_B}{H}\,. \tag{9.20}$$

Here

$$H(R) = Z_A - Z_B = 1 + \frac{1+\beta}{2\beta}R^2$$

is the gap between the upper and lower surfaces at R, neglecting terms of $O(\epsilon)$.

We solve Equation 9.14 for $\partial W_0/\partial Z$ and integrate from $Z = Z_B$ to $Z = Z_A$. Since $v_z = 0$ on the sphere surfaces, we obtain the result

$$\begin{aligned} W_0(R, Z_A) - W_0(R, Z_B) &= 0 \\ &= -\frac{1}{R}\int_0^H \frac{\partial}{\partial R}(RU_0)d\zeta - \frac{1}{R}\int_0^H V_0\,d\zeta \\ &= -\frac{1}{R}\left[\frac{\partial}{\partial R}\int_0^H (RU_0)d\zeta - R\frac{dZ_A}{dR}\right] - \frac{1}{R}\int_0^H V_0\,d\zeta\ , \end{aligned}$$

with $\zeta = Z - Z_B$. Since Z_A depends on R, we applied the Leibniz rule to change the order of integration and differentation.

We use $dZ_A/dR = R$, $dH/dR = R(1+\beta)/\beta$ and the following results for the velocity integrals,

$$\int_0^H U_0\,d\zeta = -\frac{H^3}{12}P_0' + \frac{H}{2} \qquad \int_0^H V_0\,d\zeta = \frac{H^3}{12}\frac{P_0}{R} - \frac{H}{2}\ ,$$

to obtain the *Reynolds equation,*

$$R^2 P_0'' + \left[R + 3(1+\frac{1}{\beta})\frac{R^3}{H}\right]P_0' - P_0 = -6(1-\frac{1}{\beta})\frac{R^3}{H^3}\ . \tag{9.21}$$

The exact expressions for the contribution to the force and torque from an infinitesimal surface element dS on S_A are

$$\begin{aligned} dF_x &= -p(\boldsymbol{n}\cdot\boldsymbol{\delta}_x) + [(\boldsymbol{\delta}_r\tau_{rr} + \boldsymbol{\delta}_\theta\tau_{\theta r})(\boldsymbol{\delta}_r\cdot\boldsymbol{n}) + (\boldsymbol{\delta}_r\tau_{rz} + \boldsymbol{\delta}_\theta\tau_{\theta z})(\boldsymbol{\delta}_z\cdot\boldsymbol{n})]\cdot\boldsymbol{\delta}_x dS \\ dT_y &= a(n_z dF_x - n_x dF_z) \\ &= n_z\left[(\boldsymbol{\delta}_r\tau_{rr} + \boldsymbol{\delta}_\theta\tau_{\theta r})(\boldsymbol{\delta}_r\cdot\boldsymbol{n}) + (\boldsymbol{\delta}_r\tau_{rz} + \boldsymbol{\delta}_\theta\tau_{\theta z})(\boldsymbol{\delta}_z\cdot\boldsymbol{n})\right]\cdot\boldsymbol{\delta}_x dS \\ &\quad + n_x\left[\tau_{zr}(\boldsymbol{\delta}_r\cdot\boldsymbol{n}) + \tau_{zz}(\boldsymbol{\delta}_z\cdot\boldsymbol{n})\right]dS\ . \end{aligned}$$

We employ spherical polar coordinates (ρ, χ, θ), noting the following identities:

$$\boldsymbol{\delta}_r\cdot\boldsymbol{n} = \sin\chi\ , \qquad \boldsymbol{\delta}_z\cdot\boldsymbol{n} = \cos\chi\ ,$$

$$\boldsymbol{\delta}_r\cdot\boldsymbol{\delta}_x = \cos\theta\ , \qquad \boldsymbol{\delta}_\theta\cdot\boldsymbol{\delta}_x = -\sin\theta\ , \qquad \boldsymbol{n}\cdot\boldsymbol{\delta}_x = \sin\chi\cos\theta\ .$$

We also use the stress tensor expressions in cylindrical coordinates and the velocity representation to obtain the following dimensionless expression for dF_x and dT_y:

$$\begin{aligned} dF_x &= \sin\chi\left[-P\cos^2\theta + 2\frac{\partial U}{\partial r}\cos^2\theta - \frac{\partial V}{\partial r}\sin^2\theta\right] \\ &+ \cos\chi\left[\frac{\partial U}{\partial z}\cos^2\theta - \frac{\partial V}{\partial z}\sin^2\theta + \frac{\partial W}{\partial r}\cos^2\theta\right] \end{aligned}$$

$$
\begin{aligned}
dT_y &= \sin\chi\cos\chi\left[+4\frac{\partial U}{\partial r}\cos^2\theta - \frac{\partial V}{\partial r}\sin^2\theta\right] \\
&+ \cos^2\chi\left[\frac{\partial U}{\partial z}\cos^2\theta - \frac{\partial V}{\partial z}\sin^2\theta + \frac{\partial W}{\partial r}\cos^2\theta\right] \\
&- \sin^2\chi\left[\frac{\partial U}{\partial z}\cos^2\theta + \frac{\partial W}{\partial r}\cos^2\theta\right] .
\end{aligned}
$$

The boundary conditions $U = -V = 1$ and $W = 0$ and the equation of continuity were used to simplify these expressions.

We may integrate the $\cos^2\theta$ and $\sin^2\theta$ factors over $0 \le \theta \le 2\pi$ and convert to the stretched variables to arrive at the following expressions for the force and torque:

$$
\begin{aligned}
\frac{F_x}{\pi\mu a U_A} &= \int_0^\pi \left\{\left[-P + \epsilon^{-1/2}\left(2\frac{\partial U}{\partial R} - \frac{\partial V}{\partial R}\right)\right]\sin\chi \right. \\
&\left. + \left[\epsilon^{-1}\left(\frac{\partial U}{\partial Z} - \frac{\partial V}{\partial Z}\right) + \epsilon^{-1/2}\frac{\partial W}{\partial R}\right]\cos\chi\right\}\sin\chi\, d\chi \\
\frac{T_y}{\pi\mu a^2 U_A} &= \int_0^\pi \left\{\left[\epsilon^{-1}\left(\frac{\partial U}{\partial Z} - \frac{\partial V}{\partial Z}\right) + \epsilon^{-1/2}\frac{\partial W}{\partial R}\right]\cos^2\chi \right. \\
&-\epsilon^{-1/2}\left(4\frac{\partial U}{\partial R} - \frac{\partial V}{\partial R}\right)\sin\chi\cos\chi \\
&\left. - \left[\epsilon^{-1}\frac{\partial U}{\partial Z} + \epsilon^{-1/2}\frac{\partial W}{\partial R}\right]\sin^2\chi\right\}\sin\chi\, d\chi .
\end{aligned}
$$

The extent of the gap region is denoted by the angle χ_0, with $\pi - \chi_0 \ll 1$; now change variables as

$$
\begin{aligned}
R &= \epsilon^{-1/2}\sin\chi \\
d\chi &= \frac{\epsilon^{1/2} dR}{\sqrt{1-\epsilon R^2}} .
\end{aligned}
$$

The resulting integrals for the force and torque are then expanded in powers of ϵ as

$$
\begin{aligned}
\frac{F_x}{\pi\mu a U_A} &= \int_0^{R_0}\left[-P_0 R + \left(\frac{\partial V_0}{\partial Z} - \frac{\partial U_0}{\partial Z}\right)\right] R dR \\
&+ \epsilon\int_0^{R_0}\left[\frac{P_0}{2}R^3 + P_1 R - \left(2\frac{\partial U_0}{\partial R} - \frac{\partial V_0}{\partial R}\right) R \right. \\
&\left. + \left(\frac{\partial U_1}{\partial Z} - \frac{\partial V_1}{\partial Z}\right) + \frac{\partial W_0}{\partial R}\right] RdR \\
&= -\frac{16\beta(2+\beta+2\beta^2)}{5(1+\beta)^3}\ln R_0 \qquad (9.22) \\
&- \frac{16(16-45\beta+58\beta^2-45\beta^3+16\beta^4)}{125(1+\beta)^4}\epsilon \ln R_0 + \ldots
\end{aligned}
$$

$$\begin{aligned}
\frac{T_y}{\pi\mu a^2 U_A} &= \int_0^{R_0}\left(\frac{\partial U_0}{\partial Z}-\frac{\partial V_0}{\partial Z}\right)RdR \\
&\quad + \epsilon\int_0^{R_0}\left[-\frac{3}{2}R^2\frac{\partial U_0}{\partial Z}-\frac{1}{2}R^2\frac{\partial V_0}{\partial Z}+R\left(4\frac{\partial U_0}{\partial R}-\frac{\partial V_0}{\partial R}\right)\right. \\
&\qquad\left.(1-R^2)\frac{\partial W_0}{\partial R}+\frac{\partial U_1}{\partial Z}-\frac{\partial V_1}{\partial Z}\right]RdR \\
&= \frac{8\beta(4+\beta)}{5(1+\beta)^2}\ln R_0 \\
&\quad + \frac{8(32-33\beta+83\beta^2+43\beta^3)}{125(1+\beta)^3}\epsilon\ln R_0+\dots\ .
\end{aligned} \tag{9.23}$$

Here, $R_0 = r_0/\epsilon^{1/2}$ corresponds to the limit of the lubrication region. As discussed by ONeill and Stewartson [59], a precise assignment of R_0 is not required; instead, a matching procedure is used. The above "inner" solution is matched in the limit of large R_0 with the corresponding entity from the "outer" solution. The principal result is that the leading term may be obtained by replacing $\ln R_0$ with $-(1/2)\ln\epsilon$, with corrections of $O(1)$ and $O(\epsilon)$ determined from the details of the matching with the outer solution. The lubrication results for the force and torque on sphere A are thus [39]

$$\begin{aligned}
\frac{F_x}{6\pi\mu a U_A} &= -\frac{4\beta(2+\beta+2\beta^2)}{15(1+\beta)^3}\ln\frac{1}{\epsilon}+A(\beta) \\
&\quad -\frac{4(16-45\beta+58\beta^2-45\beta^3+16\beta^4)}{375(1+\beta)^4}\epsilon\ln\frac{1}{\epsilon}+O(\epsilon)
\end{aligned} \tag{9.24}$$

$$\begin{aligned}
\frac{T_y}{8\pi\mu a^2 U_A} &= \frac{\beta(4+\beta)}{10(1+\beta)^2}\ln\frac{1}{\epsilon}+B(\beta) \\
&\quad +\frac{(32-33\beta+83\beta^2+43\beta^3)}{250(1+\beta)^3}\epsilon\ln\frac{1}{\epsilon}+O(\epsilon)\ .
\end{aligned} \tag{9.25}$$

In principle, the $O(1)$ terms $A(\beta)$ and $B(\beta)$ should be determined by matching with the outer solution. In practice, values for $A(\beta)$ and $B(\beta)$ may be obtained by a curve fit of numerical results for small ϵ.

The companion problem of calculating forces and torques produced by the rotation of sphere A about the y-axis (see Figure 9.2) can be solved with almost identical steps, with U_A replaced by $\omega_A a$. The final result for the force and torque are [39]

$$\begin{aligned}
\frac{F_x}{8\pi\mu a^2\omega_A} &= \frac{\beta(4+\beta)}{10(1+\beta)^2}\ln\frac{1}{\epsilon}+B(\beta) \\
&\quad +\frac{(32-33\beta+83\beta^2+43\beta^3)}{250(1+\beta)^3}\epsilon\ln\frac{1}{\epsilon}+O(\epsilon)
\end{aligned} \tag{9.26}$$

$$\begin{aligned}
\frac{T_y}{8\pi\mu a^3\omega_A} &= -\frac{2\beta}{5(1+\beta)}\ln\frac{1}{\epsilon}+C(\beta) \\
&\quad -\frac{2(8+6\beta+33\beta^2)}{125(1+\beta)^2}\epsilon\ln\frac{1}{\epsilon}+O(\epsilon)\ .
\end{aligned} \tag{9.27}$$

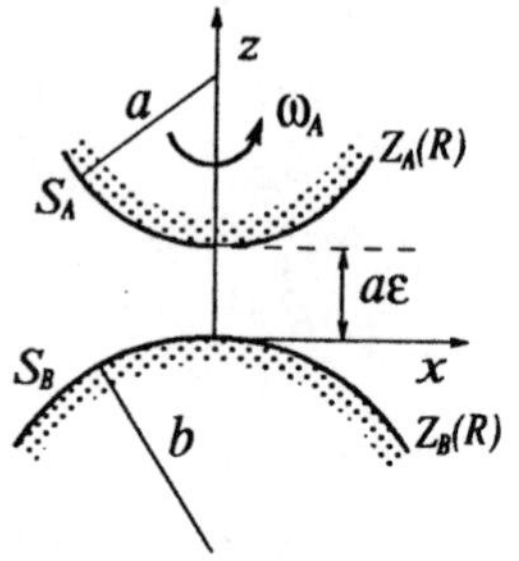

Figure 9.2: Flow produced by a shearing rotation of one sphere near another.

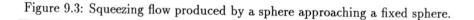

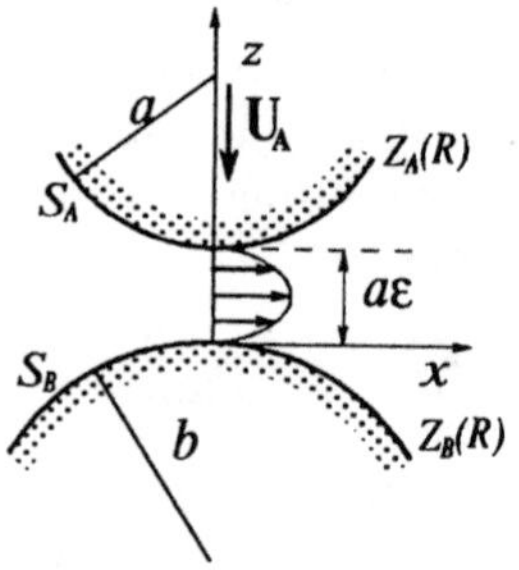

Figure 9.3: Squeezing flow produced by a sphere approaching a fixed sphere.

Note that the same coefficient appears in Equations 9.26 and 9.25 as required by the Lorentz reciprocal theorem.

9.3 Squeezing Motions of Rigid Surfaces

Here we follow closely the development in [37]. Consider a moving sphere (again denoted by S_A) of radius a and a stationary sphere (S_B) of radius $b = \beta a$. As shown in Figure 9.3, we choose a Cartesian coordinate system in which sphere S_A moves in the negative z-direction, with velocity U_A, and in which the sphere-sphere axis is coincident with the z-axis. The stationary sphere lies in the lower half space with the xy-plane as a tangent plane at the origin. The clearance between the two spheres is denoted by $a\epsilon$. In this case, the motion of one sphere toward the other results in a squeezing motion of the interstitial fluid and produces a stronger singularity in the hydrodynamic force.

Since this flow is axisymmetric, we find it convenient to use the Stokes streamfunction ψ, with

$$\begin{aligned} v_r &= U_A \frac{1}{r}\frac{\partial \psi}{\partial z} \\ v_z &= -U_A \frac{1}{r}\frac{\partial \psi}{\partial r} \\ v_\theta &= 0 \,. \end{aligned}$$

The governing equations then become

$$\left(\frac{\partial^2}{\partial z^2} + \frac{\partial^2}{\partial r^2} - \frac{1}{r}\frac{\partial}{\partial r}\right)^2 \psi = 0 \,, \tag{9.28}$$

with the boundary conditions

$$\psi = \frac{1}{2}r^2 \,, \qquad \frac{\partial \psi}{\partial z} = 0 \qquad \text{on } z = z_A,$$

$$\psi = \frac{\partial \psi}{\partial z} = 0 \qquad \text{on } z = z_B,$$

and with z_A and z_B as in the previous subsection on shearing motions. We convert again to the "stretched" coordinates,

$$R = \epsilon^{-1/2}\frac{r}{a} \qquad Z = \epsilon^{-1}\frac{z}{a} \,.$$

In these new coordinates, the sphere surfaces are given by

$$Z_A = 1 + \frac{1}{2}R^2 + \frac{1}{8}\epsilon R^4 + O(\epsilon^2) \tag{9.29}$$

$$Z_B = -\frac{1}{2\beta}R^2 - \frac{1}{(2\beta)^3}\epsilon R^4 + O(\epsilon^2) \,, \tag{9.30}$$

and the governing equation becomes

$$\left[\frac{\partial^2}{\partial Z^2} + \epsilon\left(\frac{\partial^2}{\partial R^2} - \frac{1}{R}\frac{\partial}{\partial R}\right)\right]^2 \psi = 0 \,. \tag{9.31}$$

Now the boundary conditions imply that v_z/U_A is O(1), so the streamfunction must scale as ϵ. We insert a regular expansion,

$$\psi(R,Z) = a^2\epsilon(\psi_0(R,Z) + \epsilon\psi_1(R,Z) + \psi_2(R,Z) + \ldots) \,, \tag{9.32}$$

into the governing equation to obtain a hierarchy of equations,

$$\begin{aligned} \frac{\partial^4 \psi_0}{\partial Z^4} &= 0 \\ \frac{\partial^4 \psi_1}{\partial Z^4} &= -2\Upsilon \frac{\partial^2 \psi_0}{\partial Z^2} \\ \frac{\partial^4 \psi_2}{\partial Z^4} &= -2\Upsilon \frac{\partial^2 \psi_1}{\partial Z^2} - \Upsilon^2 \psi_0 \,, \end{aligned}$$

where

$$\Upsilon = \frac{\partial^2}{\partial R^2} - \frac{1}{R}\frac{\partial}{\partial R} .$$

The boundary conditions are obtained by taking the Taylor expansions at $Z = Z_A$ and $Z = Z_B$. For ψ_0, they are

$$\psi_0 = \frac{1}{2}R^2 \qquad \frac{\partial\psi_0}{\partial Z} = 0 , \qquad \text{on } Z = Z_A$$

$$\psi_0 = 0 \qquad \frac{\partial\psi_0}{\partial Z} = 0 , \qquad \text{on } Z = Z_B .$$

For ψ_1, they are

$$\psi_1 = -\frac{1}{8}R^4\frac{\partial\psi_0}{\partial Z} = 0 , \qquad \frac{\partial\psi_1}{\partial Z} = -\frac{1}{8}R^4\frac{\partial^2\psi_0}{\partial Z^2} , \qquad \text{on } Z = Z_A$$

$$\psi_1 = \frac{1}{8\beta^3}R^4\frac{\partial\psi_0}{\partial Z} = 0 , \qquad \frac{\partial\psi_1}{\partial Z} = \frac{1}{8\beta^3}R^4\frac{\partial^2\psi_0}{\partial Z^2} , \qquad \text{on } Z = Z_B .$$

For ψ_2, they are

$$\begin{aligned}
\psi_2 &= -\frac{1}{8}R^4\frac{\partial\psi_1}{\partial Z} - \frac{1}{16}R^6\frac{\partial\psi_0}{\partial Z} - \frac{1}{128}R^8\frac{\partial^2\psi_0}{\partial Z^2} \\
&= \frac{1}{128}R^8\frac{\partial^2\psi_0}{\partial Z^2} \qquad \text{on } Z = Z_A \\
\frac{\partial\psi_2}{\partial Z} &= -\frac{1}{8}R^4\frac{\partial^2\psi_1}{\partial Z^2} - \frac{1}{16}R^6\frac{\partial^2\psi_0}{\partial Z^2} - \frac{1}{128}R^8\frac{\partial^3\psi_0}{\partial Z^3} \qquad \text{on } Z = Z_A \\
\psi_2 &= \frac{1}{128\beta^6}R^8\frac{\partial^2\psi_0}{\partial Z^2} \qquad \text{on } Z = Z_B \\
\frac{\partial\psi_2}{\partial Z} &= \frac{1}{8\beta^3}R^4\frac{\partial^2\psi_1}{\partial Z^2} + \frac{1}{16\beta^5}R^6\frac{\partial^2\psi_0}{\partial Z^2} - \frac{1}{128\beta^6}R^8\frac{\partial^3\psi_0}{\partial Z^3} \qquad \text{on } Z = Z_B .
\end{aligned}$$

We now turn our attention to the solution of these equations and the associated result for the force on the spheres. The exact expression for the contribution to the force is

$$\frac{F_z}{\pi\mu a U_A} = \int_0^\pi r^3\frac{\partial}{\partial n}\frac{E^2\psi}{r^2}\,ds ,$$

where ds is an element of arc length (in radians) along the meridian. On the moving sphere,

$$\begin{aligned}
ds &= \epsilon^{1/2}(1 + \frac{\epsilon}{2}R^2 + \frac{3\epsilon^2}{8}R^4)dR + O(\epsilon^{5/2}) \\
\frac{\partial}{\partial n} &= -\epsilon^{-1}\frac{\partial}{\partial Z} + \frac{R^2}{2}\frac{\partial}{\partial Z} + R\frac{\partial}{\partial R} + \epsilon\frac{R^4}{8}\frac{\partial}{\partial Z} + O(\epsilon^2) .
\end{aligned}$$

On the stationary sphere,

$$\begin{aligned}
ds &= \epsilon^{1/2}(1 + \frac{\epsilon}{2\beta^2}R^2 + \frac{3\epsilon^2}{8\beta^4}R^4)dR + O(\epsilon^{5/2}) \\
\frac{\partial}{\partial n} &= \epsilon^{-1}\frac{\partial}{\partial Z} - \frac{R^2}{2\beta^2}\frac{\partial}{\partial Z} - \frac{R}{\beta}\frac{\partial}{\partial R} - \epsilon\frac{R^4}{8\beta^4}\frac{\partial}{\partial Z} + O(\epsilon^2) .
\end{aligned}$$

The extent of the gap region where lubrication effects dominate is denoted by R_0 as before, and the integral for the force expands in powers of ϵ as

$$\frac{F_x}{\pi\mu a U_A} = \epsilon^{-1} I_0 + I_1 + \epsilon I_2 ,$$

with

$$\begin{aligned}
I_0 &= \int_0^{R_0} \left[R\frac{\partial^3\psi_0}{\partial Z^3} \right]_{Z=Z_A} dR \\
I_1 &= \int_0^{R_0} \left[R\frac{\partial^3\psi_1}{\partial Z^3} - \frac{\partial^2\psi_0}{\partial Z\partial R} + 2\frac{\partial^2\psi_0}{\partial Z^2} + R\frac{\partial^3\psi_0}{\partial Z\partial R^2} - R^2\frac{\partial^3\psi_0}{\partial Z^2\partial R} \right]_{Z=Z_A} dR \\
I_2 &= \int_0^{R_0} \left[R\frac{\partial^3\psi_2}{\partial Z^3} - \frac{\partial^2\psi_1}{\partial Z\partial R} + R\frac{\partial^2\psi_1}{\partial Z^2} + 2R\frac{\partial^3\psi_1}{\partial Z\partial R^2} - R^2\frac{\partial^3\psi_1}{\partial Z^2\partial R} \right. \\
&\quad \left. -3\frac{\partial\psi_0}{\partial R} + 3\frac{\partial^2\psi_0}{\partial R^2} + R^3\frac{\partial^2\psi_0}{\partial Z^2} - R^2\frac{\partial^3\psi_0}{\partial R^3} - \frac{R^2}{4}\frac{\partial^3\psi_0}{\partial Z^2\partial R} \right]_{Z=Z_A} dR .
\end{aligned}$$

These expressions indicate the specific information required from the asymptotic expansion of the streamfunction.

From the equation for ψ_0, we get

$$\psi_0 = A_0(R)Z^3 + B_0(R)Z^2 + C_0(R)Z + D_0(R) ,$$

with

$$A_0 = R^2/H^3 , \qquad B_0 = \frac{3}{2}R^2(Z_A + Z_B)/H^3 ,$$

$$C_0 = -3R^2 Z_A Z_B/H^3 , \qquad D_0 = \frac{1}{2}R^2(3Z_A - Z_B)Z_B^2/H^3 ,$$

determined from the boundary conditions.

The solution for ψ_1 is

$$\psi_1 = A_1(R)Z^3 + B_1(R)Z^2 + C_1(R)Z + D_1(R) - \frac{Z^5}{10}\Upsilon A_0 - \frac{Z^4}{6}\Upsilon B_0 ,$$

with

$$\begin{aligned}
A_1 &= (3Z_A^2 + 4Z_A Z_B + 3Z_B^2)\frac{1}{10}\Upsilon A_0 + (Z_A + Z_B)\frac{1}{3}\Upsilon B_0 \\
&\quad + \frac{3(1+\beta^3)}{8\beta^3}\frac{R^6}{H^4} \\
B_1 &= -(Z_A + Z_B)(Z_A^2 + 3Z_A Z_B + Z_B^2)\frac{1}{5}\Upsilon A_0 - (Z_A^2 + 4Z_A Z_B + Z_B^2)\frac{1}{6}\Upsilon B_0 \\
&\quad - \frac{3}{8}\frac{R^6}{H^4}[Z_A + 2Z_B + \frac{1}{\beta^3}(Z_B + 2Z_A)] \\
C_1 &= Z_A Z_B(4Z_A^2 + 7Z_A Z_B + 4Z_B^2)\frac{1}{10}\Upsilon A_0 + Z_A Z_B(Z_A + Z_B)\frac{1}{3}\Upsilon B_0 \\
&\quad + \frac{3}{8}\frac{R^6}{H^4}[Z_B(2Z_A + Z_B) + \frac{1}{\beta^3}Z_A(2Z_B + Z_A)]
\end{aligned}$$

$$
\begin{aligned}
D_1 &= -Z_A^2 Z_B^2 (Z_A + Z_B)\frac{1}{5}\Upsilon A_0 - Z_A^2 Z_B^2 \frac{1}{6}\Upsilon B_0 \\
&\quad - \frac{3}{8}\frac{R^6}{H^4} Z_A Z_B (Z_B + \frac{1}{\beta^3} Z_A) .
\end{aligned}
$$

The solution for ψ_2 is

$$
\begin{aligned}
\psi_2 &= A_2(R)Z^3 + B_2(R)Z^2 + C_2(R)Z + D_2(R) \\
&\quad + \frac{Z^7}{280}\Upsilon^2 A_0 + \frac{Z^6}{120}\Upsilon^2 B_0 - \frac{Z^5}{10}(\Upsilon A_1 + \frac{1}{12}\Upsilon^2 C_0 \\
&\quad - \frac{Z^4}{6}(\Upsilon B_1 + \frac{1}{4}\Upsilon^2 D_0 .
\end{aligned}
$$

From the expression for the force, we see that only A_2 is needed and this turns out to be

$$
\begin{aligned}
A_2 &= -(5Z_A^4 + 8Z_A^3 Z_B + 9Z_A^2 Z_B^2 + 8Z_A Z_B^3 + 5Z_B^4)\frac{1}{280}\Upsilon^2 A_0 \\
&\quad - (2Z_A^3 + 3Z_A^2 Z_B + 3Z_A Z_B^2 + 2Z_B^3)\frac{1}{60}\Upsilon^2 B_0 \\
&\quad + \frac{1}{10}(3Z_A^2 + 4Z_A Z_B + 3Z_B^2)(\Upsilon A_1 + \frac{1}{12}\Upsilon^2 C_0) \\
&\quad + \frac{1}{3}(Z_A + Z_B)(\Upsilon B_1 + \frac{1}{4}\Upsilon^2 D_0) \\
&\quad + \frac{3}{32}\frac{(1+\beta^6)}{\beta^6}\frac{R^{10}}{H^5} + \frac{3}{16}\frac{(1+\beta^5)}{\beta^5}\frac{R^8}{H^4} \\
&\quad \frac{1}{4}\frac{R^4}{H^2}[(Z_A^3 - \frac{1}{\beta^3}Z_B^3)\Upsilon A_0 + (Z_A^2 - \frac{1}{\beta^3}Z_B^2)\Upsilon B_0 \\
&\quad - 3A_1(Z_A - \frac{1}{\beta^3}Z_B) - B_1(1 - \frac{1}{\beta^3})] .
\end{aligned}
$$

The integrand for I_0 is

$$
6A_0 = -\frac{6R^3}{H^3} ,
$$

and the asymptotic behavior of those for I_1 and I_2 are

$$
-\frac{2(1 + 7\beta + \beta^2)}{5(1+\beta)^3 R}
$$

and

$$
\frac{24 + 61\beta - 206\beta^2 - 499\beta^3 - 256}{525(1+\beta)^3} R - \frac{2(1 + 18\beta - 29\beta^2 + 18\beta^3 + \beta^4)}{21(1+\beta)^4 R} .
$$

As discussed in O'Neill and Stewartson [59], the $O(R^{-1})$ terms will produce the $\ln \epsilon$ type singularity, while the $O(R)$ terms will be canceled during the matching process with the outer solution. Thus the lubrication result for the

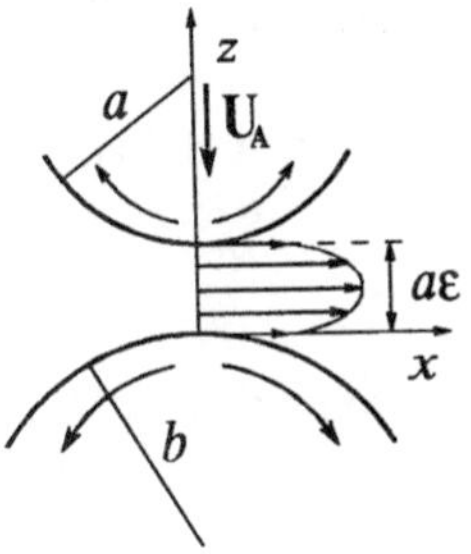

Figure 9.4: Squeezing flow produced by a drop approaching another drop.

force on sphere A is

$$\begin{aligned}\frac{F_x}{6\pi\mu a U_A} &= -\frac{\beta^2}{(1+\beta)^2}\epsilon^{-1} - \frac{(1+7\beta+\beta^2)}{5(1+\beta)^3}\ln\frac{1}{\epsilon} + K(\beta) \\ &\quad - \frac{(1+18\beta-29\beta^2+18\beta^3+\beta^4)}{21(1+\beta)^4}\epsilon\ln\frac{1}{\epsilon} + O(\epsilon)\ .\end{aligned} \quad (9.33)$$

In principle, the $K(\beta)$ and $L(\beta)$ should be determined by matching with the outer solution. In practice, these values may be obtained by a curve fit of numerical results for small ϵ. It should also be pointed out that a direct asymptotic analysis of the exact solution in bispherical coordinates has been done for two equal spheres ($\beta = 1$) [23, 26] and for the sphere and plane ($\beta = \infty$) [8], and that these results are consistent with the form proposed in Equation 9.33.

9.4 Squeezing Flow Between Viscous Drops

Two viscous drops approaching each other produce a squeezing flow that is different from that produced by the motion of rigid particles, as shown by Davis *et al.* [12]. Because of the mobility of the fluid-fluid interface, there is a slip velocity superposed on top of the parabolic profile (see Figure 9.4), and the pressure buildup is much less than that of the rigid case. The force singularity then scales as $\epsilon^{-1/2}$ instead of ϵ^{-1}, so that two drops pushed toward each other by a constant attractive force can meet in a finite time.

We write the radial component of the velocity as the sum of a uniform slip velocity, $u_t(r)$, and the familiar parabolic profile, $u_p(r,z)$, of the previous section:

$$v_r(r,z) = u_t(r) + u_p(r,z)\ ,$$

with

$$u_p(r,z) = \frac{1}{2\mu}\frac{\partial p}{\partial r}(z - z_A)(z - z_B)\ .$$

To find the expression for this unknown slip velocity, we examine the tractions at the fluid-fluid interface. To leading order, we may neglect surface curvature in the gap region so that the surface normals point in the z-direction. We denote the relevant component of the surface traction by f_t, and

$$f_t = -\mu \frac{\partial v_r}{\partial z}|_{z=z_A} = \mu \frac{\partial v_r}{\partial z}|_{z=z_B} = -(z_A - z_B)\frac{\partial p}{\partial r} \ . \tag{9.34}$$

The mass balance provides one relation between the velocities and the traction:

$$\pi r^2 U_A = 2\pi r \int_{z_A}^{z_B} v_r \, dz = 2\pi r \left((z_A - z_B)u_t + \frac{(z_A - z_B)^2}{6\mu} f_t \right) \ . \tag{9.35}$$

We obtain a second equation from the integral representation for the flow *inside* the drop:

$$\boldsymbol{v}(\boldsymbol{x}) = \frac{1}{8\pi\lambda\mu} \oint_S (\boldsymbol{\sigma} \cdot \boldsymbol{n}) \cdot \mathcal{G}(\boldsymbol{x} - \boldsymbol{\xi}) \, dS(\boldsymbol{\xi}) + \oint_S \boldsymbol{v}(\boldsymbol{\xi}) \cdot \boldsymbol{\Sigma}(\boldsymbol{x} - \boldsymbol{\xi}) \cdot \boldsymbol{n} \, dS(\boldsymbol{\xi}) \ .$$

Here, $\lambda\mu$ is the viscosity of the drop and $\boldsymbol{n}$ points from the drop interior to the exterior. Now consider the interfacial point $\boldsymbol{\eta}$ in the gap region. We know that in the gap (where the dominant contribution originates) the interface is essentially flat and thus the double layer term vanishes. Recalling the jump condition, we replace the LHS of the velocity representation with $\boldsymbol{v}(\boldsymbol{\eta})/2$ and insert the appropriate expressions for the surface tractions to simplify the integral equation. As in the rigid case, we extend the gap region to infinity with the understanding that this approximation does not affect the leading order term. The final result is

$$\begin{aligned} u_t(r) &= \frac{1}{\lambda\mu} \int_0^\infty K(r,s) f_t(s) \, ds \\ K(r,s) &= \frac{1}{2\pi} \frac{s}{\sqrt{r^2+s^2}} \int_0^\pi \frac{\cos\phi \, d\phi}{\sqrt{1-k^2\cos\phi}} \\ k^2 &= \frac{2rs}{r^2+s^2} \ . \end{aligned} \tag{9.36}$$

The kernel $K(r,s)$ involves an elliptic integral, and at $k=1$ ($r=s$) it possesses a logarithmic singularity that we readily attribute to the behavior of the Oseen tensor $\mathcal{G}(\boldsymbol{\eta} - \boldsymbol{\xi})$ for $\boldsymbol{\eta}$ near $\boldsymbol{\xi}$.

A scaling analysis is required at this point to delineate different regimes that differ in the degree of surface mobility. The most important parameter is the ratio u_t/u_p. When this parameter is small, we expect essentially the same behavior as in the rigid case (where this parameter is identically zero). On the other hand, if u_t/u_p is large, then the interface is *mobile* and we may obtain solutions quite different from the rigid case.

At the fluid-fluid interface, we have the traction boundary condition,

$$\mu \frac{\partial u_p}{\partial z} = \lambda\mu \frac{\partial v_r}{\partial z} \ .$$

In the exterior fluid, we scale $\partial u_p/\partial z \sim u_p/(a\epsilon)$. Inside the drop, $v_r \sim u_t$ and *both* the radial and axial lengths scale as $\sqrt{\bar{a}(a\epsilon)}$, where $\bar{a} = ab/(a+b)$ is the "reduced radius." This then implies $\partial u/\partial z \sim u_t/\sqrt{\bar{a}a\epsilon}$. This motivates Davis *et al.*'s [12] definition of the *interfacial mobility* parameter, m:

$$\frac{u_t}{u_p} \sim \lambda^{-1}\sqrt{\frac{\bar{a}}{a\epsilon}} = m \ . \tag{9.37}$$

Drops with nearly immobile interfaces have $m \ll 1$ (in other words, $u_t \ll u_p$) and behave essentially as rigid particles. On the other hand, under conditions where $m \gg 1$ ($u_t \gg u_p$), we obtain results that are quite different from those seen previously for rigid particles.

9.4.1 Nearly Rigid Drops

For nearly rigid drops ($m \ll 1$), the system of equations may be solved by a regular perturbation in m, with the leading term of each expression equal to the corresponding term from the rigid case [12]. The lubrication result for the force may be written, to leading order in ϵ, as

$$\frac{F_z}{6\pi\mu\bar{a}U_A} = \frac{\bar{a}}{a\epsilon}(1 - 1.31m + 1.78m^2 - 2.46m^3 + 3.44m^4 - 4.83m^5 + \cdots) \ .$$

9.4.2 Fully Mobile Interfaces

For fully mobile interfaces ($m \gg 1$) the drops offer no resistance to the outflow, so we simply neglect the parabolic profile, and obtain $u_t(r) = U_A r/(a\epsilon)$. We put this into the integral representation, Equation 9.36, and solve for f_t. This inversion is greatly simplified by the exact formula from Jansons and Lister [35],

$$f_t(r) = 8\lambda\mu \int_0^\infty K(r,s)\left(\frac{u_t}{s^2} - \frac{1}{s}\frac{du_t}{ds} - \frac{d^2u_t}{ds^2}\right) ds \ . \tag{9.38}$$

Now that f_t is known, the pressure distribution $p(r)$ can be obtained by integrating Equation 9.34, with $p(\infty) = 0$. The force follows as

$$F_z = 2\pi \int_0^\infty p(r) r dr = 6\pi\mu\bar{a}(2U_A)\left(\frac{\bar{a}}{a\epsilon}\right)\left(\frac{0.876}{m}\right) \approx 33.0\lambda\mu\bar{a}U_A\sqrt{\frac{\bar{a}}{a\epsilon}} \ .$$

Thus for a fully mobile interface, the force behaves as the *inverse square root* of the dimensionless gap width.

9.4.3 Solution in Bispherical Coordinates

The problem of two viscous drops translating along their line of centers can be solved by separating the equation for the streamfunction, $E^4\psi = 0$, in bispherical coordinates, in a manner analogous to the Stimson–Jeffrey solution for rigid spheres. Haber *et al.* [24] consider a very general situation with unequal drop sizes, velocities, and viscosities. For the case of two identical drops moving toward each other with equal speed, their solution reduces to

$$\frac{F_1}{6\pi\mu a U_1} = \frac{2}{3}\sinh\alpha \sum_{n=1}^{\infty} C_n \frac{N_n^{(0)}(\alpha,\lambda) + \lambda N_n^{(1)}(\alpha,\lambda)}{D_n^{(0)}(\alpha,\lambda) + \lambda D_n^{(1)}(\alpha,\lambda)}, \tag{9.39}$$

with

$$\begin{aligned}
N_n^{(0)}(\alpha,\lambda) &= 2[(2n+1)\sinh 2\alpha + 2\cosh 2\alpha - 2e^{-(2n+1)\alpha}] \\
N_n^{(1)}(\alpha,\lambda) &= (2n+1)^2\cosh 2\alpha - 2(2n+1)\sinh 2\alpha \\
&\quad - (2n+3)(2n-1) + 4e^{-(2n+1)\alpha} \\
D_n^{(0)}(\alpha,\lambda) &= 4\sinh(n-\tfrac{1}{2})\alpha \sinh(n+\tfrac{3}{2})\alpha \\
D_n^{(1)}(\alpha,\lambda) &= 2\sinh(2n+1)\alpha - (2n+1)\sinh 2\alpha \\
C_n &= \frac{n(n+1)}{(2n-1)(2n+3)} .
\end{aligned}$$

Here, F_1 and U_1 denote the force and speed of drop 1, and α is related to the drop-drop separation by $R = 2a\cosh\alpha$. Beshkov *et al.* [3] have considered the same problem, but their result for the force contains an error (effectively, their expression for C_n is $(n+1)/((2n+2)(2n-1))$ instead of the correct result above), so that their asymptotic expressions for the force between two nearly touching drops is also in error — by a factor of exactly $\sqrt{2}$, as inferred correctly by Davis *et al.* after a comparison with their numerical solution. For large λ, this expression above also reduces to the result for rigid spheres.

For small gaps, $\alpha \sim \epsilon^{1/2}$ is also small, and therefore many terms are needed in the bispherical solution. Following Cox and Brenner [8], we break the sum into an "inner sum" $\sum_{n=1}^{N}$ and an "outer sum" $\sum_{n=N+1}^{\infty}$. The breakpoint N is determined by requiring $\alpha N \sim 1$. Then in the inner sum, $\alpha n \ll 1$ for $\alpha \to 0$, and the leading expression in the asymptotic expansion for small α simplifies to

$$\begin{aligned}
\frac{F_1^{(inner)}}{6\pi\mu a U_1} &= \frac{2\lambda}{3\alpha}\sum_{n=1}^{N}\frac{8n(n+1)}{(2n-1)^2(2n+3)^2} \\
&= \frac{2\lambda}{3\alpha}\left\{\frac{5}{16}\sum_{n=1}^{N}\left[\frac{1}{2n-1}-\frac{1}{2n+3}\right]+\frac{3}{8}\sum_{n=1}^{N}\left[\frac{1}{(2n-1)^2}-\frac{1}{(2n+3)^2}\right]\right\} \\
&= \frac{2\lambda}{3\alpha}\left\{\frac{5}{16}\left[\frac{4}{3}+O(N^{-1})\right]+\frac{3}{8}\left[-\frac{10}{9}+O(N^{-1})\right]\right\} \\
&= \frac{2\lambda}{3\alpha}\left\{\frac{3\pi^2}{32}+O(N^{-1})\right\} .
\end{aligned}$$

As $\alpha \to 0$, the outer sum is bounded, so the leading term comes solely from the inner sum. For two equal spheres, the reduced radius is given by $\bar{a} = a/2$, and the final result for the leading behavior of the force may be written as

$$F_1 = \frac{3}{4}\pi^3\sqrt{2}\lambda\mu\bar{a}U_1\sqrt{\frac{\bar{a}}{a\epsilon}} \approx 32.887\lambda\mu\bar{a}U_1\sqrt{\frac{\bar{a}}{a\epsilon}} ,$$

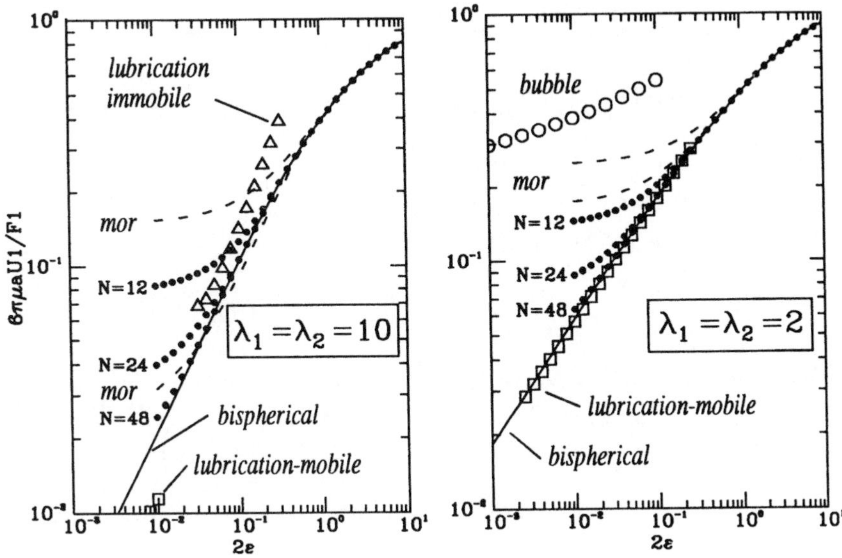

Figure 9.5: Mobility of two equal drops approaching each other: a comparison of different methods.

which is in excellent agreement with the result of Davis *et al.* The mobility functions obtained from lubrication theory, separation in bispherical coordinates, and the image method (Chapter 10) are shown in the figure. We compare the asymptotic solutions with the exact result for $\lambda = 0.1$, 2, and 10. The lubrication and image solutions are in reasonable agreement with the exact solution over their respective domain of validity and overlap quite nicely at moderate drop-drop gaps.

9.4.4 Nearly Touching Bubbles

When $\lambda \ll 1$ (bubble), the flow in the gap no longer gives the dominant contribution to the hydrodynamic force on the drop, and on intuitive grounds we would expect a much weaker singularity than $\epsilon^{-1/2}$ — perhaps even a regular solution, meaning that the bubbles do not offer any significant resistance to coalescense. The truth is actually somewhere in between, since the force is in fact singular, but it is the weak logarithmic singularity.

We set $\lambda = 0$ in Equation 9.39 and consider the behavior of the inner and outer sums. The inner sum simplifies to

$$\frac{F_1^{(inner)}}{6\pi\mu aU_1} = \frac{2}{3}\sum_{n=1}^{N}\left\{\frac{1}{2n-1} + \frac{1}{2n+3} + \frac{3}{4(2n-1)^2} - \frac{3}{4(2n+3)^2}\right\} .$$

The sums may be expanded for large N as

$$\sum_{n=1}^{N} \frac{2}{2n-1} \sim \ln N + \gamma + 2\ln 2 + O(N^{-2}) ,$$

where $\gamma = 0.577216\ldots$ is Euler's constant. Therefore, one set of sums simplify as

$$\sum_{n=1}^{N} \left\{ \frac{1}{2n+3} + \frac{1}{2n-1} \right\} \sim \ln N + \gamma + 2\ln 2 - \frac{4}{3} + o(1) ,$$

while the other can telescoped to

$$\frac{3}{4}\sum_{n=1}^{N} \left\{ \frac{1}{(2n-1)^2} - \frac{1}{(2n+3)^2} \right\} \sim \frac{5}{6} + O(N^{-2}) .$$

We define (see [8]) $X = N\alpha$ so that the inner sum may be written as

$$\frac{F_1^{(inner)}}{6\pi\mu a U_1} \sim \frac{2}{3}\left\{ \ln \alpha^{-1} + \gamma + 2\ln 2 - \frac{1}{2} + \ln X \right\} . \tag{9.40}$$

In the outer sum, we fix $\nu = n\alpha$ and let $\alpha \to 0$. The outer sum then simplifies to

$$\frac{F_1^{(outer)}}{6\pi\mu a U_1} \sim \frac{2}{3}\sum_{n=N+1}^{\infty} \Delta\nu \frac{(2\nu + 1 - e^{-2\nu})\Delta\nu}{4\sinh^2 \nu} .$$

When $n = N+1$, we have $\nu = (N+1)\alpha \sim X$, and the sum may be approximated using the Euler–MacLaurin formula, so that

$$\frac{F_1^{(outer)}}{6\pi\mu a U_1} \sim \frac{2}{3}\int_X^{\infty} f(\nu)\, d\nu - \frac{\alpha}{2}[f(\infty) - f(X)] ,$$

with

$$f(\nu) = \frac{2\nu + 1 - e^{-2\nu}}{4\sinh^2 \nu} .$$

We match with the inner sum by looking at small X. The integral has the asymptotic expansion,

$$\begin{aligned} \int_X^{\infty} f(\nu)\, d\nu &= \int_{2X}^{\infty} \frac{t + 1 - e^{-t}}{4(\cosh t - 1)} dt \\ &\sim -\ln X + \frac{1}{2} - \ln 2 + o(1) , \end{aligned}$$

so that

$$\frac{F_1^{(outer)}}{6\pi\mu a U_1} = -\ln X + \frac{1}{2} - \ln 2 + o(1) . \tag{9.41}$$

Combining Equations 9.40 and 9.41, we obtain the result for the drag on the bubble,

$$\begin{aligned} \frac{F_1}{6\pi\mu a U_1} &= \frac{2}{3}\left[\ln \alpha^{-1} + \gamma + \ln 2 \right] + o(1) \\ &= \frac{1}{3}\ln\left(\frac{a}{h}\right) + \frac{2}{3}(\gamma + \ln 2) + o(1) . \end{aligned} \tag{9.42}$$

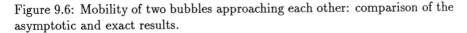

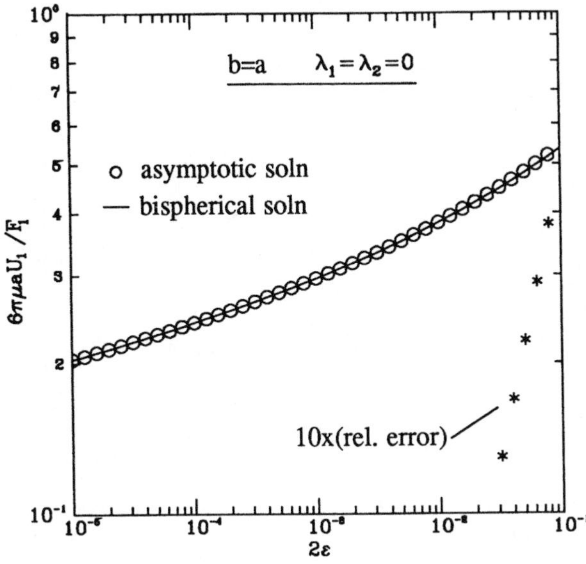

Figure 9.6: Mobility of two bubbles approaching each other: comparison of the asymptotic and exact results.

In the figure, we compare this asymptotic solution with the exact solution obtained by separation in bispherical coordinates. As shown in the figure, the $O(1)$ correction gives a dramatic improvement in accuracy.

9.5 Shearing Flow Between Viscous Drops

The flow between drops in transverse translations is not dominated by the gap region and thus lubrication theory does not apply. The solution must be obtained by numerical methods, such as the boundary collocation method of Chapter 13, but we present the results here to complete the picture of drops in relative motion. In the figure, we show the mobility for two equal drops (at various drop viscosities) for geometries ranging from near contact to large separations (the Hadamard–Rybczynski solution).[2] These results are consistent with those obtained by Zinchenko by bispherical coordinates (see the series of translated articles [72] to [76]).

[2]We acknowledge the assistance of Mr. John Geisz and Dr. Osman Basaran of Oak Ridge National Laboratory in the development of these results.

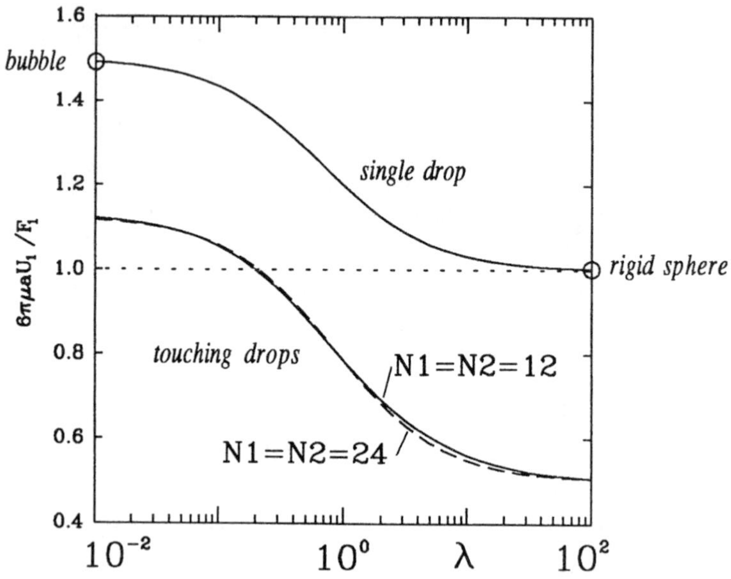

Figure 9.7: Mobility of two drops in transverse translation.

Exercises

Exercise 9.1 Crossed-Cylinders Force Apparatus.
The "crossed-cylinders" force apparatus is used to measure the colloidal forces between surfaces [32, 33, 34]. The colloidal force is deduced from the squeezing motion of the two surfaces — hence the solution for the leading term of $O(\epsilon^{-1})$ is required. Use lubrication theory to calculate this result and compare your answer with that given in the references. Also find the logarithmic singularity for one cylinder sliding over another.

Exercise 9.2 Contact Times for Two Drops.
Consider two equal-sized drops, one heavy and the other neutrally buoyant. If the heavy drop is released above the neutrally buoyant drop, with a small eccentric displacement, estimate the "contact time" between the two surfaces as a function of eccentricity and drop viscosity. Compare the predictions with the experimental data in the article by Yoon Luttrell [70]

Chapter 10

Interactions Between Large and Small Particles

10.1 Multiple Length Scales

The interactions between large and small particles involve multiple length scales. Let a and b denote the size of the large and small particles, respectively, and let h denote a measure of the gap between them. This chapter is concerned with the case where $b \ll h \ll a$ (see Figure 10.1). In essence, whereas the larger particle is in the far field of the smaller one, the smaller particle is in the near field of the larger one, whence a special form of the method of reflections is required. The key idea is that over lengths scales associated with the larger particle, the disturbance fields produced by the smaller one may be approximated by fields produced by equivalent a small number of Stokes singularities. From Chapter 3, we known that the higher order singularities (higher order terms in the multipole expansion) diminish in influence as powers of b/h.

The resistance and mobility functions are obtained by the method of reflections, as in Chapter 8 (see Figure 8.1), but with one crucial difference. For reflections at the larger particle, we retain the entire multipole solution, or

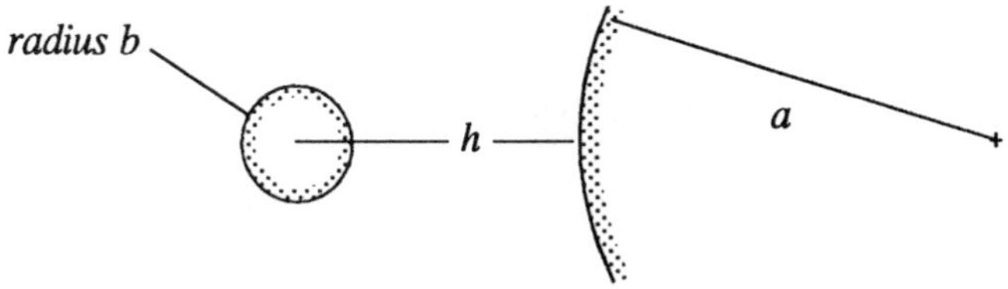

Figure 10.1: The geometry of large-small hydrodynamic interactions.

equivalently, use the exact result for image singularities induced by Stokes singularities near the particle. In practice, such information is available only for simple shapes — such as the sphere, cylinder, and plane wall — so there is a restriction on the shape of the larger particle (or wall-shape). The smaller particle can be of arbitrary shape, since it is represented by a collection of low-order Stokes singularities. We illustrate this point in Section 10.6 and 10.7 by examining the interactions between spheres of radii a and b with $\beta = b/a \ll 1$.

In the next four sections, we derive images for the Stokes singularities near a rigid sphere and spherical drop. In the next section, we consider the image system for a Stokeslet. The analysis differs for the axisymmetric case (Stokeslet directed along the line of centers) and the transverse case. In Section 10.3, these ideas are extended to spherical drops. The images for the higher order singularities (Stokes dipoles) are derived in Section 10.4 by superposition of images of two Stokeslets. The images for the stresslet derived therein are of particular importance in the calculation of resistance and mobility functions. In Section 10.5, we derive the image for the degenerate quadrupole, since it is an important entity in the singularity representation for spheres.

10.2 Image System for the Stokeslet Near a Rigid Sphere

Consider a Stokeslet at $\boldsymbol{x}_2$ and a sphere of radius a centered at $\boldsymbol{x}_1$, as shown in Figure 10.2. We simplify the notation by scaling lengths with the sphere radius a. Our task is to find the solution to the following problem:

$$-\nabla p + \mu \nabla^2 \boldsymbol{v} = -\boldsymbol{F}\delta(\boldsymbol{x} - \boldsymbol{x}_2), \qquad \nabla \cdot \boldsymbol{v} = 0, \tag{10.1}$$

with $|\boldsymbol{x}_2 - \boldsymbol{x}_1| > 1$ and the boundary conditions,

$$\boldsymbol{v} = 0 \qquad \text{for } r = a\ , \qquad\qquad \boldsymbol{v} \to 0 \qquad \text{as } r \to \infty,$$

with $r = |\boldsymbol{x} - \boldsymbol{x}_1|$.

We first write the solution as

$$\boldsymbol{v} = \boldsymbol{F} \cdot \mathcal{G}(\boldsymbol{x} - \boldsymbol{x}_2)/(8\pi\mu) + \boldsymbol{v}^*\ , \tag{10.2}$$

i.e., $\boldsymbol{v}^*$ is the image field. The essential steps in the solution procedure are as follows:

1. The Stokeslet at $\boldsymbol{x}_2$ is first written in terms of spherical harmonics, which are then transformed into harmonics based at $\boldsymbol{x}_1$ by an application of the addition theorem [30].

2. The image system is expanded as a multipole series about $\boldsymbol{x}_1$, and it too is rewritten in terms of the spherical harmonics at $\boldsymbol{x}_1$.

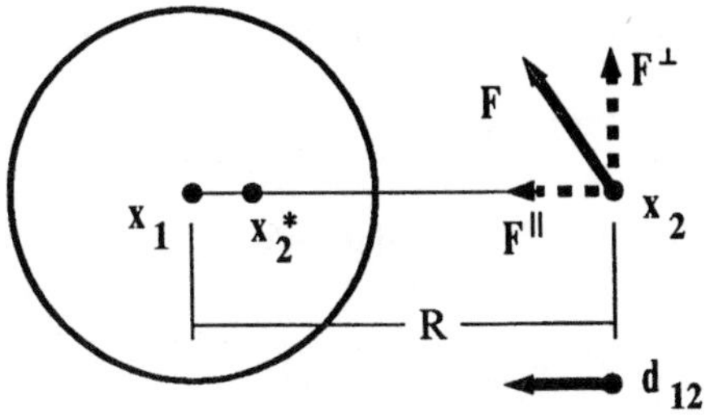

Figure 10.2: The Stokeslet-sphere geometry.

3. The coefficients in the image system may be determined by matching the boundary conditions at the sphere surface (term by term). In this case, we have the no-slip condition, while for viscous drops we use the more general boundary conditions described in Chapter 4.

If desired, the multipole expansion can be converted (by solving the appropriate set of moment equations) into a singularity solution. It turns out that the image system is the line between $\boldsymbol{x}_1$ and $\boldsymbol{x}_2^*$.

We start with the representation for the Stokeslet:

$$F_j \mathcal{G}_{ij} = F_i \frac{1}{r_2} + F_j (x - x_2)_i \frac{\partial}{\partial x_{2j}} \frac{1}{r_2} \,, \tag{10.3}$$

where we define $r_\alpha = |\boldsymbol{x} - \boldsymbol{x}_\alpha|$. Now harmonics about $\boldsymbol{x}_2$ may be recast in terms of those at $\boldsymbol{x}_1$ by the Legendre expansion,

$$\frac{1}{r_2} = \sum_{n=0}^{\infty} \frac{r_1^{2n+1}}{R^{n+1}} \frac{(\boldsymbol{d}_{12} \cdot \nabla)^n}{n!} \frac{1}{r_1} \,, \tag{10.4}$$

with the vector $\boldsymbol{d}_{12}$ defined by[1] $\boldsymbol{d}_{12} = (\boldsymbol{x}_1 - \boldsymbol{x}_2)/R$, $R = |\boldsymbol{x}_2 - \boldsymbol{x}_1| > 1$. Thus the Stokeslet at $\boldsymbol{x}_2$ may be written in terms of harmonics centered at $\boldsymbol{x}_1$,

$$\begin{aligned} F_j \mathcal{G}_{ij} &= F_i \sum_{n=0}^{\infty} \frac{r_1^{2n+1}}{R^{n+1}} \frac{(\boldsymbol{d}_{12} \cdot \nabla)^n}{n!} \frac{1}{r_1} \\ &+ (\boldsymbol{x} - \boldsymbol{x}_1)_i \, F_j \frac{\partial}{\partial x_{2j}} \sum_{n=0}^{\infty} \frac{r_1^{2n+1}}{R^{n+1}} \frac{(\boldsymbol{d}_{12} \cdot \nabla)^n}{n!} \frac{1}{r_1} \\ &- (\boldsymbol{x}_2 - \boldsymbol{x}_1)_i \, F_j \frac{\partial}{\partial x_{2j}} \sum_{n=0}^{\infty} \frac{r_1^{2n+1}}{R^{n+1}} \frac{(\boldsymbol{d}_{12} \cdot \nabla)^n}{n!} \frac{1}{r_1} \,, \end{aligned} \tag{10.5}$$

where $\partial/\partial x_{2j}$ denotes the nabla operator with respect to $\boldsymbol{x}_2$. The nabla operation with respect to $\boldsymbol{x}_2$ may be performed by noting that

$$\nabla_2 R^{-(n+1)} = -(n+1) R^{-(n+2)} (\boldsymbol{x}_2 - \boldsymbol{x}_1)/R = (n+1) R^{-(n+2)} \boldsymbol{d}_{12} \tag{10.6}$$

[1] When written in index notation, the subscripts (12) in $\boldsymbol{d}_{12}$ will be omitted.

and that $(\boldsymbol{d}_{12} \cdot \nabla_2)\boldsymbol{d}_{12} = \boldsymbol{0}$.

At this point, it is convenient to consider separately the two cases: $\boldsymbol{F}$ parallel to $\boldsymbol{d}_{12}$ (which is an axisymmetric problem) and $\boldsymbol{F}$ perpendicular to $\boldsymbol{d}_{12}$.

10.2.1 The Axisymmetric Stokeslet

For this case, we write $\boldsymbol{F} = F\boldsymbol{d}_{12}$ and Equation 10.6 becomes

$$\begin{aligned} Fd_j\mathcal{G}_{ij} &= Fd_i \sum_{n=0}^{\infty} (n+2)\frac{r_1^{2n+1}}{R^{n+1}} \frac{(\boldsymbol{d}_{12}\cdot\nabla)^n}{n!}\frac{1}{r_1} \\ &+ F(\boldsymbol{x}-\boldsymbol{x}_1)_i \sum_{n=0}^{\infty} (n+1)\frac{r_1^{2n+1}}{R^{n+2}} \frac{(\boldsymbol{d}_{12}\cdot\nabla)^n}{n!}\frac{1}{r_1} . \end{aligned} \tag{10.7}$$

Thus the Stokeslet at $\boldsymbol{x}_2$ has been expressed in terms of harmonics centered at $\boldsymbol{x}_1$.

We write the image system, $\boldsymbol{v}^*$, as a multipole expansion:

$$\boldsymbol{v}^* = \frac{F\boldsymbol{d}_{12}}{8\pi\mu} \sum_{n=0}^{\infty} \left\{ A_n^{\|}\frac{(\boldsymbol{d}_{12}\cdot\nabla)^n}{n!}\mathcal{G}_{ij}(\boldsymbol{x}-\boldsymbol{x}_1) + B_n^{\|}\frac{(\boldsymbol{d}_{12}\cdot\nabla)^n}{n!}\nabla^2\mathcal{G}_{ij}(\boldsymbol{x}-\boldsymbol{x}_1) \right\} .$$

The assumed form follows as a consequence of axisymmetry in the flow field (see Exercise 4.2).

These Stokes multipoles may be written as

$$\frac{(\boldsymbol{d}_{12}\cdot\nabla)^n}{n!}\mathcal{G}_{ij}(\boldsymbol{x}-\boldsymbol{x}_1) = \frac{(\boldsymbol{d}_{12}\cdot\nabla)^n}{n!}\left\{\delta_{ij}\frac{1}{r_1} - (\boldsymbol{x}-\boldsymbol{x}_1)_i\frac{\partial}{\partial x_j}\frac{1}{r_1}\right\} ,$$

which, since only two terms are nonzero when the Leibniz product rule is applied, simplifies to

$$\begin{aligned} \frac{(\boldsymbol{d}_{12}\cdot\nabla)^n}{n!}\mathcal{G}_{ij}(\boldsymbol{x}-\boldsymbol{x}_1) &= \delta_{ij}\frac{(\boldsymbol{d}_{12}\cdot\nabla)^n}{n!}\frac{1}{r_1} \\ &- (\boldsymbol{x}-\boldsymbol{x}_1)_i\frac{\partial}{\partial x_j}\frac{(\boldsymbol{d}_{12}\cdot\nabla)^n}{n!}\frac{1}{r_1} \\ &- d_i\frac{\partial}{\partial x_j}\frac{(\boldsymbol{d}_{12}\cdot\nabla)^{n-1}}{(n-1)!}\frac{1}{r_1} . \end{aligned} \tag{10.8}$$

The $B_n^{\|}$ terms may be rewritten in terms of vector harmonics by recalling that $\nabla^2\mathcal{G}(\boldsymbol{x}) = -2\nabla\nabla r^{-1}$. Thus the image system may be written as

$$\begin{aligned} v_i^* &= \frac{F}{8\pi\mu} \sum_{n=0}^{\infty} \left\{ A_n^{\|}\left[(1-n)d_i\frac{(\boldsymbol{d}_{12}\cdot\nabla)^n}{n!}\frac{1}{r_1} - (\boldsymbol{x}-\boldsymbol{x}_1)_i\frac{(\boldsymbol{d}_{12}\cdot\nabla)^{n+1}}{8\pi\mu n!}\frac{1}{r_1}\right] \right. \\ &\left. - 2B_n^{\|}\frac{(\boldsymbol{d}_{12}\cdot\nabla)^{n+1}}{n!}\frac{\partial}{\partial x_i}\frac{1}{r_1} \right\} . \end{aligned} \tag{10.9}$$

We can determine the coefficients $A_n^{\|}$ and $B_n^{\|}$ from the boundary conditions by collecting terms in harmonics of each order in Equations 10.8 and 10.9.

However, these harmonics are not independent; for example, we may eliminate terms involving $\boldsymbol{d}_{12}(\boldsymbol{d}_{12} \cdot \nabla)^n (1/r_1)$ by using the identity

$$\begin{aligned}
\boldsymbol{d}_{12}(\boldsymbol{d}_{12} \cdot \nabla)^n \frac{1}{r_1} &= \frac{-r_1^2}{(2n+1)(n+1)} \nabla (\boldsymbol{d}_{12} \cdot \nabla)^{n+1} \frac{1}{r_1} \\
&+ \frac{n}{2n+1} \nabla (\boldsymbol{d}_{12} \cdot \nabla)^{n-1} \frac{1}{r_1} \\
&- \frac{(2n+3)}{(2n+1)(n+1)} (\boldsymbol{x} - \boldsymbol{x}_1)(\boldsymbol{d}_{12} \cdot \nabla)^{n+1} \frac{1}{r_1} .
\end{aligned}$$

Thus at $r_1 = 1$, the expansion for the Stokeslet becomes

$$\begin{aligned}
F d_j \mathcal{G}_{ij} &= F \sum_{n=0}^{\infty} \frac{(n+3)}{(2n+3) R^{n+2}} \frac{\partial}{\partial x_i} \frac{(\boldsymbol{d}_{12} \cdot \nabla)^n}{n!} \frac{1}{r_1} \\
&- F \sum_{n=1}^{\infty} \frac{(n+1)}{(2n-1) R^n} \frac{\partial}{\partial x_i} \frac{(\boldsymbol{d}_{12} \cdot \nabla)^n}{n!} \frac{1}{r_1} \\
&+ F(\boldsymbol{x} - \boldsymbol{x}_1)_i \sum_{n=0}^{\infty} \frac{(n+1)}{R^{n+2}} \frac{(\boldsymbol{d}_{12} \cdot \nabla)^n}{n!} \frac{1}{r_1} \\
&- F(\boldsymbol{x} - \boldsymbol{x}_1)_i \sum_{n=1}^{\infty} \frac{(n+1)(2n+1)}{(2n-1) R^n} \frac{(\boldsymbol{d}_{12} \cdot \nabla)^n}{n!} \frac{1}{r_1} .
\end{aligned}$$

Note that the two terms with $n = 0$ cancel, so we may start all four summations from $n = 1$. Similarly, the expansion for the image system becomes

$$\begin{aligned}
v_i^* &= -\frac{F}{8\pi\mu} \sum_{n=1}^{\infty} \frac{2(n+1)}{(2n-1)} A_{n-1}^{\|} (\boldsymbol{x} - \boldsymbol{x}_1)_i \frac{(\boldsymbol{d}_{12} \cdot \nabla)^n}{n!} \frac{1}{r_1} \\
&+ \frac{F}{8\pi\mu} \sum_{n=1}^{\infty} \Big[\frac{(n-2)}{(2n-1)} A_{n-1}^{\|} - \frac{n}{(2n+3)} A_{n+1}^{\|} \\
&\qquad - 2n B_{n-1}^{\|} \Big] \frac{\partial}{\partial x_i} \frac{(\boldsymbol{d}_{12} \cdot \nabla)^n}{n!} \frac{1}{r_1} .
\end{aligned}$$

We collect terms in $(\boldsymbol{x} - \boldsymbol{x}_1)(\boldsymbol{d}_{12} \cdot \nabla)^n (1/r_1)$ and $\nabla(\boldsymbol{d}_{12} \cdot \nabla)^n (1/r_1)$ and set each equal to zero. Thus for $n \geq 1$, we have

$$\begin{aligned}
\frac{2(n+1)}{2n-1} A_{n-1}^{\|} &= (n+1) R^{-(n+2)} - \frac{(n+1)(2n+1)}{2n-1} R^{-n} \\
\frac{n A_{n+1}^{\|}}{2n+3} - \frac{(n-2)}{(2n-1)} A_{n-1}^{\|} - 2n B_{n-1}^{\|} &= \frac{n+3}{2n+3} R^{-(n+2)} - \frac{n+1}{2n-1} R^{-n} ,
\end{aligned}$$

which lead to

$$A_n^{\|} = -(n + \frac{3}{2}) R^{-(n+1)} + (n + \frac{1}{2}) R^{-(n+3)} \tag{10.10}$$

$$B_n^{\|} = -\frac{1}{4} (1 - R^2)^2 R^{-(n+5)} . \tag{10.11}$$

In the multipole expansion, the coefficient of the monopole field is the hydrodynamic force exerted on the sphere, *i.e.*,

$$\boldsymbol{F}^{\mathrm{Hyd}} = -A_0^{\|} F \boldsymbol{d}_{12} = \left(\frac{3}{2}R^{-1} - \frac{1}{2}R^{-3}\right) F \boldsymbol{d}_{12} \,, \tag{10.12}$$

while the coefficient of the Stokes dipole, which in this case is symmetric, must be the stresslet, so that

$$\boldsymbol{S} = A_1^{\|} F \boldsymbol{d}_{12} = \left(-\frac{5}{2}R^{-2} + \frac{3}{2}R^{-4}\right) F \boldsymbol{d}_{12}\boldsymbol{d}_{12} \,. \tag{10.13}$$

These results are consistent with that obtained by applying the Faxén laws for the force and stresslet to the ambient field, $\boldsymbol{F} \cdot \mathcal{G}(\boldsymbol{x} - \boldsymbol{x}_2)/(8\pi\mu)$.

10.2.2 The Transverse Stokeslet

We now consider the transverse case with $\boldsymbol{F} \cdot \boldsymbol{d}_{12} = 0$. Equation 10.6 becomes

$$\begin{aligned} F_j \mathcal{G}_{ij} &= F_i \sum_{n=0}^{\infty} \frac{r_1^{2n+1}}{R^{n+1}} \frac{(\boldsymbol{d}_{12} \cdot \nabla)^n}{n!} \frac{1}{r_1} \\ &- (\boldsymbol{x} - \boldsymbol{x}_1)_i \sum_{n=1}^{\infty} \frac{r_1^{2n+1}}{R^{n+2}} (\boldsymbol{F} \cdot \nabla) \frac{(\boldsymbol{d}_{12} \cdot \nabla)^{n-1}}{(n-1)!} \frac{1}{r_1} \\ &- d_i \sum_{n=1}^{\infty} \frac{r_1^{2n+1}}{R^{n+1}} (\boldsymbol{F} \cdot \nabla) \frac{(\boldsymbol{d}_{12} \cdot \nabla)^{n-1}}{(n-1)!} \frac{1}{r_1} \,. \end{aligned} \tag{10.14}$$

The Stokeslet at $\boldsymbol{x}_2$ has been expressed in terms of harmonics centered at $\boldsymbol{x}_1$.

We write the image system for the transverse case as follows:

$$\begin{aligned} v_i^* &= \frac{F_j}{8\pi\mu} \sum_{n=0}^{\infty} \left\{ A_n^{\perp} \frac{(\boldsymbol{d}_{12} \cdot \nabla)^n}{n!} \mathcal{G}_{ij}(\boldsymbol{x} - \boldsymbol{x}_1) + B_n^{\perp} \frac{(\boldsymbol{d}_{12} \cdot \nabla)^n}{n!} \nabla^2 \mathcal{G}_{ij}(\boldsymbol{x} - \boldsymbol{x}_1) \right\} \\ &+ \frac{\epsilon_{ijk} t_j}{8\pi\mu} \sum_{n=0}^{\infty} C_n^{\perp} \frac{(\boldsymbol{d}_{12} \cdot \nabla)^n}{n!} \frac{\partial}{\partial x_k} \frac{1}{r_1} \,. \end{aligned} \tag{10.15}$$

Here, $\boldsymbol{t} = \boldsymbol{F} \times \boldsymbol{d}_{12}$ and the $C_n^{\perp}$ terms originate because the Stokeslet at $\boldsymbol{x}_2$ is not pointed along the axis of symmetry. These terms may also be expressed as derivatives of the rotlet using the identity

$$\begin{aligned} \epsilon_{ijk} t_j \frac{\partial}{\partial x_k} \frac{1}{r_1} &= \frac{1}{2} \epsilon_{jkl} t_j \mathcal{G}_{il,k}(\boldsymbol{x} - \boldsymbol{x}_1) \\ &= \frac{1}{2} (F_k d_j - F_j d_k) \mathcal{G}_{ij,k}(\boldsymbol{x} - \boldsymbol{x}_1) \\ &= (F_k d_i - F_i d_k) \frac{\partial}{\partial x_k} \frac{1}{r_1} \,. \end{aligned} \tag{10.16}$$

Equation 10.16 may be used to convert harmonics in Equation 10.14 of the form $\boldsymbol{d}_{12}(\boldsymbol{F} \cdot \nabla)(\boldsymbol{d}_{12} \cdot \nabla)^{n-1}(1/r_1)$ to terms of the form $\boldsymbol{F}(\boldsymbol{d}_{12} \cdot \nabla)^n(1/r_1)$ plus derivatives of the rotlet. The image system may then be written as

$$
\begin{aligned}
v_i^* &= \sum_{n=0}^{\infty} A_n^{\perp} \Big[F_i(1-n)\frac{(\boldsymbol{d}_{12}\cdot\nabla)^n}{8\pi\mu\, n!}\frac{1}{r_1} - (\boldsymbol{x}-\boldsymbol{x}_1)_i(\boldsymbol{F}\cdot\nabla)\frac{(\boldsymbol{d}_{12}\cdot\nabla)^n}{8\pi\mu\, n!}\frac{1}{r_1} \\
&\qquad - n(\boldsymbol{t}\times\nabla)_i\frac{(\boldsymbol{d}_{12}\cdot\nabla)^{n-1}}{8\pi\mu\, n!}\frac{1}{r_1} \Big] \\
&- \sum_{n=0}^{\infty} 2B_n^{\perp}\frac{\partial}{\partial x_i}(\boldsymbol{F}\cdot\nabla)\frac{(\boldsymbol{d}_{12}\cdot\nabla)^n}{8\pi\mu\, n!}\frac{\partial}{\partial x_i}\frac{1}{r_1} \\
&+ \sum_{n=0}^{\infty} C_n^{\perp}(\boldsymbol{t}\times\nabla)_i\frac{(\boldsymbol{d}_{12}\cdot\nabla)^n}{8\pi\mu\, n!}\frac{1}{r_1}\ .
\end{aligned}
\tag{10.17}
$$

Since we have "three" coefficients — $A_n^{\perp}$, $B_n^{\perp}$, and $C_n^{\perp}$ — we should collect terms in three linearly independent vector functions. We have already mentioned how (using Equation 10.16) one of the vector functions may be eliminated. An examination of Equations 10.14 and 10.17 reveals that this still leaves four types. However, for $\boldsymbol{F}\cdot\boldsymbol{d}_{12}=0$ we have on the surface $r_1=1$ the identity,

$$
\begin{aligned}
&(\boldsymbol{x}-\boldsymbol{x}_1)(\boldsymbol{F}\cdot\nabla)(\boldsymbol{d}_{12}\cdot\nabla)^n\,\frac{1}{r_1} \\
&= \frac{-1}{(2n+3)}\nabla(\boldsymbol{F}\cdot\nabla)(\boldsymbol{d}_{12}\cdot\nabla)^n\frac{1}{r_1} + \frac{n(n-1)}{(2n+3)}\nabla(\boldsymbol{F}\cdot\nabla)(\boldsymbol{d}_{12}\cdot\nabla)^{n-2}\frac{1}{r_1} \\
&- \frac{(2n+1)(n+1)}{(2n+3)}\boldsymbol{F}(\boldsymbol{d}_{12}\cdot\nabla)^n\frac{1}{r_1} \\
&- \frac{n(2n+1)}{(2n+3)}(\boldsymbol{t}\times\nabla)(\boldsymbol{d}_{12}\cdot\nabla)^{n-1}\frac{1}{r_1}\ .
\end{aligned}
$$

Therefore, the final expressions for the Stokeslet and the image system involve three basis functions. For example, the Stokeslet may be written as

$$
\begin{aligned}
F_j\mathcal{G}_{ij} &= F_i\sum_{n=0}^{\infty}\Big[(1-n)R^{-(n+1)} + \frac{(2n+1)(n+1)}{(2n+3)R^{n+3}}\Big]\frac{(\boldsymbol{d}_{12}\cdot\nabla)^n}{n!}\frac{1}{r_1} \\
&+ \sum_{n=0}^{\infty}\Big[\frac{R^{-(n+3)}}{(2n+3)} - \frac{R^{-(n+5)}}{(2n+7)}\Big]\frac{\partial}{\partial x_i}(\boldsymbol{F}\cdot\nabla)\frac{(\boldsymbol{d}_{12}\cdot\nabla)^n}{n!}\frac{1}{r_1} \\
&- \sum_{n=0}^{\infty}\Big[R^{-(n+2)} - \frac{(2n+3)}{(2n+5)}R^{-(n+4)}\Big](\boldsymbol{t}\times\nabla)_i\frac{(\boldsymbol{d}_{12}\cdot\nabla)^n}{n!}\frac{1}{r_1}\ .
\end{aligned}
$$

As before, we determine the coefficients $A_n^{\perp}$, $B_n^{\perp}$, and $C_n^{\perp}$ from the boundary conditions by collecting terms in like harmonics. The terms in $\boldsymbol{F}(\boldsymbol{d}_{12}\cdot\nabla)^n(1/r_1)$, $\nabla(\boldsymbol{F}\cdot\nabla)(\boldsymbol{d}_{12}\cdot\nabla)^n(1/r_1)$, and $(\boldsymbol{t}\times\nabla)(\boldsymbol{d}_{12}\cdot\nabla)^n(1/r_1)$ yield the following set of equations:

$$
\begin{aligned}
\Big[1-n+\frac{(2n+1)(n+1)}{(2n+3)}\Big]A_n^{\perp} &= \frac{(n-1)}{R^{n+1}} + \frac{(2n+1)(n+1)}{(2n+3)R^{n+3}} \\
\frac{A_n^{\perp}}{(2n+3)} - \frac{A_{n+2}^{\perp}}{(2n+7)} - 2B_n^{\perp} &= -\frac{R^{-(n+3)}}{(2n+3)} + \frac{R^{-(n+5)}}{(2n+7)}
\end{aligned}
$$

$$\left[-1+\frac{(2n+3)}{(2n+5)}\right]A^{\perp}_{n+1}+C^{\perp}_{n} \;=\; R^{-(n+2)}-\frac{(2n+3)}{(2n+5)}R^{-(n+4)}\,,$$

which lead to

$$A^{\perp}_{n} \;=\; \frac{(n-1)(2n+3)}{2(n+2)}R^{-(n+1)}-\frac{(n+1)(2n+1)}{2(n+2)}R^{-(n+3)} \qquad (10.18)$$

$$B^{\perp}_{n} \;=\; \frac{n-1}{4(n+2)}R^{-(n+1)}-\frac{n^2+3n-1}{2(n+2)(n+4)}R^{-(n+3)} + \frac{n+1}{4(n+4)}R^{-(n+5)} \qquad (10.19)$$

$$C^{\perp}_{n} \;=\; \frac{2n+3}{n+3}R^{-(n+2)}-\frac{2n+3}{n+3}R^{-(n+4)}\,. \qquad (10.20)$$

The force, torque, and stresslet exerted by the fluid on the sphere are

$$\boldsymbol{F}^{\text{Hyd}} \;=\; -A^{\perp}_{0}\boldsymbol{F}=\left(\frac{3}{4}R^{-1}+\frac{1}{4}R^{-3}\right)\boldsymbol{F} \qquad (10.21)$$

$$\boldsymbol{T}^{\text{Hyd}} \;=\; (C^{\perp}_{0}-A^{\perp}_{1})\boldsymbol{t}=R^{-2}\boldsymbol{F}\times\boldsymbol{d}_{12} \qquad (10.22)$$

$$\boldsymbol{S} \;=\; A^{\perp}_{1}\frac{1}{2}(\boldsymbol{F}\boldsymbol{d}_{12}+\boldsymbol{d}_{12}\boldsymbol{F})=-\frac{1}{2}R^{-4}(\boldsymbol{F}\boldsymbol{d}_{12}+\boldsymbol{d}_{12}\boldsymbol{F})\,. \qquad (10.23)$$

Again, these results are consistent with the Faxén laws (see Exercise 10.1).

10.2.3 Singularity Solutions for the Image of a Stokeslet

We have obtained analytic expressions for the image of a Stokeslet near a sphere. As the Stokeslet is placed nearer to the sphere, the contribution from the higher order singularities becomes important. On the other hand, it would be nice to have the option of not dealing with the higher order Stokes singularities; these are awkward to implement in numerical schemes and are not as readily identified with physical quantities. Indeed, a Stokeslet placed very close to the sphere surface corresponds to, in the limit of vanishingly small separations, a Stokeslet near a plane wall. This limiting behavior is difficult to attain as long as the sphere center remains as a parameter in the problem. We would be trying to perform a multipole expansion about a point at infinity!

Here, we show that the image is indeed just a line distribution of a small collection of the lower order Stokes singularities. The line segment runs along the sphere-Stokeslet axis from the sphere center to the inverse point, $\boldsymbol{x}_2^*$, defined by $|\boldsymbol{x}_2^*-\boldsymbol{x}_1| = |\boldsymbol{x}_2-\boldsymbol{x}_1|^{-1}$. In Chapter 12, we will show that in the limit of vanishingly small separations, the contribution from the inverse point dominates and the image for the sphere reduces to the Lorentz image for the plane wall.

The Axisymmetric Problem

Consider a representation of the image system of the form

$$\boldsymbol{v}^* = -\boldsymbol{d}_{12}\cdot\int_0^{1/R} f(\xi)\frac{\mathcal{G}(\boldsymbol{x}-\boldsymbol{\xi})}{8\pi\mu}\,d\xi \;-\; \boldsymbol{d}_{12}\cdot\int_0^{1/R} g(\xi)\nabla^2\frac{\mathcal{G}(\boldsymbol{x}-\boldsymbol{\xi})}{8\pi\mu}\,d\xi\,,$$

where $\boldsymbol{\xi}$ lies on the line between $\boldsymbol{x}_1$ and $\boldsymbol{x}_2$. Now for any ξ in this interval, we have the Taylor expansion about $\boldsymbol{\xi} = \boldsymbol{x}_1$,

$$\mathcal{G}(\boldsymbol{x}-\boldsymbol{\xi}) = \sum_{n=0}^{\infty} \xi^n \frac{(\boldsymbol{d}_{12}\cdot\nabla)^n}{n!}\mathcal{G}(\boldsymbol{x}-\boldsymbol{x}_1) \; .$$

Therefore, the line distribution can also be written as a multipole expansion,

$$\begin{aligned} \boldsymbol{v}^* &= -\boldsymbol{d}_{12}\cdot \sum_{n=0}^{\infty} \Big[\int_0^{1/R} f(\xi)\xi^n d\xi \Big] \frac{(\boldsymbol{d}_{12}\cdot\nabla)^n}{n!} \frac{\mathcal{G}(\boldsymbol{x}-\boldsymbol{\xi})}{8\pi\mu}\, d\xi \\ &- \boldsymbol{d}_{12}\cdot \sum_{n=0}^{\infty} \Big[\int_0^{1/R} g(\xi)\xi^n d\xi \Big] \frac{(\boldsymbol{d}_{12}\cdot\nabla)^n}{n!} \nabla^2 \frac{\mathcal{G}(\boldsymbol{x}-\boldsymbol{\xi})}{8\pi\mu}\, d\xi \; . \end{aligned}$$

The line representation and the multipole expansion form derived earlier are equivalent if and only if

$$-A_n^{\|} = (n+\frac{3}{2})R^{-(n+1)} - (n+\frac{1}{2})R^{-(n+3)} = \int_0^{1/R} f(\xi)\xi^n \, d\xi$$

and

$$-B_n^{\|} = \frac{1}{4}(1-R^2)^2 R^{-(n+5)} = \int_0^{1/R} g(\xi)\xi^n \, d\xi \; ,$$

for all $n \geq 0$.

These equations yield the following solution for $f(\xi)$ and $g(\xi)$:

$$\begin{aligned} f(\xi) &= (\frac{3}{2}R^{-1} - \frac{1}{2}R^{-3})\delta(\xi - R^{-1}) - (R^{-2} - R^{-4})\frac{\partial}{\partial\xi}\delta(\xi - R^{-1}) \\ g(\xi) &= \frac{1}{4}R^{-5}(1-R^2)^2\delta(\xi - R^{-1}) \; , \end{aligned}$$

when we exploit the following identities for generalized functions:

$$\begin{aligned} \int_0^{1/R} \delta(\xi - R^{-1})\xi^n \, d\xi &= R^{-n} \\ \int_0^{1/R} \frac{\partial}{\partial\xi}\delta(\xi - R^{-1})\xi^n \, d\xi &= -nR^{-(n-1)} \; . \end{aligned}$$

We insert the result for $f(\xi)$ and $g(\xi)$ to obtain the following image for a Stokeslet oriented along the axis:

$$\begin{aligned} \boldsymbol{v}^* &= -(\frac{3}{2}R^{-1} - \frac{1}{2}R^{-3})F\boldsymbol{d}_{12}\cdot \frac{\mathcal{G}(\boldsymbol{x}-\boldsymbol{x}_2^*)}{8\pi\mu} \\ &- [(R^{-2} - R^{-4})F(\boldsymbol{d}_{12}\boldsymbol{d}_{12} - \frac{1}{3}\boldsymbol{\delta})\cdot\nabla]\cdot \frac{\mathcal{G}(\boldsymbol{x}-\boldsymbol{x}_2^*)}{8\pi\mu} \\ &- \frac{1}{4}R^{-1}(1-R^{-2})^2 F\boldsymbol{d}_{12}\cdot\nabla^2 \frac{\mathcal{G}(\boldsymbol{x}-\boldsymbol{x}_2^*)}{8\pi\mu} \; . \end{aligned} \quad (10.24)$$

So for the axisymmetric problem the "line distribution" is actually confined to the inverse point, and the collection of singularities consists of a Stokeslet of strength $(\frac{3}{2}R^{-1} - \frac{1}{2}R^{-3})$, a stresslet of strength $(R^{-2} - R^{-4})$, and a degenerate quadrupole of strength $\frac{1}{4}R^{-1}(1-R^{-2})^2$.

The Transverse Problem

Consider a representation of the image system of the form

$$\begin{aligned} \boldsymbol{v}^* &= -\boldsymbol{F} \cdot \int_0^{1/R} f(\xi) \frac{\mathcal{G}(\boldsymbol{x} - \boldsymbol{\xi})}{8\pi\mu}\, d\xi - \boldsymbol{F} \cdot \int_0^{1/R} g(\xi) \nabla^2 \frac{\mathcal{G}(\boldsymbol{x} - \boldsymbol{\xi})}{8\pi\mu}\, d\xi \\ &+ \frac{\boldsymbol{F} \times \boldsymbol{d}_{12}}{8\pi\mu} \times \nabla \int_0^{1/R} \frac{h(\xi)\, d\xi}{|\boldsymbol{x} - \boldsymbol{\xi}|}, \end{aligned}$$

where $\boldsymbol{\xi}$ lies on the $\boldsymbol{d}_{12}$ axis as before.

We rewrite the solution for $A_n^\perp$, $B_n^\perp$, and $C_n^\perp$, which was derived earlier as

$$\begin{aligned} A_n^\perp &= (R^{-2} - R^{-4}) n R^{-(n-1)} - (\frac{3}{2} R^{-1} - \frac{1}{2} R^{-3}) R^{-n} + \frac{3}{2}(R - R^{-1}) \frac{R^{-(n+2)}}{(n+2)} \\ B_n^\perp &= \frac{1}{4}(R^{-1} - 2R^{-3} + R^{-5}) R^{-n} - \frac{3}{4}(R - R^{-1}) \frac{R^{-(n+2)}}{(n+2)} \\ &\quad + \frac{3}{4}(R - R^{-1}) \frac{R^{-(n+4)}}{(n+4)} \\ C_n^\perp &= 2(R^{-2} - R^{-4}) R^{-n} - 3(R - R^{-1}) \frac{R^{-(n+3)}}{(n+3)}, \end{aligned}$$

to facilitate the identification of the density functions. Terms of the form $R^{-(n+k)}/(n+k)$ appearing above correspond to a true line distribution along $0 \le \xi \le R^{-1}$ with density ξ^{k-1}.

Following the experience from the axisymmetric problem, we make assignments of the form

$$\begin{aligned} -A_n^\perp &= \int_0^{1/R} f(\xi) \xi^n\, d\xi \\ -B_n^\perp &= \int_0^{1/R} g(\xi) \xi^n\, d\xi \\ C_n^\perp &= \int_0^{1/R} h(\xi) \xi^n\, d\xi, \end{aligned}$$

for all $n \ge 0$. The solutions for $f(\xi)$, $g(\xi)$, and $h(\xi)$ are

$$\begin{aligned} f(\xi) &= (R^{-2} - R^{-4}) \frac{\partial}{\partial \xi} \delta(\xi - R^{-1}) + (\frac{3}{2} R^{-1} - \frac{1}{2} R^{-3}) \delta(\xi - R^{-1}) \\ &\quad - \frac{3}{2}(R - R^{-1}) \xi \\ g(\xi) &= -\frac{1}{4}(R^{-1} - 2R^{-3} + R^{-5}) \delta(\xi - R^{-1}) + \frac{3}{4}(R - R^{-1})(\xi - \xi^3) \\ h(\xi) &= 2(R^{-2} - R^{-4}) \delta(\xi - R^{-1}) - 3(R - R^{-1}) \xi^2, \end{aligned}$$

and the image may be written as

$$\boldsymbol{v}^* = -(\frac{3}{2} R^{-1} - \frac{1}{2} R^{-3}) \boldsymbol{F} \cdot \frac{\mathcal{G}(\boldsymbol{x} - \boldsymbol{x}_2^*)}{8\pi\mu} + [(R^{-2} - R^{-4}) \boldsymbol{F} (\boldsymbol{d}_{12} \cdot \nabla)] \cdot \frac{\mathcal{G}(\boldsymbol{x} - \boldsymbol{x}_2^*)}{8\pi\mu}$$

$$
\begin{aligned}
&+ \frac{1}{4}R^{-1}(1-R^{-2})^2\boldsymbol{F}\cdot\frac{\nabla^2\mathcal{G}(\boldsymbol{x}-\boldsymbol{x}_2^*)}{8\pi\mu} + 2R^{-2}(1-R^{-2})\frac{\boldsymbol{F}\times\boldsymbol{d}_{12}}{8\pi\mu}\times\nabla\frac{1}{|\boldsymbol{x}-\boldsymbol{x}_2^*|} \\
&- \frac{3}{4}(R-R^{-1})\int_0^{1/R}(\xi-\xi^3)\boldsymbol{F}\cdot\nabla^2\frac{\mathcal{G}(\boldsymbol{x}-\boldsymbol{\xi})}{8\pi\mu}\,d\xi \\
&+ \frac{3}{2}(R-R^{-1})\int_0^{1/R}\xi\boldsymbol{F}\cdot\frac{\mathcal{G}(\boldsymbol{x}-\boldsymbol{\xi})}{8\pi\mu}\,d\xi \\
&- 3(R-R^{-1})\int_0^{1/R}\xi^2\frac{\boldsymbol{F}\times\boldsymbol{d}_{12}}{8\pi\mu}\times\nabla\frac{d\xi}{|\boldsymbol{x}-\boldsymbol{\xi}|}\,. \qquad (10.25)
\end{aligned}
$$

In conclusion, a point force acting transverse to the axis requires an image system consisting of a Stokeslet of strength $(\frac{3}{2}R^{-1}-\frac{1}{2}R^{-3})$, a Stokes dipole of strength $(R^{-2}-R^{-4})$ (which in turn can be further decomposed into a stresslet and rotlet), a rotlet of strength $2R^{-2}(1-R^{-2})$, and a degenerate quadrupole of strength $\frac{1}{4}R^{-1}(1-R^{-2})^2$ — all located at the inverse point, plus an additional distribution of Stokeslets, degenerate quadrupoles, and rotlets along the axis from the sphere center to the inverse point. These integals may be evaluated analytically to give the Oseen's expression for the image [62]. The solution is also given by Higdon [28].[2] In numerical calculations, these integrals over the line segment may be evaluated by Gaussian quadratures, thus effectively reducing the image to a finite number of Stokes singularities at a finite number of quadrature points. As the disturbance Stokeslet is placed further away, *i.e.*, as $R \to \infty$, the inverse point moves towards the sphere center and the line segment collapses onto the sphere center. This then leaves us with the usual situation for the reaction of a sphere to a distant disturbance. On the other hand, if the Stokeslet is very close to the sphere, *i.e.*, $R \approx 1+\epsilon$, the inverse point also moves to a point almost at the sphere surface, *i.e.*, $1/R \approx 1-\epsilon$, and the line distribution extends almost the entire distance from the sphere center to the surface.

Finally, in the limit of vanishingly small separations, the problem reduces to that of a Stokeslet near a plane wall. We start with the line distribution above and expand in a Taylor series *about the inverse point.* We find, as shown Chapter 12, that in the limit of a plane wall the image reduces to lower order Stokes singularities at the inverse point, which are then located exactly at the mirror image of $\boldsymbol{x}_2$.

10.2.4 Image for the Stokeslet Near a Viscous Drop

The image system for a Stokeslet near a spherical viscous drop has been derived by Fuentes *et al.* [16, 17]. Consider a spherical drop of viscosity μ_i immersed in a fluid of viscosity μ_e. As usual, we denote the viscosity ratio by $\lambda = \mu_i/\mu_e$. The derivation of the image system is similar to that given earlier in this chapter for the rigid sphere. The boundary conditions for the rigid sphere are replaced by those for a fluid-fluid interface, *viz.*, the kinematic condition for the radial

[2] There are typographical errors in Higdon's expression for the image but these are easily corrected by dimensional arguments.

velocity, continuity of tangential velocities, and continuity of the tangential component of the surface tractions (see Chapter 4).

For the *axisymmetric* problem of a Stokeslet pointing along the axis, we write the image field, $\boldsymbol{v}^*$, as a multipole expansion,

$$\boldsymbol{v}^* = \frac{F^{\|}\boldsymbol{d}_{12}}{8\pi\mu_e}\sum_{n=0}^{\infty}\left\{ A_n^{\|}\frac{(\boldsymbol{d}_{12}\cdot\nabla)^n}{n!}\mathcal{G}(\boldsymbol{x}-\boldsymbol{x}_1) + B_n^{\|}\frac{(\boldsymbol{d}_{12}\cdot\nabla)^n}{n!}\nabla^2\mathcal{G}(\boldsymbol{x}-\boldsymbol{x}_1)\right\},$$

outside the drop. The interior solution can be written using Lamb's general solution:

$$\begin{aligned}\boldsymbol{v}^{(i)} &= \frac{F^{\|}}{8\pi\mu_e}\sum_{n=1}^{\infty}\left\{a_n^{\|}\left[\frac{(n+3)}{2}r_1^{2n+3}\nabla\frac{(\boldsymbol{d}_{12}\cdot\nabla)^n}{n!}\frac{1}{r_1}\right.\right.\\ &\qquad\left.+\frac{(n+1)(2n+3)}{2}r_1^{2n+1}(\boldsymbol{x}-\boldsymbol{x}_1)\frac{(\boldsymbol{d}_{12}\cdot\nabla)^n}{n!}\frac{1}{r_1}\right]\\ &+ \left. b_n^{\|}\left[r_1^{2n+1}\nabla\frac{(\boldsymbol{d}_{12}\cdot\nabla)^n}{n!}\frac{1}{r_1}+(2n+1)r_1^{2n-1}(\boldsymbol{x}-\boldsymbol{x}_1)\frac{(\boldsymbol{d}_{12}\cdot\nabla)^n}{n!}\frac{1}{r_1}\right]\right\}.\end{aligned}$$

From the boundary conditions, we obtain a set of four coupled difference equations, which may be solved to yield

$$a_n^{\|} = (1-\Lambda)\left[R^{-(n+2)}-R^{-n}\right] \tag{10.26}$$

$$b_n^{\|} = -(1-\Lambda)\frac{(n+1)}{2}\left[R^{-(n+2)}-R^{-n}\right] \tag{10.27}$$

$$A_n^{\|} = \Lambda\left[\frac{-(2n+3)}{2R^{(n+1)}}+\frac{(2n+1)}{2R^{(n+3)}}\right]-(1-\Lambda)R^{-(n+1)} \tag{10.28}$$

$$B_n^{\|} = -\frac{1}{4}\Lambda(1-R^2)^2R^{-(n+5)}. \tag{10.29}$$

The parameter $\Lambda = \lambda/(\lambda+1)$ has been introduced to simplify the notation. Note that $0 \le \Lambda \le 1$, with $\Lambda = 0$ corresponding to a bubble and $\Lambda = 1$ to a rigid sphere. We make the following observations about this solution. For $\Lambda = 1$, the expressions above simplify to Equations 10.10 and 10.11, as expected. Note also that the image field for a bubble is particularly simple (we will say more about this later when we examine the singularity form of the image system). Finally, since the solution is linear in Λ, the general solution can be obtained by linear interpolation of the results for a bubble and a rigid sphere!

For the *transverse* problem, outside the drop, we use a multipole expansion,

$$\begin{aligned}\boldsymbol{v}^* &= \frac{\boldsymbol{F}^{\perp}}{8\pi\mu_e}\cdot\sum_{n=0}^{\infty}\left\{ A_n^{\perp}\frac{(\boldsymbol{d}_{12}\cdot\nabla)^n}{n!}\mathcal{G}(\boldsymbol{x}-\boldsymbol{x}_1) + B_n^{\perp}\frac{(\boldsymbol{d}_{12}\cdot\nabla)^n}{n!}\nabla^2\mathcal{G}(\boldsymbol{x}-\boldsymbol{x}_1)\right\}\\ &+\sum_{n=0}^{\infty}\left\{ C_n^{\perp}\frac{(\boldsymbol{d}_{12}\cdot\nabla)^n}{n!}(\boldsymbol{t}\times\nabla)\frac{1}{r_1}\right\}-(C_0^{\perp}-A_1^{\perp})(\boldsymbol{t}\times\nabla)\frac{1}{r_1},\end{aligned}$$

while inside we have Lamb's solution,

$$\boldsymbol{v}^{(i)} = \sum_{n=1}^{\infty}\left\{ a_n^{\perp}\left[\frac{(n+3)}{2}r_1^{2n+3}\nabla(\boldsymbol{F}\cdot\nabla)\frac{(\boldsymbol{d}_{12}\cdot\nabla)^{n-1}}{n!}\frac{1}{r_1}\right.\right.$$

$$+ \frac{(n+1)(2n+3)}{2} r_1^{2n+1} (\boldsymbol{x} - \boldsymbol{x}_1)(\boldsymbol{F}^{\perp} \cdot \nabla) \frac{(\boldsymbol{d}_{12} \cdot \nabla)^{n-1}}{n!} \frac{1}{r_1} \Bigg]$$

$$+ b_n^{\perp} \nabla \left[r_1^{2n+1} (\boldsymbol{F} \cdot \nabla) \frac{(\boldsymbol{d}_{12} \cdot \nabla)^{n-1}}{n!} \frac{1}{r_1} \right]$$

$$+ c_n^{\perp} \Bigg[r_1^{2n-1} (\boldsymbol{t} \times \nabla) \frac{(\boldsymbol{d}_{12} \cdot \nabla)^{n-1}}{(n-1)!} \frac{1}{r_1}$$

$$+ (2n-1) r_1^{2n-3} \boldsymbol{t} \times (\boldsymbol{x} - \boldsymbol{x}_1) \frac{(\boldsymbol{d}_{12} \cdot \nabla)^{n-1}}{(n-1)!} \frac{1}{r_1} \Bigg] .$$

From the boundary conditions, we obtain a set of four coupled difference equations, which may be solved to yield

$$a_n^{\perp} = \frac{1}{(1+\lambda)} \left[\frac{(n-2)}{(n+1)} R^{-n} - \frac{n}{(n+1)} R^{-(n+2)} \right] \tag{10.30}$$

$$b_n^{\perp} = -\frac{(n+1)}{2} a_n^{\perp} + c_{n+1}^{\perp} \tag{10.31}$$

$$c_n^{\perp} = \frac{2(2n-3)}{(n-2)(n+(n-3)\lambda)} R^{-(n-1)} \tag{10.32}$$

$$A_n^{\perp} = \Lambda \frac{(2n+1)}{2(n+2)} \left[(n-1) R^{-(n+1)} - (n+1) R^{-(n+3)} \right] + \frac{(n-1)}{(n+2)} R^{-(n+1)} \tag{10.33}$$

$$B_n^{\perp} = \Lambda \left[\frac{(n+1)}{4(n+4)} R^{-(n+5)} - \frac{(n+1)}{2(n+2)} R^{-(n+3)} + \frac{(n-1)}{4(n+2)} R^{-(n+1)} \right] + \Lambda \frac{(2n+5)(n+1)}{2(n+2)(n+3)(n+4)} \frac{(n+6-3\Lambda)}{(n+4-3\Lambda)} R^{-(n+3)} \tag{10.34}$$

$$C_n^{\perp} = \Lambda \frac{(2n+3)}{(n+2)(n+3)} \left[n R^{-(n+2)} - (n+2) R^{-(n+4)} \right] + \Lambda \frac{2n(2n+3)}{(n+2)(n+3)} \frac{1}{(n+3-3\Lambda)} R^{-(n+2)} . \tag{10.35}$$

Using the same procedure as that employed for the rigid sphere, we may convert the multipole expansion for the image into line distributions. The motivation for this action is as before. For a Stokeslet aligned with $\boldsymbol{d}_{12}$ (the axisymmetric problem), the image is again confined to the inverse point and is given by

$$\begin{aligned} \boldsymbol{v}^* &= \left[-\Lambda \left(\frac{3}{2} R^{-1} - \frac{1}{2} R^{-3} \right) - (1-\Lambda) R^{-1} \right] \boldsymbol{F}^{\|} \cdot \frac{\mathcal{G}(\boldsymbol{x} - \boldsymbol{x}_2^*)}{8\pi\mu_e} \\ &- \Lambda (R^{-2} - R^{-4}) F^{\|} [(\boldsymbol{d}_{12}\boldsymbol{d}_{12} - \frac{1}{3}\boldsymbol{\delta}) \cdot \nabla] \cdot \frac{\mathcal{G}(\boldsymbol{x} - \boldsymbol{x}_2^*)}{8\pi\mu_e} \\ &- \frac{1}{4} \Lambda R^{-1} (1 - R^{-2})^2 \boldsymbol{F}^{\|} \cdot \nabla^2 \frac{\mathcal{G}(\boldsymbol{x} - \boldsymbol{x}_2^*)}{8\pi\mu_e} \\ &= \Lambda \boldsymbol{v}_{rigid}^* + (1-\Lambda) \boldsymbol{v}_{bubble}^* . \end{aligned} \tag{10.36}$$

The image for the spherical drop is simply a linear combination of the results for the rigid sphere and the bubble ($\Lambda = 0$). As we shall see below, this nice result does not hold in the transverse problem.

For the transverse problem, the image is

$$\begin{aligned}
\boldsymbol{v}^* &= \Lambda \boldsymbol{v}^*_{Rigid} + (1-\Lambda)\boldsymbol{v}^*_{Bubble} \\
&- \frac{3}{2}\Lambda(1-\Lambda)R^{-2}\int_0^{1/R}\Big[R\xi\, L(\xi, 2-3\Lambda) + 2R^2\xi^2 L(\xi, 1-3\Lambda) \\
&\qquad -3R^3\xi^3 L(\xi, -3\Lambda)\Big]\, \boldsymbol{F}^{\perp}\cdot\nabla^2\frac{\mathcal{G}(\boldsymbol{x}-\boldsymbol{\xi})}{8\pi\mu_e}\, d\xi \\
&+ 6\Lambda(1-\Lambda)R^{-1}\int_0^{1/R}\Big[R\xi\, L(\xi, 1-3\Lambda) - 3R^2\xi^2 L(\xi, -3\Lambda)\Big] \\
&\qquad \times \frac{\boldsymbol{F}^{\perp}\times \boldsymbol{d}_{12}}{8\pi\mu_e}\times\nabla\frac{1}{|\boldsymbol{x}-\boldsymbol{\xi}|}\, d\xi\ ,
\end{aligned} \tag{10.37}$$

where

$$L(\xi, \epsilon) = \frac{1-(R\xi)^{\epsilon}}{\epsilon}\ ,$$

and $\boldsymbol{v}^*_{Rigid}$ and $\boldsymbol{v}^*_{Bubble}$ are given by

$$\begin{aligned}
\boldsymbol{v}^*_{Rigid} &= -(\frac{3}{2}R^{-1} - \frac{1}{2}R^{-3})\boldsymbol{F}^{\perp}\cdot\frac{\mathcal{G}(\boldsymbol{x}-\boldsymbol{x}_2^*)}{8\pi\mu_e} \\
&+ (R^{-2}-R^{-4})\,\boldsymbol{F}^{\perp}(\boldsymbol{d}_{12}\cdot\nabla)\cdot\frac{\mathcal{G}(\boldsymbol{x}-\boldsymbol{x}_2^*)}{8\pi\mu_e} \\
&+ \frac{1}{4}R^{-1}(1-R^{-2})^2\boldsymbol{F}^{\perp}\cdot\nabla^2\frac{\mathcal{G}(\boldsymbol{x}-\boldsymbol{x}_2^*)}{8\pi\mu_e} \\
&+ 2R^{-2}(1-R^{-2})\frac{\boldsymbol{F}^{\perp}\times\boldsymbol{d}_{12}}{8\pi\mu_e}\times\nabla\frac{1}{|\boldsymbol{x}-\boldsymbol{x}_2^*|} \\
&+ \frac{3}{2}(R-R^{-1})\int_0^{1/R}\xi\,\boldsymbol{F}^{\perp}\cdot\frac{\mathcal{G}(\boldsymbol{x}-\boldsymbol{\xi})}{8\pi\mu_e}\, d\xi \\
&- \frac{3}{4}(R-R^{-1})\int_0^{1/R}(\xi-\xi^3)\boldsymbol{F}^{\perp}\cdot\nabla^2\frac{\mathcal{G}(\boldsymbol{x}-\boldsymbol{\xi})}{8\pi\mu_e}\, d\xi \\
&- 3(R-R^{-1})\int_0^{1/R}\xi^2\frac{\boldsymbol{F}^{\perp}\times\boldsymbol{d}_{12}}{8\pi\mu_e}\times\nabla\frac{d\xi}{|\boldsymbol{x}-\boldsymbol{\xi}|}
\end{aligned} \tag{10.38}$$

$$\boldsymbol{v}^*_{Bubble} = R^{-1}\boldsymbol{F}^{\perp}\cdot\frac{\mathcal{G}(\boldsymbol{x}-\boldsymbol{x}_2^*)}{8\pi\mu_e} - 3\int_0^{1/R}R\xi\,\boldsymbol{F}^{\perp}\cdot\frac{\mathcal{G}(\boldsymbol{x}-\boldsymbol{\xi})}{8\pi\mu_e}\, d\xi\ . \tag{10.39}$$

To summarize, when the point force is directed along the axis, the flow outside the drop is that due to the Stokeslet plus that of the image, with the image consisting of a Stokeslet of strength $\Lambda(\frac{3}{2}R^{-1} - \frac{1}{2}R^{-3}) + (1-\Lambda)R^{-1}$, a stresslet of strength $\Lambda(R^{-2} - R^{-4})$, and a degenerate quadrupole of strength $\frac{1}{4}\Lambda R^{-1}(1-R^{-2})^2$ — all located at the inverse point. When the point force is perpendicular to the z-axis, the image consists of a Stokeslet, a stresslet, a degenerate quadrupole, and a rotlet — all located at the inverse point, plus

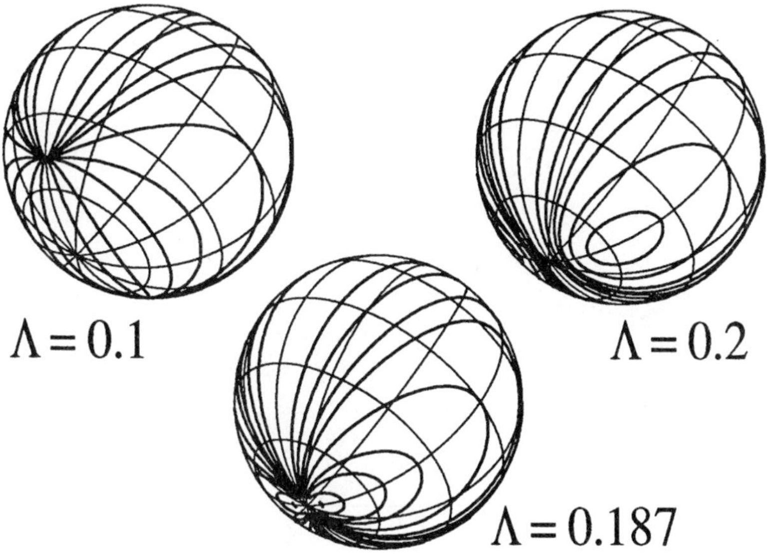

Figure 10.3: Surface flows due to a nearby Stokeslet.

line distributions of Stokeslets, degenerate quadrupoles, and rotlets, from the drop center to the inverse point. The solution for the transverse case is *not* a linear combination of the solutions for the rigid sphere and the bubble. In constrast to the result for rigid spheres, *non-integral* powers of ξ appear in the density functions, the power depending on the viscosity ratio. For $\Lambda = 1/3$ and $\Lambda = 2/3$, *i.e.*, $\lambda = 1/2$ and $\lambda = 2$, we obtain a logarithmic term, because the function $L(\xi, \epsilon)$ becomes the indeterminate form $\lim\{1 - (R\xi)^\epsilon/\epsilon\}$.

Figure 10.3 shows flow lines for the surface velocities induced by the action of a Stokeslet for three different viscosity ratios ($\Lambda = 0.1,\ 0.187,\ 0.2$). The image field was evaluated using the singularity form with Gauss–Jacobi quadratures, with the order of the Jacobi polynomial dictated by Λ. In all three cases, the (transverse) Stokeslet is placed on the z-axis, at $z = 1.2$. These plots show a sequence in which the surface flow field undergoes a change in topology. For relatively inviscid drops, the surface flow field contains two stagnation points that migrate toward the "south pole" with increasing drop viscosity. For $R = 1.2$, at the critical viscosity ratio corresponding to $\Lambda = 0.187$, the stagnation points coalesce. Figure 10.4 shows these critical values as a function of R. For Λ greater than this critical value, the surface of the drop undergoes a "rolling" motion, with centers shifting away from the south pole with increasing viscosity. At values of Λ near unity, this rolling motion is essentially identical to rigid-body rotation, as expected. However, rather curiously, as Λ approached unity, the locus of centers first overshoots the equator, then drifts back. The surface

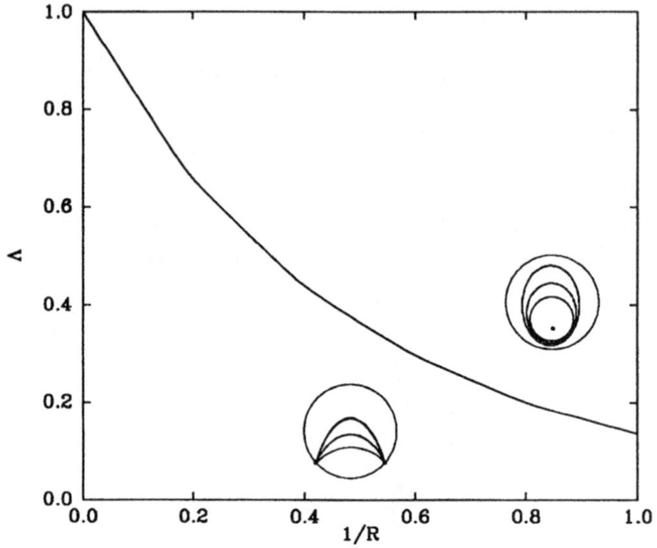

Figure 10.4: Transition to rolling surface motions for the transverse Stokeslet.

flow topology, in particular the presence or absence of the stagnation points, dictates whether surface convection can focus contaminants or surfactants into certain regions.

10.3 Image Systems for Stokes Dipoles

The image system for Stokes dipoles may be obtained by superposition of neighboring Stokeslets of opposite signs and taking the usual limit for a dipole. We effect this mathematically by taking derivatives of the image system for the Stokeslet with respect to $\boldsymbol{x}_2$, the location of the singularity.

Just as $\boldsymbol{F}$ may be decomposed, for axisymmetric geometries, into $\boldsymbol{F}^{\|}$ and $\boldsymbol{F}^{\perp}$, the stresslet $\boldsymbol{S}$ too can be written in a special form that exploits the symmetry. Explicitly, we write the stresslet as

$$\begin{aligned}
\boldsymbol{S} &= \boldsymbol{S}^{(0)} + \boldsymbol{S}^{(1)} + \boldsymbol{S}^{(2)} \\
&= \frac{3}{2}(\boldsymbol{d}_{12}\boldsymbol{d}_{12} - \frac{1}{3}\boldsymbol{\delta})\, \boldsymbol{d}_{12}\boldsymbol{d}_{12} : \boldsymbol{S} \qquad (10.40) \\
&+ \ \boldsymbol{d}_{12}\,(\boldsymbol{S}\cdot\boldsymbol{d}_{12}) + (\boldsymbol{S}\cdot\boldsymbol{d}_{12})\,\boldsymbol{d}_{12} - 2\boldsymbol{d}_{12}\boldsymbol{d}_{12}\,\boldsymbol{d}_{12}\boldsymbol{d}_{12} : \boldsymbol{S} \\
&+ \ \boldsymbol{S} + \frac{1}{2}\boldsymbol{\delta}\,\boldsymbol{d}_{12}\boldsymbol{d}_{12} : \boldsymbol{S} - \boldsymbol{d}_{12}(\boldsymbol{S}\cdot\boldsymbol{d}_{12}) - (\boldsymbol{S}\cdot\boldsymbol{d}_{12})\boldsymbol{d}_{12} + \frac{1}{2}\boldsymbol{d}_{12}\boldsymbol{d}_{12}\,\boldsymbol{d}_{12}\boldsymbol{d}_{12} : \boldsymbol{S} \ ,
\end{aligned}$$

where the labelling indices, $m = 0, 1, 2$ in $\boldsymbol{S}^{(m)} : \boldsymbol{xx}$, correspond to the azimuthal dependence, $e^{im\phi}$, of each stresslet type in polar spherical coordinates about the axis $\boldsymbol{d}_{12}$ (see Chapter 4 or 5). We consider the three cases separately because they correspond to three distinct operations on the Stokeslet image solution.

10.3.1 The Axisymmetric Stresslet

It may be helpful to keep in mind that the axisymmetric case corresponds to a stresslet whose components may be written as

$$\begin{pmatrix} -1 & 0 & 0 \\ 0 & -1 & 0 \\ 0 & 0 & 2 \end{pmatrix}$$

when the z-axis is taken as the axis of symmetry. The field $(\boldsymbol{S}^{(0)} \cdot \nabla) \cdot \mathcal{G}(\boldsymbol{x} - \boldsymbol{x}_2)$ is equivalent to

$$-(\boldsymbol{d}_{12} \cdot \nabla_2) \frac{3}{2} (\boldsymbol{S} : \boldsymbol{d}_{12}\boldsymbol{d}_{12})\, \boldsymbol{d}_{12} \cdot \mathcal{G}(\boldsymbol{x} - \boldsymbol{x}_2) \ ,$$

i.e., a dipole formed by two axisymmetric Stokeslets of strength $\frac{3}{2}|\boldsymbol{S}^{(0)} \cdot \boldsymbol{d}_{12}|$ separated along a line parallel to the $\boldsymbol{d}_{12}$ axis, as shown in Figure 10.5. Therefore, the image system for the axisymmetric stresslet, $(\boldsymbol{S}^{(0)} \cdot \nabla) \cdot \mathcal{G}(\boldsymbol{x} - \boldsymbol{x}_2)$, may be obtained by operating on the image system of the axisymmetric Stokeslet with $-(\boldsymbol{d}_{12} \cdot \nabla_2)$. The result is

$$\boldsymbol{v}^* = \sum_{n=0}^{\infty} \left\{ A_n^{(0)} \frac{(\boldsymbol{d}_{12} \cdot \nabla)^n}{n!} \boldsymbol{f} \cdot \frac{\mathcal{G}(\boldsymbol{x} - \boldsymbol{x}_1)}{8\pi\mu} + B_n^{(0)} \frac{(\boldsymbol{d}_{12} \cdot \nabla)^n}{n!} \boldsymbol{f} \cdot \nabla^2 \frac{\mathcal{G}(\boldsymbol{x} - \boldsymbol{x}_1)}{8\pi\mu} \right\} , \quad (10.41)$$

with $\boldsymbol{f} = \boldsymbol{S}^{(0)} \cdot \boldsymbol{d}_{12}$ and

$$\begin{aligned} A_n^{(0)} &= \frac{3}{2} \frac{\partial A_n^{\|}}{\partial R} && (10.42) \\ &= \frac{3}{2}(n + \frac{3}{2})(n+1)R^{-(n+2)} - \frac{3}{2}(n + \frac{1}{2})(n+3)R^{-(n+4)} \\ B_n^{(0)} &= \frac{3}{2} \frac{\partial B_n^{\|}}{\partial R} && (10.43) \\ &= \frac{3}{8}(n+1)R^{-(n+2)} - \frac{3}{4}(n+3)R^{-(n+4)} + \frac{3}{8}(n+5)R^{-(n+6)} \ . \end{aligned}$$

10.3.2 The Stresslet in Hyperbolic Flow ($m = 1$)

We now consider the stresslet $\boldsymbol{S}^{(1)}$. If the z-axis is taken as the axis of symmetry, the components are proportional to

$$\begin{pmatrix} 0 & 0 & 1 \\ 0 & 0 & 0 \\ 1 & 0 & 0 \end{pmatrix} \quad \text{and} \quad \begin{pmatrix} 0 & 0 & 0 \\ 0 & 0 & 1 \\ 0 & 1 & 0 \end{pmatrix} .$$

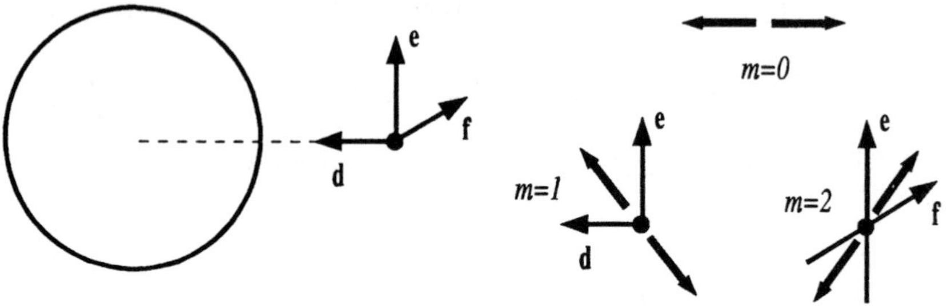

Figure 10.5: Construction of Stokes dipoles from Stokeslets near a sphere.

The two matrices represent the same physical problem if the geometry is axisymmetric about the z-axis. The velocity field $(\boldsymbol{S}^{(1)} \cdot \nabla) \cdot \mathcal{G}(\boldsymbol{x}-\boldsymbol{x}_2)$ is equivalent to

$$-[(\boldsymbol{e}\boldsymbol{d}_{12} + \boldsymbol{d}_{12}\boldsymbol{e}) \cdot \nabla_2] \cdot \mathcal{G}(\boldsymbol{x}-\boldsymbol{x}_2) ,$$

where $\boldsymbol{e} = (\boldsymbol{S} \cdot \boldsymbol{d}_{12}) \cdot (\boldsymbol{\delta} - \boldsymbol{d}_{12}\boldsymbol{d}_{12})$ is the component of $\boldsymbol{S} \cdot \boldsymbol{d}_{12}$ that is orthogonal to $\boldsymbol{d}_{12}$. Thus the case $m = 1$ consists of a symmetric sum of two situations: the first corresponding to a Stokes dipole formed by two Stokeslets (directed along $\boldsymbol{d}_{12}$) separated along a direction $\boldsymbol{e}$ perpendicular to $\boldsymbol{d}_{12}$; the second corresponding to a Stokes dipole formed with the roles of $\boldsymbol{d}_{12}$ and $\boldsymbol{e}$ reversed. An antisymmetric combination of these two dipoles would give the image system for the rotlet oriented in the direction $\boldsymbol{e} \times \boldsymbol{d}_{12}$, transverse to the symmetry axis.

Compared to the $m = 0$ case, the $m = 1$ case is "twice the work." The derivations of the image systems for the two Stokes dipole that make up the stresslet $\boldsymbol{S}^{(1)}$ are quite different and we shall examine each in turn.

The Stokes Dipole $(\boldsymbol{e} \cdot \nabla)\, \boldsymbol{d}_{12} \cdot \mathcal{G}(\boldsymbol{x}-\boldsymbol{x}_2)$

This dipole represents the difference between two axisymmetric Stokeslets, the difference taken in a direction orthogonal to the line of centers, as shown in Figure 10.5. Since

$$(\boldsymbol{e} \cdot \nabla_2)\boldsymbol{d}_{12} = -\frac{1}{R}\boldsymbol{e} ,$$

the exact relation between the Stokes dipole and the ∇_2 operation follows as

$$\begin{aligned}(\boldsymbol{e} \cdot \nabla)\boldsymbol{d}_{12} \cdot \mathcal{G}(\boldsymbol{x}-\boldsymbol{x}_2) &= -\boldsymbol{d}_{12} \cdot (\boldsymbol{e} \cdot \nabla_2)\mathcal{G}(\boldsymbol{x}-\boldsymbol{x}_2) \\ &= -(\boldsymbol{e} \cdot \nabla_2)\boldsymbol{d}_{12} \cdot \mathcal{G}(\boldsymbol{x}-\boldsymbol{x}_2) - \frac{1}{R}\boldsymbol{e} \cdot \mathcal{G}(\boldsymbol{x}-\boldsymbol{x}_2) ,\end{aligned}$$

and the desired image system is the sum of two contributions: $-(\boldsymbol{e} \cdot \nabla_2)$ applied to the image system of the axisymmetric Stokeslet and $-\frac{1}{R}$ times the image system of the transverse Stokeslet.

For the first operation, we will need to know that

$$\begin{aligned} & -(\boldsymbol{e}\cdot\nabla_2)(\boldsymbol{d}_{12}\cdot\nabla)^n\boldsymbol{d}_{12}\cdot\mathcal{G} \\ &= \frac{1}{R}\left[(\boldsymbol{d}_{12}\cdot\nabla)^n\boldsymbol{e}\cdot\mathcal{G}+n(\boldsymbol{d}\cdot\nabla)^{n-1}(\boldsymbol{e}\cdot\nabla)\boldsymbol{d}_{12}\cdot\mathcal{G}\right] \\ &= \frac{1}{R}\left[(n+1)(\boldsymbol{d}_{12}\cdot\nabla)^n\boldsymbol{e}\cdot\mathcal{G}+2n(\boldsymbol{d}_{12}\cdot\nabla)^{n-1}(\boldsymbol{e}\times\boldsymbol{d}_{12})\nabla\frac{1}{r}\right], \end{aligned}$$

where the second step follows from Equation 10.16. The image system for the Stokes dipole $(\boldsymbol{e}\cdot\nabla)\,\boldsymbol{d}_{12}\cdot\mathcal{G}(\boldsymbol{x}-\boldsymbol{x}_2)$ is eventually obtained as

$$\begin{aligned} \boldsymbol{v}^* &= \sum_{n=0}^{\infty}\left\{A_n\frac{(\boldsymbol{d}_{12}\cdot\nabla)^n}{n!}\boldsymbol{e}\cdot\frac{\mathcal{G}(\boldsymbol{x}-\boldsymbol{x}_1)}{8\pi\mu}+B_n\frac{(\boldsymbol{d}_{12}\cdot\nabla)^n}{n!}\boldsymbol{e}\cdot\nabla^2\frac{\mathcal{G}(\boldsymbol{x}-\boldsymbol{x}_1)}{8\pi\mu}\right. \\ &\quad\left.+\,C_n\frac{(\boldsymbol{d}_{12}\cdot\nabla)^n}{8\pi\mu n!}\boldsymbol{t}\times\nabla\frac{1}{r}\right\}, \end{aligned} \tag{10.44}$$

with $\boldsymbol{t}=\boldsymbol{e}\times\boldsymbol{d}_{12}$ and

$$A_n = \frac{(n+1)}{R}A_n^{\parallel}-\frac{1}{R}A_n^{\perp} \tag{10.45}$$

$$B_n = \frac{(n+1)}{R}B_n^{\parallel}-\frac{1}{R}B_n^{\perp} \tag{10.46}$$

$$C_n = \frac{2}{R}A_{n+1}^{\parallel}-\frac{1}{R}C_n^{\perp}\,. \tag{10.47}$$

The Stokes Dipole $(\boldsymbol{d}_{12}\cdot\nabla)\,\boldsymbol{e}\cdot\mathcal{G}(\boldsymbol{x}-\boldsymbol{x}_2)$

This dipole represents the difference between two transverse Stokeslets, the difference taken in the direction of the line of centers as shown in Figure 10.5, whence its image system is obtained by applying $-(\boldsymbol{d}_{12}\cdot\nabla_2)$ on the transverse Stokeslet $\boldsymbol{e}\cdot\mathcal{G}(\boldsymbol{x}-\boldsymbol{x}_2)$. The result is

$$\begin{aligned} \boldsymbol{v}^* &= \sum_{n=0}^{\infty}\left\{A_n\frac{(\boldsymbol{d}_{12}\cdot\nabla)^n}{n!}\boldsymbol{e}\cdot\frac{\mathcal{G}(\boldsymbol{x}-\boldsymbol{x}_1)}{8\pi\mu}+B_n\frac{(\boldsymbol{d}_{12}\cdot\nabla)^n}{n!}\boldsymbol{e}\cdot\nabla^2\frac{\mathcal{G}(\boldsymbol{x}-\boldsymbol{x}_1)}{8\pi\mu}\right. \\ &\quad\left.+\,C_n\frac{(\boldsymbol{d}_{12}\cdot\nabla)^n}{8\pi\mu\, n!}\boldsymbol{t}\times\nabla\frac{1}{r}\right\}, \end{aligned} \tag{10.48}$$

with $\boldsymbol{t}=\boldsymbol{e}\times\boldsymbol{d}_{12}$ and

$$A_n = \frac{\partial A_n^{\perp}}{\partial R} \tag{10.49}$$

$$B_n = \frac{\partial B_n^{\perp}}{\partial R} \tag{10.50}$$

$$C_n = \frac{\partial C_n^{\perp}}{\partial R}\,. \tag{10.51}$$

Final Result for Hyperbolic Flow ($m = 1$)

The image system for the stresslet, $(\boldsymbol{S}^{(1)} \cdot \nabla) \cdot \mathcal{G}(\boldsymbol{x} - \boldsymbol{x}_2)$, follows from the sum of the previous two results. We group all terms in the forms $(\boldsymbol{d}_{12}\boldsymbol{e} + \boldsymbol{e}\boldsymbol{d}_{12}) = \boldsymbol{S}^{(1)}$, where $\boldsymbol{e} = \boldsymbol{S}^{(1)} \cdot \boldsymbol{d}_{12}$. It is also helpful to keep in mind that some of the singularities, such as the degenerate quadrupole, are already symmetric with respect to $\{\boldsymbol{d}_{12}, \boldsymbol{e}\}$. The image system is

$$\begin{aligned} \boldsymbol{v}^* &= \sum_{n=0}^{\infty} \left\{ A_n^{(1)} \frac{(\boldsymbol{d}_{12} \cdot \nabla)^n}{n!} \boldsymbol{e} \cdot \frac{\mathcal{G}(\boldsymbol{x} - \boldsymbol{x}_1)}{8\pi\mu} + B_n^{(1)} \frac{(\boldsymbol{d}_{12} \cdot \nabla)^n}{n!} \boldsymbol{e} \cdot \nabla^2 \frac{\mathcal{G}(\boldsymbol{x} - \boldsymbol{x}_1)}{8\pi\mu} \right. \\ &\quad \left. + C_n^{(1)} \frac{(\boldsymbol{d}_{12} \cdot \nabla)^n}{8\pi\mu\, n!} \boldsymbol{t} \times \nabla \frac{1}{r} \right\} , \end{aligned} \qquad (10.52)$$

with $\boldsymbol{e} = \boldsymbol{S}^{(1)} \cdot \boldsymbol{d}_{12}$, $\boldsymbol{t} = \boldsymbol{e} \times \boldsymbol{d}_{12}$ and

$$A_n^{(1)} = \frac{(n+1)}{R} A_n^{\|} + R \frac{\partial}{\partial R} \left(\frac{A_n^{\perp}}{R} \right) \qquad (10.53)$$

$$B_n^{(1)} = \frac{(n+1)}{R} B_n^{\|} + R \frac{\partial}{\partial R} \left(\frac{B_n^{\perp}}{R} \right) \qquad (10.54)$$

$$C_n^{(1)} = \frac{2}{R} A_{n+1}^{\|} + R \frac{\partial}{\partial R} \left(\frac{C_n^{\perp}}{R} \right) . \qquad (10.55)$$

10.3.3 The Stresslet in Hyperbolic Flow ($m = 2$)

We now consider the last portion, $\boldsymbol{S}^{(2)}$, of the stresslet. When the z-axis is taken as the axis of symmetry, the components are proportional to

$$\begin{pmatrix} 0 & 1 & 0 \\ 1 & 0 & 0 \\ 0 & 0 & 0 \end{pmatrix} \quad \text{and} \quad \begin{pmatrix} 1 & 0 & 0 \\ 0 & -1 & 0 \\ 0 & 0 & 0 \end{pmatrix} .$$

The two matrices represent the same physical problem if the geometry is axisymmetric about the z-axis (one follows from the other by a rotation of 45° about the z-axis). The velocity field $(\boldsymbol{S}^{(2)} \cdot \nabla) \cdot \mathcal{G}(\boldsymbol{x} - \boldsymbol{x}_2)$ is thus equivalent to -$(\boldsymbol{e}\boldsymbol{f} + \boldsymbol{f}\boldsymbol{e}) \cdot \nabla_2) \cdot \mathcal{G}(\boldsymbol{x} - \boldsymbol{x}_2)$, where $\boldsymbol{e}$ and $\boldsymbol{f}$ can be any vectors that form an orthogonal basis for the plane normal to $\boldsymbol{d}_{12}$ (see Figure 10.5). Thus the case $m = 2$ also consists of a symmetric sum of two Stokes-dipoles, but here the role played by $\boldsymbol{e}$ and $\boldsymbol{f}$ are equivalent, so that the solution for the dipole $(\boldsymbol{e} \cdot \nabla)\boldsymbol{f} \cdot \mathcal{G}$ may be obtained by switching $\boldsymbol{e}$ and $\boldsymbol{f}$ in the image system for the dipole $(\boldsymbol{f} \cdot \nabla)\boldsymbol{e} \cdot \mathcal{G}$.

The image system for the Stokes dipole, $(\boldsymbol{f} \cdot \nabla)\boldsymbol{e} \cdot \mathcal{G}(\boldsymbol{x} - \boldsymbol{x}_2)$, can be obtained by applying $(-\boldsymbol{f} \cdot \nabla_2)$ to the image system for the transverse Stokeslet, $\boldsymbol{e} \cdot \mathcal{G}(\boldsymbol{x} - \boldsymbol{x}_2)$. Here, the useful relations are

$$(\boldsymbol{f} \cdot \nabla_2)\boldsymbol{e} = 0 , \qquad (\boldsymbol{f} \cdot \nabla_2)\boldsymbol{d}_{12} = -\frac{1}{R}\boldsymbol{f} , \qquad (\boldsymbol{f} \cdot \nabla_2)R = 0 .$$

After these preliminaries, we obtain the following result for the image system for the Stokes dipole $(\boldsymbol{f}\cdot\nabla)\boldsymbol{e}\cdot\mathcal{G}(\boldsymbol{x}-\boldsymbol{x}_2)$:

$$\begin{aligned}\boldsymbol{v}^* &= \sum_{n=1}^{\infty}\Big\{\frac{A_n^{\perp}}{R}\frac{(\boldsymbol{d}_{12}\cdot\nabla)^{n-1}}{(n-1)!}(\boldsymbol{f}\cdot\nabla)\boldsymbol{e}\cdot\frac{\mathcal{G}(\boldsymbol{x}-\boldsymbol{x}_1)}{8\pi\mu} \\ &\quad+\frac{B_n^{\perp}}{R}\frac{(\boldsymbol{d}_{12}\cdot\nabla)^{n-1}}{(n-1)!}(\boldsymbol{f}\cdot\nabla)\boldsymbol{e}\cdot\nabla^2\frac{\mathcal{G}(\boldsymbol{x}-\boldsymbol{x}_1)}{8\pi\mu}\Big\} \\ &+\sum_{n=0}^{\infty}\frac{C_n^{\perp}}{R}\left[n\frac{(\boldsymbol{d}_{12}\cdot\nabla)^{n-1}}{8\pi\mu n!}(\boldsymbol{f}\cdot\nabla)\boldsymbol{t}\times\nabla\frac{1}{r}+\frac{(\boldsymbol{d}_{12}\cdot\nabla)^{n}}{8\pi\mu n!}(e\times\boldsymbol{f})\times\nabla\frac{1}{r}\right]. \end{aligned} \tag{10.56}$$

If we symmetrize this with respect to $\boldsymbol{e}$ and $\boldsymbol{f}$, then the image system for the stresslet $\boldsymbol{S}^{(2)}=\boldsymbol{ef}+\boldsymbol{fe}$ follows as

$$\begin{aligned}\boldsymbol{v}^* &= \sum_{n=0}^{\infty}\Big\{A_n^{(2)}\frac{(\boldsymbol{d}_{12}\cdot\nabla)^{n}}{n!}(\boldsymbol{S}^{(2)}\cdot\nabla)\cdot\frac{\mathcal{G}(\boldsymbol{x}-\boldsymbol{x}_1)}{8\pi\mu} \\ &\quad+B_n^{(2)}\frac{(\boldsymbol{d}_{12}\cdot\nabla)^{n}}{n!}(\boldsymbol{S}^{(2)}\cdot\nabla)\cdot\nabla^2\frac{\mathcal{G}(\boldsymbol{x}-\boldsymbol{x}_1)}{8\pi\mu} \\ &\quad+C_n^{(2)}\frac{(\boldsymbol{d}_{12}\cdot\nabla)^{n}}{8\pi\mu\, n!}[(\boldsymbol{S}^{(2)}\cdot\nabla)\times\boldsymbol{d}_{12}]\times\nabla\frac{1}{r}\Big\}, \end{aligned} \tag{10.57}$$

with

$$A_n^{(2)} = \frac{A_{n+1}^{\perp}}{R} \tag{10.58}$$

$$B_n^{(2)} = \frac{B_{n+1}^{\perp}}{R} \tag{10.59}$$

$$C_n^{(2)} = \frac{C_{n+1}^{\perp}}{R}. \tag{10.60}$$

Note that no hydrodynamic force and torque are exerted on the sphere.

10.4 Image System for the Degenerate Stokes Quadrupole

The degenerate quadrupole, $\boldsymbol{F}\cdot\nabla^2\mathcal{G}(\boldsymbol{x}-\boldsymbol{x}_2)$, is an important entity, because it appears in the singularity solution for a sphere (Chapter 3), *i.e.*, the disturbance velocity of a translating sphere of radius b, center at $\boldsymbol{x}_2$ is

$$\boldsymbol{F}\cdot(1+\frac{b^2}{6}\nabla^2)\frac{\mathcal{G}(\boldsymbol{x}-\boldsymbol{x}_2)}{8\pi\mu}. \tag{10.61}$$

We could try a solution approach analogous to the one employed to derive the image system for the Stokes dipole: differentiate the image system of the Stokeslet with respect to $\boldsymbol{x}_2$, the location of the singular point. It turns out however, that this is more work than the direct method, because we have to differentiate twice. In fact, since the degenerate quadrupole is a vector harmonic, the direct analysis is much easier than that for the Stokeslet.

We start with the representation for the degenerate quadrupole:

$$\boldsymbol{F} \cdot \nabla^2 \mathcal{G}(\boldsymbol{x} - \boldsymbol{x}_2) = -2\nabla \boldsymbol{F} \cdot \nabla \frac{1}{r_2} . \qquad (10.62)$$

We differentiate the Legendre expansion, Equation 10.4, twice with respect to $\boldsymbol{x}_2$ to obtain the following relations for the degenerate quadrupole:

$$\begin{aligned} \nabla \boldsymbol{F}^{\|} \cdot \nabla \frac{1}{r_2} &= -F^{\|} \sum_{n=0}^{\infty} \frac{(n+1)(2n+1)}{R^{n+2}} r_1^{2n-1} (\boldsymbol{x} - \boldsymbol{x}_1) \frac{(\boldsymbol{d}_{12} \cdot \nabla)^n}{n!} \frac{1}{r_1} \\ &- F^{\|} \sum_{n=0}^{\infty} \frac{(n+1)}{R^{n+2}} r_1^{2n+1} \nabla \frac{(\boldsymbol{d}_{12} \cdot \nabla)^n}{n!} \frac{1}{r_1} \end{aligned} \qquad (10.63)$$

and

$$\begin{aligned} \nabla \boldsymbol{F}^{\perp} \cdot \nabla \frac{1}{r_2} &= \sum_{n=0}^{\infty} \frac{(2n+3)}{R^{n+3}} r_1^{2n+1} (\boldsymbol{x} - \boldsymbol{x}_1)(\boldsymbol{F}^{\perp} \cdot \nabla) \frac{(\boldsymbol{d}_{12} \cdot \nabla)^n}{n!} \frac{1}{r_1} \\ &+ \sum_{n=0}^{\infty} \frac{r_1^{2n+3}}{R^{n+3}} \nabla \, (\boldsymbol{F}^{\perp} \cdot \nabla) \frac{(\boldsymbol{d}_{12} \cdot \nabla)^n}{n!} \frac{1}{r_1} . \end{aligned} \qquad (10.64)$$

As before, we examine the axisymmetric and transverse cases separately.

10.4.1 The Axisymmetric Problem

We write $\boldsymbol{F} = F\boldsymbol{d}_{12}$ and use Equation 10.63 to express the degenerate quadrupole at $\boldsymbol{x}_2$ in terms of harmonics centered at $\boldsymbol{x}_1$. The image system, $\boldsymbol{v}^*$, is written as a multipole expansion,

$$\boldsymbol{v}^* = \frac{F\boldsymbol{d}_{12}}{8\pi\mu} \cdot \sum_{n=0}^{\infty} \left\{ \tilde{A}_n^{\|} \frac{(\boldsymbol{d}_{12} \cdot \nabla)^n}{n!} \mathcal{G}(\boldsymbol{x} - \boldsymbol{x}_1) + \tilde{B}_n^{\|} \frac{(\boldsymbol{d}_{12} \cdot \nabla)^n}{n!} \nabla^2 \mathcal{G}(\boldsymbol{x} - \boldsymbol{x}_1) \right\} .$$

From here on, the analysis is similar to that performed for the Stokeslet problem. The only difference is that when matching the boundary conditions we use terms originating from Equation 10.63 in place of the expression for the Stokeslet. Thus the equations for $\tilde{A}_n^{\|}$ and $\tilde{B}_n^{\|}$ $(n \geq 1)$ are

$$\begin{aligned} \frac{1}{(2n-1)} \tilde{A}_{n-1}^{\|} &= \frac{(2n+1)}{R^{n+2}} \\ \frac{(n-2)}{(2n-1)} \tilde{A}_{n-1}^{\|} - \frac{n}{(2n+3)} \tilde{A}_{n+1}^{\|} - 2n\tilde{B}_{n-1}^{\|} &= -\frac{2(n+1)}{R^{n+2}} , \end{aligned}$$

so that

$$\begin{aligned} \tilde{A}_n^{\|} &= (2n+1)(2n+3)R^{-(n+3)} & (10.65) \\ \tilde{B}_n^{\|} &= \frac{1}{2}(2n+1)R^{-(n+3)} - \frac{1}{2}(2n+7)R^{-(n+5)} . & (10.66) \end{aligned}$$

The hydrodynamic force exerted on the large sphere is

$$\boldsymbol{F}^{\text{Hyd}} = -\tilde{A}_0^{\|} F \boldsymbol{d}_{12} = -3R^{-3} F \boldsymbol{d} ,$$

consistent with the Faxén law. This image system is equivalent to a Stokeslet, stresslet, both degenerate and nondegenerate quadrupoles, and a degenerate octupole — all located at the inverse point.

10.4.2 The Transverse Problem

We now consider the transverse case with $\boldsymbol{F}\cdot\boldsymbol{d}_{12} = 0$. Equation 10.64 will be used to express the singularity at $\boldsymbol{x}_2$ in terms of harmonics at $\boldsymbol{x}_1$. The image system for the transverse case will be the same as that used in the transverse Stokeslet problem: The "three" coefficients — $\tilde{A}_n^{\perp}$, $\tilde{B}_n^{\perp}$, and $\tilde{C}_n^{\perp}$ — are determined from the boundary conditions, which ultimately yields the following set of equations:

$$\begin{aligned}
\left[n-1-\frac{(2n+1)(n+1)}{(2n+3)}\right]\tilde{A}_n^{\perp} &= 2(2n+1)(n+1)R^{-(n+3)} \\
\frac{\tilde{A}_n^{\perp}}{(2n+3)} - \frac{\tilde{A}_{n+2}^{\perp}}{(2n+7)} - 2\tilde{B}_n^{\perp} &= 2R^{-(n+5)} \\
\left[1-\frac{(2n+3)}{(2n+5)}\right]\tilde{A}_{n+1}^{\perp} - \tilde{C}_n^{\perp} &= 2(2n+3)R^{-(n+4)} ,
\end{aligned}$$

so that

$$\tilde{A}_n^{\perp} = -\frac{(2n+3)(2n+1)(n+1)}{(n+2)}R^{-(n+3)} \tag{10.67}$$

$$\tilde{B}_n^{\perp} = -\frac{(2n+1)(n+1)}{2(n+2)}R^{-(n+3)} + \frac{(n+1)(2n+7)}{2(n+4)}R^{-(n+5)} \tag{10.68}$$

$$\tilde{C}_n^{\perp} = -\frac{2(2n+3)(2n+5)}{(n+3)}R^{-(n+4)} . \tag{10.69}$$

The hydrodynamic force and torque exerted by the fluid on the sphere are

$$\boldsymbol{F}^{\text{hyd}} = -\tilde{A}_0^{\perp}\boldsymbol{F} = \frac{3}{2}R^{-3}\boldsymbol{F} \tag{10.70}$$

$$\boldsymbol{T}^{\text{hyd}} = (\tilde{C}_0^{\perp} - \tilde{A}_1^{\perp})\boldsymbol{F}\times\boldsymbol{d}_{12} = 0 , \tag{10.71}$$

consistent with the Faxén laws for the force and torque.

This image system may be rewritten as a distribution of singularities consisting of a Stokeslet, a Stokes dipole (which in turn can be further decomposed into a stresslet and rotlet), a rotlet, degenerate and nondegenerate quadrupoles and a degenerate octupole — all located at the inverse point, plus an additional line distribution of Stokeslets, degenerate quadrupoles, and rotlets along the axis from the sphere center to the inverse point.

10.5 Hydrodynamic Interactions Between Large and Small Spheres

The image solutions for Stokes singularites may be employed in method of reflections calculations for the interactions between a large sphere and an arbitrary small particle. The key idea is that over lengths scales associated with the large sphere, the disturbance fields produced by the small particle may be approximated by those produced by equivalent Stokes singularities. We illustrate this point by examining the interactions between spheres of radii a and b with $\beta = b/a \ll 1$. In the reflections (see Figure 8.1) at the small sphere, we truncate the multipole expansion at the desired order in b/R. For reflections at the large sphere, we retain the entire multipole solution, which of course is the image systems of the Stokes singularities. In the rest of this section, we derive expressions for the mobility functions $x^a_{\alpha\beta}$ to $O(\beta^5)$.

10.5.1 Mobility Functions x_{12} and x^a_{22}

We assume no net force on sphere 1 and a hydrodynamic force $\boldsymbol{F}_2$ on sphere 2, with $\boldsymbol{F}_2 \parallel \boldsymbol{d}_{12}$. The zeroth-order solution is

$$\begin{aligned} 6\pi\mu b \boldsymbol{U}_2^{(0)} &= -\boldsymbol{F}_2 , \\ \boldsymbol{v}_2 &= -\boldsymbol{F}_2 \cdot (1 + \frac{b^2}{6}\nabla^2) \frac{\mathcal{G}(\boldsymbol{x}-\boldsymbol{x}_2)}{8\pi\mu} . \end{aligned}$$

At sphere 1, the resulting translational motion induced by $\boldsymbol{v}_2$ is, from the Faxén law,

$$\boldsymbol{U}_1^{(1)} = (1 + \frac{a^2}{6}\nabla^2) \boldsymbol{v}_2 |_{x=x_1}$$

or

$$6\pi\mu b \boldsymbol{U}_1^{(1)} = -\boldsymbol{F}_2 \left[\frac{3}{2}\left(\frac{b}{R}\right) - \frac{1}{2}\left(\frac{b}{R}\right)\left(\frac{a}{R}\right)^2 - \frac{1}{2}\left(\frac{b}{R}\right)^3 \right] .$$

Up to this point, the solution procedure is identical to that employed earlier. However, for $\boldsymbol{v}_{12}$ we use the image system for a Stokeslet of strength $\boldsymbol{F}_2$ and a degenerate quadrupole of strength $(b^2/6)\boldsymbol{F}_2$, *for a force-free sphere* at $\boldsymbol{x}_1$. Using results derived earlier in this chapter, we have

$$\begin{aligned} \boldsymbol{v}_{12} &= \boldsymbol{F}_2 \cdot \sum_{n=0}^{\infty} \left\{ A_n \frac{(\boldsymbol{d}_{12}\cdot\nabla)^n}{n!} \frac{\mathcal{G}(\boldsymbol{x}-\boldsymbol{x}_1)}{8\pi\mu} + B_n \frac{(\boldsymbol{d}_{12}\cdot\nabla)^n}{n!} \nabla^2 \frac{\mathcal{G}(\boldsymbol{x}-\boldsymbol{x}_1)}{8\pi\mu} \right\} \\ &- \boldsymbol{F}_2 A_0 \cdot \left\{ 1 + \frac{a^2}{6}\nabla^2 \right\} \frac{\mathcal{G}(\boldsymbol{x}-\boldsymbol{x}_1)}{8\pi\mu} , \end{aligned}$$

where

$$A_n = A_n^{\|} + \frac{b^2}{6}\tilde{A}_n^{\|} , \qquad B_n = B_n^{\|} + \frac{b^2}{6}\tilde{B}_n^{\|} ,$$

with $A_n^{\|}$, $B_n^{\|}$, $\tilde{A}_n^{\|}$, and $\tilde{B}_n^{\|}$ given by Equations 10.10, 10.11, 10.65, and 10.66. The image for a force-free sphere is obtained from that of the stationary sphere by placing an additional Stokeslet of opposite sign (and the associated degenerate quadrupole) at $\boldsymbol{x}_1$.

The second reflection result is given by the Faxén law as

$$\boldsymbol{U}_2^{(2)} = (1 + \frac{b^2}{6}\nabla^2) \boldsymbol{v}_{12} |_{x=x_2} ,$$

and since

$$\begin{aligned}\frac{(\boldsymbol{d}_{12}\cdot\nabla)^n}{n!}\boldsymbol{F}_2\cdot\mathcal{G}(\boldsymbol{x}-\boldsymbol{x}_1)|_{x=x_2} &= 2R^{-(n+1)}\boldsymbol{F}_2\\ \frac{(\boldsymbol{d}_{12}\cdot\nabla)^n}{n!}\boldsymbol{F}_2\cdot\nabla^2\mathcal{G}(\boldsymbol{x}-\boldsymbol{x}_1)|_{x=x_2} &= -2(n+1)(n+2)R^{-(n+3)}\boldsymbol{F}_2\ ,\end{aligned}$$

we obtain

$$\begin{aligned}6\pi\mu b\boldsymbol{U}_2^{(2)} &= -\boldsymbol{F}_2\left\{\left(\frac{b}{R}\right)\left[-\frac{3}{2}x^3+\frac{1}{4}x^5-\frac{9}{4}\sum_{n=1}^{\infty}x^{2n+1}\right]\right.\\ &+\left(\frac{b}{R}\right)^3\left[\frac{1}{2}x^3+\frac{1}{2}\sum_{n=1}^{\infty}(n+1)(4n+3)x^{2n+1}\right]\\ &-\left.\left(\frac{b}{R}\right)^5\left[\frac{1}{24}\sum_{n=1}^{\infty}(n+1)(n+2)(2n+1)(2n+3)x^{2n+1}\right]\right\}\ ,\end{aligned}$$

with the notation x for a/R.

For the third reflection,

$$\boldsymbol{U}_1^{(3)} = (1+\frac{a^2}{6}\nabla^2)\boldsymbol{v}_{212}|_{x=x_1}\ , \tag{10.72}$$

with

$$\boldsymbol{v}_{212}\sim(\boldsymbol{S}_2^{(2)}\cdot\nabla)\cdot\frac{\mathcal{G}(\boldsymbol{x}-\boldsymbol{x}_2)}{8\pi\mu}\ .$$

Only the axisymmetric component of the stresslet appears in this problem, so it suffices to compute only $\boldsymbol{S}:\boldsymbol{d}_{12}\boldsymbol{d}_{12}$. We apply the Faxén law for the stresslet to the incident field $\boldsymbol{v}_{21}$ and find that

$$\begin{aligned}\boldsymbol{S}_2^{(2)}:\boldsymbol{d}_{12}\boldsymbol{d}_{12} &= \frac{20}{3}\pi\mu b^3\boldsymbol{e}_{21}:\boldsymbol{d}_{12}\boldsymbol{d}_{12}\\ &= \frac{5}{6}b^3\frac{\boldsymbol{F}_2\cdot\boldsymbol{d}_{12}}{R^2}\left[\sum_{n=1}^{\infty}[2(n+1)A_n^{\|}x^n-2(n+1)(n+2)(n+3)B_n^{\|}x^{n+2}]\right.\\ &\quad\left.-\ 12(B_0^{\|}-\frac{1}{6}A_0^{\|})x^2\right]\ .\end{aligned}$$

Recall that the last two terms (the $A_0^{\|}$ and $B_0^{\|}$ terms) are present because the reflection at sphere 1 is for the *force-free* condition. We may now write the stresslet field as

$$(\boldsymbol{S}_2^{(2)}\cdot\nabla)\cdot\mathcal{G}(\boldsymbol{x}-\boldsymbol{x}_2) = \frac{3}{2}[\boldsymbol{S}_2^{(2)}:\boldsymbol{d}_{12}\boldsymbol{d}_{12}](\boldsymbol{d}_{12}\cdot\nabla)\boldsymbol{d}_{12}\cdot\mathcal{G}(\boldsymbol{x}-\boldsymbol{x}_2)\ ,$$

with

$$[\boldsymbol{S}_2^{(2)}:\boldsymbol{d}_{12}\boldsymbol{d}_{12}]\boldsymbol{d}_{12} = \frac{5}{6}\left(\frac{b}{R}\right)^3\left[-4x^2+x^4-3\sum_{n=1}^{\infty}(n+1)x^{2n}\right]a\boldsymbol{F}_2\ .$$

The third reflection contribution, as given by Equation 10.72, thus reduces to the evaluation of the following dipole and octupole fields:

$$\begin{aligned}(\boldsymbol{d}_{12}\cdot\nabla)\boldsymbol{d}_{12}\cdot\mathcal{G}(\boldsymbol{x}-\boldsymbol{x}_2)|_{x=x_1} &= -2R^{-2}\boldsymbol{d}_{12}\\ (\boldsymbol{d}_{12}\cdot\nabla)\boldsymbol{d}_{12}\cdot\nabla^2\mathcal{G}(\boldsymbol{x}-\boldsymbol{x}_2)|_{x=x_1} &= 12R^{-4}\boldsymbol{d}_{12}\ .\end{aligned}$$

Collecting all of the above together, we find that the contribution from the third reflection is

$$6\pi\mu b\boldsymbol{U}_1^{(3)} = -\boldsymbol{F}_2\left(\frac{b}{R}\right)^4\left[\frac{105}{8}x^3-\frac{75}{8}x^5+\frac{15}{8}x^7+\frac{45}{8}\sum_{n=1}^{\infty}x^{2n+1}\right].$$

At the fourth reflection, we have a contribution to the translational velocity of sphere 2, given by

$$\boldsymbol{U}_2^{(4)} = \boldsymbol{v}_{2121}\ ,$$

where $\boldsymbol{v}_{2121}$ is the image for $\boldsymbol{v}212$, which to the desired order of accuracy may be represented by the stresslet field $(\boldsymbol{S}_2^{(2)}\cdot\nabla)\cdot\mathcal{G}(\boldsymbol{x}-\boldsymbol{x}_2)/8\pi\mu$. The Faxén correction, $(b^2/6)\nabla^2\boldsymbol{v}_{2121}$, has been dropped because, as we shall see shortly, it gives a correction of $O(\beta^6)$. From the first part of this chapter we know that for a force-free sphere the image for an axisymmetric stresslet at $\boldsymbol{x}_2$, evaluated at $\boldsymbol{x}_2$, is

$$\begin{aligned}&\frac{3}{2}\left[\frac{\boldsymbol{S}_2^{(2)}}{8\pi\mu}:\boldsymbol{d}_{12}\boldsymbol{d}_{12}\right]\frac{\boldsymbol{d}_{12}}{R^2}\left[\sum_{n=1}^{\infty}[2A_n^{(0)}x^{n-1}-2(n+1)(n+2)B_n^{(0)}x^{n+1}]\right.\\ &\qquad\left.-\,4x(b_0^{(0)}-\frac{1}{6}a_0^{(0)})\right]\\ &= \frac{3}{2}\left[\frac{\boldsymbol{S}_2^{(2)}}{8\pi\mu}:\boldsymbol{d}_{12}\boldsymbol{d}_{12}\right]\frac{\boldsymbol{d}_{12}}{R^2}\left[4x^3-x^5+3\sum_{n=1}^{\infty}(n+1)x^{2n+1}\right].\end{aligned}$$

We have the expression for $\boldsymbol{S}_2^{(2)}$, and thus an $O(\beta^4)$ contribution from the fourth reflection follows as

$$6\pi\mu b\boldsymbol{U}_2^{(4)} = \boldsymbol{F}_2\left(\frac{b}{R}\right)^4\left(\frac{15}{16}x^2\right)\left[4x^2-x^4+3\sum_{n=1}^{\infty}(n+1)x^{2n}\right]^2.$$

Looking back at the steps in the derivation, we see that the occurrence of the perfect square is a consequence of the Lorentz reciprocal theorem. The following expressions for x_{22}^a and x_{12}^a may be obtained by summing contributions from the even and odd reflections, respectively:

$$\begin{aligned}6\pi b x_{12}^a &= \beta\left[\frac{3}{2}x-\frac{1}{2}x^3\right]-\beta^3\left[\frac{1}{2}x^3\right]\\ &+\ \beta^4\left[\frac{15}{8}\left(7x^7-5x^9+x^{11}+3\sum_{n=1}^{\infty}x^{2n+5}\right)\right]+O(\beta^6) \qquad (10.73)\\ 6\pi b x_{22}^a &= 1-\beta\left[\frac{3}{2}x^4-\frac{1}{4}x^6+\frac{9}{4}\sum_{n=1}^{\infty}x^{2n+2}\right]\end{aligned}$$

$$+\ \beta^3\left[\frac{1}{2}x^6+\frac{1}{2}\sum_{n=1}^{\infty}(n+1)(4n+3)x^{2n+4}\right]$$
$$-\ \beta^4\left[\frac{15}{16}\left(4x^5-x^7+3\sum_{n=1}^{\infty}(n+1)x^{2n+3}\right)^2\right]$$
$$-\ \beta^5\left[\frac{1}{24}\sum_{n=1}^{\infty}(n+1)(n+2)(2n+1)(2n+3)x^{2n+6}\right]+O(\beta^6)\ . \quad (10.74)$$

10.5.2 Mobility Functions x_{11} and x^a_{21}

We consider a hydrodynamic force $\boldsymbol{F}_1$ on the larger sphere and none on the smaller sphere (labelled 2). From the reciprocal theorem, $x^a_{21} = x^a_{12}$, and x^a_{12} was derived in the preceding example. Consequently, we consider only x^a_{11}. The leading order contribution is

$$6\pi\mu a\boldsymbol{U}_1^{(0)} = -\boldsymbol{F}_1\ ,$$

while the next correction is given by $\boldsymbol{U}_1^{(2)}$.

Now, terms of the form

$$\left(\frac{b}{R}\right)^N\sum_{n=1}^{\infty}f_n\left(\frac{a}{R}\right)^n$$

will appear only for $N \geq 6$. The $N = 6$ term appears at the fourth reflection and comes from the velocity field of $\boldsymbol{S}_2^{(3)}$, which in turn is proportional to the rate-of-strain of the *image* of the velocity field of $\boldsymbol{S}_2^{(1)}$. Since the solution is desired accurate to $O(\beta^5)$, we will stop after the second reflection.

The following expression for $\boldsymbol{U}_1^{(2)}$ can be obtained from the the expression for $\boldsymbol{U}_2^{(2)}$ of the preceding subsection by switching the labels for the spheres:

$$\begin{aligned}6\pi\mu\boldsymbol{U}_1^{(2)} &= -\boldsymbol{F}_1\left(\frac{a}{R}\right)\left\{\left[-\frac{3}{2}\left(\frac{b}{R}\right)^3+\frac{1}{4}\left(\frac{b}{R}\right)^5-\frac{9}{4}\sum_{n=1}^{\infty}\left(\frac{b}{R}\right)^{2n+1}\right]\right.\\ &+\ \left(\frac{a}{R}\right)^3\left[\frac{1}{2}\left(\frac{b}{R}\right)^3+\frac{1}{2}\sum_{n=1}^{\infty}(n+1)(4n+3)\left(\frac{b}{R}\right)^{2n+1}\right]\\ &-\ \left.\left(\frac{a}{R}\right)^5\left[\frac{1}{24}\sum_{n=1}^{\infty}(n+1)(n+2)(2n+1)(2n+3)\left(\frac{b}{R}\right)^{2n+1}\right]\right\}\ .\end{aligned}$$

We may truncate this expression, retaining terms up to $(b/R)^5$, and the result for x^a_{11} is

$$\begin{aligned}6\pi a x^a_{11} &= 1-\beta^3\left[\frac{15}{4}x^4-\frac{15}{2}x^6+\frac{15}{4}x^8\right]\\ &-\ \beta^5\left[2x^6-\frac{33}{2}x^8+\frac{35}{2}x^{10}\right]+O(\beta^6)\ . \quad (10.75)\end{aligned}$$

The translational velocity that results from the application of a net external force on the large sphere is given by Stokes law with a small correction of $O(\beta^3)$ due to the presence of the small, force-free sphere.

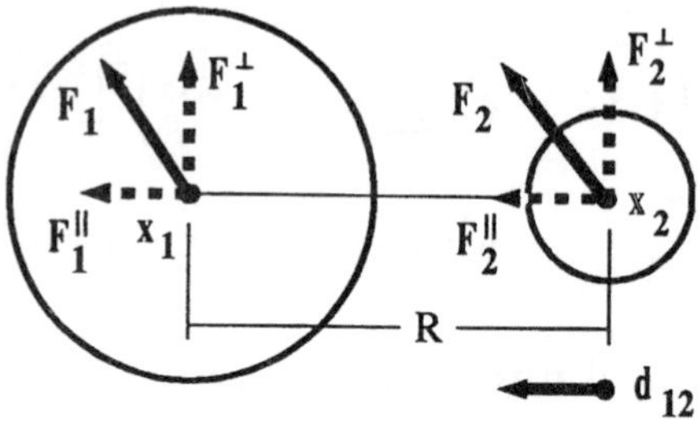

Figure 10.6: Forces and motions for the two-drop geometry.

10.6 Hydrodynamic Interactions Between Large and Small Drops

The preceding discussion on the interactions between large and small *rigid* spheres can be extended to the case of viscous drops, as shown by Fuentes *et al.* [16, 17], the essential modification being the use of the image solutions for a viscous drop in place of that for the rigid sphere. The forces acting on the drops are decomposed into components along and transverse to the axis, and we employ the same notation as used earlier for the rigid spheres (see Figure 10.6).

The final results for the axisymmetric mobility functions $x^a_{\alpha\beta}$ and the asymmetric mobility functions $y^a_{\alpha\beta}$, defined by

$$\begin{pmatrix} \boldsymbol{U}_1^{\|} \\ \boldsymbol{U}_2^{\|} \end{pmatrix} = \mu^{-1} \begin{pmatrix} x^a_{11} & x^a_{12} \\ x^a_{21} & x^a_{22} \end{pmatrix} \begin{pmatrix} \boldsymbol{F}_1^{\|} \\ \boldsymbol{F}_2^{\|} \end{pmatrix}$$

$$\begin{pmatrix} \boldsymbol{U}_1^{\perp} \\ \boldsymbol{U}_2^{\perp} \end{pmatrix} = \mu^{-1} \begin{pmatrix} y^a_{11} & y^a_{12} \\ y^a_{21} & y^a_{22} \end{pmatrix} \begin{pmatrix} \boldsymbol{F}_1^{\perp} \\ \boldsymbol{F}_2^{\perp} \end{pmatrix},$$

are

$$\begin{aligned} 6\pi a x^a_{11} &- \frac{3}{2+\Lambda_1} \\ &= -\beta^3\left[\frac{9\Lambda_2+6}{4}x^4 - \left(\frac{3\Lambda_1}{2+\Lambda_1}\right)\left(\frac{9\Lambda_2+6}{2}\right)x^6 + \frac{3}{4}\left(\frac{3\Lambda_1}{2+\Lambda_1}\right)^2(7\Lambda_2-2)x^8\right] \\ &- \beta^5\left[\frac{3(1+\Lambda_2)}{2+\Lambda_2}x^6 - \frac{3}{2}\left(\frac{3\Lambda_1}{2+\Lambda_1}\right)(7\Lambda_2+4)x^8\right. \\ &\qquad \left.+\frac{5}{2}\left(\frac{3\Lambda_1}{2+\Lambda_1}\right)^2(9\Lambda_2-2)x^{10}\right] \\ &+ O(\beta^6) \end{aligned} \qquad (10.76)$$

$$\begin{aligned}6\pi b x_{12}^a &= 6\pi b x_{21}^a \\ &= \beta\left[\frac{3}{2}x - \frac{3\Lambda_1}{2(2+\Lambda_1)}x^3\right] - \beta^3\left[\frac{3\Lambda_2}{2(2+\Lambda_2)}x^3\right] \\ &+ \beta^4\Bigg[\frac{3}{8}(2+3\Lambda_2)\Bigg(7\Lambda_1 x^7 - \frac{15\Lambda_1^2}{2+\Lambda_1}x^9 + \frac{9\Lambda_1^3}{(2+\Lambda_1)^2}x^{11} \\ &\qquad + \sum_{n=1}^{\infty}[3\Lambda_1 + 2(1-\Lambda_1)(n+1)]x^{2n+5}\Bigg)\Bigg] + O(\beta^6) \qquad (10.77)\end{aligned}$$

$$\begin{aligned}&6\pi b x_{22}^a - \frac{3}{2+\Lambda_2} \\ &= -\beta\left[\frac{3}{2}\Lambda_1 x^4 - \frac{3\Lambda_1^2 x^6}{4(2+\Lambda_1)} + \frac{3}{4}(2+\Lambda_1)\sum_{n=1}^{\infty}x^{2n+2}\right] \qquad (10.78) \\ &+ \beta^3\left(\frac{3\Lambda_2}{2+\Lambda_2}\right)\left[\frac{\Lambda_1}{2}x^6 + \frac{1}{2}\sum_{n=1}^{\infty}[\Lambda_1(4n+3) + (1-\Lambda_1)(n+2)](n+1)x^{2n+4}\right] \\ &- \beta^4\left[\frac{3}{16}(2+3\Lambda_2)\left(4\Lambda_1 x^5 - \frac{3\Lambda_1^2}{2+\Lambda_1}x^7 + (2+\Lambda_1)\sum_{n=1}^{\infty}(n+1)x^{2n+3}\right)^2\right] \\ &- \beta^5\left(\frac{3\Lambda_2}{2+\Lambda_2}\right)^2\sum_{n=1}^{\infty}\left[\frac{\Lambda_1(2n+3) - 2(1-\Lambda_1)}{24}\right](n+1)(n+2)(2n+1)x^{2n+6} \\ &+ O(\beta^6)\end{aligned}$$

$$\begin{aligned}&6\pi a y_{11}^a - \frac{3}{2+\Lambda_1} \\ &= -\beta^3\frac{1}{4}\left(\frac{3\Lambda_1}{2+\Lambda_1}\right)^2(3\Lambda_2+2)x^8 \qquad (10.79) \\ &+ \beta^5\Bigg[\frac{3}{8}\left(\frac{\Lambda_2^2}{2(2+\Lambda_2)} - \frac{3\Lambda_2 - 2(1-\Lambda_2) + 3\Lambda_2(1-\Lambda_2)}{4-3\Lambda_2}\right)x^6 \\ &\quad + \frac{3}{8}\left(\frac{3\Lambda_1}{2+\Lambda_1}\right)(\Lambda_2+2)x^8 - \frac{15}{16}\left(\frac{3\Lambda_1}{2+\Lambda_1}\right)^2(5\Lambda_2+2)x^{10}\Bigg] + O(\beta^6)\,.\end{aligned}$$

$$\begin{aligned}6\pi b y_{12}^a &= 6\pi b y_{21}^a \\ &= \beta\left[\frac{3}{4}x + \frac{3\Lambda_1}{4(2+\Lambda_1)}x^3\right] \qquad (10.80) \\ &+ \beta^3\left[\frac{3\Lambda_2}{4(2+\Lambda_2)}x^3\right] \\ &+ \beta^4\Bigg[\frac{3}{16}(2+3\Lambda_2)\left(\frac{3\Lambda_1}{2+\Lambda_1}\right)\Bigg(-\frac{\Lambda_1^2}{(2+\Lambda_1)}x^{11} \\ &\qquad + \sum_{n=1}^{\infty}(2n+3)\frac{\Lambda_1 n + \Lambda_1(1-\Lambda_1)}{n+3-3\Lambda_1}x^{2n+9}\Bigg)\Bigg] + O(\beta^6)\end{aligned}$$

$$
\begin{aligned}
6\pi b y_{22}^{a} & - \frac{3}{2+\Lambda_2} \\
&= \beta\left[\frac{3\Lambda_1^2 x^6}{16(2+\Lambda_1)} - \frac{3}{8}\sum_{n=1}^{\infty}\frac{\Lambda_1 3n-(1-\Lambda_1)2n+3\Lambda_1(1-\Lambda_1)}{n+3-3\Lambda_1}x^{2n+4}\right] \\
&+ \beta^3\left(\frac{3\Lambda_2}{2+\Lambda_2}\right)\left[\frac{1}{8}\sum_{n=1}^{\infty}[\Lambda_1(4n^2+6n-1)+(1-\Lambda_1)2n(n+2)]x^{2n+6}\right] \\
&- \beta^4\left[\frac{9}{64}(2+3\Lambda_2)\left(\frac{\Lambda_1^2}{2+\Lambda_1}x^7-\sum_{n=1}^{\infty}(2n+3)\frac{\Lambda_1 n+\Lambda_1(1-\Lambda_1)}{n+3-3\Lambda_1}x^{2n+5}\right)^2\right] \\
&- \beta^5\left(\frac{3\Lambda_2}{2+\Lambda_2}\right)^2\sum_{n=1}^{\infty}\left[\frac{\Lambda_1(2n+3)+2(1-\Lambda_1)}{48}\right](n+1)^2(2n+1)x^{2n+6} \\
&+ O(\beta^6)\,. \qquad (10.81)
\end{aligned}
$$

Exercises

Exercise 10.1 The Moments on a Rigid Sphere Induced by a Nearby Stokeslet.
Consider a sphere at $\boldsymbol{x}_1$ with a Stokeslet nearby at $\boldsymbol{x}_2$. The force, torque, and stresslet may be obtained by applying the Faxén laws to the ambient field, $\boldsymbol{F}\cdot\mathcal{G}(\boldsymbol{x}-\boldsymbol{x}_2)/(8\pi\mu)$. Show that for the axisymmetric Stokeslet,

$$
\begin{aligned}
\boldsymbol{F}^{\mathrm{Hyd}} &= \left(\frac{3}{2}R^{-1}-\frac{1}{2}R^{-3}\right)F^{\parallel}\boldsymbol{d}_{12} \\
\boldsymbol{S} &= \left(-\frac{5}{2}R^{-2}+\frac{3}{2}R^{-4}\right)F^{\parallel}\boldsymbol{d}_{12}\boldsymbol{d}_{12}\,,
\end{aligned}
$$

and for the transverse Stokeslet,

$$
\begin{aligned}
\boldsymbol{F}^{\mathrm{Hyd}} &= \left(\frac{3}{4}R^{-1}+\frac{1}{4}R^{-3}\right)\boldsymbol{F}^{\perp} \\
\boldsymbol{T}^{\mathrm{Hyd}} &= R^{-2}\boldsymbol{F}^{\perp}\times\boldsymbol{d}_{12} \\
\boldsymbol{S} &= -\frac{1}{2}R^{-4}(\boldsymbol{F}^{\perp}\boldsymbol{d}_{12}+\boldsymbol{d}_{12}\boldsymbol{F}^{\perp})\,.
\end{aligned}
$$

Exercise 10.2 Derivation of the General Faxén Law for the Moments on a Sphere.
In Chapter 3, we derived the Faxén laws for a particle of arbitary shape by first deriving expressions for the moments for the special ambient flow produced by a nearby Stokeslet. Now that we know the exact result for a sphere in the Stokeslet ambient field, use the developments in Chapter 3 in the reverse order to derive the Faxén Law for the n-th moment,

$$\oint_S(\boldsymbol{\sigma}\cdot\hat{\boldsymbol{n}})\boldsymbol{\xi}\boldsymbol{\xi}\ldots\boldsymbol{\xi}\,dS(\boldsymbol{\xi})$$

on a sphere.

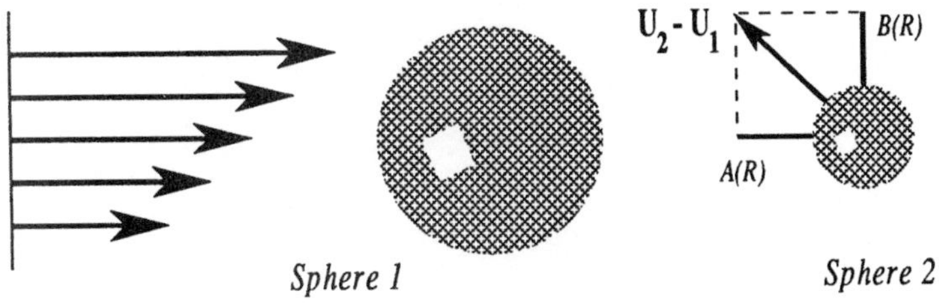

Figure 10.7: Large and small spheres in a linear field.

Exercise 10.3 Two Spheres in a Linear Field.
Consider two neutrally buoyant spheres of radius a and b in the linear ambient field, $\boldsymbol{\Omega}^\infty \times \boldsymbol{x} + \boldsymbol{E}^\infty \cdot \boldsymbol{x}$. Keeping with the theme of this chapter, we assume $\beta = b/a \ll 1$. Use the image system of the stresslet near a rigid sphere to calculate to $O(\beta^5)$ the A and B mobility functions of Batchelor and Green. These were defined in Section 8.4 (see also Exercise 8.6) by the expression

$$\boldsymbol{U}_2 - \boldsymbol{U}_1 = \boldsymbol{\Omega}^\infty \times (\boldsymbol{x}_2 - \boldsymbol{x}_1) + \boldsymbol{E}^\infty \cdot (\boldsymbol{x}_2 - \boldsymbol{x}_1) - [A\boldsymbol{dd} + B(\boldsymbol{\delta} - \boldsymbol{dd})] \cdot \boldsymbol{E}^\infty \cdot (\boldsymbol{x}_2 - \boldsymbol{x}_1) \, .$$

Now consider simple shear flow with sphere centers in the plane of shear. Compute the trajectories of the small sphere for various values of β. How closely do these trajectories follow the streamlines for shear flow past the large sphere in isolation?

Exercise 10.4 Stokeslet Near a Spherical Drop: Surface Deformation.
With a Stokeslet nearby, the shape of a drop is spherical only in the limit of large surface tension γ. For finite surface tensions, we must examine the normal component of the surface tractions and perform a balance between surface tension and hydrodynamic forces. If we take the drop shape to be spherical, what is the error in this force balance? Define an effective capillary number for this problem using the Stokeslet strength F, viscosity μ_e, drop size a, distance between Stokeslet and the drop surface h and surface tension γ. Find the leading corrections to the drop shape (for a history and review of this method, see [9]).

Chapter 11

The Complete Set of Resistance and Mobility Functions for Two Rigid Spheres

11.1 Regimes of Interaction

In this section, we present a summary of all known analytical results for interactions between two rigid spheres. The parameter space for two spheres of radii a and b with $a \geq b$ is shown in Figure 11.1. We delineate three overlapping regions of interest: *widely separated spheres*, *nearly touching spheres*, and *large-small interactions*. For the first case, we have $R \gg a \geq b$, and Jeffrey and Onishi's [38] twin-multipole expansions (Chapter 8) is the method of choice (the results given in the tables in this chapter were obtained by this method). The leading order terms in the far field results may be obtained from the standard method of reflections, but the algebraic manipulations become increasingly difficult for the higher order terms.

For nearly touching spheres, the singular terms may be obtained by lubrication theory (see Chapter 9) and there is a large body of literature on this subject. (Readers interested in the original literature for the various resistance functions are directed to [23, 39, 41, 55, 59, 60, 61].) Unfortunately, at this time we do not have the complete set of analytical results for *all* resistance and mobility functions. A complete set (almost-touching to far field) of information is available for the functions $\boldsymbol{A}$, $\boldsymbol{B}$, $\boldsymbol{C}$, $\boldsymbol{a}$, $\boldsymbol{b}$, $\boldsymbol{c}$ from Jeffrey and Onishi [38] and is reproduced here. Other results in the tables were obtained by combining known lubrication analyses for the resistance functions $\boldsymbol{G}$, $\boldsymbol{H}$, and $\boldsymbol{M}$, as summarized by Jeffrey and coworkers [7, 40, 41] (see also Chapter 9) with boundary collocation results of Kim and coworkers [45, 69] for the $O(1)$ and $o(1)$ terms of the resistance functions in almost-touching geometries.[1] The results for the

[1]Boundary collocation is described in Chapter 13. We thank Mr. Gary Huber for contributing to the numerical analysis of the boundary collocation results. We also thank Dr. David Jeffrey of the University of Western Ontario for providing us with a complete set of results obtained from the twin multipole expansion algorithm.

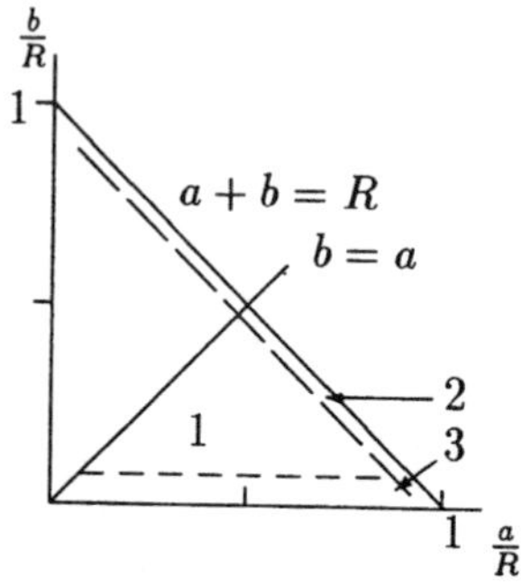

Figure 11.1: Schematic diagram of the parameter space for two spheres. Far field (1); near field (2); disparate spheres (3).

interaction between large and small spheres (disparate-sized interactions) are described in Chapter 10, and thus not repeated here. In the next section, we illustrate the use of these tables.

11.2 Examples of the Usage of Resistance and Mobility Functions

We illustrate how these results can be used to solve a number of resistance and mobility problems. We consider five examples (see Figure 11.2): two spheres moving in tandem along their line of centers; two spheres approaching each other (squeezing flow); sedimentation of a heavy sphere sliding past a neutrally buoyant sphere; two almost-touching spheres undergoing rigid-body rotation about a center located at the midpoint of the gap region; and two force-free and torque-free spheres in a linear field.

11.2.1 Two Spheres Moving in Tandem Along Their Line of Centers

We set $\boldsymbol{U}_2 = \boldsymbol{U}_1 = \boldsymbol{U}$, with $\boldsymbol{U}$ directed along the sphere-sphere axis. The relevant resistance relations for sphere 1 are

$$F = -\mu(X_{11}^A + X_{12}^A)U \ , \qquad S = -\mu(X_{11}^G + X_{12}^G)U \ .$$

At large separations, we simply insert the expansions in inverse powers of R^{-1} from the tables for the resistance functions X_{11}^A, X_{12}^A, X_{11}^G, and X_{12}^G. The result for near-contact is more interesting. We define the small parameter,

$$\xi = \frac{2(R-a-b)}{(a+b)} = \frac{2\epsilon}{1+\beta} \ .$$

This is just the gap divided by a length scale that is symmetric with respect to both sphere labels. The singular terms cancel exactly and we are left with[2]

$$\begin{aligned}\frac{-F}{6\pi\mu aU} &= A^X_{11}(\beta) + \frac{1}{2}(1+\beta)A^X_{12}(\beta) \\ &\quad + (L^X_{11}(\beta) + \frac{1}{2}(1+\beta)L^X_{12}(\beta))\xi + O(\xi^2 \ln \xi) \\ \frac{-S}{4\pi\mu a^2 U} &= G^X_{11}(\beta) + \frac{1}{2}(1+\beta)G^X_{12}(\beta) + O(\xi) \ .\end{aligned}$$

The finite terms at $\xi = 0$ correspond to the resistance functions for the "dumbbell" with zero connector length.

11.2.2 Two Spheres Approaching Each Other

Consider two spheres approaching each other, so that $\boldsymbol{U}_2 = -\boldsymbol{U}_1 = \boldsymbol{U}$ with $\boldsymbol{U}$ parallel to the sphere-sphere axis. The force on sphere 1 is then given by

$$F = \mu(X^A_{11} - X^A_{12})U \ ,$$

and at near-contact the singular terms do not cancel. Instead we have

$$\begin{aligned}\frac{F}{6\pi\mu aU} &= \frac{4\beta^2}{(1+\beta)^3}\xi^{-1} + \frac{2\beta(1+7\beta+\beta^2)}{5(1+\beta)^3}\ln \xi^{-1} \\ &+ A^X_{11}(\beta) - \frac{1}{2}(1+\beta)A^X_{12}(\beta) \\ &+ \frac{1+18\beta-29\beta^2+18\beta^3+\beta^4}{21(1+\beta)^3}\xi \ln \xi^{-1} \\ &+ (L^X_{11}(\beta) - \frac{1}{2}(1+\beta)L^X_{12}(\beta))\xi + O(\xi^2 \ln \xi) \ .\end{aligned}$$

This result is consistent with that derived in Chapter 9.

11.2.3 Heavy Sphere Falling Past a Neutrally Buoyant Sphere

Consider a heavy sphere falling past a second, neutrally buoyant sphere, at the instant when the line between the sphere centers is horizontal. The drag on sphere 1, $\boldsymbol{F}_1$, is equal and opposite to the known external force (gravity), while $\boldsymbol{F}_2 = 0$. Under these circumstances, the mobility functions y^a_{11} and $y^a_{21} = y^a_{12}$[3] describe the motions of sphere 1 and sphere 2, as given by the following relations:

$$U_1 = y^a_{11}(-F_1/\mu) \ , \qquad U_2 = y^a_{21}(-F_1/\mu) \ .$$

Here, both $\boldsymbol{U}_1$ and $\boldsymbol{U}_2$ point in the direction of gravity.

---

[2] In the tables, we denote the combination $L^X_{11}(\beta) + (1+\beta)L^X_{12}(\beta) + \beta L^X_{22}(\beta)$ as W^X.

[3] Why are these two functions equal?

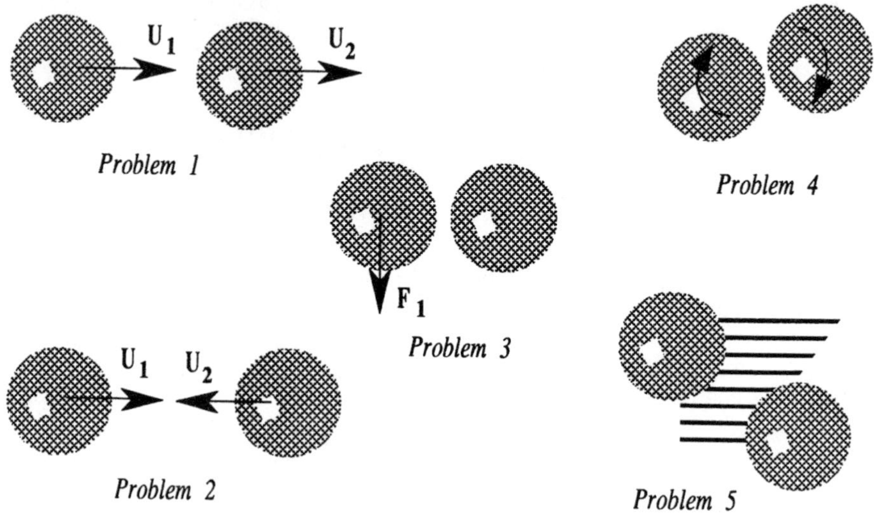

Figure 11.2: Resistance and mobility problems.

An examination of the results for $y^a_{\alpha\beta}$ given in the tables of this chapter reveal that the far field behavior is as expected: The heavy sphere falls with essentially the Stokes sedimentation velocity, whilst the neutrally buoyant sphere barely moves in reaction to its distant neighbor. Indeed, this far field behavior persists until the partices are almost touching. For the near-contact regime, we again employ the gap parameter, ξ. To capture the sharp transition from the lubrication solution to the far field solution, we use $1/\ln\xi^{-1}$ on the abscissa in Figure 11.3. These results of the continuum theory can be interpreted as follows: Unless the spheres are practically touching, lubrication stresses are too weak to keep the spheres together. The heavier one falls; the neutrally-buoyant one does not. On the other hand, at contact (a heavy and a neutrally-buoyant sphere fused together) the two must fall with a common sedimentation velocity, plus the correction for the rigid-body rotation caused by the torque caused by the asymmetric loading (verify this: use the translation theorems of Chapter 5 and the results in this chapter for the $y^c_{\alpha\beta}$ functions). Note that the transitions occur at gap thicknesses for which the continuum theory may be at best a framework for mathematical model building.

The solid lines in Figure 11.3 correspond to the lubrication solutions, while the discrete points are those obtained from the boundary collocation solutions. The positive role of osculation points at the poles is highlighted in this plot. The solid circles (•) and pluses (+) are 48-point (per sphere) and 96-point (per sphere) boundary collocation solutions with equidistant spacing between

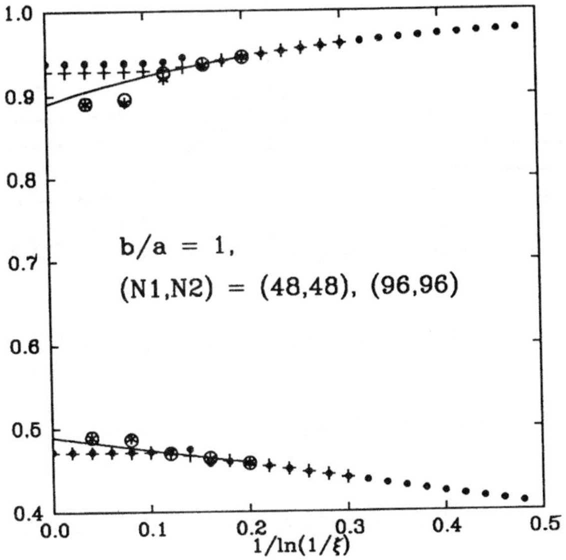

Figure 11.3: Mobility functions y^a_{11} and y^a_{21}.

collocation points, and no points at the poles. The asterisk (∗) and open circle (∘) are 48-point and 96-point solutions, with equidistant spacing and osculation points at the poles. Without osculation at the poles, the system becomes ill-conditioned and diverges at some critical ξ. Placement of osculation points at the poles circumvents this problem. *Note also that at some small value for ξ, the discrete system (with osculation) cannot resolve the thin film separating the spheres, but this in fact can be used to our advantage: It provides a fairly simple numerical procedure for obtaining values of the mobility functions for touching spheres.*

11.2.4 Two Almost-Touching Spheres in Rigid-Body Rotation

For two almost-touching spheres rotating about the center of sphere 1, the relevant velocities are

$$\omega_1 = \omega_2 = \omega \ ,$$

$$U_1 = 0 \ , \qquad U_2 = a(1+\beta)(1+\xi)\omega \times d \ .$$

The force, torque, and stresslet on sphere 1 are given by

$$\begin{aligned}
F_1 &= -\mu Y^A_{12} U_2 + \mu (Y^B_{11} + Y^B_{21}) \omega \times d \\
T_1 &= -\mu Y^B_{12} U_2 \times d - \mu (Y^C_{11} + Y^C_{12}) \omega \\
S_1 &= -\mu Y^G_{12} (dU_2 + U_2 d) - \mu (Y^H_{11} + Y^H_{12}) (\omega \times dU_2 + U_2 \omega \times d) \ .
\end{aligned}$$

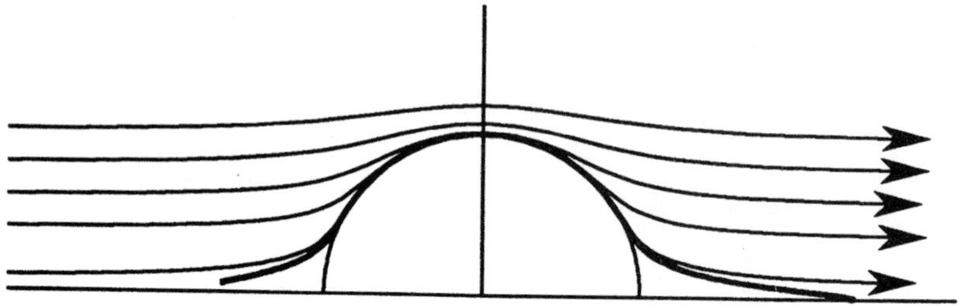

Figure 11.4: Two equal spheres in a shear flow.

The singular terms cancel exactly, and for $\xi = 0$ (touching spheres) we are left with

$$\begin{aligned}
\frac{F_1}{4\pi\mu a^2\omega} &= -\frac{3}{4}(1+\beta)^2 A_{12}^Y(\beta) + B_{11}^Y(\beta) - \frac{1}{4}(1+\beta)^2 B_{12}^Y(\beta^{-1}) \\
\frac{T_1}{8\pi\mu a^3\omega} &= \frac{1}{8}(1+\beta)B_{12}^Y - C_{11}^Y - \frac{1}{8}(1+\beta)^3 C_{12}^Y \\
\frac{S}{8\pi\mu a^3\omega} &= -\frac{1}{8}(1+\beta)^3 G_{12}^Y - H_{11}^Y - \frac{1}{8}(1+\beta)^3 H_{12}^Y .
\end{aligned}$$

11.2.5 Two Force-Free and Torque-Free Spheres in a Linear Field

In the ambient field $\boldsymbol{\Omega}^\infty \times \boldsymbol{x} + \boldsymbol{E}^\infty \cdot \boldsymbol{x}$, the translational velocity of spheres 1 and 2 are given by the mobility relations

$$\begin{aligned}
\boldsymbol{\Omega}^\infty \times \boldsymbol{x}_1 + \boldsymbol{E}^\infty \cdot \boldsymbol{x}_1 - \boldsymbol{U}_1 &= \tilde{\boldsymbol{g}}_1 : \boldsymbol{E}^\infty \\
\boldsymbol{\Omega}^\infty \times \boldsymbol{x}_2 + \boldsymbol{E}^\infty \cdot \boldsymbol{x}_2 - \boldsymbol{U}_2 &= \tilde{\boldsymbol{g}}_2 : \boldsymbol{E}^\infty .
\end{aligned}$$

The relative velocity then follows as

$$\begin{aligned}
\boldsymbol{U}_2 - \boldsymbol{U}_1 &= \boldsymbol{\Omega}^\infty \times (\boldsymbol{x}_2 - \boldsymbol{x}_1) + \boldsymbol{E}^\infty \cdot (\boldsymbol{x}_2 - \boldsymbol{x}_1) - (\tilde{\boldsymbol{g}}_2 - \tilde{\boldsymbol{g}}_1) : \boldsymbol{E}^\infty \\
&= \boldsymbol{\Omega}^\infty \times (\boldsymbol{x}_2 - \boldsymbol{x}_1) + \boldsymbol{E}^\infty \cdot (\boldsymbol{x}_2 - \boldsymbol{x}_1) \\
&\quad - [A\boldsymbol{dd} + B(\boldsymbol{\delta} - \boldsymbol{dd})] \cdot \boldsymbol{E}^\infty \cdot (\boldsymbol{x}_2 - \boldsymbol{x}_1) ,
\end{aligned}$$

where

$$\begin{aligned}
A &= R^{-1}(x_{11}^g + x_{21}^g - x_{12}^g - x_{22}^g) \\
B &= 2R^{-1}(y_{11}^g + y_{21}^g - y_{12}^g - y_{22}^g) .
\end{aligned}$$

Using the tables for $x^g_{\alpha\beta}$ and $y^g_{\alpha\beta}$ (read down the column below f_k to pick out the coefficients for the polynomial, $f_k(\beta)$), we will compute A and B accurate to terms of R^{-8}.

$$\begin{aligned} x^g_{11} &= 2a\left[200\beta^3\left(\frac{a}{2R}\right)^5 + (-1760\beta^3 + 640\beta^5)\left(\frac{a}{2R}\right)^7 + \cdots\right] \\ &= 2a\left[\frac{25a^2b^3}{R^5} + \frac{-55a^4b^3 + 20a^2b^5}{4R^7} + \cdots\right] \\ x^g_{22} &= -2b\left[\frac{25a^3b^2}{R^5} + \frac{-55a^3b^4 + 20a^5b^2}{4R^7} + \cdots\right], \end{aligned}$$

and similarly for x^g_{12}, x^g_{21}, and $y^g_{\alpha\beta}$. Note that the expression for x^g_{22} has been obtained from that for x^g_{11} by switching the labels on the spheres $a \to b$, $b \to a$, and $\boldsymbol{d} \to -\boldsymbol{d}$. The final results for A and B are

$$\begin{aligned} A &= \frac{5(a^3+b^3)}{2R^3} - \frac{3(a^5+b^5)+5a^2b^2(a+b)}{2R^5} + \frac{25a^3b^3}{R^6} - \frac{35a^3b^3(a^2+b^2)}{2R^8} \\ &\quad + O\left(\left(\frac{a+b}{R}\right)^9\right), \\ B &= \frac{3(a^5+b^5)+5a^2b^2(a+b)}{3R^5} - \frac{25a^3b^3(a^2+b^2)}{6R^8} + O\left(\left(\frac{a+b}{R}\right)^{10}\right). \end{aligned}$$

In Figure 11.4, we show trajectories traced by one of two equal spheres in a shear field, as viewed from the center-of-mass frame. Although all sphere pairs starting from infinity sweep past each other, a critical trajectory (darker solid line) delineates a region with doublets that orbit each other. The hydrodynamic force on its own cannot disrupt the doublet; over a very long time other cumulative effects, such as Brownian motion, may break up the doublet.

11.3 Tables of the Resistance and Mobility Functions

In these tables we present the asymptotic expressions for the resistance and mobility functions. The notation is as follows:

$$\begin{aligned} R &= |\boldsymbol{x}_2 - \boldsymbol{x}_1| \\ \boldsymbol{d} &= (\boldsymbol{x}_2 - \boldsymbol{x}_1)/R \\ \beta &= b/a \\ \xi &= 2(R-a-b)/(a+b). \end{aligned}$$

Only the "11" and "12" functions are provided; the reader should use the following symmetries to obtain the "21" and "22" functions:

$$X^A_{\alpha\beta}\left(\frac{2R}{a+b}, \beta\right) = X^A_{(3-\alpha)(3-\beta)}\left(\frac{2R}{a+b}, \beta^{-1}\right)$$

$$
\begin{aligned}
Y^A_{\alpha\beta}\left(\frac{2R}{a+b},\beta\right) &= Y^A_{(3-\alpha)(3-\beta)}\left(\frac{2R}{a+b},\beta^{-1}\right) \\
Y^B_{\alpha\beta}\left(\frac{2R}{a+b},\beta\right) &= -Y^B_{(3-\alpha)(3-\beta)}\left(\frac{2R}{a+b},\beta^{-1}\right) \\
X^C_{\alpha\beta}\left(\frac{2R}{a+b},\beta\right) &= X^C_{(3-\alpha)(3-\beta)}\left(\frac{2R}{a+b},\beta^{-1}\right) \\
Y^C_{\alpha\beta}\left(\frac{2R}{a+b},\beta\right) &= Y^C_{(3-\alpha)(3-\beta)}\left(\frac{2R}{a+b},\beta^{-1}\right) \\
X^G_{\alpha\beta}\left(\frac{2R}{a+b},\beta\right) &= -X^G_{(3-\alpha)(3-\beta)}\left(\frac{2R}{a+b},\beta^{-1}\right) \\
Y^G_{\alpha\beta}\left(\frac{2R}{a+b},\beta\right) &= -Y^G_{(3-\alpha)(3-\beta)}\left(\frac{2R}{a+b},\beta^{-1}\right) \\
Y^H_{\alpha\beta}\left(\frac{2R}{a+b},\beta\right) &= Y^H_{(3-\alpha)(3-\beta)}\left(\frac{2R}{a+b},\beta^{-1}\right) \\
X^M_{\alpha\beta}\left(\frac{2R}{a+b},\beta\right) &= X^M_{(3-\alpha)(3-\beta)}\left(\frac{2R}{a+b},\beta^{-1}\right) \\
Y^M_{\alpha\beta}\left(\frac{2R}{a+b},\beta\right) &= Y^M_{(3-\alpha)(3-\beta)}\left(\frac{2R}{a+b},\beta^{-1}\right) \\
Z^M_{\alpha\beta}\left(\frac{2R}{a+b},\beta\right) &= Z^M_{(3-\alpha)(3-\beta)}\left(\frac{2R}{a+b},\beta^{-1}\right) ,
\end{aligned}
$$

with a similar set results for the mobility functions. Also, because of the reciprocal theorem, we have the symmetry relations for functions from the diagonal tensors (Chapter 7):

$$
\begin{aligned}
X^A_{\alpha\beta}\left(\frac{2R}{a+b},\beta\right) &= X^A_{\beta\alpha}\left(\frac{2R}{a+b},\beta\right) \\
Y^A_{\alpha\beta}\left(\frac{2R}{a+b},\beta\right) &= Y^A_{\beta\alpha}\left(\frac{2R}{a+b},\beta\right) \\
X^C_{\alpha\beta}\left(\frac{2R}{a+b},\beta\right) &= X^C_{\beta\alpha}\left(\frac{2R}{a+b},\beta\right) \\
Y^C_{\alpha\beta}\left(\frac{2R}{a+b},\beta\right) &= Y^C_{\beta\alpha}\left(\frac{2R}{a+b},\beta\right) \\
X^M_{\alpha\beta}\left(\frac{2R}{a+b},\beta\right) &= X^M_{\beta\alpha}\left(\frac{2R}{a+b},\beta\right) \\
Y^M_{\alpha\beta}\left(\frac{2R}{a+b},\beta\right) &= Y^M_{\beta\alpha}\left(\frac{2R}{a+b},\beta\right) \\
Z^M_{\alpha\beta}\left(\frac{2R}{a+b},\beta\right) &= Z^M_{\beta\alpha}\left(\frac{2R}{a+b},\beta\right)
\end{aligned}
$$

The analogous set for the mobility functions may be deduced from expressions given in Chapter 7.

$$
\begin{aligned}
X_{11}^A &= 6\pi a\left(\frac{2\beta^2}{(1+\beta)^3}\xi^{-1}+\frac{\beta(1+7\beta+\beta^2)}{5(1+\beta)^3}\ln\xi^{-1}+A_{11}^X(\beta)\right.\\
&\quad\left.+\frac{1+18\beta-29\beta^2+18\beta^3+\beta^4}{42(1+\beta)^3}\xi\ln\xi^{-1}+L_{11}^X(\beta)\xi+O(\xi^2\ln\xi)\right)\\
X_{12}^A &= -6\pi a\left(\frac{2\beta^2}{(1+\beta)^3}\xi^{-1}+\frac{\beta(1+7\beta+\beta^2)}{5(1+\beta)^3}\ln\xi^{-1}-\frac{1}{2}(1+\beta)A_{12}^X(\beta)\right.\\
&\quad+\frac{1+18\beta-29\beta^2+18\beta^3+\beta^4}{42(1+\beta)^3}\xi\ln\xi^{-1}\\
&\quad\left.-\frac{1}{2}(1+\beta)L_{12}^X(\beta)\xi+O(\xi^2\ln\xi)\right)\\
Y_{11}^A &= 6\pi a\left(\frac{4\beta(2+\beta+2\beta^2)}{15(1+\beta)^3}\ln\xi^{-1}+A_{11}^Y(\beta)\right.\\
&\quad\left.+\frac{2(16-45\beta+58\beta^2-45\beta^3+16\beta^4)}{375(1+\beta)^3}\xi\ln\xi^{-1}\right)\\
Y_{12}^A &= -6\pi a\left(\frac{4\beta(2+\beta+2\beta^2)}{15(1+\beta)^3}\ln\xi^{-1}-\frac{1}{2}(1+\beta)A_{12}^Y(\beta)\right.\\
&\quad\left.+\frac{2(16-45\beta+58\beta^2-45\beta^3+16\beta^4)}{375(1+\beta)^3}\xi\ln\xi^{-1}\right)
\end{aligned}
$$

β	A_{11}^X	A_{12}^X	A_{22}^X	W^X	A_{11}^Y	A_{12}^Y	A_{22}^Y
0.1	1.0398	−0.0787	0.4692	0.0011	0.9869	−0.0072	0.2438
0.125	1.0496	−0.0993	0.5001	0.0022	0.9907	−0.0268	0.3637
0.2	1.0730	−0.1556	0.5789	0.0084	1.0015	−0.0844	0.5930
0.25	1.0836	−0.1880	0.6253	0.0146	1.0073	−0.1181	0.6871
0.5	1.0881	−0.2957	0.8083	0.0575	1.0193	−0.2246	0.9009
1.0	0.9954	−0.3502	0.9954	0.1163	0.9983	−0.2737	0.9983

Table 11.1: Near field forms of the **A** resistance functions.

$$X_{11}^A = 6\pi a \sum_{k=0}^{\infty} f_{2k}(\beta) \left(\frac{a}{2R}\right)^{2k}$$

$$X_{12}^A = -6\pi a \sum_{k=0}^{\infty} f_{2k+1}(\beta) \left(\frac{a}{2R}\right)^{2k+1}$$

	f_0	f_2	f_4	f_6	f_8	f_{10}
1	1					
β		9	−24	16		
β^2			81	108	576	2304
β^3			36	281	4848	20736
β^4				648	5409	42804
β^5				144	4524	115849
β^6					3888	76176
β^7					576	39264
β^8						20736
β^9						2304

	f_1	f_3	f_5	f_7	f_9	f_{11}
1	0					
β	3	−4				
β^2		27	72	288	1152	4608
β^3		−4	243	1620	9072	46656
β^4			72	1515	14752	108912
β^5				1620	26163	269100
β^6				288	14752	319899
β^7					9072	269100
β^8					1152	108912
β^9						46656
β^{10}						4608

Table 11.2: Far field forms of the resistance functions $X_{\alpha\beta}^A$.

$$Y_{11}^A = 6\pi a \sum_{k=0}^{\infty} f_{2k}(\beta) \left(\frac{a}{2R}\right)^{2k}$$

$$Y_{12}^A = -6\pi a \sum_{k=0}^{\infty} f_{2k+1}(\beta) \left(\frac{a}{2R}\right)^{2k+1}$$

	f_0	$4f_2$	$16f_4$	$64f_6$	$256f_8$	$1024f_{10}$
1	1					
β		9	96	256		
β^2			81	3456	71424	1179648
β^3			288	1241	136352	2011392
β^4				5184	126369	6303168
β^5				4608	−3744	10548393
β^6					165888	8654976
β^7					73728	−179712
β^8						3981312
β^9						1179648

	$2f_1$	$8f_3$	$32f_5$	$128f_7$	$512f_9$	$2048f_{11}$
1	0					
β	3	16				
β^2		27	1008	18432	294912	4718592
β^3		16	243	16848	580608	14598144
β^4			1008	19083	967088	22600704
β^5				16848	766179	43912080
β^6				18432	967088	95203835
β^7					580608	43912080
β^8					294912	22600704
β^9						14598144
β^{10}						4718592

Table 11.3: Far field forms of the resistance functions $Y_{\alpha\beta}^A$.

$$Y_{11}^{B} = 4\pi a^2 \sum_{k=0}^{\infty} f_{2k+1}(\beta) \left(\frac{a}{2R}\right)^{2k+1}$$

$$Y_{12}^{B} = -4\pi a^2 \sum_{k=0}^{\infty} f_{2k}(\beta) \left(\frac{a}{2R}\right)^{2k}$$

	f_0	f_2	$2f_4$	$8f_6$	$32f_8$	$128f_{10}$
1	0					
β		−6				
β^2			−27	−864	−13824	−221184
β^3				−243	−15552	−497664
β^4				−576	−77451	−1905696
β^5					−12960	−1125603
β^6					−9216	−2816256
β^7						−373248
β^8						−147456

	f_1	f_3	$4f_5$	$16f_7$	$64f_9$	$256f_{11}$
1	0					
β		−9	−48			
β^2			−81	−3024	−55296	−884736
β^3			−144	−8409	−50544	−1741824
β^4				−3888	−283041	−9831696
β^5				−2304	−488400	−4579497
β^6					−103680	−8586864
β^7					−36864	−18941184
β^8						−2322432
β^9						−589824

Table 11.4: Far field forms of the resistance functions $Y_{\alpha\beta}^{B}$.

$$X_{11}^C = 8\pi a^3 \left(\frac{\beta^3}{(1+\beta)^3}\zeta(3,\beta/(1+\beta)) - \frac{\beta^2}{4(1+\beta)}\xi \ln \xi^{-1} \right)$$

$$X_{12}^C = -8\pi a^3 \left(\frac{\beta^3}{(1+\beta)^3}\zeta(3,1) - \frac{\beta^2}{4(1+\beta)}\xi \ln \xi^{-1} \right) ,$$

where the Riemann zeta function is defined as

$$\zeta(z,x) = \sum_{k=0}^{\infty}(k+x)^{-z} .$$

In Jeffrey and Onishi, $\zeta(3,1)$ is denoted by the usual notation as just $\zeta(3)$.

$$\begin{aligned}
Y_{11}^B &= 4\pi a^2 \left(-\frac{\beta(4+\beta)}{5(1+\beta)^2} \ln \xi^{-1} + B_{11}^Y(\beta) \right. \\
&\quad \left. - \frac{(32 - 33\beta + 83\beta^2 + 43\beta^3)}{250(1+\beta)^2}\xi \ln \xi^{-1} \right) \\
Y_{12}^B &= 4\pi a^2 \left(\frac{\beta(4+\beta)}{5(1+\beta)^2} \ln \xi^{-1} + \frac{1}{4}(1+\beta)^2 B_{12}^Y(\beta) \right. \\
&\quad \left. + \frac{(32 - 33\beta + 83\beta^2 + 43\beta^3)}{250(1+\beta)^2}\xi \ln \xi^{-1} \right) \\
Y_{11}^C &= 8\pi a^3 \left(\frac{2\beta}{5(1+\beta)} \ln \xi^{-1} + C_{11}^Y(\beta) \right. \\
&\quad \left. + \frac{(8 + 6\beta + 33\beta^2)}{125(1+\beta)}\xi \ln \xi^{-1} \right) \\
Y_{12}^C &= 8\pi a^3 \left(\frac{\beta^2}{10(1+\beta)} \ln \xi^{-1} + \frac{1}{8}(1+\beta)^3 C_{12}^Y(\beta) \right. \\
&\quad \left. + \frac{\beta(43 - 24\beta + 43\beta^2)}{250(1+\beta)}\xi \ln \xi^{-1} \right)
\end{aligned}$$

β	C_{11}^Y	C_{12}^Y	C_{22}^Y	B_{11}^Y	B_{12}^Y	B_{21}^Y	B_{22}^Y
0.1	0.9683	−0.0165	−0.1909	0.0408	−0.0560	0.0214	−0.7559
0.125	0.9625	−0.0210	−0.0918	0.0433	−0.0306	0.0288	−0.7124
0.25	0.9280	−0.0349	0.2097	0.0620	0.0592	0.0594	−0.5664
0.5	0.8489	−0.0349	0.4839	0.1201	0.0817	0.0686	−0.4016
1.0	0.7028	−0.0274	0.7028	0.2390	−0.0017	0.0017	−0.2390

Table 11.5: Near field forms of the **B** and **C** resistance functions.

$$X^C_{11} = 8\pi a^3 \sum_{k=0}^{\infty} f_{2k}(\beta) \left(\frac{a}{2R}\right)^{2k}$$

$$X^C_{12} = -8\pi a^3 \sum_{k=0}^{\infty} f_{2k+1}(\beta) \left(\frac{a}{2R}\right)^{2k+1}$$

	f_0	f_2	f_4	f_6	f_8	f_{10}
1	1					
β		0				
β^2			0			
β^3				64		
β^4						
β^5					768	
β^6						
β^7						6144

	f_1	f_3	f_5	f_7	f_9	f_{11}
β^3	0	8				
β^4			0			
β^5				0		
β^6					512	6144
β^7						
β^8						6144

Table 11.6: Far field forms of the resistance functions $X^C_{\alpha\beta}$.

$$Y_{11}^C = 8\pi a^3 \sum_{k=0}^{\infty} f_{2k}(\beta) \left(\frac{a}{2R}\right)^{2k}$$
$$Y_{12}^C = 8\pi a^3 \sum_{k=0}^{\infty} f_{2k+1}(\beta) \left(\frac{a}{2R}\right)^{2k+1}$$

	f_0	f_2	f_4	f_6	$4f_8$	$16f_{10}$
1	1					
β		0	12			
β^2				27	864	13824
β^3				256	243	15552
β^4					864	151179
β^5					9984	15552
β^6						20736
β^7						294912

	f_1	f_3	f_5	$2f_7$	$8f_9$	$32f_{11}$
β^2	0					
β^3		4				
β^4			18	144	2304	36864
β^5				81	3888	103680
β^6				144	−6439	−175152
β^7					3888	518049
β^8					2304	−175152
β^9						103680
β^{10}						36864

Table 11.7: Far field forms of the resistance functions $Y_{\alpha\beta}^C$.

$$X_{11}^G = 4\pi a^2 \left(\frac{3\beta^2}{(1+\beta)^3}\xi^{-1} + \frac{3\beta(1+12\beta-4\beta^2)}{10(1+\beta)^3}\ln\xi^{-1} + G_{11}^X(\beta) + \frac{5+181\beta-453\beta^2+566\beta^3-65\beta^4}{140(1+\beta)^3}\xi\ln\xi^{-1} + O(\xi) \right)$$

$$X_{12}^G = -4\pi a^2 \left(\frac{3\beta^2}{(1+\beta)^3}\xi^{-1} + \frac{3\beta(1+12\beta-4\beta^2)}{10(1+\beta)^3}\ln\xi^{-1} - \frac{1}{4}(1+\beta)^2 G_{12}^X(\beta) + \frac{5+181\beta-453\beta^2+566\beta^3-65\beta^4}{140(1+\beta)^3}\xi\ln\xi^{-1} + O(\xi) \right)$$

$$Y_{11}^G = 4\pi a^2 \left(\frac{\beta(4-\beta+7\beta^2)}{10(1+\beta)^3}\ln\xi^{-1} + G_{11}^Y(\beta) + \frac{32-179\beta+532\beta^2-356\beta^3+221\beta^4}{500(1+\beta)^3}\xi\ln\xi^{-1} + O(\xi) \right)$$

$$Y_{12}^G = -4\pi a^2 \left(\frac{\beta(4-\beta+7\beta^2)}{10(1+\beta)^3}\ln\xi^{-1} - \frac{1}{4}(1+\beta)^2 G_{12}^Y(\beta) + \frac{32-179\beta+532\beta^2-356\beta^3+221\beta^4}{500(1+\beta)^3}\xi\ln\xi^{-1} + O(\xi) \right)$$

$$Y_{11}^H = 8\pi a^3 \left(\frac{\beta(2-\beta)}{10(1+\beta)^2}\ln\xi^{-1} + H_{11}^Y(\beta) + \frac{16-61\beta+180\beta^2+2\beta^3}{500(1+\beta)^2}\xi\ln\xi^{-1} + O(\xi) \right)$$

$$Y_{12}^H = 8\pi a^3 \left(\frac{\beta^2(1+7\beta)}{20(1+\beta)^2}\ln\xi^{-1} + \frac{1}{8}(1+\beta)^3 H_{12}^Y(\beta) + \frac{\beta(43+147\beta-185\beta^2+221\beta^3)}{1000(1+\beta)^2}\xi\ln\xi^{-1} + O(\xi) \right)$$

β	G_{11}^X	G_{12}^X	G_{11}^Y	G_{12}^Y	H_{11}^Y	H_{12}^Y
0.125	0.059	−0.217	−0.025	0.040	−0.018	−0.010
0.25	0.065	−0.292	−0.040	0.039	−0.032	−0.015
0.5	−0.078	−0.137	−0.072	0.068	−0.056	−0.011
1.0	−0.469	0.195	−0.142	0.103	−0.074	−0.030
2.0	−0.780	0.229	−0.331	0.125	−0.056	−0.086
4.0	−0.472	0.051	−0.736	0.108	0.013	−0.098
8.0	0.487	−0.027	−1.318	0.063	0.116	−0.061

Table 11.8: Near field forms of the **G** and **H** resistance functions.

$$X_{11}^{G} = 4\pi a^2 \sum_{k=0}^{\infty} f_{2k+1}(\beta) \left(\frac{a}{2R}\right)^{2k+1}$$

$$X_{12}^{G} = -4\pi a^2 \sum_{k=0}^{\infty} f_{2k}(\beta) \left(\frac{a}{2R}\right)^{2k}$$

	f_0	f_2	f_4	f_6	f_8	f_{10}
1	0					
β		15	−36			
β^2			135	216	864	3456
β^3			−60	1215	6804	34992
β^4				900	8679	66384
β^5					12960	155007
β^6					5040	75900
β^7						97200
β^8						25920
β^9						

	f_1	f_3	f_5	f_7	f_9	f_{11}
1	0					
β		45	−168	144		
β^2			405	108	1728	6912
β^3			360	−1251	27072	77760
β^4				4860	25893	178260
β^5				2160	−3888	790845
β^6					38880	500580
β^7					11520	14880
β^8						259200
β^9						57600
β^{10}						

Table 11.9: Far field forms of the resistance functions $X_{\alpha\beta}^{G}$.

$$Y^G_{11} = 4\pi a^2 \sum_{k=0}^{\infty} f_{2k+1}(\beta) \left(\frac{a}{2R}\right)^{2k+1}$$
$$Y^G_{12} = -4\pi a^2 \sum_{k=0}^{\infty} f_{2k}(\beta) \left(\frac{a}{2R}\right)^{2k}$$

	f_0	f_2	f_4	f_6	f_8	f_{10}
1	0					
β		0	12			
β^2				27	216	864
β^3			20		243/4	972
β^4				135	−1008	42123/16
β^5					1215/4	648
β^6					1080	−96073/16
β^7						4860
β^8						6240
β^9						

	f_1	f_3	f_5	f_7	f_9	f_{11}
1	0					
β		0	18	24		
β^2				81/2	378	1728
β^3			90	−336	21209/8	3158/2
β^4				405/2	−972	103329/32
β^5				600	−62147/8	69387
β^6					2970	763173/32
β^7					3360	−166737/2
β^8						24840
β^9						17280
β^{10}						

Table 11.10: Far field forms of the resistance functions $Y^G_{\alpha\beta}$.

$$Y_{11}^H = 8\pi a^3 \sum_{k=0}^{\infty} f_{2k}(\beta) \left(\frac{a}{2R}\right)^{2k}$$
$$Y_{12}^H = 8\pi a^3 \sum_{k=0}^{\infty} f_{2k+1}(\beta) \left(\frac{a}{2R}\right)^{2k+1}$$

	f_0	f_2	f_4	f_6	f_8	f_{10}
1	0					
β		0	0	24		
β^2					54	432
β^3				−120	1280	243/2
β^4					270	−1872
β^5					−1280	46015/2
β^6						2880
β^7						−9600
β^8						
β^9						

	f_1	f_3	f_5	f_7	f_9	f_{11}
1	0					
β						
β^2						
β^3		10	0			
β^4				36	144	576
β^5					81	972
β^6				180	5248	153049/4
β^7					405	−864
β^8					1200	186173/4
β^9						5940
β^{10}						6720

Table 11.11: Far field forms of the resistance functions $Y_{\alpha\beta}^H$.

$$X_{11}^M = \frac{20}{3}\pi a^3 \left(\frac{6\beta^2}{5(1+\beta)^3}\xi^{-1} + \frac{3\beta(1+17\beta-9\beta^2)}{25(1+\beta)^3}\ln\xi^{-1} + M_{11}^X(\beta) \right.$$
$$\left. + \frac{5+272\beta-831\beta^2+1322\beta^3-415\beta^4}{350(1+\beta)^3}\xi\ln\xi^{-1} + O(\xi) \right)$$
$$X_{12}^M = \frac{20}{3}\pi a^3 \left(\frac{6\beta^3}{5(1+\beta)^3}\xi^{-1} - \frac{3\beta^2(4-17\beta+4\beta^2)}{25(1+\beta)^3}\ln\xi^{-1} + \frac{1}{8}(1+\beta)^3 M_{12}^X(\beta) \right.$$
$$\left. - \frac{\beta(65-832\beta+1041\beta^2-832\beta^3+65\beta^4)}{350(1+\beta)^3}\xi\ln\xi^{-1} + O(\xi) \right)$$

$$Y_{11}^M = \frac{20}{3}\pi a^3 \left(\frac{6\beta(1-\beta+4\beta^2)}{25(1+\beta)^3}\ln\xi^{-1} + M_{11}^Y(\beta) \right.$$
$$\left. + \frac{3(8-67\beta+294\beta^2-394\beta^3+197\beta^4)}{625(1+\beta)^3}\xi\ln\xi^{-1} + O(\xi) \right)$$
$$Y_{12}^M = \frac{20}{3}\pi a^3 \left(\frac{3\beta^2(7-10\beta+7\beta^2)}{50(1+\beta)^3}\ln\xi^{-1} + \frac{1}{8}(1+\beta)^3 M_{12}^Y(\beta) \right.$$
$$\left. - \frac{3\beta(221-748\beta+1902\beta^2-748\beta^3+221\beta^4)}{2500(1+\beta)^3}\xi\ln\xi^{-1} + O(\xi) \right)$$

Z^M is also a mobility function, and therefore displayed as z^m.

β	$M_{11}^X + \frac{1}{8}(1+\beta)^3 M_{12}^X$	$M_{11}^Y + \frac{1}{8}(1+\beta)^3 M_{12}^Y$
0.125	1.0194	0.9679
0.25	0.9908	0.9414
0.5	0.8438	0.8819
1.0	0.5712	0.6760
2.0	0.3219	0.0577
4.0	0.5503	−1.6130
8.0	3.1595	−6.3412

Table 11.12: Near field forms of the ***M*** resistance functions.

$$X_{11}^M = \frac{20}{3}\pi a^3 \sum_{k=0}^{\infty} f_{2k}(\beta)\left(\frac{a}{2R}\right)^{2k}$$
$$X_{12}^M = \frac{20}{3}\pi a^3 \sum_{k=0}^{\infty} f_{2k+1}(\beta)\left(\frac{a}{2R}\right)^{2k+1}$$

	f_0	f_2	f_4	f_6	f_8	f_{10}
1	1					
β		0	60	−288	345.6	
β^2				540	−432.0	1382.4
β^3				1120	−9348.0	52416.0
β^4					8640.0	28380.0
β^5					10560.0	−151632.0
β^6						86400.0
β^7						76800.0

	f_1	f_3	f_5	f_7	f_9	f_{11}
1	0					
β						
β^2						
β^3		40	−192			
β^4			180	1008	5184.0	25344.0
β^5			−192	1620	15552.0	108864.0
β^6				1008	24128.8	169257.6
β^7					15552.0	266695.2
β^8					5184.0	169257.6
β^9						108864.0
β^{10}						25344.0

Table 11.13: Far field forms of the resistance functions $X_{\alpha\beta}^M$.

$$Y_{11}^M = \frac{20}{3}\pi a^3 \sum_{k=0}^{\infty} f_{2k}(\beta)\left(\frac{a}{2R}\right)^{2k}$$
$$Y_{12}^M = \frac{20}{3}\pi a^3 \sum_{k=0}^{\infty} f_{2k+1}(\beta)\left(\frac{a}{2R}\right)^{2k+1}$$

	f_0	f_2	f_4	f_6	f_8	f_{10}
1	1					
β		0			115.2	
β^2			0			259.2
β^3				640	−4736.0	16384.0
β^4						2592.0
β^5					6720.0	−110592.0
β^6						6480.0
β^7						51200.0

	f_1	f_3	f_5	f_7	f_9	f_{11}
1	0					
β						
β^2						
β^3		−20	128			
β^4				0	864.0	5760.0
β^5			128			1944.0
β^6					−14067.2	−75825.6
β^7						10108.8
β^8					864.0	−75825.6
β^9						1944.0
β^{10}						5760.0

Table 11.14: Far field forms of the resistance functions $Y_{\alpha\beta}^M$.

$$3\pi(a_\alpha + a_\beta)x^a_{\alpha\beta} = d^{(1)}_{\alpha\beta}(\beta) + d^{(2)}_{\alpha\beta}(\beta)\xi + d^{(3)}_{\alpha\beta}(\beta)\xi^2 \ln \xi + d^{(4)}_{\alpha\beta}(\beta)\xi^2$$

β	$d^{(1)}_{11}$	$d^{(2)}_{11}$	$d^{(3)}_{11}$	$d^{(4)}_{11}$
0.125	0.9997	$*-0.0002$	0.003	0.008
0.25	0.9951	0.009	0.026	0.013
0.5	0.9537	0.152	0.194	-0.322
1.0	0.7750	0.930	0.900	$*-2.685$
2.0	0.4768	2.277	2.188	$*-6.236$
4.0	0.2488	3.610	4.061	$*-9.165$
8.0	0.1250	5.620	8.500	$*-16.26$

β	$d^{(1)}_{12}$	$d^{(2)}_{12}$	$d^{(3)}_{12}$	$d^{(4)}_{12}$
0.125	0.5623	-0.170	-0.256	$*-0.114$
0.25	0.6219	-0.372	-0.408	$*0.360$
0.5	0.7152	-0.766	-0.691	$*1.566$
1.0	0.7750	-1.070	-0.900	$*2.697$

* These entries differ from the original work of Jeffrey and Onishi.

Table 11.15: Near field forms of the mobility functions $x^a_{\alpha\beta}$.

$$x^a_{11} = (6\pi a)^{-1} \sum_{k=0}^{\infty} f_{2k}(\beta) \left(\frac{a}{2R}\right)^{2k}$$

$$x^a_{12} = -(6\pi a)^{-1} \sum_{k=0}^{\infty} f_{2k+1}(\beta) \left(\frac{a}{2R}\right)^{2k+1}$$

	f_0	f_2	f_4	f_6	f_8	f_{10}
1	1					
β		0				
β^2						
β^3			−60	480	−960	
β^4						
β^5				−128	4224	−17920
β^6						−96000
β^7					−576	30720
β^8						
β^9						−2304

	f_1	f_3	f_5	f_7	f_9	f_{11}
1	−3	4				
β						
β^2		4	0			
β^3				−2400	1920	−15360
β^4						
β^5					1920	231936
β^6						
β^7						−15360

Table 11.16: Far field forms of the mobility functions $x^a_{\alpha\beta}$.

$$3\pi(a_\alpha + a_\beta)y^a_{\alpha\beta} = \frac{a^{(1)}_{\alpha\beta}(\ln\xi^{-1})^2 + a^{(2)}_{\alpha\beta}\ln\xi^{-1} + a^{(3)}_{\alpha\beta}}{(\ln\xi^{-1})^2 + e^{(1)}\ln\xi^{-1} + e^{(2)}} + O(\xi(\ln\xi)^3)$$

$$\pi(a_\alpha + a_\beta)^2 y^b_{\alpha\beta} = \frac{b^{(1)}_{\alpha\beta}(\ln\xi^{-1})^2 + b^{(2)}_{\alpha\beta}\ln\xi^{-1} + b^{(3)}_{\alpha\beta}}{(\ln\xi^{-1})^2 + e_{(1)}\ln\xi^{-1} + e^{(2)}} + O(\xi\ln\xi)\ .$$

β	$a^{(1)}_{11}$	$a^{(2)}_{11}$	$a^{(3)}_{11}$	$a^{(1)}_{12}$	$a^{(2)}_{12}$	$a^{(3)}_{12}$
0.125	0.99415	1.53362	−1.54846	0.55315	0.45721	−0.64730
0.25	0.97292	3.84204	0.33945	0.57085	1.53280	−0.06410
0.5	0.92729	5.61052	4.40223	0.53482	2.50225	1.23963
1.0	0.89056	5.77196	7.06897	0.48951	2.80545	1.98174
2.0	0.76425	5.01983	5.60143	0.53482	2.50225	1.23963
4.0	0.47343	3.71016	1.88922	0.57085	1.53280	−0.06410
8.0	0.23775	2.70340	−0.84740	0.55315	0.45721	−0.64730

β	$b^{(1)}_{11}$	$b^{(2)}_{11}$	$b^{(3)}_{11}$	$b^{(1)}_{12}$	$b^{(2)}_{12}$	$b^{(3)}_{12}$
0.125	0.00638	−0.01937	0.00922	−0.17225	−0.02980	0.13570
0.25	0.03176	−0.05584	−0.01788	−0.20424	−0.31661	0.07668
0.5	0.09519	−0.03922	−0.23881	−0.20385	−0.68106	−0.10094
1.0	0.13368	0.19945	−0.79238	−0.13368	−0.92720	−0.18805
2.0	0.09060	0.52663	−1.45080	−0.05355	−0.99178	0.04487
4.0	0.03268	0.61847	−1.74796	−0.01241	* − 0.86793	*0.47544
8.0	0.00851	0.53163	−1.74774	−0.00202	−0.71569	0.82610

$\beta,\ \beta^{-1}$	$e^{(1)}$	$e^{(2)}$
1.0	6.04250	6.32549
2.0	5.59906	4.17702
4.0	3.79489	0.32014
8.0	1.51572	−1.54010

* These entries differ from the original work of Jeffrey and Onishi.

Table 11.17: Near field forms of the mobility functions $y^a_{\alpha\beta}$ and $y^b_{\alpha\beta}$.

$$y_{11}^a = (6\pi a)^{-1} \sum_{k=0}^{\infty} f_{2k}(\beta) \left(\frac{a}{2R}\right)^{2k}$$

$$y_{12}^a = (6\pi a)^{-1} \sum_{k=0}^{\infty} f_{2k+1}(\beta) \left(\frac{a}{2R}\right)^{2k+1}$$

	f_0	f_2	f_4	f_6	f_8	f_{10}
1	1					
β		0	0			
β^2						
β^3					−320	
β^4						
β^5				−68	288	−6720
β^6						
β^7					−288	3456
β^8						
β^9						−1152

	f_1	f_3	f_5	f_7	f_9	f_{11}
1	3/2	2				
β						
β^2		2	0			
β^3				0		8960
β^4					0	
β^5						−8848
β^6						
β^7						8960

Table 11.18: Far field forms of the mobility functions $y^a_{\alpha\beta}$.

$$y_{11}^b = (4\pi a^2)^{-1} \sum_{k=0}^{\infty} f_{2k+1}(\beta) \left(\frac{a}{2R}\right)^{2k+1}$$

$$y_{12}^b = (4\pi a^2)^{-1} \sum_{k=0}^{\infty} f_{2k}(\beta) \left(\frac{a}{2R}\right)^{2k}$$

	f_0	f_2	f_4	f_6	f_8	f_{10}
1	0	−2				
β			0			
β^2				0		
β^3					0	−4480
β^4						
β^5						−3200

	f_1	f_3	f_5	f_7	f_9	f_{11}
β^3	0	0	0	160		
β^4						
β^5				48	2240	
β^6						
β^7					192	21504
β^8						
β^9						768

Table 11.19: Far field forms of the mobility functions $y_{\alpha\beta}^b$.

(Near field forms easily obtained by inversion of X^C functions)

$$x^c_{11} = (8\pi a^3)^{-1} \sum_{k=0}^{\infty} f_{2k}(\beta) \left(\frac{a}{R}\right)^{2k}$$

$$x^c_{12} = -(8\pi a^3)^{-1} \sum_{k=0}^{\infty} f_{2k+1}(\beta) \left(\frac{a}{R}\right)^{2k+1}$$

	f_0	f_2	f_4	f_6	f_8	f_{10}
1	1	0	0	0		
β						
β^2						
β^3						
β^4						
β^5					-3	
β^6						
β^7						-6

	f_1	f_3	f_5	f_7	f_9	f_{11}
1	0	1	0	0	0	0
β						
β^2						
β^3						
β^4						
β^5						
β^6						
β^7						

Table 11.20: Far field forms of the mobility functions $x^c_{\alpha\beta}$.

$$\pi(a_\alpha + a_\beta)^3 y^c_{\alpha\beta} = \frac{c^{(1)}_{\alpha\beta}(\ln \xi^{-1})^2 + c^{(2)}_{\alpha\beta} \ln \xi^{-1} + c^{(3)}_{\alpha\beta}}{(\ln \xi^{-1})^2 + e_{(1)} \ln \xi^{-1} + e^{(2)}} .$$

β	$c^{(1)}_{11}$	$c^{(2)}_{11}$	$c^{(3)}_{11}$	$c^{(1)}_{12}$	$c^{(2)}_{12}$	$c^{(3)}_{12}$
0.125	0.97916	1.57872	−1.57064	0.17427	−1.07599	1.17174
0.25	0.88738	3.99132	0.37078	0.21664	−1.09737	*0.78409
0.5	0.61012	5.80416	4.98252	0.25740	−1.09618	0.45026
1.0	0.26736	5.60896	9.28111	0.26736	−1.05770	0.29981
2.0	0.07626	4.62489	9.87487	0.25740	* − 1.09618	*0.45026
4.0	0.01386	3.75550	6.51310	0.21664	* − 1.09737	*0.78409
8.0	0.00191	3.20768	2.68635	0.17427	−1.07599	1.17174

$\beta,\ \beta^{-1}$	$e^{(1)}$	$e^{(2)}$
1.0	6.04250	6.32549
2.0	5.59906	4.17702
4.0	3.79489	0.32014
8.0	1.51572	−1.54010

* These entries differ from the original work of Jeffrey and Onishi.

Table 11.21: Near field forms of the mobility functions $y^c_{\alpha\beta}$.

$$y_{11}^c = (8\pi a^3)^{-1} \sum_{k=0}^{\infty} f_{2k}(\beta) \left(\frac{a}{2R}\right)^{2k}$$

$$y_{12}^c = (8\pi a^3)^{-1} \sum_{k=0}^{\infty} f_{2k+1}(\beta) \left(\frac{a}{2R}\right)^{2k+1}$$

	f_0	f_2	f_4	f_6	f_8	f_{10}
1	1					
β		0	0			
β^2						
β^3				−240		
β^4						
β^5					−2496	
β^6						
β^7						−18432

	f_1	f_3	f_5	f_7	f_9	f_{11}
1	0	−4				
β			0			
β^2				0		
β^3					4800	30720
β^4						
β^5						30720

Table 11.22: Far field forms of the mobility functions $y_{\alpha\beta}^c$.

$$x^g_{\alpha\beta} = (a_\alpha + a_\beta)\left(g^{(1)}_{\alpha\beta}(\beta) + g^{(2)}_{\alpha\beta}(\beta)\xi\ln\xi^{-1} + g^{(3)}_{\alpha\beta}(\beta)\xi\right)$$

β	$g^{(1)}_{11}$	$g^{(2)}_{11}$	$g^{(3)}_{11}$	$g^{(1)}_{12}$	$g^{(2)}_{12}$	$g^{(3)}_{12}$
0.125	$2.0e-05$	0.0014	−0.0016	−0.8889	−0.3863	0.6726
0.25	0.0031	0.0072	−0.0128	−0.7950	−0.2885	0.9797
0.5	0.0379	0.0217	−0.1598	−0.6162	−0.1379	1.3129
1.0	0.1792	0.0000	−0.8703	−0.3208	0.0000	0.9184
2.0	0.3709	−0.2065	−1.7555	−0.0861	0.0290	0.2184
4.0	0.4679	−0.7210	−1.7835	−0.0128	0.0116	0.0132
8.0	0.4941	−1.7384	−0.5125	−0.0013	0.0026	−0.0032

Table 11.23: Near field forms of the mobility functions $x^g_{\alpha\beta}$.

$$x^m_1 = \frac{20}{3}\pi a^3\left(1 + m^{(1)}_1(\beta) + m^{(2)}_1(\beta)\xi\ln\xi^{-1} + m^{(3)}_1(\beta)\xi\right)$$

β	$m^{(1)}_1$	$m^{(2)}_1$	$m^{(3)}_1$
0.125	0.00011	0.0067	−0.0026
0.25	0.017	0.034	−0.046
0.5	0.199	0.115	−0.810
1.0	0.910	0.0000	−3.85
2.0	1.77	−0.720	−7.04
4.0	1.81	−1.27	−8.59
8.0	1.14	−2.44	−2.23

Table 11.24: Near field forms of the mobility functions x^m_1.

$$x^g_{11} = (2a)\sum_{k=0}^{\infty} f_{2k+1}(\beta)\left(\frac{a}{2R}\right)^{2k+1}$$

$$x^g_{12} = -(2a)\sum_{k=0}^{\infty} f_{2k}(\beta)\left(\frac{a}{2R}\right)^{2k}$$

	f_0	f_2	f_4	f_6	f_8	f_{10}
1	0	5	−12			
β						
β^2			−20	0		
β^3					8000	−12800
β^4						
β^5						16000

	f_1	f_3	f_5	f_7	f_9	f_{11}
β^3	0	0	200	−1760	3840	−19200
β^4						
β^5				640	−20736	89600
β^6						320000
β^7					3840	−189440
β^8						
β^9						19200

Table 11.25: Far field forms of the mobility functions $x^g_{\alpha\beta}$.

$$y^g_{\alpha\beta} = (a_\alpha + a_\beta)\frac{g^{(1)}_{\alpha\beta}(\ln\xi^{-1})^2 + g^{(2)}_{\alpha\beta}\ln\xi^{-1} + g^{(3)}_{\alpha\beta}}{(\ln\xi^{-1})^2 + e_{(1)}\ln\xi^{-1} + e^{(2)}} + O(\xi\ln\xi)\ .$$

$$y^h_{\alpha\beta} = \frac{h^{(1)}_{\alpha\beta}(\ln\xi^{-1})^2 + h^{(2)}_{\alpha\beta}\ln\xi^{-1} + h^{(3)}_{\alpha\beta}}{(\ln\xi^{-1})^2 + e_{(1)}\ln\xi^{-1} + e^{(2)}} + O(\xi\ln\xi)\ .$$

β	$g^{(1)}_{11}$	$g^{(2)}_{11}$	$g^{(3)}_{11}$	$g^{(1)}_{12}$	$g^{(2)}_{12}$	$g^{(3)}_{12}$
0.125	0.0021	−0.0074	0.0035	−0.3784	−0.0937	0.3488
0.25	0.0070	−0.0176	−0.0100	−0.2650	−0.5043	0.2345
0.5	0.0086	−0.0059	−0.0988	−0.1337	−0.5547	0.1588
1.0	0.0145	0.0786	−0.3193	−0.0869	−0.2956	0.1584
2.0	0.1278	0.0586	−0.6096	−0.0611	−0.0435	0.0888
4.0	0.3281	−0.5175	−0.4652	−0.0213	0.0248	0.0105
8.0	0.4446	−1.4007	0.4763	−0.0043	0.0118	−0.0054

β	$h^{(1)}_{11}$	$h^{(2)}_{11}$	$h^{(3)}_{11}$	$h^{(1)}_{12}$	$h^{(2)}_{12}$	$h^{(3)}_{12}$
0.125	0.0068	−0.0239	0.0116	0.5068	3.2733	−4.1376
0.25	0.0239	−0.0635	−0.0272	0.5239	3.4701	−2.0760
0.5	0.0214	−0.0713	−0.2932	0.5214	2.6560	−0.6409
1.0	−0.1014	0.0764	−0.7905	0.3986	1.0762	−0.3510
2.0	−0.3130	0.5024	−1.0825	0.1870	0.1161	−0.1902
4.0	−0.4525	1.1984	−1.3297	0.0475	−0.0533	−0.0180
8.0	−0.4920	1.9231	−2.1161	0.0080	−0.0214	0.0099

$\beta,\ \beta^{-1}$	$e^{(1)}$	$e^{(2)}$
1.0	6.04250	6.32549
2.0	5.59906	4.17702
4.0	3.79489	0.32014
8.0	1.51572	−1.54010

Table 11.26: Near field forms of the mobility functions $y^g_{\alpha\beta}$ and $y^h_{\alpha\beta}$.

$$y^g_{11} = (2a)\sum_{k=0}^{\infty} f_{2k+1}(\beta)\left(\frac{a}{2R}\right)^{2k+1}$$

$$y^g_{12} = -(2a)\sum_{k=0}^{\infty} f_{2k}(\beta)\left(\frac{a}{2R}\right)^{2k}$$

	f_0	f_2	f_4	f_6	f_8	f_{10}
1	0	0	4			
β						
β^2			20/3	0		
β^3					0	−11200/3
β^4						
β^5						8000/3

	f_1	f_3	f_5	f_7	f_9	f_{11}
β^3	0	0	0	−400/3	2560/3	
β^4						
β^5				560/3	−8224/3	22400
β^6						
β^7					1120	−31232
β^8						
β^9						5760

Table 11.27: Far field forms of the mobility functions $y^g_{\alpha\beta}$.

$$y_{11}^h = \sum_{k=0}^{\infty} f_{2k}(\beta)\left(\frac{a}{2R}\right)^{2k}$$

$$y_{12}^h = \sum_{k=0}^{\infty} f_{2k+1}(\beta)\left(\frac{a}{2R}\right)^{2k+1}$$

	f_0	f_2	f_4	f_6	f_8	f_{10}
1	0	0	0			
β						
β^2						
β^3				−200	1280	
β^4						
β^5					−1280	22400
β^6						
β^7						−9600

	f_1	f_3	f_5	f_7	f_9	f_{11}
1	0	10				
β			0			
β^2				0		
β^3					4000	
β^4						
β^5						64000
β^6						

Table 11.28: Far field forms of the mobility functions $y_{\alpha\beta}^h$.

$$x_1^m = \frac{20}{3}\pi a^3 \sum_{k=0}^{\infty} f_k(\beta)\left(\frac{a}{2R}\right)^k$$

	f_0	f_2	f_4	f_6	f_8	f_{10}
1	1					
β		0				
β^2			0			
β^3				1600	−15360	36864
β^4						
β^5					9600	−248832
β^6						
β^7						76800

	f_1	f_3	f_5	f_7	f_9	f_{11}
1	0					
β						
β^2						
β^3		40	−192			
β^4						
β^5			−192	0		
β^6					64000	76800
β^7						
β^8						76800
β^9						
β^{10}						

Table 11.29: Far field forms of the mobility function x_1^m.

$$y_1^m = \frac{20}{3}\pi a^3 \frac{m_1^{(1)}(\ln \xi^{-1})^2 + m_1^{(2)} \ln \xi^{-1} + m_1^{(3)}}{(\ln \xi^{-1})^2 + e^{(1)} \ln \xi^{-1} + e^{(2)}} + O(\xi(\ln \xi)^3)$$

β	$m_1^{(1)}$	$m_1^{(2)}$	$m_1^{(3)}$
0.125	1.0118	1.4746	−1.5212
0.25	1.0377	3.7041	0.2573
0.5	1.0473	5.5678	3.5328
1.0	1.1456	6.1694	3.7112
2.0	2.2657	4.2124	−1.2260
4.0	4.7033	−4.4574	−2.5413
8.0	7.0021	−18.9350	8.4871

β, β^{-1}	$e^{(1)}$	$e^{(2)}$
1.0	6.04250	6.32549
2.0	5.59906	4.17702
4.0	3.79489	0.32014
8.0	1.51572	−1.54010

Table 11.30: Near field forms of the mobility function y_1^m.

$$z_1^m = \frac{20}{3}\pi a^3 \left(m_1^{(1)}(\beta) + m_1^{(2)}(\beta)\xi + m_1^{(3)}(\beta)\xi^2\right)$$

β	$m_1^{(1)}$	$m_1^{(2)}$	$m_1^{(3)}$
0.125	0.9995	0.00008	0.0017
0.25	0.9967	0.00385	0.0014
0.5	0.9851	0.02584	−0.0156
1.0	0.9527	0.0914	−0.081
2.0	0.8689	0.266	−0.26
4.0	0.6979	0.632	−0.57
8.0	0.4783	1.12	−0.94

Table 11.31: Near field forms of the mobility function z_1^m.

$$y_1^m = \frac{20}{3}\pi a^3 \sum_{k=0}^{\infty} f_k(\beta)\left(\frac{a}{2R}\right)^k$$

	f_0	f_2	f_4	f_6	f_8	f_{10}
1	1					
β		0				
β^2			0			
β^3				400	−5120	16384
β^4						
β^5					6400	−110592
β^6						
β^7						51200

	f_1	f_3	f_5	f_7	f_9	f_{11}
1	0					
β						
β^2						
β^3		−20	128			
β^4						
β^5			128	0		
β^6					−39998/5	−383994/5
β^7						
β^8						−383994/5

Table 11.32: Far field forms of the mobility functions y_1^m.

$$z_1^m = \frac{20}{3}\pi a^3 \sum_{k=0}^{\infty} f_k(\beta) \left(\frac{a}{2R}\right)^k$$

	f_0	f_2	f_4	f_6	f_8	f_{10}
1	1					
β		0				
β^2			0			
β^3				0		1024
β^4						
β^5					800	−7040
β^6						
β^7						9088

	f_1	f_3	f_5	f_7	f_9	f_{11}
1	0					
β		0				
β^2						
β^3			−32			
β^4				0		
β^5			−32		0	
β^6						0
β^7						

Table 11.33: Far field forms of the mobility functions z_1^m.

Chapter 12

Particle-Wall Interactions

12.1 The Lorentz Image

In previous chapters we have considered interactions between particles. We now address another important class of hydrodynamic interaction problems, namely, the influence of nearby walls on the motion of small particles. In many respects, the analysis for particle-wall interactions parallels that for particle-particle interactions. Here again, the method of choice depends on the ratio of two length scales: the particle size and the separation between the particle and wall. When the particle is far away from the wall, a suitably modified form of the method-of-reflections technique is appropriate — as in the previous chapter, the reflections off the wall are conveniently represented by image singularities. This approach has a long history dating back to the works of Lorentz [52] and Faxén [13]. An excellent summary of work up to 1965 on particle-wall interactions by the method of reflections (for particles of arbitrary shape and various wall geometries) may be found in Chapter 7 of Happel and Brenner [27], along with analytical solutions for special geometries (*e.g.*, sphere near a plane). More recent works by Liron and Mochon [51] and Hackborn [25] consider flows produced by Stokeslets and rotlets in channels bounded by two parallel plane walls; these works provide analytic expressions for the Green's function in such geometries and thus form the basis for further numerical work.

For particles and walls near contact, a combination of lubrication and numerical methods is required. The resistance functions for rigid particles and walls in relative motion are singular at touching and, as was the case for interactions between two particles, the singular leading order terms originate from the flow in the gap region, whence lubrication methods apply. Contributions to the higher order, nonsingular terms come from all regions and thus, except for a limited number of fortuitous geometries such as the sphere-plane problem [4, 56], a numerical approach is unavoidable. Also, for particles sliding near a wall, the leading order singular term is only logarithmic (see Chapter 9), and thus higher order terms really are essential for most applications.

For moderate separations, we again require numerical solutions. We will present two such methods in this book. The boundary-multipole collocation

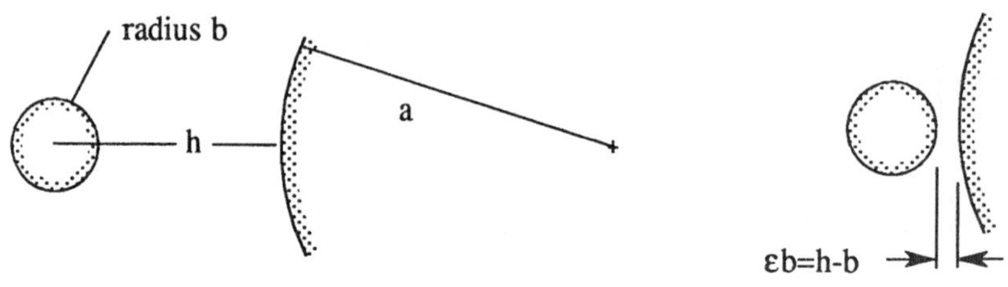

Figure 12.1: Particle-wall interactions: widely-separated and near-contact.

method of the following chapter provides an efficient method for particles of simple shape (for a representative sample, see [19, 20, 31, 65]). For particles and walls of complex shape and many-body problems, all of which lead to large system of equations, we recommend the boundary integral method described in Part IV.

The focus of this chapter will be primarily on the analysis of the image systems of the Stokeslet and other lower order singularities for two pedagogical reasons. These singularities may be used to represent the particle as in Chapter 3, and their images form the basis for the method of reflections. Furthermore, most numerical methods for particulate Stokes involve velocity representations that are distributions of Stokeslets and higher order singularities on or inside the particle. Therefore, given the image system for a Stokeslet near a wall, we have a straightforward generalization of the numerical method to the case of particles near a wall.

12.2 Stokeslet Near a Rigid Wall

Consider the image solution for a point force near a stationary sphere, which we derived in Section 10.2.3. From it, we can obtain the image for the plane wall by increasing the radius a while keeping constant the distance between the singularity and the nearest point on the sphere (see Figure 12.2). We write

$$\frac{R}{a} = 1 + \frac{h}{a} ,$$

and in the limit of small h/a, the image inside the sphere reduce to the following solution:

$$\begin{aligned} \boldsymbol{v}^* &= -(\boldsymbol{F}^{\perp} + \boldsymbol{F}^{\parallel}) \cdot \mathcal{G}(\boldsymbol{x} - \boldsymbol{y}^*) - 2h[(\boldsymbol{F}^{\perp} - \boldsymbol{F}^{\parallel}) \cdot \nabla]\boldsymbol{n} \cdot \mathcal{G}(\boldsymbol{x} - \boldsymbol{y}^*) \\ &\quad + h^2(\boldsymbol{F}^{\perp} - \boldsymbol{F}^{\parallel}) \cdot \nabla^2 \mathcal{G}(\boldsymbol{x} - \boldsymbol{y}^*) , \end{aligned} \tag{12.1}$$

where $\boldsymbol{n}$ (pointing into the fluid) is the normal for the plane and $\boldsymbol{y}$ and $\boldsymbol{y}^*$ are the new designations for the singular point $\boldsymbol{x}_2$ and the inverse point $\boldsymbol{x}_2^*$.

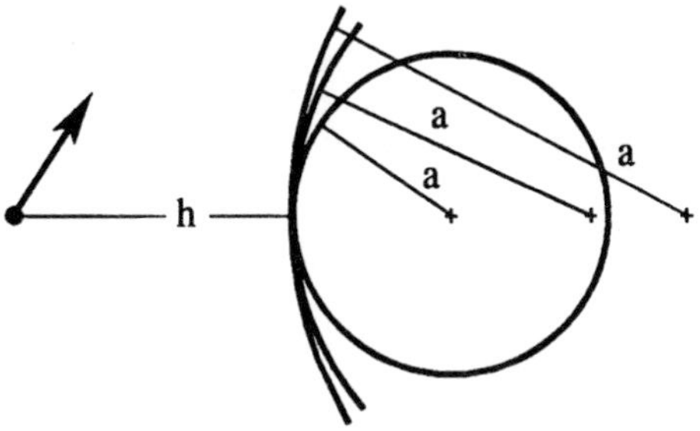

Figure 12.2: Stokeslet-wall interaction from a limiting procedure using large spheres.

Note that $\boldsymbol{y}^*$ is the mirror image of $\boldsymbol{y}$, whence if we choose our coordinate system so that the Stokeslet is on the z-axis and the plane is at $z = 0$, then $\boldsymbol{y}^* = -\boldsymbol{y}$. In contrast to the solution for the sphere, here the image systems for both axisymmetric and transverse cases are degenerate, consisting only of the inverse point, $\boldsymbol{y}^*$. Furthermore, the various singularities possess equal strengths in the axisymmetric ($\boldsymbol{F}^{||}$) and transverse ($\boldsymbol{F}^{\perp}$) cases.[1] We shall exploit these properties presently.

Although Equation 12.1 is a valid expression for the image system (and as we shall see later, quite useful in its own right), we display the following alternate expression in which the geometric parameter h does not appear explicitly:

$$\boldsymbol{v}^* = -(\boldsymbol{\delta} - 2\boldsymbol{nn}) \cdot \boldsymbol{u} - 2\boldsymbol{n} \cdot \boldsymbol{x}\nabla(\boldsymbol{n} \cdot \boldsymbol{u}) + (\boldsymbol{n} \cdot \boldsymbol{x})^2 \nabla^2 \boldsymbol{u} \ , \tag{12.2}$$

with

$$\boldsymbol{u} = (\boldsymbol{F}^{\perp} - \boldsymbol{F}^{||}) \cdot \mathcal{G}(\boldsymbol{x} - \boldsymbol{y}^*) \ . \tag{12.3}$$

Equation 12.2 is the image solution of Lorentz [52] and the equivalence of Equations 12.1 and 12.2 is readily demonstrated.

According to Equation 12.3, $\boldsymbol{u}$ is the field generated by a Stokeslet located at the image point. Since this image Stokeslet ($\boldsymbol{F}^{\perp} - \boldsymbol{F}^{||}$) is the mirror reflection of the original (see Figure 12.3), its vector field, $\boldsymbol{u}$, is also obtained by mirror reflection of $\boldsymbol{F} \cdot \mathcal{G}(\boldsymbol{x} - \boldsymbol{y})$, as can be verified directly. The mirror reflection operation may be defined formally by introducing the operator $\mathcal{M}$ defined by[2] $\boldsymbol{u} = \mathcal{M}\boldsymbol{v}$, with

$$\boldsymbol{u}(x, y, z) = \mathcal{M}(\boldsymbol{v}) = (\boldsymbol{\delta} - 2\boldsymbol{e}_3\boldsymbol{e}_3) \cdot \boldsymbol{v}(x, y, -z) \ , \tag{12.4}$$

for a plane wall coincident with the xy-plane.

[1]The $||$ and $\perp$ designations are relative to the symmetry axis. Some authors reverse this notation for the plane, *i.e.*, $\perp$ designates orientation normal to the plane.

[2]Lee [47] denotes this operation by an asterisk and the Lorentz operation with a caret.

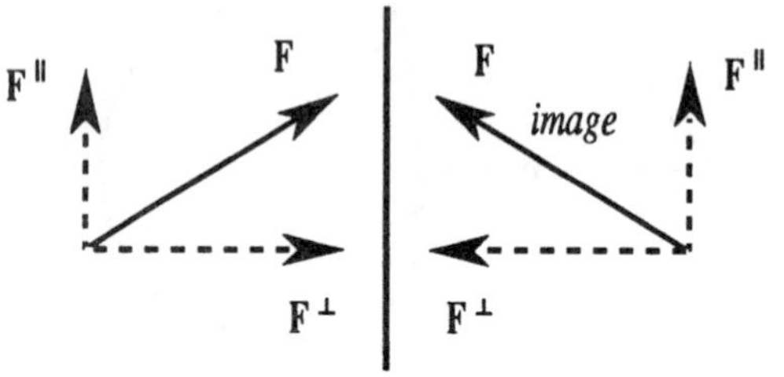

Figure 12.3: The mirror reflection operation for the Stokeslet field.

Going back to Equation 12.2, we see that the image field for the Stokeslet may be written as

$$\boldsymbol{v}^* = \widehat{\boldsymbol{u}} \ , \qquad \boldsymbol{u} = \mathcal{M}(\boldsymbol{v}) \ , \tag{12.5}$$

where the "hat" operator is defined by

$$\widehat{\boldsymbol{u}} = -(\boldsymbol{\delta} - 2\boldsymbol{nn}) \cdot \boldsymbol{u} - 2z\nabla u_z + z^2\nabla^2\boldsymbol{u} \ . \tag{12.6}$$

Equations 12.5 and 12.6 transform the Stokeslet into its image solution. By the linearity of the Stokes equation (or direct verification), we see that the *Lorentz reflection* for the wall, defined by Equations 12.5 and 12.6, is in fact a general result that applies to any solution of the Stokes equations.

The mirror reflection and Lorentz ("hat") operators share two important properties. First, if $\boldsymbol{v}$ is a solution of the Stokes equation, then $\mathcal{M}(\boldsymbol{v}(x,y,z))$ and $\widehat{\boldsymbol{v}}$ are also solutions of the Stokes equations. The associated pressure fields are

$$\mathcal{M}(p(x,y,z)) = p(x,y,-z)$$

and

$$\widehat{p} = p + 2z\frac{\partial p}{\partial z} - 4\frac{\partial v_z}{\partial z} \ .$$

Second, the operators commute, *i.e.*, $\mathcal{M}(\widehat{\boldsymbol{v}}) = \widehat{\boldsymbol{u}}$ where $\boldsymbol{u} = \mathcal{M}(\boldsymbol{v})$.

In the following example, we use the preceding result to calculate the mobility of a sphere translating near a plane wall.

Example 12.1 Mobility of a Sphere Far Away from a Plane Wall. Consider a torque-free sphere near a plane wall. We will derive expressions for the translational and rotational velocities by the method of reflections, when the sphere is subject to a net external force $\boldsymbol{F}^{ext}$.

We center the sphere of radius b at a distance h from the plane. We will correct Stokes' law to $O(h^{-3})$ by incorporating the contributions from the images for the Stokeslet and degenerate quadrupole. The velocity field due to an isolated, translating sphere, subject to an external force $\boldsymbol{F}^{ext}$, in a quiescent fluid is

$$\boldsymbol{v}_1 = \boldsymbol{F}^{ext} \cdot (1 + \frac{b^2}{6}\nabla^2)\frac{\mathcal{G}(\boldsymbol{x}-\boldsymbol{y})}{(8\pi\mu)} ,$$

with $\boldsymbol{F}^{ext} = 6\pi\mu b\boldsymbol{U}$ and $\boldsymbol{y} = h\boldsymbol{n}$.

At the first reflection on the wall (surface 2), $\boldsymbol{v}_{12}$ is simply the image system for the Stokeslet and the degenerate quadrupole. If we split the external force into components along and transverse to the plane normal, $\boldsymbol{F}^{ext} = \boldsymbol{F}^{\|} + \boldsymbol{F}^{\perp}$, then the image systems for the Stokeslet and degenerate Stokes quadrupole coefficient $\boldsymbol{Q} = (b^2/6)\boldsymbol{F}^{ext}$ may be written as

$$\begin{aligned}
\boldsymbol{v}_{12} \;=\; & -(\boldsymbol{F}^{\|} + \boldsymbol{F}^{\perp}) \cdot \frac{\mathcal{G}(\boldsymbol{x}-\boldsymbol{y}^*)}{8\pi\mu} \\
& + 2h[(\boldsymbol{F}^{\|} - \boldsymbol{F}^{\perp}) \cdot \nabla]\boldsymbol{n} \cdot \frac{\mathcal{G}(\boldsymbol{x}-\boldsymbol{y}^*)}{8\pi\mu} \\
& - h^2(\boldsymbol{F}^{\|} - \boldsymbol{F}^{\perp}) \cdot \nabla^2\frac{\mathcal{G}(\boldsymbol{x}-\boldsymbol{y}^*)}{8\pi\mu} \\
& + 4(\boldsymbol{n}\cdot\nabla)^2(\boldsymbol{Q}^{\|} + \boldsymbol{Q}^{\perp}) \cdot \frac{\mathcal{G}(\boldsymbol{x}-\boldsymbol{y}^*)}{8\pi\mu} \\
& - 2h(\boldsymbol{n}\cdot\nabla)(\boldsymbol{Q}^{\|} - \boldsymbol{Q}^{\perp}) \cdot \frac{\mathcal{G}(\boldsymbol{x}-\boldsymbol{y}^*)}{8\pi\mu} \\
& - (3\boldsymbol{Q}^{\|} - \boldsymbol{Q}^{\perp}) \cdot \nabla^2\frac{\mathcal{G}(\boldsymbol{x}-\boldsymbol{y}^*)}{8\pi\mu} \\
& - 8(\boldsymbol{n}\cdot\nabla)(\boldsymbol{Q}^{\perp} \times \boldsymbol{n}) \times \nabla\frac{1}{8\pi\mu|\boldsymbol{x}-\boldsymbol{y}^*|} ,
\end{aligned}$$

with $\boldsymbol{y}^* = -h\boldsymbol{n}$.

The second reflection is at the sphere, and here the translational and rotational velocities are obtained by applying the Faxén laws for a force-free and torque-free sphere. Note that we do not need the Faxén correction for the image field of the degenerate quadrupole, since this image is already $O(h^{-3})$. Thus the terms of interest are

$$\begin{aligned}
\boldsymbol{v}_{12}|_{x=y} \;=\; & -\left(\frac{3}{2h}\right)\frac{\boldsymbol{F}^{\|}}{8\pi\mu} - \left(\frac{3}{4h}\right)\frac{\boldsymbol{F}^{\perp}}{8\pi\mu} \\
& + \left(\frac{2}{h^3}\right)\frac{\boldsymbol{Q}^{\|}}{8\pi\mu} + \left(\frac{1}{2h^3}\right)\frac{\boldsymbol{Q}^{\perp}}{8\pi\mu}
\end{aligned}$$

and

$$\nabla^2\boldsymbol{v}_{12}|_{x=y} \;=\; \left(\frac{2}{h^3}\right)\frac{\boldsymbol{F}^{\|}}{8\pi\mu} + \left(\frac{1}{2h^3}\right)\frac{\boldsymbol{F}^{\perp}}{8\pi\mu} .$$

The equality between the $O(h^{-3})$ coefficients in $\boldsymbol{v}_{12}$ and $\nabla^2\boldsymbol{v}_{12}$ is a consequence of the Lorentz reciprocal theorem (see Exercise 12.4). The translational velocity follows immediately as

$$\begin{aligned} 6\pi\mu b\boldsymbol{U} &= \boldsymbol{F}^{ext} + 6\pi\mu b(1 + \frac{b^2}{6}\nabla^2)\boldsymbol{v}_{12}(\boldsymbol{x})|_{x=y} \\ &= \left[1 - \frac{9}{8}\frac{b}{h} + \frac{1}{2}\left(\frac{b}{h}\right)^3\right]\boldsymbol{F}^{\parallel} + \left[1 - \frac{9}{16}\frac{b}{h} + \frac{1}{8}\left(\frac{b}{h}\right)^3\right]\boldsymbol{F}^{\perp} , \end{aligned}$$

with an error of $O((b/h)^5)$.

The rotational velocity is obtained in an analogous fashion, by applying the Faxén law to the image fields. The vorticity of the image for the Stokeslet is identically zero at $\boldsymbol{y}$, so the leading term for the rotational velocity comes from the image of the degenerate quadrupole. Thus, at the second reflection,

$$\begin{aligned} \boldsymbol{\omega}^{(2)} &= \frac{1}{2}\nabla \times \boldsymbol{v}_{12}(\boldsymbol{x})|_{x=y} \\ &= -\frac{3}{4h^4}\frac{\boldsymbol{Q}^{\perp}}{8\pi\mu} \times \boldsymbol{n} \\ &= -\frac{b^2}{8h^4}\frac{\boldsymbol{F}^{\perp} \times \boldsymbol{n}}{8\pi\mu} . \end{aligned}$$

Thus a torque-free sphere dragged near a wall by an external force rotates weakly in a rolling motion.

The fourth reflection contributes the next term for the rotational velocity, and this contribution comes from the vorticity field of the image of the stresslet,

$$\boldsymbol{S} = \frac{5b^3}{16h^2}(\boldsymbol{F}^{\perp}\boldsymbol{n} + \boldsymbol{n}\boldsymbol{F}^{\perp}) , \qquad (12.7)$$

induced at the second reflection. This stresslet is of type $m = 1$ defined in Chapter 10 in the discussion on image solutions for the sphere. If we take the limit of small h/a, then the vorticity of this image system becomes

$$\begin{aligned} \nabla \times \Bigg[(\boldsymbol{S} \cdot \nabla) \cdot \frac{\mathcal{G}(\boldsymbol{x} - \boldsymbol{y}^*)}{8\pi\mu} + 2h(\boldsymbol{n} \cdot \nabla)(\boldsymbol{S} \cdot \nabla) \cdot \frac{\mathcal{G}(\boldsymbol{x} - \boldsymbol{y}^*)}{8\pi\mu} \\ + 4h(\boldsymbol{n} \cdot \nabla)[(\boldsymbol{S} \cdot \boldsymbol{n}) \times \boldsymbol{n}] \times \nabla \frac{1}{8\pi\mu|\boldsymbol{x} - \boldsymbol{y}^*|}\Bigg] , \end{aligned}$$

which at $\boldsymbol{x} = \boldsymbol{y}$, with Equation 12.7 for the stresslet, reduces to

$$\frac{15b^3}{64h^5}\frac{\boldsymbol{F}^{\perp} \times \boldsymbol{n}}{8\pi\mu} .$$

Combining the second and fourth reflection contributions, we obtain

$$\boldsymbol{\omega} = -\frac{b^2}{8h^4}\left(1 - \frac{15}{16}\frac{b}{h}\right)\frac{\boldsymbol{F}^{\perp} \times \boldsymbol{n}}{8\pi\mu} .$$

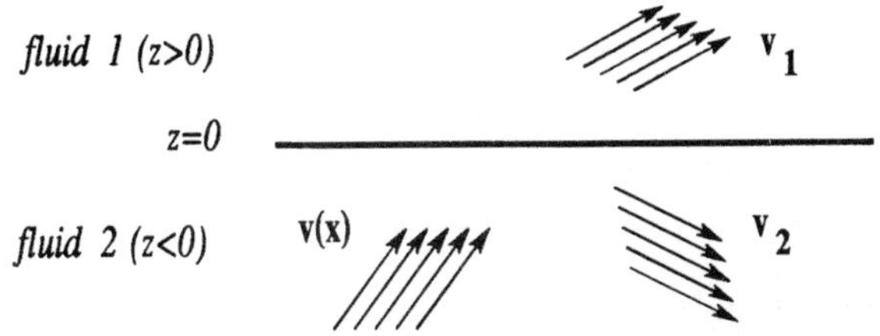

Figure 12.4: Geometry for the reflection from a fluid-fluid interface.

We may replace $\boldsymbol{F}^{\perp}$ with the more readily observable translational velocity, and our expression for the rotational velocity now becomes

$$\begin{aligned}\boldsymbol{\omega} &= -\frac{b^2}{8h^4}\left(1-\frac{15}{16}\frac{b}{h}\right)\frac{3b}{4}\left(1-\frac{9}{16}\frac{b}{h}\right)^{-1}\boldsymbol{U}\times\boldsymbol{n} \\ &= -\frac{3}{32b}\left(\frac{b}{h}\right)^4\left(1-\frac{3}{8}\frac{b}{h}\right)\boldsymbol{U}\times\boldsymbol{n}\ ,\end{aligned}$$

a result given in Faxén's thesis and cited in [27]. ◇

12.3 A Drop Near a Fluid-Fluid Interface

In the previous section, we described Lorentz' reflection for a rigid plane wall. The generalization to a planar fluid-fluid interface has been derived by Lee *et al.* [47]. We place the fluid-fluid interface on the plane $z = 0$, and designate the region $z > 0$ as "fluid 1" and $z < 0$ as "fluid 2," as shown in Figure 12.4. We denote the viscosity ratio as $\lambda = \mu_1/\mu_2$.

Consider a velocity field $\boldsymbol{v}$ in fluid 2 that satisfies the Stokes equations. Then the following velocity fields in fluids 1 and 2,

$$\begin{aligned}\boldsymbol{v}_1 &= \frac{1}{1+\lambda}(\boldsymbol{v}-\widehat{\boldsymbol{v}}) \\ \boldsymbol{v}_2 &= (\boldsymbol{v}+\mathcal{M}(\boldsymbol{v}))-\frac{\lambda}{1+\lambda}\mathcal{M}(\boldsymbol{v}+\widehat{\boldsymbol{v}})\ ,\end{aligned}$$

with the pressure fields,

$$\begin{aligned}p_1 &= \frac{\lambda}{1+\lambda}(p-\widehat{p}) \\ p_2 &= (p+\mathcal{M}(p))-\frac{\lambda}{1+\lambda}\mathcal{M}(p+\widehat{p})\ ,\end{aligned}$$

satisfy the Stokes equations in the two regions, as well as the boundary conditions at the fluid-fluid interface:

$$\begin{aligned} \boldsymbol{n}\cdot\boldsymbol{v}_1 &= 0 \\ \boldsymbol{n}\cdot(\boldsymbol{v}+\boldsymbol{v}_2) &= 0 \\ (\boldsymbol{\delta}-\boldsymbol{nn})\cdot\boldsymbol{v}_1 &= (\boldsymbol{\delta}-\boldsymbol{nn})\cdot(\boldsymbol{v}+\boldsymbol{v}_2) \\ (\boldsymbol{\delta}-\boldsymbol{nn})\cdot(\boldsymbol{\sigma}_1\cdot\boldsymbol{n}) &= (\boldsymbol{\delta}-\boldsymbol{nn})\cdot(\boldsymbol{\sigma}_2\cdot\boldsymbol{n})\ . \end{aligned}$$

We refer to this result as *Lee's reflection lemma.* Note that in the limit of the rigid wall, $\lambda\to\infty$, $\boldsymbol{v}_2$ is the Lorentz reflection from the rigid wall, as required.

Although this lemma can be verified directly by substitution, an easier proof follows from the results of Section 10.2.4 concerning the image solution for a point force near a stationary spherical drop. Analogous to the derivation of the Lorentz reflection principle for the rigid wall, we increase the drop radius a while keeping constant the distance h between the singularity and the nearest point on the sphere. It can be shown that all line integrals appearing in the image field vanish in the limit of small h/a at least as fast as $(h/a)\log(h/a)$ (see Exercise 12.1). But all terms nonlinear in $\Lambda=\lambda/(1+\lambda)$ appear with the line integrals, so *the image of a Stokeslet near a fluid-fluid interface must be a linear function in* Λ of the form

$$\Lambda\boldsymbol{v}^w+(1-\Lambda)\boldsymbol{v}^b\ ,$$

where $\boldsymbol{v}^w$ is the image field for the rigid wall, as described in Section 12.2, and $\boldsymbol{v}^b$ is the image field for a shear-free interface (the limit obtained from a large bubble). The latter is given by the expression,

$$\boldsymbol{v}^b=(\boldsymbol{F}^{\perp}-\boldsymbol{F}^{\parallel})\cdot\mathcal{G}(\boldsymbol{x}-\boldsymbol{y}^*)\ . \tag{12.8}$$

According to this equation, the image of a Stokeslet in fluid 2 when the interface is shear-free is simply its mirror reflection in (inviscid) fluid 1, which of course could have been derived directly by symmetry arguments that apply to any Stokes velocity field in fluid 2.

We now put together the results for the rigid wall and shear-free interface; *the reflection at the fluid-fluid interface is a linear combination (in Λ) of the results for a rigid plane wall and a shear-free surface, with the latter obtained simply by a mirror reflection*, and this is precisely the restatement of *Lee's Reflection Lemma.*

The lemma may be used to derive analytical expressions for the resistance and mobility functions for a particle or drop far away from a fluid-fluid interface, in a manner reminiscent of the preceding section for the rigid wall. Examples may be found in the literature ranging from spheres to slender bodies [47, 66, 67]. Furthermore, it can also be used as the basis of numerical schemes, with Stokeslets and other singularities distributed on the particle and drops, and the interface treated with the appropriate images [48].

For large but finite surface tensions, the fluid-fluid interface must deform because of particle motion. The effect of small deviations in the shape of the interface can be analyzed by linearization of the boundary condition at the interface, and new effective boundary conditions are obtained. These, however, can be treated exactly by the Lorentz reciprocal theorem so that the first effects of interfacial deformation can be handled entirely within the context of particulate motion near a planar interface (see Yang and Leal [68]).

Exercises

Exercise 12.1 Images for a Stokeslet Near a Rigid Plane Wall and for a Stokeslet Near a Planar Fluid-Fluid Interface.
Derive the images for a Stokeslet near a rigid wall and for a Stokeslet near a fluid-fluid interface, from the image solutions for a Stokeslet at a distance h from a sphere of radius a, by taking the limit of small h/a. Show that the line distributions (the terms nonlinear in Λ) in the image for a spherical drop vanish in the limit of small h/a.

Exercise 12.2 The Lorentz Image for a Stokeslet Near a Rigid Plane Wall.
Show that the image for a Stokeslet near a rigid plane wall derived by the above limiting procedure is identical to that of Lorentz, Equation 12.2.
Comment: The rigid wall has a dissipative effect on the flow produced by the Stokeslet; we should expect a much weaker flow in the far field, in comparison to the flow produced by a Stokeslet in unbounded flow. For a Stokeslet in the channel bounded by *two* plane walls, the far field is yet weaker. For example, the paper by Liron and Mochon [51] shows that the far field of a Stokeslet parallel to the plane walls is that of a two-dimensional source-sink doublet, *i.e.*,

$$v_i \sim \frac{D_j}{2\pi\mu}\left(-\frac{\delta_{ij}}{r^2}+\frac{2x_i x_j}{r^4}\right) ,$$

where the strength of the source doublet $\boldsymbol{D}$ depends on the problem geometry (location of the Stokeslet and distance between walls). On the other hand, a Stokeslet directed normal to the plane walls has a far field velocity that vanishes exponentially.

Exercise 12.3 Properties of the Lorentz and Mirror Reflection Operations.
Show that the Lorentz and mirror reflection operations commute, *i.e.*, $\mathcal{M}(\widehat{\boldsymbol{v}}) = \widehat{\boldsymbol{u}}$ where $\boldsymbol{u} = \mathcal{M}(\boldsymbol{v})$. Show that if $\boldsymbol{v}$ is satisfies the Stokes equations, then $\widehat{\boldsymbol{v}}$ is also a solution.

Exercise 12.4 A Consequence of the Lorentz Reciprocal Theorem.
Let $\boldsymbol{v}^F$ be the image of the Stokeslet field, $\boldsymbol{F}^{ext}\cdot\mathcal{G}(\boldsymbol{x}-\boldsymbol{y})$, and let $\boldsymbol{v}^Q$ be the image of the Stokes quadrupole field, $\boldsymbol{F}^{ext}\cdot\nabla^2\mathcal{G}(\boldsymbol{x}-\boldsymbol{y})$. Use the Lorentz reciprocal theorem to prove that

$$\nabla^2\boldsymbol{v}^F(\boldsymbol{x})|_{x=y} = \boldsymbol{v}^Q(\boldsymbol{y}) .$$

The extension of this result to the image fields of a viscous drop is given in Fuentes *et al.* [17].

Exercise 12.5 Image of a Rotlet Near a Rigid Plane Wall.
Derive the image for a rotlet near a plane wall. For the axisymmetric rotlet with $\boldsymbol{T}^{ext} \parallel \boldsymbol{n}$, what is the nature of the image singularity?

Exercise 12.6 A Spherical Drop Translating Near a Fluid-Fluid Interface.
Generalize Example 12.1 to obtain the translational mobility of a small spherical drop moving near a planar fluid-fluid interface. Compare your result with that obtain from the two-sphere problem of Chapter 10 in the limit of small h/a [16, 17].

Exercise 12.7 Force Balance on the Fluid-Fluid Interface.
Consider a fluid-fluid interface (placed on the plane $z = 0$) with large surface tension γ. Examine σ_{33} for a Stokeslet plus its image and estimate the amount of deformation of the interface [47]. Consider both axisymmetric and transverse Stokeslets.

Exercise 12.8 Computer-Aided Method of Reflections for a Sphere Near a Plane.
Consider a sphere of radius b near a plane wall. By combining the Lorentz reflection principle and the addition theorems for the spherical harmonics, as in Jeffrey and Onishi [38], derive a recursion scheme to generate the series solution in b/h to high orders. Compare your results with the solutions obtained by separation of variables in bispherical coordinates [4, 22].

Chapter 13

Boundary-Multipole Collocation

13.1 Introduction

On several occasions, we mentioned the boundary-multipole collocation method as a tool for solving particle-particle hydrodynamic interaction problems. We now devote a chapter to explore this technique in depth: definition of the method, analysis of discretization errors, variations on the standard method, and last but not least, an exposition of the method as applied to interactions between particles. The emphasis will be on the applications to the calculation of resistance and mobility functions of spheres and ellipsoids.

Collocation methods are well established in the world of numerical methods as an important tool for solving boundary value problems. In the more familiar interior collocation (or simply collocation) method, the trial solution is expanded in a set of basis functions that trivially satisfy the boundary conditions, but not the governing differential equation. The coefficients in this expansion are determined by equations that correspond to the minimization in some sense, of the residual or error in the differential equation. In *boundary collocation*, the role played by the boundary condition and differential equation is reversed. We write

$$\boldsymbol{v}(\boldsymbol{x}) = \sum_n c_n \boldsymbol{v}_{(n)}(\boldsymbol{x}) \ ,$$

i.e., the trial solution is expanded in a complete set of basis functions $\{\boldsymbol{v}_{(n)}\}_1^\infty$. Each $\boldsymbol{v}_{(n)}$ satisfies the governing differential equation, but not the boundary conditions. In *boundary-multipole collocation*, the multipole expansion is used as the basis. The coefficients c_n are determined from the boundary conditions.

The boundary conditions actually provide us with a continuum of equations, one for each surface point. Satisfying all of these is impractical (or viewed in another way, we would then relabel it as a simple problem with an analytical solution), so the numerical procedure is to first truncate the series to finite sums and then enforce some finite set of equations obtained from the boundary conditions to get a linear system determining the unknown coefficients. If these equations are found by imposing the boundary conditions exactly at a finite number of discrete surface points only, the method is called *collocation.* If in addition derivatives of the boundary conditions with respect to the continuous variables, *i.e.*, location on the surface, are imposed also on discrete surface points, the method of discretizing the continuum of equations is called *osculation.* In *discrete least squares* the equations are formed in a manner similar to collocation, but this is continued until we have an overdetermined system of equations for the coefficients. This system is solved with the ordinary least squares procedure, originally developed by Gauss. One could also approximate a *continuous least squares* procedure numerically with integration quadratures, or try to minimize the maximum residual (*min-max* procedure). In the *method of weighted residuals* the continuous variables are integrated away from the boundary conditions with some weights to get a set of discrete equations for the unknowns. The reader may compare this to determining the Fourier coefficients of a known function, where the resulting linear system is particularly simple. When the weights of integration are chosen to be the same basis functions that the unknowns are multiplying, the procedure is called the *Galerkin method*; this is the most popular of the actual weighted residual methods. It should be clear from the above that once the series expansion has been chosen and truncated, all that remains is a standard curve-fitting problem: We have to find a combination of parameters so that the resulting series approximates the boundary conditions well in some sense. The combination of series expansion and collocation shall be called the *boundary collocation method* (BCM). In Part IV, the BCM for multiparticle cases will be theoretically justified. It will be shown that the disturbance field of a multiparticle system can be uniquely decomposed into disturbance fields of the individual particles, so that single particle basis functions can be used and the system is (analytically) determinate. In practice there are good and bad choices of collocation points, and only rules of thumb, based on generic concepts from interpolation theory, can be given about these.

The application of boundary-multipole collocation to Stokes flow was popularized by the 1971 work of Gluckman *et al.* [21], in which the method was used to solve axisymmetric streaming flow past a collection of spheres and spheroids. One can find earlier applications of the method, *e.g.*, O'Brien's (1968) work on the motion of a slightly deformed drop [58]. Over the subsequent decades, the method has been used to solve a wide range of problems in Stokes flow (a few illustrative examples include: multiple spheres in a cylinder [49]; three-sphere sedimentation problems [18]; two spheroids in a uniform stream [50]; sphere between two plane walls [19, 20]; a sphere in a circular orifice [10]; resistance and mobility functions for two spheres [45]; spheroids and other nonspherical particles settling in a channel [31]; and mobility functions for two spheroids in a linear field [71]). A fairly complete overview of the range of problems that has been successfully tackled by this method during the period 1978–1990 is available in the review article by Weinbaum *et al.* [65].

In general, the boundary-multipole collocation method is a very efficient tool for the class of Stokes flow involving interactions between particles of simple shape (spheres, spheroids, *etc.*) We illustrate the power of the technique by outlining the steps used to obtain the entire collection of resistance and mobility functions for two spheres of Chapter 11. The results compare favorably with those obtained from other methods, such as separation of variables in bispherical coordinates, method of reflections, and lubrication theory. We also study the convergence of the method, by looking at the boundary collocation solution for a Stokeslet near a sphere.

Perhaps even more impressive is the success of the boundary collocation method for nonaxisymmetric geometries, such as the one formed by two oblate spheroids at arbitrary configurations. The resistance and mobility functions for the two-spheroid problem can be obtained quite readily, even for nearly touching and touching configurations, at a fraction of the computational time required by other methods. Here again, we may gauge the discretization error by comparison with exact analytic solutions for a Stokeslet near a spheroid.

13.2 Two-Sphere Problems

From Chapter 3 we know that the velocity field in any two-sphere Stokes flow problem can be written as a twin multipole expansion about each sphere center. Thus, we are lead to a natural choice for the basis functions for the boundary collocation method: the Stokes multipoles themselves. Since these multipole fields and Lamb's general solution span the same set, we choose, mainly for the sake of ease of notation, to use the latter.

By appropriate choice of $A_{\ell m}$ in the following expression, we obtain all surface velocities associated with disturbance fields of the resistance problem:

$$\boldsymbol{v}_s = \sum_{\ell=1}^{2} \sum_{m=0}^{\ell} \nabla \left[(A_{\ell 0}\delta_{0m} + A_{\ell m} \sin m\phi) r_1^{\ell} P_{\ell}^{m}(\xi_1) \right] \; . \tag{13.1}$$

Explicitly, if we set all but the following coefficient equal to zero:

1. $A_{10} = 1$ gives a translational velocity along the sphere-sphere axis.

2. $A_{11} = 1$ gives a translational velocity perpendicular to the sphere-sphere axis.

3. $A_{20} = 1$ gives the axisymmetric rate-of-strain field about the sphere-sphere axis.

4. $A_{21} = 1$ gives the hyperbolic straining field associated with the shear flow in the xz-coordinate plane.

5. $A_{22} = 1$ gives the hyperbolic straining field associated with the shear flow in the xy-coordinate plane.

6. $B_{10} = 1$ gives the rotational velocity about the sphere-sphere axis.

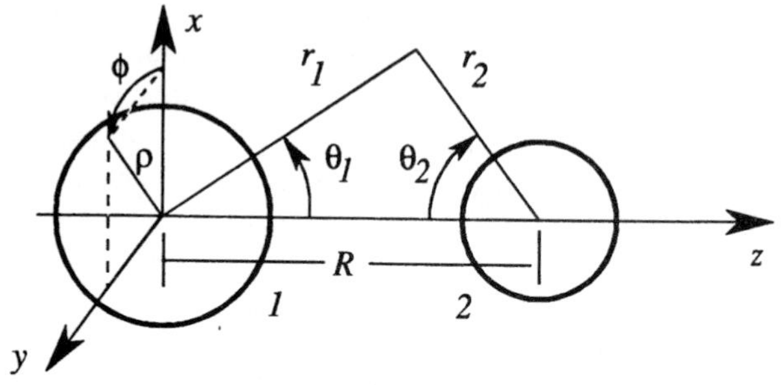

Figure 13.1: The two-sphere geometry.

7. $B_{11} = 1$ gives the rotational velocity about an axis perpendicular to the sphere-sphere axis.

The velocity field must also contain the same ϕ-dependence, so we write Lamb's solution as

$$\begin{aligned} \boldsymbol{v} - \boldsymbol{v}^{\infty} &= \sum_{n=1}^{\infty} \left\{ \frac{n+1}{n(2n-1)} \boldsymbol{r}_{\alpha} p_{-n-1}^{(\alpha)} - \frac{n-2}{2n(2n-1)} r_{\alpha}^2 \nabla p_{-n-1}^{(\alpha)} \right. \\ &\quad \left. + \nabla \Phi_{-n-1}^{(\alpha)} + \nabla \times \left[\boldsymbol{r}_1 \chi_{-n-1}^{(\alpha)} \right] \right\} , \end{aligned} \tag{13.2}$$

with $\boldsymbol{r}_{\alpha} = \boldsymbol{x} - \boldsymbol{x}_{\alpha}$, $r_{\alpha} = |\boldsymbol{r}_{\alpha}|$, for $\alpha = 1, 2$, and

$$\begin{aligned} p_{-n-1}^{(\alpha)} &= \sum_{m=0}^{n} (a_{0n}^{(\alpha)} \delta_{0m} + a_{mn}^{(\alpha)} \sin m\phi) r_{\alpha}^{-(n+1)} P_n^m(\cos\theta_{\alpha}) \\ \Phi_{-n-1}^{(\alpha)} &= \sum_{m=0}^{n} (b_{0n}^{(\alpha)} \delta_{0m} + b_{mn}^{(\alpha)} \sin m\phi) r_{\alpha}^{-(n+1)} P_n^m(\cos\theta_{\alpha}) \\ \chi_{-n-1}^{(\alpha)} &= \sum_{m=0}^{n} c_{mn}^{(\alpha)} r_{\alpha}^{-(n+1)} P_n^m(\cos\theta_{\alpha}) \cos m\phi . \end{aligned}$$

The two-sphere geometry and the spherical polar angles θ_{α} and ϕ are shown in Figure 13.1.

Further manipulation of the boundary conditions are required, and these are done more easily in cylindrical coordinates (z, ρ, ϕ) (see Figure 13.1). We may collect terms in $\sin m\phi$ and $\cos m\phi$, and obtain an independent set of equations for each Fourier mode. Thus the ϕ-dependence is eliminated from the problem! The three velocity components yield three independent conditions:

$$\begin{aligned} &\sum_{\alpha=1}^{2} (-1)^{\alpha} \sum_{n=\ell}^{\infty} \left\{ M_{mn}^{11(\alpha)} a_{mn}^{(\alpha)} + M_{mn}^{12(\alpha)} b_{mn}^{(\alpha)} + M_{mn}^{13(\alpha)} c_{mn}^{(\alpha)} \right\} \\ &\quad = A_{\ell m} \left(\ell \xi_1 P_{\ell}^m(\xi_1) + (1 - \xi_1^2) P_{\ell}^{m\prime}(\xi_1) \right) + B_{\ell m} m P_{\ell}^m(\xi_1) \end{aligned} \tag{13.3}$$

$$\sum_{\alpha=1}^{2} (-1)^{\alpha} \sum_{n=\ell}^{\infty} \left\{ M_{mn}^{21(\alpha)} a_{mn}^{(\alpha)} + M_{mn}^{22(\alpha)} b_{mn}^{(\alpha)} + M_{mn}^{23(\alpha)} c_{mn}^{(\alpha)} \right\}$$

$$= A_{\ell m}\left(\ell \sin\theta_1 P_\ell^m(\xi_1) - [\xi_1 P_\ell^{m+1}(\xi_1) + m\sin\theta_1 P_\ell^m(\xi_1)]\right) - B_{\ell m} P_\ell^{m+1}(\xi_1) \tag{13.4}$$

$$\sum_{\alpha=1}^{2}(-1)^\alpha \sum_{n=\ell}^{\infty}\left\{M_{mn}^{31(\alpha)} a_{mn}^{(\alpha)} + M_{mn}^{32(\alpha)} b_{mn}^{(\alpha)} + M_{mn}^{33(\alpha)} c_{mn}^{(\alpha)}\right\} = A_{\ell m} m P_\ell^m(\xi_1)/\sin\theta_1 + B_{\ell m}\sin\theta_1 P_\ell^{m\prime}(\xi_1)\ , \tag{13.5}$$

where

$$M_{mn}^{11(\alpha)} = r_\alpha^{-n}\left[\frac{n+1}{4n-2}\xi_\alpha P_n^m(\xi_\alpha) - \frac{n-2}{n(4n-2)}(1-\xi_\alpha^2)P_n^{m\prime}(\xi_\alpha)\right]$$

$$M_{mn}^{12(\alpha)} = r_\alpha^{-n-2}\left[-(n+1)\xi_\alpha P_n^m(\xi_\alpha) + (1-\xi_\alpha^2)P_n^{m\prime}(\xi_\alpha)\right]$$

$$M_{mn}^{13(\alpha)} = r_\alpha^{-n-1} m P_n^m(\xi_\alpha)$$

$$M_{mn}^{21(\alpha)} = r_\alpha^{-n}\left[\frac{n+1}{4n-2}\sin\theta_\alpha P_n^m(\xi_\alpha) + \frac{n-2}{n(4n-2)}[\xi_\alpha P_n^{m+1}(\xi_\alpha) + m\sin\theta_\alpha P_n^m(\xi_\alpha)]\right]$$

$$M_{mn}^{22(\alpha)} = -r_\alpha^{-n-2}\left[(n+1)\sin\theta_\alpha P_n^m(\xi_\alpha) + [\xi_\alpha P_n^{m+1}(\xi_\alpha) + m\sin\theta_\alpha P_n^m(\xi_\alpha)]\right]$$

$$M_{mn}^{23(\alpha)} = -r_\alpha^{-n-1} P_n^{m+1}(\xi_\alpha)$$

$$M_{mn}^{31(\alpha)} = -m r_\alpha^{-n}\frac{n-2}{n(4n-2)}\frac{P_n^m(\xi_\alpha)}{\sin\theta_\alpha}$$

$$M_{mn}^{32(\alpha)} = m r_\alpha^{-n-2}\frac{P_n^m(\xi_\alpha)}{\sin\theta_\alpha}$$

$$M_{mn}^{33(\alpha)} = r_\alpha^{-n-1}\sin\theta_\alpha P_n^{m\prime}(\xi_\alpha)\ ,$$

with $\xi = \cos\theta$ throughout. There is, of course, an analogous set of equations from the boundary conditions on sphere 2.

Equations 13.3 and 13.5 follow directly from the boundary conditions on the z-component and the ϕ-component of the velocity. However, Equation 13.4 is obtained by subtracting the ϕ-component equation from the ρ-component equation. By this maneuver, we ensure that in each equation each term contains the same factor of $\sin\theta_\alpha$, namely, $\sin\theta_\alpha^m$, $\sin\theta_\alpha^{m+1}$, and $\sin\theta_\alpha^{m-1}$, in Equations 13.3–13.5. The importance of this procedure follows from the fact that at $\theta_1 = 0$ and $\theta_1 = \pi$, these factors of $\sin\theta_\alpha$ cause degeneracy $(0 = 0)$ in the collocation equations, a problem that is circumvented by first removing these factors of $\sin\theta_\alpha^m$. The benefits gained by placing points at 0 and π are discussed below. Note that on the surface $r_1 = 1$, we have

$$\frac{\sin\theta_2}{\sin\theta_1} = \left[(R+\cos\theta_1)^2 + \sin^2\theta_1\right]^{-1/2}\ ,$$

so that this ratio remains between $R-1$ and $R+1$ for $0 \le \theta_1 \le \pi$.

Now if the multipole expansion is truncated after a finite number of terms, say N_1 and N_2 in the two series, then clearly Equations 13.3–13.5 cannot be satisfied exactly. The *collocation equations* are obtained by requiring these equalities at N_1 collocation points on sphere 1, *i.e.*, $r_1 = 1$, θ_{1j}, $j = 1, 2, \ldots, N_1$. We do the analogous steps for sphere 2, but with N_2 collocation points. The end result is that with $N_1 + N_2$ collocation points, we obtain $3N_1 + 3N_2$ equations. We also have this many unknowns in the truncated multipole expansions. We note degenerate cases: For axisymmetric streaming flows, the ϕ-component vanishes everywhere and we obtain a degenerate system of order $2N_1 + 2N_2$ by $2N_1 + 2N_2$; for axisymmetric swirling flows, the ϕ-component is the only nonvanishing component so we obtain a degenerate system of order $N_1 + N_2$ by $N_1 + N_2$.

13.2.1 Equal Spheres

For two equal spheres, we reduce the system of equations by exploiting the symmetry with respect to the xy-coordinate plane. An examination of each resistance problem reveals that it possesses one of the following two types of symmetry or may be decomposed into two subproblems, with a subproblem of each symmetry type. A velocity with *mirror symmetry* with respect to the xy-coordinate plane satisfies

$$\begin{aligned} v_x(x,y,z) &= v_x(x,y,-z) \\ v_y(x,y,z) &= v_y(x,y,-z) \\ v_z(x,y,z) &= -v_z(x,y,-z), \end{aligned}$$

i.e., the flow vectors in $z < 0$ and $z > 0$ are mirror images of each other. A field with *mirror anti-symmetry* satisfies

$$\begin{aligned} v_x(x,y,z) &= -v_x(x,y,-z) \\ v_y(x,y,z) &= -v_y(x,y,-z) \\ v_z(x,y,z) &= v_z(x,y,-z). \end{aligned}$$

For such fields, the flow vectors in one half-space is the negative of the mirror image of the vectors in the other half-space.

For problems with these symmetries, the coefficients for the expansion centered at $\boldsymbol{x}_2$ are related to those centered at $\boldsymbol{x}_1$ by the relations,

$$a^{(2)}_{mn} = \mathcal{S} a^{(1)}_{mn}\,, \qquad b^{(2)}_{mn} = \mathcal{S} b^{(1)}_{mn}\,, \qquad c^{(2)}_{mn} = \mathcal{S} c^{(1)}_{mn}\,,$$

where the symmetry parameter $\mathcal{S}$ is defined by

$$\mathcal{S} = \begin{cases} 1 & \text{for problems with mirror symmetry} \\ -1 & \text{for problems with mirror anti-symmetry.} \end{cases}$$

The collocation solution for these coefficients also satisfy the symmetry condition exactly, *if the collocation points on sphere 2 are placed at the mirror images of the points on sphere 1* and the system of equations reduce to two identical subsystems, namely,

$$\sum_{n=\ell}^{\infty}\{(M^{11(1)}_{mn} - \mathcal{S}M^{11(2)}_{mn})a^{(\alpha)}_{mn} + (M^{12(1)}_{mn} - \mathcal{S}M^{12(2)}_{mn})b^{(\alpha)}_{mn}$$
$$+(M^{13(1)}_{mn} - \mathcal{S}M^{13(2)}_{mn})c^{(\alpha)}_{mn}\}$$
$$= A_{\ell m}\left(\ell\xi_1 P^m_\ell(\xi_1) + (1-\xi_1^2)P^{m\prime}_\ell(\xi_1)\right) + B_{\ell m} m P^m_\ell(\xi_1) \tag{13.6}$$

$$\sum_{\alpha=1}^{2}(-1)^\alpha \sum_{n=\ell}^{\infty}\{(M^{21(1)}_{mn} + \mathcal{S}M^{21(2)}_{mn})a^{(\alpha)}_{mn} + (M^{22(1)}_{mn} + \mathcal{S}M^{22(2)}_{mn})b^{(\alpha)}_{mn}$$
$$+(M^{23(1)}_{mn} + \mathcal{S}M^{23(2)}_{mn})c^{(\alpha)}_{mn}\}$$
$$= A_{\ell m}\left(\ell \sin\theta_1 P^m_\ell(\xi_1) - [\xi_1 P^{m+1}_\ell(\xi_1) + m\sin\theta_1 P^m_\ell(\xi_1)]\right)$$
$$- B_{\ell m} P^{m+1}_\ell(\xi_1) \tag{13.7}$$

$$\sum_{\alpha=1}^{2}(-1)^\alpha \sum_{n=\ell}^{\infty}\{(M^{31(1)}_{mn} + \mathcal{S}M^{31(2)}_{mn})a^{(\alpha)}_{mn} + (M^{32(1)}_{mn} + \mathcal{S}M^{32(2)}_{mn})b^{(\alpha)}_{mn}$$
$$+(M^{33(1)}_{mn} + \mathcal{S}M^{33(2)}_{mn})c^{(\alpha)}_{mn}\}$$
$$= A_{\ell m} m P^m_\ell(\xi_1)/\sin\theta_1 + B_{\ell m}\sin\theta_1 P^{m\prime}_\ell(\xi_1)\ . \tag{13.8}$$

13.2.2 Resistance Functions

Given the boundary collocation solution, we now extract the resistance functions, as defined in Chapter 11. Using the relations given in Chapter 4 between the force, torque, and stresslet and the coefficients in Lamb's general solution, we obtain the following expressions for the resistance functions:

$$\widehat{X}^A_{11} = \frac{1}{3}[a_{01}(1,-1) + a_{01}(1,1)]$$
$$\widehat{X}^A_{12} = \frac{1}{3}[a_{01}(1,-1) - a_{01}(1,1)]$$
$$\widehat{Y}^A_{11} = \frac{1}{3}[a_{11}(1,-1) + a_{11}(1,1)]$$
$$\widehat{Y}^A_{12} = -\frac{1}{3}[a_{11}(1,-1) - a_{11}(1,1)]$$
$$\widehat{Y}^B_{11} = -[c_{11}(1,-1) + c_{11}(1,1)]$$
$$\widehat{Y}^B_{12} = c_{11}(1,-1) - c_{11}(1,1)$$
$$\widehat{X}^G_{11} = -\frac{1}{4}[a_{02}(1,-1) + a_{02}(1,1)]$$
$$\widehat{X}^G_{12} = -\frac{1}{4}[a_{02}(1,-1) - a_{02}(1,1)]$$
$$\widehat{Y}^G_{11} = -\frac{1}{4}[a_{12}(1,-1) + a_{12}(1,1)]$$
$$\widehat{Y}^G_{12} = \frac{1}{4}[a_{12}(1,-1) - a_{12}(1,1)]$$

$$\widehat{X}_{11}^{M} + \widehat{X}_{12}^{M} = \frac{1}{10}a_{02}(2,1)$$
$$\widehat{Y}_{11}^{M} + \widehat{Y}_{12}^{M} = \frac{1}{10}a_{12}(2,-1)$$
$$\widehat{Z}_{11}^{M} + \widehat{Z}_{12}^{M} = \frac{1}{10}a_{22}(2,1) .$$

In $a_{mn}(\ell, \mathcal{S})$ and $c_{mn}(\ell, \mathcal{S})$, the index m and argument ℓ denote which $A_{\ell m}$ is set to unity and $\mathcal{S}$ denotes the type of symmetry (or equivalently, a condition placed on the motion of sphere 2).

The remaining resistance functions are associated with rotational motions, and these may be expressed as

$$\widehat{X}_{11}^{C} = \frac{1}{2}[c_{01}(1,-1) + c_{01}(1,1)]$$
$$\widehat{X}_{12}^{C} = -\frac{1}{2}[c_{01}(1,-1) - c_{01}(1,1)]$$
$$\widehat{Y}_{11}^{C} = \frac{1}{2}[c_{11}(1,-1) + c_{11}(1,1)]$$
$$\widehat{Y}_{12}^{C} = \frac{1}{2}[c_{11}(1,-1) - c_{11}(1,1)]$$
$$\widehat{Y}_{11}^{H} = -\frac{1}{8}[a_{12}(1,-1) + a_{12}(1,1)]$$
$$\widehat{Y}_{12}^{H} = -\frac{1}{8}[a_{12}(1,-1) - a_{12}(1,1)] .$$

The index m and argument ℓ now denote which $B_{\ell m}$ is set to unity. The convergence of the method *vs.* the number of collocation points is described in Section 13.3.

13.2.3 Collocation Schemes

We now address the issue of "optimal location" of the collocation points. Although an *a priori* "best" collocation criterion does not exist, for two-sphere problems superior results can be obtained by placing an even number of collocation points uniformly over the meridional chord, as shown in Figure 13.2. We can rationalize this result as follows. The boundary conditions are satisfied only at the collocation points, and nonuniform distributions lead to large, oscillatory deviations from the required boundary conditions in the sparsely collocated region reminiscent of the behavior found in interpolation theory [11]. The implications for the drag calculations, as well the calculations for the higher moments of the tractions, can be seen by using the Lorentz reciprocal theorem in the following manner. We express the error in any multipole moment as an integral of the error in the boundary velocity, weighted by the stress field of the conjugate velocity problem. For example, the error in the force may be written

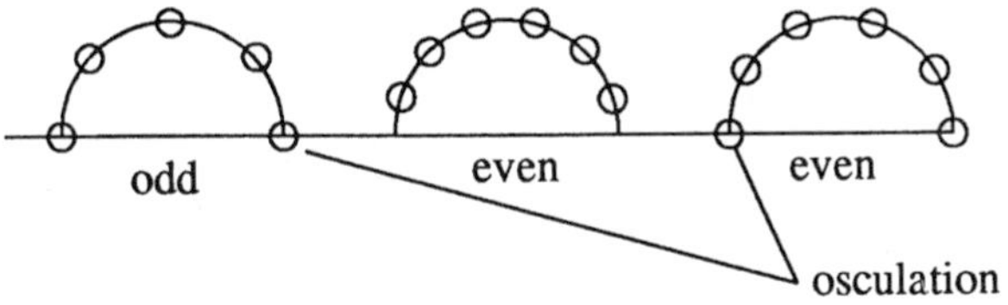

Figure 13.2: Collocation schemes for the two-sphere problem.

as

$$F^{app} - F^{exact} = \oint_S v_{error} \cdot (\sigma(;\delta) \cdot \hat{n})\, dS$$

where $\sigma(;\delta) \cdot \hat{n}$ is the surface traction induced by the constant dyadic "velocity" δ. From this line of reasoning, we conclude that large asymmetric fluctuations in the error velocity field lead to large errors in the multipole moments.

Uniform spacing with an *odd* number of collocation points yield poorer results than that obtained with comparable even numbers. This can be traced to the deterioration in the condition number in the linear system of equations. The point $\theta_1 = \pi/2$ is used when N is odd. Since $P_{2k+1}(\cos\theta_1) = 0$, for all integer k, this collocation point (and equation) places no constraints on the odd coefficients, and in some sense the system is underdetermined. In conclusion, superior results are obtained by uniform placement of an even number of collocation points.

13.3 Error Analysis for Spheres

Since the velocity field in the multisphere problem is generated by a set of multipoles expanded about each sphere center, the essential aspect of the solution is captured by the model problem consisting of a single sphere with a nearby Stokeslet. This "model problem" has an analytical solution, *viz.*, the image system for the Stokeslet near a sphere (see Chapter 10), and thus there are closed-form expressions for all multipole moments. We may compare this with the boundary collocation solution and determine the errors in the numerical method as a function of the adjustable parameters in the numerical scheme. We shall see from these comparisons that the boundary collocation method is quite robust and that the lower order moments, such as the force, torque, and stresslet, are determined quite accurately, even when the Stokeslet is near the sphere.

Throughout this section, we work with the velocity field $F \cdot \mathcal{G}(x - x_2)$ associated with a Stokeslet at x_2 *sans* the factor of $8\pi\mu$. We place a stationary,

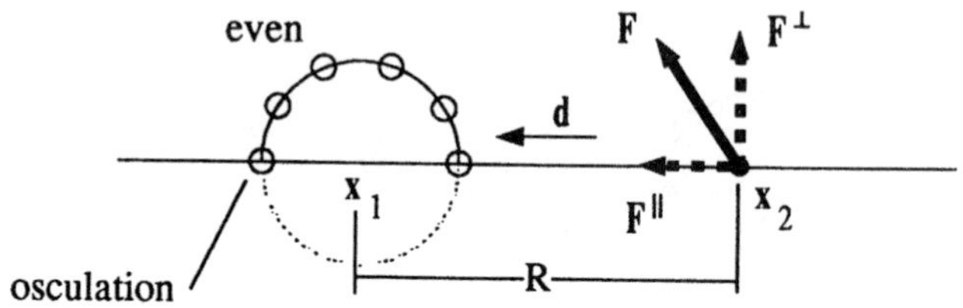

Figure 13.3: Collocation points on a sphere with a Stokeslet nearby.

nonrotating sphere of unit radius centered at $\boldsymbol{x}_1$ and let $R = |\boldsymbol{x}_2 - \boldsymbol{x}_1|$ and $\boldsymbol{d} = (\boldsymbol{x}_1 - \boldsymbol{x}_2)/R$ (see Figure 13.3). We will first consider the axisymmetric case, $\boldsymbol{F}^{\|} = \boldsymbol{F} \cdot \boldsymbol{dd}$, and then consider the asymmetric (or transverse) case, $\boldsymbol{F}^{\perp} = \boldsymbol{F} - \boldsymbol{F} \cdot \boldsymbol{dd}$.

13.3.1 Axisymmetric Stokeslet

From Section 10.2, we know that the image system, $\boldsymbol{v}^*$, can be expressed as a multipole expansion,

$$\boldsymbol{v}^* = Fd_j \sum_{n=0}^{\infty} \left\{ A_n^{\|} \frac{(\boldsymbol{d} \cdot \nabla)^n}{n!} \mathcal{G}_{ij}(\boldsymbol{x} - \boldsymbol{x}_1) + B_n^{\|} \frac{(\boldsymbol{d} \cdot \nabla)^n}{n!} \nabla^2 \mathcal{G}_{ij}(\boldsymbol{x} - \boldsymbol{x}_1) \right\} , \qquad (13.9)$$

with the coefficients given by

$$\begin{aligned} A_n^{\|} &= -(n + \frac{3}{2}) R^{-(n+1)} + (n + \frac{1}{2}) R^{-(n+3)} \\ B_n^{\|} &= -\frac{1}{4}(1 - R^2)^2 R^{-(n+5)} . \end{aligned}$$

From Chapter 4, we also have the relation between the multipole expansion and Lamb's general solution. For the latter, we express $\boldsymbol{v}^*$ as

$$\boldsymbol{v}^* = \sum_{n=1}^{\infty} \left\{ -\frac{n-2}{2n(2n-1)} r^2 \nabla p_{-n-1} + \nabla \Phi_{-n-1} \right\} , \qquad (13.10)$$

with $p_{-n-1} = a_{0n} r^{-(n+1)} P_n(\cos\theta)$ and $\Phi_{-n-1} = b_{0n} r^{-(n+1)} P_n(\cos\theta)$, and with the relations

$$a_{0n} = -2n A_{n-1}^{\|} \qquad b_{0n} = -\frac{n}{2n+3} A_{n+1}^{\|} - 2n B_{n-1}^{\|} .$$

Combining these relations together, we arrive at the analytical expression for a_{0n} and b_{0n}:

$$\begin{aligned} a_{0n} &= n(2n+1) R^{-n} - n(2n-1) R^{-(n+2)} \\ b_{0n} &= \frac{n}{2} R^{-n} - \frac{n(2n+1)}{2(2n+3)} R^{-(n+2)} . \end{aligned}$$

The boundary collocation solution for the image of the Stokeslet can be obtained with only minor modifications of the collocation equations from the two-sphere problem. We retain the multipole expansion about sphere 1, but for the surface velocity field on sphere 1, we use the velocity field produced by a Stokeslet at $\boldsymbol{x}_2$. The collocation equations are obtained from

$$\sum_{n=1}^{\infty}\left\{a_{0n}\left[\frac{n+1}{4n-2}\xi_1 P_n(\xi_1)-\frac{n-2}{n(4n-2)}(1-\xi_1^2)P_n'(\xi_1)\right]\right.$$
$$\left.+\,b_{0n}\left[-(n+1)\xi_1 P_n(\xi_1)+(1-\xi_1^2)P_n'(\xi_1)\right]\right\} = r_2^{-1}(1+\xi_2^2)$$
$$\sum_{n=1}^{\infty}\left\{a_{0n}\left[\frac{n+1}{4n-2}\sin\theta_1 P_n(\xi_1)+\frac{n-2}{n(4n-2)}[\xi_1 P_n^1(\xi_1)\right]\right.$$
$$\left.-\,b_{0n}\left[(n+1)\sin\theta_1 P_n(\xi_1)+\xi_1 P_n^1(\xi_1)\right]\right\} = r_2^{-1}\xi_2\sin\theta_2\ .$$

In the accompanying tables, the boundary collocation results for a_{0n} and b_{0n} are compared with the analytical results. We examine the influence of the location of the Stokeslet ($R = 1.1, 2.0, 10.0$) and the accuracy obtained with increasing number of collocation points. As expected, fewer collocation points are required with increasing R. For a given R, the solution accuracy degrades with increasing order of the multipole moment. Thus the quantities of greatest interest, such as the force, torque, and stresslet on the particle, may be portrayed quite faithfully with relatively few basis elements. For example, for $R = 2$ and $N = 12$, the relative error in the force (which is proportional to a_{01}) is less than 0.01%. At $R = 1.1$, the 12-point solution still yields the force coefficient accurate to 2%. Similar conclusions may be drawn for the stresslet, which is proportional to a_{02}.

In the tables for $R = 2$, we also show typical results for N odd. As discussed earlier, the collocation equation for the point at $\theta = \pi/2$ do not impose any conditions on the odd coefficients, and the system of equations become ill-conditioned. Consequently, gross errors are produced for the odd coefficients. The even coefficients, nevertheless, are reproduced accurately.

In Figure 13.4, we examine the convergence of a_{01}, *i.e.*, the drag on the sphere, as a function of N. Note that the asymptotic rate of convergence is in line with that expected from interpolation theory.

n	a_{0n} $(exact)$	$N = 12$	$N = 6$
1	0.2990 0000	0.2990 0000	0.2989 9998
2	0.0994 0000	0.0994 0000	0.0994 0013
3	0.0208 5000	0.0208 5000	0.0208 4892
4	0.0035 7200	0.0035 7200	0.0035 6980
5	0.0005 4550	0.0005 4550	0.0005 5489
6	0.0000 7734	0.0000 7734	0.0000 8136
7	0.0000 1041	0.0000 1041	
8	0.0000 0135	0.0000 0135	
9	0.0000 0017	0.0000 0017	
	$\vdots$	$\vdots$	
12	2.97×10^{-10}	3.06×10^{-10}	

n	b_{0n} $(exact)$	$N = 12$	$N = 6$
1	0.0497 0000	0.0497 0000	0.0497 0004
2	0.0099 2857	0.0099 2857	0.0099 2855
3	0.0014 8833	0.0014 8833	0.0014 8814
4	0.0001 9836	0.0001 9836	0.0001 9829
5	0.0000 2479	0.0000 2479	0.0000 2528
6	0.0000 0297	0.0000 0297	0.0000 0310
7	0.0000 0035	0.0000 0035	
8	0.0000 0004	0.0000 0004	
	$\vdots$	$\vdots$	
12	5.94×10^{-12}	6.11×10^{-12}	

Table 13.1: Axisymmetric Stokeslet at $R = 10$ for 6 and 12 collocation points.

n	a_{0n} (*exact*)	$N = 13$	$N = 12$	$N = 6$
1	1.3750 000	1.50×10^7	1.3749 948	1.3728 351
2	2.1250 000	2.1249 949	2.1250 092	2.1265 069
3	2.1562 500	3.28×10^8	2.1561 516	2.0732 118
4	1.8125 000	1.8123 653	1.8126 081	1.6942 577
5	1.3671 875	1.61×10^9	1.3667 642	2.1913 532
6	0.9609 3750	0.9606 1693	0.9617 2449	1.6017 759
7	0.6425 7813	7.33×10^9	0.6409 8340	
8	0.4140 6250	0.4128 0571	0.4240 3929	
9	0.2592 7734	6.45×10^{10}	0.2489 1388	
10	0.1586 9141	0.1582 7188	0.0929 1618	
11	0.0953 3691	-3.50×10^{11}	0.1507 9108	
12	0.0563 9648	0.0825 1236	0.1412 5853	

n	b_{0n} (*exact*)	$N = 13$	$N = 12$	$N = 6$
1	0.2125 0000	-7.49×10^6	0.2125 0359	0.2127 2613
2	0.2053 5714	0.2053 6018	0.2053 5866	0.1995 4268
3	0.1510 4167	1.09×10^7	0.1510 4105	0.1363 7757
4	0.0994 3182	0.0994 2728	0.0994 3845	0.0980 2443
5	0.0615 9856	5.38×10^7	0.0615 8859	0.1047 1251
6	0.0367 1875	0.0367 1154	0.0367 4256	0.0603 6923
7	0.0213 1204	2.01×10^8	0.0212 6457	
8	0.0121 2993	0.0120 9749	0.0123 1510	
9	0.0068 0106	1.48×10^9	0.0064 1732	
10	0.0037 6826	0.0036 8745	0.0024 1398	
11	0.0020 6787	-6.81×10^9	0.0033 5303	
12	0.0011 2576	0.0016 9952	0.0027 2114	

Table 13.2: Axisymmetric Stokeslet at $R = 2$ for 6, 12, and 13 collocation points.

n	a_{0n} (*exact*)	$N = 48$	$N = 12$
1	1.9759 579	1.9758 626	1.9978 184
2	4.1663 821	4.1663 929	3.9983 077
3	6.4637 910	6.4621 287	6.6397 658
4	8.7832 144	8.7833 387	6.7645 759
5	11.0585 57	11.0533 37	10.6850 73
	⋮	⋮	⋮
12	22.9098 19	22.9173 03	-3.00×10^2
	⋮	⋮	
48	9.1468 796	1.02×10^3	

n	b_{0n} (*exact*)	$N = 48$	$N = 12$
1	0.2291 5101	0.2291 9937	0.2047 3955
2	0.3385 7952	0.3385 8059	0.3138 8465
3	0.4025 6398	0.4025 1006	0.3729 7097
4	0.4423 4230	0.4423 4850	0.3161 6323
5	0.4667 7651	0.4666 0456	0.3827 8957
	⋮	⋮	⋮
12	0.4488 3352	0.4489 6889	−4.8118 295
	⋮	⋮	
48	0.0470 6371	5.1354 072	

Table 13.3: Axisymmetric Stokeslet at $R = 1.1$ for 12 and 48 collocation points.

13.3.2 The Transverse Stokeslet

We now consider the transverse case $\boldsymbol{F} = \boldsymbol{F}^\perp$ with $\boldsymbol{F}^\perp \cdot \boldsymbol{d} = 0$. With $\boldsymbol{t} = \boldsymbol{F}^\perp \times \boldsymbol{d}$, we have the two velocity representations:

$$\begin{aligned} v_i^* &= F_j \sum_{n=0}^{\infty} \left\{ A_n^\perp \frac{(\boldsymbol{d}\cdot\nabla)^n}{n!} \mathcal{G}_{ij}(\boldsymbol{x}-\boldsymbol{x}_1) + B_n^\perp \frac{(\boldsymbol{d}\cdot\nabla)^n}{n!} \nabla^2 \mathcal{G}_{ij}(\boldsymbol{x}-\boldsymbol{x}_1) \right\} \\ &\quad + \epsilon_{ijk} t_j \sum_{n=0}^{\infty} C_n^\perp \frac{(\boldsymbol{d}\cdot\nabla)^n}{n!} \frac{\partial}{\partial x_k} \frac{1}{r_1} \end{aligned} \tag{13.11}$$

and

$$\begin{aligned} \boldsymbol{v}^* &= \sum_{n=1}^{\infty} \left\{ -\frac{n-2}{2n(2n-1)} r^2 \nabla p_{-n-1} + \nabla \Phi_{-n-1} \right. \\ &\quad \left. + \nabla \times [(\boldsymbol{x}-\boldsymbol{x}_1)\chi_{-n-1}] \right\} , \end{aligned} \tag{13.12}$$

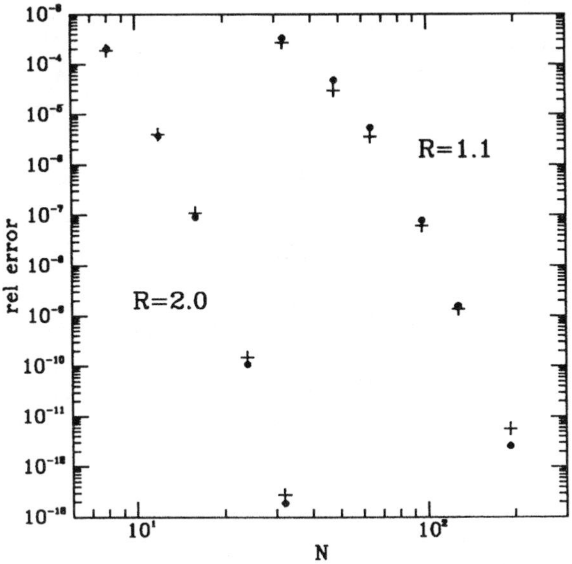

Figure 13.4: Convergence of a_{01} (•) and a_{11} (+) *vs.* number of points.

with

$$\begin{aligned} p_{-n-1} &= a_{1n} r^{-(n+1)} P_n^1(\cos\theta)\sin\phi \\ \Phi_{-n-1} &= b_{1n} r^{-(n+1)} P_n^1(\cos\theta)\sin\phi \\ \chi_{-n-1} &= c_{1n} r^{-(n+1)} P_n^1(\cos\theta)\cos\phi \ . \end{aligned}$$

From Section 10.2, the multipole expansion coefficients are:

$$A_n^\perp = \frac{(n-1)(2n+3)}{2(n+2)} R^{-(n+1)} - \frac{(n+1)(2n+1)}{2(n+2)} R^{-(n+3)} \quad (13.13)$$

$$B_n^\perp = \frac{(n-1)}{4(n+2)} R^{-(n+1)} - \Big[\frac{(n-1)}{4(n+2)} + \frac{(n+1)}{4(n+4)}\Big] R^{-(n+3)} \quad (13.14)$$

$$C_n^\perp = \frac{2n+3}{n+3} R^{-(n+2)} - \frac{2n+3}{n+3} R^{-(n+4)} \ . \quad (13.15)$$

These coefficients are related to those in Lamb's solution by the following:

$$a_{1n} = 2A_{n-1}^\perp \quad (13.16)$$

$$c_{1n} = \frac{1}{n} C_{n-1}^\perp - \frac{2}{n(n+1)} A_n^\perp \quad (13.17)$$

$$b_{1n} = 2B_{n-1}^\perp - \frac{1}{n+1} C_n^\perp + \frac{(n+3)}{(n+1)(2n+3)} A_{n+1}^\perp \ , \quad (13.18)$$

so that

$$a_{1n} = \frac{(n-2)(2n+1)}{(n+1)}R^{-n} - \frac{n(2n-1)}{(n+1)}R^{-(n+2)} \tag{13.19}$$

$$b_{1n} = \frac{(n-2)}{2(n+1)}R^{-n} - \frac{n(2n+1)}{2(n+1)(2n+3)}R^{-(n+2)} \tag{13.20}$$

$$c_{1n} = \frac{2}{n(n+1)}R^{-(n+1)} \; . \tag{13.21}$$

For the transverse Stokeslet, the collocation equations are

$$\begin{aligned}
&\sum_{n=1}^{\infty}\left\{a_{1n}\left[\frac{n+1}{4n-2}\xi_1 P_n^1(\xi_1) - \frac{n-2}{n(4n-2)}(1-\xi_1^2)P_n^{1\prime}(\xi_1)\right]\right. \\
&\qquad \left. + b_{1n}\left[-(n+1)\xi_1 P_n^1(\xi_1) + (1-\xi_1^2)P_n^{1\prime}(\xi_1)\right] + c_{1n}P_n^1(\xi_1)\right\} \\
&= r_2^{-1}\xi_2 \sin\theta_2 \\
&\sum_{n=1}^{\infty}\left\{a_{1n}\left[\frac{n+1}{4n-2}\sin\theta_1 P_n(\xi_1) + \frac{n-2}{n(4n-2)}[\xi_1 P_n^2(\xi_1) + \sin\theta_1 P_n^1(\xi_1)]\right]\right. \\
&\qquad \left. - b_{1n}\left[(n+2)\sin\theta_1 P_n^1(\xi_1) + \xi_1 P_n^2(\xi_1)\right] - c_{1n}P_n^2(\xi_1)\right\} \\
&= r_2^{-1}\sin^2\theta_2 \\
&\sum_{n=1}^{\infty}\left\{-a_{1n}\frac{n-2}{n(4n-2)}\frac{P_n^1(\xi_1)}{\sin\theta_1}\right. \\
&\qquad \left. + b_{1n}\frac{P_n^1(\xi_1)}{\sin\theta_1} + c_{1n}\sin\theta_1 P_n^{1\prime}(\xi_1)\right\} = r_2^{-1} \; .
\end{aligned}$$

In the accompanying tables, these exact results for the coefficients are compared with the corresponding boundary collocation results. Again, we conclude that the numerical results remain accurate even when the Stokeslet is quite near the sphere and that the lower order moments, such as the force, torque, and stresslet, are obtained with higher accuracy than the overall solution.

In Figure 13.4, we examine the convergence of a_{11}, which is proportional to the drag on the sphere, as a function of N, and, as in the axisymmetric problem, the rate of convergence is reminiscent of that found in interpolation theory.

n	a_{1n} (*exact*)	$N = 12$	$N = 6$
1	−0.1505 0000	−0.1505 0000	−0.1505 0001
2	−0.0002 0000	−0.0002 0000	−0.0001 9999
3	0.0017 1250	0.0017 1250	0.0017 1227
4	0.0003 5440	0.0003 5440	0.0003 5425
5	0.0000 5425	0.0000 5425	0.0000 5542
6	0.0000 0733	0.0000 0733	0.0000 0759
	⋮	⋮	
12	1.90×10^{-11}	1.96×10^{-11}	

n	b_{1n} (*exact*)	$N = 12$	$N = 6$
1	−0.0251 5000	−0.0251 5000	−0.0251 5001
2	−0.0000 2381	−0.0000 2381	−0.0000 2380
3	0.0001 2208	0.0001 2208	0.0001 2208
4	0.0000 1967	0.0000 1967	0.0000 1966
5	0.0000 0246	0.0000 0246	0.0000 0251
6	0.0000 0028	0.0000 0028	0.0000 0029
	⋮	⋮	
12	3.80×10^{-13}	3.92×10^{-13}	

n	c_{1n} (*exact*)	$N = 12$	$N = 6$
1	0.0100 0000	0.0100 0000	0.0100 0000
2	0.0003 3333	0.0003 3333	0.0003 3335
3	0.0000 1667	0.0000 1667	0.0000 1666
4	0.0000 0100	0.0000 0100	0.0000 0097
5	6.67×10^{-8}	6.67×10^{-8}	6.78×10^{-8}
6	4.76×10^{-9}	4.76×10^{-9}	1.69×10^{-8}
	⋮	⋮	
12	1.28×10^{-15}	8.46×10^{-15}	

Table 13.4: Transverse Stokeslet at $R = 10$ for 6 and 12 collocation points.

n	a_{1n} (*exact*)	$N = 12$	$N = 6$
1	−0.8125 0000	−0.8125 0321	−0.8138 0688
2	−0.1250 0000	−0.1249 9819	−0.1194 9522
3	0.1015 6250	0.1015 4305	0.0709 9711
4	0.1375 0000	0.1375 1330	0.0528 2180
5	0.1132 8125	0.1132 3093	0.2268 4604
6	0.0792 4107	0.0793 0014	0.2113 7383
	⋮	⋮	
12	0.0033 9919	0.0077 6035	

n	b_{1n} (*exact*)	$N = 12$	$N = 6$
1	−0.1437 5000	−0.1437 5201	−0.1440 5275
2	−0.0148 8095	−0.0148 8105	−0.0126 8696
3	0.0065 1042	0.0065 0795	0.0053 0967
4	0.0073 8636	0.0073 8688	0.0009 5563
5	0.0050 5809	0.0050 5478	0.0097 0791
6	0.0030 1339	0.0030 1613	0.0086 1188
	⋮	⋮	
12	6.78×10^{-5}	1.59×10^{-4}	

n	c_{1n} (*exact*)	$N = 12$	$N = 6$
1	0.2500 0000	0.2500 0013	0.2492 3892
2	0.0416 6667	0.0416 6886	0.0436 6289
3	0.0104 1667	0.0104 1668	0.0096 3527
4	0.0031 2500	0.0031 2670	−0.0009 2969
5	0.0010 4167	0.0010 4160	0.0016 8348
6	0.0003 7202	0.0003 7438	0.0020 8919
	⋮	⋮	
12	1.56×10^{-6}	1.60×10^{-5}	

Table 13.5: Transverse Stokeslet at $R = 2$ for 6 and 12 collocation points.

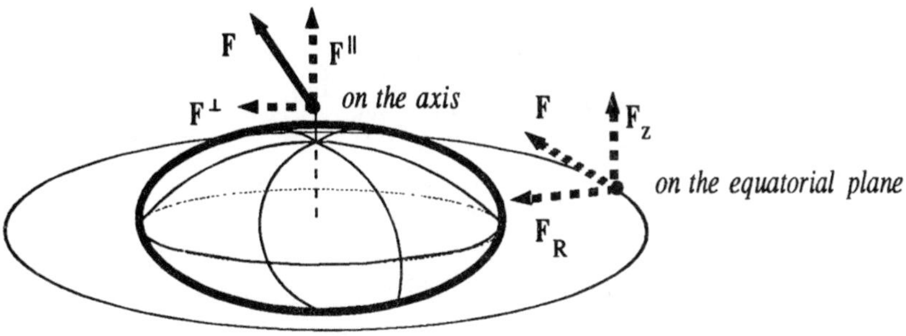

Figure 13.5: The Stokeslet near a stationary oblate spheroid.

13.4 Error Analysis for Spheroids

We examine the problem of a Stokeslet placed near an oblate spheroid to gauge the accuracy of the boundary collocation solution. The principles involved are the same as those discussed previously for the sphere.

From Chapter 3, we have the following expression (Faxén law) for the force on a stationary oblate spheroid in the field produced by the Stokeslet:

$$\begin{aligned} \boldsymbol{F}^S &= 6\pi\mu a[X^A \boldsymbol{dd} + Y^A(\boldsymbol{\delta} - \boldsymbol{dd})] \cdot \left\{ \frac{\sin D_z}{D_z} \right. \\ &\quad \left. - \frac{a^2}{2}\left(\frac{1}{D_z}\frac{\partial}{\partial D_z}\right)\frac{\sin D_z}{D_z}\nabla^2 \right\} \boldsymbol{F} \cdot \frac{\mathcal{G}(\boldsymbol{x} - \boldsymbol{x}_2)}{8\pi\mu}\Big|_{x=x_1} , \qquad (13.22) \end{aligned}$$

where $D_z^2 = (a^2 - c^2)\partial^2/\partial z^2$ and the resistance functions X^A and Y^A are as defined in Chapter 5. Where expedient, we will write the radius of the focal disk, $(a^2 - c^2)^{1/2} = ae$ as a_E. We will also use e to denote the eccentricity of the spheroid.

In comparison with the analysis for spheres, the spheroid problem inevitably requires more geometric parameters. Note also that if the Stokeslet is off the symmetry axis of the spheroid, then the spheroid-Stokeslet geometry is not axisymmetric. Thus the complete solution will depend on the numerous parameters that specify the location of the Stokeslet relative to the spheroid. However, our interest here is only in the exposition of the accuracy of the collocation solutions, and this information may be extracted by examining two special configurations: a Stokeslet placed on the axis of symmetry and a Stokeslet placed on the "equatorial plane," as shown in Figure 13.5. The collocation scheme is also depicted in the figure.

n	a_{1n} (*exact*)	$N = 48$	$N = 12$
1	−1.7392 938	−1.7393 460	−1.7960 163
2	−1.3660 269	−1.3659 905	−1.1717 432
3	−1.0136 541	−1.0139 570	−1.4739 896
4	−0.7022 0558	−0.7021 0205	0.0332 0390
5	−0.4336 1861	−0.4341 8951	−1.7801 860
6	−0.2052 6349	−0.2050 7080	2.8300 659
	⋮	⋮	⋮
12	0.5367 9068	0.5374 4096	1.51×10^2
	⋮	⋮	
48	0.1458 6194	0.1440 3522	

n	b_{1n} (*exact*)	$N = 48$	$N = 12$
1	−0.3399 6995	−0.3399 9637	−0.3625 6759
2	−0.1626 2225	−0.1626 1649	−0.1232 1870
3	−0.0871 8770	−0.0872 1832	−0.1256 7090
4	−0.0481 3422	−0.0481 2714	0.0135 7631
5	−0.0256 9080	−0.0257 2281	−0.0904 0912
6	−0.0119 9590	−0.0119 8742	0.1399 9798
	⋮	⋮	⋮
12	0.0100 1559	0.0100 2978	3.2409 243
	⋮	⋮	
48	0.0007 5013	0.0755 4548	

n	c_{1n} (*exact*)	$N = 48$	$N = 12$
1	0.8264 4628	0.8264 3625	0.7679 1789
2	0.2504 3827	0.2504 6764	0.2860 8967
3	0.1138 3558	0.1138 3157	0.0885 8993
4	0.0620 9213	0.0621 0825	0.0965 2831
5	0.0376 3160	0.0376 2896	0.0177 1595
6	0.0244 3610	0.0244 4747	0.0853 5150
	⋮	⋮	⋮
12	0.0037 1365	0.0037 2040	0.5079 2543
	⋮	⋮	
48	7.97×10^{-6}	2.89×10^{-3}	

Table 13.6: Transverse Stokeslet at $R = 1.1$ for 12 and 48 collocation points.

13.4.1 Stokeslet on the Axis

We now show that a Stokeslet of strength $\boldsymbol{F}$ placed on the axis of symmetry exerts a force $\boldsymbol{F}^S$ given by

$$\begin{aligned}\boldsymbol{F}^S &= \boldsymbol{F}\cdot\boldsymbol{dd}\frac{3X^A}{4e^3}\left[(2e^2-1)\cot^{-1}(R/ae)+\frac{eR/a}{(R/a)^2+e^2}\right]\\ &\quad+(\boldsymbol{F}-\boldsymbol{F}\cdot\boldsymbol{dd})\frac{3Y^A}{8e^3}\left[(2e^2+1)\cot^{-1}(R/ae)-\frac{eR/a}{(R/a)^2+e^2}\right]\end{aligned}$$

on the stationary spheroid. Since $R=|\boldsymbol{x}_2-\boldsymbol{x}_1|$ is the separation between the spheroid center and the Stokeslet, we assume $c<R<\infty$. The two lines on the RHS of the preceding equation correspond to Stokeslet orientations along and perpendicular to the symmetry axis.

Stokeslet Aligned with the Axis

If we have $\boldsymbol{F}=\boldsymbol{F}^{||}$, *i.e.*, the Stokeslet is aligned with the symmetry axis, then only the axial component is nonzero. We employ the following expressions for $\partial^n\mathcal{G}/\partial z^n$ from Chapter 3:

$$\begin{aligned}\frac{(\boldsymbol{d}\cdot\nabla)^n}{n!}\boldsymbol{F}^{||}\cdot\mathcal{G}(\boldsymbol{x}-\boldsymbol{x}_2)|_{x=x_1} &= 2\boldsymbol{F}^{||}R^{-(n+1)}\\ \frac{(\boldsymbol{d}\cdot\nabla)^n}{n!}\boldsymbol{F}^{||}\cdot\nabla^2\mathcal{G}(\boldsymbol{x}-\boldsymbol{x}_2)|_{x=x_1} &= -2(n+1)(n+2)\boldsymbol{F}^{||}R^{-(n+3)}\end{aligned}$$

to reduce Equation 13.22 to

$$F^S=F^{||}\frac{3}{2}X^A\frac{a}{R}\sum_{k=0}^{\infty}(-1)^k\left[\frac{1}{2k+1}-\frac{a^2}{R^2}\frac{k+1}{2k+3}\right]\left(\frac{a_E}{R}\right)^{2k}. \qquad (13.23)$$

Note that the infinite series in R^{-1} converges only if $R^{-1}\le a_E$. We may visualize this constraint with the help of a "hemispheric dome" of radius a_E over the focal circle of the spheroid. If we move the Stokeslet along the axis of symmetry, the infinite series will converge only if the Stokeslet lies outside the hemisphere. In general, when the Faxén law is written as an infinite series in differential operators, the resulting expression will have a finite radius of convergence in R^{-1}, and, furthermore, this expression will fail to converge when the Stokeslet is too close to the particle. This example also shows that the exact value of the radius of convergence depends on the details of the particle geometry. *In particular, this value corresponds neither to the particle surface nor the surface of the "effective sphere" that circumscribes the particle.*

In Chapter 3, we showed other forms for the Faxén relation for the ellipsoid that did not require an infinite series expansion in differential operators. The final expression for the force obtained from these alternate forms will be an analytic function of R that converges for all physical values of R. In principle, we may obtain these force expressions starting from the appropriate Faxén relation. However, it is easier to use analytic continuation of Equation 13.23 to reproduce these results. We replace the infinite series by using the identities

$$\sum_{k=0}^{\infty} \frac{(-1)^k}{2k+1} x^{2k} = x^{-1} \cot^{-1} x^{-1}$$

$$\sum_{k=0}^{\infty} (-1)^k \frac{2k+2}{2k+1} x^{2k} = \frac{1}{1+x^2} - x^{-2} + x^{-3} \cot^{-1} x^{-1}$$

to obtain

$$F^S = F^{\|} \frac{3X^A}{4e^3} \left[(2e^2 - 1) \cot^{-1}(R/ae) + \frac{eR/a}{(R/a)^2 + e^2} \right] . \qquad (13.24)$$

In the limiting case $R \to c$, this reduces to $F^S \to F^{\|}$, and when $e \to 0$ we recover the result for the sphere,

$$F^S = F^{\|} \left[\frac{3}{2} \frac{a}{R} - \frac{1}{2} \left(\frac{a}{R} \right)^3 \right] .$$

Stokeslet Perpendicular to the Axis

If $\boldsymbol{F} = \boldsymbol{F}^{\perp}$ with $\boldsymbol{F}^{\perp} \cdot \boldsymbol{d} = 0$, then only the component parallel to $\boldsymbol{F}^{\perp}$ is nonzero, and with the help of

$$\begin{aligned} \frac{(\boldsymbol{d} \cdot \nabla)^n}{n!} \boldsymbol{F}^{\perp} \cdot \mathcal{G}(\boldsymbol{x} - \boldsymbol{x}_2)|_{x=x_1} &= \boldsymbol{F}^{\perp} R^{-(n+1)} \\ \frac{(\boldsymbol{d} \cdot \nabla)^n}{n!} \boldsymbol{F}^{\perp} \cdot \nabla^2 \mathcal{G}(\boldsymbol{x} - \boldsymbol{x}_2)|_{x=x_1} &= (n+1)(n+2) \boldsymbol{F}^{\perp} R^{-(n+3)} , \end{aligned}$$

Equation 13.22 reduces to

$$F^S = F^{\perp} \frac{3}{4} Y^A \frac{a}{R} \sum_{k=0}^{\infty} (-1)^k \left[\frac{1}{2k+1} + \frac{a^2}{R^2} \frac{k+1}{2k+3} \right] \left(\frac{a_E}{R} \right)^{2k} . \qquad (13.25)$$

Again, the radius of convergence is at $R^{-1} = a_E$. The expression that is valid for all $R^{-1} \geq c$ is

$$F^S = F^{\perp} \frac{3Y^A}{8e^3} \left[(2e^2 + 1) \cot^{-1}(R/ae) - \frac{eR/a}{(R/a)^2 + e^2} \right] . \qquad (13.26)$$

The limiting cases give consistent results. When $R \to c$, we obtain $F^S \to F^{\perp}$, and when $e \to 0$ we recover the result for the sphere,

$$F^S = F^{\perp} \left[\frac{3}{4} \frac{a}{R} + \frac{1}{4} \left(\frac{a}{R} \right)^3 \right] .$$

In Figure 13.6 the exact solutions for the force are compared with those obtained by boundary collocation. We see that even when the solution is expanded only up to (distributed) quadrupoles, the collocation solutions reproduce the net force accurately for all Stokeslets, except those placed very close to the surface.

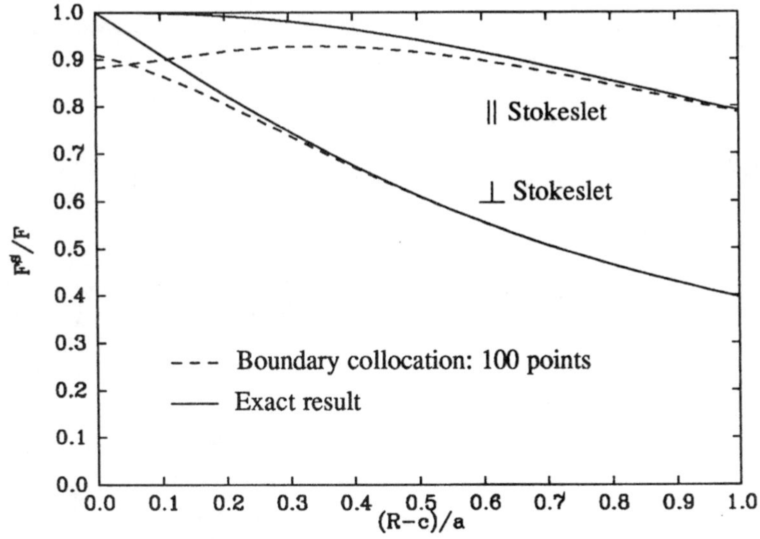

Figure 13.6: Force on an oblate spheroid due to a nearby Stokeslet: comparison of exact and collocation results.

13.4.2 Stokeslet on the Equatorial Plane

If the Stokeslet is in the equatorial plane, it is convenient to decompose the problem into three special cases in which the Stokeslet is aligned with the basis vectors $\boldsymbol{e}_z$, $\boldsymbol{e}_R$, and $\boldsymbol{e}_\phi$ of the cylindrical coordinate system. Keeping in mind that $a < R < \infty$, the final result for the force on the spheroid may be written as

$$
\begin{aligned}
\boldsymbol{F}^S &= \boldsymbol{F}\cdot\boldsymbol{dd}\frac{3X^A}{4e^3}\left[(2e^2-1)\chi+(1-e^2)\frac{ae}{(R^2-a^2e^2)^{1/2}}\right] \\
&\quad + \boldsymbol{F}\cdot\boldsymbol{e}_R\boldsymbol{e}_R\frac{3Y^A}{8e^3}\left[(2e^2+1)\chi+\frac{(2e^2-1)ae-a^3e^3/R^2}{(R^2-a^2e^2)^{1/2}}\right] \\
&\quad + \boldsymbol{F}\cdot\boldsymbol{e}_\phi\boldsymbol{e}_\phi\frac{3Y^A}{8e^3}\left[(2e^2+1)\chi-e\frac{a}{R}[1-e^2(a/R)^2]^{1/2}\right] ,
\end{aligned}
$$

where

$$
\chi = \cot^{-1}\left(\frac{\sqrt{(R/a)^2-e^2}}{e}\right) . \tag{13.27}
$$

We now consider the three cases.

Stokeslet Aligned with the Axis

If the Stokeslet is aligned with the symmetry axis, then only the axial component of the force on the spheroid is nonzero. We use

$$\begin{aligned}
\frac{(\boldsymbol{d}\cdot\nabla)^n}{n!}F_z\boldsymbol{e}_z\cdot\mathcal{G}(\boldsymbol{x}-\boldsymbol{x}_2)|_{x=x_1} &= -(n-1)F_z\boldsymbol{e}_z\frac{P_n(0)}{R^{n+1}} \\
\frac{(\boldsymbol{d}\cdot\nabla)^n}{n!}F_z\boldsymbol{e}_z\cdot\nabla^2\mathcal{G}(\boldsymbol{x}-\boldsymbol{x}_2)|_{x=x_1} &= -2(n+1)(n+2)F_z\boldsymbol{e}_z\frac{P_{n+2}(0)}{R^{n+3}}
\end{aligned}$$

to reduce Equation 13.22 to

$$F^S = F_z\frac{3}{4}X^A\frac{a}{R}\sum_{k=0}^{\infty}(-1)^k\left[P_{2k}(0)\frac{1-2k}{2k+1}-\frac{a^2}{R^2}P_{2k+2}(0)\frac{2k+2}{2k+3}\right]\left(\frac{a_E}{R}\right)^{2k} \tag{13.28}$$

Now, identities such as

$$\sum_{k=0}^{\infty}(-1)^kP_{2k}(0)x^{2k} = (1-x^2)^{-1/2}$$

$$\sum_{k=0}^{\infty}\frac{(-1)^kP_{2k}(0)}{2k+1}x^{2k+1} = \cot^{-1}\left(\frac{(1-x^2)^{1/2}}{x}\right)$$

may be used to replace Equation 13.28 with

$$F^S = F_z\frac{3X^A}{4e^3}\left[(2e^2-1)\chi+(1-e^2)\frac{ae}{(R^2-a^2e^2)^{1/2}}\right], \tag{13.29}$$

with χ as defined previously in Equation 13.27. For $R\to a$, $F^S\to F_z$, and $e\to 0$ we recover the result for the transverse Stokeslet near a sphere.

Stokeslet in the Radial Direction

For $\boldsymbol{F} = F_R\boldsymbol{e}_R$ (see Figure 13.5) only the R-component of the force on the spheroid is nonzero. We use

$$\begin{aligned}
\boldsymbol{e}_R\cdot\frac{(\boldsymbol{d}\cdot\nabla)^n}{n!}\boldsymbol{F}\cdot\mathcal{G}(\boldsymbol{x}-\boldsymbol{x}_2)|_{x=x_1} &= (n+2)F_R\frac{P_n(0)}{R^{n+1}} \\
\boldsymbol{e}_R\cdot\frac{(\boldsymbol{d}\cdot\nabla)^n}{n!}\boldsymbol{F}\cdot\nabla^2\mathcal{G}(\boldsymbol{x}-\boldsymbol{x}_2)|_{x=x_1} &= -2(n+1)(n+2)F_R\frac{P_n(0)}{R^{n+3}}
\end{aligned}$$

to reduce Equation 13.22 to

$$F^S = F_R\frac{3}{4}Y^A\frac{a}{R}\sum_{k=0}^{\infty}(-1)^kP_{2k}(0)\left[\frac{2k+2}{2k+1}-\frac{a^2}{R^2}\frac{2k+2}{2k+3}\right]\left(\frac{a_E}{R}\right)^{2k}. \tag{13.30}$$

We may rewrite Equation 13.30 as

$$F^S = F_R\frac{3Y^A}{8e^3}\left[(2e^2+1)\chi+\frac{(2e^2-1)ae-a^3e^3/R^2}{(R^2-a^2e^2)^{1/2}}\right], \tag{13.31}$$

with χ as defined previously in Equation 13.27. In the limiting case $R\to a$ this reduces to $F^S\to F_R$, and when $e\to 0$ we recover the result for the axisymmetric Stokeslet near a sphere.

Stokeslet in the Azimuthal Direction

For $\boldsymbol{F} = F_\phi \boldsymbol{e}_\phi$ only the ϕ-component of the force on the spheroid is nonzero. We use

$$\begin{aligned} \boldsymbol{e}_\phi \cdot \frac{(\boldsymbol{d}\cdot\nabla)^n}{n!}\boldsymbol{F}\cdot\mathcal{G}(\boldsymbol{x}-\boldsymbol{x}_2)|_{x=x_1} &= F_\phi R^{-(n+1)} P_n(0) \\ \boldsymbol{e}_\phi \cdot \frac{(\boldsymbol{d}\cdot\nabla)^n}{n!}\boldsymbol{F}\cdot\nabla^2\mathcal{G}(\boldsymbol{x}-\boldsymbol{x}_2)|_{x=x_1} &= 2(n+1) F_\phi R^{-(n+3)} P_n(0) \end{aligned}$$

to reduce Equation 13.22 to

$$F^S = F_\phi \frac{3}{4} Y^A \frac{a}{R} \sum_{k=0}^{\infty} (-1)^k P_{2k}(0) \left[\frac{1}{2k+1} + \frac{a^2}{R^2}\frac{1}{2k+3}\right] \left(\frac{a_E}{R}\right)^{2k} . \tag{13.32}$$

We may rewrite Equation 13.32 as

$$F^S = F_\phi \frac{3Y^A}{8e^3}\left[(2e^2+1)\chi - e\frac{a}{R}[1-e^2(a/R)^2]^{1/2}\right] . \tag{13.33}$$

In the limiting case $R \to a$, this reduces to $F^S \to F_\phi$, and when $e \to 0$ we recover the result for the transverse Stokeslet near a sphere.

In Figure 13.6, the exact solutions for the force are compared with those obtained by boundary collocation. Here also, the collocation solutions reproduce the net force accurately for all Stokeslets except those placed very close to the spheroid.

We conclude this chapter on boundary-multipole collocation by stating that the basis functions are the key to the success of the method. In the extreme limit where the exact analytical solution is known (let us denote it as $c_1\boldsymbol{v}_{(1)}$), then a one-point collocation of the one-term expansion, $\boldsymbol{v} = c_1\boldsymbol{v}_{(1)}$, would work. More generally, a good basis set gives a rapidly converging expansion so that accurate solutions are obtained with a small number of unknown coefficients. In the case of ellipsoids and spheroids, the use of Lamb's general solution or any other expansion based on the spherical harmonics is a bad idea; from Chapter 3 we know that even the translating ellipsoid has a slowly converging infinite series expansion of this type. On the other hand, it should be apparent from the discussion in this section that expansions based on the singularity solutions of Chapter 3 should work quite nicely. The recent work by Yoon and Kim [71] demonstrate that this is indeed the case.

Exercises

Exercise 13.1 Stokeslet Near a Spheroid.
Show that the results in Section 13.4 for the force on a spheroid in the field of a Stokeslet — Equations 13.24, 13.26, 13.29, 13.31, and 13.33 — are consistent with those obtained from the velocity field of a translating spheroid.
Hint: Use of the Lorentz reciprocal theorem.

Exercise 13.2 Error Analysis for Spherical Drops.
Extend the error analysis for rigid spheres (Section 13.3) to the case of the spherical viscous drop. How does the viscosity ratio λ affect the convergence behavior?

Exercise 13.3 Boundary Collocation for the Oblate Spheroid with Lamb's General Solution.
Consider an oblate spheroid in an axisymmetric streaming flow. Expand the velocity with Lamb's general solution and examine the convergence behavior of the collocation solution. At what aspect ratio does the method deteriorate?

Exercise 13.4 Convergence of Torque and Stresslet Coefficients.
Consider a stationary, nonrotating sphere in an ambient field produced by a nearby Stokeslet. Examine the convergence of a_{02}, a_{12}, and c_{11} as a function of the number of collocation points, in a manner analogous to the discussion on a_{01} and a_{11} (see Figure 13.4).
Comment: The analysis of these coefficients reveals the rate at which the torque and stresslet resistance functions converge with number of collocations points.

Exercise 13.5 Collocation Schemes for Two Unequal Spheres.
It seems reasonable that for two unequal spheres, the larger sphere should receive a greater number of collocation points, given that the total number of points $N_1 + N_2$ is fixed. In this problem, we offer additional guidelines for distributing the points between the two spheres. Given spheres sizes and center-to-center separation, examine each sphere in turn, with the other sphere replaced by an "equivalent" Stokeslet. Use the error analysis from the sphere-Stokeslet problem to determine an acceptable value for the number of collocation points.
Hint: Consider the following options for the equivalent Stokeslet:

1. For lubrication-type problems, large stresses develop in the gap region, so the Stokeslet is located at the nearest point on the surface of the other sphere.
2. For surfaces moving in tandem as a rigid body, the gap region does not dominate the flow, so the equivalent Stokeslet is located at the center of the other sphere.

Do these strategies provide optimal rates of convergence?
Note: An example program for the two unequal spheres is available on `flossie`, in the subdirectory `chapter13`. (See the instructions in the preamble of this book for copying computer programs over Internet.)

References

[1] G. K. Batchelor and J. T. Green. The hydrodynamic interaction of two small freely-moving spheres in a linear flow field. *J. Fluid Mech.*, 56:375–400, 1972.

[2] G. K. Batchelor and J. T. Green. The determination of the bulk stress in a suspension of spherical particles to order c^2. *J. Fluid Mech.*, 56:401–427, 1972.

[3] V. N. Beshkov, B.P. Radoev, and I. B. Ivanov. Slow motion of two droplets and a droplet towards a fluid or solid interface. *Intl. J. Multiphase Flow*, 4:563–570, 1978.

[4] H. Brenner. The slow motion of a sphere through a viscous fluid towards a plane surface. *Chem. Eng. Sci.*, 16:242–251, 1961.

[5] H. Brenner and M. E. O'Neill. On the Stokes resistance of multiparticle systems in a linear shear field. *Chem. Eng. Sci.*, 27:1421–1439, 1972.

[6] S. B. Chen and H. J. Keh. Electrophoresis in a dilute suspension of colloidal spheres. *AIChE J.*, 34:1075–1085, 1988.

[7] R. M. Corless and D. J. Jeffrey. Stress moments of nearly touching spheres in low Reynolds number flow. *Z. Angew. Math. Phys.*, 39:874–884, 1988.

[8] R. G. Cox and H. Brenner. The slow motion of a sphere through a viscous fluid towards a plane surface. Part II. Small gap widths, including inertial effects. *Chem. Eng. Sci.*, 22:1753–1777, 1967.

[9] R. G. Cox. The deformation of a drop in a general time-dependent fluid flow. *J. Fluid Mech.*, 37:601–623, 1969.

[10] Z. Dagan, S. Weinbaum, and R. Pfeffer. General theory for the creeping motion of a finite sphere along the axis of a circular orifice. *J. Fluid Mech.*, 117:143–170, 1982.

[11] G. Dahlquist and A. Bjorck. *Numerical Methods.* Prentice-Hall, Englewood Cliffs, 1974. Translated by Ned Anderson.

[12] R. H. Davis, J. A. Schonberg, and J. M. Rallison. The lubrication force between two viscous drops. *Phys. Fluids*, A1:77–81, 1989.

[13] H. Faxén. Der Widerstand gegen die Bewegung einer starren Kugel in einer zähen Flüssigkeit, die zwischen zwei parallelen Ebenen Wänden eingeschlossen ist (The resistance against the movement of a rigid sphere in a viscous fluid enclosed between two parallel planes). *Arkiv fur Matematik, Astronomi och Fysik*, 18(29):1–52, 1924.

[14] D. L. Feke and W. R. Schowalter. The effect of Brownian diffusion on shear-induced coagulation of colloidal dispersion. *J. Fluid Mech.*, 133:17–35, 1983.

[15] D. L. Feke and W. R. Schowalter. The influence of Brownian diffusion on binary flow-induced collision rates in colloidal dispersions. *J. Colloid Interface Sci.*, 106:203–214, 1985.

[16] Y. O. Fuentes, S. Kim, and D. J. Jeffrey. Mobility functions for two unequal viscous drops in Stokes flow I. Axisymmetric motions. *Phys. Fluids*, 31:2445–2455, 1988.

[17] Y. O. Fuentes, S. Kim, and D. J. Jeffrey. Mobility functions for two unequal viscous drops in Stokes flow II. Asymmetric motions. *Phys. Fluids*, A1:61–76, 1989.

[18] P. Ganatos, R. Pfeffer, and S. Weinbaum. A numerical-solution technique for three-dimensional Stokes flows, with application to the motion of strongly interacting spheres in a plane. *J. Fluid Mech.*, 84:79–111, 1978.

[19] P. Ganatos, S. Weinbaum, and R. Pfeffer. A strong interaction theory for the creeping motion of a sphere between plane parallel boundaries. Part 1. Perpendicular motion. *J. Fluid Mech.*, 99:739–753, 1980.

[20] P. Ganatos, R. Pfeffer, and S. Weinbaum. A strong interaction theory for the creeping motion of a sphere between plane parallel boundaries. Part 2. Parallel motion. *J. Fluid Mech.*, 99:755–783, 1980.

[21] M. J. Gluckman, R. Pfeffer, and S. Weinbaum. A new technique for treating multiparticle slow viscous flow: axisymmetric flow past spheres and spheroids. *J. Fluid Mech.*, 50:705–740, 1971.

[22] A. J. Goldman, R. G. Cox, and H. Brenner. Slow viscous motion of a sphere parallel to a plane wall — I. Motion through a quiescent fluid. *Chem. Eng. Sci.*, 22:637–651, 1967.

[23] J. T. Green. *Properties of Suspensions of Rigid Spheres.* Ph. D. Dissertation, University of Cambridge, 1971.

[24] S. Haber, G. Hetsroni, and A. Solan. On the low Reynolds number motion of two droplets. *Intl. J. Multiphase Flow*, 1:57–71, 1973.

[25] W. W. Hackborn. Asymmetric Stokes flow between parallel planes due to a rotlet. *J. Fluid Mech.*, 000:000–000, 1990.

[26] R. E. Hansford. On converging solid spheres in a highly viscous fluid. *Mathematika*, 17:250–254, 1970.

[27] J. Happel and H. Brenner. *Low Reynolds Number Hydrodynamics.* Martinus Nijhoff, The Hague, 1983.

[28] J. J. L. Higdon. A hydrodynamic analysis of flagellar propulsion. *J. Fluid Mech.*, 90:685–711, 1979.

[29] E. J. Hinch. An averaged-equation approach to particle interactions in a fluid suspension. *J. Fluid Mech.*, 83:695–720, 1977.

[30] E. W. Hobson. *The Theory of Spherical and Ellipsoidal Harmonics.* Chelsea, New York, 1955.

[31] R. Hsu and P. Ganatos. The motion of a rigid body in a viscous fluid bounded by a plane wall. *J. Fluid Mech.*, 207:29–72, 1989.

[32] J. N. Israelachvili. The calculation of van der Waals dispersion forces between macroscopic bodies. *Proc. R. Soc.*, A331:39–55, 1972.

[33] J. N. Israelachvili. Van der Waals forces in biological systems. *Q. Rev. Biophys.*, 6:341–387, 1974.

[34] J. N. Israelachvili and B. W. Ninham. Intermolecular forces — the long and short of it. *J. Colloid Interface Sci.*, 58:14–25, 1977.

[35] K. M. Jansons and J. R. Lister. The general solution of Stokes flow in a half-space as an integral of the velocity on the boundary. *Phys. Fluids*, 31:1321–1323, 1988.

[36] D. J. Jeffrey. Group expansions for the bulk properties of a statistically homogeneous, random suspension. *Proc. R. Soc. (London) A*, 338:503–516, 1974.

[37] D. J. Jeffrey. Low-Reynolds-number flow between converging spheres. *Mathematika*, 29:58–66, 1982.

[38] D. J. Jeffrey and Y. Onishi. Calculation of the resistance and mobility functions for two unequal rigid spheres in low-Reynolds-number flow. *J. Fluid Mech.*, 139:261–290, 1984.

[39] D. J. Jeffrey and Y. Onishi. The forces and couples acting on two nearly touching spheres in low-Reynolds-number flow. *Z. Angew. Math. Phys.*, 35:634–641, 1984.

[40] D. J. Jeffrey and R. M. Corless. Forces and stresslets for the axisymmetric motion of nearly touching unequal spheres. *PCH PhysicoChem. Hydrodynamics*, 10:461–470, 1988.

[41] D. J. Jeffrey. Stresslet resistance functions for low Reynolds number flow using deformable spheres. *Z. Angew. Math. Phys.*, 40:1–8, 1989.

[42] H. J. Keh and J. L. Anderson. Boundary effects on electrophoretic motion of colloidal spheres. *J. Fluid Mech.*, 153:417–439, 1985.

[43] S. Kim. *Modeling of Porous Media via Renormalization of the Stokes Equations.* Ph.D. Dissertation, Princeton University, 1983.

[44] S. Kim. Sedimentation of two arbitrarily oriented spheroids in a viscous fluid. *Intl. J. Multiphase Flow*, 11(5):699–712, 1985.

[45] S. Kim and R. T. Mifflin. The resistance and mobility functions of two equal spheres in low-Reynolds-number flow. *Phys. Fluids*, 28:2033–2045, 1985.

[46] S. Kim. Stokes flow past three spheres: an analytic solution. *Phys. Fluids*, 30(8):2309–2314, 1987.

[47] S. H. Lee, R. S. Chadwick, and L. G. Leal. Motion of a sphere in the presence of a plane interface. Part 1. An approximate solution by generalization of the method of Lorentz. *J. Fluid Mech.*, 93:705–726, 1979.

[48] S. H. Lee and L. G. Leal. The motion of a sphere in the presence of a deformable interface II. A numerical study of the translation of a sphere normal to an interface. *J. Colloid Interface Sci.*, 87(1):81–106, 1982.

[49] S. Leichtberg, S. Weinbaum, R. Pfeffer, and M. J. Gluckman. A study of unsteady forces at low Reynolds number: a strong interaction theory for the coaxial settling of three or more spheres. *Phil. Trans. R. Soc. (London)*, A282:585–613, 1976.

[50] W. H. Liao and D. Krueger. Multipole expansion calculation of slow viscous flow about spheroids of different sizes. *J. Fluid Mech.*, 96:223–241, 1980.

[51] N. Liron and S. Mochon. Stokes flow for a Stokeslet between two parallel flat plates. *J. Eng. Math.*, 10:287–303, 1976.

[52] H. A. Lorentz. Ein allgemeiner Satz, die Bewegung einer reibenden Flüssigkeit betreffend, nebst einigen Anwendungen desselben (A general theorem concerning the motion of a viscous fluid and a few applications from it). *Versl. Kon. Akad. Wetensch. Amsterdam*, 5:168–174, 1896. Also in *Abhandlungen über Theoretische Physik*, 1:23–42 (1907) and *Collected Papers*, 4:7–14, Martinus Nijhoff, The Hague, 1937.

[53] S. Y. Lu and S. Kim. The general solution and Faxén laws for the temperature fields in and outside an isolated ellipsoid. *J. Eng. Math.*, 21(3):179–200, 1987.

[54] J. H. C. Luke. Convergence of a multiple reflection method for calculating Stokes flow in a suspension. *SIAM J. Appl. Math.*, 49:1635–1651, 1989.

[55] S. R. Majumdar. Slow motion of an incompressible viscous liquid generated by the rotation of two spheres in contact. *Mathematika*, 14:43–46, 1967.

[56] A. D. Maude. End effects in a falling-sphere viscometer. *Br. J. Appl. Phys.*, 12:293–295, 1961.

[57] R. W. O'Brien. A method for the calculation of the effective transport properties of suspensions of interacting particles. *J. Fluid Mech.*, 91(Part 1):17–39, 1979.

[58] V. O'Brien. Form factors for deformed spheroids in Stokes flow. *AIChE J.*, 14:870–875, 1968.

[59] M. E. O'Neill and K. Stewartson. On the slow motion of a sphere parallel to a nearby wall. *J. Fluid Mech.*, 27:705–724, 1967.

[60] M. E. O'Neill. On asymmetrical slow viscous flows caused by the motion of two equal spheres almost in contact. *Proc. Camb. Phil. Soc.*, 65:543–556, 1969.

[61] M. E. O'Neill and S. R. Majumdar. Asymmetrical slow viscous fluid motions caused by the translation or rotation of two spheres. Part II: Asymptotic forms of the solutions when the minimum clearance between the spheres approaches zero. *Z. Angew. Math. Phys.*, 21:180–187, 1970.

[62] C. W. Oseen. *Hydrodynamik.* Akad. Verlagsgesellschaft, Leipzig, 1927.

[63] O. Pinkus and B. Sternlicht. *Lubrication Theory.* McGraw-Hill, New York, 1961.

[64] M. Smoluchowski. On the mutual action of spheres which move in a viscous liquid. *Bull. Acad. Sci. Cracovie A*, 1:28–39, 1911.

[65] S. Weinbaum, P. Ganatos, and Z. Yan. Numerical multipole and boundary integral equation techniques in Stokes flow. *Ann. Rev. Fluid Mech.*, 22:275–316, 1990.

[66] S. -M. Yang and L. G. Leal. Particle motion in Stokes flow near a plane fluid-fluid interface. Part 1. Slender body in a quiescent fluid. *J. Fluid Mech.*, 136:393–421, 1983.

[67] S. -M. Yang and L. G. Leal. Particle motion in Stokes flow near a plane fluid-fluid interface. Part 1. Linear shear and axisymmetric straining flows. *J. Fluid Mech.*, 149:275–304, 1984.

[68] S. -M. Yang and L. G. Leal. Particle motion in Stokes flow near a deformed fluid-fluid interface. *Intl. J. Multiphase Flow*, 16:597–616, 1990.

[69] B. J. Yoon and S. Kim. Note on the direct calculation of mobility functions for two equal-sized spheres in Stokes flow. *J. Fluid Mech.*, 185:437–446, 1987.

[70] R. H. Yoon and G. H. Luttrell. The effect of bubble size on fine particle flotation. *Mineral Proc. and Extrac. Metall. Rev.*, 5:101–122, 1989.

[71] B. J. Yoon and S. Kim. A boundary collocation method for the motion of two spheroids in Stokes flow: hydrodynamic and colloidal interactions. *Intl. J. Multiphase Flow*, 16:639–650, 1990.

[72] A. Z. Zinchenko. Calculation of hydrodynamic interaction between drops at low Reynolds number. *PMM Applied Math. and Mech.* 42:1046–1051, 1978.

[73] A. Z. Zinchenko. The slow asymmetric motion of two drops in a viscous medium. *PMM Applied Math. and Mech.* 44:30–37, 1981.

[74] A. Z. Zinchenko. Calculation of close interaction between two drops, with internal circulation and slip effect taken into account. *PMM Applied Math. and Mech.* 45:564–567, 1982.

[75] A. Z. Zinchenko. Hydrodynamic interaction of two identical liquid spheres in linear flow field. *PMM Applied Math. and Mech.* 47:37–43, 1984.

[76] A. Z. Zinchenko. Effect of hydrodynamic interactions between the particles on the rheological properties of dilute emulsions. *PMM Applied Math. and Mech.* 48:198–206, 1984.

Part IV

Foundations of Parallel Computational Microhydrodynamics

Chapter 14

The Boundary Integral Equations for Stokes Flow

14.1 The Setting for Computational Microhydrodynamics

Despite the simplicity of the Stokes equations, the extraction of quantitative information for complex geometries, in which the resolution of the particle and boundary shape is an important part of the mathematical model, requires substantial computational resources. Situations where large-scale computations may be avoided were described in Parts II and III: If the particle geometry is simple, *e.g.*, the boundaries fit a coordinate system in which the Laplacian separates, closed-form analytic solutions can be obtained; on the other hand, even for particles of fairly complex shape, the disturbance far away from the particles can be described by a multipole expansion, although the multipole moments in this expansion must be derived from a numerical solution. With regard to the numerical solution, if each particle fits a convenient coordinate system, the multipole boundary collocation scheme provides an accurate solution, with only moderate demands on the computational facility. The basis functions can be represented as distributions of Stokes singularities (inside the particle), which, in some sense, capture the shape of the particle, so that the number of functions needed in the truncated expansion is greatly reduced. (The ultimate limit of this argument would be the use of the exact solution as a basis element, with an adjustable scalar coefficient determined by setting the boundary condition at one point.)

With more complex geometries, such as those shown in Figure 14.1, the only way to capture the details of the particle shape is to bring the Stokes singularities to the particle surface. The resulting class of numerical techniques are known as the *boundary integral methods* (or some variations of this name with the word *equation* added). Their distinct advantage over spatial methods such as finite elements or finite difference is *reduction of dimensionality*. That is, instead of a three-dimensional PDE, we solve a two-dimensional (boundary) integral equation, in which the unknowns are densities of the Stokes singularities

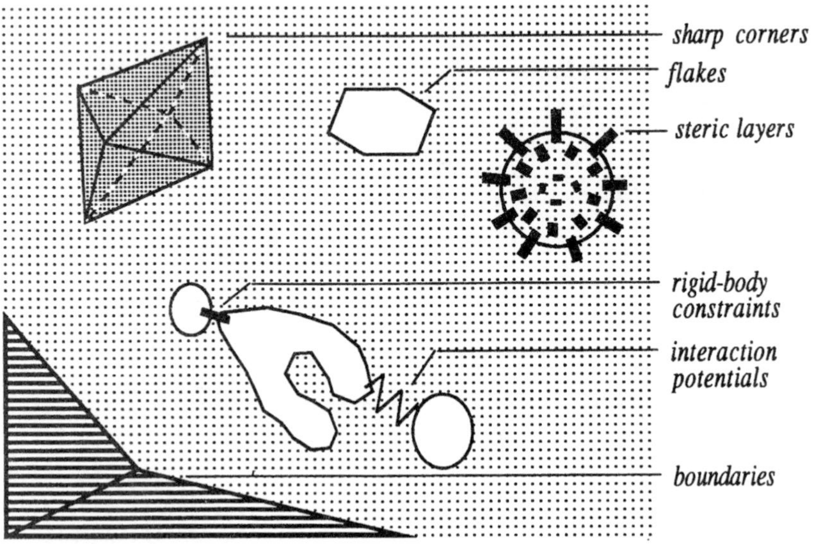

Figure 14.1: Complex microstructures of microhydrodynamics.

distributed over the boundary of the fluid domain, and even infinite fluid domains without a container boundary cause no problems. Furthermore, although the mathematical theory places a smoothness requirement on the surface, this restriction can be loosened in the actual numerical procedure.

By going to two-dimensional formulation, we reduce both the amount of work for discretization and the number of variables needed for getting some required numerical resolution. Especially in time-dependent simulations where the mutual configuration of individual rigid boundaries is changing, no new discretization may be needed after each time step. This is in contrast to spatial methods such as finite elements. Even so, when boundary element methods are applied to problems of interest, we tax even the "state of the art" computers to the fullest. These are, after all, three-dimensional fluid flows with complex geometries and time evolution. Consequently, we would gladly exploit orders of magnitude speedups in computing power.

At first glance, it may appear that our emphasis of boundary methods goes against current trends in computational technology. For example, the emergence of fast, inexpensive processors suggest that the most interesting innovations will occur in parallel computational architectures. Indeed, fundamental physical barriers such as the speed of light and the size of an atom place a limitation on the speed and integration of a single processor. It is unlikely that in the future the speed attainable by using a single processor would be orders of magnitude faster than today. The obvious answer to this dilemma is parallel processing.

It is easy to visualize how some computational problems in areas such as molecular dynamics and operations research can exploit parallel processing. Computational tasks are readily divided into independent pieces. Similarly, in computational fluid dynamics one intuitively senses that the equations from spatial methods, such as the finite element method and the finite difference method, as relations connecting only near neighbors, are also attractive candidates for parallel computers. Boundary methods, on the other hand, lead to dense systems, for the following reason. The (flow) fields emanating from Stokes singularities on one boundary element propagate in all directions with a fairly persistent algebraic decay, so that in some sense, all elements interact with all other elements. The mapping of such dense systems to parallel architectures is a nontrivial problem.

With this environment in mind, the emphasis of Part IV will be the construction of a new boundary integral method, which we call the Completed Double Layer Boundary Integral Equation Method (CDL-BIEM). The name is derived from the fact that Stokes double layer densities are employed, but since the double layers alone do not form a complete basis, a completion procedure is necessary. Using techniques from linear operator theory, we shall prove that the solution of the general mobility problem of N particles in either a bounded or unbounded domain can be cast in terms of a fixed point problem in which the linear operator (or some finite power of it) is a contraction mapping, since the spectral radius is less that one. Furthermore, the three steps in the computational procedure (pre-processing or creation of the system and contraction mapping, iterative solution of the discretized problem, and post-processing to extract the physical variables of interest) can be formulated as parallel algorithms. We will draw upon the developments of Parts II and III to show that for many-body suspension problems, the interactions between double layer elements on different particles are very small, in the sense that the eigenvalue(s) determining the spectral radius of the operator are perturbed only slightly by hydrodynamic interactions. Thus we raise the possibility of attacking large-scale suspension simulations by a network of parallel processor, with each processor powerful enough to handle all boundary elements for a given particle. *In essence we recover the nearest neighbor property of the spatial methods — only now the fundamental entity is not an element or node, but the collection of elements for the particle.* For many-body problems, the resulting iterative strategy will involve asynchronous iterations accelerating the convergence relative to synchronous iterations similarly as the Gauss–Seidel iterations do in comparison with Gauss–Jacobi iterations.

This chapter provides a sufficient background for comprehension of the material covered, to the extent that the reader should be able to apply the theory and algorithms presented for obtaining numerical results. We go over Fredholm integral equations of the first and second kind, and some smoothness requirements for the kernels, which for our application imply smoothness requirements for the surface. The kernels encountered are *weakly singular*, so that an extension of the Fredholm theory, namely, the *Fredholm–Riesz–Schauder* theory for compact operators, is necessary. The fact that the integral operators are compact will be exploited in two ways. First, it implies that CDL-BIEM is a well-posed problem. Second, the spectrum of a compact operator is discrete, which in our case will be used to show that, even before discretization, the integral equations of CDL-BIEM can be solved by Neumann series expansion, in practice replaced by direct (Picard) iterations (also called successive substitution) corresponding to truncating the Neumann series (see the exercises).

14.2 Integral Operators and Integral Equations

14.2.1 Motivation

Although integral equations seem to be unduly slighted in courses of advanced calculus, they have turned out to be very useful both for theoretical and numerical purposes. In this chapter we shall give a brief introduction (with references for more detailed study) to both the theory and practical numerical solution of integral equations. The theory, roughly speaking, is such that anyone familiar with ordinary matrices should feel comfortable with it.

We summarize here only the main properties of those types of IEs that we shall encounter. These properties may seem abstract and "too mathematical" for engineers, but this should be only a temporary impression. The existence and uniqueness of a solution are significant for practical computation also. It would be a waste of time to look for a solution that does not exist, and multiple solutions can make a linear system hard to treat numerically.

14.2.2 First- and Second-Kind Equations

In an integral equation the unknown function appears within an integral. We shall only be concerned with linear integral equations. The Fredholm equations of the first and second kind are of the form

$$\lambda x(s) - \int_I K(s,t)x(t)\,dt = g(s)\ , \quad \text{for } s \in I, \tag{14.1}$$

where λ is a given constant, g and K are given functions (usually, the *kernel* K is assumed to be continuous), x is to be solved for, and I is a finite interval for the one-dimensional case. If $\lambda = 0$, the equation is of the first kind, otherwise it is of the second kind. In the latter case the equation can be divided by λ, so that we see the identity operator acting on x; this identity operator is the main reason for distinguishing between the first- and second-kind equations. Usually the functions g and x are assumed to be continuous (see Whittaker and Watson [69]). We shall need an extension of this theory to *weakly singular* kernels, and this is provided by the *Fredholm–Riesz–Schauder* theory. It covers the abstract second-kind equation

$$\lambda x - \mathcal{K}x = g\ , \quad (\lambda \neq 0)\ , \tag{14.2}$$

where $\mathcal{K}$ is a *compact* (also called *completely continuous*) linear operator. Henceforth it will always be tacitly assumed that any *operator* under discussion is linear.

We shall limit our attention to the case of operators acting on a Hilbert space, which will be that of square summable functions L_2 in the application of the theory. More general results for Banach spaces, *etc.*, can be found in the references.

14.2.3 Weakly Singular Kernels

Let Ω be a bounded measurable m-dimensional set in Euclidean space, like a region in m-dimensional space or a surface in $m+1$-dimensional space. A *weakly singular kernel* is of the form

$$K(s,t) = \frac{A(s,t)}{r^{\alpha}}, \text{ for } s,t \in \Omega, \tag{14.3}$$

where r is the Euclidean distance $|s-t|$, A is continuous, and $0 \leq \alpha < m$. Such kernels are also called *polar* or *potential* kernels. Mikhlin [51, 52] shows that the corresponding integral operator, mapping a function x to

$$\int_{\Omega} K(s,t)x(t)\, dt, \tag{14.4}$$

is compact on $L_2(\Omega)$ and also on $C(\Omega)$ (the space of continuous functions equipped with sup-norm). In the integral above, dt signifies integration with respect to the measure on Ω, and in the case of a surface we have a surface integral. Note that by setting $\alpha = 0$, we recover the usual continuous kernels, and so the theory of weakly singular kernels extends the ordinary theory. Also Petrovskii [55] covers weakly singular kernels in a similar manner as Mikhlin. A more advanced and more demanding presentation is given by Kantorovich and Akilov [36].

The weakly singular operator has the following properties not implied by its compactness. It maps bounded functions to continuous, and for continuous data g, any solution $x \in L_2(\Omega)$ of the second-kind equation is also continuous. The latter also implies that all eigenfunctions in $L_2(\Omega)$ are continuous.

14.2.4 Ill- and Well-Posedness

The definition of this classification for linear problems dates back to Hadamard (1932). He stated that a linear problem $Ax = b$ is well-posed if it is uniquely solvable for any data b, and the solution x depends continuously on the data. More concisely, the operator A must have a well-defined bounded inverse. Linear problems that are not well-posed are called ill-posed.

Originally it was envisioned that any physical problem should have a well-posed formulation, but physically reasonable ill-posed problems have been encountered quite frequently. From a philosophical point of view, well-posed problems are such that the mathematical model has no serious difficulties in dealing with them. Ill-posed problems, on the other hand, are somewhat "incompatible" with the model used, the solution being excessively sensitive to any variations in the data. As in numerical computations, we may be relying on approximations or measured quantities, and in any case there will be some round-off errors in floating-point numbers; such excessive sensitivity can disturb the calculations badly. Typically the computed result must be considered only intermediate in character — only some projection of it is reliable. Some post-processing that uses just this projection may remove the effects of the "indeterminate part," and give quite accurate results.

There are two standard ways for dealing with ill-posedness. The first is to impose some constraints on the solution x and look for the (best approximate) solution satisfying the constraints. This corresponds to using some extra information not previously utilized to get rid of the indeterminate part of the solution. The second method is to construct a sequence of well-posed problems that approximate, in some sense, better and better the original ill-posed problem. Typically this is done with so-called Phillips–Tikhonov regularization, by adding a "regularizing functional" and thus changing the equations.

In summary, we would like our problems to be well-posed, or at least be aware of the ill-posedness of a formulation, since some difficulties can be expected. An ill-posed problem may suggest that we are computing results that are only intermediate in character, or that the model used is inappropriate.

14.2.5 Compact Operators

In addition to the references in the previous sections, the reader may wish to consult the books by Dieudonné [14], Friedman [21], Ramkrishna and Amundson [60], and Rudin [62]. Both Friedman and Dieudonné give self-contained expositions starting from the fundamentals. In addition, the book by Ramkrishna and Amundson contains applications of linear operator theory to transport phenomena and reactor design problems.

Several properties of compact operators on a Hilbert space H shall be listed here. The set of compact operators is a subspace of the bounded operators, closed with respect to the operator norm. Multiplication with a bounded operator from left or right preserves compactness, as does the involutive operation of taking the adjoint. The adjoint of a bounded operator $\mathcal{K}$ is the unique operator $\mathcal{K}^*$ satisfying the following "flipping rule" within the inner product:

$$\langle \mathcal{K}x, y\rangle = \langle x, \mathcal{K}^* y\rangle, \text{ for all } x, y \in H. \tag{14.5}$$

Inner products, if complex, will be conjugate linear with respect to their *second* argument. In linear algebra the adjoint of a matrix is the complex conjugate of the transpose matrix, and this should not be confused with the *classical adjoint* whose elements are cofactors of the elements in the original matrix.

The identity operator I is compact, iff (if and only if) H is finite-dimensional. From this it follows that a compact $\mathcal{K}$ will not have a bounded inverse if $\dim(H) = \infty$, for assuming the converse, $I = \mathcal{K}\mathcal{K}^{-1}$ would be compact as

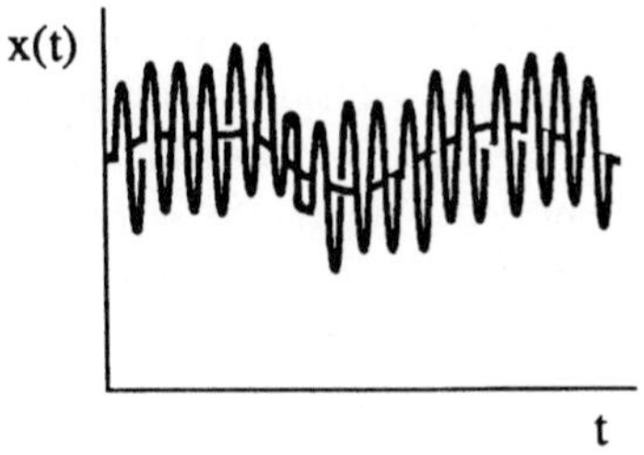

Figure 14.2: Fredholm integral equations of the first kind are ill-posed.

the product of a compact and a bounded operator. This shows why usually little space is devoted to discussing the first-kind equation $\mathcal{K}x = g$. Even if this relation can be inverted to $x = \mathcal{K}^{-1}g$, the unboundedness of $\mathcal{K}^{-1}$ implies that a small change in the data g can result in an arbitrarily large change in the solution x. As this type of excessive sensitivity of the solution to variations in the data is commonly called ill-posedness, *first-kind equations are ill-posed.* The term *compact* describes well the "smallness" of these operators, corresponding to the "largeness" of the inverse operators. In practice the ill-posedness of the first-kind equations manifests itself when a numerical solution is attempted using the discretized linear system. The linear system will be ill-conditioned, increasingly so with finer discretizations.

The situation is illustrated in Figure 14.2. Assume that the smooth curve is the desired solution of the integral equation. Now superimpose on this curve a high-frequency oscillation. The integral operator smooths out such oscillations so both functions get mapped to similar outputs. The inverse problem, of determining the input given the output, must be ill-posed. When very fine discretizations are employed, these high-frequency oscillations can exist in the numerical solution and the corresponding symptom in the discretized problem, *ill-conditioning*, is observed. Later on (Chapter 17) we shall derive explicit examples of this effect as applied to boundary integral equations. On the surface of a sphere, oscillatory inputs of the form $P_n(\cos\theta)$ will map to small outputs that vanish for large n as $O(n^{-1})$.

For second-kind equations the situation is dramatically different. Here the smallness of compact operators is an asset, implying that they only somewhat perturb the identity operator in the equation. As a result, the second-kind equations share several properties with ordinary square matrices, which is the essence of the Fredholm–Riesz–Schauder theory. Often some of the results listed below are termed *Fredholm theorems.* The *Fredholm alternative* states that either the second-kind equation is uniquely solvable for any data g, or else the corresponding homogeneous equation (found by setting $g = 0$) has nontrivial solutions. In the latter case these nontrivial solutions are called *eigenvectors* or *eigenfunctions* corresponding to the *eigenvalue* λ, and if g happens to allow a solution x, other solutions are found by adding an eigenvector to x; thus solutions are not unique. This alternative is sometimes stated as "1-1 implies onto," meaning that uniqueness of solutions implies solvability for any data. Moreover, if the second-kind operator is invertible, or equivalently λ is not an eigenvalue, the inverse is bounded. Therefore *second-kind equations are well-posed*, although special treatment is needed when an eigenvalue is encountered. Later on we shall introduce a *bordering* scheme to handle these situations.

Suppose now that $\lambda \neq 0$ is an eigenvalue; the statements that follow are not necessarily valid for the case of a first-kind operator with $\lambda = 0$. The corresponding eigenvectors form the *null space* $N(\lambda I - \mathcal{K})$ and are also called *null vectors* (of the second-kind operator). The second-kind equation is solvable iff g can be represented as $\lambda x - \mathcal{K}x$, in other words, iff g is in the *range* $R(\lambda I - \mathcal{K})$. The null space is always finite dimensional, so that there is a finite basis of eigenvectors for each eigenvalue. Just like matrices may be nondiagonalizable, having nontrivial Jordan blocks and *principal vectors* other than eigenvectors (in the terminology used by Householder [31]), a compact operator also can have these; also the term *principal function* may be used. In case a matrix is diagonalizable, its Jordan blocks being trivial, the Jordan canonical form actually gives the diagonal matrix achievable with similarity transformations. Principal vectors are such that repeated application of the second-kind operator on them will at some step give a zero result. All the nonzero vectors that come up during this process form a *Jordan chain*, in which the last vector is an eigenvector. Thus to each eigenvector there corresponds a Jordan chain and Jordan block (in the matrix representation with the principal vectors chosen as the basis), and if the Jordan chain consists of the eigenvector alone, the Jordan block is trivial. The Jordan blocks of a compact operator are all finite, and for a fixed eigenvalue there is only a finite number of these, as the number of linearly independent eigenvectors is finite.

The null spaces of the adjoint operators $\lambda I - \mathcal{K}$ and $\lambda^* I - \mathcal{K}^*$ (λ^* is the complex conjugate of λ) are of equal dimension (the same goes for the Jordan blocks in general), and further they contain all the information about the ranges that are closed subspaces:

$$R(\lambda I - \mathcal{K}) = N(\lambda^* I - \mathcal{K}^*)^{\perp}. \tag{14.6}$$

The second-kind operator and its adjoint can be switched here, since applying the operation of taking the adjoint twice to a bounded operator recovers the original (this, together with conjugate linearity and preserving the norm, makes the adjoint operation involutive), just as with matrices. From the equation above it follows that $g \in R(\lambda I - \mathcal{K})$ and a solution x exists, iff g is orthogonal to the (finite basis of the) null functions of the adjoint $\lambda^* I - \mathcal{K}^*$. This also shows that the *codimension* of the range, meaning the dimension of its orthogonal complement, equals the dimension of the null space, for these second-kind operators. We could also say that the range has a deficiency equal to the dimension of the null space.

The observations above are crucial for the applications to be presented. Firstly we know that even if the second-kind operator is not invertible due to us having encountered an eigenvalue, there will only be a finite number of independent eigenvectors that we need to consider, and the number of these will indicate the deficiency in the range. Secondly, solving the homogeneous adjoint problem will completely describe "the shape of this deficiency," *i.e.*, what is missing from the range.

The term *Jordan block* is used in none of the references listed, but we use it here in order to show the similarity with ordinary matrix algebra and to help interpret the results. The Jordan canonical form of a matrix is explained in Gantmacher [23]. Familiarity with the Jordan form is also useful later on, when it is used to help visualize Wielandt's deflation.

As a compact operator is bounded, so is the set of its eigenvalues. A much stronger statement can be made, namely, there is at most a denumerable set of eigenvalues, and the only possible accumulation point of them is zero. Therefore each nonzero eigenvalue is isolated: It has a neighborhood within which no other eigenvalues are found. Knowledge of this will be useful on considering the spectral radius, which determines whether an iterative solution of a linear system of equations is possible. The set of eigenvalues of a compact operator will also be called the *spectrum* of the operator; for operators in general the spectrum is so defined that it may contain numbers other than eigenvalues. The result above, characterizing the spectrum of a compact operator, can be understood in the following way. Suppose we had a nonzero eigenvalue repeated an infinite number of times. This would mean that in an infinite-dimensional subspace the operator acts as a multiple of the identity, which is not compact. Thus the operator itself would not be compact. The situation remains essentially the same if we have an infinite number of eigenvalues in any neighborhood of one accumulation point; this also is not allowed for a compact operator. Point zero is an exception, as zero multiple of the identity is just zero.

14.2.6 Numerical Solution Methods

In an IE, basically three discretizing approximations are needed to get an algorithmic method ready for computer implementation. Approximate the solution function with another that depends on a finite set of parameters, replace the infinite process of integration with a finite computable approximation, and finally the continuum of equations that resulted has to be dealt with. In practice the first two discretizations here are coupled and, to some extent, performed "in one step."

In general for approximating a function with another of fixed form, say, a polynomial, different criteria may be used to determine the free parameters, such as the coefficients of a polynomial. In *collocation*, a functional equation is enforced exactly at a (finite) set of (collocation) points. In *osculation*, not only are the values of the function enforced, but also some derivatives at some points. The *least squares method* minimizes a number describing the error somehow on an average, and the *min-max* method strives to minimize the largest error. In accordance with this terminology, the following procedure shall be called *the quadrature collocation method.* Consider a linear IE of the first or second kind, and approximate the integration (infinite process) by a finite quadrature (*e.g.*, Gaussian), as in

$$\int_0^1 u(t)\,dt \approx \sum_{i=1}^{N} \omega_i u(t_i) , \tag{14.7}$$

where ω_i and t_i are the quadrature weights and points, respectively. Now the discretization of the first-kind equation is

$$\sum_{i=1}^{N} \omega_i K(s,t_i) X(t_i) = g(s) , \tag{14.8}$$

where X is an approximation for x. As only a finite set of values of X is involved, the quadrature has discretized also the unknown function simultaneously with the integration. There are only N values $X(t_i)$ to be determined, so 14.8 cannot be valid for all s (in general). Instead one can enforce the equation exactly at N collocation points:

$$\sum_{i=1}^{N} \omega_i K(s_j,t_i) X(t_i) = g(s_j) \quad , \text{for } j = 1, \ldots, N . \tag{14.9}$$

This is a typical linear system. With similar treatment of the second-kind equation it would be necessary to choose the collocation points to be the quadrature points, $s_j = t_j$, since X comes up both alone with arguments s_j and inside the quadrature with t_j. The choice of collocation points to be the quadrature points is also beneficial if the kernel K is symmetric, $K(t,s) = K(s,t)$, since then the discrete equation system (for the unknowns $\omega_i X(t_i)$) will also be symmetric. Utilization of this symmetry will clearly bring savings in numerical computations, both in storage space and through the use of more efficient routines for this special case.

Consider now an expansion of the solution in terms of some trial (basis) functions, which might be, *e.g.*, the eigensolutions of a related problem, or piecewise polynomials. As numerical integration is fairly expensive computationally, a set of basis functions that allows doing the integrations with the kernel analytically would be preferable. If such basis functions are not known, numerical integrations are necessary, but the way in which these are carried out has no particular effect on the next step here. The discretization of the function leaves us (after the integrations) still with the discretization of the "continuum" of equations, as in going from 14.8 to 14.9 above. This may be done by any of the methods mentioned for approximating a function with another, depending on a finite set of parameters. The "quadrature step" 14.8 above (which discretized both the function and the integral) has been replaced with

$$x(t) \approx \sum_{i=1}^{N} c_i \phi_i(t) = X(t) . \tag{14.10}$$

Substitution into the Fredholm equation of the first kind gives

$$\int_0^1 K(s,t) \sum_{i=1}^{N} c_i \phi_i(t)\, dt = g(s) \ . \tag{14.11}$$

Choosing collocation points s_j would not preserve the symmetry of the kernel K in the discretized equations, so the *Galerkin method* shall be introduced here. This means that s is integrated out of Equation 14.11 with weight functions, and, moreover, often the weight functions are chosen to be those in the basis function set, *i.e.*, the ϕ_is. Then

$$\int_0^1 \int_0^1 K(s,t) \sum_{i=1}^{N} c_i \phi_i(t) \phi_j(s)\, dt\, ds = \int_0^1 g(s)\phi_j(s)\, ds \ , \tag{14.12}$$

which, in case K is symmetric, is a symmetric linear system for the unknowns c_i. For unsymmetric kernels the advantage of Galerkin method is lost as far as the symmetry of the resulting system is regarded; it will not create symmetry that was not there to start with. Furthermore, this method requires an extra integration compared with collocation. Thus it may be anticipated that collocation methods will in general be superior to Galerkin methods computationally, unless the choice of basis functions has made the required integrations numerically less difficult.

When one has no prior knowledge of the shape of the solution x, it seems natural to use piecewise polynomial approximation in the basis function approach. Then each basis function will get nonzero values only within some subinterval of the whole range of the integral equation, and this division into subintervals may be compared with the division into elements for a FEM solution; the basis functions can also naturally be called shape functions. The simplest type of approximation would be to use piecewise constant shape functions, and collocation at the centers of the subintervals (or elements). This approach is also the easiest to implement numerically, but by using so crude an approximation to the solution some accuracy will be lost. For a more detailed discussion on this subject, we refer the reader to the papers by Cruse [10, 11].

In fact, a piecewise constant approximation will be used later on with the three-dimensional codes. The reason in that case is that there are three significant advantages in using planar elements for approximating the boundary surfaces. As the approximation of the boundaries is "rough," restricting the accuracy of the solutions, it would be pointless to use very accurate representation of the functions solved for. In addition, matrix generation with such elements is more readily mapped to parallel architectures. Lastly, post-processing for physical quantities such as surface tractions is more straightforward.

14.2.7 Applications to Boundary Integral Equations

In general there will be more than one integration variable in the surface integrals that come up in hydrodynamic problems, although, as will be shown later, some problems with axisymmetric boundaries can be reduced to systems of integral equations with just one integration variable.

The quadrature collocation scheme introduced for one-variable problems could still be applied in the multi-dimensional case. For surface integrals there is not much choice in choosing the quadrature rule; typically for two-dimensional integrations one uses a product rule found by applying one-dimensional quadratures to the iterated integral. Furthermore, it is not very wise to try using, say, very high order Gaussian quadratures. Instead one should split the interval of integration to subintervals and apply a "decent" order quadrature on each of these separately. However, with the weakly singular kernels here, a simple quadrature collocation scheme would run into difficulties, since at any fixed collocation point there is a singularity that prevents the direct evaluation of the kernel. This situation can be remedied by using the *singularity subtraction* technique.

The basis function approach has so far been overwhelmingly more popular in applications. For surface integral equations, this leads to what is now called the *boundary element method*, a term initially coined by C. Brebbia in applications to solid structures. The procedure is to divide the surface(s) of interest to small elements, over each of which some low-order approximation for the solution can be applied with good conscience; the elements will also give a natural approach to calculating the surface integrals by the use of (perhaps modified) product rule integration on each of these. A modification of the quadrature will be necessary in particular for the element in which the singularity corresponding to the collocation point is located. Naturally it is not necessary to use piecewise polynomials for the basis functions, but use of some "global" approximating functions can only be justified when the shape of the results to be expected is known *a priori*, and still generating a quadrature rule for the surface would have to be accomplished probably by subdivision into elements. Thus the boundary element formulation seems to be the most generally applicable approach in sight.

In one-dimensional cases the subdivision of the integration interval is not a big problem, whereas for surfaces the division into elements would require some algorithmic approach. This is probably the reason why, in spite of its generality, the boundary element method, at least in fluid mechanics, seems to have been applied only to combinations of geometrically simple topologies where some natural coordinate systems allow for easy grid generation. Actually triangulations of surfaces can be refined fairly easily by using a subroutine that generates a surface point approximately "midway between" two given ones; with such a routine available it is straightforward to transform one triangle into four smaller ones fitting the surface more closely by forming new edges that connect the approximate midpoints of the original edges and "twisting" the original edges. Starting from an initial coarse triangulation, it is then possible to generate triangulations with desired resolution of the surface.

14.3 Notation and Definitions

In the following the Stokes equations in fluid domains that are regions in the three-dimensional Euclidean space shall be considered. A *region* is, as usual, an open connected set, meaning that each point of a region has some neighborhood wholly contained within the region, and any two points of the region can be joined by a curve that also lies within the region. The boundary of the fluid domain is denoted by S, and it is further required that each connected component of S be a closed bounded *Lyapunov-smooth surface* (see Günter [27]); this ensures, for example, that the (Gauss) divergence theorem can be applied to bounded fluid domains.

In practice surfaces without edges or corners usually are Lyapunov-smooth. Without going into great detail (see the reference above), it is just required that the surface normal changes its direction in a certain uniformly continuous fashion as we move along the surface, and further that "opposing parts" of the surface are uniformly separated from each other. Sometimes, we will omit the "-smooth" part of the terminology.

The fluid domain will also be called the *interior* and the complement of it and its boundary the *exterior*, although this conflicts with the mathematical term *interior* denoting the largest open set within *any* given set. Each single closed Lyapunov surface divides the space into a bounded and an unbounded region; these are called the *inside* and the *outside*, respectively. A boundary surface restricting the fluid domain to its outside (inside) will sometimes be called a *particle* (*container*). It is clear that for a connected fluid domain there can be at most one container boundary.

The unit normal on S pointing *into the interior fluid domain* will be denoted by $\hat{\boldsymbol{n}}$, and it has the Cartesian components n_i; optionally the surface point determining $\hat{\boldsymbol{n}}$ will be indicated as its argument. To keep the notation explicit, the symbols $\boldsymbol{\xi}$ and $\boldsymbol{\eta}$ shall be used for the position vectors of surface points on S, with corresponding components ξ_i and η_i. Vectors corresponding to points not on S will be denoted by $\boldsymbol{x}$, $\boldsymbol{y}$, *etc.* The Euclidean distance of two points, say, $\boldsymbol{x}$ and $\boldsymbol{y}$, is denoted by r_{xy} or $|\boldsymbol{x}-\boldsymbol{y}|$, and the corresponding vector is $\boldsymbol{r}_{xy} = \boldsymbol{y} - \boldsymbol{x}$ (read "vector from x to y"). The Euclidean norm of a single vector $\boldsymbol{x}$ will also be denoted by x.

The *natural inner product* is defined for square integrable vector surface fields on S as the over S integrated pointwise dot product of these fields. With respect to this inner product, the adjoint kernel to any given kernel $\boldsymbol{K}$ is $\boldsymbol{K}^*$ defined by

$$\boldsymbol{K}^*(\boldsymbol{\eta},\boldsymbol{\xi}) = \boldsymbol{K}^t(\boldsymbol{\xi},\boldsymbol{\eta}) \ , \tag{14.13}$$

so that transformation with $\boldsymbol{K}$ inside an inner product may be flipped to the other side of the inner product by changing the transformation to $\boldsymbol{K}^*$. This corresponds to the transpose of a matrix and its relation to the ordinary dot product (a real inner product) of finite vectors. For complex inner products one also needs to complex conjugate the transpose, to create the adjoint. As with matrices, the flipping rule is verified by exchanging the order of finite summations (Exercise 16.1); with the natural inner product here, one needs to exchange the order of the implied integrations. From now on an inner product will be understood to be the natural inner product; deviations from this practice will be clear from the context.

In Chapter 2 it was shown that the rate of work done by a moving solid particle equals the energy dissipation in the fluid domain. More generally the equations used there show that the dissipation equals the rate of work done on the boundaries of the fluid domain, the latter being defined as the pointwise inner product of the velocity with the surface tractions integrated over the boundary surfaces with sign change:

$$\int_V \boldsymbol{\sigma} : \boldsymbol{e} dV = \oint_S \boldsymbol{v} \cdot (-\boldsymbol{\sigma} \cdot \hat{\boldsymbol{n}}) dS \, . \tag{14.14}$$

Recall that $(\boldsymbol{\sigma} \cdot \hat{\boldsymbol{n}})dS$ is the force by the fluid on the surface element, while here we use the reaction force on the fluid; this causes the sign change. This result, equating the dissipation with a natural inner product of the velocity and traction fields, will be referred to as the *energy relation*, and will be frequently used in the following sections.

14.4 The Boundary Integral Equation in the Primary Variables

Let us now consider the homogeneous Stokes solutions for problems without a body-force field. In Chapter 2 the following integral representations for the Stokes field pair $(\boldsymbol{u}, p)$ in the interior $Q_{(i)}$ were obtained:

$$\begin{aligned} \boldsymbol{u}(\boldsymbol{y}) &= -\frac{1}{8\pi\mu} \oint_S \mathcal{G}(\boldsymbol{y}-\boldsymbol{\xi}) \cdot (\boldsymbol{\sigma}(\boldsymbol{\xi};\boldsymbol{u}) \cdot \hat{\boldsymbol{n}}(\boldsymbol{\xi})) \; dS(\boldsymbol{\xi}) \\ &\quad - \oint_S \{\hat{\boldsymbol{n}}(\boldsymbol{\xi}) \cdot \boldsymbol{\Sigma}(\boldsymbol{y},\boldsymbol{\xi})\} \cdot \boldsymbol{u}(\boldsymbol{\xi}) \; dS(\boldsymbol{\xi}) \; , \text{ for } \boldsymbol{y} \in Q_{(i)} \qquad (14.15) \\ p(\boldsymbol{y}) &= -\frac{1}{8\pi\mu} \oint_S \mathcal{P}(\boldsymbol{y}-\boldsymbol{\xi}) \cdot (\boldsymbol{\sigma}(\boldsymbol{\xi};\boldsymbol{u}) \cdot \hat{\boldsymbol{n}}(\boldsymbol{\xi})) \; dS(\boldsymbol{\xi}) \\ &\quad - \frac{1}{4\pi} \oint_S \{\hat{\boldsymbol{n}}(\boldsymbol{\xi}) \cdot \nabla_y \mathcal{P}(\boldsymbol{y},\boldsymbol{\xi})\} \cdot \boldsymbol{u}(\boldsymbol{\xi}) \; dS(\boldsymbol{\xi}) \; , \text{ for } \boldsymbol{y} \in Q_{(i)} \; , (14.16) \end{aligned}$$

and it was noted that these representations give zeroes in the exterior $Q_{(e)}$. We mentioned the jump discontinuity across the boundary, and in the velocity representation 14.15 this must come from the last term. Indeed, $\mathcal{G}(\boldsymbol{y}-\boldsymbol{\xi}) = O(1/r_{y\xi})$ implies that in local polar coordinates about the surface point $\boldsymbol{\eta}$ that $\boldsymbol{y}$ passes through, the first term on the RHS has no essential singularity; so the jump must be from the last term. The analogy with electrostatics has been mentioned before. The electrostatic potential of a surface charge distribution is continuous across the boundary, while that of a surface dipole distribution (the electric double layer) suffers a jump (the idea behind a dry cell battery).

Assume now that the last term exists as an improper integral when we substitute $\boldsymbol{y} \leftarrow \boldsymbol{\eta}$. This is true for sufficiently smooth surfaces such as Lyapunov

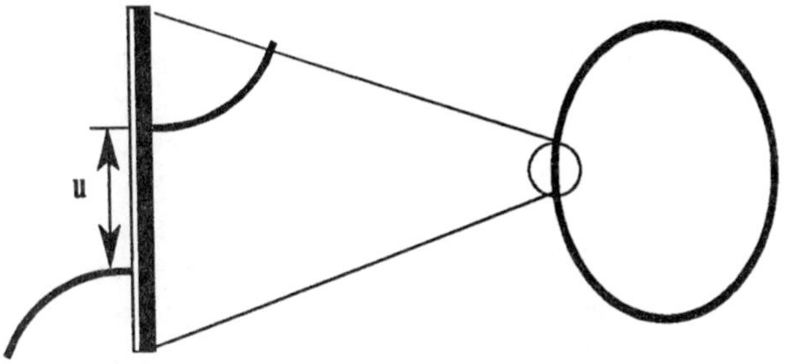

Figure 14.3: Jump in $\boldsymbol{v}$ for a double layer distribution.

surfaces, even though $\boldsymbol{\Sigma}(\boldsymbol{\eta},\boldsymbol{\xi}) = \mathrm{O}(r_{\xi\eta}^{-2})$, because $\hat{\boldsymbol{n}}(\boldsymbol{\xi}) \cdot \boldsymbol{r}_{\xi\eta}^{0} = \mathrm{O}(r_{\xi\eta}^{\alpha})$ as $\boldsymbol{\xi} \to \boldsymbol{\eta}$, with $\alpha \in (0,1]$. Assuming $\boldsymbol{u}$ to be smooth, we see that the integrand is $\mathrm{O}(r_{\xi\eta}^{\alpha-2})$, which is the reason why the kernel in this term is weakly singular and integrable in the local polar coordinates about the singular point.

As $\boldsymbol{y}$ approaches $\boldsymbol{\eta}$ from either side, the surface will locally look planar, and since the jump is due to the immediate neighborhood of $\boldsymbol{\eta}$, it can be claimed by symmetry that the jump is equal in magnitude whether $\boldsymbol{y}$ comes to sit on the surface from $Q_{(i)}$ or $Q_{(e)}$. Therefore, on substituting $\boldsymbol{y} \leftarrow \boldsymbol{\eta}$ in Equation 14.15, the RHS will attain the value symmetrically halfway, as shown in Figure 14.3. This leads to the equation

$$\begin{aligned} \frac{1}{2}\boldsymbol{u}(\boldsymbol{\eta}) &= -\frac{1}{8\pi\mu}\oint_S \mathcal{G}(\boldsymbol{\eta}-\boldsymbol{\xi}) \cdot (\boldsymbol{\sigma}(\boldsymbol{\xi};\boldsymbol{u}) \cdot \hat{\boldsymbol{n}}(\boldsymbol{\xi}))\, dS(\boldsymbol{\xi}) \\ &\quad - \oint_S \{\hat{\boldsymbol{n}}(\boldsymbol{\xi}) \cdot \boldsymbol{\Sigma}(\boldsymbol{\eta},\boldsymbol{\xi})\} \cdot \boldsymbol{u}(\boldsymbol{\xi}) dS(\boldsymbol{\xi}) \text{ , for } \boldsymbol{\eta} \in S \, . \end{aligned} \tag{14.17}$$

This is a *boundary integral equation* (BIE) for the surface tractions when the velocity field is known on the surface, or for the velocity field on the surface when the tractions there are known. The kernels, explicitly $\mathcal{G}(\boldsymbol{\eta}-\boldsymbol{\xi})$ and $\hat{\boldsymbol{n}}(\boldsymbol{\xi}) \cdot \boldsymbol{\Sigma}(\boldsymbol{\eta},\boldsymbol{\xi})$, are weakly singular and integrable. For the tractions the integral equation is an *equation of the first kind*, since these are only integrated with the kernel as a weight. For the velocity surface field we have an *equation of the second kind*, since in addition to being integrated with a kernel, $\boldsymbol{u}$ comes up also as such in the equation.

A rigorous derivation of the equation above, for Lyapunov surfaces and $\boldsymbol{u}$ continuous, can be given by considering about $\boldsymbol{\eta}$ a Lyapunov sphere, which in the limit is shrunk to a point. This type of technique is applied to problems of potential theory in Günter [27], and also Ladyzhenskaya [43, Chapter 3] employs it. On a spherical or hemispherical surface S about $\boldsymbol{\eta} = \mathbf{0}$ one can write

$$\begin{aligned} &\oint_S \hat{\boldsymbol{n}}(\boldsymbol{\xi}) \cdot \boldsymbol{\Sigma}(\boldsymbol{\eta},\boldsymbol{\xi})\, dS(\boldsymbol{\xi}) = \oint_S \frac{3}{4\pi}\hat{\boldsymbol{n}}(\boldsymbol{\xi})\boldsymbol{\xi} R^{-3}\, dS(\boldsymbol{\xi}) \\ &= R^{-3}\int_Q \frac{3}{4\pi}\nabla\boldsymbol{x}\, dQ(\boldsymbol{x}) = R^{-3}\int_Q \frac{3}{4\pi}\boldsymbol{\delta}\, dQ(\boldsymbol{x}) = V/\left(\frac{4}{3}\pi R^3\right)\boldsymbol{\delta}\,, \end{aligned}$$

where R is the radius and V is the volume of the region Q enclosed by this surface. This calculation tacitly utilized the fact that the planar part of the hemispherical surface does not contribute to the surface integrals. So the result is one (half) for a (hemi)sphere. This is the key result required in the limiting procedure explained, and further details shall not be given here.

An extension to infinite fluid domains is possible by assuming that we consider the *disturbance field* of a particle or particles that decays to zero at a sufficient rate as infinity is approached. From Lamb's general solution, valid for a fluid domain outside a sphere (see Happel and Brenner [29]) enclosing the particles, it is seen that a decaying velocity field will have at least $O(|\boldsymbol{x}|^{-1})$ as the decay rate. The corresponding stresses behave as the derivatives of the velocity field and will decay at least as $O(|\boldsymbol{x}|^{-2})$, except for an arbitrary constant base pressure, which is at first assumed so chosen that this decay rate is obeyed. Consider now another larger sphere enclosing the particles, and as the fluid domain use the volume enclosed in between. A simple order of magnitude argument shows that on the particle surfaces the contribution to the integrals from the larger enclosing sphere S_c will become negligible as we let it expand to infinity. Letting R be the radius of this sphere, observe that the surface S_c is $O(R^2)$, and use the estimates above together with similar estimates for $\mathcal{G}$ and $\boldsymbol{\Sigma}$ in 14.17 with $\boldsymbol{\eta}$ on the particles and $\boldsymbol{\xi}$ on S_c. In summary *if a homogeneous Stokes solution is decaying as infinity is approached, then its decay rate is sufficient for neglecting the "boundary at infinity" in the integral equation above.* Since in a bounded domain, like the inside of the larger sphere, we can arbitrarily set the base pressure without affecting the result from the integral involving the tractions in 14.15, we can in fact allow any base pressure with the disturbance fields and still use the above "enclosing sphere expanded to infinity argument" with 14.15 and 14.17.

Later on also, the energy relation 14.14 shall be used in unbounded domains. For this case the base pressure needs more careful consideration. If the pressure is chosen to become zero at infinity, a similar order of magnitude estimate as above shows that the "enclosing sphere expanded to infinity argument" allows us to again neglect the "surface at infinity." Now adding a base pressure to these tractions on the enclosing sphere will result in a term

$$\oint \boldsymbol{u} \cdot \hat{\boldsymbol{n}}\, dS \tag{14.18}$$

evaluated over this container surface, which will vanish only in a particular case. It vanishes iff the same integral evaluated over the totality of particle surfaces vanishes, due to the velocity field being mass-conserving in the fluid domain in between. In summary *the energy equation 14.14 can be applied to decaying homogeneous Stokes solutions in infinite fluid domains, neglecting the "surface at infinity," when either the pressure vanishes at infinity, or the total generation of fluid within the particle surfaces is zero.* The Lorentz reciprocal theorem can be extended to decaying Stokes fields in infinite domains with similar restrictions; now the nonvanishing of the pressure of one flow field at infinity is allowed if *the other* flow field has no net sources or sinks within the particle surfaces.

We need not consider only disturbance fields in infinite fluid domains, but we can also assume that before any particles were inserted into the space there was some *ambient flow* satisfying the Stokes equations. The change needed to produce the flow field after inserting some particles will then serve as the decaying disturbance field to which the theory above is applicable.

Let $\boldsymbol{u}$ now be the disturbance field cancelling an ambient field $\boldsymbol{v}$ on the particle surfaces. Then $\boldsymbol{u}$ satisfies Equation 14.17 with the normal pointing into the fluid domain as usual, *i.e.*, outwards from the particles. By using the same equation for $\boldsymbol{v}$ *inside* the particles, it is found by observing the sign change for the normal and that $\boldsymbol{v}$ is the negative of $\boldsymbol{u}$ on the surface, that

$$\begin{aligned} -\frac{1}{2}\boldsymbol{u}(\boldsymbol{\eta}) &= \frac{1}{8\pi\mu}\oint_S \mathcal{G}(\boldsymbol{\eta}-\boldsymbol{\xi})\cdot(\boldsymbol{\sigma}(\boldsymbol{\xi};\boldsymbol{v})\cdot\hat{\boldsymbol{n}}(\boldsymbol{\xi}))\,dS(\boldsymbol{\xi}) \\ &\quad -\oint_S \{\hat{\boldsymbol{n}}(\boldsymbol{\xi})\cdot\boldsymbol{\Sigma}(\boldsymbol{\eta},\boldsymbol{\xi})\}\cdot\boldsymbol{u}(\boldsymbol{\xi})dS(\boldsymbol{\xi})\ ,\ \text{for } \boldsymbol{\eta}\in S\ . \end{aligned} \tag{14.19}$$

By subtracting this from 14.17 one obtains another boundary integral equation where the integration of $\boldsymbol{u}$ over the surface has dropped out.

$$\boldsymbol{u}(\boldsymbol{\eta}) = -\frac{1}{8\pi\mu}\oint_S \mathcal{G}(\boldsymbol{\eta}-\boldsymbol{\xi})\cdot(\{\boldsymbol{\sigma}(\boldsymbol{\xi};\boldsymbol{v})+\boldsymbol{\sigma}(\boldsymbol{\xi};\boldsymbol{u})\}\cdot\hat{\boldsymbol{n}}(\boldsymbol{\xi}))\,dS(\boldsymbol{\xi})\ . \tag{14.20}$$

This equation, which shall be called the *single layer equation* for reasons that become clear in Section 15.2, is useful when the ambient field is given in such a form that its stress field is readily calculable, and we are considering the disturbance created by immersing a finite number of particles into this given flow field. It is obvious that in coding a numerical procedure we benefit by having fewer surface integrals to consider, assuming that we want to use any of the formulations presented above. This equation also allows a simple physical interpretation. The tractions of the final flow field (sum of $\boldsymbol{u}$ and $\boldsymbol{v}$) are such that they prevent any penetration or slip at the particle boundaries, cancelling the ambient velocity there. Here the simple superposition of point forces on the surfaces is visible.

The integral equations shown so far each involve only physical variables of *primary* interest, namely, velocity and surface tractions, which explains the title of this section.

14.4.1 Physically Occurring Boundary Conditions

The two main categories of problems with solid particles in a fluid are mobility and resistance problems. In mobility problems we need to determine the translation and rotation velocities of the particles from the given forces and torques acting on them, while in the resistance problems the roles of the given and solved quantities are inverted. As the Stokes equations are linear, these problems are closely related, and the resistance and mobility matrices are simply each other's inverses. However, construction of such a matrix, say, the resistance matrix, requires solving many resistance problems. Therefore it is desired to be able to solve mobility problems directly. Moreover, most of the physically occurring problems are of the mobility type, and resistance problems only come up somewhat artificially.

Why is it then that mobility problems seldom are solved directly? The data given in a mobility problem does not really provide us with an ordinary set of boundary conditions for solving partial differential equations. Rather, what we are given are quantities found from the tractions upon integration over the particle surfaces, and that the particles move as rigid bodies. As opposed to this, in resistance problems we know the disturbance velocity field on the particle surfaces, and ordinary solution methods for the Stokes equations can be applied.

Mathematically it is clear that a problem need not fit either of the categories above; we could start from constraints involving both surface velocities and the total forces and torques. Physically such problems are very rare.

Other problems outside this classification include free surface problems and multiphase flows with different fluids. In these cases the boundary shapes, together with the surface tensions, give boundary conditions for the tractions. As an example of such a problem, see the work of Rallison and Acrivos [59] and Pozrikidis [58].

14.4.2 Difficulties with the BIE in the Primary Variables

The boundary integral equation in the primary variables is of the first kind for the tractions, given velocity boundary conditions. This is something of a problem for two reasons.

Firstly there is no general theory for the existence or uniqueness of solutions to integral equations of the first kind. In fact, the BIE is valid only if a Stokes solution with *smooth* tractions exists for the given velocity BCs. Granted the existence of such a solution, multiple solutions, in fact, will occur. This is due to the indeterminacy of the base pressure in a Stokes solution, allowing the addition of a constant multiple of the surface normal to the tractions.

Secondly, first-kind equations are ill-posed. Thus *numerical* solution attempts for the tractions may encounter some problems. Recall that the ill-posedness means excessive sensitivity of the solved tractions for variations in the given velocity data on the boundary. Physically we can understand this, as the tractions should behave as the derivatives of the velocity field, and therefore we are in some sense doing numerical differentiation here. In the next section it is shown that the ill-posedness cannot be removed by restricting the solution space to continuous functions.

14.5 On Solving Problems with Velocity BCs

The boundary integral equation (BIE) 14.17 is of the first kind for the tractions. Assuming that a velocity surface field $\boldsymbol{u}$ has been given, such that allows a Stokes solution in the whole fluid domain, one can try to solve for the tractions. Since all the integral operators in the BIE are weakly singular and therefore compact, their inverses are unbounded. So $\boldsymbol{u}$ is first affected by a second-kind operator, and the result is mapped to the tractions by an unbounded operator. From this arises the question, whether this unboundedness is an artifact of the formulation used, so that on restricting the surface fields $\boldsymbol{u}$ to "allowed fields" (in the sense of corresponding to a Stokes solution in the interior) one would get a bounded mapping to the tractions. Unfortunately this is not the case.

To give an example of this unboundedness, one can again use Lamb's general solution outside a sphere, which previously was utilized in inspecting the decay rates of disturbance fields. Recall from Chapter 3 (see also Happel and Brenner [29]) that the Stokes velocity field $\nabla\Phi_n$ has the traction field $2(n-1)\mu\nabla\Phi_n/r$ on the surface of a sphere, where r is the radius and Φ_n is a solid spherical harmonic of order n. For a decaying solution any negative integer values are allowed for n, so that these allowed velocity surface fields are mapped to multiples of themselves as tractions, the multiplier attaining arbitrarily large values (see also Exercise 17.3).

Thus, even for a sphere, it is impossible to find a bounded mapping from velocity surface fields to corresponding tractions, provided that the same norm is used both in the domain (the velocities) and the range (the tractions). The reason for this is that tractions are related to the stresses, which essentially are derivatives of the velocity field, and differentiation is an unbounded operation; a small variation in a function can give strong peaks in the derivative. One could use different norms in the domain and range, using also the derivatives of the surface velocity in constructing its norm. This, however, seems to lead to some complications with the integral equation approach adopted here.

Because of the unboundedness observed, it can be stated that *it is impossible to have a second-kind equation for the tractions in which allowed surface velocities would be operated on with a bounded operator.* If, however, we are able to formulate the same problem so that we start from given tractions (different from those of the final solution) instead of velocity BCs, it is possible to have well-posed formulations where the final tractions are obtained from a second-kind equation with no unbounded operators involved.

After these notes we are ready to go on to the secondary variable formulations, in which one of the primary variables of physical interest is replaced by another surface field, namely, an aphysical surface density. This approach will allow a well-posed formulation also for problems with velocity BCs, and it will turn out that a specific set of tractions can be found similarly from a well-posed equation. The reason for this is that ill-posedness is related to the formulation used, and in particular to the information used. For example, although numerical differentiation is not a nice procedure, if we have a recursion formula relating consecutive derivatives, we can differentiate fairly accurately. The basic idea is doing some analytical work before doing numerics. *In the well-posed formulation for tractions given later, the actual velocity field is not used explicitly at all.*

Exercises

Exercise 14.1 Neumann Series and Picard Iterations.
The equation $(1-\mathcal{K})x = b$ has the solution given by the Neumann series

$$x = b + \mathcal{K}b + \mathcal{K}^2 b + \mathcal{K}^3 b + \dots ,$$

provided that the series on the RHS converges. This is analogous to the geometric series

$$1/(1-k) = 1 + k + k^2 + \dots ,$$

which converges if $|k| < 1$; the Neumann series converges for arbitrary b if the spectral radius of $\mathcal{K}$ satisfies $\rho(\mathcal{K}) < 1$. Show that if S_n is the n-th partial sum of the Neumann series, the recursion relation

$$S_{n+1} = b + \mathcal{K}S_n$$

holds. Here $S_n \to x$ as $n \to \infty$, and if we have a good approximation $x_0 \approx x$ to start with, it makes sense to apply the recursion

$$x_{n+1} = b + \mathcal{K}x_n \ , \ \text{for } n = 0, 1, 2, \dots .$$

Thus we are led to Picard iterations.

Exercise 14.2 Systems of Integral Equations.
Let $\boldsymbol{A}$, $\boldsymbol{B}$, $\boldsymbol{C}$, and $\boldsymbol{D}$ be matrices, and $\boldsymbol{x}$, $\boldsymbol{y}$, $\boldsymbol{a}$, and $\boldsymbol{b}$ vectors, such that the operations in the following equation system are well defined:

$$\begin{aligned} \boldsymbol{Ax} + \boldsymbol{By} &= \boldsymbol{a} \\ \boldsymbol{Cx} + \boldsymbol{Dy} &= \boldsymbol{b} \ . \end{aligned}$$

How would you write this as a single matrix equation? A corresponding system of integral equations, with unknown functions $\boldsymbol{x}$ and $\boldsymbol{y}$, can analogously be transformed to a single integral equation; this just requires accepting a piecewise definition of the kernel function.
Note: For this reason systems of integral equations, with unknown "vector functions," are not treated separately in books on integral equations, even though their occurrence in physical models is quite natural.

Chapter 15

Odqvist's Approach for a Single Particle Surface

The integral representation in terms of boundary velocity and tractions was already known to Lorentz. Odqvist realized that the terms in the integral representation could be separately inspected and used to create Stokes solutions. He called these terms the single and the double layer, and established the properties of these for single closed boundary surfaces. Much of the remainder of this and the next chapter is just an extension of his ideas to fluid domains with multiple boundaries and inspection of numerical methods, although the fairly new idea of completing the range [57] plays a significant role in the recent developments.

The formulations presented in this section can be called secondary variable formulations, since instead of the physical variables of primary interest we employ nonphysical density distributions of secondary importance by themselves.

15.1 Smoothness of the Boundary Surfaces

So far we have restricted the boundary surfaces to be Lyapunov-smooth. There are two reasons for the necessity of this restriction. Firstly we need the Fredholm–Riesz–Schauder theory of compact operators, and this is applicable to weakly singular kernels. Without the smoothness restriction the kernels that come up would not be weakly singular any more. Secondly we know that a Stokeslet is nonphysical in the sense that it gives infinite dissipation. In the single layer term Stokeslets are distributed over the surface, and for this to result in a nice Stokes flow with finite (physical) dissipation it is necessary to require some smoothness of both the surface and the distribution density. Thus this second reason is concerned with the relation to Stokes solutions in the three-dimensional domain, instead of just with the solution of boundary integral equations.

The discussion in Ladyzhenskaya shows that for the single layer term it suffices to require Lyapunov-smoothness of the boundary surfaces and continuity of single layer density for the resulting tractions to be well defined and smooth. In this connection Odqvist defined surfaces of class Ah for similar purposes, meaning that the surface can piecewise be represented by one Cartesian coordinate as a function of the other two, so that this function has Hölder-continuous first derivatives.

For the double layer somewhat stronger restrictions are necessary, because the singularity distributed over the boundary is stronger (of higher order). While Ladyzhenskaya does not consider this case, Odqvist showed the following sufficient conditions. The surface must be of class Bh, meaning that Hölder-continuity is required of also the second derivatives in the Cartesian piecewise representation. Furthermore, the double layer density must have Hölder-continuous first derivatives. With these conditions satisfied, the tractions due to the double layer will again exist as the surface is approached, having a continuous limit on the surface. We shall not discuss these mathematically elaborate deductions in detail here.

In order to avoid complications we shall in the central discussion of this chapter always assume that the surfaces are both Lyapunov-smooth and of class Bh. However, the theory can be developed for just Lyapunov-smooth surfaces [37] (see also Section 16.2), but then the mathematical complications may hide the essential physical ideas from the reader to some extent.

Odqvist has shown further that the eigenfunctions of the double layer operator will be not only continuous (as implied by the standard theory of weakly singular IEs), but also their first derivatives will be Hölder-continuous. This follows by noting that an eigenfunction is given by the integral operator acting on this function, and the integral operator has a "smoothing effect." The same conclusion holds for other solutions of the second kind equation than eigenfunctions, provided that the data used are as smooth as we want the solution to be. As a result, we can use the eigenfunctions (or other solution densities with the provision above) to generate nice smooth Stokes flows, such that the tractions are well defined and continuous. The significance of this is that now energy dissipation arguments can be utilized, expressing the dissipation as the rate of work done on the boundaries of the fluid domain (the energy relation), and also the uniqueness theorem for Stokes solutions with velocity boundary conditions is valid.

15.2 Single and Double Layer Potentials, and Some of Their Properties

Consider the terms on the RHSs of the velocity representation 14.15 and the pressure representation 14.16. They will be called the *single layer potentials* and the *double layer potentials*, respectively. The reason for this terminology is, that the single layer potentials are just superpositions of the hydrodynamic potentials of a point force, whereas the double layer potentials are those caused by a layer of sources (or sinks) and doublets of point forces.

Let us look at the double layer a bit more precisely. In 14.15 singularities of the type $\hat{\boldsymbol{n}} \cdot \boldsymbol{\Sigma} \cdot \boldsymbol{u}$ are distributed over the surface; in the following the vectors

$\hat{\boldsymbol{n}}$ and $\boldsymbol{u}$ are considered constant. Due to the symmetry of $\boldsymbol{\Sigma}$ the same result is obtained from $\hat{\boldsymbol{n}}\boldsymbol{u}{:}\boldsymbol{\Sigma}$. Using the definition of $\boldsymbol{\Sigma}$ in terms of the fundamental solution, this is the scaled sum of

$$\hat{\boldsymbol{n}}\boldsymbol{u}{:}(-\boldsymbol{\delta}\mathcal{P}) = -(\hat{\boldsymbol{n}} \cdot \boldsymbol{u})\mathcal{P} \ , \tag{15.1}$$

$$\mu\hat{\boldsymbol{n}}\boldsymbol{u}{:}(\nabla\mathcal{G}) = \mu(\boldsymbol{u} \cdot \nabla)(\hat{\boldsymbol{n}} \cdot \mathcal{G}) \ , \tag{15.2}$$

and

$$\mu\hat{\boldsymbol{n}}\boldsymbol{u}{:}({}^t\nabla\mathcal{G}) = \mu(\hat{\boldsymbol{n}} \cdot \nabla)(\boldsymbol{u} \cdot \mathcal{G}) \ . \tag{15.3}$$

The first of these represents a source or a sink depending on the sign of the dot product on its RHS, and the two remaining terms are clearly doublets formed by opposing point forces. The torques due to these doublets are given by $\mu\boldsymbol{u} \times \hat{\boldsymbol{n}}$ and $\mu\hat{\boldsymbol{n}} \times \boldsymbol{u}$, which cancel each other. Thus *the double layer singularities can give no total force or torque.*

To examine the effects of arbitrary *density distributions* given on S for these potentials, substitute $\boldsymbol{\sigma} \cdot \hat{\boldsymbol{n}} \leftarrow -2\boldsymbol{\psi}$ and $\boldsymbol{u} \leftarrow 2\boldsymbol{\varphi}$ to get the following contributions to the velocity and pressure fields:

$$\boldsymbol{V}(\boldsymbol{y}) = \frac{1}{4\pi\mu} \oint_S \frac{1}{r_{\xi y}}(\boldsymbol{\delta} + \overset{0}{\boldsymbol{r}}_{\xi y}\overset{0}{\boldsymbol{r}}_{\xi y}) \cdot \boldsymbol{\psi}(\boldsymbol{\xi})\, dS(\boldsymbol{\xi}) \tag{15.4}$$

$$\begin{aligned} \Omega(\boldsymbol{y}) &= \frac{1}{2\pi} \oint_S \frac{\boldsymbol{r}_{\xi y}}{r_{\xi y}^3} \cdot \boldsymbol{\psi}(\boldsymbol{\xi})\, dS(\boldsymbol{\xi}) \\ &= -\frac{1}{2\pi}\nabla_y \cdot \oint \frac{\boldsymbol{\psi}(\boldsymbol{\xi})}{r_{\xi y}}\, dS(\boldsymbol{\xi}) \end{aligned} \tag{15.5}$$

for the single layer density $\boldsymbol{\psi}$, and correspondingly,

$$\boldsymbol{W}(\boldsymbol{y}) = \frac{3}{2\pi} \oint_S \hat{\boldsymbol{n}}(\boldsymbol{\xi}) \cdot \frac{\boldsymbol{r}_{\xi y}\boldsymbol{r}_{\xi y}\boldsymbol{r}_{\xi y}}{r_{\xi y}^5} \cdot \boldsymbol{\varphi}(\boldsymbol{\xi})\, dS(\boldsymbol{\xi}) \tag{15.6}$$

$$\begin{aligned} \Pi(\boldsymbol{y}) &= -\frac{\mu}{\pi} \oint_S \nabla_y \left\{ \frac{\boldsymbol{r}_{\xi y}}{r_{\xi y}^3} \cdot \hat{\boldsymbol{n}}(\boldsymbol{\xi}) \right\} \cdot \boldsymbol{\varphi}(\boldsymbol{\xi})\, dS(\boldsymbol{\xi}) \\ &= -\frac{\mu}{\pi}\nabla_y \cdot \oint \left\{ \hat{\boldsymbol{n}}(\boldsymbol{\xi})\frac{\boldsymbol{r}_{\xi y}}{r_{\xi y}^3} \right\} \cdot \boldsymbol{\varphi}(\boldsymbol{\xi})\, dS(\boldsymbol{\xi}) \end{aligned} \tag{15.7}$$

for the double layer density $\boldsymbol{\varphi}$. Here the base pressures are so chosen that as infinity is approached the pressure decays towards zero. The alternative forms of $\Pi(\boldsymbol{y})$ are due to the symmetry of $\nabla_y\mathcal{P}$, mentioned when the integral representation for the pressure was introduced:

$$\hat{\boldsymbol{n}}(\boldsymbol{\xi}) \cdot \nabla_y \boldsymbol{P}(\boldsymbol{y} - \boldsymbol{\xi}) = \nabla_y \mathcal{P}(\boldsymbol{y} - \boldsymbol{\xi}) \cdot \hat{\boldsymbol{n}}(\boldsymbol{\xi}) \ . \tag{15.8}$$

The kernels giving the velocity and pressure fields of a double layer potential will be denoted by

$$\boldsymbol{K}(\boldsymbol{y}, \boldsymbol{\xi}) = -2\hat{\boldsymbol{n}}(\boldsymbol{\xi}) \cdot \boldsymbol{\Sigma}(\boldsymbol{y}, \boldsymbol{\xi}) = \frac{3}{2\pi}\hat{\boldsymbol{n}}(\boldsymbol{\xi}) \cdot \frac{\boldsymbol{r}_{\xi y}\boldsymbol{r}_{\xi y}\boldsymbol{r}_{\xi y}}{r_{\xi y}^5} \tag{15.9}$$

and

$$\boldsymbol{k}(\boldsymbol{y},\boldsymbol{\xi}) = -\frac{1}{2\pi}\hat{\boldsymbol{n}}(\boldsymbol{\xi}) \cdot \nabla_y \mathcal{P}(\boldsymbol{y}-\boldsymbol{\xi}) = -\frac{\mu}{\pi}\hat{\boldsymbol{n}}(\boldsymbol{\xi}) \cdot \nabla_y \frac{\boldsymbol{r}_{\xi y}}{r_{\xi y}^3} \, . \tag{15.10}$$

Thus the density distribution is premultiplied with $\boldsymbol{K}$ or $\boldsymbol{k}$ and integrated over the surface, so that the *first* argument of each kernel will remain, to get the corresponding velocity and pressure fields.

The natural inner product was defined for square integrable vector surface fields on S as the over S integrated pointwise dot product of these fields. With respect to this inner product, the adjoint kernel to any given kernel $\boldsymbol{K}$ is $\boldsymbol{K}^*$ defined by

$$\boldsymbol{K}^*(\boldsymbol{\eta},\boldsymbol{\xi}) = \boldsymbol{K}^t(\boldsymbol{\xi},\boldsymbol{\eta}) \, , \tag{15.11}$$

so that transformation with $\boldsymbol{K}$ inside an inner product may be flipped to the other side of the inner product by changing the transformation to $\boldsymbol{K}^*$. This corresponds to the transpose of a matrix and its relation to the ordinary dot product (a real inner product) of finite vectors. For complex inner products one also needs to complex conjugate the transpose, to create the adjoint. The "operator notation" $\boldsymbol{W} = \mathcal{K}\varphi$, where $\mathcal{K}$ is the linear operator corresponding to kernel $\boldsymbol{K}$, shall be used as a shorthand for Equation 15.6; similarly $\mathcal{K}^*$ corresponds to $\boldsymbol{K}^*$. Since these operators, corresponding to weakly singular kernels, are compact on L_2, square integrable densities are mapped to square integrable surface fields. The inner product is denoted by $\langle \bullet, \bullet \rangle$, so that the flipping rule looks like

$$\langle \mathcal{K}\varphi_1, \varphi_2 \rangle = \langle \varphi_1, \mathcal{K}^*\varphi_2 \rangle \, , \tag{15.12}$$

which is easy to verify just by interchanging the order of the implied surface integrations. To indicate an argument position vector at which an image (function) given by, say, $\mathcal{K}$ is to be evaluated, notation like $\mathcal{K}\varphi(\boldsymbol{\xi})$ shall be used.

The surface jumps for $\boldsymbol{V}$ and $\boldsymbol{W}$ were discussed when the BIE in the primary variables was derived for smooth (continuous) density distributions. Later on density distributions in L_2 shall be considered with the integral equations for secondary variables, but the smoothness assumption is essential for the following "jump conditions." Explicitly, $\boldsymbol{V}$ is continuous through the surface, whereas the double layer potential has the jump (see Figure 14.3)

$$\left.\begin{aligned} \boldsymbol{W}(\boldsymbol{\eta})_{(i)} &= \varphi(\boldsymbol{\eta}) + \mathcal{K}\varphi(\boldsymbol{\eta}) \\ \boldsymbol{W}(\boldsymbol{\eta})_{(e)} &= -\varphi(\boldsymbol{\eta}) + \mathcal{K}\varphi(\boldsymbol{\eta}) \end{aligned}\right\} , \tag{15.13}$$

where the obvious notation

$$\boldsymbol{W}(\boldsymbol{\eta})_{(i)} = \lim_{\epsilon \to 0+} \boldsymbol{W}\left(\boldsymbol{\eta} + \epsilon\hat{\boldsymbol{n}}(\boldsymbol{\eta})\right) \tag{15.14}$$

is used for the limiting value on the interior side (to which the surface normal points), with similar notation for the exterior. The one-dimensional limit on the RHS here is just representative and clearly shows the direction of approach; the limits exists in the ordinary three-dimensional sense as the surface point is approached from either side. The surface tractions for the single layer potential $\boldsymbol{V}$ will also show a similar jump, as is shown shortly.

Consider now the stress field (with a particular choice of base pressure) for the single layer potential 15.4. On interpreting the surface integral as a superposition of fundamental solutions $\mathcal{G}$, and superposing the corresponding stress fields $\boldsymbol{\Sigma}$, we get

$$\boldsymbol{\sigma}(\boldsymbol{y};\boldsymbol{V}) = 2\oint_S \boldsymbol{\Sigma}(\boldsymbol{y},\boldsymbol{\xi})\cdot\boldsymbol{\psi}(\boldsymbol{\xi})\,dS(\boldsymbol{\xi})\ . \tag{15.15}$$

The kernel here is $O(r_{\xi y}^{-2})$, and is no more weakly singular on substituting $\boldsymbol{y} \leftarrow \boldsymbol{\eta} \in S$. Multiply with the surface normal at $\boldsymbol{\eta}$:

$$\hat{\boldsymbol{n}}(\boldsymbol{\eta})\cdot\boldsymbol{\sigma}(\boldsymbol{y};\boldsymbol{V}) = 2\oint_S \hat{\boldsymbol{n}}(\boldsymbol{\eta})\cdot\boldsymbol{\Sigma}(\boldsymbol{y};\boldsymbol{\xi})\cdot\boldsymbol{\psi}(\boldsymbol{\xi})\,dS(\boldsymbol{\xi})\ . \tag{15.16}$$

Now the integrand on the RHS is weakly singular as one substitutes $\boldsymbol{y} \leftarrow \boldsymbol{\eta}$, and the kernel $\boldsymbol{K}^*(\boldsymbol{\eta},\boldsymbol{\xi})$ results. Adding the double layer velocity field $\mathcal{K}\boldsymbol{\psi}(\boldsymbol{y})$ to the RHS of 15.16 gives a function that is continuous as $\boldsymbol{y}$ passes through S at $\boldsymbol{\eta}$. This shows that the tractions of the single layer potential (with suitable choice of base pressures on both sides of S) have a jump similar to the velocity jump of the double layer. Explicitly,

$$\left.\begin{aligned} \hat{\boldsymbol{n}}(\boldsymbol{\eta})\cdot\boldsymbol{\sigma}(\boldsymbol{\eta};\boldsymbol{V})_{(i)} &= -\boldsymbol{\psi}(\boldsymbol{\eta}) + \mathcal{K}^*\boldsymbol{\psi}(\boldsymbol{\eta}) \\ \hat{\boldsymbol{n}}(\boldsymbol{\eta})\cdot\boldsymbol{\sigma}(\boldsymbol{\eta};\boldsymbol{V})_{(e)} &= \boldsymbol{\psi}(\boldsymbol{\eta}) + \mathcal{K}^*\boldsymbol{\psi}(\boldsymbol{\eta}) \end{aligned}\right\}\ , \tag{15.17}$$

As stated earlier, the results above will be referred to as "jump conditions." Verbally, *on passing from the exterior to the interior, the double layer velocity field will have a jump equal to twice the double layer density, and the single layer tractions (with suitable base pressures) jump by minus twice the single layer density.*

From the physical picture of the double layer, having just sources and pairs of opposing forces, one can infer that there will be no traction jump through it. On any surface element of S there is nothing to generate a total force, and so the tractions $\times$ surface element area must be equal on the two sides. Mathematically the situation is more complicated.

For a Lyapunov surface in general, the kernel that comes up on forming the stresses of the double layer will not be weakly singular, even after dotting with the surface normal to get the tractions. There is no guarantee that a finite limit will exist for the stresses as the surface is approached. Odqvist [53] has shown that for this limit to exist it is sufficient that the double layer density and its first derivatives along the surface are *Hölder continuous* with a similar requirement for the second derivatives of one Cartesian coordinate of points on the surface in terms of the other two coordinates (the coordinate system may be suitably chosen for each surface patch). For this reason the article by Karrila and Kim [38] is not completely rigorous; there tractions due to a double layer were used without requiring more than Lyapunov-smoothness of the surfaces. The results presented there are correct though, and they have been re-established with different arguments in the thesis of Karrila [37]. To keep the theoretical considerations simple, we shall here restrict the boundary surfaces to be both Lyapunov-smooth and of class Bh as defined by Odqvist. Then the null functions shall be sufficiently smooth for the generated Stokes fields to always have well-defined tractions on these surfaces, and the energy relation and jump-conditions hold.

The energy relation 14.14 shall be applied to unbounded domains with the single layer generated velocity field. A single layer on a closed surface generates a Stokes velocity field both inside and outside the surface, and since mass is conserved on the inside and the velocity is continuous through the surface, the total generation of fluid as observed from the outside is zero. (Physically the single layer simply has no sources.) Therefore the energy equation can be applied, neglecting the "surface at infinity," regardless of how the base pressure is chosen. Choosing the base pressure for the single layer so that its pressure field is decaying (as was done on defining Ω in this section) will enable the use of Lorentz reciprocal theorem with any other decaying velocity field, regardless of how the base pressure for this is chosen or whether this has sources within the particle surfaces.

15.3 Results for a Single Closed Surface

Assume now that a (homogeneous) Stokes solution is given by the double layer alone. Then according to 15.13

$$\boldsymbol{u}(\boldsymbol{\eta}) = -\lambda\boldsymbol{\varphi}(\boldsymbol{\eta}) + \mathcal{K}\boldsymbol{\varphi}(\boldsymbol{\eta}) , \qquad (15.18)$$

where λ is -1 or $+1$ according as we are considering the interior or the exterior. Assuming that the velocity surface field $\boldsymbol{u}$ has Hölder-continuous first derivative and admits a solution, the solution $\boldsymbol{\varphi}$ will also be similarly smooth. Since the jump condition 15.13 holds for such densities, the corresponding double layer potential $\boldsymbol{W}$ will be a Stokes solution having $\boldsymbol{u}$ as its pointwise limit as the surface S is approached, and further the corresponding tractions are continuous. Thus for smooth enough data $\boldsymbol{u}$ on S it suffices to solve the BIE 15.18 to get the Stokes solution in the fluid domain corresponding to these velocity boundary conditions, assuming that the boundary S consists of closed Lyapunov surfaces of class Bh. Soon it shall be investigated when, given $\boldsymbol{u} \in L_2$, one can solve for $\boldsymbol{\varphi} \in L_2$. In doing that, the adjoint single layer problem shall be used:

$$\boldsymbol{T}(\boldsymbol{\eta}) = -\lambda^*\boldsymbol{\psi}(\boldsymbol{\eta}) + \mathcal{K}^*\boldsymbol{\psi}(\boldsymbol{\eta}) . \qquad (15.19)$$

Comparing this with 15.17 it is easy to see that, with some smoothness restrictions, this corresponds to the interior (exterior) problem with $\lambda = \lambda^*$ set to $+1$ (-1) and $\boldsymbol{T}$ set equal to the surface tractions $\hat{\boldsymbol{n}} \cdot \boldsymbol{\sigma}(;\boldsymbol{V})$. The stresses corresponding to these tractions, given by 15.15, are decaying. So a double layer representation corresponds to velocity boundary conditions, while a single layer representation corresponds to traction boundary conditions, and on switching from one formulation above to the adjoint (keeping the same $\lambda \in \{-1, 1\}$) interior problems are switched to exterior problems and *vice versa*.

In the following, "1" is also used for the identity operator, since the effect of multiplying with one is just the identity mapping. Furthermore, the main interest here is with the double layer equation for the interior and therefore mostly the eigenvalue $\lambda = -1$ is considered; the theorems will be labeled according to the interior.

Theorem 1 (A single particle) *Let S be a single particle surface, defining the interior as its outside. Then* $\dim N(1+\mathcal{K}) = 6$, *and these null double layer densities coincide with RBM velocities on S, giving zero flow in the interior. The null functions of the adjoint second-kind operator, forming $N(1+\mathcal{K}^*)$, are single layer densities that correspond to RBM velocities on S and are proportional to the resulting tractions outside S. The corresponding Jordan blocks are trivial.*

Proof. Choose $\boldsymbol{u}$ to be an RBM velocity inside the particle surface (*i.e.*, in the exterior) with the corresponding stresses set to zero in 14.17. Rewriting the equation using the double layer operator (noting the change in the direction of the normal) we find:

$$\frac{1}{2}\boldsymbol{u} = -\frac{1}{2}\mathcal{K}(;\boldsymbol{u}) \ , \tag{15.20}$$

which shows that $\boldsymbol{u}$ is a null function of $1+\mathcal{K}$. The integral representation also shows that the zero flow is generated in the interior. To show that there are no more null functions, the homogeneous adjoint problem 15.19 with $\lambda = -1$ is considered. This corresponds to a single layer generated flow $\boldsymbol{V}$, for which the tractions vanish on the exterior side of S. The energy equation 14.14 implies that there is no dissipation in the exterior, and so only RBM is possible there. Since the velocity $\boldsymbol{V}$ is continuous through S, on the outside (interior) there is also single layer generated flow with RBM on S. Because of the uniqueness of Stokes solutions with velocity BCs, the flow field in the interior has maximally six degrees of freedom (dof). Because the base pressure in the interior is set to zero at infinity in 15.19, the stress and outside traction fields have no more dof. The single layer density is proportional to the jump in tractions over S, so that it also has maximally six dof. Since for the adjoint double layer problem six linearly independent null functions are already known, the equal dimension of the null spaces must be exactly six.

Now we show that the Jordan blocks corresponding to these null functions are trivial, *i.e.*, no other density is mapped onto a null function by the same second-kind operator. As the Jordan block structures of an operator and its adjoint are similar, we only consider the null functions in $N(1+\mathcal{K})$, namely, RBM velocities on the surface. Let φ be one such RBM null function. If the Jordan block were not trivial, then $\varphi \in R(1+\mathcal{K}) = N(1+\mathcal{K}^*)^{\perp}$. But then this RBM velocity φ would be orthogonal to any RBM tractions, in particular to the tractions corresponding to φ as boundary velocity. This means that with that boundary velocity, no work would be done on the boundary and the dissipation should be zero. However, the fluid motion with such boundary velocity clearly is not RBM, and thus there is deformation and nonzero dissipation. This contradiction shows the triviality of the Jordan blocks. ◇

For a proof of a part of this theorem without invoking the single layer, see the exercises. The triviality of the Jordan blocks is where the single layer results are needed here. This theorem has some immediate corollaries. Since the null functions of second-kind operators are continuous, and RBM tractions of a single particle (with a specific choice of the base pressure) are null functions of 15.19, these RBM tractions exist and are continuous. Also the orthogonality of a null space to the range of the adjoint second kind operator can be utilized. The RBM densities as null functions in $N(1+\mathcal{K})$ are the orthogonal complement of $R(1+\mathcal{K}^*)$, and so the single layer in 15.19 can represent exactly such inside flow fields whose tractions are orthogonal to RBM velocities on S, or equivalently correspond to no total force or torque on the inside of S. This is not much of a restriction, since according to Lorentz reciprocal theorem any Stokes flow inside satisfies these conditions. On the other hand, $N(1+\mathcal{K}^*)$ consists of the RBM tractions on the outside of S. By the Lorentz reciprocal theorem (assuming that the integrals in it exist) the natural inner product on S of any outside velocity field $\boldsymbol{u}$ with these RBM tractions gives components of the total force and torque corresponding to $\boldsymbol{u}$ on S. If the limiting tractions of a Stokes solution do not exist and allow integration to get the total force and torque, components of the total force and torque are defined by the inner products of the surface velocity field and suitable RBM tractions. Since the RBM tractions are smooth and so in L_2, these inner products are well defined for any $\boldsymbol{u} \in L_2$. The outside velocity fields $\boldsymbol{W} \in L_2$ generated by a double layer are exactly those that in this sense correspond to no total force or torque on S. Physically this result is understood, since the force doublets and sources in a double layer can give no total force or torque on a surface enclosing the particle surface, and the total force and torque are transmitted unchanged by Stokes flow. The Lorentz reciprocal theorem could be applied above, neglecting the surface at infinity, because the single layer generated velocity field has no generation of fluid inside the particle surface, and the base pressure corresponding to the RBM tractions given by 15.19 is such that the stress field is decaying.

Theorem 2 (A container) *Let S be a single closed container surface, defining the interior as its inside. Then* $\dim N(1+\mathcal{K}) = 1$*, and these null double layer densities are nonorthogonal to the surface normal on S and generate zero flow in the interior. The null functions of the adjoint second kind operator, forming $N(1+\mathcal{K}^*)$, are single layer densities proportional to the surface normal. The corresponding Jordan blocks are trivial.*

Proof. Consider the single layer outside problem given by 15.19 with $\lambda = -1$. The null functions are such that the tractions on the outside vanish. Then the dissipation on the outside is zero, and there can only be RBM. Since the single layer generated velocity field $\boldsymbol{V}$ is decaying, it must be zero on the outside. The velocity $\boldsymbol{V}$ is continuous through S, and so vanishes also on the inside. Then, by the uniqueness theorem (which can be used with single layer generated flow fields), there is zero motion everywhere inside.

The corresponding stress field is a constant pressure, and so the tractions on the inside are a constant multiple of the surface normal. Now the single layer null density must also be a multiple of the surface normal, since it is proportional to the traction jump over S. That such single layer density really is a null density can be seen by applying 14.15 to zero velocity field with constant pressure and $\boldsymbol{y}$ in the exterior. This shows that the generated velocity field is zero in the exterior, and since the chosen single layer stresses in Equation 15.19 are decaying, the stresses and tractions on the outside really are zero. Thus it has been shown that the equal dimensions of the null spaces are one.

As the null double layer density gives zero velocity on the container boundary and has continuous tractions, the uniqueness theorem for Stokes flows implies that it gives zero flow in all of the interior. (This is where we require the class Bh smoothness of the boundaries, as defined by Odqvist. Without it the uniqueness theorem would be inapplicable.)

To show that the corresponding Jordan block is trivial (since the null space has dimension one, there is only one Jordan block), assume to the contrary that there exists a single layer density that gives the null function found. This density should give outside tractions that are a multiple of the surface normal. Since the velocity field $\boldsymbol{V}$ is mass-conserving on the inside and continuous through S, it is orthogonal to the surface normal. Thus again from Equation 14.14 zero dissipation is found in the outside. The rest of the argument is as above, and it is found that principal functions other than the null function should coincide with the null function. This contradiction shows that the Jordan block is trivial.

Since the surface normal is not in $R(1+\mathcal{K}^*)$, it necessarily has a component in the orthogonal complement of this closed subspace, namely, in $N(1+\mathcal{K}) = R(1+\mathcal{K}^*)^{\perp}$. The null space $N(1+\mathcal{K})$ has dimension one, and therefore all (nonzero) functions in it are nonorthogonal to the surface normal. $\diamond$

For another way of showing that the container null functions in $N(1+\mathcal{K})$ are nonorthogonal to the surface normal, see the exercises.

This theorem again has a corollary following similarly as with the previous theorem. The double layer can represent any inside velocity field that is orthogonal to the surface normal (*i.e.*, mass-conserving) and only these. Again this restriction is not too severe, since all Stokes flows in a container with rigid walls satisfy mass-conservation. The single layer necessarily has a deficiency of similar dimension and cannot represent arbitrary tractions, its restriction being the decay condition. From this it can be immediately inferred that the surface normal (on the outside of a surface) is not in the range of a single layer.

15.4 The Completion Method of Power and Miranda for a Single Particle

Although Odqvist's work was published as early as 1930, it has received little attention to date. This holds particularly in connection with numerical applications. In fact as his work has partly been summarized by Ladyzhenskaya, and the original was published in German, sometimes his results have been attributed to Ladyzhenskaya.

It was only as recently as 1987 that Power and Miranda showed how the weaknesses of the double layer, namely, inability to represent net force and torque, and indeterminacy of the density, could be utilized beneficially. As Odqvist had restricted his attention to single closed boundaries, Power and Miranda also considered just a single particle. Further, as they chose to couple the total force and torque acting on the particle with the known null functions, they did not consider the direct solution of mobility problems. In fact for a single particle it does not make a whole lot of difference if one is able to solve mobility problems directly, but with multiparticle systems forming the whole resistance matrix needed to solve a mobility problem becomes prohibitively expensive computationally. Later on it will be seen how the mobility problems actually are computationally easier than the resistance problems, in that they allow the iterative solution of the discretized system. This single fact accounts for significant computational savings with large linear systems. In the following the main results of Power and Miranda are briefly explained, in a somewhat different fashion from their original derivation.

Consider the outside problem for a single surface, the outside now being the interior fluid domain where the surface normal points. Power and Miranda [57] observed that although the double layer representation is able to represent only those flow fields that correspond to a force- and torque-free surface, the representation may be completed by adding terms that give arbitrary total force and torque in suitable linear combination. They chose to use a Stokeslet $\boldsymbol{F} \cdot \mathcal{G}$ and a rotlet $-\frac{1}{2}(\boldsymbol{T} \times \nabla) \cdot \mathcal{G}$ positioned at the origin chosen inside the particle. Thus their equation is

$$\boldsymbol{u}(\boldsymbol{\eta}) = \boldsymbol{\varphi}(\boldsymbol{\eta}) + \mathcal{K}\boldsymbol{\varphi}(\boldsymbol{\eta}) + (\boldsymbol{F} - \frac{1}{2}\boldsymbol{T} \times \nabla) \cdot \mathcal{G}(\boldsymbol{\eta})/(8\pi\mu) \quad \text{for } \boldsymbol{\eta} \in S. \tag{15.21}$$

Note again that, although this is the outside problem, $\boldsymbol{W}_{(i)}$ of Equation 15.13 is used, since the fluid domain of consideration is called the interior and the surface normal points in that direction. The total force and torque are fully determined by the square integrable $\boldsymbol{u}$ on S so that $\boldsymbol{F}$ and $\boldsymbol{T}$ are fixed, see the discussion after Theorem 1. Since the double layer part here has null functions (six linearly independent), the resulting system is indeterminate. Power and Miranda chose to associate these null functions with the added six parameters, namely, three components of force and torque each, thus making the density fully determined. Denoting the null functions by $\boldsymbol{\varphi}_i$, where $i = 1, \ldots, 6$ and using the natural inner product

$$\langle \boldsymbol{a}, \boldsymbol{b} \rangle = \oint_S \boldsymbol{a}(\boldsymbol{\xi}) \cdot \boldsymbol{b}(\boldsymbol{\xi})\, dS(\boldsymbol{\xi}) \tag{15.22}$$

for vector surface fields, the association was done by

$$\left. \begin{array}{rcl} F_i & = & \langle \boldsymbol{\varphi}, \boldsymbol{\varphi}_i \rangle \\ T_i & = & \langle \boldsymbol{\varphi}, \boldsymbol{\varphi}_{i+3} \rangle \end{array} \right\} \text{ for } i \in \{1, 2, 3\}. \tag{15.23}$$

Substituting these in the equation for $\boldsymbol{\varphi}$ above gives an ordinary IE of the second kind. Since the double layer is now fully determined, the *Fredholm alternative* implies that a solution always exists for any given $\boldsymbol{u}$. This corresponds to the result of linear algebra, that a square linear system either always has a solution and it is unique, or that solutions exist only sometimes but then they are not unique.

The point to note here is that the association of the null space with the components of force and torque is completely arbitrary (the basis for the null space can be chosen and permuted freely), although Power and Miranda did not mention this. Their method is mathematically nice because it leads to an ordinary IE of the second kind without any extra terms, but after solving for $\boldsymbol{\varphi}$, $\boldsymbol{F}$ and $\boldsymbol{T}$ have to be separately computed. Their mathematical argument, however, is much more elaborate than the one given here and does not easily allow extension to multiparticle systems or bounded domains. It may be observed that this method is related to Wielandt's deflation, see Bodewig [5], in that the original integral operator is modified by adding some tensor products. In fact, we will show in Chapter 17 that such deflations can be carried out for the mobility problem. The end result is a fast iterative algorithm for the solution of mobility problems.

Finally, we note that the flow fields of the Stokeslet and rotlet are infinitely differentiable on the boundary surface, and therefore subtraction of these terms from given sufficiently smooth boundary conditions gives again equally smooth boundary conditions, representable by the double layer alone when $\boldsymbol{F}$ and $\boldsymbol{T}$ are chosen correctly.

Exercises

Exercise 15.1 Regarding Theorem 1.

Show by using the double layer operator $\mathcal{K}$ (without invoking the adjoint operator) that $\dim N(1 + \mathcal{K}) \leq 6$.

Hint: The chain of deduction could be [the double layer null functions give zero outside flow] — [the tractions on both sides are zero] — [the inside flow is RBM] — [apply the jump condition].

Note: The advantage of the earlier proof in this chapter is that it can be used when requiring only Lyapunov-smoothness.

Exercise 15.2 Regarding Theorem 2.

Show directly with the double layer operator that the container null function is nonorthogonal to the surface normal.

Hint: [zero flow on the inside implies constant pressure stress field] — [also on the outside the tractions are a multiple of the surface normal] — [orthogonality would imply no dissipation in the outside due to energy relation and jump condition] — [zero flow on the outside due to decay condition] — [jump condition gives zero double layer density, a contradiction].

Exercise 15.3 Double Layer Representation and Solution of an Exterior Traction Problem.

Suppose $\boldsymbol{u}$ is a velocity field representable by a double layer. Show how to construct its double layer representation assuming that the tractions $\hat{\boldsymbol{n}} \cdot \boldsymbol{\sigma}(;\boldsymbol{u})$ are known.

Hint: In the exterior construct a single layer generated field $\boldsymbol{V}$ with tractions equal to the interior tractions of $\boldsymbol{u}$. This is possible when the tractions of $\boldsymbol{u}$ correspond to no total force or torque. Use the integral representation 14.15 in the interior both for $\boldsymbol{u}$ and for $\boldsymbol{V}$ (the latter having zero LHS), noting that the single layer terms are equal.

Note: We know that the completed double layer equations (either bordered or deflated) have a unique solution. But is this solution the one we are seeking? We have a uniqueness theorem for Stokes flows with nice smooth tractions, but that cannot be applied to flows generated by a double layer without smoothness restrictions for both the surface shapes and the density. Using the technique above it can be shown that if a solution with nice smooth tractions exists, then the completed equations will give just that solution [37] .

Exercise 15.4 The Container Null Function on a Sphere.

Show that on a sphere the container null function is the surface normal.

Hint: The dimension of that null space is one. Because of symmetry any rotation about the center of the sphere, or reflection through an equatorial plane, would transform a null function to a null function again. The only null function must be invariant with respect to these, which easily implies the result desired.

Exercise 15.5 Green's Functions Expressed with Single or Double Layers.

In general a Green's function (in our case tensor) is such that it satisfies some homogeneous boundary conditions, so that several such functions can be superposed while the boundary conditions remain the same. Also the Green's function is an inhomogeneous solution of the governing linear partial differential equations, corresponding to a delta function as the driving force term that makes the equations inhomogeneous. In our ordinary Green's function $\mathcal{G}$, the homogeneous boundary conditions are satisfied at infinity. Replace these by BCs requiring zero velocity on a container surface S_c. Show that the resulting Green's function can be expressed as the sum of $\mathcal{G}$ and a double layer generated flow field in the interior (inside S_c). As a more difficult problem, show that the double layer "correction" can be replaced by a single layer "correction" to $\mathcal{G}$.

Hint: All you really need to do is cancel $\mathcal{G}$ on the boundary S_c. With the single layer first generate an exterior flow field that matches the tractions of $\mathcal{G}$ (with some choice of the base pressure). Deduce that the velocity fields are the same in the exterior, and use continuity of the single layer generated velocity field through the surface.
Note: The Green's functions and their representations with single and double layers can be generalized to more complicated geometries [37].

Exercise 15.6 Second-Kind IEs from Other Green's Functions.
Let $\boldsymbol{G} = \mathcal{G} + \mathcal{K}\varphi$ be the Green's function relative to some fixed finite measurable boundaries S_0, over which the ordinary double layer integral operator $\mathcal{K}$ is defined. In the interior fluid domain insert some new closed finite boundaries, denoted by S, and denote the new smaller fluid domain between S and S_0 by Q. For $\boldsymbol{u}$ vanishing on S_0 derive an integral representation similar to 14.15 by using the Green's formula (*i.e.*, the Lorentz reciprocal theorem with the volume integrals retained) with $\boldsymbol{G}$ instead of $\mathcal{G}$. Inspect the resulting terms (integrals over S) separately. Clearly the new single layer term gives a Stokes velocity field vanishing on S_0, with no velocity jump over S. Show that the new double layer term also generates a Stokes velocity field that vanishes on S_0 (provided, in case Q is bounded, that the double layer density is orthogonal to $\hat{\boldsymbol{n}}$ on S).
Hint: With the completed double layer representation on $S+S_0$ we can generate a Stokes field $\boldsymbol{u}$ vanishing on S_0 and coinciding with the on S given double layer density (used with the "new" kernel). Then the new integral representation for $\boldsymbol{u}$ verifies the claim.
Exercise continued: Note that the new integral representation gets value zero outside Q, and use a symmetry argument with the new double layer to get an IE of the second kind for velocity BCs on S (with zero velocity on S_0 assumed).
Note: In some cases the Green's function is available in analytic form. Such is the case of half-space restricted by an infinite plane boundary, as shown in Chapter 12. The new Green's function $\boldsymbol{G}$ is found by adding the Lorentz image to $\mathcal{G}$. In this way some problems with infinite boundaries can be treated with compact integral operators (this in spite of the fact that in the exercise above we require S_0 to be finite).

Exercise 15.7 Constant Pressure Solutions.
Take the pressure representation corresponding to the completed double layer representation for a single particle. By order of magnitude analysis of the decay rates show that if the pressure is constant, the total force has to vanish.
Note: This exercise was inspired by [16]. Also Lamb's general solution could be used to solve this problem.

Exercise 15.8 Another Variational Principle for Mobility Problems; Towards a Convergence Proof of the Method of Reflections.
Consider a mobility problem with rigid particles (forces and torques given, translation and rotation or angular velocities to be determined). All the flow fields

complying with the supplied data (but not necessarily corresponding to rigid particles) constitute the set

$$\{\sum_{\alpha}(F_{\alpha} - \frac{1}{2}\boldsymbol{T}_{\alpha} \times \nabla) \cdot \frac{\mathcal{G}(\boldsymbol{x} - \boldsymbol{x}_{\alpha})}{8\pi\mu} + (\mathcal{K}\varphi)(\boldsymbol{x}) \mid \varphi \text{ smooth.}\}$$

Show that the mobility solution, corresponding to rigid-body motion and found within this set, minimizes the energy dissipation among all the flow fields considered here, *i.e.*, having the specified forces and torques.

Hint: This again can be considered an application of the Pythagorean theorem, as in Chapter 2. To prove orthogonality, just observe that the variations in the flow field must be orthogonal to RBM tractions, since they correspond to no total force or torque on any of the particles.

Note: A similar result holds for the "suboptimization problem" where the double layer on just one of the particles (at a time) is adjusted so as to make the surface velocity on that particle an RBM. Performing such suboptimizations cyclically, stepping through each of the particles at each cycle, is precisely the method of reflections for mobility problems, as described in Part III. Using results from functional analysis, Luke [46] has shown that for *mobility problems* this type of method of reflections always converges (provided the particles are smooth and nontouching). Luke's proof is not completely rigorous for unbounded domains, and he actually suggests using integral equation formulations for extending the validity to the cases without a container.

Chapter 16

Multiparticle Problems in Bounded and Unbounded Domains

In this chapter the work of Power and Miranda is extended in two directions. The completion method is extended to multiparticle problems, also allowing a container surface, in such a way that the direct solution of mobility problems becomes possible. As a preliminary step the properties of the double layer are examined in the multisurface setting, and this will in essence show just that no new null functions arise from the interactions of the individual connected components of the boundary of the fluid domain. Then we show how tractions corresponding to the RBM of particles can be computed from a well-posed formulation.

For mobility problems the real strength of the double layer lies in that it cannot represent a total force or torque; these variables stand separately in the equations. In fact the double layer has other nice properties too, as will be noted when iterative solutions are considered.

16.1 The Double Layer on Multiple Surfaces

We shall directly start with a theorem that shows the behavior of the double layer on multiple surfaces. The triviality of the Jordan blocks will be necessary when iterative solutions are considered. Again the boundary surfaces are assumed to be similarly smooth, as in the previous sections.

Theorem 3 (Particles with a container) *Let S_i, $i = 1, \ldots, M$, be particle surfaces enclosed by a container surface S_c, the interior fluid domain being bound between these. Then* $\dim N(1 + \mathcal{K}) = 6M + 1$, *and double layer densities that coincide with RBM velocities on the particle surfaces are in this null space. Any other null functions in $N(1 + \mathcal{K})$ are nonorthogonal to the surface normal on S_c, and the null function of the container alone (see Theorem 2) is one of these. All null functions in $N(1 + \mathcal{K})$ generate zero velocity field in the interior,* *when used as double layer densities. The null functions of the adjoint second kind operator, forming $N(1+\mathcal{K}^*)$, are single layer densities that generate RBM velocities on the particle surfaces and zero velocity on S_c, the resulting tractions in the interior being proportional to these null functions; in particular a constant multiple of the surface normal on the totality of boundary surfaces is such a null function and it also generates zero velocity field. The $6M+1$ Jordan blocks are trivial.*

Proof. First consider the interior double layer problem. For each particle surface separately we know six linearly independent null functions, given by RBM-velocities. For the container boundary, if present, we know that there exists a null function. These all generate zero velocity on the other boundary components, and therefore are null functions for the multiboundary case as well. Thus the dimension of the null space is at least $6M+1$.

Consider now the single layer null functions, generating zero tractions in the exterior. As the tractions vanish outside the container, no work is done on the imaginary fluid there, and due to the decay condition the velocity field must vanish. Similarly inside each particle there is no dissipation, and only RBM is allowed. As the single layer generated velocity field is continuous through the boundary surfaces, in the interior we also have RBM on the particle surfaces and zero velocity on the container. This gives $6M$ degrees of freedom (dof) for the velocity boundary conditions, and due to the indeterminacy in the base pressure the interior tractions have maximally $6M+1$ dof. Due to the jump condition the single layer densities are proportional to these tractions.

Since the dimensions of the null spaces for the two adjoint problems above must coincide, this common dimension is exactly $6M+1$. Thus the double layer null functions are exactly those that are found for each connected component of the boundary separately (and linear combinations of these), while the single layer null functions follow from the jump condition applied to the analysis carried out above for this adjoint problem.

Showing that the Jordan blocks are trivial is done in very much the same way as with a single boundary surface. As a multiple of the surface normal on all of the bounding surfaces corresponds to constant pressure, it is one of the RBM tractions in $N(1+\mathcal{K}^*)$. To show that the corresponding Jordan block is trivial, again try to find a single layer density that gives exterior tractions equaling this null function. The tractions outside S_c must then be a multiple of the surface normal, and, due to the single layer having no sources, no work is done in that component of the exterior. Thus the velocity field there is RBM, and being a decaying single layer generated velocity it is identically zero. Then the stress field must be just a constant pressure, and again being decaying it must vanish. This gives zero tractions outside S_c, showing that a nonzero multiple of the surface normal cannot be generated there. Thus neither the surface normal of the container alone nor the normal on all the boundary surfaces is in the range $R(1+\mathcal{K}^*)$.

Now it shall be proven that the Jordan blocks corresponding to the other single layer null functions than a multiple of the surface normal are also trivial.

Choose one such single layer null function ψ generating RBM velocities on the particle surfaces and zero velocity on the container, the null function being proportional to the tractions in the interior. The flow field $\boldsymbol{V}$ in the interior is then not RBM, and so there is nonzero dissipation; this is the reason why the surface normal had to be inspected separately. The energy equation 14.14 implies that the null function ψ, being proportional to the tractions, is not orthogonal to the corresponding surface velocity $\boldsymbol{V}$. But this surface velocity is a null function of the adjoint double layer equation, $\boldsymbol{V} \in N(1+\mathcal{K})$, as shown above. If it were true that $\psi \in R(1+\mathcal{K}^*) = N(1+\mathcal{K})^{\perp}$, then ψ would have to be orthogonal to $\boldsymbol{V}$, which is a contradiction.

That the double layer null functions generate zero velocity in all of the interior follows from the uniqueness of Stokes flows (with smooth tractions) that satisfy given velocity boundary conditions. ◊

Again corollaries follow. The tractions corresponding to RBMs of the particles with zero velocity on the container are null functions and therefore are continuous. Exactly such continuous velocity BCs that are orthogonal to these RBM tractions and orthogonal to the surface normal can be satisfied by a double layer generated Stokes flow. The former condition can again be interpreted as the total force and torque on each particle having to vanish, and the latter as requiring mass conservation.

A theorem similar to the previous holds for unbounded domains, and since the proof differs little from that above, it shall not be elaborated on.

Theorem 4 (Particles without a container) *Let S_i, $i = 1, \ldots, M$, be particle surfaces, the interior fluid domain being outside these. Then* $\dim N(1+\mathcal{K}) = 6M$, *and double layer densities that coincide with RBM velocities on the particle surfaces form this null space, giving zero flow fields in the interior. The null functions of the adjoint second kind operator, forming $N(1+\mathcal{K}^*)$, are single layer densities that generate RBM velocities on the particle surfaces, the resulting tractions in the interior being proportional to these null functions. The* $6M$ *Jordan blocks are trivial.*

The following theorem will be utilized on doing the 'mathematical deflations' in the case of particles with a container, when Wielandt deflations are applied to our integral equations to make iterative solutions converge. Physically the theorem has not much content: since we can deal with different connected components of fluid separately, only connected fluid domains are of interest. The theorem could also be useful on solving a problem with traction BCs, as the adjoint single layer problem deals with boundary conditions in the interior. (The theorem is again labelled according to the double layer BCs.)

Theorem 5 (Exterior problem for particles with a container.)
Let S_i, $i = 1, \ldots, M$, be particle surfaces enclosed by a container surface S_c, the interior fluid domain being bound between these. Then $\dim N(-1+\mathcal{K}) = 6+M$, *and double layer densities that coincide with an RBM velocity on the totality*

of surfaces are in this null space. The null functions of the adjoint second kind operator, forming $N(-1+\mathcal{K}^*)$, are single layer densities on S_c that generate an RBM velocity outside it, the resulting tractions outside S_c being proportional to these null functions, and multiples of the surface normal on each particle surface. The $6+M$ Jordan blocks are trivial.

Proof. It is easy to show that the particle surface normals ψ_i, $i=1,\ldots,M$ are in $N(-1+\mathcal{K}^*)$. Similarly RBMs on the totality of surfaces φ_j, $j=1,\ldots,6$, are in $N(-1+\mathcal{K})$. Since $\langle\psi_i,\varphi_j\rangle=0$ for any i and j, these null functions do not correspond to each other in the biorthogonal sense, but rather complement each other. Therefore $\dim N(-1+\mathcal{K})\geq 6+M$.

Consider a single layer null function in $N(-1+\mathcal{K}^*)$, generating zero interior tractions. The interior velocity field must then be RBM, and due to continuity of the single layer generated velocity field the same RBM prevails inside the particles and on S_c viewed from the outside. Now inside each particle the stress field is a constant pressure, and the traction jump over that surface is a constant multiple of the particle surface normal; on the particle surfaces the single layer density must be proportional to these according to the jump condition. The tractions outside S_c correspond to RBM on S_c and vanish at infinity, therefore having six dof. Thus in all $\dim N(-1+\mathcal{K}^*)\leq 6+M$. Together with the previous inequality this implies that $\dim N(-1+\mathcal{K}^*)=6+M$, and all the single layer densities considered above must be in this null space.

Now we show the triviality of the Jordan blocks corresponding to the ψ_i's. Assume to the contrary, that, for example,

$$\psi_1=\begin{cases}\hat{\boldsymbol{n}} & \text{on } S_1\\ \mathbf{0} & \text{otherwise}\end{cases}\tag{16.1}$$

were in $R(-1+\mathcal{K}^*)$, *i.e.*, could be generated as interior tractions by a single layer. Then no work is done on the fluid in the interior, as these tractions are orthogonal to single layer generated flows without sources. Thus there is RBM in the interior, and the stress field is just a constant pressure. But then the tractions are a constant multiple of $\hat{\boldsymbol{n}}$ on the totality of surfaces and cannot coincide with ψ_1 — a contradiction.

Finally we verify the triviality of the Jordan blocks corresponding to the φ_js. Assume to the contrary that, for example, $\varphi_1\in R(-1+\mathcal{K})=N(-1+\mathcal{K}^*)^{\perp}$. Then we could generate a nonzero RBM surface velocity outside S_c with a double layer, corresponding to nonzero dissipation outside S_c. But the corresponding tractions outside S_c are in $N(-1+\mathcal{K}^*)$, and due to orthogonality by the energy relation, there should be zero dissipation — a contradiction. ◊

All the previous theorems before this last one have dealt with connected fluid domains for the interior double layer problem. For those the results can be summarized simply by stating that *the interactions of boundary components generate no new double layer null functions.*

The integral equations 15.18 and 15.19 can be called *secondary variable formulations*, since the function solved does not *in general* represent any physically measurable quantity of interest to us, although in the theorems above the null functions were sometimes associated with RBM velocities and tractions. The jump conditions show the physical meanings of these solutions.

16.2 The Lyapunov-Smooth Container

The double layer eigenfunction with $\lambda = 1$ is not known explicitly for a container of arbitrary shape. Odqvist's proof that this eigenfunction generates zero flow field inside the container (when used as a double layer density) requires Hölder-continuous second derivatives in the parametric representation of the surface. When there are particles inside, the null function for the container alone will remain a null function for this multisurface problem only if it generates zero flow.

On the other hand, by inspection of the velocity representation 14.15, we see that RBMs as double layer densities generate no flow outside a Lyapunov (surface) particle. This asymmetry is removed in the following stronger version of Theorem 3 involving *Lyapunov-smooth* particle and container surfaces.

Theorem 6 (Particles with a container) *Let S_i, $i = 1, \ldots, M$, be Lyapunov-smooth particle surfaces enclosed by a Lyapunov-smooth container surface S_c, the interior fluid domain being bound between these. Then* $\dim N(1 + \mathcal{K}) = 6M + 1$, *and double layer densities that coincide with RBM velocities on the particle surfaces are in this null space. Any other null functions in $N(1+\mathcal{K})$ are nonorthogonal to the surface normal on S_c, and the null function of the container alone (see Theorem 2) is one of these. All null functions in $N(1 + \mathcal{K})$ generate zero velocity field in the interior, when used as double layer densities. The null functions of the adjoint second-kind operator, forming $N(1 + \mathcal{K}^*)$, are single layer densities that generate RBM velocities on the particle surfaces and zero velocity on S_c, the resulting tractions in the interior being proportional to these null functions; in particular a constant multiple of the surface normal on the totality of boundary surfaces is such a null function, and it also generates zero velocity field. The $6M + 1$ Jordan blocks are trivial.*

Proof. Using 14.15 with zero velocity and constant pressure in the interior, we see that a multiple of the surface normal as single layer density gives zero velocity fields in the components of the exterior. The velocity field, being continuous through all the surfaces, vanishes identically everywhere. Therefore in each connected component of the exterior and the interior there is just a constant pressure. Outside S_c the chosen stress field is decaying, and being a constant pressure it must vanish. Due to the jump condition 15.17 the pressure jumps over S_c and over any particle surface are equal in magnitude but opposite in direction, on passing from outside S_c through the interior to inside a particle. Therefore inside the particles the pressure is again zero, and a multiple of the surface normal really is a null function.

To show that the corresponding Jordan block is trivial, again try to find a single layer density that gives exterior tractions equaling this null function.

The tractions outside S_c must then be a multiple of the surface normal, and, due to the single layer having no sources, no work is done in that component of the exterior. Thus the velocity field there is RBM, and being a decaying single layer generated velocity it is identically zero. Then the stress field must be just a constant pressure, and again being decaying it must vanish. This gives zero tractions outside S_c, showing that a nonzero multiple of the surface normal cannot be generated there. Thus neither the surface normal of the container alone nor the normal on all the boundary surfaces is in the range $R(1+\mathcal{K}^*)$.

Consider now the single layer null functions in general. Since the tractions are zero inside the particles and outside the container, the energy equation 14.14 implies that dissipation in each connected component of the exterior is zero. Thus in each of these only RBM velocity fields are allowed, and outside the container the velocity field must vanish due to the decay condition. Since the single layer velocities are continuous through the surfaces, in the interior the velocity field has $6M$ dofs, because of the uniqueness of Stokes velocity fields in terms of the velocity boundary conditions. These dofs come from the three components of translation and rotation each, on each particle surface, the velocity vanishing on the container. The corresponding tractions have one more dof, because of the pressure indeterminacy in the interior. Thus the traction jumps giving the single layer densities that are null functions have at most $6M+1$ dofs.

Using the velocity representation 14.15 within any one of the particle surfaces with an RBM velocity $\boldsymbol{u}$ and zero tractions, it is seen that RBM densities are not only null functions for these individual surfaces, but they generate a zero velocity field everywhere outside the particles. Therefore these RBM densities on particle surfaces are null functions of the multisurface problem also. Here we have $6M$ linearly independent null functions in $N(1+\mathcal{K})$. If these were all the null functions, the surface normal of the container being orthogonal to these would be in the range $R(1+\mathcal{K}^*)$. Since this is not the case, there must be exactly $6M+1$ independent null functions, as it is known that this is the maximal number of them. In addition the null functions in $N(1+\mathcal{K})$ other than RBM on particle surfaces must be nonorthogonal to the surface normal on the container.

Now it shall be proven that the Jordan blocks corresponding to the other single layer null functions than a multiple of the surface normal are also trivial. Choose one such single layer null function $\boldsymbol{\psi}$ generating RBM velocities on the particle surfaces and zero velocity on the container, the null function being proportional to the tractions in the interior. The flow field $\boldsymbol{V}$ in the interior is then not RBM, and so there is nonzero dissipation. The energy equation 14.14 implies that the null function $\boldsymbol{\psi}$ is not orthogonal to the corresponding surface velocity $\boldsymbol{V}$. But this surface velocity is a null function of the adjoint double layer equation, $\boldsymbol{V} \in N(1+\mathcal{K})$, as shown above. If it were true that $\boldsymbol{\psi} \in R(1+\mathcal{K}^*) = N(1+\mathcal{K})^{\perp}$, then $\boldsymbol{\psi}$ would have to be orthogonal to $\boldsymbol{V}$, which is a contradiction.

Choose an orthonormal basis for $N(1+\mathcal{K})$, in which $6M$ of these null functions coincide with RBM velocities on the particle surfaces and vanish on the container. It shall be shown that the remaining basis element φ is a null function of the container alone, and generates zero flow field inside the container. First one can show that any such tractions $\boldsymbol{T}$ that are orthogonal to RBM velocities on the particle surfaces and vanish on the container are in the range $R(1+\mathcal{K}^*)$. If this were not the case, then necessarily $\langle \boldsymbol{T}, \varphi \rangle \neq 0$, since orthogonality to the remaining RBM null functions of the adjoint is satisfied by assumption. But also $\langle \hat{\boldsymbol{n}}, \varphi \rangle \neq 0$, whereas $\hat{\boldsymbol{n}}$ is orthogonal to the RBM null functions. Then there exists a constant p such that $\boldsymbol{T} + p\hat{\boldsymbol{n}}$ is orthogonal to $N(1+\mathcal{K})$ and therefore is in $R(1+\mathcal{K}^*)$. But if $p \neq 0$, a constant multiple of the surface normal on the exterior side of the container surface results, and as noted before this is not possible for single layer generated tractions. Therefore $p = 0$ and $\boldsymbol{T} \in R(1+\mathcal{K}^*)$.

Now φ is orthogonal to both the RBM null functions and all $\boldsymbol{T}$ of the type above; these in turn span linearly all surface fields that vanish on the container. Therefore φ can attain no nonzero values on the particle surfaces, and so is nonvanishing only on the container. But now $\varphi \in N(1+\mathcal{K})$ is also in $N(1+\mathcal{K})$ for the container surface alone, since it gives zero inside velocity on this surface and vanishes itself on the other components of the boundary — so it is the container null function in Theorem 2. As this container null function gives zero velocity on any enclosed particle surfaces, however these are chosen, the corresponding flow field is zero flow inside the container. This completes the proof of this theorem. $\diamond$

Again, we emphasize that even though the container null function generates a Stokes velocity field that vanishes in the limit as S_c is approached, one cannot apply the uniqueness theorem to infer that this velocity field is zero everywhere inside the container. This is because the proof of uniqueness utilized the tractions on the bounding surfaces. With a double layer on a Lyapunov surface there is no guarantee that the tractions exist as finite-valued functions, even if the double layer density is continuous. This situation was remedied in Theorem 3 by requiring the container surface to have Hölder-continuous second derivatives in suitably chosen Cartesian representation of the surface patchwise. Then the corresponding null function and its first derivatives will also be Hölder-continuous (*c.f.*, Odqvist), and the tractions are well defined.

In this discussion, these difficulties have been circumvented, and Odqvist's theory has been extended to Lyapunov surfaces in general, in a multisurface setting. Of course, for a Lyapunov-smooth container without particles we now see the behavior of the container null function: For any point in the fluid, use a virtual particle whose surface runs through this point. This mathematical trickery makes the proofs look fairly complicated, but the general idea of the proofs can be easily summarized. One needs to use the energy relation (which is allowed only with the single layer since the tractions are involved), continuity and jump properties through the surfaces, decay conditions in the unbounded component regions, and the special properties of RBM velocities and constant pressures.

16.3 The Canonical Equations

16.3.1 The Bordering Method in General

The method of Power and Miranda can be simply visualized in terms of *square* matrices. If a matrix is not of full rank its null space has the same dimension as the "deficiency" of its range space, *i.e.*, the orthogonal complement of its range. The dimension of the orthogonal complement of a subspace (here the range) is also called the *codimension* of the subspace. The result mentioned is often expressed as the row and column ranks of a matrix being equal. Now the matrix can be "completed" to an invertible matrix of full rank by linearly associating the null space with the missing part of the range space. As an example, the matrix

$$\begin{bmatrix} 0 & 1 & 0 \\ 1 & 0 & 1 \\ 0 & 0 & 0 \end{bmatrix} \tag{16.2}$$

has its null space and the orthogonal complement of its range spanned by $(1,0,-1)$ and $(0,0,1)$, respectively. Associating the component in null space with the missing part of the range space gives

$$\begin{bmatrix} 0 & 1 & 0 \\ 1 & 0 & 1 \\ 0 & 0 & 0 \end{bmatrix} + a \begin{bmatrix} 0 \\ 0 \\ 1 \end{bmatrix} \begin{bmatrix} 1 & 0 & -1 \end{bmatrix} = \begin{bmatrix} 0 & 1 & 0 \\ 1 & 0 & 1 \\ a & 0 & -a \end{bmatrix} \quad \text{where } a \neq 0, \tag{16.3}$$

which is invertible. The added part here could be written in inner product notation when operating on a vector

$$\begin{bmatrix} 0 \\ 0 \\ 1 \end{bmatrix} \begin{bmatrix} 1 & 0 & -1 \end{bmatrix} \begin{bmatrix} x \\ y \\ z \end{bmatrix} = \begin{bmatrix} 0 \\ 0 \\ 1 \end{bmatrix} \langle \begin{bmatrix} 1 & 0 & -1 \end{bmatrix}^t, \begin{bmatrix} x & y & z \end{bmatrix}^t \rangle \tag{16.4}$$

to see the similarity with the equation of Power and Miranda. This association is arbitrary to some degree, as is here demonstrated by the freedom in choosing a. Also instead of the null vector, any vector with a nonzero component in the null space could be used, and there is similar freedom in choosing the completing range vector.

In the case at hand the multipliers of the completing vectors, namely, the total force and torque, are exactly those values that are wanted. Therefore we take these as actual new variables, and remove the indeterminacy in the double layer density by requiring it to be orthogonal to the original null space. Thus instead of adding something to the original matrix to complete it, here it is preferred to expand the matrix by bordering it with range vectors corresponding to added variables as new columns and orthogonality conditions to the original null space as added rows. With the example above the bordered matrix is

$$\left[\begin{array}{ccc|c} 0 & 1 & 0 & 0 \\ 1 & 0 & 1 & 0 \\ 0 & 0 & 0 & 1 \\ \hline 1 & 0 & -1 & 0 \end{array}\right] \tag{16.5}$$

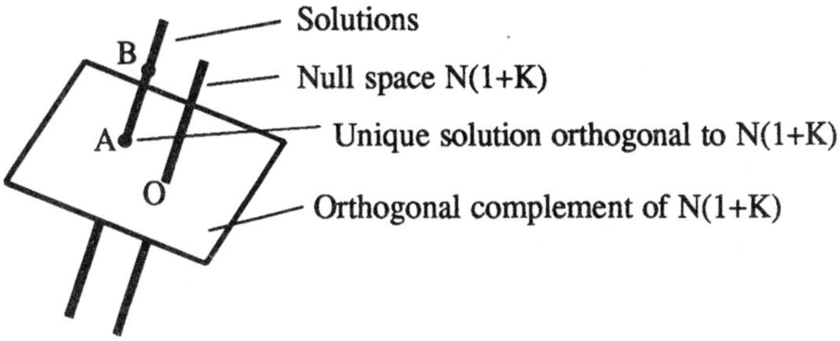

Figure 16.1: Solutions obtained from completion procedures.

and a 0 is added as the last element to the given vector to enforce the orthogonality condition. Since the dimension of the null space equals the codimension of the range (the number of completing range vectors needed), bordering will always keep the matrix square. This situation is also illustrated in Figure 16.1, where point A represents the canonical solution (the solution that is orthogonal to the null space of the matrix operator), point B represents some other solution obtained by the completion procedure, and the distance between A and B corresponds to the free parameter a.

Second-kind operators with compact integral operators behave analogously. The dimension of the null space coincides with the codimension of the range, and the bordering procedure can be applied. In the case of a single particle in infinite fluid the equations become

$$\begin{aligned} \boldsymbol{u}(\boldsymbol{\eta}) &= \varphi(\boldsymbol{\eta}) + \mathcal{K}\varphi(\boldsymbol{\eta}) + (\boldsymbol{F} - \frac{1}{2}\boldsymbol{T} \times \nabla) \cdot \mathcal{G}(\boldsymbol{\eta})/(8\pi\mu) \\ &\quad \text{for } \boldsymbol{\eta} \in S \qquad (16.6) \\ \langle \varphi, \varphi_i \rangle &= 0 \text{ for } i = 1, \ldots, 6 \,, \qquad (16.7) \end{aligned}$$

from which $\boldsymbol{F}$ and $\boldsymbol{T}$ can be directly solved. Note again that the origin can be freely chosen within the particle as we please, or actually the Stokeslet can be replaced with a cluster of Stokeslets whose contribution to the total torque must be separately considered. The equation system above is again of the second kind, although this is not obvious from the way the equations are written above, the operator now acting on the product space $L_2(S) \times R^6$.

16.3.2 On Removing the Null Functions

Admittedly the choice of the orthogonality conditions is again somewhat arbitrary. Actually the use of arbitrarily chosen orthogonality conditions—just having the right number of them—would almost surely remove the indeterminacy. This happens if the projections of the chosen vectors to the null space are linearly independent, *i.e.*, if the volume of the corresponding parallelepiped in the null space is nonzero. This volume being small although nonzero can, however, lead to numerical difficulties through an ill-conditioned linear system. Therefore it is better to use the null functions when they are known, or otherwise try to make a good choice for removing the indeterminacy caused by nontrivial null space.

The double layer representation can now be made fully determined for a multiparticle system within a container, since the exact number of null densities is known. In summary, the complete set of null functions is those given by RBMs on the particle surfaces plus the extra one for the container (if present). The last one, although not explicitly known, is denied by requiring orthogonality to the surface normal of the container. These are immediate consequences of Theorems 3 and 4.

16.3.3 Additions to the Range Space

Consider again a multiparticle system with or without a container. A double layer on the totality of surfaces can induce no total force or torque on any of the particles. Therefore one can add a Stokeslet and rotlet (or some fixed distribution of them) to the inside of each particle in sequence, knowing that from within one particle the flow fields of these will induce no force or torque on the surface of another — this argument shows that the codimension of the range is reduced at each step by the number of the functions added. For the container add a multiple of its surface normal (zero on the particle surfaces). According to Theorem 3 the range $R(1+\mathcal{K})$ is orthogonal to the surface normal on the totality of surfaces S, and so are the previous additions to the range, whereas this last one is not; therefore the codimension of the range is reduced again. Now the number of these additions to the range exactly matches the number of the original linearly independent null densities, which in turn equals the "deficiency" of the range, so it is known that the range is fully completed. *Thus a simple counting argument is used, such that it easily allows the treatment of multiparticle systems, even with a container.*

16.3.4 Summary of the Canonical Equations for Resistance and Mobility Problems

The following notation shall be used. There are n particles with surfaces S_i, $i = 1, \ldots, n$, with or without a containing surface S_c. Each solid particle is cancelling the ambient velocity field $\boldsymbol{u}$ and in addition moving with some RBM velocity $\boldsymbol{U}_i + \boldsymbol{\omega}_i \times \boldsymbol{\eta}$, exerting force $\boldsymbol{F}_i$ and torque $\boldsymbol{T}_i$ on the fluid. On the containing surface, if present, some velocity boundary conditions are imposed for the flow field, given by $\boldsymbol{u}_c$; in this case the ambient velocity field $\boldsymbol{u}$ on the particle surfaces is set to zero. The whole equation system is

$$-\boldsymbol{u}(\boldsymbol{\eta}) + \boldsymbol{U}_i + \boldsymbol{\omega}_i \times \boldsymbol{\eta} =$$
$$\varphi(\boldsymbol{\eta}) + \mathcal{K}\varphi(\boldsymbol{\eta}) + \sum_{j=1}^{n} \left\{ (\boldsymbol{F}_j - \frac{1}{2}\boldsymbol{T}_j \times \nabla) \cdot \mathcal{G}(\boldsymbol{\eta} - \boldsymbol{c}_j)/(8\pi\mu) \right\}$$

$$\text{for } \boldsymbol{\eta} \in S_i,\ i = 1, \ldots, n \tag{16.8}$$

$$\oint_{S_i} \boldsymbol{\varphi}(\boldsymbol{\xi}) \cdot \boldsymbol{\varphi}_{i,k}(\boldsymbol{\xi})\, dS(\boldsymbol{\xi}) = 0 \quad \text{for } i = 1, \ldots, n \text{ and } k = 1, \ldots, 6 \tag{16.9}$$

$$\begin{aligned} \boldsymbol{u}_c(\boldsymbol{\eta}) &= \boldsymbol{\varphi}(\boldsymbol{\eta}) + \mathcal{K}\boldsymbol{\varphi}(\boldsymbol{\eta}) + \sum_{j=1}^{n} \left\{ (\boldsymbol{F}_j - \frac{1}{2}\boldsymbol{T}_j \times \nabla) \cdot \mathcal{G}(\boldsymbol{\eta} - \boldsymbol{c}_j)/(8\pi\mu) \right\} \\ &+ \ V\hat{\boldsymbol{n}} \qquad \text{for } \boldsymbol{\eta} \in S_c \end{aligned} \tag{16.10}$$

$$\oint_{S_c} \boldsymbol{\varphi}(\boldsymbol{\xi}) \cdot \hat{\boldsymbol{n}}(\boldsymbol{\xi})\, dS(\boldsymbol{\xi}) = 0 \ . \tag{16.11}$$

Here $\boldsymbol{\varphi}_{i,k}$ are the RBM null densities for each of the particle surfaces, $\boldsymbol{c}_i$ the points inside each particle where the inside singularities are placed, and V the added variable for the container surface (usually zero). These shall be called the *canonical equations* for resistance and mobility problems, because the boundary conditions (for resistance problems) or force/torque constraints (for mobility problems) can be directly imposed and the results of interest directly solved numerically. In a resistance problem $\boldsymbol{F}_i$ and $\boldsymbol{T}_i$ are solved for, whereas these are given in a mobility problem and one solves for $\boldsymbol{U}_i$ and $\boldsymbol{\omega}_i$. With most methods the total force and torque are not explicit variables in the equations and so mobility problems cannot be directly attacked.

Several exceptions are listed below. The exterior singularity distribution method of Dabros [12] provides an approximate method, but currently there is no proof that it would always allow the pursuit of high accuracy and lead to well-posed systems. The method of Dabros is actually a variant of the general method explained by Mathon and Johnston [48] (he does not refer to this earlier work), and a more recent application to elasticity problems is given by Han and Olson [28]. For two equal spheres Yoon and Kim [71] have examined the direct solution of mobility problems, based on boundary collocation with an expansion of the velocity field in terms of Lamb's general solution for each sphere individually. The results indicate that the direct solution of mobility problems provides us with accurate results over a wide range of surface separations, including surfaces almost touching. Jeffrey and Onishi [35] have considered both resistance and mobility problems for two unequal spheres, and Ganatos *et al.* [22] for multiple spheres. With most methods it is necessary to first solve a general resistance problem and then to invert the linear relationship connecting all the forces and torques with all the RBM velocities of the particles. Especially in problems where the surfaces are almost touching, this inversion may be ill-conditioned.

The traction surface field is almost never known *a priori* for a physical flow problem. An exception here is the drop deformation problem studied by Rallison and Acrivos [59], where surface tension together with surface shape gives a boundary condition for the tractions, and the IE in primary variables is of the second kind. Some problems with velocity BCs can be transformed to *exterior* problems with traction BCs, but this transformation is mainly of theoretical interest [37]. The remaining problems that are well suited for BIE are either resistance or mobility problems for a collection of rigid particles. For these problems the canonical equations provide a natural setting, with full mathematical rigor for any smooth surface shapes and guaranteed well-posedness of the discretized linear systems. Corners and edges change the IE so that the Fredholm–Riesz–Schauder theory seems not to be applicable any more. Through numerical examples the complications caused by edges on the surface of a single particle in infinite fluid shall be studied: It seems that the smoothness restriction for surfaces, although necessary for the mathematical development here, can be neglected in numerical applications to some extent.

The bordered double layer equations can naturally be applied to any Stokes flow problems with velocity boundary conditions, not just those involving solid particles. The integral representation for the disturbance flow field in the interior fluid domain is found from the solved double layer density and the total forces and torques by dropping the $\boldsymbol{\varphi}$-terms and the $V\hat{\boldsymbol{n}}$-term from the RHSs of the canonical equations and substituting $\boldsymbol{\eta} \leftarrow \boldsymbol{y} \in \mathcal{Q}_{(i)}$. Recall that the $\boldsymbol{\varphi}$-term — giving a second-kind instead of a first-kind equation — came from a limiting procedure, when the double layer integral was evaluated *on the boundary surface* and this was used to replace the double layer representation as the surface is approached. In the interior fluid domain the double layer generated velocity field is $\boldsymbol{W}(\boldsymbol{y})$, as in Equation 15.6, and may also be denoted by $\mathcal{K}\boldsymbol{\varphi}(\boldsymbol{y})$. (This notation is *not* consistent with the interpretation of $\mathcal{K}$ as a linear operator mapping surface fields to *surface* fields.)

The completed double layer representation finds greatest utility in numerical applications, but the following analytic examples for a single sphere illustrate the general properties of the double layer representation. Note that the double layer density $\boldsymbol{\varphi}(\boldsymbol{\xi})$ can be determined from the jump properties of the double layer operator.

Example 16.1 The Translating Sphere

For translation induced by an external force, $\boldsymbol{F}$, we know that the Stokes solution may be written in the singularity form,

$$\boldsymbol{v}(\boldsymbol{y}) \;=\; \boldsymbol{F}\cdot\frac{\mathcal{G}(\boldsymbol{y})}{8\pi\mu} \;+\; \frac{a^2}{6}\boldsymbol{F}\cdot\frac{\nabla^2\mathcal{G}(\boldsymbol{y})}{8\pi\mu}\;.$$

Thus the problem reduces to finding a double layer density $\boldsymbol{\varphi}(\boldsymbol{\xi})$ that satisfies

$$\frac{a^2}{6}\boldsymbol{F}\cdot\nabla^2\frac{\mathcal{G}(\boldsymbol{y})}{8\pi\mu} \;=\; \oint_S \boldsymbol{K}(\boldsymbol{y},\boldsymbol{\xi})\cdot\boldsymbol{\varphi}(\boldsymbol{\xi})\;dS(\boldsymbol{\xi})\;.$$

We know that the velocity field induced by a double layer has a traction field that is continuous across the surface, while the velocity itself has a jump equal to twice the double layer density. So the solution strategy is to derive first the expression for the traction field of the degenerate Stokes quadrupole, and then find an interior solution that matches this traction. The inner and outer velocity fields will differ at the surface, this difference being exactly twice the desired density.

The degenerate quadrupole $(a^2/48\pi\mu)\boldsymbol{F}\cdot\nabla^2\mathcal{G}(\boldsymbol{y})$ has the stress field

$$\boldsymbol{\sigma} = -a^2|4\pi y|^5(\boldsymbol{Fy}+\boldsymbol{yF}+\boldsymbol{F}\cdot\boldsymbol{y\delta}) + \frac{5a^2}{4\pi|y|^7}\boldsymbol{F}\cdot\boldsymbol{yyy}\,,$$

so at the sphere surface $|\boldsymbol{y}| = a$ the traction is

$$\boldsymbol{\sigma}\cdot\hat{\boldsymbol{n}} = -\frac{\boldsymbol{F}}{4\pi a^4}(\boldsymbol{\delta}-3\hat{\boldsymbol{n}}\hat{\boldsymbol{n}})\,.$$

The interior velocity fields may be constructed from Lamb's general solution. However, from our experience with the Hadamard–Rybczynski drop problem, we recognize that the above matches the traction field of the Stokeson,

$$\begin{aligned}\boldsymbol{v} &= 2|\boldsymbol{y}|^2\boldsymbol{U}-\boldsymbol{U}\cdot\boldsymbol{yy}\\ p &= 10\mu\boldsymbol{U}\cdot\boldsymbol{y}\\ \boldsymbol{\sigma}\cdot\hat{\boldsymbol{n}} &= 3\mu a\boldsymbol{U}(\boldsymbol{\delta}-3\hat{\boldsymbol{n}}\hat{\boldsymbol{n}})\,.\end{aligned}$$

From a comparison of the two traction fields, we see that the required velocity field inside the sphere is

$$\boldsymbol{v}^{(i)} = \frac{\boldsymbol{F}}{12\pi\mu a^3}(\boldsymbol{yy}-2|\boldsymbol{y}|^2\boldsymbol{\delta})\,.$$

The desired solution for the double layer density follows:

$$\begin{aligned}\boldsymbol{\varphi}(\boldsymbol{\xi}) &= \frac{1}{2}(\boldsymbol{v}^{(o)}-\boldsymbol{v}^{(i)})\\ &= \frac{a^2}{96\pi\mu}\boldsymbol{F}\cdot\nabla^2\mathcal{G}(\boldsymbol{y})|_{y=a} - \frac{1}{24\pi\mu a}\boldsymbol{F}\cdot(\hat{\boldsymbol{n}}\hat{\boldsymbol{n}}-2\boldsymbol{\delta})\\ &= \frac{1}{48\pi\mu a}\boldsymbol{F}\cdot(\boldsymbol{\delta}-3\hat{\boldsymbol{n}}\hat{\boldsymbol{n}}) - \frac{1}{24\pi\mu a}\boldsymbol{F}\cdot(\hat{\boldsymbol{n}}\hat{\boldsymbol{n}}-2\boldsymbol{\delta})\\ &= -\frac{5\boldsymbol{F}\cdot\hat{\boldsymbol{n}}\hat{\boldsymbol{n}}}{48\pi\mu a} + \frac{5\boldsymbol{F}}{48\pi\mu a}\,.\end{aligned}$$

To this particular solution, we may add any null function. The solutions of interest are all of the form

$$\boldsymbol{\varphi} = -\frac{5\boldsymbol{F}\cdot\hat{\boldsymbol{n}}\hat{\boldsymbol{n}}}{48\pi\mu a} + C\frac{\boldsymbol{F}}{8\pi\mu a}\,.$$

We determine C of the *canonical solution* from the orthogonality condition, $\langle\boldsymbol{\varphi},\boldsymbol{F}\rangle = 0$ as $C = 5/18$ so that

$$\boldsymbol{\varphi}(\boldsymbol{\xi}) = -\frac{5\boldsymbol{F}}{48\pi\mu a}\cdot(\hat{\boldsymbol{n}}\hat{\boldsymbol{n}}-\frac{1}{3}\boldsymbol{\delta}) \quad \text{(Canonical solution)}$$

The solution approach of Power and Miranda, with the force set proportional to the projection on the null space,

$$\boldsymbol{F}\cdot\boldsymbol{U} = \frac{\mu}{a} < \boldsymbol{U},\boldsymbol{\varphi} >\,,$$

leads to $C = 2 + 5/18$ and the solution,

$$\varphi(\xi) = -\frac{5\boldsymbol{F}}{48\pi\mu a}\cdot(\hat{\boldsymbol{n}}\hat{\boldsymbol{n}} - \frac{1}{3}\boldsymbol{\delta}) + \frac{\boldsymbol{F}}{4\pi\mu a}\,.$$

Note that for the canonical solution the density is proportional to the degenerate quadrupole. We will have more to say on this matter when we examine the spectrum of $\mathcal{K}$ acting on a single sphere in Section 17.2.

We conclude this example with a comment on the use of an off-center Stokeslet, thus illustrating the point that range completion is possible with any vector that contains a nonzero projection in the missing part of the range. If we place the Stokeslet at a distance $R < a$ from the origin, or $R\boldsymbol{F}/F$, instead of at the origin, we must add the following terms:

$$\begin{aligned}\tilde{\varphi}(\xi) &= \frac{F}{8\pi\mu a}\sum_{n=2}^{\infty}\left\{\frac{(R/a)^{n-1}}{2(n-1)}\left[\boldsymbol{e}_r n P_n(\cos\theta) + \boldsymbol{e}_\theta\frac{\partial P_n}{\partial\theta}\right]\right.\\ &\quad + \left(\frac{(R/a)^{n-1}}{2n+1} - \frac{(R/a)^{n+1}}{2n+3}\right)\frac{n(2n+1)(2n+3)}{2(2n^2+4n+3)}\\ &\quad \left.\times\left[-\boldsymbol{e}_r(n+1)P_n(\cos\theta) + \boldsymbol{e}_\theta\frac{\partial P_n}{\partial\theta}\right]\right\}\end{aligned}$$

to the double layer density. The techniques for deriving this expression are described in Chapter 4. For small R/a, the dominant, $O(R/a)$, part of the correction is that due to a double layer representation for a Stokes dipole. As $R/a \to 1$, the contributions from the higher order terms become significant and the highly oscillatory nature of φ is the price paid for the poor choice of range completion.◇

Example 16.2 The Rotating Sphere

This is a trivial example, because a rotlet placed at the sphere center produces the exact solution for the rotating sphere. Therefore, the canonical solution is $\varphi(\xi) = 0$.◇

Example 16.3 Fixed Sphere in a Constant Rate-of-Strain Field

The disturbance field is

$$v_i(\boldsymbol{y}) = S_{jk}\mathcal{G}_{ij,k}(\boldsymbol{y})/8\pi\mu + \frac{a^2}{10}S_{jk}\frac{\nabla^2\mathcal{G}_{ij,k}(\boldsymbol{y})}{8\pi\mu}\,,$$

and the double layer distribution must represent both the dipole and degenerate octupole terms. At $r = a$, the disturbance field generates the traction

$$\boldsymbol{\sigma}\cdot\hat{\boldsymbol{n}} = \frac{9\boldsymbol{S}\cdot\hat{\boldsymbol{n}}}{20\pi a^3}\,.$$

This traction is also generated by the linear interior velocity field,

$$\boldsymbol{v}^{(i)} = \frac{9\boldsymbol{S}\cdot\boldsymbol{y}}{40\pi\mu a^3},$$

and the double layer density follows as

$$\varphi(\boldsymbol{\xi}) = \frac{1}{2}(\boldsymbol{v}^{(o)} - \boldsymbol{v}^{(i)}) = \frac{-3\boldsymbol{S}\cdot\hat{\boldsymbol{n}}}{40\pi\mu a^2} - \frac{9\boldsymbol{S}\cdot\hat{\boldsymbol{n}}}{80\pi\mu a^2} = \frac{-3\boldsymbol{S}\cdot\hat{\boldsymbol{n}}}{16\pi\mu a^2}. \diamond$$

16.4 RBM-Tractions from the Riesz Representation Theorem

16.4.1 The Riesz Representation Theorem

In finite dimensional inner product spaces it is easy to see that any linear functional L can be uniquely represented by a corresponding vector $\boldsymbol{v}$, the functional being the inner product (dot product) with this vector. For this purpose let $\boldsymbol{x} = \sum_{i=1}^m x_i\hat{e}_i$ be a general vector expressed in terms of the orthonormal basis vectors. Then

$$\begin{aligned} L(\boldsymbol{x}) &= \sum_{i=1}^m x_i L(\hat{e}_i) \\ &= \sum_{i=1}^m \langle \boldsymbol{x}, \hat{e}_i\rangle \; L(\hat{e}_i) \\ &= \langle\, \boldsymbol{x}, \sum_{i=1}^m \hat{e}_i L(\hat{e}_i)\,\rangle \\ &= \langle \boldsymbol{x}, \boldsymbol{v}\rangle \end{aligned}$$

shows the existence of $\boldsymbol{v}$. (The equations above require complex conjugation of the scalars absorbed into the second arguments of the inner products, in case the inner product is complex.) The uniqueness follows on substituting $\boldsymbol{x} = \boldsymbol{v}_1 - \boldsymbol{v}_2$ into $\langle \boldsymbol{x}, \boldsymbol{v}_1\rangle = \langle \boldsymbol{x}, \boldsymbol{v}_2\rangle$. This unique representation of bounded linear functionals as inner products is extended to certain infinite dimensional spaces, the *Hilbert spaces*, by the *Riesz representation theorem* (see any of the references cited when discussing compact operators in the beginning of this part). For our purposes it suffices to know that the set of infinitely differentiable functions, and therefore also the larger class of functions with Hölder-continuous first derivatives, is a dense subspace of the Hilbert space whose elements are quadratically summable functions. Then any bounded linear functional on either of these subspaces is also uniquely representable by a quadratically summable function (which may or may not be continuous). This is because such a functional can uniquely be extended to a bounded functional on the whole Hilbert space of functions.

16.4.2 Force and Torque from Lorentz Reciprocal Theorem

Let again S be the totality of bounding surfaces for a fluid volume. According to Lorentz reciprocal theorem

$$\oint_S \boldsymbol{u}(\boldsymbol{\xi}) \cdot \boldsymbol{\sigma}(\boldsymbol{\xi}; \boldsymbol{v}) \cdot \hat{\boldsymbol{n}}(\boldsymbol{\xi})\, dS(\boldsymbol{\xi}) = \oint_S \boldsymbol{v}(\boldsymbol{\xi}) \cdot \boldsymbol{\sigma}(\boldsymbol{\xi}; \boldsymbol{u}) \cdot \hat{\boldsymbol{n}}(\boldsymbol{\xi})\, dS(\boldsymbol{\xi}) \ , \tag{16.12}$$

or in inner product notation,

$$\langle \boldsymbol{u}, \boldsymbol{\sigma}(;\boldsymbol{v}) \cdot \hat{\boldsymbol{n}} \rangle = \langle \boldsymbol{v}, \boldsymbol{\sigma}(;\boldsymbol{u}) \cdot \hat{\boldsymbol{n}} \rangle \tag{16.13}$$

for any Stokes velocity fields $\boldsymbol{u}$ and $\boldsymbol{v}$ in the fluid, with smooth tractions. Choose $\boldsymbol{v}$ so that on one particle surface it equals $\boldsymbol{U} + \boldsymbol{\omega} \times \boldsymbol{\eta}$ and is zero on the rest of S. This corresponds to moving one particle while holding the rest of the surfaces fixed. Then the equations above become

$$\langle \boldsymbol{u}, \boldsymbol{\sigma}(;\boldsymbol{v}) \cdot \hat{\boldsymbol{n}} \rangle = -\left(\boldsymbol{U} \cdot \boldsymbol{F} + \boldsymbol{\omega} \cdot \boldsymbol{T}\right) \ , \tag{16.14}$$

where $\boldsymbol{F}$ and $\boldsymbol{T}$ are the force and torque exerted by the particle surface on the fluid with flow field $\boldsymbol{u}$. From the theorems proven in this chapter we know that the tractions corresponding to $\boldsymbol{v}$ are single layer null functions and therefore are smooth. The same holds for $\boldsymbol{u}$ at least if this has Hölder-continuous first derivatives, since then the bordered double layer representation can be applied with a density of similar smoothness, and this results in smooth tractions. By choosing $\boldsymbol{v}$ suitably we can pick out components of $\boldsymbol{F}$ and $\boldsymbol{T}$ in the form of an inner product acting on the given field $\boldsymbol{u}$, *i.e.*, here we have the linear functionals that associate the surface velocity field to a corresponding force and torque, on a dense subspace of the underlying Hilbert space. These functionals are bounded since the tractions are smooth and thus quadratically summable, and we already have their representations in the form given by the Riesz representation theorem. This connection of RBM tractions to force and torque in a general flow field has been noted by Brenner (1964), but the *uniqueness* seems to have escaped notice so far. The uniqueness means in practice that these equations can be used to solve for the RBM tractions, provided that we already have a linear functional that maps a surface velocity field to a component of corresponding force or torque. Such functionals are supplied (at least as numerical approximations) by any method that can solve the general resistance problem, in particular by a numerical implementation of the canonical equations.

On applying the Lorentz reciprocal theorem to infinite fluid domains, we must again assume that the fields obey a decay condition enabling an "enclosing sphere expanded to infinity" argument. Then the pressure field corresponding to the RBM tractions is uniquely determined by this decay condition, and no undetermined constants arise. The case of a finite fluid domain with a containing surface is discussed below and in the next section.

We shall also present here a somewhat different view on applying the Lorentz and Riesz theorems to get RBM tractions. Consider the canonical equations, which are well-posed equations of the second kind. The linear functionals mapping the data to components of the solution are bounded, which is just another way of expressing the well-posedness. The Riesz representation theorem can be applied to the bounded linear functionals mapping the LHSs of the equation system to components of the forces and torques. Let $\boldsymbol{v}$ be the surface field that represents, in the sense of Riesz theorem, the mapping to the component F_1 acting from within some specified particle. If the data is in $R(1+\mathcal{K})$, *i.e.*, representable by a double layer alone, the total forces and torques must vanish. Thus $\boldsymbol{v}$ must be orthogonal to the range above, or equivalently it must be in $N(1+\mathcal{K}^*)$, and so it is the sum of an RBM traction field and the tractions due to constant pressure p in the interior. With the Lorentz reciprocal theorem it is found that the RBM traction field must be that corresponding to translation of the particle in question in the "1-direction," while on the other boundaries this velocity field vanishes. This is most easily accomplished by choosing the "test field" $\boldsymbol{u}$ to be that due to the inside singularities (point forces and torques). The surface normal on the container alone corresponds to a solution where only V is nonzero of all the solved quantities, and so must be orthogonal to $\boldsymbol{v}$; this condition determines the constant multiple of the surface normal in the RBM tractions, due to base pressure. Therefore again the RBM tractions correspond through the Riesz theorem to the functionals mapping velocity boundary conditions to components of forces and torques. If there is no container surface, the tractions correspond to a decaying stress field, as in Equation 15.19. If a container is present, RBM tractions are indeterminate up to a constant pressure, but on using Riesz theorem with the canonical equations this constant is determined, as was just shown. Later on, this connection of the RBM tractions to the functionals giving the total force and torque shall be applied in the numerical computations.

16.4.3 Tractions for Rigid-Body Motion

We now show how the existence and uniqueness considerations above can be utilized in practice. On solving our canonical equations we find the double layer density and the strengths of the inside singularities. Getting the surface tractions from these requires, in general, elaborate integrations. We should take the stress of the bordered double layer representation, and use this representation of the stress field in the fluid with the known double layer density distribution. However, for the particular case of RBM tractions on the particle surfaces, we can use the theory above to create an efficient numerically applicable procedure. On discretizing the IE we replace the infinite process of integration with some finite approximation to get a numerically solvable linear system. This same discretization replaces our original natural inner product with a discrete version $\langle \boldsymbol{a}, \boldsymbol{b} \rangle \approx [\boldsymbol{A}, \boldsymbol{B}]$. Let the discrete linear system be $\boldsymbol{U} = \boldsymbol{K}\boldsymbol{X}$, where $\boldsymbol{U}$ contains the discretized velocity surface field and the zeroes for orthogonality conditions. Consider one component F_1 of one of the forces, a scalar that is also one of the variables in $\boldsymbol{X}$. We first extend the discrete inner product so that the added variables are accounted for. Choosing $\boldsymbol{C}$ properly we have

$$\begin{aligned} F_1 &= [\boldsymbol{C}, \boldsymbol{X}] \\ &= [\boldsymbol{C}, \boldsymbol{K}^{-1}\boldsymbol{U}] \\ &= [\boldsymbol{K}^{-t}\boldsymbol{C}, \boldsymbol{U}] \\ &= [\boldsymbol{D}, \boldsymbol{U}], \end{aligned}$$

where $\boldsymbol{D}$ is found by solving

$$\boldsymbol{K}^t\boldsymbol{D} = \boldsymbol{C}. \tag{16.15}$$

Because the extended part of $\boldsymbol{U}$ contains only zeroes for the orthogonality conditions, we can drop the extension also from $\boldsymbol{D}$ and consider the inner product with $\boldsymbol{U}$ as the discretized approximation alone. Since for a suitable RBM traction field $\boldsymbol{d}$, *and only for this*, we have

$$F_1 = \langle \boldsymbol{d}, \boldsymbol{u} \rangle, \text{ for any smooth } \boldsymbol{u}, \tag{16.16}$$

we must have a discrete approximation of this traction surface field in the current form of $\boldsymbol{D}$. *So we use the reciprocal theorem in exactly the opposite direction from how it conventionally is used and justify this with Riesz theorem.* Thus the discretized RBM tractions can be found by solving the transposed linear system, and after this computing the strengths of singularities reduce to computing dot products. If the forces and torques are to be found for several given flow fields, we save some time by just once solving for the vectors $\boldsymbol{D}$ and applying these to all of the flow fields — this being just the conventional way of applying Brenner's result. The line of reasoning above shows that any method that provides the forces and torques in *arbitrary* flow fields necessarily contains all the information about the RBM traction fields. The uniqueness part of Riesz representation theorem is essential to this deduction.

The particular case where we have a container deserves a further comment. Since the Stokes velocity field in a bounded volume is mass conserving on the totality of surfaces, the velocity surface fields $\boldsymbol{u}$ used with the reciprocal theorem are restricted by this one condition $\langle \boldsymbol{u}, \hat{\boldsymbol{n}} \rangle = 0$. Then the vector $\boldsymbol{d}$ is also unique only up to a constant multiple of $\hat{\boldsymbol{n}}$. This is in complete agreement with the fact that in a bounded volume the flow field determines the pressure only up to a constant. We have, however, added the new variable V to complete the range of the container double layer, and this makes F_1 fully determined for *any* surface field $\boldsymbol{u}$ (given velocity BCs), including those that are not mass conserving. Therefore there is no further indeterminacy in our equations, but the pressure is automatically set to some fixed value.

Later on a modification of the canonical equations will be presented, which allows the efficient solution of mobility problems. For that case it can be shown, quite similarly as has been done above for resistance problems, that the functionals mapping ambient-velocity boundary conditions to translation and rotation velocities that keep the particles force- and torque-free are again represented by RBM tractions. The difference is that now the tractions are "mobility-based," corresponding to given forces and torques, instead of the "resistance-based" tractions corresponding to given translation and rotation velocities. These ideas are also brought out in the exercises.

16.5 The Stresslet

Above it was shown how the tractions corresponding to rigid-body motion *without any ambient field* can be numerically found from a well-posed formulation. Naturally for this case other weighted averages of the tractions can then be computed also, such as the stresslet. However, it turns out that the double layer singularity is closely related to the stresslet, so that for *solid particles* a completely general formula can be derived that relates the stresslet to the double layer density. This of course implies that, in complete analogy to the previous section, the surface tractions for a particle in a rate-of-strain field can be obtained directly by the combined use of the Riesz and Lorentz theorems.

As the reader may recall from Part II, the first terms in the multipole expansion, before the stresslet, involve the total force and torque, and these are directly solvable from the canonical equations. As these terms do not interact with the stresslet, we shall drop them from the completed (bordered) double layer representation, and inspect the double layer term alone.

Consider the multipole expansion of the double layer term:

$$\begin{aligned}\int_S K_{ij}(\boldsymbol{y},\boldsymbol{\xi})\varphi_j(\boldsymbol{\xi})\,dS(\boldsymbol{\xi}) &= \frac{\mathcal{P}_i(\boldsymbol{y})}{4\pi\mu}\oint_S \boldsymbol{\varphi}(\boldsymbol{\xi})\cdot\hat{\boldsymbol{n}}\,dS(\boldsymbol{\xi}) \\ &- \frac{1}{4\pi}\oint_S \hat{n}_k\varphi_j(\boldsymbol{\xi})\,dS(\boldsymbol{\xi})\ \left(\mathcal{G}_{ij,k}(\boldsymbol{y})+\mathcal{G}_{ik,j}(\boldsymbol{y})\right)\ +\ \cdots\ .\end{aligned}$$

The first term on the RHS vanishes because $2\boldsymbol{\varphi} = \boldsymbol{v}^{(o)} - \boldsymbol{v}^{(i)}$ and both velocity fields have no net flux through the particle surface. Since the stresslet is the coefficient of the symmetric Stokes dipole, we conclude that the stresslet is given exactly by the expression

$$\boldsymbol{S} \;=\; -2\mu\oint_S (\hat{\boldsymbol{n}}\boldsymbol{\varphi}+\boldsymbol{\varphi}\hat{\boldsymbol{n}})\ dS(\boldsymbol{\xi})\ .$$

This result for the stresslet should be compared with that obtained in Part II for the mobile interface; there the surface velocity played the same role as the double layer density (or more precisely, the density $2\boldsymbol{\varphi}$). Finally, we note that this more general result concerning the stresslet is consistent with the result for the sphere in a rate-of-strain field (Example 16.3).

Exercises

Exercise 16.1 The Adjoint of a Matrix.
Note: In this exercise we exceptionally denote the complex conjugation of a scalar with the superscript c for clarity.
Let $\boldsymbol{A} = (a_{ij})_{N\times N}$ be a complex matrix, and let the inner product between

two complex N-vectors be $\langle \boldsymbol{x}, \boldsymbol{y} \rangle = x_i y_i^c$. Show that the matrix $\boldsymbol{A}^*$ satisfying $\langle \boldsymbol{A}\boldsymbol{x}, \boldsymbol{y} \rangle = \langle \boldsymbol{x}, \boldsymbol{A}^* \boldsymbol{y} \rangle$, for all $\boldsymbol{x}$ and $\boldsymbol{y}$, is given by $\boldsymbol{A}^* = (a_{ij}^*)_{N \times N}$ with $a_{ij}^* = a_{ji}^c$. Deduce that $(\boldsymbol{A}^*)^* = \boldsymbol{A}$.

Exercise 16.2 The Relation Between Null Space and Range.
Show that for an ordinary square matrix $\boldsymbol{A}$ the relation $R(\boldsymbol{A}) = N(\boldsymbol{A}^*)^\perp$ holds. By substituting $\boldsymbol{A}^*$ for $\boldsymbol{A}$ derive the complementary relation $N(\boldsymbol{A}) = R(\boldsymbol{A}^*)^\perp$.
Hint: See the previous exercise.

Exercise 16.3 The RBM Tractions for a Mobility Problem.
Let $\boldsymbol{u}$ be a given velocity surface field, $\boldsymbol{v}_{RBM}$ the velocity field corresponding to total force $\boldsymbol{F}$ and torque $\boldsymbol{T}$ (mobility solution), and $\boldsymbol{u}_{RBM} = \boldsymbol{U} + \boldsymbol{\omega} \times \boldsymbol{r}$ such RBM velocity that $\boldsymbol{u} - \boldsymbol{u}_{RBM}$ is force- and torque-free. Show that

$$\boldsymbol{U} \cdot \boldsymbol{F} + \boldsymbol{\omega} \cdot \boldsymbol{T} = -\langle \boldsymbol{u}, \hat{\boldsymbol{n}} \cdot \boldsymbol{\sigma}(;\boldsymbol{v}_{RBM}) \rangle \ .$$

Hint: Explain the following steps:

$$\begin{aligned} 0 &= \langle \hat{\boldsymbol{n}} \cdot \boldsymbol{\sigma}(;\boldsymbol{u} - \boldsymbol{u}_{RBM}), \boldsymbol{v}_{RBM} \rangle \\ &= -\langle \boldsymbol{u}_{RBM}, \hat{\boldsymbol{n}} \cdot \boldsymbol{\sigma}(;\boldsymbol{v}_{RBM}) \rangle + \langle \boldsymbol{u}, \hat{\boldsymbol{n}} \cdot \boldsymbol{\sigma}(;\boldsymbol{v}_{RBM}) \rangle \\ &= (\boldsymbol{U} \cdot \boldsymbol{F} + \boldsymbol{\omega} \cdot \boldsymbol{T}) + \langle \boldsymbol{u}, \hat{\boldsymbol{n}} \cdot \boldsymbol{\sigma}(;\boldsymbol{v}_{RBM}) \rangle \ . \end{aligned}$$

Note: This shows the physical significance of mobility-based tractions; they map a given velocity field to (components of) such an RBM that absorbs the total force and torque. Another view of the same fact is presented in the section on iterative solution of RBM tractions, in the next chapter, using the completed (deflated) double layer representation.

Exercise 16.4 Unique Decomposition of Disturbance Fields.
The integral representation of Stokes fields, involving both single and double layers, can be utilized to show that the disturbance field of a multisurface problem can be decomposed to the sum of disturbance fields of each individual surface. Show that such a decomposition is unique in case we have no container. State the smoothness conditions involved.
Hint: The integral over totality of surfaces is the sum of integrals over individual surfaces. For uniqueness use the completed double layer representation, known to have a unique solution.
Note: The significance of this result is that multiparticle problems are sometimes handled by matching single-particle disturbance fields, for example, a two-sphere problem can be handled by superposing Lamb's general solution for each sphere (see Chapter 13). Now we know *a priori* that the resulting system has a unique solution; we have not introduced any indeterminacies by artificial decomposition of the problem.

Chapter 17

Iterative Solutions for Mobility Problems

The solution of large linear systems is best accomplished by some iterative method. Such methods are available if an approximate inverse for the coefficient matrix can be (inexpensively) constructed, or if the matrix satisfies certain conditions so that direct iteration is possible.

The spectrum of the double layer integral operator will be inspected, and we shall show how the operator can be modified so that this modified analytic operator satisfies the conditions for feasibility of direct iteration. If the discretized linear system closely enough approximates the analytical equations, we may expect it to share the same properties, thus allowing direct iterative solution.

17.1 Conditions for Successful Direct Iteration

Here we shall briefly discuss the direct, or Picard, iteration. Let $\mathcal{K}$ be a linear operator on a Hilbert space and b a given vector. On trying to solve the equation $x = \mathcal{K}x + b$ for x, it seems natural to try the iteration $x_{i+1} = \mathcal{K}x_i + b$ — note that the second kind IEs are directly of this form. Suppose the iterations are started with the approximate solution x_1, and consider the errors $z_i = x_i - x$. These errors satisfy the recursion relation $z_{i+1} = \mathcal{K}z_i$ from which by induction $z_i = \mathcal{K}^{i-1}z_1$. For the errors to vanish in the limit of a large number of iterations, regardless of how the initial guess x_1 was chosen, it is necessary and sufficient that the powers of $\mathcal{K}$ vanish in the limit of large exponents. For compact operators $\mathcal{K}$ this condition is equivalent to requiring the *spectral radius* of $\mathcal{K}$ to be strictly less than one. The spectral radius is by definition the largest, of all absolute values of the eigenvalues of $\mathcal{K}$. Finite matrices are compact operators, so that in particular the result mentioned applies to ordinary matrices as well.

Assume for the moment that $\mathcal{K}$ would have an eigenvalue larger or equal to one in absolute magnitude. Then it is easy to see that the iterations will not converge if the initial error vector z_1 is the corresponding eigenvector, as on each iteration this is just multiplied with the eigenvalue. This shows the necessity of the condition on the spectral radius.

Assume now that all the eigenvalues of $\mathcal{K}$ are less than one in absolute magnitude. If z_1 is an eigenvector, clearly the error will go to zero in the limit. Suppose z_1 is a principal vector such that $(\lambda 1 - \mathcal{K})z_1$ is an eigenvector z. Then $\mathcal{K}z_1 = \lambda z_1 - z$ and by induction $\mathcal{K}^i z_1 = \lambda^i z_1 - i\lambda^{i-1}z$, which again tends to zero since $|\lambda| < 1$. This argument can be extended to principal vectors further back in the Jordan chain with induction. Finally, as any initial guess vector z_1 can be represented in the basis formed by the eigenvectors and other principal vectors, it is sufficient to inspect these separately, as was done above. This shows (readers desiring complete rigor are referred to textbooks on mathematics) the sufficiency of the condition on the spectral radius. Also please note how the errors in the directions of the proper vectors decrease with iterations as the powers of the corresponding eigenvalue — the spectral radius gives the dominating eigenvalue, and therefore characterizes the rate of convergence. In our case the double layer operator in fact is diagonalizable (see the exercises), so that discussion of other principal vectors than eigenvectors is just for the sake of generality.

The direct iteration scheme is naturally not the only possible one, but it is the simplest and suffices for the application to be presented. Sometimes an *approximate inverse* H is known for the system matrix $1 - K$, such that $H(1 - K)$ is very nearly identity. Then the original equation can be written as $x = [1 - H(1 - K)]x + Hb$, by a small rearrangement after premultiplication with H, and now the bracketed "iteration operator" here is small enough (in terms of the spectral radius) to allow direct iteration. As a final note we observe that any operator norm derived from a vector norm by $|\mathcal{K}| = \sup |\mathcal{K}x|/|x|$ gives an upper bound for the spectral radius and thus can be used to show the feasibility of iterative solutions.

17.2 The Spectrum of the Double Layer Operator

In this section two results given by Odqvist [53] for single bounding surfaces shall be established in the multiboundary setting, namely, that the spectra of the operators $\mathcal{K}$ and $\mathcal{K}^*$ are real and therefore coincide, and this common spectrum is within the interval $[-1, 1]$. Since Odqvist omits the details, our exposition more closely follows that given in Karrila [37]. As the reader may recall, in general the spectra of an operator and its adjoint are mapped to each other by complex conjugation (see Section 14.2.5). The eigenvalues of $\mathcal{K}$ at the endpoints of this interval were already examined in the theorems within this chapter, since these are the physically relevant ones.

In the following complex conjugation shall be denoted by the superscript c.

For complex surface fields in L_2 the natural inner product is

$$\langle \boldsymbol{f}, \boldsymbol{g} \rangle = \oint_S \boldsymbol{f} \cdot \boldsymbol{g}^c \, dS \ . \tag{17.1}$$

A complex Stokes solution is such that its real and imaginary parts are ordinary Stokes solutions. The stress field is defined as before and will also be complex. By using the Lorentz reciprocal theorem for real homogeneous Stokes solutions, and the linearity of the operation of taking the stresses, one can verify that the Lorentz reciprocal theorem is valid for complex Stokes fields in the form:

$$\langle \boldsymbol{v}, \hat{\boldsymbol{n}} \cdot \boldsymbol{\sigma}(; \boldsymbol{u}) \rangle = \langle \hat{\boldsymbol{n}} \cdot \boldsymbol{\sigma}(; \boldsymbol{v}), \boldsymbol{u} \rangle \ . \tag{17.2}$$

Also the dissipation of a complex velocity field is the sum of the dissipations of its real and imaginary parts, and is given by $\langle \boldsymbol{v}, \hat{\boldsymbol{n}} \cdot \boldsymbol{\sigma}(; \boldsymbol{v}) \rangle$. Thus the dissipation is always non-negative and is zero only if there is complex RBM.

Let $\boldsymbol{\psi}$ be a complex single layer eigenfunction on the totality of bounding surfaces S. *If the total dissipation* (sum of dissipations in the interior and the exterior) *is zero*, then there can be only RBM in each connected component of the interior and exterior. Thus on each of these there is a constant pressure, and the single layer density being proportional to the traction jumps over surfaces is a multiple of the surface normal on each surface. A constant multiple of the surface normal $p\hat{\boldsymbol{n}}$ on a container surface will give zero stress on the outside, $-p\hat{\boldsymbol{n}}$ on S_c (Theorem 2), and the constant pressure field $-2p\boldsymbol{\delta}$ as the stress field on the inside. On a particle surface the main difference is the direction of the surface normal, now pointing outwards. Then one gets zero stress field outside, $p\hat{\boldsymbol{n}}$ on S_i, and the constant pressure field $2p\boldsymbol{\delta}$ as the stress field on the inside. If the fluid domain is bounded, having a container boundary surface, the single layers on the particles will not affect on S_c according to the above, and so the eigenvalue must be -1. If there is no container, the eigenvalue is similarly deduced to be 1, since the particles do not affect each other with these densities (see Exercise 17.13). Thus *an eigenfunction with $\lambda \neq \pm 1$ necessarily induces nonzero (positive) dissipation in the whole space.*

Let $\boldsymbol{\psi}$ now be a complex eigenfunction of $\mathcal{K}^*$, satisfying

$$0 = -\lambda \boldsymbol{\psi} + \mathcal{K}^* \boldsymbol{\psi} \ , \tag{17.3}$$

where $\lambda \neq \pm 1$. On denoting the LHSs of Equations 15.17 by $\boldsymbol{T}_{(i)}$ and $\boldsymbol{T}_{(e)}$, it is found that

$$2\boldsymbol{\psi} = \boldsymbol{T}_{(e)} - \boldsymbol{T}_{(i)} \tag{17.4}$$

$$2\mathcal{K}^* \boldsymbol{\psi} = \boldsymbol{T}_{(e)} + \boldsymbol{T}_{(i)} \ . \tag{17.5}$$

Substitute these into 17.3 to get

$$0 = (1 - \lambda) \boldsymbol{T}_{(e)} + (1 + \lambda) \boldsymbol{T}_{(i)} . \tag{17.6}$$

Note also that the energy equation 14.14 gives the dissipations

$$\left. \begin{aligned} D_{(i)} &= 2\mu \textstyle\int_{Q_{(i)}} \boldsymbol{e}{:}\boldsymbol{e}^c \, dQ = -\oint_S \boldsymbol{T}_{(i)} \cdot \boldsymbol{V}^c \, dS \\ D_{(e)} &= 2\mu \textstyle\int_{Q_{(e)}} \boldsymbol{e}{:}\boldsymbol{e}^c \, dQ = \oint_S \boldsymbol{T}_{(e)} \cdot \boldsymbol{V}^c \, dS \end{aligned} \right\} , \tag{17.7}$$

where the continuity of $\boldsymbol{V}$ across the boundary was used. The sign change is due to using the normal $\hat{\boldsymbol{n}}$ pointing into $Q_{(i)}$ in all equations. Multiply the first equation by $-(1+\lambda)$ and the second by $1-\lambda$ before adding; then the surface integrals cancel according to Equation 17.6 and

$$-(1+\lambda)D_{(i)} + (1-\lambda)D_{(e)} = 0 . \tag{17.8}$$

Taking the imaginary part of this equation gives

$$\Im(\lambda)\left(D_{(i)} + D_{(e)}\right) = 0 . \tag{17.9}$$

Since the total dissipation is a positive real number, λ must be real. Going back to the previous equation, solving for λ gives

$$\lambda = \left(D_{(e)} - D_{(i)}\right) / \left(D_{(e)} + D_{(i)}\right) , \tag{17.10}$$

where the division is allowed for $\lambda \neq \pm 1$, and this shows that $\lambda \in [-1, 1]$, since the dissipations are non-negative.

Odqvist referred to Plemelj's work, presumably on electrostatic problems, in connection with the proof of the spectrum being real. He did not consider complex Stokes solutions, as is done here, and also otherwise the treatment here is more detailed, but it is most likely that the proof he had in mind was along the spirit of the one presented here. In the next two subsections, we derive the spectrum for the single-sphere and for the two-sphere system. In both cases, we obtain analytic results for the spectrum that illustrate the salient points of the general theory.

17.2.1 The Spectrum for the Sphere

For a sphere, the eigenfunctions can be obtained by using Lamb's solution for the interior and exterior flows, and applying the jump conditions across a double layer distribution. Suppose that φ is an eigenfunction of the double layer operator, *i.e.*,

$$\int_S \boldsymbol{K}(\boldsymbol{\eta}, \boldsymbol{\xi}) \cdot \boldsymbol{\varphi}(\boldsymbol{\xi})\, dS(\boldsymbol{\xi}) \;=\; \lambda \boldsymbol{\varphi}(\boldsymbol{\eta}) .$$

The flow fields outside and inside the particle then satisfy the following conditions at the particle surface:

$$\begin{aligned} \boldsymbol{\varphi} + \mathcal{K}(\boldsymbol{\varphi}) &= (\lambda+1)\boldsymbol{\varphi} = \boldsymbol{v}^{(o)} \\ -\boldsymbol{\varphi} + \mathcal{K}(\boldsymbol{\varphi}) &= (\lambda-1)\boldsymbol{\varphi} = \boldsymbol{v}^{(i)} . \end{aligned}$$

Therefore, it follows that φ is an eigenfunction of the double layer operator if and only if it satisfies the jump conditions,

$$\boldsymbol{\sigma}^{(o)} \cdot \hat{\boldsymbol{n}} = \boldsymbol{\sigma}^{(i)} \cdot \hat{\boldsymbol{n}} \qquad \text{and} \qquad \boldsymbol{v}^{(i)} = \kappa \boldsymbol{v}^{(o)} ,$$

with κ defined as $\kappa = (\lambda - 1)/(\lambda + 1)$. The first condition is the usual statement that tractions are continuous across a double layer distribution; the second condition says that the velocity discontinuity takes a special form for eigenfunctions.

It turns out that the toroidal field, $\nabla \times (\boldsymbol{x}\chi)$, decouples from the other problems, so that we may break the task at hand into two smaller subproblems. On the unit sphere, the velocity field $\nabla \times (\boldsymbol{x}\chi_n) = -\boldsymbol{x} \times \nabla\chi_n$ has the surface traction (Chapter 4)

$$\boldsymbol{\sigma} \cdot \hat{\boldsymbol{n}} \;=\; \mu(n-1)\nabla \times (\boldsymbol{x}\chi_n)$$

for both positive n (interior flows) and negative n (exterior flows). The exterior and interior surface tractions match if we apply the usual correspondence between n and $-n-1$. For example, for axisymmetric problems $m = 0$, if the harmonics χ_n are written as $c_n r^n P_n$ and $C_n r^{-n-1} P_n$, we have

$$\begin{aligned}\boldsymbol{\sigma}^{(i)} \cdot \hat{\boldsymbol{n}} &= -c_n \mu(n-1)\boldsymbol{e}_\phi \frac{\partial P_n}{\partial \theta} \\ \boldsymbol{\sigma}^{(o)} \cdot \hat{\boldsymbol{n}} &= C_n \mu(n+2)\boldsymbol{e}_\phi \frac{\partial P_n}{\partial \theta} \;.\end{aligned}$$

Matching of surface tractions implies

$$c_n(n-1) \;+\; C_n(n+2) \;=\; 0 \;,$$

while the jump condition for the velocity fields requires

$$\boldsymbol{x} \times \nabla\chi_n \;=\; \kappa \boldsymbol{x} \times \nabla\chi_{-n-1}$$

or

$$c_n \;=\; \kappa_n C_n \;.$$

The system of equations for c_n and C_n has a nontrivial solution if and only if the determinant vanishes, which implies that

$$\kappa_n \;=\; -\frac{n+2}{n-1} \;.$$

The eigenvalues follow from the definition of κ_n as

$$\lambda_n \;=\; \frac{1+\kappa_n}{1-\kappa_n} \;=\; \frac{-3}{2n+1}\;, \qquad n = 1, 2, 3, \ldots \;.$$

The first few eigenvalues are -1, $-3/5$, $-3/7, \ldots$, with -1 corresponding to rigid-body rotation about the z-axis, as required by the general theory. It is also easy to show that eigenfunctions are also obtained starting from the more general form for χ_n involving $P_n^m e^{im\phi}$, $0 \le |m| \le n$, so that λ_n has multiplicity $2n+1$, with the corresponding eigenfunctions,

$$(1+\lambda_n)^{-1}\nabla \times (\boldsymbol{x}\chi_n)|_{r=1} \;.$$

The factor of $(1+\lambda_n)^{-1}$ was introduced here to make the velocity generated by the double layer equal to $\nabla \times (\boldsymbol{x}\chi_n)$. Note also that this velocity field is proportional to the orthogonal basis function, $\boldsymbol{\varphi}^{(3)}$, of Chapter 4.

We now turn our attention to the pressure and $\nabla\Phi$ terms in Lamb's solution. Again, consider axisymmetric fields first. Suppose that the interior and exterior velocity fields are given by

$$\begin{aligned} \boldsymbol{v}^{(i)} &= b_n \nabla\Phi_n + a_n \left\{ \frac{(n+3)r^2\nabla p_n}{2(n+1)(2n+3)} - \frac{n\boldsymbol{x}p_n}{(n+1)(2n+3)} \right\} \\ \boldsymbol{v}^{(o)} &= B_n \nabla\Phi_{-n-1} + A_n \left\{ -\frac{(n-2)r^2\nabla p_{-n-1}}{2n(2n-1)} + \frac{(n+1)\boldsymbol{x}p_{-n-1}}{n(2n-1)} \right\} . \end{aligned}$$

Then on the unit sphere, the surface tractions are (Chapter 4)

$$\begin{aligned} \frac{\boldsymbol{\sigma}^{(i)}\cdot\hat{\boldsymbol{n}}}{\mu} &= b_n 2(n-1)\nabla\Phi_n + a_n \left\{ -\frac{(2n^2+4n+3)}{(n+1)(2n+3)}\boldsymbol{x}p_n + \frac{n(n+2)r^2\nabla p_n}{(n+1)(2n+3)} \right\} \\ &= \boldsymbol{e}_r P_n \left\{ b_n 2n(n-1) + a_n \frac{(n^2-n-3)}{(2n+3)} \right\} \\ &\quad + \boldsymbol{e}_\theta \frac{\partial P_n}{\partial\theta} \left\{ b_n 2(n-1) + a_n \frac{n(n+2)}{(n+1)(2n+3)} \right\} \\ \frac{\boldsymbol{\sigma}^{(o)}\cdot\hat{\boldsymbol{n}}}{\mu} &= -B_n 2(n+2)\nabla\Phi_{-n-1} \\ &\quad + A_n \left\{ -\frac{(2n^2+1)}{n(2n-1)}\boldsymbol{x}p_{-n-1} + \frac{(n+1)(n-1)}{n(2n-1)} r^2 \nabla p_{-n-1} \right\} \\ &= \boldsymbol{e}_r P_n \left\{ B_n 2(n+1)(n+2) - A_n \frac{(n^2+3n-1)}{(2n-1)} \right\} \\ &\quad + \boldsymbol{e}_\theta \frac{\partial P_n}{\partial\theta} \left\{ -B_n 2(n+2) + A_n \frac{(n+1)(n-1)}{n(2n-1)} \right\} \end{aligned}$$

and matching implies that

$$2n(n-1)b_n + \frac{(n^2-n-3)}{(2n+3)}a_n - 2(n+1)(n+2)B_n + \frac{(n^2+3n-1)}{(2n-1)}A_n = 0 \quad (17.11)$$

$$2(n-1)b_n + \frac{n(n+2)}{(n+1)(2n+3)}a_n + 2(n+2)B_n - \frac{(n+1)(n-1)}{n(2n-1)}A_n = 0 \,. \quad (17.12)$$

The relation $\boldsymbol{v}^{(i)} = \kappa\boldsymbol{v}^{(o)}$ provides two more equations (from the r and θ components):

$$nb_n + \frac{n}{2(2n+3)}a_n + \kappa(n+1)B_n - \kappa\frac{(n+1)}{2(2n-1)}A_n = 0 \quad (17.13)$$

$$b_n + \frac{(n+3)}{2(n+1)(2n+3)}a_n - \kappa B_n + \kappa\frac{(n-2)}{2n(2n-1)}A_n = 0 \,. \quad (17.14)$$

The problem thus reduces to finding κ such that

$$\begin{vmatrix} 2n(n-1) & \frac{n^2-n-3}{2n+3} & -2(n+2)(n+1) & \frac{n^2+3n-1}{2n-1} \\ 2(n-1) & \frac{n(n+2)}{(n+1)(2n+3)} & 2(n+2) & -\frac{(n+1)(n-1)}{n(2n-1)} \\ n & \frac{n}{2(2n+3)} & (n+1)\kappa & -\frac{(n+1)}{2(2n-1)}\kappa \\ 1 & \frac{n+3}{2(n+1)(2n+3)} & -\kappa & \frac{(n-2)}{2n(2n-1)}\kappa \end{vmatrix} = 0 \,.$$

This is a quadratic equation for κ. Note that the determinant is invariant with respect to the operation of replacing n and κ by $-n-1$ and $1/\kappa$, because of the corresponding relationship between the interior and exterior fields in Lamb's general solution. Therefore, if $\kappa(n)$ is a root, then so is $1/\kappa(-n-1)$.

After expansion of the determinant, we obtain the quadratic equation explicitly,

$$\kappa^2 + \frac{8n^4+16n^3+4n^2-4n+3}{2(n^2-1)(2n^2+4n+3)}\kappa + \frac{n(n+2)(2n^2+1)}{(n^2-1)(2n^2+4n+3)} = 0$$

or

$$\kappa^2 + \left[2 + \frac{3(4n+5)}{2(n^2-1)(2n^2+4n+3)}\right]\kappa + \left[1 + \frac{3(2n+1)}{(n^2-1)(2n^2+4n+3)}\right] = 0\,.$$

The discriminant of this quadratic equation,

$$\frac{16n^4+32n^3+24n^2+8n+1}{4(n^2-1)^2(2n^2+4n+3)^2} = \frac{(2n+1)^4}{4(n^2-1)^2(2n^2+4n+3)^2}\,,$$

is a perfect square, and the roots turn out to be

$$\kappa_n^{\pm} = -1 + \frac{3}{4}\left[\frac{-(4n+5) \pm (2n+1)^2}{(n^2-1)(2n^2+4n+3)}\right]\,.$$

We may simplify this further as

$$\kappa_n^- = -\frac{2n^2+1}{2(n^2-1)}$$
$$\kappa_n^+ = -\frac{2n(n+2)}{2n^2+4n+3}\,,$$

and we see that the two branches do indeed satisfy the relation,

$$\kappa_{-(n+1)}^- = \kappa_{n+1}^- = \frac{1}{\kappa_n^+}\,,$$

as stated earlier. The corresponding eigenvalues are found to be

$$\lambda_n^- = \frac{1+\kappa_n^-}{1-\kappa_n^-} = \frac{-3}{(2n-1)(2n+1)}\,, \qquad \text{for } n = 1, 2, 3, \ldots$$
$$\lambda_n^+ = \frac{1+\kappa_n^+}{1-\kappa_n^+} = \frac{3}{(2n+1)(2n+3)}\,, \qquad \text{for } n = 0, 1, 2, \ldots\,.$$

The two eigenvalue branches satisfy the relation,

$$\lambda_{n+1}^- = -\lambda_n^+\,, \qquad \text{for } n = 0, 1, 2, \ldots\,.$$

The first few are ± 1, $\pm 1/5$, $\pm 3/35$, $\pm 1/21$, $\pm 1/33$, $\ldots$.

As before, the nonaxisymmetric case follows the same structure, so that the degeneracy of the eigenvalue $\lambda_n^\pm$ is $(2n+1)$. Note that the matched pair λ_{n+1}^- and λ_n^+ have different multiplicities, *viz.*, $2n+3$ and $2n+1$, respectively. For $n=0$, we already know from the general theory that there are three translational null functions ($\lambda=-1$) and one container null function ($\lambda=1$). This analysis also confirms the obvious result that for a sphere, the container null function is in fact the surface normal.

The five eigenfunctions with $\lambda_2^- = -1/5$ correspond to a sphere in a rate-of-strain field, $\boldsymbol{v}^\infty = \boldsymbol{E}\cdot\boldsymbol{x}$, of which there are five linearly independent fields. These, in fact, were encountered in Section 16.3 in the discussion of the sphere in a rate-of-strain field. The three eigenfunctions with $\lambda_2^+ = 1/5$ correspond to the degenerate quadrupole, $\boldsymbol{F}\cdot\nabla^2\mathcal{G}(\boldsymbol{\xi})$, which we also encountered in Section 16.3.

We turn our attention to the general description of the eigenfunctions. We assume that κ is appropriately chosen so that the system of equations, 17.11–17.14, has a nontrivial solution. We discard 17.11 and use pairings 17.13 and 17.14, as well as 17.12 and 17.14, to obtain two independent equations without the b_ns:

$$(2n+1)\kappa B_n - \frac{\kappa}{2}A_n = \frac{na_n}{(n+1)(2n+3)}$$

$$\{2\kappa(n-1)+2(n+2)\}\,B_n - \frac{(n-1)\,(\kappa(n-2)+n+1)}{n(2n-1)}A_n = \frac{-3a_n}{(n+1)(2n+3)}\ .$$

We may eliminate a_n from these equations to obtain

$$\left[\kappa(2n^2+4n+3)+2n(n+2)\right]B_n = \frac{\kappa(2n^2+1)+2(n-1)(n+1)}{2(2n-1)}A_n\ . \tag{17.15}$$

If we use $\kappa=\kappa_n^+$, then the nontrivial solution can be chosen as $A_n=0$ and $B_n=1$, and the eigenfunctions φ_n^+, proportional to

$$\nabla\Phi_{-n-1}|_{r=1}\ ,\qquad \text{with}\quad \lambda_n^+ = \frac{3}{(2n+1)(2n+3)}\ ,$$

for $n\ge 0$ are obtained. On the other hand, if we use $\kappa=\kappa_n^-$, then Equation 17.15 simplifies to

$$B_n = \frac{1}{2(2n+1)}A_n\ .$$

These eigenfunctions, which we denote as φ_n^-, are of the form,

$$\left\{-\frac{(n-2)r^2\nabla p_{-n-1}}{2n(2n-1)}+\frac{(n+1)\boldsymbol{x}p_{-n-1}}{n(2n-1)}\right\}\Big|_{r=1}+\frac{\nabla p_{-n-1}|_{r=1}}{2(2n+1)}\ ,$$

and the associated eigenvalues are

$$\lambda_n^- = -\frac{3}{(2n-1)(2n+1)}\ .$$

The eigenfunctions may also be written in terms of interior fields. With the above relation between A_n and B_n, we find that $a_n = 0$, so that only the b_n term survives. Therefore, the eigenfunction φ_n^- is also proportional to $\nabla(r^n P_n)|_{r=1}$.

Using the relations between Lamb's general solution and the multipole expansion (Chapter 4), we may *define* φ_n^+ and φ_n^- as

$$(1+\lambda_n^+)\varphi_n^+ = \boldsymbol{P}^{(n-1)} \cdot \nabla^2 \mathcal{G}|_{r=1} \tag{17.16}$$

$$(1+\lambda_n^-)\varphi_n^- = \boldsymbol{P}^{(n-1)} \cdot \left\{1 + \frac{1}{2(2n+1)}\nabla^2\right\} \mathcal{G}|_{r=1} - \frac{\boldsymbol{P}^{(n-3)} \cdot \mathcal{G}|_{r=1}}{2(2n-1)} , \tag{17.17}$$

where $\boldsymbol{P}^{(n)}$ is the vector differential operator that gives an n-th multipole. As double layer densities, the eigenfunctions $\varphi_n^{\pm}$ generate outside the sphere the velocity fields on the RHS of Equations 17.16 and 17.17. For $\lambda_1^- = -1$, Equation 17.17 should be interpreted as the RBM translational null function generating zero velocity outside the sphere.

The multipole expansion for axisymmetric flows (symmetry axis $\boldsymbol{d} = -\boldsymbol{e}_z$) past a sphere,

$$\boldsymbol{v} = \sum_{n=0}^{\infty} \left\{ A_n \frac{(\boldsymbol{d}\cdot\nabla)^n}{n!}\boldsymbol{d}\cdot\mathcal{G} + B_n \frac{(\boldsymbol{d}\cdot\nabla)^n}{n!}\boldsymbol{d}\cdot\nabla^2\mathcal{G} \right\} ,$$

may also be written as a completed double layer representation,

$$\boldsymbol{v} = A_0 \boldsymbol{d}\cdot\mathcal{G} + \oint_S \boldsymbol{K}(\boldsymbol{x},\boldsymbol{\xi})\cdot\boldsymbol{\varphi}(\boldsymbol{\xi})\, dS ,$$

with the double layer density on the sphere surface expanded as

$$\boldsymbol{\varphi}(\boldsymbol{\xi}) = \sum_{n=1}^{\infty} \alpha_n \varphi_{n+1}^-(\boldsymbol{\xi}) + \sum_{n=0}^{\infty} \beta_n \varphi_{n+1}^+(\boldsymbol{\xi}) .$$

Using this result, we may express many of the results from Parts II and III with the completed double layer representation, thereby expanding our knowledge of analytic solutions for CDL-BIEM. In particular, the image for a Stokeslet near a fixed rigid sphere may be expressed as a combination of a Stokeslet inside the sphere plus a double layer density on the sphere surface. The "smoothness" of the double layer density depends on the location of the Stokeslet inside the sphere, thus providing an instructive example of the importance of selecting a good basis for completing the range (see the exercises).

We conclude the discussion by noting that $\mathcal{K}$ is self-adjoint for the sphere, *i.e.*, $K_{ij}(\boldsymbol{\eta},\boldsymbol{\xi}) = K_{ji}(\boldsymbol{\xi},\boldsymbol{\eta})$. This result follows from, among other things,

$$\hat{\boldsymbol{n}}(\boldsymbol{\xi})\cdot(\boldsymbol{\eta}-\boldsymbol{\xi}) = -\hat{\boldsymbol{n}}(\boldsymbol{\eta})\cdot(\boldsymbol{\eta}-\boldsymbol{\xi}) ,$$

a condition satisfied for points $\boldsymbol{\xi}$ and $\boldsymbol{\eta}$ on a sphere. It then follows that the eigenfunctions of the double layer operator are orthogonal. This is the underlying principle behind the orthogonal properties of the orthonormalized basis of

Chapter 4:

$$\begin{aligned}
\varphi_{nm}^{(1)}(\boldsymbol{\xi}) &= \sqrt{n(2n+1)}\frac{(2n-1)}{\eta_{nm}(n+1)} \\
&\quad \times \left\{ -\frac{(n-2)r^2\nabla}{2n(2n-1)} + \frac{(n+1)\boldsymbol{x}}{n(2n-1)} + \frac{\nabla}{2(2n+1)} \right\} \left\{ r^{-n-1}\tilde{Y}_n^m \right\} \\
&= \frac{\sqrt{n(2n+1)}}{\eta_{nm}} \left[\frac{r^{-n}\boldsymbol{A}_{nm}}{n(2n+1)} + \frac{(2n-1)}{(2n+1)} \frac{(r^{-2}-1)\, r^{-n}\boldsymbol{B}_{nm}}{2(n+1)} \right]
\end{aligned} \tag{17.18}$$

$$\varphi_{nm}^{(2)} = \frac{\nabla\left\{ r^{-n-1}\tilde{Y}_n^m \right\}}{\eta_{nm}\sqrt{(n+1)(2n+1)}} = \frac{r^{-n-2}\boldsymbol{B}_{nm}}{\eta_{nm}\sqrt{(n+1)(2n+1)}} \tag{17.19}$$

$$\varphi_{nm}^{(3)} = \frac{i}{\eta_{nm}\sqrt{n(n+1)}} \nabla \times \left\{ \boldsymbol{x} r^{-n-1}\tilde{Y}_n^m \right\} = \frac{i r^{-n-1}\boldsymbol{C}_{nm}}{\eta_{nm}\sqrt{n(n+1)}}\ , \tag{17.20}$$

all evaluated at $r = 1$. The associated eigenvalues are $-3/(2n-1)(2n+1)$ $(n \geq 1)$, $3/(2n+1)(2n+3)$ $(n \geq 0)$, and $-3/(2n+1)$ $(n \geq 1)$, respectively.

It is also now clear why the canonical solution in the translation problem (Section 16.3) was simply a degenerate quadrupole evaluated at the sphere surface – the degenerate quadrupole "generates itself," since it is an eigenfunction, and it is orthogonal to the null functions since those are eigenfunctions too.

17.2.2 Double Layer Eigenfunctions for Ellipsoids in Rate-of-Strain Fields

This section emphasizes the fact that only the extreme eigenvalues at $\lambda = \pm 1$ (associated with the null functions) are universal (independent of particle shape). The specific analytical example considered here is the eigenvalues/eigenfunctions corresponding to the disturbance fields for an ellipsoid in the rate-of-strain fields, $\boldsymbol{E} \cdot \boldsymbol{x}$. For the sphere we just saw that only the eigenvalue $-1/5$ comes up in this way, so that there are five independent eigenfunctions corresponding to it. For the general ellipsoid this degeneracy disappears, and we shall discover five eigenfunctions with, in general, different eigenvalues. We will inspect how these eigenvalues change continuously with the shape parameters of the ellipsoid.

As before, we pick our coordinate system so that the ellipsoid is described by the equation,

$$\frac{x^2}{a^2} + \frac{y^2}{b^2} + \frac{z^2}{c^2} = 1\ ,$$

with $a \geq b \geq c > 0$. We will show that the linear field $\boldsymbol{E} \cdot \boldsymbol{\xi}$, $\xi \in S$ is an eigenfunction of the double layer operator on the ellipsoidal surface, with $\boldsymbol{E}$ corresponding physically to a constant rate-of-strain tensor.

We decompose the rate-of-strain tensor as follows:

$$\begin{pmatrix} E_{11} & E_{12} & E_{13} \\ E_{21} & E_{22} & E_{23} \\ E_{31} & E_{32} & E_{33} \end{pmatrix} = \begin{pmatrix} 0 & E_{12} & 0 \\ E_{21} & 0 & 0 \\ 0 & 0 & 0 \end{pmatrix} + \begin{pmatrix} 0 & 0 & E_{13} \\ 0 & 0 & 0 \\ E_{31} & 0 & 0 \end{pmatrix}$$

$$+\begin{pmatrix} 0 & 0 & 0 \\ 0 & 0 & E_{23} \\ 0 & E_{32} & 0 \end{pmatrix} + \begin{pmatrix} E_{11} & 0 & 0 \\ 0 & E_{22} & 0 \\ 0 & 0 & E_{33} \end{pmatrix} ,$$

with $E_{11} + E_{22} + E_{33} = 0$. In the following, we denote these rate-of-strain fields as $\boldsymbol{E}^{(1)}, \ldots, \boldsymbol{E}^{(4)}$. We first consider the "off-diagonal" cases involving ($\boldsymbol{E}^{(1)}$, $\boldsymbol{E}^{(2)}$, $\boldsymbol{E}^{(3)}$).

Off-Diagonal Rate-of-Strain Fields

Our strategy is to construct velocity fields outside and inside the ellipsoid, with matching traction fields. The double layer density is one-half of the velocity jump, and if it is also proportional to the outer and inner fields evaluated at the ellipsoid surface, then we have an eigenfunction.

Consider the rate-of-strain field,

$$\boldsymbol{E}^{(1)} = \begin{pmatrix} 0 & E_{12} & 0 \\ E_{21} & 0 & 0 \\ 0 & 0 & \end{pmatrix} ,$$

immerse in it a torque-free ellipsoid rotating about the z-axis with the rotational velocity,

$$\omega_3 = \left(\frac{a^2 - b^2}{a^2 + b^2}\right) E_{12} .$$

The disturbance field of this force-free, torque-free ellipsoid can be represented by the double layer alone. From Jeffery's classical paper (see also Chapter 3), we have the following simple result for the tractions *on the ellipsoid surface*:

$$\boldsymbol{\sigma}^{(o)} \cdot \hat{\boldsymbol{n}} = \left[\frac{4\mu}{\gamma_0'(a^2 + b^2)} - 2\mu\right] \boldsymbol{E}^{(1)} \cdot \hat{\boldsymbol{n}} .$$

We have followed the definition,

$$\gamma_0' = abc \int_0^\infty \frac{dt}{(a^2 + t)(b^2 + t)\Delta(t)} ,$$

which differs from Jeffery's by the factor of (abc).

This surface traction is matched inside the ellipsoid by the linear velocity field,

$$\boldsymbol{v}^{(i)} = \left[\frac{2}{\gamma_0'(a^2 + b^2)} - 1\right] \boldsymbol{E}^{(1)} \cdot \boldsymbol{x} ,$$

and the double layer density can be determined immediately from

$$2\boldsymbol{\varphi} = \boldsymbol{v}^{(o)} - \boldsymbol{v}^{(i)} = \left[-\boldsymbol{E}^{(1)} \cdot \boldsymbol{\xi} + \boldsymbol{\omega} \times \boldsymbol{\xi}\right] - \left[\frac{2}{\gamma_0'(a^2 + b^2)} - 1\right] \boldsymbol{E}^{(1)} \cdot \boldsymbol{\xi} .$$

The rotatonal motion is a null solution and generates no flow, so we may drop it. Or equivalently, we could put the same RBM rotation inside the particle, a flow that generates no traction, and then the expression for the velocity jump (= 2φ) would not involve angular velocities at all. We thus arrive at the conclusion that the double layer density is given by

$$\varphi = -\frac{\boldsymbol{E}^{(1)} \cdot \boldsymbol{\xi}}{\gamma_0'(a^2+b^2)} ,$$

and is proportional to $\boldsymbol{v}^{(i)}$ and $\boldsymbol{v}^{(o)}$. Therefore, $\boldsymbol{E}^{(1)} \cdot \boldsymbol{\xi}$ is an eigenfunction. The eigenvalue follows from

$$\lambda = \frac{1+\kappa}{1-\kappa} , \qquad \text{with} \qquad \boldsymbol{v}^{(i)} = \kappa \boldsymbol{v}^{(o)} .$$

Since this eigenvalue is associated with E_{12} and E_{21}, we denote it as $\lambda^{(12)}$.

We determine κ from the expression for the inner and outer velocities, and the expressions for the eigenvalues follow from the preceding equations as

$$\begin{aligned}
\lambda^{(12)} &= (a^2+b^2)\gamma_0' - 1 \\
\lambda^{(23)} &= (b^2+c^2)\alpha_0' - 1 \\
\lambda^{(31)} &= (c^2+a^2)\beta_0' - 1 ,
\end{aligned}$$

with the last two following by permutative symmetry. All three eigenvalues become equal to $-1/5$ when $a = b = c$, but for the general ellipsoid, they are nondegenerate. This completes the discussion of the eigenvalues associated with the off-diagonal rate-of-strain fields.

Diagonal Rate-of-Strain Fields

Consider the rate-of-strain field,

$$\begin{pmatrix} E_{11} & 0 & 0 \\ 0 & E_{22} & 0 \\ 0 & 0 & E_{33} \end{pmatrix} ,$$

with $E_{11} + E_{22} + E_{33} = 0$. Jeffery defines the constants, A, B, and C by

$$\begin{aligned}
A &= \frac{1}{6}\frac{2\alpha_0''E_{11} - \beta_0''E_{22} - \gamma_0''E_{33}}{\beta_0''\gamma_0'' + \gamma_0''\alpha_0'' + \alpha_0''\beta_0''} \\
B &= \frac{1}{6}\frac{2\beta_0''E_{22} - \gamma_0''E_{33} - \alpha_0''E_{11}}{\beta_0''\gamma_0'' + \gamma_0''\alpha_0'' + \alpha_0''\beta_0''} \\
C &= \frac{1}{6}\frac{2\gamma_0''E_{33} - \alpha_0''E_{11} - \beta_0''E_{22}}{\beta_0''\gamma_0'' + \gamma_0''\alpha_0'' + \alpha_0''\beta_0''}
\end{aligned}$$

and shows that on the ellipsoid surface the tractions of the disturbance field simplify to

$$\begin{aligned}
(\boldsymbol{\sigma}\cdot\hat{\boldsymbol{n}})_x &= [8\mu A - 2\mu E_{11} - 4\mu(\alpha_0 A + \beta_0 B + \gamma_0 C)]\, n_x \\
(\boldsymbol{\sigma}\cdot\hat{\boldsymbol{n}})_y &= [8\mu B - 2\mu E_{22} - 4\mu(\alpha_0 A + \beta_0 B + \gamma_0 C)]\, n_y \\
(\boldsymbol{\sigma}\cdot\hat{\boldsymbol{n}})_z &= [8\mu C - 2\mu E_{33} - 4\mu(\alpha_0 A + \beta_0 B + \gamma_0 C)]\, n_z .
\end{aligned}$$

Again, Jeffery's expression for the traction differs from ours (the first terms on the RHSs of the above equations) because of the factor of (abc) difference in the definitions of the elliptic integrals.

The traction field is of the form,

$$\boldsymbol{\sigma}^{(o)} \cdot \hat{\boldsymbol{n}} = 2\mu \boldsymbol{D} \cdot \hat{\boldsymbol{n}} - 2\mu \boldsymbol{E}^{(4)} \cdot \hat{\boldsymbol{n}} \; - \; 4\mu(\alpha_0 A + \beta_0 B + \gamma_0 C)\hat{\boldsymbol{n}} \; ,$$

with

$$\boldsymbol{D} = \begin{pmatrix} 4A & 0 & 0 \\ 0 & 4B & 0 \\ 0 & 0 & 4C \end{pmatrix} .$$

The constant pressure may be dropped from further consideration, because it is not used at all in the expression for the interior field,

$$\boldsymbol{v}^{(i)} \; = \; \boldsymbol{D} \cdot \boldsymbol{x} - \boldsymbol{E}^{(4)} \cdot \boldsymbol{x} \; .$$

We now set the condition, $\boldsymbol{v}^{(i)} = \kappa \boldsymbol{v}^{(o)}$, which leads to

$$4A \; = \; (1-\kappa)E_{11} \; , \qquad 4B \; = \; (1-\kappa)E_{22} \; , \qquad 4C \; = \; (1-\kappa)E_{33} \; .$$

We insert the expressions for A, B, and C to arrive at the following system of equations:

$$\begin{aligned} \left[(\kappa-1) + \frac{4\alpha_0''}{3d}\right] E_{11} - \frac{2\beta_0''}{3d}E_{22} - \frac{2\gamma_0''}{3d}E_{33} \; &= \; 0 \\ -\frac{2\alpha_0''}{3d}E_{11} \left[(\kappa-1) + \frac{4\beta_0''}{3d}\right] E_{22} - \frac{2\gamma_0''}{3d}E_{33} \; &= \; 0 \\ -\frac{2\alpha_0''}{3d}E_{11} - \frac{2\beta_0''}{3d}E_{22} + \left[(\kappa-1) + \frac{4\gamma_0''}{3d}\right] E_{33} \; &= \; 0 \; . \end{aligned}$$

Note, however, that $E_{11}+E_{22}+E_{33}=0$, and so these three variables and the three equations are not independent. Under these circumstances, κ is determined by taking any two equations from the preceding set of three, plus the zero-trace condition, $E_{11}+E_{22}+E_{33}=0$. The problem thus reduces to finding κ such that

$$\begin{vmatrix} (\kappa-1) + \frac{4\alpha_0''}{3d} & -\frac{2\beta_0''}{3d} & -\frac{2\gamma_0''}{3d} \\ -\frac{2\alpha_0''}{3d} & (\kappa-1) + \frac{4\beta_0''}{3d} & -\frac{2\gamma_0''}{3d} \\ 1 & 1 & 1 \end{vmatrix} = 0 \; .$$

After expansion of the determinant, we obtain the quadratic equation,

$$(\kappa-1)^2 \; + \; \frac{4}{3d}(\alpha_0''+\beta_0''+\gamma_0'')(\kappa-1) \; + \; \frac{4}{3d} \; = 0 \; ,$$

with roots

$$\kappa^{\pm} \; = \; 1 - \frac{2}{3d}(\alpha_0''+\beta_0''+\gamma_0'') \pm \frac{2}{3d}\sqrt{[\alpha_0'']^2 + [\beta_0'']^2 + [\gamma_0'']^2 - d} \; ,$$

and the *two* eigenvalues follow as

$$\lambda^{\pm} \; = \; \frac{1+\kappa^{\pm}}{1-\kappa^{\pm}} \; .$$

We now consider further simplifications of these results as applied to ellipsoids of revolution.

Prolate Spheroids

We set $a \geq b = c$, then the eigenvalues from the off-diagonal case simplify to

$$\begin{aligned} \lambda^{(12)} = \lambda^{(31)} &= (a^2 + c^2)\gamma_0' - 1 \\ \lambda^{(23)} &= 2c^2\alpha_0' - 1 \, , \end{aligned}$$

while the eigenvalues from the diagonal rate-of-strain field reduce to

$$\begin{aligned} \lambda^- &= -1 + 3\gamma_0'' \\ \lambda^+ &= -1 + 2\alpha_0'' + \gamma_0'' \, . \end{aligned}$$

The rate-of-strain field for the eigenvalue λ^- corresponds to uniaxial extension along the x-axis (an axisymmetric flow about the x-axis), while that for $\lambda^+ = \lambda^{(23)}$ corresponds to two-dimensional hyperbolic flow in the yz-plane. Thus as expected, for a prolate spheroid there are three distinct eigenvalues. If the shape of the spheroid is specified by the eccentricity, $e = \sqrt{a^2 - c^2}/a$, of the generating ellipse, then the relevant elliptic integrals simplify to (see Chapter 3):

$$\begin{aligned} \alpha_0' &= \frac{1}{a^2 8e^5(1-e^2)}\left[2(5e^3 - 3e) + 3(1-e^2)^2 \log\left(\frac{1+e}{1-e}\right)\right] \\ \gamma_0' &= \frac{1}{a^2e^5}\left[3e - 2e^3 - \frac{3}{2}(1-e^2)\log\left(\frac{1+e}{1-e}\right)\right] \\ \alpha_0'' &= \frac{3-e^2}{4e^4} - \frac{(1-e^2)(e^2+3)}{8e^5}\log\left(\frac{1+e}{1-e}\right) \\ \gamma_0'' &= \frac{(1-e^2)}{e^5}\left[-3e + \frac{1}{2}(3-e^2)\log\left(\frac{1+e}{1-e}\right)\right] \, . \end{aligned}$$

The eigenvalues have the explicit expressions,

$$\begin{aligned} \lambda^{(12)} = \lambda^{(31)} &= -1 + \frac{(2-e^2)}{e^5}\left[3e - 2e^3 - \frac{3}{2}(1-e^2)\log\left(\frac{1+e}{1-e}\right)\right] \\ \lambda^{(23)} = \lambda^+ &= -1 + \frac{5e^2-3}{2e^4} + \frac{3(1-e^2)^2}{4e^5}\log\left(\frac{1+e}{1-e}\right) \\ \lambda^- &= -1 + \frac{3(1-e^2)}{e^5}\left[-3e + \frac{1}{2}(3-e^2)\log\left(\frac{1+e}{1-e}\right)\right] \, . \end{aligned}$$

The asymptotic behavior in the limit for near-spheres ($e \to 0$) are

$$\begin{aligned} \lambda^{(12)} = \lambda^{(31)} &\sim -\frac{1}{5} - \frac{2}{35}e^2 + \cdots \\ \lambda^{(23)} = \lambda^+ &\sim -\frac{1}{5} + \frac{4}{35}e^2 + \cdots \\ \lambda^- &\sim -\frac{1}{5} - \frac{4}{35}e^2 + \cdots \, . \end{aligned}$$

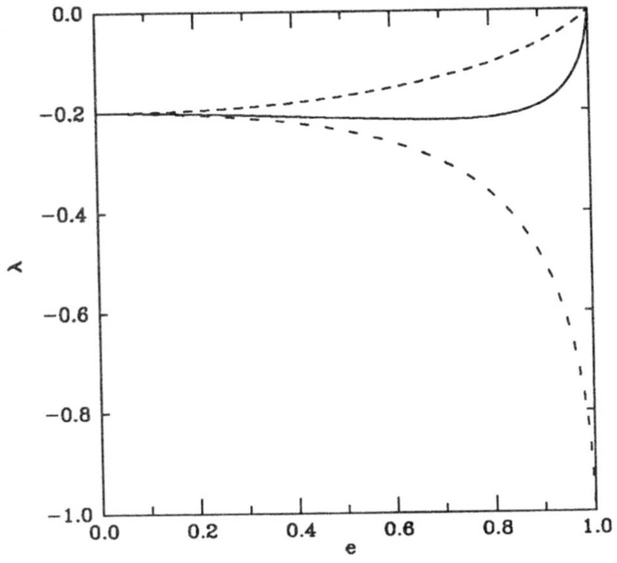

Figure 17.1: Eigenvalues for the prolate spheroid in a rate-of-strain field.

The asymptotic behavior in the slender-body limit ($e \to 1$) are

$$\begin{aligned}
\lambda^{(12)} = \lambda^{(31)} &\sim \{-3\log(2\epsilon) + 5\}\,\epsilon^2 + \cdots \\
\lambda^{(23)} = \lambda^{+} &\sim -\frac{1}{2}\epsilon^2 + \cdots \\
\lambda^{-} &\sim -1 + \{6\log(2\epsilon) - 9\}\,\epsilon^2 + \cdots ,
\end{aligned}$$

with $\epsilon = c/a = \sqrt{1-e^2}$, the inverse of the aspect ratio. We see that the spectral radius of the deflated system must approach unity with increasing aspect ratio. The accompanying figure provides a plot of these eigenvalues *vs.* e.

Oblate Spheroids

We set $a = b > c$, then the eigenvalues from the off-diagonal case simplify to

$$\begin{aligned}
\lambda^{(12)} &= 2a^2\gamma_0' - 1 \\
\lambda^{(23)} = \lambda^{(31)} &= (a^2 + c^2)\alpha_0' - 1 ,
\end{aligned}$$

while the eigenvalues from the diagonal rate-of-strain field reduce to

$$\begin{aligned}
\lambda^{-} &= -1 + 3\alpha_0'' \\
\lambda^{+} &= -1 + \alpha_0'' + 2\gamma_0'' .
\end{aligned}$$

The rate-of-strain field for the eigenvalue λ^- corresponds to uniaxial extension along the z-axis (an axisymmetric flow about the x-axis), while that for $\lambda^+ =$

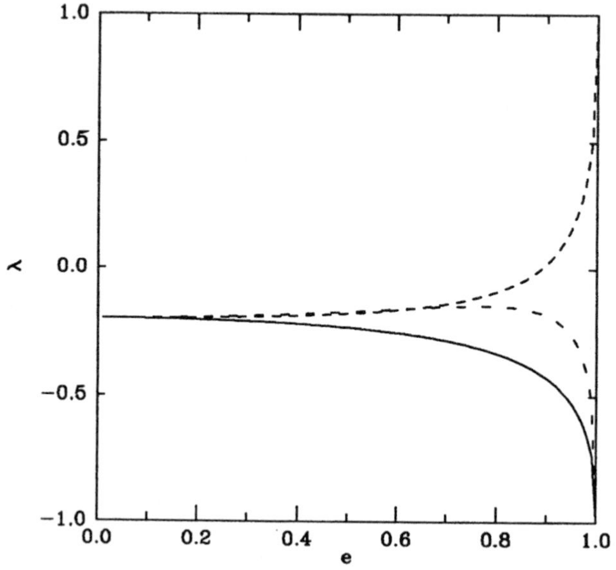

Figure 17.2: Eigenvalues for the oblate spheroid in a rate-of-strain field.

$\lambda^{(12)}$ corresponds to two-dimensional hyperbolic flow in the xy-plane, so again, as expected, there are three distinct eigenvalues. If the shape of the spheroid is specified by the eccentricity, $e = \sqrt{a^2 - c^2}/a$, of the generating ellipse, then the relevant elliptic integrals simplify to

$$\begin{aligned}
\alpha_0' &= \frac{1}{a^2}\left[\frac{3-e^2}{e^4} - \frac{3\sqrt{1-e^2}}{e^5}\cot^{-1}\left(\frac{\sqrt{1-e^2}}{e}\right)\right] \\
\gamma_0' &= \frac{1}{a^2}\left[\frac{(1-e^2)(e^2-3)}{4e^4} + \frac{3\sqrt{1-e^2}}{4e^5}\cot^{-1}\left(\frac{\sqrt{1-e^2}}{e}\right)\right] \\
\alpha_0'' &= -\frac{3(1-e^2)}{e^4} + \frac{(3-2e^2)\sqrt{1-e^2}}{e^5}\cot^{-1}\left(\frac{\sqrt{1-e^2}}{e}\right) \\
\gamma_0'' &= \frac{(1-e^2)(3-2e^2)}{4e^4} - \frac{(3-4e^2)\sqrt{1-e^2}}{4e^5}\cot^{-1}\left(\frac{\sqrt{1-e^2}}{e}\right) .
\end{aligned}$$

The eigenvalues have the explicit expressions

$$\begin{aligned}
\lambda^{(12)} = \lambda^+ &= -1 - \frac{(1-e^2)(2e^2+3)}{2e^4} + \frac{3\sqrt{1-e^2}}{2e^5}\cot^{-1}\left(\frac{\sqrt{1-e^2}}{e}\right) \\
\lambda^{(23)} = \lambda^{(31)} &= -1 + (2-e^2)\left[\frac{3-e^2}{e^4} - \frac{3\sqrt{1-e^2}}{e^5}\cot^{-1}\left(\frac{\sqrt{1-e^2}}{e}\right)\right] \\
\lambda^- &= -1 - \frac{9(1-e^2)}{e^4} + \frac{3(3-2e^2)\sqrt{1-e^2}}{e^5}\cot^{-1}\left(\frac{\sqrt{1-e^2}}{e}\right) .
\end{aligned}$$

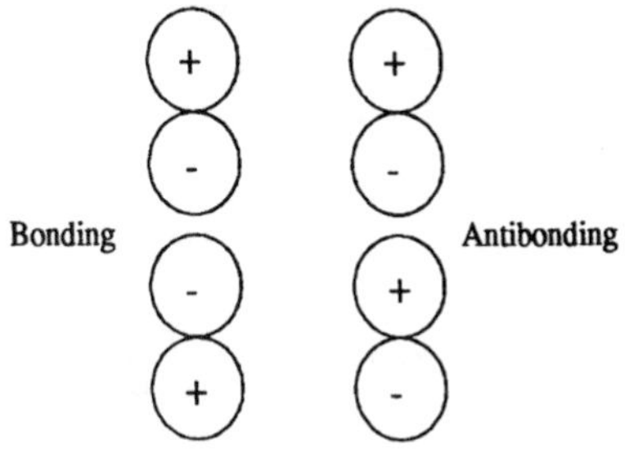

Figure 17.3: Bonding and anti-bonding combinations of p_z orbitals.

The asymptotic behavior in the limit for near-spheres ($e \to 0$) are

$$\begin{aligned}
\lambda^{(12)} = \lambda^{+} &\sim -\frac{1}{5} - \frac{4}{35}e^2 + \cdots \\
\lambda^{(23)} = \lambda^{(31)} &\sim -\frac{1}{5} + \frac{2}{35}e^2 + \cdots \\
\lambda^{-} &\sim -\frac{1}{5} + \frac{4}{35}e^2 + \cdots .
\end{aligned}$$

The asymptotic behavior in the thin-body limit ($e \to 1$) are

$$\begin{aligned}
\lambda^{(12)} = \lambda^{+} &\sim -1 + \frac{3\pi}{4}\epsilon - 4\epsilon^2 + \cdots \\
\lambda^{(23)} = \lambda^{(31)} &\sim 1 - \frac{3\pi}{2}\epsilon + 10\epsilon^2 + \cdots \\
\lambda^{-} &\sim -1 + \frac{3\pi}{2}\epsilon - 12\epsilon^2 + \cdots ,
\end{aligned}$$

with $\epsilon = c/a = \sqrt{1-e^2}$, as before. We see that the spectral radius of the deflated system approaches unity, as the oblate spheroid is flattened, with eigenvalues approaching both ± 1, as shown in the figure. As expected, thin particles present a greater challenge.

17.2.3 The Spectrum for Two Spheres: HI-Induced Spectral Splitting

Consider two equal spheres with their centers $\boldsymbol{x}_1$, $\boldsymbol{x}_2$ separated by a distance $R = |\boldsymbol{x}_1 - \boldsymbol{x}_2|$. If R is very large, we may neglect interactions between the two and the eigensystem would simply be that obtained by superposition of the results for each sphere. The question is what happens as we decrease R.

At this point, it may be helpful to draw an analogy with a similar problem from quantum mechanics, *viz.*, the mixing of atomic orbitals to create molecular orbitals. For example, we can mix the p_z-orbitals of two widely separated atoms in either a *bonding* $(\psi_1-\psi_2)$ or *anti-bonding* $(\psi_1+\psi_2)$ fashion. The former builds up electron densities in the overlap region, while the later creates an electron-deficient region near the newly created node.

With these ideas in mind, we write the eigenfunctions for the two-sphere system as

$$\varphi_n(1) = \varphi_n^{(0)}(1) + \sum_{j\neq n} c_{nj}\varphi_j^{(0)}(1) \qquad \text{on } S_1 \tag{17.21}$$

$$\varphi_n(2) = \varphi_n^{(0)}(2) + \sum_{k\neq n} c_{nk}\varphi_k^{(0)}(2) \qquad \text{on } S_2, \tag{17.22}$$

where $\varphi_n^{(0)}$s are the eigenfunctions of the single-sphere problem. The eigenfunction $\varphi_k^{(0)}(2)$ as a double layer density generates a velocity field "$\boldsymbol{v}_k$" outside sphere 2. The inner product between this velocity field and the basis elements on the surface of sphere 1 is an important quantity. We introduce the notation,

$$\langle\, \varphi_j^{(0)}(1), \mathcal{K}\varphi_k^{(0)}(2)\,\rangle_1 = \int_{S_1} \varphi_j^{(0)}(1)\cdot \boldsymbol{v}_k \, dS \ , \tag{17.23}$$

so that *on* S_1

$$\int_{S_2} \boldsymbol{K}(\boldsymbol{x},\boldsymbol{\xi}_2)\cdot\varphi_k^{(0)}(\boldsymbol{\xi}_2)\, dS_2 = \sum_j \varphi_j^{(0)}(1)\left\langle \varphi_j^{(0)}(1), \mathcal{K}\varphi_k^{(0)}(2)\right\rangle_1 \ . \tag{17.24}$$

On the surface of sphere 1, the eigenvalue problem, $\mathcal{K}(\varphi_n) = \lambda_n\varphi_n$ becomes

$$\begin{aligned} \lambda_n^{(0)}\varphi_n^{(0)} \ &+ \ \sum_{j\neq n} c_{nj}\lambda_j^{(0)}\varphi_j^{(0)}(1) \ \pm \ \sum_j \left\langle \varphi_j^{(0)}(1), \mathcal{K}\varphi_k^{(0)}(2)\right\rangle_1 \varphi_j^{(0)}(1) \\ = \ \lambda_n\varphi_n^{(0)} \ &+ \ \lambda_n \sum_{j\neq n} c_{nj}\varphi_j^{(0)}(1) \ . \end{aligned} \tag{17.25}$$

The first and second terms on the left-hand side of this equation result from the fact that the $\varphi_j^{(0)}$s are eigenfunctions of $\mathcal{K}$ restricted to S_1, with eigenvalues $\lambda_j^{(0)}$. Looking at each mode, we find the following set of equations:

$$\text{For } j=n: \quad \lambda_n = \lambda_n^{(0)} \pm \left\langle \varphi_j^{(0)}(1), \mathcal{K}\varphi_k^{(0)}(2)\right\rangle_1 \tag{17.26}$$

$$\text{For } j\neq n: \quad c_j = \frac{\pm\left\langle \varphi_j^{(0)}(1), \mathcal{K}\varphi_k^{(0)}(2)\right\rangle_1}{\lambda_n^{(0)} - \lambda_j^{(0)} \pm \left\langle \varphi_j^{(0)}(1), \mathcal{K}\varphi_k^{(0)}(2)\right\rangle_1} \ . \tag{17.27}$$

Here, $\pm$ denotes the sign used to construct the base solution, $\varphi_n^{(0)}(1) \pm \varphi_n^{(0)}(2)$.

We have arrived at the desired result. The eigenvalues of the two-sphere system are given by those of the single-sphere problem, plus a small perturbation proportional to the interaction factor, $\left\langle \varphi_j^{(0)}(1), \mathcal{K}\varphi_k^{(0)}(2)\right\rangle_1$, which we evaluate presently with an addition theorem. But first, we may estimate the R-dependence of the spectral radius with this limited information. From the discussion for the single sphere, we know that the dominant eigenfunctions (the ones with eigenvalue $-3/5$) generate Stokes quadrupole fields, which decay as

$|\boldsymbol{x} - \boldsymbol{x}_2|^{-3}$. The fact that $j = n$ determines the eigenvalue shift implies that we need the quadrupole moment induced on sphere 1, which from the Faxén relation is proportional to the *second* derivative of the incident field. Therefore, we expect the shift in the spectral radius to scale as $|\boldsymbol{x}_1 - \boldsymbol{x}_2|^{-5}$, which we now verify *via* the exact formula.

We require the following addition theorem for Stokes flow [35, 63]:

$$\begin{aligned}\left\langle \varphi^{(3)}_{n'm'}(1), \mathcal{K}\varphi^{(3)}_{nm}(2)\right\rangle &= (1+\lambda^{(0)}_{n'})(-1)^{m+n}\delta_{mm'}\sqrt{n(n+1)n'(n'+1)} \\ &\quad \times \frac{\eta_{nm}}{\eta_{n'm}}\begin{pmatrix} n+n' \\ n+m \end{pmatrix}\frac{-R^{-(n+n'+1)}}{(n+1)(n'+1)} .\end{aligned}$$

The factor of $(1+\lambda^{(0)}_{n'})$ takes into account the jump between the surface density $\varphi(\boldsymbol{\xi}_2)$ on S_2 and the associated velocity field emanating from sphere 2. *This factor also ensures that the eigenvalues at* -1 *do not get shifted,* which we know has to be the case, since the eigenvalues corresponding to the null densities are always equal to -1, independent of the (multiparticle) geometry.

We now apply these results with the eigenvalue as $-3/(2n+1)$. The perturbation result for the two-sphere system becomes

$$\begin{aligned}\lambda_{2m} &= -\frac{3}{5} \pm \frac{2}{5}\left\langle \varphi^{(3)}_{2m}(1), \mathcal{K}\varphi^{(3)}_{2m}(2)\right\rangle \\ &= -\frac{3}{5} \pm (-1)^{m+1}\frac{4}{15}\begin{pmatrix} 4 \\ 2+m \end{pmatrix} R^{-5} \\ &= -\frac{3}{5} \pm \frac{(-1)^{m+1}32R^{-5}}{5(m+2)!\,(2-m)!}\end{aligned}$$

We may write these results explicitly as

$$\begin{aligned} m=0\text{:} \quad \lambda &= -\frac{3}{5}\left[1 \pm \frac{8}{3}R^{-5}\right] \\ |m|=1\text{:} \quad \lambda &= -\frac{3}{5}\left[1 \pm \frac{-16}{9}R^{-5}\right] \\ |m|=2\text{:} \quad \lambda &= -\frac{3}{5}\left[1 \pm \frac{4}{9}R^{-5}\right] .\end{aligned}$$

There are five pairs, but two of these pairs are degenerate, so that there are three *distinct pairs,* with each pair consisting of a "bonding" and "anti-bonding" split. As with molecular orbitals, the value of m determines whether the "+" or "-" shifts the eigenvalue downwards.

In summary, the ten degenerate eigenvalues at $-3/5$ of the decoupled system splits into a set of six distinct eigenvalues with multiplicities: 1,2,2,2,2, and 1. The magnitude of the splits, as computed numerically and analytically, are in excellent agreement, as shown in Figure 17.4. (We defer until Chapter 19 the discussion on the discretization schemes for the numerical solution.) More importantly, in contrast with the more familiar situation in quantum mechanics, the shift in the spectrum as a result of hydrodynamic interactions is actually

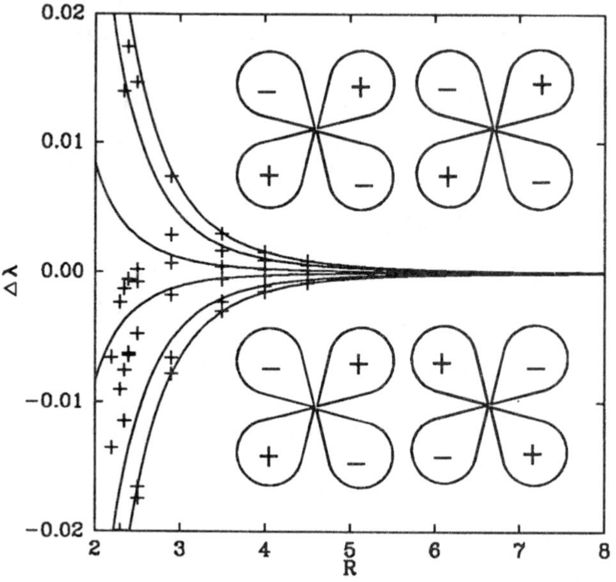

Figure 17.4: Comparison of numerical and asymptotic results for λ.

quite small. This has important ramifications for computations. We show in the following sections a deflate-and-iterate solution scheme, which consists of a deflation procedure for removing all eigenvalues at ± 1, and an iterative solution of the resulting set of equations. The iterations converge since the spectral radius of the deflated system is less than one. But now we know that the number of iterations is essentially the same as in the single-particle problem. For the sphere, according to the discussion above, the spectral radius for the two-sphere system remains approximately 0.6 (the single-sphere result) at all configurations except almost-touching spheres. We will in fact observe this property directly in multisphere problems, where we discover that the iterative solutions converge with essentially the same number of iterations as in the single-sphere problem.

17.3 Wielandt's Deflation

Consider the double layer IE of the second kind without any completion. We know now that the integral operator ("iteration operator" for direct iteration) there has spectral radius exactly one. Thus direct iteration would not be successful here, as spectral radius strictly less than one is required. However, if we can remove the endpoints of the spectrum, for example, by moving those eigenvalues to zero, *without affecting the rest of the eigenvalues*, the new modified spectrum will allow direct iterative solution. Such a modification is possible by using *Wielandt's deflation.* (For a compact operator eigenvalues other than zero are isolated and removing the extreme eigenvalues will strictly diminish the spectral radius — if, to the contrary, the extreme eigenvalues were possibly accumulation points, there still could remain eigenvalues arbitrarily close to these, and the spectral radius would have to be defined as the *supremum* of the magnitudes of the eigenvalues, possibly not attained.)

Consider the Jordan form of a matrix. If the matrix happens to be diagonalizable, having only eigenvectors and no other principal vectors, the Jordan form is just the diagonal form of the matrix. Otherwise to each nontrivial Jordan chain there corresponds a Jordan block, in which the diagonal elements coincide with the eigenvalue in question and the upper bidiagonal is filled with ones. (Equivalently the lower bidiagonal could be used for the Jordan blocks; this corresponds to reversal of the order of the principal vectors in the basis used). So the Jordan form is just an "almost diagonal representation" of the original matrix, obtained by using the principal vectors as the basis for the vector space. This similarity transformation can be represented as $\boldsymbol{K} = \boldsymbol{B}\boldsymbol{J}\boldsymbol{B}^{-1}$, where $\boldsymbol{K}$ is the original matrix, $\boldsymbol{B}$ contains the principal vectors as its columns, and $\boldsymbol{J}$ is the Jordan form. On taking the adjoint of this equation the order of matrix multiplications is reversed, and we find $\boldsymbol{K}^* = \boldsymbol{B}^{-*}\boldsymbol{J}^*\boldsymbol{B}^*$; recall that for matrices the adjoint is the complex conjugated transpose. By inspection, $\boldsymbol{B}^{-*}$ must have the principal vectors of the adjoint as its columns. As $\boldsymbol{B}$ is obtained from the principal vectors of the original matrix, this shows that the two sets of principal vectors are *biorthogonal* bases for the vector space, meaning that the (complex) inner product of a fixed principal vector is zero with all but one of the adjoint's principal vectors.

Let $\boldsymbol{x}_i$ be the principal vectors of $\boldsymbol{K}$, and $\boldsymbol{y}_i$ be those of the adjoint $\boldsymbol{K}^*$. Represent the operator $\boldsymbol{x}_{\tilde{i}}\langle\cdot,\boldsymbol{y}_{\tilde{i}}\rangle$ (*no summation over* $\tilde{i}$) as a matrix relative to the basis given by the $\boldsymbol{x}_i$s. Due to the biorthogonality we have $\langle\boldsymbol{x}_j,\boldsymbol{y}_{\tilde{i}}\rangle = \langle\boldsymbol{x}_{\tilde{i}},\boldsymbol{y}_{\tilde{i}}\rangle\delta_{\tilde{i}j}$, so that only elements in the $\tilde{i}$-column are nonzero. Moreover, this operator gives only multiples of $\boldsymbol{x}_{\tilde{i}}$, so that in that column only the $\tilde{i}$-element is nonzero. Thus in fact this operator is represented by the $\tilde{i}$-th diagonal element being $\langle\boldsymbol{x}_{\tilde{i}},\boldsymbol{y}_{\tilde{i}}\rangle$ with all other elements being zeroes. Addition of this operator to the original operator $\boldsymbol{K}$ will correspond to modifying just one of the diagonal elements in the Jordan form; the result is not necessarily a Jordan form any more (namely, if $\boldsymbol{x}_{\tilde{i}}$ is a principal vector in a nontrivial Jordan chain), but, as it is still triangular, all the eigenvalues are visible on the diagonal. Therefore one of the eigenvalues has been modified (if the corresponding Jordan block was nontrivial the original eigenvalue has not been removed completely, just its multiplicity has changed), while the rest of these remain unchanged. Operators of the type above, here used for modifying the original matrix, are sometimes called *rank-one* operators or *tensor products.*

Let us now concentrate on eigenvectors, as a special case of principal vectors. The procedure above can be used when both an eigenvector of the original operator and the corresponding eigenvector of the adjoint are known. We can, however, get essentially the same effect with less information; this leads to actual

Wielandt's deflations. Consider first the case where only $x_{\bar{i}}$ is known. Instead of $y_{\bar{i}}$ we use just some vector y within the inner product part of the rank-one operator. When represented in the basis of principal vectors of the operator of interest K, this corresponds to one nonzero row. By choosing $x_{\bar{i}}$ to be the first vector in the basis, this is the first row, and therefore addition to the original matrix K modifies its Jordan form again to a triangular form. This modification only changes one of the diagonal elements, therefore retaining the rest of the eigenvalues unchanged. The first diagonal element has been changed by adding the number $\langle x_{\bar{i}}, y\rangle$ to it; this shows explicitly the change in that eigenvalue. Note that the order in which $x_{\bar{i}}$ and y are used in the tensor product is crucial: Switching them gives a rank-one operator without the desired properties in general. For the second case assume that $y_{\bar{i}}$ is known, while $x_{\bar{i}}$ is not. In place of $x_{\bar{i}}$ use some vector x to get the rank-one operator $x\langle\cdot, y_{\bar{i}}\rangle$. This, when represented in the same basis as the original Jordan form, gives just one nonzero column. By rearranging the basis as necessary, this column can be chosen to be the last. Again, addition of such a tensor product keeps the original Jordan form triangular, and the eigenvalues remain unchanged, except for the last one. The shift in the last eigenvalue is explicitly given by $\langle x, y_{\bar{i}}\rangle$.

In summary, Wielandt's deflation can be carried out when either an eigenvector of the operator itself or that of its adjoint is known. The former must be used outside the inner product in the rank-one modification, and the latter inside the inner product. The rank-one modification will modify only the eigenvalue corresponding to the eigenvector used, leaving the rest unchanged. If the eigenvector used corresponds to a trivial Jordan block, no new eigenvectors with the relevant eigenvalue are created; otherwise, the situation gets more complicated, but the Jordan form still provides some insight as to what happens.

Now we consider multiple Wielandt's deflations done at once. Suppose we know two eigenvectors x_1 and x_2 of K. In the basis used for the Jordan canonical form of K, the operators $x_1\langle\cdot, z_1\rangle$ and $x_2\langle\cdot, z_2\rangle$ are represented each by a single nonzero row. Here the z-vectors are arbitrary to start with. Addition of both these operators to K will modify just two rows in the original Jordan form, and by permutation of the basis used these can be chosen as the first two rows. Now the resulting matrix is *block-triangular*, with the lower left-hand side $(n-2)\times 2$ block being zero, and the diagonal blocks being square. For this situation it is known from linear algebra (theory of determinants) that the eigenvalues of the total matrix are given by those of the diagonal blocks. The latter of these diagonal blocks is again triangular, unchanged from the original Jordan form, and therefore the corresponding eigenvalues visible on the diagonal have remained unchanged, as with simple deflation. The modification to the first 2×2 diagonal block is given by the matrix

$$\begin{bmatrix} \langle x_1, z_1\rangle & \langle x_2, z_1\rangle \\ \langle x_1, z_2\rangle & \langle x_2, z_2\rangle \end{bmatrix}, \tag{17.28}$$

and in general it would be difficult to give a *nice* formula for the change in the original eigenvalues due to this modification. The situation is considerably simplified by requiring that the off-diagonal elements in this 2×2 matrix are zeroes, while the diagonal elements are not; such orthogonality requirements are easy to accommodate in practice. If the 2×2 matrix above is invertible, there exist linear combinations of the $\boldsymbol{z}$-vectors such that when used in place of these the 2×2 identity matrix results. This procedure can be applied to as many vectors as is necessary, before carrying out the deflation; here two vectors were used just to keep the notation simple and easy to follow.

17.4 Deflation for a Single Particle

From Theorem 1 it is known that -1 is an eigenvalue of $\mathcal{K}$, corresponding to six linearly independent eigenfunctions that are given by rigid-body motions. In addition, the corresponding Jordan blocks are trivial, so that deflation with these eigenfunctions really removes the eigenvalue -1 completely; recall that in the case of nontrivial Jordan blocks the same eigenvalue would remain on the diagonal of the original Jordan form after deflation using only the eigenvectors.

The original operator in the unmodified double layer equation is $1 + \mathcal{K}$. On deflation using the known RBM null functions, we need to add six rank-one modifications, namely,

$$\sum_{i=1}^{3}\{\boldsymbol{U}_i\langle\varphi, \boldsymbol{z}_i\rangle + (\boldsymbol{\omega}_i \times \boldsymbol{r})\langle\varphi, \boldsymbol{Z}_i\rangle\} \ , \tag{17.29}$$

where the functions $\boldsymbol{z}_i$ and $\boldsymbol{Z}_i$ are suitably chosen. Note that the known eigenfunctions of the operator are used for output, not within the inner product, as Wielandt's deflation requires. Furthermore, it seems to be easiest to choose the set of $\boldsymbol{U}_i$s and $\boldsymbol{\omega}_i \times \boldsymbol{r}$s to form an orthonormal set with respect to the natural inner product, and further choose the $\boldsymbol{z}_i$s and $\boldsymbol{Z}_i$s to equal these functions, respectively. This can be accomplished by using the Gram–Schmidt orthonormalization procedure (normalizing to unit length after the orthogonalization) to some translations and rotations, and ensures the orthonormality of the functions used within the inner products, which is useful on doing controlled multiple deflation. Since all the eigenvectors correspond to the same eigenvalue, the orthogonalization forming linear combinations again gives eigenvectors.

Thus the added part now looks like

$$\sum_{i=1}^{3}\{\boldsymbol{U}_i\langle\varphi, \boldsymbol{U}_i\rangle + (\boldsymbol{\omega}_i \times \boldsymbol{r})\langle\varphi, (\boldsymbol{\omega}_i \times \boldsymbol{r})\rangle\}. \tag{17.30}$$

It is easy to see that translations in dot-product orthogonal directions are mutually orthogonal also with respect to the natural inner product over the boundary surface, and the rotational eigenfunctions are orthogonal to these iff the origin (center of rotation) is chosen to be the center of mass of the boundary:

$$\oint_S \boldsymbol{r} \, dS = 0 \ . \tag{17.31}$$

Then on orthonormalization the first rotational eigenfunction is just normalized and therefore it remains a rotation, while the second and third of these get rotational additions only, since they already are orthogonal to translational eigenfunctions. Therefore orthogonalization just modifies the rotational velocities ω_i. For example, for a sphere it is easy to verify directly that the ω_is can be chosen in the directions of the coordinate axes, but in general this need not be the case.

Typically the Gram–Schmidt procedure will give axes of rotation that are *not* orthogonal with respect to the *dot product.* Another way to perform the orthogonalization is to choose the orthogonal coordinate axes so that the moment of inertia tensor, for the surface with uniform mass-density, is diagonalized, keeping the origin at the center of mass. In practice this only requires diagonalizing a 3×3 symmetric matrix, and is easily incorporated into the numerical codes. The major advantage of the latter procedure is keeping the rotation axes mutually orthogonal, and these axes can be chosen to be the coordinate axes. This simplifies the writing of codes to some extent (see Chapter 19). Also we see that for most particle shapes (*i.e.*, those whose moment of inertia tensor has three distinct eigenvalues) there will be a unique coordinate system, determined by the geometry, that automatically orthogonalizes the "natural basis" for RBMs. An important point raised in Chapter 5 should be re-emphasized here: *This natural choice of the origin and coordinate axes, as motivated by computational efficiency, differs from that dictated by classical hydrodynamics (diagonalization of the resistance/mobility tensors).* In general, one should not expect superior computational techniques to result from the use of physical insight alone, and it is precisely in these instances that mathematical analysis plays an important role.

Now the rank-one operators in the sum above each modify one row of the original Jordan form, the diagonal modification being $+1$, with other elements being zero in the corresponding diagonal block. This modification shifts the original eigenvalue -1 of $\mathcal{K}$ to 0.

The modification above resembles that used by Power and Miranda. The main difference is that instead of using total forces and torques on adding the missing part of the range, one now uses translation and rotation velocities. These are naturally connected with the total forces and torques in a linear fashion, through either the resistance or the mobility matrix. In any case a deduction similar to that used on explaining the bordering method shows that the range completion is carried out by the deflation above. Therefore two things have been accomplished at once here: the shift of one endpoint of the spectrum, and range completion of the double layer operator.

It is noteworthy that, although traditionally resistance problems have been favored over mobility problems, Wielandt's deflation here is only possible for problems of the latter kind. The reason for the traditional trend has been that for mobility problems the given data (total forces and torques) do not supply boundary conditions for use with ordinary methods of solving partial differential equations. For integral equation formulations the situation is different, as was already seen when the canonical equations were presented.

One of the endpoints of the spectrum of $\mathcal{K}$ still remains, namely, that at 1. In order to make the direct (Picard) iteration possible, also this has to be moved towards the origin. Whether a single surface is considered a particle or a container affects only the direction of the surface normal in the definition of the integral operator $\mathcal{K}$; this change corresponds to multiplication of the operator with -1. Thus for "particle-$\mathcal{K}$" the eigenvalue 1 corresponds to -1 for a "container-$\mathcal{K}$," and the eigenfunction is the container null function of Theorem 2. This eigenfunction is not available for a general surface shape, but the adjoint eigenfunction is. According to Theorem 2 the adjoint eigenfunction is the surface normal, and the corresponding Jordan block is again trivial. In Wielandt's deflation the adjoint eigenfunction is used within the inner product, and the addition must be of the form

$$\boldsymbol{z}\langle \boldsymbol{\varphi}, \hat{\boldsymbol{n}} \rangle \ , \tag{17.32}$$

where $\boldsymbol{z}$ is a suitable vector not orthogonal to $\hat{\boldsymbol{n}}$. This condition is most certainly satisfied by the choice $\boldsymbol{z} = a\hat{\boldsymbol{n}}$ with $a \neq 0$ a constant. It seems optimal for the convergence rate of the iterations to shift the original eigenvalue 1 of $\mathcal{K}$ to 0, and the shift caused by addition of the rank-one operator above is $a\langle \hat{\boldsymbol{n}}, \hat{\boldsymbol{n}} \rangle = a|S|$, where $|S|$ denotes the surface area. Thus the proper choice is $a = -1/|S|$. Please note that although $\hat{\boldsymbol{n}}$ is a unit vector in the norm induced by the ordinary dot product for vectors, the same clearly need not hold with respect to the natural inner product on the surface S.

The effect of this deflation differs somewhat from the previous ones, as this is not related to range completion. The role of this latter deflation is purely mathematical, making direct iteration possible. Thus the effect of the deflation, *i.e.*, changing the original operator, has to be compensated for. Luckily this is also easy to accomplish. First note that here the adjoint eigenvector satisfies the relation $\mathcal{K}^*\hat{\boldsymbol{n}} = \hat{\boldsymbol{n}}$. Denote by $\mathcal{H}$ the operator resulting from the previous multiple deflations using the RBM null functions of $(1 + \mathcal{K})$. Then $(1 + \mathcal{H})$ is invertible and $(1 + \mathcal{H})^*\hat{\boldsymbol{n}} = 2\hat{\boldsymbol{n}}$. Therefore we first take the adjoint of the deflated operator

$$(1 + \mathcal{H} + a\hat{\boldsymbol{n}}\langle \cdot, \hat{\boldsymbol{n}} \rangle)^* = 1 + \mathcal{H}^* + a^*\hat{\boldsymbol{n}}\langle \cdot, \hat{\boldsymbol{n}} \rangle) \ . \tag{17.33}$$

(In general, on taking the adjoint the two vectors defining a tensor product are switched, while scalars are complex-conjugated.) On multiplying from left with $(1 + \mathcal{H}^*)^{-1}$ we find

$$(1 + \mathcal{H}^*)^{-1}(1 + \mathcal{H} + a\hat{\boldsymbol{n}}\langle \cdot, \hat{\boldsymbol{n}} \rangle)^* = 1 + (a^*\hat{\boldsymbol{n}}/2)\langle \cdot, \hat{\boldsymbol{n}} \rangle \ . \tag{17.34}$$

Taking the adjoint of this equation reveals that

$$(1 + \mathcal{H} + a\hat{\boldsymbol{n}}\langle \cdot, \hat{\boldsymbol{n}} \rangle)(1 + \mathcal{H})^{-1} = 1 + (a\hat{\boldsymbol{n}}/2)\langle \cdot, \hat{\boldsymbol{n}} \rangle \tag{17.35}$$

or

$$(1 + \mathcal{H})^{-1} = (1 + \mathcal{H} + a\hat{\boldsymbol{n}}\langle \cdot, \hat{\boldsymbol{n}} \rangle)^{-1}(1 + (a\hat{\boldsymbol{n}}/2)\langle \cdot, \hat{\boldsymbol{n}} \rangle) \ . \tag{17.36}$$

Therefore, instead of solving $(1+\mathcal{H})\boldsymbol{x} = \boldsymbol{b}$ for $\boldsymbol{x}$, we can equivalently solve

$$(1+\mathcal{H}+a\hat{\boldsymbol{n}}\langle\cdot,\hat{\boldsymbol{n}}\rangle)\,\boldsymbol{x} = (1+(a\hat{\boldsymbol{n}}/2)\langle\cdot,\hat{\boldsymbol{n}}\rangle)\boldsymbol{b}\ . \tag{17.37}$$

Thus all that is needed is a simple premultiplication of the data vector, giving $\tilde{\boldsymbol{b}}$, before the iterative solution $\boldsymbol{x}_{i+1} = -\mathcal{H}\boldsymbol{x}_i - a\hat{\boldsymbol{n}}\langle\boldsymbol{x}_i,\hat{\boldsymbol{n}}\rangle + \tilde{\boldsymbol{b}}$ is carried out.

17.5 Deflation for a Container

Handling of the container problem is similar in most essential features to that of the handling of a single particle. The major difference is that now the deficiency in the range is only of dimension one, while at the other endpoint of the spectrum a sixfold deflation is needed with compensation. This is just because on switching from a particle to a container the operator has experienced a sign change, and the roles of the endpoints of the spectrum have been reversed.

The container null function is not known in general, but the adjoint null function is again the surface normal. Thus the natural deflation is of the form

$$a\hat{\boldsymbol{n}}\langle\cdot,\hat{\boldsymbol{n}}\rangle\ , \tag{17.38}$$

where $\hat{\boldsymbol{n}}$ is as usual the unit surface normal, and a is a scalar that normalizes this rank-one modification. As in the previous section, the choice $a = 1/|S|$, where $|S|$ is the surface area, shifts the eigenvalue to 0. Recall that this normalization comes from using the natural inner product over the surface. Mathematically this deflation obviously works; physically the addition is in the direction of $\hat{\boldsymbol{n}}$, and this is because we need a non-mass-conserving addition to the double layer representation on dealing with a bounded volume to be able to comply to arbitrary (possibly non-mass-conserving) data.

Now the range completion is done, and the other endpoint of the spectrum, namely 1, is dealt with. At this endpoint the eigenfunctions are the rigid-body motions, the dimension of the eigenspace being six and the corresponding Jordan blocks trivial. Therefore there are no serious mathematical difficulties in carrying out the deflations. Let φ_i, $i = 1,\ldots,6$, be the RBM eigenfunctions, *orthonormal* with respect to the natural inner product. Then the modification necessary is

$$\boldsymbol{M} = -\sum_{i=1}^{6}\varphi_i\langle\cdot,\varphi_i\rangle\ , \tag{17.39}$$

where the minus sign is necessary, since this time we are shifting eigenvalues that were originally 1 to 0. Denote again the operator after the previous (range completing) deflation by $\mathcal{H} = \mathcal{K} + a\hat{\boldsymbol{n}}\langle\cdot,\hat{\boldsymbol{n}}\rangle$. Then $(1+\mathcal{H})$ is invertible, and $(1+\mathcal{H})^{-1}\varphi_i = (1/2)\varphi_i$ for the RBM eigenfunctions. From this it easily follows that

$$(1+\mathcal{H})^{-1} = (1+(1/2)\boldsymbol{M})\,(1+\mathcal{H}+\boldsymbol{M})^{-1}\ . \tag{17.40}$$

This means that, for a container, we should first iteratively solve the equation

$$(1+\mathcal{H}+\boldsymbol{M})\,\boldsymbol{x} = \boldsymbol{b} \tag{17.41}$$

for $\boldsymbol{x}$, this being done with the Picard iteration $\boldsymbol{x}_{i+1} = -(\mathcal{H} + \boldsymbol{M})\,\boldsymbol{x}_i + \boldsymbol{b}$, and then premultiply the intermediate result $\boldsymbol{x}$ to get *the final result* $(1 + (1/2)\boldsymbol{M})\,\boldsymbol{x}$. Note the difference with the case of a particle, where the data were premultiplied before the iterations, and these directly gave the final result.

17.6 Multiparticle Problems in Bounded and Unbounded Domains

The deflations for a particle and a container surface have been discussed at length in the previous two sections. In this section we indicate how to deal with more complex cases, involving multiple particles and possibly also a container. The case of multiple containers is not of practical interest; we only deal with connected fluid domains, and these have at most one container boundary. The development of the theory here will utilize orthogonal projections, and the previous cases with a single boundary could have been handled similarly, as they are only special cases. We hope that the previous approaches complement the one given here, making it easier to follow the presentation.

Let S_i, $i = 1, \ldots, N$, be disjoint particle surfaces and S_c a surrounding container surface. The totality of surfaces is denoted by S. First we carry out the "physical deflations" necessary for range completion and removal of indeterminacies in the double layer density. Then the "mathematical deflations" are discussed at length. The purpose of the latter deflations is again to make the spectrum of the "iteration operator" in the second-kind equation small enough, *i.e.*, such that all its eigenvalues are strictly less than one in magnitude, and this mathematical trick needs to be compensated for in the solution procedure, while the physical deflations need no compensation. For analytical purposes the physical deflations would suffice, and the shift of the other endpoint of the spectrum serves only numerical purposes.

Let φ_{ij} be the "j-th RBM of particle i" ($j = 1, \ldots, 6$); each of these double layer densities attains nonzero values only on one particle surface S_i and $\mathcal{K}\varphi_{ij} = -\varphi_{ij}$. Assume these densities are orthonormal

$$\langle \varphi_{ij}, \varphi_{kl} \rangle = \delta_{ik}\delta_{jl}. \tag{17.42}$$

Here the δ_{ik} -part is trivially satisfied, and in practice the Gram–Schmidt orthonormalization needs to be applied separately for each i over the indices $j = 4, 5, 6$ if for each particle surface its center of mass is used as the center of rotations. These are $6N$ of the $6N + 1$ linearly independent eigenvectors with eigenvalue -1. Again the remaining container null function is not explicitly known in general, but instead the adjoint eigenfunction $\psi_t = -\mathcal{K}^*\psi_t$ is known to be

$$\psi_t = \alpha_t \hat{\boldsymbol{n}}, \tag{17.43}$$

where $\hat{\boldsymbol{n}}$ is the unit normal pointing into the fluid on all of S and the normalization is $\alpha_t = |S|^{-1/2}$. Now it is natural to carry out the Wielandt's deflations and define

$$\mathcal{H} = \mathcal{K} + \varphi_{ij}\langle\cdot,\varphi_{ij}\rangle + \psi_t\langle\cdot,\psi_t\rangle \ , \tag{17.44}$$

where *the summation convention was used, as is frequently done in the rest of this section.*

Now the physical deflations have been carried out and $(1+\mathcal{H})$ is invertible, as the eigenvalues at -1 of $\mathcal{K}$ have been shifted to 0 for $\mathcal{H}$. Otherwise the spectra of $\mathcal{H}$ and $\mathcal{K}$ coincide. For any smooth surface field $\boldsymbol{u}$ (velocity BCs) there exists a unique φ such that

$$(1+\mathcal{H})\varphi = \boldsymbol{u} \ . \tag{17.45}$$

Let us now interpret the definition of $\mathcal{H}$ physically. Any nonzero RBMs of the particles (except when the same RBM prevails on all the particles *and* on the container) would necessarily correspond to some nonzero forces or torques acting on them, and are thus not represented by a double layer alone. These particle RBMs are added separately. The additions so far are mass conserving, and only BCs, similarly mass conserving, could be satisfied without the final addition (deflation with ψ_t), which removes this restriction.

Consider now a mobility problem, with given forces $\boldsymbol{F}_i$ and torques $\boldsymbol{T}_i$ acting at points $\boldsymbol{x}_i$ within each particle. Define

$$\boldsymbol{u} = \boldsymbol{F}_i \cdot \mathcal{G}(\boldsymbol{x}-\boldsymbol{x}_i)/(8\pi\mu) - \frac{1}{2}(\boldsymbol{T}_i \times \nabla)\cdot \mathcal{G}(\boldsymbol{x}-\boldsymbol{x}_i)/(8\pi\mu) \ . \tag{17.46}$$

Using these BCs we solve φ from $(1+\mathcal{H})\varphi = \boldsymbol{u}$. The non-mass-conserving term in $\mathcal{H}$ must vanish, as it is the only such term in the whole equation: $\langle\varphi,\psi_t\rangle = 0$. The remaining equation is rewritten as

$$\varphi_{ij}\langle\varphi,\varphi_{ij}\rangle = -(1+\mathcal{K})\varphi + \boldsymbol{u}. \tag{17.47}$$

The first term on the RHS here gives no total force or torque, and so the RBMs on the LHS must correspond to the forces and torques used on forming $\boldsymbol{u}$. Thus the mobility problem can be directly solved.

The spectrum of $\mathcal{H}$ now lies in the real interval $(-1,1]$, and in order to enable iterative solution, deflations are needed so that the modified spectrum is within $(-1,1)$. For both $\mathcal{K}$ and $\mathcal{H}$ the multiplicity of the eigenvalue 1 is $6+N$. Explicitly six linearly independent eigenfunctions are given by the RBMs on the totality of surfaces S, for which

$$\mathcal{K}\varphi_{tj} = \varphi_{tj}, \ \text{for } j = 1,\ldots,6 \tag{17.48}$$

(by invoking Gram–Schmidt, if necessary, these are assumed orthonormal with respect to the natural inner product), and N orthonormal adjoint eigenfunctions are given by the scaled particle surface normals

$$\psi_i = \mathcal{K}^*\psi_i = \begin{cases} \alpha_i\hat{\boldsymbol{n}} & \text{on } S_i \\ 0 & \text{otherwise.} \end{cases} \tag{17.49}$$

For normalization choose $\alpha_i = |S_i|^{-1/2}$. These were orthonormalized eigenfunctions of $\mathcal{K}$ and its adjoint, and while for $\mathcal{H}$ and its adjoint each eigenvalue and its multiplicity are the same as for $\mathcal{K}$, the corresponding eigenfunctions are different.

So far everything here has been very similar to the cases with single connected boundaries, *i.e.*, a single particle or a container. It is here in the mathematical deflations that complications arise, as these need to be done and compensated for simultaneously both for eigenfunctions and adjoint eigenfunctions when particles and a container coexist. Suppose we have orthonormal sets of eigenfunctions of $\mathcal{H}$ and its adjoint, such that

$$\mathcal{H}\varphi_j' = \varphi_j', \text{ for } j = 1, \ldots, 6 \tag{17.50}$$

$$\mathcal{H}^*\psi_i' = \psi_i', \text{ for } i = 1, \ldots, N. \tag{17.51}$$

Define the "left" projection

$$\mathcal{P}_L = \psi_i'\langle\cdot, \psi_i'\rangle \tag{17.52}$$

and the "right" projection

$$\mathcal{P}_R = \varphi_j'\langle\cdot, \varphi_j'\rangle \; ; \tag{17.53}$$

the motivation for the names will be apparent presently.

Immediately one can verify that

$$\mathcal{H}\mathcal{P}_R = \mathcal{P}_R \tag{17.54}$$

$$\mathcal{H}^*\mathcal{P}_L = \mathcal{P}_L \; , \tag{17.55}$$

and taking the adjoint of the latter equation gives

$$\mathcal{P}_L\mathcal{H} = \mathcal{P}_L. \tag{17.56}$$

Recall that for projections $\mathcal{P}$ in general, $\mathcal{P} = \mathcal{P}^* = \mathcal{P}^2$. Note also that due to mutual orthogonality

$$\mathcal{P}_R\mathcal{P}_L = \mathcal{P}_L\mathcal{P}_R = 0. \tag{17.57}$$

Now the natural deflated operator is

$$\tilde{\mathcal{H}} = \mathcal{H} - \mathcal{P}_R - \mathcal{P}_L, \tag{17.58}$$

in which the eigenvalue 1 has been shifted to 0. Using the special properties of the projections when multiplied with $\mathcal{H}$, the following simple connection between $(1 + \mathcal{H})$ and $(1 + \tilde{\mathcal{H}})$ is found:

$$(1 - \mathcal{P}_L/2)(1 + \mathcal{H})(1 - \mathcal{P}_R/2) = 1 + \tilde{\mathcal{H}}. \tag{17.59}$$

Operators of the type $(1 - \mathcal{P}/2)$ are invertible, the inverse being $(1 + \mathcal{P})$, so that the relation above can easily be inverted.

The equation $(1 + \mathcal{H})\varphi = \boldsymbol{u}$ is equivalent to

$$\tilde{u} = (1 - \mathcal{P}_L/2)\boldsymbol{u} \tag{17.60}$$

$$(1 + \tilde{\mathcal{H}})\tilde{\varphi} = \tilde{u} \tag{17.61}$$

$$\varphi = (1 - \mathcal{P}_R/2)\tilde{\varphi}, \tag{17.62}$$

and this gives the steps for the computational procedure. First transform the data. Then solve the second-kind equation, which now allows direct iterative solution. Finally modify the found solution with a simple linear transformation.

In practice one should not form $\tilde{\mathcal{H}}$ explicitly at all, as this would be a computationally expensive matrix operation. Rather, always compute

$$(1+\tilde{\mathcal{H}})\tilde{\varphi} = (1+\mathcal{H}-\mathcal{P}_L-\mathcal{P}_R)\tilde{\varphi}, \tag{17.63}$$

where vectors are multiplied by matrices or other vectors on forming the projections. Similarly $\mathcal{H}$ should not be explicitly formed from $\mathcal{K}$, but rather have it "operationally defined" in a subroutine.

Now an algorithmic approach to finding the orthonormal eigenvectors φ'_j and ψ'_i will be given. First note that

$$(1+\mathcal{H})\varphi_{tj} = 2\varphi_{tj} + \varphi_{ik}\langle\varphi_{tj},\varphi_{ik}\rangle \tag{17.64}$$

and

$$(1+\mathcal{H})\varphi_{ik} = \varphi_{ik}. \tag{17.65}$$

Then for

$$\varphi''_j = \varphi_{tj} + \varphi_{ik}\langle\varphi_{tj},\varphi_{ik}\rangle \tag{17.66}$$

we have

$$(1+\mathcal{H})\varphi''_j = 2\varphi''_j\ . \tag{17.67}$$

These found six eigenfunctions of $\mathcal{H}$ can numerically easily be orthonormalized with the Gram–Schmidt procedure to get the φ'_j vectors that define the projection $\mathcal{P}_R$ for numerical purposes. In fact

$$\varphi_{ik}\langle\varphi_{tj},\varphi_{ik}\rangle = \begin{cases} 0 & \text{on } S_c \\ \varphi_{tj} & \text{otherwise,} \end{cases} \tag{17.68}$$

and no cumbersome double sums need be evaluated. Thus

$$\varphi''_j = \begin{cases} \varphi_{tj} & \text{on } S_c \\ 2\varphi_{tj} & \text{otherwise.} \end{cases} \tag{17.69}$$

Using the general identity $(\boldsymbol{a}\langle\cdot,\boldsymbol{b}\rangle)^* = \boldsymbol{b}\langle\cdot,\boldsymbol{a}\rangle$ it is easy to verify that

$$(1+\mathcal{H})^*\psi_i = 2\psi_i + \frac{\alpha_t}{\alpha_i}\psi_t, \tag{17.70}$$

and since $(1+\mathcal{H})^*\psi_t = \psi_t$ we have

$$(1+\mathcal{H})^*\psi''_i = 2\psi''_i \tag{17.71}$$

for

$$\psi''_i = \psi_i + \frac{\alpha_t}{\alpha_i}\psi_t. \tag{17.72}$$

Without the use of Gram–Schmidt these found N eigenvectors are nonorthogonal. In fact,

$$\langle\psi''_i,\psi''_j\rangle = \delta_{ij} + 3\frac{\alpha_t^2}{\alpha_i\alpha_j}. \tag{17.73}$$

Using these known inner products one could choose to carry out the Gram–Schmidt procedure numerically. The form of the inner products here is, however, simple enough to allow doing the orthonormalization analytically, as will now be shown.

Define the column vector $\boldsymbol{y} \in R^N$ by $y_i = \alpha_t/\alpha_i$, and the matrix $\boldsymbol{A} = (a_{ij})_{N\times N}$ by $a_{ij} = \langle \psi_i'', \psi_j'' \rangle$. Then $\boldsymbol{A} = 1 + 3\boldsymbol{y}\boldsymbol{y}^t$. Let

$$\psi_i' = b_{ij}\psi_j'' \tag{17.74}$$

and require orthogonality

$$\langle \psi_i', \psi_j' \rangle = \delta_{ij}. \tag{17.75}$$

This requirement is equivalent to the matrix equation

$$\boldsymbol{B}\boldsymbol{A}\boldsymbol{B}^t = 1, \tag{17.76}$$

where $\boldsymbol{B} = (b_{ij})_{N\times N}$. Assume that we can find a solution of the form $\boldsymbol{B} = 1 + \beta\boldsymbol{y}\boldsymbol{y}^t = \boldsymbol{B}^t$. Then $\boldsymbol{B}$ and $\boldsymbol{A}$ commute, and we find

$$\boldsymbol{B} = \boldsymbol{A}^{-1/2}. \tag{17.77}$$

Let $\mathcal{P} = |\boldsymbol{y}|^{-2}\boldsymbol{y}\boldsymbol{y}^t$ be the orthogonal projection in the direction of $\boldsymbol{y}$. Write

$$\boldsymbol{A} = [1 - \mathcal{P}] + (3|\boldsymbol{y}|^2 + 1)\mathcal{P}, \tag{17.78}$$

which expresses $\boldsymbol{A}$ as the weighted sum of two mutually orthogonal projections. In this form functions of $\boldsymbol{A}$ can be formed analogously to corresponding operations on diagonal matrices. Thus

$$\boldsymbol{A}^{-1/2} = [1 - \mathcal{P}] + \mathcal{P}/\sqrt{3|\boldsymbol{y}|^2 + 1} = 1 + \left((3|\boldsymbol{y}|^2 + 1)^{-1/2} - 1\right)\frac{\boldsymbol{y}\boldsymbol{y}^t}{|\boldsymbol{y}|^2}. \tag{17.79}$$

Since $|S| = |S_c| + \sum_{i=1}^N |S_i|$,

$$|\boldsymbol{y}|^2 = \sum_{i=1}^N \frac{\alpha_t^2}{\alpha_i^2} = \left(\sum_{i=1}^N |S_i|\right)/|S| = 1 - |S_c|/|S|, \tag{17.80}$$

and computing $|\boldsymbol{y}|^2$ is easy.

We found that $\boldsymbol{B} = 1 + \beta\boldsymbol{y}\boldsymbol{y}^t$ with

$$\beta = |\boldsymbol{y}|^{-2}\left(\frac{1}{\sqrt{3|\boldsymbol{y}|^2 + 1}} - 1\right), \tag{17.81}$$

so that

$$\psi_i' = b_{ij}\psi_j'' = \psi_i'' + \beta y_i(y_j\psi_j''). \tag{17.82}$$

Numerically the orthonormalization is now in a very efficient form, as the sum $y_j\psi_j''$ is calculated just once, and each ψ_i'' is corrected with a multiple of this.

As a check let us inspect what happens when the container boundary becomes very large. Firstly, when $|S_c| \approx |S|$, $\frac{\alpha_t}{\alpha_i} \approx 0$ and $\boldsymbol{\psi}_i'' \approx \boldsymbol{\psi}_i$. Secondly $|\boldsymbol{y}|^2 \approx 0$ implies $\beta|\boldsymbol{y}|^2 \approx 0$, and thus $\boldsymbol{B} \approx 1$. Then also $\boldsymbol{\psi}_i' \approx \boldsymbol{\psi}_i''$. Therefore $\boldsymbol{\psi}_i' \approx \boldsymbol{\psi}_i$, and the N orthonormal eigenfunctions of the adjoint — used for the mathematical deflations — are approximated quite closely by the N scaled particle surface normals. This is reasonable, since we expect the effect of the container surface to diminish as it "expands to infinity."

Also we note that if there is no container $\mathcal{P}_R = \mathbf{0}$, and the mathematical deflations assume the form found when a single particle was considered. Similarly for a container alone $\mathcal{P}_L = \mathbf{0}$, and the previously found result is recovered.

In summary this section has shown how the original double layer operator in discretized form can be so modified that efficient iterative solution of *mobility problems* is possible. Currently the authors are not aware of a deflation procedure that would enable similar modification of the second-kind integral equations for the iterative solution of resistance problems, except in special cases such as a spherical particle. Thus it seems that on using the second-kind IE formulations mobility problems will be favored over resistance problems, while the use of conventional solution methods leads to favoring resistance problems.

17.7 Iterative Solution of the Tractions for a Mobility Problem

Numerically the most important properties of the deflated double layer equations are the well-posedness and the feasibility of efficient iterative solutions. The latter especially is important since the linear systems are large and dense, and ordinary solution methods would be too expensive. Another useful feature of the iterative solution method is that good initial guesses can be utilized. An application where good initial guesses are available (except for the first time step) is the simulation of some system of particles over some period of time. With the quasi-steady approximation, at each time step one then solves the steady Stokes equations and uses the found RBM velocities to determine the particle configuration after a small increment in time, at which point a new steady Stokes solution is computed. When the changes of configuration are small, the previous solution provides a good approximation to the current one, and a small number of iterations will suffice.

In this section we show how the tractions can also be found iteratively for mobility problems. In fact we want the tractions that actually correspond to the given external forces and torques acting on the rigid particles, with as few calculations as possible, and not just tractions corresponding to some RBM. In some sense our mobility-based tractions complement Brenner's result relating the "resistance-based" tractions to mappings on total force and torque. The Lorentz reciprocal theorem can be used with the Riesz theorem (see below) to show that the "mobility-based" RBM tractions correspond to linear functionals that map a given surface velocity to components of that RBM that absorbs the total forces and torques, leaving the rest of the disturbance field representable by a double layer alone. The same result can be reached, based on our knowledge about the double layer operator, without using the Lorentz reciprocal theorem; the linear algebra steps required will also be shown in the following paragraphs.

Let $\boldsymbol{u}$ be a given velocity surface field, $\boldsymbol{v}_{RBM}$, the velocity field corresponding to total force $\boldsymbol{F}$ and torque $\boldsymbol{T}$ (mobility solution), and $\boldsymbol{u}_{RBM} = \boldsymbol{U} + \boldsymbol{\omega} \times \boldsymbol{r}$ such RBM velocity that $\boldsymbol{u} - \boldsymbol{u}_{RBM}$ is force- and torque-free. We will use the fact that

$$\langle \boldsymbol{u}, \hat{\boldsymbol{n}} \cdot \boldsymbol{\sigma}(; \boldsymbol{v}_{RBM}) \rangle = -(\boldsymbol{U} \cdot \boldsymbol{F} + \boldsymbol{\omega} \cdot \boldsymbol{T}) . \tag{17.83}$$

(The Lorentz reciprocal theorem was implicitly used here.) This shows the physical significance of mobility-based tractions: They map a given velocity field, $\boldsymbol{u}$, to (components of) such an RBM that absorbs the total force and torque, as claimed above. To derive this, we reason as follows:

$$\begin{aligned} 0 &= \langle \hat{\boldsymbol{n}} \cdot \boldsymbol{\sigma}(; \boldsymbol{u} - \boldsymbol{u}_{RBM}), \boldsymbol{v}_{RBM} \rangle \\ &= -\langle \boldsymbol{u}_{RBM}, \hat{\boldsymbol{n}} \cdot \boldsymbol{\sigma}(; \boldsymbol{v}_{RBM}) \rangle + \langle \boldsymbol{u}, \hat{\boldsymbol{n}} \cdot \boldsymbol{\sigma}(; \boldsymbol{v}_{RBM}) \rangle \\ &= (\boldsymbol{U} \cdot \boldsymbol{F} + \boldsymbol{\omega} \cdot \boldsymbol{T}) + \langle \boldsymbol{u}, \hat{\boldsymbol{n}} \cdot \boldsymbol{\sigma}(; \boldsymbol{v}_{RBM}) \rangle . \end{aligned}$$

It may be helpful to see how this works out for a single sphere. According to the Faxén law (Chapter 3), a force-free sphere, subject to the surface field $\boldsymbol{u}$, must translate as

$$\boldsymbol{U} = (\boldsymbol{u} + \frac{a^2}{6} \nabla^2 \boldsymbol{u})|_{x=0} = \frac{1}{4\pi a^2} \oint_{S_p} \boldsymbol{u} \, dS , \tag{17.84}$$

so that

$$\boldsymbol{U} \cdot \boldsymbol{F} = \frac{1}{4\pi a^2} \oint_{S_p} \boldsymbol{u} \cdot \boldsymbol{F} \, dS = \frac{1}{4\pi a^2} \langle \boldsymbol{u}, \boldsymbol{F} \rangle , \tag{17.85}$$

which correctly identifies $\boldsymbol{\sigma} \cdot \hat{\boldsymbol{n}} = -\boldsymbol{F}/4\pi a^2$ as the surface tractions on a sphere translating while exerting a force $\boldsymbol{F}$ on the fluid. Again, to repeat the main theme: The Riesz theorem guarantees both the existence and uniqueness of the RBM surface tractions in this role.

We are now ready to consider the iterative solution of the tractions in a mobility problem. Consider again N particle surfaces S_i surrounded by a container surface S_c, the interior fluid domain being bound between these. As in the previous section φ_{ij} denote the orthonormalized particle RBMs, explicitly given by

$$\varphi_{ij}(\boldsymbol{\xi}) = \boldsymbol{U}_{ij} + \boldsymbol{\omega}_{ij} \times (\boldsymbol{r} - \boldsymbol{r}_i) \tag{17.86}$$

on S_i and vanishing elsewhere. It is natural to choose $\boldsymbol{r}_i$ as the center of mass of the surface S_i, as has been discussed previously. The adjoint eigenfunction ψ_t is also defined as in the previous section. Given $\boldsymbol{u}$ we can solve the equation $(1 + \mathcal{H})\varphi = \boldsymbol{u}$ and the resulting RBM components ("absorbing" the total forces and torques not representable by the double layer) are given by $-\varphi_{ij}\langle \varphi, \varphi_{ij} \rangle$, as is seen by expanding $\mathcal{H}$:

$$\varphi_{ij}\langle \varphi, \varphi_{ij} \rangle = -(1 + \mathcal{K})\varphi + (\boldsymbol{u} - \psi_t \langle \varphi, \psi_t \rangle) . \tag{17.87}$$

The effect of the $\boldsymbol{\psi}_t$-term is to correct $\boldsymbol{u}$ to mass conserving in case it was not that to start with. The resulting RBM is thus completely characterized by the inner products on the LHS here. Consider one such inner product. The mapping $\boldsymbol{u} \mapsto \langle \varphi, \varphi_{ij} \rangle$ is a linear functional, and thus by the Riesz representation theorem this image is of the form $\langle \boldsymbol{u}, \boldsymbol{w} \rangle$ for some fixed $\boldsymbol{w}$.

Now fixing $\boldsymbol{w}$, we may insert various $\boldsymbol{u}$ in Equation 17.87 and inspect the result, $\langle \boldsymbol{u}, \boldsymbol{w} \rangle$, to identify $\boldsymbol{w}$. If $\boldsymbol{u} \in R(1+\mathcal{K})$, the inner products must all vanish in Equation 17.87. Thus $\boldsymbol{w} \perp R(1+\mathcal{K})$, or $\boldsymbol{w} \in N(1+\mathcal{K}^*)$. The latter shows that $\boldsymbol{w} = \hat{\boldsymbol{n}} \cdot \boldsymbol{\sigma}(;\boldsymbol{v})$ for some RBM $\boldsymbol{v}$, according to Theorem 3. Similarly testing with $\boldsymbol{u} = \boldsymbol{\psi}_t$ shows that $\boldsymbol{w} \perp \boldsymbol{u} = \boldsymbol{\psi}_t$, which, in physical terms, removes the pressure indeterminacy in the tractions within a bounded fluid domain. An easy way to show the last result is to note that the solution for φ is the container null function $\boldsymbol{\psi}_t \in N(1+\mathcal{K})$, so that $\langle \boldsymbol{u}, \boldsymbol{w} \rangle = \langle \varphi, \varphi_{ij} \rangle = \langle \boldsymbol{\psi}_t, \varphi_{ij} \rangle = 0$.

Let $\boldsymbol{w} \perp \boldsymbol{\psi}_t$ (and also $\boldsymbol{w} \perp R(1+\mathcal{K})$) be the tractions of some RBM $\boldsymbol{v}$. Applying $\langle \cdot, \boldsymbol{w} \rangle$ to Equation 17.87 above now gives

$$\begin{aligned} \langle \boldsymbol{u}, \boldsymbol{w} \rangle &= \langle \varphi, \varphi_{ij} \rangle \langle \varphi_{ij}, \boldsymbol{w} \rangle \\ &= \langle \varphi, \varphi_{ij} \rangle \langle \varphi_{ij}, \hat{\boldsymbol{n}} \cdot \boldsymbol{\sigma}(;\boldsymbol{v}) \rangle \\ &= -\langle \varphi, \varphi_{ij} \rangle (\boldsymbol{F}_i \cdot \boldsymbol{U}_{ij} + \boldsymbol{T}_i \cdot \boldsymbol{\omega}_{ij}) \ , \end{aligned} \tag{17.88}$$

where $\boldsymbol{F}_i$ *and* $\boldsymbol{T}_i$ *are the external total forces and torques affecting the particles relative to the points* $\boldsymbol{r}_i$ *(exerted by the particles on the fluid), corresponding to the mobility solution* $\boldsymbol{v}$. This discussion provided us another view to the significance of the "mobility-based" RBM-tractions, *avoiding the use of the Lorentz theorem and illustrating the functional analysis viewpoint.*

Now we look at things in the reversed way. Given a mobility problem for which we want to calculate the resulting RBM tractions, we start by defining

$$\varphi_g = -\varphi_{ij}(\boldsymbol{F}_i \cdot \boldsymbol{U}_{ij} + \boldsymbol{T}_i \cdot \boldsymbol{\omega}_{ij}) \ , \tag{17.89}$$

inspired by the RHS of the previous equation. Now for *any* φ

$$\langle \varphi, (1+\mathcal{H}^*)\boldsymbol{w} \rangle = \langle (1+\mathcal{H})\varphi, \boldsymbol{w} \rangle = \langle \boldsymbol{u}, \boldsymbol{w} \rangle = \langle \varphi, \varphi_g \rangle \ , \tag{17.90}$$

and $\boldsymbol{w} = \hat{\boldsymbol{n}} \cdot \boldsymbol{\sigma}(;\boldsymbol{v})$ can be solved from

$$(1+\mathcal{H}^*)\boldsymbol{w} = \varphi_g \ . \tag{17.91}$$

The (real) spectrum of $\mathcal{H}^*$ coincides with that of $\mathcal{H}$, and iterative solution is similarly possible. In fact

$$(1-\mathcal{P}_R/2)(1+\mathcal{H}^*)(1-\mathcal{P}_L/2) = 1+\tilde{\mathcal{H}}^* \ , \tag{17.92}$$

and the sequential steps of the solution are

$$\begin{aligned} \tilde{\varphi} &= (1-\mathcal{P}_R/2)\varphi_g && (17.93) \\ (1+\tilde{\mathcal{H}}^*)\tilde{\boldsymbol{w}} &= \tilde{\varphi} && (17.94) \\ \boldsymbol{w} &= (1-\mathcal{P}_L/2)\tilde{\boldsymbol{w}} \ . && (17.95) \end{aligned}$$

Thus we are now able to find the RBM tractions corresponding to a given mobility problem with just one iterative solution of a linear system. Moreover, the linear system is just the adjoint of that used for solving the RBM velocities, meaning in practice that the transposed real discrete system is used. Thus some savings are possible by solving "simultaneously" these two sets of quantities, velocities, and tractions, as the elements of the system matrices are the same and need to be generated but once.

We conclude this subsection with an illustration of how these ideas apply to a single sphere. Our illustration for the sphere, which we chose for pedagogical reasons (an analytic solution is feasible), should not obscure the fact that the iterative approach is most attractive for *numerical solution* of more complex geometries.

For the RBM translation of a single, torque-free sphere in unbounded flow, we set $\mathcal{P}_R = \mathbf{0}$ (no container), and drop the particle index i above. The sequential steps of the solution simplify to

$$\tilde{\varphi} = \varphi_g = -\boldsymbol{U}_j(\boldsymbol{F}\cdot\boldsymbol{U}_j) \tag{17.96}$$
$$(1+\tilde{\mathcal{H}}^*)\tilde{\boldsymbol{w}} = \tilde{\varphi} \tag{17.97}$$
$$\boldsymbol{w} = (1-\mathcal{P}_L/2)\tilde{\boldsymbol{w}}\ . \tag{17.98}$$

The orthonormalized translational velocities $\boldsymbol{U}_j$ will eventually be replaced with $\boldsymbol{U}_j = \boldsymbol{e}_j/|S|^{1/2}$.

For the single sphere, we fortunately know the eigensystem completely and may spectrally decompose the double layer operator as

$$1+\mathcal{K} = 1+\sum_k \lambda_k\mathcal{P}_k\ . \tag{17.99}$$

Here, $\{\mathcal{P}_k\}$ is an orthogonal family of self-adjoint projections onto the eigenspace N_{λ_k}. The physical deflation simply removes N_{-1}, so that

$$1+\mathcal{H} = 1+\sum \lambda_k\mathcal{P}_k\ , \quad \lambda_k \in (-1,1]\ . \tag{17.100}$$

The mathematical deflation at the other end of the spectrum is accomplished with $\mathcal{P}_L = \hat{\boldsymbol{n}}\langle\cdot,\hat{\boldsymbol{n}}\rangle/|S|$, and

$$\begin{aligned}(1-\frac{1}{2}\mathcal{P}_L)(1+\mathcal{H}) &= (1+\tilde{\mathcal{H}})\\ &= 1+\sum \lambda_k\mathcal{P}_k\ , \quad \lambda_k \in (-1,1)\ .\end{aligned}$$

Since the projection operators in question are self-adjoint, so is $\mathcal{H}$ and the decomposition

$$(1+\tilde{\mathcal{H}}^*) = 1+\sum \lambda_k\mathcal{P}_k\ , \quad \lambda_k \in (-1,1) \tag{17.101}$$

is obtained. The iterative procedure reduces to

$$\tilde{\boldsymbol{w}} = -\left(\sum_k \lambda_k\mathcal{P}_k\right)\tilde{\boldsymbol{w}} - \boldsymbol{U}_j(\boldsymbol{F}\cdot\boldsymbol{U}_j)\ , \quad \lambda_k \in (-1,1)\ , \tag{17.102}$$

with the solution, $\tilde{\boldsymbol{w}} = -\boldsymbol{U}_j(\boldsymbol{F} \cdot \boldsymbol{U}_j)$, obtained immediately after just the first iteration from the fact that $\boldsymbol{U}_j \in N_{-1} \perp N_{\lambda_k}$.

Finally, the RBM traction, $\boldsymbol{w} = \boldsymbol{\sigma} \cdot \hat{\boldsymbol{n}}$, is obtained as follows:

$$\begin{aligned} \boldsymbol{w} &= (1 - \frac{1}{2}\mathcal{P}_L)\tilde{\boldsymbol{w}} \;=\; -(1 - \frac{1}{2}\mathcal{P}_L)\boldsymbol{U}_j(\boldsymbol{F} \cdot \boldsymbol{U}_j) \\ &= -\boldsymbol{U}_j(\boldsymbol{F} \cdot \boldsymbol{U}_j) \;=\; -\frac{1}{4\pi a^2}\boldsymbol{e}_j(\boldsymbol{F} \cdot \boldsymbol{e}_j) \;=\; -\frac{\boldsymbol{F}}{4\pi a^2} \,, \end{aligned}$$

and this is indeed the correct result.

Exercises

Exercise 17.1 The Double Layer Kernel on a Sphere.
Show that $\mathcal{K}$ is self-adjoint on a single sphere. How does this imply the result of Exercise 15.4?

Exercise 17.2 Complex Stokes Solutions.
Verify the complex form of the Lorentz reciprocal theorem, given when discussing the spectrum of the double layer operator.
Hint: Note that while $\langle \imath a, b \rangle = \imath \langle a, b \rangle$, the conjugate linearity with respect to the second argument implies that $\langle a, \imath b \rangle = -\imath \langle a, b \rangle$.

Exercise 17.3 Ill-Posedness of the First-Kind Equation.
In Chapter 1, we gave a heuristic argument, using highly oscillatory inputs, on why integral equations of the first kind were ill-posed problems. However, since the Oseen tensor $\mathcal{G}(\boldsymbol{x})$ is singular at the origin, we may ask whether this singular behavior somehow saves us, *i.e.*, the well-conditioned property of the second-kind equation is mimicked to some extent. The answer, of course, is no, as this exercise with a sphere shall illustrate.

Consider high-frequency velocity fields of the form $\nabla\Phi_{-n-1}$, the exterior vector harmonic in Lamb's general solution. Consider the integral representation for this velocity field in the region outside a unit sphere, use the fact that the tractions on the sphere surface are given by $-2(n+2)\mu\nabla\Phi_{-n-1}$, and that $\nabla\Phi_{-n-1}|_{r=1}$ is an eigenfunction of the double layer operator to derive the following relation:

$$\frac{1}{2}\nabla\Phi_{-n-1} = \frac{(n+2)}{4\pi}\oint \nabla\Phi_{-n-1} \cdot \mathcal{G}\, dS + \frac{1}{2}\lambda_n^+ \nabla\Phi_{-n-1}$$

or

$$\oint \nabla\Phi_{-n-1} \cdot \mathcal{G} dS = \frac{8\pi n \nabla\Phi_{-n-1}}{(2n+1)(2n+3)} \,.$$

This shows that for large n, the output becomes small. This also explains why in practice we may use the single layer representation for spheres and other smooth particles of moderate aspect ratios, since the higher oscillations are suppressed by the use of coarse meshes. For oblate spheroids, we find that the behavior is essentially the same, until the aspect ratio approaches 10; then the large tractions near the rim imply that the factor of n gets amplified. Thus the ill-conditioning is a significant problem, especially when fine meshes are used at the rim.

Exercise 17.4 The Spectral Radius of the Single Layer Operator.
Consider the single layer operator mapping $\boldsymbol{\psi}$ to

$$\frac{1}{4\pi\mu}\oint_S \mathcal{G}\cdot\boldsymbol{\psi}\,dS \ .$$

Consider the effect of rescaling a particle surface, say by factor a. Find the power of a by which the spectrum of the single layer operator gets scaled. Plot some of the eigenvalues implicitly given for a unit sphere in Exercise 17.3.
Note: This is the reason why no bounds can be given, without some restrictions to particle size, for the spectral radius of the single layer operator.

Exercise 17.5 Diagonalizability of a Square Matrix.
Show that a square matrix can be diagonalized if and only if it can be expressed as the product of a symmetric positive definite (SPD) and a symmetric matrix.
Solution: Assume diagonalizability, $A = P^{-1}DP$. Then also

$$A = (P^{-1}P^{-T})(P^TDP) \ ,$$

where we have an SPD matrix multiplying a symmetric one. Assume $A = BC$, where B is SPD and C symmetric. Then $A = B^{1/2}(B^{1/2}CB^{1/2})B^{-1/2}$, and diagonalizing the symmetric matrix in the parentheses to $Q^{-1}DQ$ gives the diagonal representation of A with $P = QB^{-1/2}$.
Note: According to the Lorentz reciprocal theorem the mapping of velocity BCs to corresponding tractions is symmetric (just write the theorem in inner product notation), and it is also positive *semi*definite due to the energy relation. Thus the operator product is almost of the form given in the exercise above, when this operator is applied after the symmetric mapping of a single layer density to corresponding surface velocity. (The symmetry of this mapping is visible from the symmetric kernel, namely, the Green's function.) From this we may conjecture that the operator $\mathcal{K}^*$ should be diagonalizable, the same equivalently holding for $\mathcal{K}$. In fact Odqvist has shown that the Jordan blocks for $|\lambda| \neq 1$ are trivial, and this (together with our theorems regarding the extreme eigenvalues) implies the diagonalizability of $\mathcal{K}$. Therefore the discussion of nontrivial Jordan blocks in connection with Wielandt's deflation was somewhat superfluous for our application and only served the purpose of generality.

Exercise 17.6 Tractions via the Reciprocal Theorem: Rate-of-Strain Fields.
Let $\boldsymbol{E}$ be a symmetric traceless 3×3 matrix, generating the Stokes velocity field $\boldsymbol{v} = \boldsymbol{E}\cdot\boldsymbol{x}$. For convenience assume that the viscosity $\mu = 1$. First recall that the rate-of-strain tensor and the stress tensor of this field both coincide

with $\boldsymbol{E}$. Derive a system of well-posed equations from which the tractions of the disturbance field cancelling $\boldsymbol{v}$ on a particle surface could be solved numerically.
Solution: Denote the disturbance field by $\boldsymbol{v}_d$ and its traction surface field by $\boldsymbol{w} = \hat{\boldsymbol{n}} \cdot \boldsymbol{\sigma}(;\boldsymbol{v}_d)$. An arbitrary disturbance field can be represented on S by $\boldsymbol{u} = (1+\mathcal{K})\boldsymbol{\varphi} + (\boldsymbol{F} - \frac{1}{2}\boldsymbol{T} \times \nabla) \cdot \mathcal{G}/(8\pi\mu)$. Applying Lorentz reciprocal theorem to the double layer part of $\boldsymbol{u}$, one finds $\langle(1+\mathcal{K})\boldsymbol{\varphi}, \boldsymbol{w}\rangle = \langle\hat{\boldsymbol{n}} \cdot \boldsymbol{\sigma}(;\mathcal{K}\boldsymbol{\varphi}), \boldsymbol{v}_d\rangle = \langle\hat{\boldsymbol{n}} \cdot \boldsymbol{\sigma}(;\mathcal{K}\boldsymbol{\varphi}), -\boldsymbol{v}\rangle = \langle(-1+\mathcal{K})\boldsymbol{\varphi}, \hat{\boldsymbol{n}} \cdot \boldsymbol{\sigma}(;-\boldsymbol{v})\rangle = -\langle(-1+\mathcal{K})\boldsymbol{\varphi}, \hat{\boldsymbol{n}} \cdot \boldsymbol{E}\rangle$, where the second application of the Lorentz reciprocal theorem was on the double layer generated velocity field and $\boldsymbol{v}$ *inside* the particle. Since $\boldsymbol{\varphi}$ is arbitrary, this gives the equation $(1+\mathcal{K}^*)\boldsymbol{w} = -(-1+\mathcal{K}^*)(\hat{\boldsymbol{n}} \cdot \boldsymbol{E})$. The operator acting on $\boldsymbol{w}$ is not invertible, and bordering will have to be used here. The added equations come from the force and torque terms: $\langle\hat{\boldsymbol{e}}_i \cdot \mathcal{G}, \boldsymbol{w}\rangle = \langle\hat{\boldsymbol{n}} \cdot \boldsymbol{\sigma}(;\hat{\boldsymbol{e}}_i \cdot \mathcal{G}), -\boldsymbol{v}\rangle$ and $\langle(\hat{\boldsymbol{e}}_i \times \nabla) \cdot \mathcal{G}, \boldsymbol{w}\rangle = \langle\hat{\boldsymbol{n}} \cdot \boldsymbol{\sigma}(;(\hat{\boldsymbol{e}}_i \times \nabla) \cdot \mathcal{G}), -\boldsymbol{v}\rangle$, for $i \in \{1,2,3\}$. Since the range of $(1+\mathcal{K}^*)$ is orthogonal to RBM velocities, the six added variables can be defined as components of translation and rotation velocities, modifying the equation to $(1+\mathcal{K}^*)\boldsymbol{w} + \boldsymbol{U} + \boldsymbol{\omega} \times \boldsymbol{\xi} = -(-1+\mathcal{K}^*)(\hat{\boldsymbol{n}} \cdot \boldsymbol{E})$. In fact, the added variables should get value zero, but in a numerical approximation this will not be exactly so.

Exercise 17.7 Deflations Over Bordering.
From your favorite text on numerical methods, check the number of operations required for the solution of a linear system of dimension m by Gaussian elimination. It should be $O(m^3)$. Consider now the iterative solution of a system of the same size. Convince yourself that each iterative step takes $O(m^2)$ operations. Since the rate of convergence depends on the largest eigenvalue, close to that of the continuous system approximated, it will stay about the same as the discretization is refined. Therefore the iterative solution with different discretizations but the same initial guess will take also $O(m^2)$ operations. This shows a definite advantage for the iterative solution when accurate results are desired.

Consider now a single particle, the deflated system size being m and the bordered system size $m+6$. For solving a *resistance problem* you have the following choices:

1. Solve iteratively six mobility problems and construct the mobility matrix, then solve by elimination one problem with the mobility matrix as the system matrix.
2. Solve directly by elimination the bordered system.

Which one is more advantageous? Show that if the iterative solutions above are changed to elimination solutions, the bordering scheme becomes more efficient for large m.

Exercise 17.8 Double Layer Eigenfunctions for the Ellipsoid.
In this chapter, we saw that the double layer density in the solution for a sphere in the constant rate-of-strain field $\boldsymbol{E} \cdot \boldsymbol{x}$ is an eigenfunction of $\mathcal{K}$, with eigenvalue $-1/5$. Show that the solutions for a torque-free ellipsoid in the constant rate-of-strain field are eigenfunctions of $\mathcal{K}$, and determine the expression for the

eigenvalues in terms of elliptic integrals. Look at the degenerate case of ellipsoids of revolution and examine what happens to the various families of eigenfunctions with increasing nonsphericity. Do your results suggest that numerical difficulties will be encountered in the numerical solution for flow past nearly flat oblate spheroids and very slender prolate spheroids?

Exercise 17.9 Perturbation Results for the Two-Sphere Spectrum. In this chapter, we computed the spectral radius of the two-sphere system, at large R, by a perturbation solution on the single-sphere eigenvalue $\lambda = -3/5$. Generalize this approach to determine the splitting induced by hydrodynamic interactions on the entire spectrum of the single sphere. Do the eigenvalues cross as R is reduced?

Exercise 17.10 Double Layer Representation for Axisymmetric Flow. For axisymmetric problems (symmetry axis $\boldsymbol{d} = -\boldsymbol{e}_z$), show that a velocity field $\boldsymbol{v}$ with the multipole representation

$$\boldsymbol{v} = \sum_{n=0}^{\infty} \left\{ A_n \frac{(\boldsymbol{d}\cdot\nabla)^n}{n!} \boldsymbol{d}\cdot\mathcal{G} + B_n \frac{(\boldsymbol{d}\cdot\nabla)^n}{n!} \boldsymbol{d}\cdot\nabla^2\mathcal{G} \right\}$$

may also be written in terms of a double layer density on a unit sphere,

$$\boldsymbol{v}(\boldsymbol{x}) = A_0 \boldsymbol{d}\cdot\mathcal{G}(\boldsymbol{x}) + \oint_S \boldsymbol{K}(\boldsymbol{x},\boldsymbol{\xi})\cdot\boldsymbol{\varphi}\, dS \ ,$$

with the double layer density given by

$$\boldsymbol{\varphi}(\boldsymbol{\xi}) = \sum_{n=1}^{\infty} \alpha_n \boldsymbol{\varphi}^-_{n+1}(\boldsymbol{\xi}) + \sum_{n=0}^{\infty} \beta_n \boldsymbol{\varphi}^+_{n+1}(\boldsymbol{\xi}) \ .$$

This result is used in the subsequent problems.

Exercise 17.11 Use of an Off-Center Stokeslet in Range Completion. Use the results of the previous exercise to derive the double layer density for a sphere of radius a translating with velocity $\boldsymbol{U}$ with an off-center Stokeslet completing the representation. Assume that the Stokeslet and the the displacement R are all in the direction of $\boldsymbol{U}$. The final result may be written as

$$\begin{aligned}
\boldsymbol{\varphi}(\boldsymbol{\xi}) &= -\frac{5\boldsymbol{F}}{48\pi\mu a}\cdot(\hat{\boldsymbol{n}}\hat{\boldsymbol{n}} - \frac{1}{3}\boldsymbol{\delta}) \\
&+ \frac{F}{8\pi\mu a}\sum_{n=2}^{\infty}\left\{ \frac{(R/a)^{n-1}}{2(n-1)} \left[\boldsymbol{e}_r n P_n(\cos\theta) + \boldsymbol{e}_\theta \frac{\partial P_n}{\partial\theta} \right] \right. \\
&\qquad + \left(\frac{(R/a)^{n-1}}{2n+1} - \frac{(R/a)^{n+1}}{2n+3} \right) \frac{n(2n+1)(2n+3)}{2(2n^2+4n+3)} \\
&\qquad \left. \times \left[-\boldsymbol{e}_r (n+1) P_n(\cos\theta) + \boldsymbol{e}_\theta \frac{\partial P_n}{\partial\theta} \right] \right\} \ .
\end{aligned}$$

Note that with increasing eccentricity, the double layer density becomes highly oscillatory. This is a typical symptom of a poor choice for the range completion.

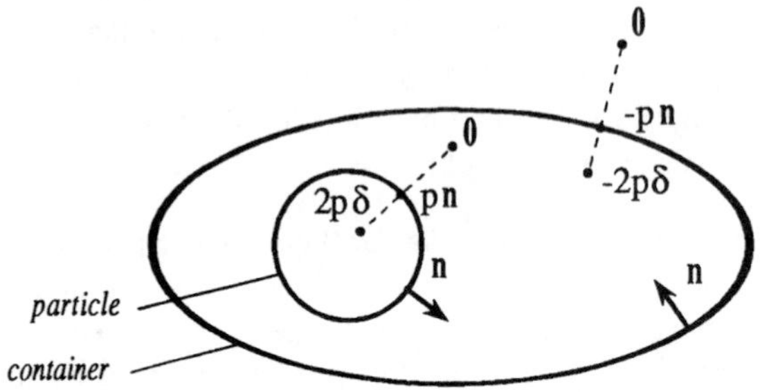

Figure 17.5: Geometry for the single layer eigenfunction with zero dissipation; stress fields of the single layer potential of density $p\hat{n}$ (on a particle surface and on a container surface).

Exercise 17.12 Oseen's Image for a Stokeslet Near a Sphere.
In Part III, we derived a multipole representation for the Oseen image for a Stokeslet near a rigid sphere. Use the result of Exercise 17.10 to express the image of an Stokeslet directed along the axis as a completed double layer representation. In this case, what is the best way of completing the range if a single Stokeslet is used? What is the best way of completing the range if a line distribution of Stokeslets is employed?

Exercise 17.13 Single Layer Eigenfunctions with Zero Total Dissipations.
Let ψ be a single layer eigenfunction (corresponding to an eigenvalue λ) on the totality of bounding surfaces S, and suppose that the velocity fields induced by ψ have zero total dissipation (sum of dissipations in the interior and the exterior domains). As discussed in Section 17.2, the condition of zero dissipation forces the single layer density to be a multiple of the surface normal on each surface, *viz.*, $p\hat{n}$. Since the traction *drops* by $2p$ across the surface and since the stresses must be zero on the "outsides" we obtain the situation shown in Figure 17.5 for the surface values of the single layer potential. Now on each surface apply the condition that ψ is an eigenfunction. Show that this forces the eigenvalue λ to be either -1 or $+1$.

Chapter 18

Fourier Analysis for Axisymmetric Boundaries

In this chapter, some axisymmetric numerical tests with the canonical equations shall be described. The idea was to verify the validity of the equations derived, including the method for finding RBM tractions through use of the Riesz theorem, with suitable example cases for which some analytical results were available. Fourier analysis was used in order to further reduce the system sizes, to make computations on ordinary computers feasible, and all the numerical solutions in this section utilized ordinary Gaussian elimination solution of the discrete linear system. The use of the deflate-and-iterate techniques of the preceding chapter will be illustrated in the next chapter, when we consider true three-dimensional geometries.

18.1 How the Components Separate in Wave-Number

Utilizing the axisymmetry of boundaries is easy when the flow is axisymmetric, *e.g.*, due to axial translation or axial rotation. For nonaxisymmetric boundary conditions the problem is somewhat more involved. In elasticity problems efforts have been made by Mayr [49], Rizzo and Shippy [61], Mayr *et al.* [50], and Shippy *et al.* [64], among others. The underlying idea is that on using cylindrical polar coordinates, with the physical components of the fields of interest multiplying the corresponding unit vectors, we may effect a separation of the Fourier components in the azimuthal variable ϕ. This is most easily seen by noting that the differential operators in the equations of interest (here the Stokes equations),

$$\mu \nabla^2 \boldsymbol{v} = \nabla p \qquad (18.1)$$

$$\nabla \cdot \boldsymbol{v} = 0\,, \qquad (18.2)$$

preserve factors of the form $e^{\imath m\phi}$ multiplying functions independent of ϕ; for a collection of formulas, see Bird *et al.* [4, Table A.7-2 in Appendix A]. Obviously a similar method may then be used for any linear differential equation systems stemming from isotropic physical problems to obtain decoupling of the mentioned Fourier components. What makes this decoupling particularly interesting to us is that all the rigid-body motion velocity fields are expressible in terms of components with wave-numbers $m \in \{0, 1\}$. The term *wave-number* for m is preferred here instead of *frequency*, reserving the latter for time-dependent problems. As a consequence of the above, mobility and resistance problems with axisymmetric boundaries require solving only problems with wave-numbers 0 and 1. Note that the separation of the ϕ-variable from the equations leaves us with two-dimensional arguments, so that reducing the dimensionality further by using the BIE approach will result in a scalar argument, *i.e.*, one gets one-dimensional integral equations. This argument will correspond to the location on a generating curve of the axisymmetric surface. Obviously this reduction can potentially bring large savings in grid generation and computational costs, while enhancing achievable accuracy. A problem on effecting this decomposition is, how to evaluate the elliptic integrals arising in the azimuthal ϕ-integrations. Dealing only with the case of axial translation, Youngren and Acrivos [72] used directly the complete elliptic integrals of the first and second kind. Since then other authors have devised different ways to deal with these in the case of higher wave-numbers, see, *e.g.*, Shippy *et al.* [64] and Hsu and Ganatos [32].

18.2 Another Symmetry Argument for the Fourier Decomposition

In this section we present an elementary symmetry argument in order to clarify why the Fourier decomposition works out the way it does in the case of axisymmetric boundaries. The argument utilizes just the rotational symmetry of the geometry and the linearity and isotropy of the Stokes and continuity equations, without any need for using the actual form of these equations. Thus the same kind of argument can be applied to any problem for which the governing equations are linear and isotropic.

On an axisymmetric surface consider the velocity field given by

$$\boldsymbol{v} = g_\rho(z)\cos(m\phi)\hat{e}_\rho + g_\phi(z)\sin(m\phi)\hat{e}_\phi + g_z(z)\cos(m\phi)\hat{e}_z \ , \tag{18.3}$$

which is its own reflection through the plane $y = 0$. This boundary velocity is assumed to determine the solution in all of the fluid domain considered (recall we have the uniqueness theorem for Stokes flows, and with smooth enough boundary conditions — satisfying mass conservation, *etc.* — we also know the existence of the Stokes solution). Rotation of this velocity surface field by α corresponds to substituting $\phi \leftarrow \phi - \alpha$ in the RHS of the equation above, while everything else there remains unchanged, and produces the velocity field $\boldsymbol{v}_\alpha$. The addition formulas for the trigonometric functions show that for a given fixed $\boldsymbol{v}$, $\boldsymbol{v}_\alpha$ depends linearly on the parameters $\sin(m\alpha)$ and $\cos(m\alpha)$. Since the equations are assumed linear, we infer

Conjecture 1 *If the boundary conditions depend linearly on a set of parameters, so does the whole flow field and the corresponding stress field.*

So the surface tractions $\boldsymbol{T}$ will also depend linearly on the two parameters, "cos($m\alpha$)" and "sin($m\alpha$)", for example,

$$\hat{e}_\rho(\phi) \cdot \boldsymbol{T}_\alpha(\phi, z) = A(\phi, z)\cos(m\alpha) + B(\phi, z)\sin(m\alpha) \ . \tag{18.4}$$

On the other hand due to the rotational symmetry,

$$\hat{e}_\rho(\phi) \cdot \boldsymbol{T}_\alpha(\phi, z) = \hat{e}_\rho(\phi - \alpha) \cdot \boldsymbol{T}(\phi - \alpha, z) = A(\phi - \alpha, z) \ , \tag{18.5}$$

and so

$$\begin{aligned} A(\phi - \alpha, z) &= A(\phi, z)\cos(m\alpha) + B(\phi, z)\sin(m\alpha) \\ \Rightarrow A(\alpha, z) &= A(0, z)\cos(m\alpha) - B(0, z)\sin(m\alpha) \ . \end{aligned} \tag{18.6}$$

Using this in the expression for the tractions gives

$$\hat{e}_\rho(\phi) \cdot \boldsymbol{T}(\phi, z) = A(\phi, z) = A(0, z)\cos(m\phi) - B(0, z) sin(m\phi) \ . \tag{18.7}$$

Requiring invariance under the reflection about $y = 0$ implies $B(0, z) = 0$, so that

$$\hat{e}_\rho(\phi) \cdot \boldsymbol{T}(\phi, z) = A(0, z)\cos(m\phi) \ . \tag{18.8}$$

Similar treatment of the other components of the surface tractions leads to $\boldsymbol{T}$ being of the same form as $\boldsymbol{v}$, with just the z-dependence modified.

Note that the only assumption needed about the tractions in the deduction above was that the surface field in question must linearly depend on the given boundary conditions. Thus any such surface vector field necessarily has the same form (*i.e.*, the same vectorial Fourier-components) as the velocity and tractions above. In particular, this applies to the double layer density.

18.3 Analytical Fourier Decomposition of the Kernel with Toroidals

Mainly cases where the flow and stress fields are mirror symmetric with respect to the plane $y = 0$ in Cartesian coordinates shall be considered. Then on the surface these fields and the surface tractions have their ϕ-dependence in the form of $\hat{e}_\rho \cos(m\phi)$, $\hat{e}_\phi \sin(m\phi)$, or $\hat{e}_z \cos(m\phi)$. At every point of the surface these are used as a local basis, and after azimuthal integrations an equation system for the ϕ-independent coefficients of these vectors is found. Axial rotation is an exception ($m = 0$ above drops $\hat{e}_\phi$), and one has to consider $\hat{e}_\phi$ alone to get the component of torque along z-axis; clearly $\hat{e}_\phi$ is not mirror symmetric. The remaining two components for $m = 0$ give an equation pair involving F_z. Finally with $m = 1$ the above-mentioned fields correspond to F_x and T_y, so that these have to be solved from an integral equation triplet. Note that the remaining Fourier components $\hat{e}_\rho \sin(m\phi)$, *etc.* can be reduced to the case above by a rotation of the coordinate axes about z; obviously one then gets F_y and T_x for $m = 1$, so that these need not be considered separately. The following formula, valid for integral m and n,

$$\int_0^{2\pi} \frac{\cos(m\phi)}{q^{n+1/2}}\, d\phi = c_n(k) Q^n_{m-1/2}(k) , \tag{18.9}$$

where $q = k - \cos(\phi)$, $k > 1$ and

$$c_n(k) = \frac{(-1)^n 2^{3/2}}{(1/2)_n (k^2 - 1)^{n/2}} \tag{18.10}$$

$$(1/2)_n = \Gamma(1/2 + n)/\Gamma(1/2) , \tag{18.11}$$

shows that one type of the elliptic integrals needed can be related to *toroidal functions*, a particular case of associated Legendre functions. Once this has been realized, trigonometric identities and recurrence relations for toroidal functions can be used to similarly relate the rest of the integrals needed to these functions. This gives a systematic method for treating the kernels in any of the integral equations that have been formulated for Stokes flow to get a Fourier decomposition. The analytical work for the double layer kernel was done for all values of the wave-number m simultaneously. It is just the numerical evaluation of these functions that gets more complicated as m is increased. However, as noted, one need not consider other values than $m = 0$ and $m = 1$ for resistance or mobility problems, as long as the configuration of boundaries is axisymmetric.

18.4 Numerical Computation of the Toroidal Functions

Recognizing the elliptic integrals as toroidal functions not only gives us a number of representations, recurrence relations, and asymptotic formulae, but also different algorithms for their computation and tabulated values. Tabulated values and some numerical methods can be found in Briggs and Lowan [8], and backward recursion has been studied by Gautschi [24]. Asymptotic formulae and other properties are available in Gradshteyn and Ryzhik [25] and Magnus and Oberhettinger [47], and in particular Snow [65] contains a wealth of information. A generalization of the arithmetic geometric mean method for complete elliptic integrals, called Bartky's lemma, can be found in Wimp [70], and the algorithm of Fettis [18] is analytically equivalent to this, but numerically easier to implement. The algorithm of Fettis was chosen, with a minor modification. Since the functions computed are singular at argument value one ($k = 1$), the deviation from one ($k - 1$) was used in all the iteration formulae. After this modification the computed values are accurate to double precision over all of the range tested.

In the computations to form the kernel functions, some very small and very large numbers do come up, so that it has been necessary to use the extended number range provided by g_floating in VAX Fortran. To further enhance accuracy, the computer algebra program MACSYMA[1] was used to factor coefficients in the kernel functions, so that catastrophic cancellation could be avoided. This factoring was also necessary for assessing the behavior of the coefficients near the singularities. Asymptotic analysis reveals, then, that only one of the kernel functions is singular. The diagonal element corresponding to the azimuthal angle ϕ has a logarithmic singularity, whereas all the others remain bounded. This knowledge is essential in performing singularity subtraction with vector functions, where only one component can be cancelled at a time.

18.5 The Numerical Solution Procedure

18.5.1 The Choice of the Numerical Method

The typical method for the solution of BIE systems is to use boundary elements. This means that the surfaces are subdivided into elements over which the solutions are approximated by typically a linear combination of some basis functions, just as in finite element methods, and appropriate quadratures are used for the integrations over elements. In the current case the kernel functions are costly to compute, and therefore it is no longer economical to use a large number of kernel function evaluations for each degree of freedom in the resulting linear system. A method that completely avoids the element aspect is combining a global quadrature with collocation at the quadrature points. In the numerical examples the fact that the integrations are in only one variable makes this choice easy to implement, whereas for two-dimensional surface integrals generating the quadratures in a flexible way could require the use of element approach again. Some quadratures for surface integrals are known though (see Stroud [67]). A difficulty one faces immediately after this choice is how to deal with the singularities at the collocation points. Also for boundary element methods the way that integrations over the singular elements are carried out has a great influence on the accuracy of the results, and even "nearly singular elements" are considered separately. It has been shown that for weakly singular integral equations, the so-called singularity subtraction technique leads to convergent algorithms (Fenyö and Stolle [17]). Using the known outside null functions of the operator, namely densities that coincide with the velocity fields of rigid-body motions on the surface, this technique can be implemented. Power and Miranda had also noticed this possibility, although the singularity subtraction was disguised in the form of an operator identity in connection with their full surface implementation using elements. They utilized the three null functions related to translations, whereas here also those related to rotations are used. Actually with a full surface method one cannot utilize more than three of the null functions for singularity subtraction, but it should be pointed out

[1]MACSYMA was originally developed by the Mathlab Group of the MIT Laboratory of Computer Science, formerly Project MAC, and is currently being enhanced, supported, and distributed by Symbolics, Inc. of Cambridge, Massachusetts. See [68].

that there is some choice. *Also, essential to the success of the singularity subtraction in the codes is that most of the kernel functions are nonsingular.* This observation was facilitated by recognizing the appearance of toroidal harmonics as the elliptic integrals to be calculated, so that the asymptotic behavior of the kernel elements near the singularity could be analyzed. For a general presentation of the various methods for solving integral equations of the second kind see Atkinson [2].

18.5.2 Theory of Singularity Subtraction

Suppose we want to solve an equation of the second kind, $x + Kx = b$, for x. Approximating the integration by a quadrature we get

$$x(s) + \sum_{j=1}^{N} w_j K(s,t_j)x(t_j) = b(s), \tag{18.12}$$

where w_j and t_j ($j = 1,\ldots,N$) represent the quadrature weights and points, respectively. On using collocation to solve these equations, it is necessary to choose the quadrature points as the collocation points, *i.e.*, to substitute t_i for s above. When the kernel is singular at the collocation point, we cannot evaluate $K(t_j,t_j)$, and this method seems to break down. If, however, we know any nontrivial identity $Kh = g$, this may be subtracted with a suitable multiplier to get

$$x(s)\left(1+\frac{g(s)}{h(s)}\right) + \int K(s,t)\left(x(t) - \frac{x(s)}{h(s)}h(t)\right)\,dt = b(s)\;, \tag{18.13}$$

and on discretizing this as above, the function on which the kernel is acting will be zero at the singular point s. Thus the difficulty is avoided. The null functions that are known in the application here were used in exactly the way described to remove singularities. A formal justification for the method can be found in Fenyö and Stolle [17], based on the theory of Kantorovich, for the case where h is identically one. This seems to be the only case usually mentioned (see also Delves and Walsh [13]), probably because the modification here can be easily transformed to this particular case. All one needs to do is to consider h as part of the kernel and to modify the IE correspondingly to

$$\frac{x(s)}{h(s)} + \int \frac{K(s,t)h(t)}{h(s)}\frac{x(t)}{h(t)}\,dt = \frac{b(s)}{h(s)}\;, \tag{18.14}$$

for which the integral identity now reads

$$\int \frac{K(s,t)h(t)}{h(s)}\,dt = \frac{g(s)}{h(s)}\;. \tag{18.15}$$

Naturally it must be required that h is nonzero almost everywhere.

18.6 Axial Torque as an Example

18.6.1 The Azimuthal Integrations

The Fourier analysis and singularity subtraction techniques shall be illustrated here with the simplest case of one integral equation for the axial torque on an axisymmetric particle. As mentioned earlier, axial force leads to two and transverse force and torque to three coupled one-dimensional integral equations, due to the vector nature of Stokes flow problems. Let an axisymmetric surface be generated by the curve $(\rho(t), z(t))$, $t_{min} \leq t \leq t_{max}$, and consider two surface points (here in cylindrical coordinates) $\boldsymbol{\xi} = (\rho, \phi, z)$ and $\boldsymbol{\eta} = (\tilde{\rho}, \tilde{\phi}, \tilde{z})$ determined by (t, ϕ) and $(\tilde{t}, \tilde{\phi})$. Let the double layer density be $\boldsymbol{\varphi}(\boldsymbol{\xi}) = f(t)\hat{\mathbf{e}}_\phi(\phi)$. The goal here is to calculate $\hat{\mathbf{e}}_\phi(\tilde{\phi}) \cdot \mathcal{K}\boldsymbol{\varphi}(\boldsymbol{\eta})$. First note that

$$dS = J(t)\, dt\, d\phi \ , \tag{18.16}$$

where

$$J(t) = \rho(t)\sqrt{\rho'(t)^2 + z'(t)^2} \tag{18.17}$$

and the surface normal is

$$\hat{\mathbf{n}}(\boldsymbol{\xi}) = a(t)\hat{\mathbf{e}}_\rho(\phi) + b(t)\hat{\mathbf{e}}_z \ . \tag{18.18}$$

When the generating curve is traversed in the positive (anticlockwise) direction, the coefficients are

$$a = z'/\sqrt{\rho'^2 + z'^2} \tag{18.19}$$

$$b = -\rho'/\sqrt{\rho'^2 + z'^2} \tag{18.20}$$

for the *outward* unit normal. Then

$$\hat{\mathbf{e}}_\phi(\tilde{\phi}) \cdot \mathcal{K}\boldsymbol{\varphi}(\boldsymbol{\eta}) = \int_{t_{min}}^{t_{max}} \left\{ \int_0^{2\pi} \hat{\mathbf{e}}_\phi(\tilde{\phi}) \cdot \frac{\mathbf{rrr} \cdot \hat{\mathbf{n}}(\boldsymbol{\xi})}{r^5} \cdot \hat{\mathbf{e}}_\phi(\phi)\, d\phi \right\} \frac{3}{2\pi} f(t)J(t)\, dt \ , \tag{18.21}$$

where $\mathbf{r} = \boldsymbol{\eta} - \boldsymbol{\xi}$. Consider the inner integrand only. Writing $\boldsymbol{\xi} = \rho\hat{\mathbf{e}}_\rho(\phi) + z\hat{\mathbf{e}}_z$ with a similar expression for $\boldsymbol{\eta}$ leads to

$$\rho \sin(\Delta\phi) \left[a(\tilde{\rho}\cos(\Delta\phi) - \rho) + b\Delta z\right] \tilde{\rho}\sin(\Delta\phi)/r^5 \ , \tag{18.22}$$

where $\Delta\phi = \tilde{\phi} - \phi$, with similar notation for other cylindrical coordinates. Now

$$r^2 = (\Delta z)^2 + \rho^2 + \tilde{\rho}^2 - 2\rho\tilde{\rho}\cos(\Delta\phi) \ , \tag{18.23}$$

so that $r = l\sqrt{k - \cos(\Delta\phi)}$ with

$$l = l(\tilde{t}, t) = \sqrt{2\rho\tilde{\rho}} \tag{18.24}$$

$$k = k(\tilde{t}, t) = 1 + \left((\Delta z)^2 + (\Delta\rho)^2\right)/(2\rho\tilde{\rho}) > 1 \quad \text{for } \tilde{t} \neq t. \tag{18.25}$$

One can replace $\Delta\phi$ everywhere by ϕ, formally by a suitable substitution followed by a shift of the integration interval (using the 2π periodicity of the integrand). Defining $q = k - \cos(\phi)$ the integrand is now

$$\frac{\rho\tilde{\rho}}{l^5}\frac{[a(\tilde{\rho}\cos(\phi) - \rho) + b\Delta z](1 - \cos^2(\phi))}{q^{5/2}} . \tag{18.26}$$

Substituting $\cos(\phi) = k - q$ in the numerator and expanding it into a polynomial in q gives integrand terms related to toroidal functions with argument k. On denoting the result from the azimuthal integration by $I(\tilde{t}, t)$,

$$\hat{\mathbf{e}}_\phi(\tilde{\phi}) \cdot \mathcal{K}(f(t)\hat{\mathbf{e}}_\phi)(\boldsymbol{\eta}) = \int_{t_{min}}^{t_{max}} I(\tilde{t}, t)\frac{3}{2\pi}f(t)J(t)\,dt , \tag{18.27}$$

for any $f(t)$. With more calculations of this type, and especially making use of the parity of the integrands, it can be shown that the Fourier components really are separated; specifically $\hat{\mathbf{e}}_\phi$-type dependence on $\tilde{\phi}$ is obtained from only the same type of double layer density, as above. This is not surprising, as we know that the mapping of velocity boundary conditions to tractions preserves the Fourier components. On the other hand, the double layer density is just a difference in surface velocities (by the jump condition) found by matching the tractions of the interior and exterior solutions. Another point of view is that the double layer density depends linearly on the velocity boundary conditions, and therefore the symmetry argument given in the beginning of this chapter applies.

18.6.2 Discretizing with Quadrature and Singularity Subtraction

In cylindrical coordinates

$$\begin{aligned} \frac{1}{2}(T_z\hat{\mathbf{e}}_z \times \nabla) \cdot \mathcal{G}(\boldsymbol{\eta}) &= \frac{T_z}{8\pi\eta^3}\hat{\mathbf{e}}_z \times \left(\tilde{\rho}\hat{\mathbf{e}}_\rho(\tilde{\phi}) + \tilde{z}\hat{\mathbf{e}}_z\right) \\ &= \frac{T_z\tilde{\rho}}{8\pi\eta^3}\hat{\mathbf{e}}_\phi(\tilde{\phi}) . \end{aligned} \tag{18.28}$$

(For convenience we are setting $\mu = 1$ here.) This is the only term from the Stokeslet and rotlet that has the kind of ϕ-dependency considered here. So only T_z comes up as an added variable in this case of a single integral equation, as expected. Considering only the relevant Fourier component of the disturbance surface velocity field, $u(t)\hat{\mathbf{e}}_\phi$, the equation is

$$u(\tilde{t}) = f(\tilde{t}) + \int_{t_{min}}^{t_{max}} I(\tilde{t}, t)\frac{3}{2\pi}f(t)J(t)\,dt + T_z\frac{\tilde{\rho}}{8\pi\eta^3} . \tag{18.29}$$

Axial rotation with the angular velocity $\hat{\mathbf{e}}_z$ causes the RBM surface velocity field given by

$$\hat{\mathbf{e}}_z \times \boldsymbol{\eta} = \tilde{\rho}\hat{\mathbf{e}}_\phi(\tilde{\phi}) , \tag{18.30}$$

which is an outside null function. When substituted in place of $f(t)$ above, $\rho(t)$ will cause the f-dependent terms to cancel:

$$0 = \rho(\tilde{t}) + \int_{t_{min}}^{t_{max}} I(\tilde{t}, t) \frac{3}{2\pi} \rho(t) J(t) \, dt \; . \tag{18.31}$$

Multiply this equation with $f(\tilde{t})/\rho(\tilde{t})$ and subtract to get

$$u(\tilde{t}) = \int_{t_{min}}^{t_{max}} I(\tilde{t}, t) \frac{3}{2\pi} \left[f(t) - \frac{\rho(t)}{\rho(\tilde{t})} f(\tilde{t}) \right] J(t) \, dt + T_z \frac{\tilde{\rho}}{8\pi\eta^3} \; . \tag{18.32}$$

Now a quadrature for the interval $[t_{min}, t_{max}]$ can be applied:

$$u(t_i) = \sum_{j \neq i} I(t_i, t_j) \frac{3}{2\pi} \left[f(t_j) - \frac{\rho(t_j)}{\rho(t_i)} f(t_i) \right] J(t_j) w_j + T_z \frac{\rho(t_i)}{8\pi\eta(t_i)^3} \; , \tag{18.33}$$

where w_i and t_i are the quadrature weights and points. To avoid difficulties at the points where the subtracted null function is zero, only such quadratures should be used that do not make use of the endpoints of the interval; for axisymmetric particles with the origin inside, these endpoints will be at the poles on the axis of symmetry. For particles like open toroids, this difficulty does not arise, but then one has to use a circular distribution of Stokeslets or rotlets inside to make use of the Fourier decomposition. To finish the bordering procedure and get a square linear system, one additional orthogonality condition is needed; here the only added variable was T_z. The only RBM density with correct type of ϕ-dependency was already used for singularity subtraction, and the orthogonality condition is

$$\langle f(t)\hat{e}_\phi, \rho\hat{e}_\phi \rangle = 0 \tag{18.34}$$

or

$$\oint_S f\rho \, dS = 0 \; . \tag{18.35}$$

In discretized form this reads

$$\sum_j w_j f(t_j) \rho(t_j) J(t_j) = 0 \; , \tag{18.36}$$

which is the final equation in the discretized system.

To get the surface tractions corresponding to axial rotation with unit angular velocity, denote these by $S(t)\hat{e}_\phi$. From the Lorentz reciprocal theorem

$$T_z = \langle S(t)\hat{e}_\phi, u(t)\hat{e}_\phi \rangle = 2\pi \int_{t_{min}}^{t_{max}} S(t) u(t) J(t) \, dt \; , \tag{18.37}$$

where the 2π came from the ϕ-integration. In discretized form,

$$T_z = 2\pi \sum_j w_j S(t_j) u(t_j) J(t_j) \; . \tag{18.38}$$

The transposing procedure, when applied to the linear system, gives a dependency of the form

$$T_z = \sum_j a_j u(t_j) \; , \tag{18.39}$$

and comparison of the coefficients of $u(t_j)$ implies

$$S(t_j) = \frac{a_j}{2\pi w_j J(t_j)} \; . \tag{18.40}$$

This concludes a fairly complete description of the method applied to a particular Fourier component.

18.7 Transverse Force and Torque

Here are some comments regarding the most complicated case on solving the resistance or mobility problem for axisymmetric surfaces, namely, three coupled one-dimensional IEs involving the transverse force and torque. A less detailed exposition than was given for the case of axial torque shall be given here. The Fourier decomposition of the double layer kernel in terms of toroidal harmonics is given at the end of this section. There are alternative ways of carrying out the Fourier decomposition and discretizing the resulting IEs. For example, it is not necessary to use the toroidal functions when considering only the wave-numbers zero and one, but instead all the integrals involved can be expressed in terms of the ordinary complete elliptic integrals. The advantage of toroidal functions is that all wave-numbers can be handled simultaneously on doing the analytic Fourier decomposition, and the asymptotic properties of the kernel functions can be generally examined.

Restricting our attention to cases mirror symmetric about the Cartesian plane $y = 0$, transverse translation must be parallel to

$$\hat{\mathbf{e}}_x = \cos(\tilde{\phi})\hat{\mathbf{e}}_\rho(\tilde{\phi}) - \sin(\tilde{\phi})\hat{\mathbf{e}}_\phi(\tilde{\phi}) \; . \tag{18.41}$$

Transverse rotation must be about the y-axis, so that this part of the velocity surface field is a multiple of

$$\begin{aligned} \hat{\mathbf{e}}_y \times \boldsymbol{\eta} &= \left(\sin(\tilde{\phi})\hat{\mathbf{e}}_\rho(\tilde{\phi}) + \cos(\tilde{\phi})\hat{\mathbf{e}}_\phi(\tilde{\phi})\right) \times (\tilde{\rho}\hat{\mathbf{e}}_\rho(\tilde{\phi}) + \tilde{z}\hat{\mathbf{e}}_z) \\ &= \tilde{z}\cos(\tilde{\phi})\hat{\mathbf{e}}_\rho(\tilde{\phi}) - \tilde{z}\sin(\tilde{\phi})\hat{\mathbf{e}}_\phi(\tilde{\phi}) - \tilde{\rho}\cos(\tilde{\phi})\hat{\mathbf{e}}_z \; . \end{aligned} \tag{18.42}$$

It is easy to verify that the Fourier components with this wave-number $m = 1$ and this symmetry are generated by the Stokeslet and rotlet only from $\boldsymbol{F} = F_x\hat{\mathbf{e}}_x$ and $\boldsymbol{T} = T_y\hat{\mathbf{e}}_y$. That the double layer kernel preserves this type of ϕ-dependency and produces it only from the same type (for axisymmetric surfaces) can be directly verified. Therefore these three vectorial Fourier components allow solving the transverse force and torque in arbitrary flow fields, and by use of the Lorentz reciprocal theorem (in the reversed way) allow finding the surface tractions corresponding to transverse translation or rotation. After separating the ϕ-dependency from the equations, similar to what was done with the axial torque above, there are three coupled IEs with one-dimensional arguments to be numerically solved. Obviously the RBM velocities for transverse translation and rotation are to be used for singularity subtraction, the integral identities coming from these being null functions as double layer densities. The subtraction was done so that the $\hat{\mathbf{e}}_\phi$- and $\hat{\mathbf{e}}_z$-components of the unknown density were cancelled at the singular point. To deal with the singularity in the kernel functions, it would be sufficient to just do the first of these cancellations, but it seems to be better to use all the available singularity subtractions to enhance accuracy, even though only one of the kernel functions is singular. The other kernel functions typically have singularities in their derivatives, which causes difficulties in their numerical evaluation; the actual collocation point cannot be used numerically, in any case, since the intermediate results go to infinity, and a nearby point is in large error when the first derivative is unbounded. Finally, the two orthogonality conditions (corresponding to the two added variables) also come from these two null functions, similarly to the case of axial torque.

18.7.1 Fourier Decomposition of the Double Layer Kernel

The following integrals are needed:

$$\mathcal{D}_m^n(k) = \int_0^{2\pi} \frac{\cos(m\phi)}{q^{n+1/2}}\,d\phi = c_n Q_{m-1/2}^n \tag{18.43}$$

$$\begin{aligned}\mathcal{E}_m^n(k) &= \int_0^{2\pi} \frac{\sin^2\phi\cos(m\phi)}{q^{n+1/2}}\,d\phi \\ &= -c_n\sqrt{k^2-1}\left\{kQ_{m-1/2}^{n-1} + \sqrt{k^2-1}(n+m^2-\frac{3}{2})Q_{m-1/2}^{n-2}\right\}\end{aligned} \tag{18.44}$$

$$\mathcal{F}_m^n(k) = \int_0^{2\pi} \frac{\sin\phi\sin(m\phi)}{q^{n+1/2}}\,d\phi = -c_n m\sqrt{k^2-1}Q_{m-1/2}^{n-1}\,. \tag{18.45}$$

The symbols given on the LHSs are introduced for the sake of brevity — they are not standard notation in mathematical literature. Here the argument of every toroidal function is k.

It can be shown that

$$\begin{aligned}\mathcal{K}(\hat{\mathbf{e}}_\rho(\phi)\cos(m\phi))(\eta) &= I_{\rho\rho}\hat{\mathbf{e}}_\rho(\tilde{\phi})\cos(m\tilde{\phi}) + I_{\phi\rho}\hat{\mathbf{e}}_\phi(\tilde{\phi})\sin(m\tilde{\phi}) \\ &\quad + I_{z\rho}\hat{\mathbf{e}}_z(\tilde{\phi})\cos(m\tilde{\phi})\,,\end{aligned} \tag{18.46}$$

with similar equations for $\hat{\mathbf{e}}_\phi(\phi)\sin(m\phi)$ and $\hat{\mathbf{e}}_z\cos(m\phi)$ used as the double layer densities defining the ϕ-independent coefficients $I_{\rho\phi}$, *etc.* These coefficients are listed below in terms of the integrals $\mathcal{D}$, $\mathcal{E}$, and $\mathcal{F}$:

$$\begin{aligned}\frac{2\pi}{3}I_{\rho\rho} = \frac{1}{l^5}\Bigg\{ \Bigg[& a\frac{(\tilde{\rho}^2-\rho^2-(\Delta z)^2)(\tilde{\rho}^2-\rho^2+(\Delta z)^2)^2}{8\rho^2\tilde{\rho}} \\ & +b\frac{\Delta z(\tilde{\rho}^2-\rho^2-(\Delta z)^2)(\tilde{\rho}^2-\rho^2+(\Delta z)^2)}{4\rho\tilde{\rho}}\Bigg]\mathcal{D}_m^2\end{aligned}$$

$$+\left[-a\frac{(\tilde{\rho}^2-\rho^2-3(\Delta z)^2)(\tilde{\rho}^2-\rho^2+(\Delta z)^2)}{4\rho}+b(\Delta z)^3\right]\mathcal{D}_m^1$$
$$+\left[-\frac{a}{2}\tilde{\rho}(\tilde{\rho}^2-\rho^2+3(\Delta z)^2)-b\Delta z\rho\tilde{\rho}\right]\mathcal{D}_m^0+\left[a\rho\tilde{\rho}^2\right]\mathcal{D}_m^{-1}\Big\} \tag{18.47}$$

$$\begin{aligned}\frac{2\pi}{3}I_{\rho\phi} &= \frac{-\tilde{\rho}}{l^5}\Bigg\{\Bigg[a\frac{(\tilde{\rho}^2-\rho^2-(\Delta z)^2)(\tilde{\rho}^2-\rho^2+(\Delta z)^2)}{4\rho\tilde{\rho}} \\ &\quad +b\frac{\Delta z(\tilde{\rho}^2-\rho^2-(\Delta z)^2)}{2\tilde{\rho}}\Bigg]\mathcal{F}_m^2 \\ &\quad +\left[a(\Delta z)^2+b\Delta z\rho\right]\mathcal{F}_m^1+\left[-a\rho\tilde{\rho}\right]\mathcal{F}_m^0\Bigg\}\end{aligned} \tag{18.48}$$

$$\begin{aligned}\frac{2\pi}{3}I_{\rho z} &= \frac{\Delta z}{l^5}\Bigg\{\Bigg[a\frac{(\tilde{\rho}^2-\rho^2-(\Delta z)^2)(\tilde{\rho}^2-\rho^2+(\Delta z)^2)}{4\rho\tilde{\rho}} \\ &\quad +b\frac{\Delta z(\tilde{\rho}^2-\rho^2-(\Delta z)^2)}{2\tilde{\rho}}\Bigg]\mathcal{D}_m^2 \\ &\quad +\left[a(\Delta z)^2+b\Delta z\rho\right]\mathcal{D}_m^1+\left[-a\rho\tilde{\rho}\right]\mathcal{D}_m^0\Bigg\}\end{aligned} \tag{18.49}$$

$$\begin{aligned}\frac{2\pi}{3}I_{\phi\rho} &= \frac{\rho}{l^5}\Bigg\{\left[a\frac{(\tilde{\rho}^2-\rho^2+(\Delta z)^2)^2}{4\rho^2}+b\frac{\Delta z(\tilde{\rho}^2-\rho^2+(\Delta z)^2)}{2\rho}\right]\mathcal{F}_m^2 \\ &\quad +\left[-a\frac{\tilde{\rho}(\tilde{\rho}^2-\rho^2+(\Delta z)^2)}{\rho}-b\Delta z\tilde{\rho}\right]\mathcal{F}_m^1+\left[a\tilde{\rho}^2\right]\mathcal{F}_m^0\Bigg\}\end{aligned} \tag{18.50}$$

$$\begin{aligned}\frac{2\pi}{3}I_{z\rho} &= \frac{\Delta z}{l^5}\Bigg\{\left[a\frac{(\tilde{\rho}^2-\rho^2+(\Delta z)^2)^2}{4\rho^2}+b\frac{\Delta z(\tilde{\rho}^2-\rho^2+(\Delta z)^2)}{2\rho}\right]\mathcal{D}_m^2 \\ &\quad +\left[-a\frac{\tilde{\rho}(\tilde{\rho}^2-\rho^2+(\Delta z)^2)}{\rho}-b\Delta z\tilde{\rho}\right]\mathcal{D}_m^1+\left[a\tilde{\rho}^2\right]\mathcal{D}_m^0\Bigg\}\end{aligned} \tag{18.51}$$

$$\frac{2\pi}{3}I_{\phi\phi} = \frac{\rho\tilde{\rho}}{l^5}\left\{\left[a\frac{\tilde{\rho}^2-\rho^2+(\Delta z)^2}{2\rho}+b\Delta z\right]\mathcal{E}_m^2+\left[-a\tilde{\rho}\right]\mathcal{E}_m^1\right\} \tag{18.52}$$

$$\frac{2\pi}{3}I_{\phi z} = \frac{\rho\Delta z}{l^5}\left\{\left[a\frac{\tilde{\rho}^2-\rho^2+(\Delta z)^2}{2\rho}+b\Delta z\right]\mathcal{F}_m^2+\left[-a\tilde{\rho}\right]\mathcal{F}_m^1\right\} \tag{18.53}$$

$$\frac{2\pi}{3}I_{zz} = \frac{(\Delta z)^2}{l^5}\left\{\left[a\frac{\tilde{\rho}^2-\rho^2+(\Delta z)^2}{2\rho}+b\Delta z\right]\mathcal{D}_m^2+\left[-a\tilde{\rho}\right]\mathcal{D}_m^1\right\} \tag{18.54}$$

$$I_{z\phi} = -\tilde{\rho}I_{\phi z}/\rho\,. \tag{18.55}$$

The bracketed coefficients in $I_{\rho\phi}$ and $I_{\rho z}$ are the same, with a similar note for the groupings $\{I_{\phi\rho}, I_{z\rho}\}$ and $\{I_{\phi\phi}, I_{\phi z}, I_{z\phi}, I_{zz}\}$.

18.8 Other Details of Implementation

The quadrature chosen was rectangular rule with equal-length subintervals, since the generation of weights is easy and there is great flexibility in choosing the number of quadrature points. On dealing only with axisymmetric particle shapes, little input is required compared with full surface methods. The user needs to give two functions describing the generating curve for the surface in one parameter, $\rho(t)$ and $z(t)$ for $t_{min} \leq t \leq t_{max}$, and two derivatives of these. Similarly, the components of the flow field are given in terms of z and ρ. At first the choice of rectangular quadrature could seem to be restrictive, giving us somehow equidistant points only, but the density of points on the generating curve can easily be modified by using a variable transformation $t = t(\bar{t})$. Viewed differently, the rectangular rule can be used to generate any number of quadratures by changing the integration variable. Still this is not quite as flexible as element methods, but the amount of input remains small, and, after having input the functions, the density of the mesh can be easily changed by just giving one number. *A selection of Fortran programs that illustrate the above ideas are available in the subdirectory, Chapter19, on* "Flossie."

18.9 Limitations of the Fourier Analysis Approach

When the configuration of all of the boundaries involved is nonaxisymmetric, as for a sphere eccentrically inside a cylinder, the different wave-numbers are not decoupled in the equations, since the surface normals do not have a single axis of symmetry. If the Fourier series for the fields are truncated after, say, five wave-numbers, this will lead to a five times bigger linear system than the decoupled wave-numbers individually would give. Depending on the eccentricity of the problem, very high wave-numbers may be significant, so that the Fourier analysis approach has lost its merits. Then it seems more reasonable to go over to a full surface method, where one can implement small surface elements at difficult areas of the surfaces, like those where two surfaces are relatively close together. *Due to the IEs being of the second kind, this will cause no ill-conditioning of the linear system.*

It still remains to be numerically tested if a direct surface quadrature implementation can handle mobility problems with surfaces close together; experience with such problems by related methods suggests that this should be feasible, even though a similar resistance problem would lead to difficulties when the quadrature is not particularly dense around the areas of peaked surface tractions.

18.10 Results from the Axisymmetric Codes

18.10.1 Prolate Spheroids; Comparison of Surface Tractions with Known Analytical Results

The surface tractions on ellipsoids in rigid-body motion in an infinite quiescent fluid are analytically known. In particular, for spheroids the analytical expressions can be formulated in terms of elementary functions, using the results in Brenner [7] and Appendix B of Kim [40]. The same goes for the total force and torque on a spheroid in translation or rotation. A prolate spheroid with axis ratio $b \leq 1$ is generated by

$$\left.\begin{array}{rcl} \rho(t) & = & b\sin(t) \\ z(t) & = & -\cos(t) \end{array}\right\}, \quad \text{for } 0 \leq t \leq \pi. \tag{18.56}$$

Using this parametrization with the chosen quadrature gives points equidistant in the polar angle t of spheroidal polar coordinates. This seems to work a little better than using points equidistant in z. All the programs used (except that for a particle in a container) have been tested with spheroids to verify correctness of the method used and the programming. Here the dependence of the convergence on the number of collocation points with the aspect ratio $1/b$ as parameter, for the case of transverse translation, is presented in Figure 18.1. A comparison of the analytical surface tractions with the numerically found ones in a particular case is shown in Figure 18.2. Of the three cases for the resistance problem with Fourier decomposition, the largest equation system is related to transverse force and torque. With 80 points the number of equations is about 240, and a MicroVAX II requires about 83 seconds of CPU time to solve the problem, including computation of the related RBM tractions. For the two other cases with smaller linear systems, the CPU times are correspondingly shorter, and typically the accuracy achieved is somewhat better with the same mesh. Double precision is used in the computations, and specifically the computation of toroidal functions is always carried out to this precision. In this sense the programs are not optimized; instead of analyzing round-off errors and allowing inaccuracies, these have been avoided as much as possible.

18.10.2 Mesh Effects: Grooved Particles

As a fairly difficult test particle, a sphere with a groove along its equator was chosen. This was generated by using $b = 1$ for the spheroid above and modifying the radial coordinate ρ by subtracting from it a Gaussian type bell shape:

$$\rho(t) = \sin(t) - \alpha \exp[-\frac{(t-\pi/2)^2}{2\beta^2}]. \tag{18.57}$$

In this way there will be no sharp edges on the surface; arbitrary order derivatives of the generating curve are continuous.

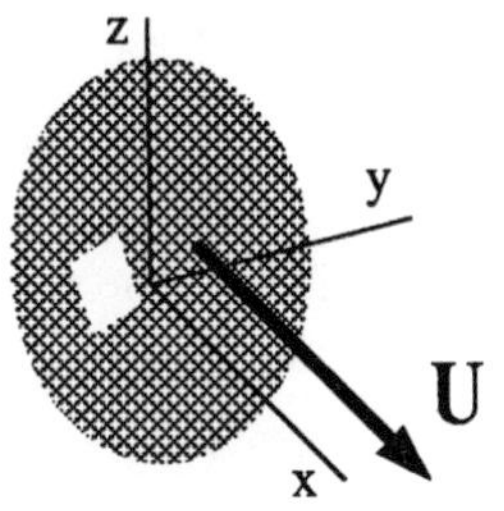

Figure 18.1: Transverse translation of a prolate spheroid.

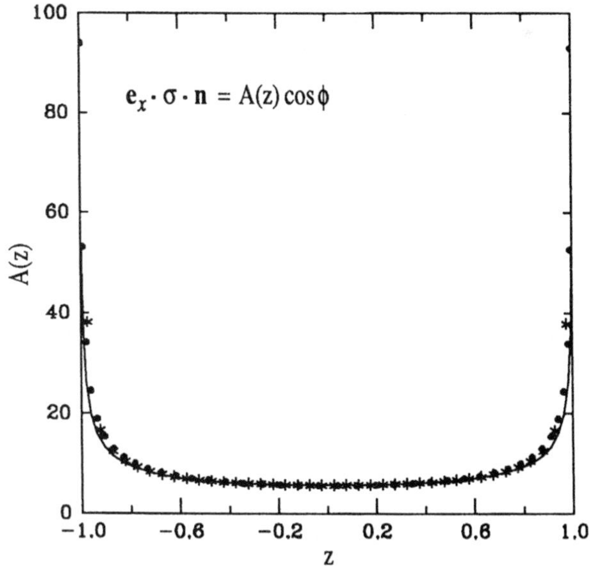

Figure 18.2: The surface tractions on a prolate spheroid (aspect ratio 10).

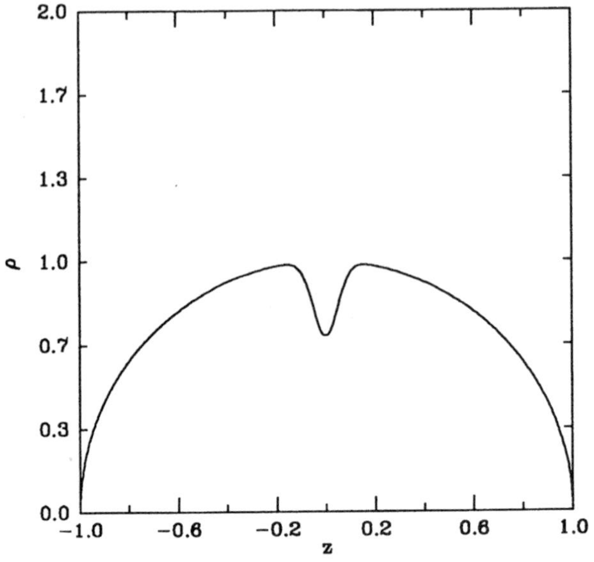

Figure 18.3: The profile of the grooved particle.

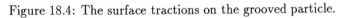

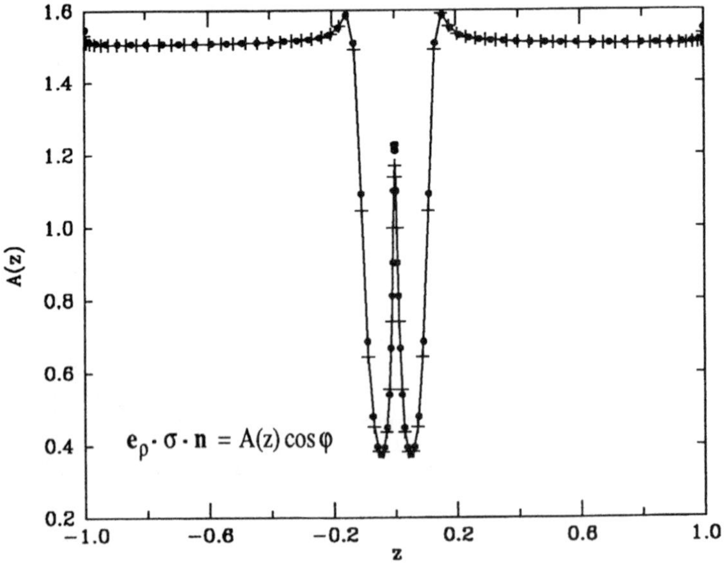

Figure 18.4: The surface tractions on the grooved particle.

In the numerical computations presented here, $\alpha = 0.3$ and $\beta = 0.05$. The resulting profile of the grooved sphere is shown in Figure 18.3. The convergence with this parametrization and the chosen quadrature was slow, so that it was decided to modify the mesh to be more dense around the equator by the following transformations. A "cubic density" is achieved by

$$t = \frac{(\bar{t} - \pi/2)^3}{(\pi/2)^2} + \pi/2, \tag{18.58}$$

and a less pronounced "quadratic density" by

$$t = \frac{(\bar{t} - \pi/2)|\bar{t} - \pi/2|}{\pi/2} + \pi/2. \tag{18.59}$$

These modifications to the parametrization of the generating curve were done in the corresponding subroutine, so that now the quadrature points are equidistant in $\bar{t}$. Use of these gives almost identical results, although the extremely high density at the equator given by the former causes slight difficulties with the surface tractions; see Figure 18.4. This can be understood since pointwise traction values with extremely small weight in the inner product used with the Lorentz reciprocal theorem do not have much effect on the total force or torque; small deviations in them are numerically allowed. It should be noted that use of these distributions of collocation points would lead to severe ill-conditioning with almost any other kind of formulation than IEs of the second kind. Here we are saved by the point evaluation of the unknown function ("the delta function added to the kernel") that distinguishes equations of the second kind from those of the first kind.

18.10.3 The Effect of Sharp Edges: Finite Circular Cylinder

It is well known that sharp corners or edges protruding into the fluid can produce singular surface tractions. One way to analytically assess the behavior of these singularities is by utilizing the Papkovich–Neuber representation in terms of harmonic functions for which the behavior near edges is known (Jaswon and Symm [33]). Chan *et al.* [9] directly transformed the problem of axial rotation to a potential problem, following the method of Jeffery [34], so they did not refer to the representation mentioned above. The numerical computations for the straight circular cylinder show that the edges can be neglected, as long as we are not explicitly interested in the type of singularity at them. The total torque on the cylinder converges fine, and getting about three-digit accuracy requires only 80 points when the cylinder has axis ratio 1 (diameter equals height). The fore-aft symmetry of the cylinder was not utilized, and the parametrization was

$$(\rho, z) = \begin{cases} (t, -1) & \text{for } t \in [0, 1] \\ (1, t-2) & \text{for } t \in [1, 3] \\ (4-t, 1) & \text{for } t \in [3, 4]. \end{cases} \tag{18.60}$$

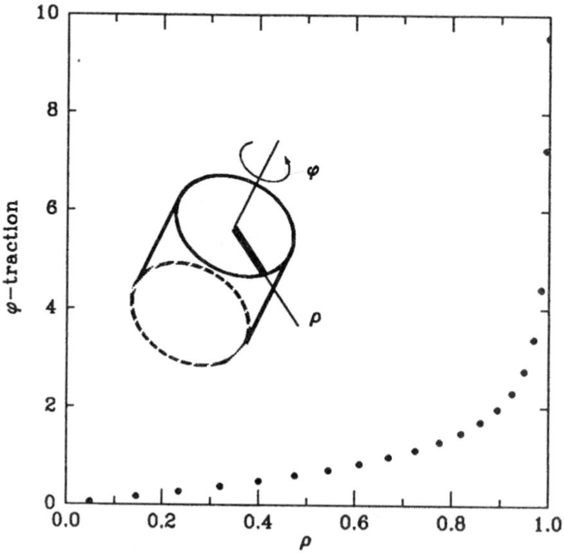

Figure 18.5: The tractions on the "lid" of a rotating cylinder.

With 80 points $T_z = 39.734$, and Chan *et al.* give 39.869. To see if better accuracy is achieved by modifying the mesh, the following "double quadratic modification" was used to get high density at both edges:

$$t = \begin{cases} (\bar{t}-1)|\bar{t}-1|+1 & \text{for } \bar{t} \in [0,2] \\ (\bar{t}-3)|\bar{t}-3|+3 & \text{for } \bar{t} \in [2,4]. \end{cases} \tag{18.61}$$

After this the torque 39.854 is found with 80 points, and the results for surface tractions are plotted along the generating curve on the top of the surface in Figure 18.5. To assess the accuracy of these the same curve was scaled so that the singularity at the edge is cancelled, and plotted also including point values of the tractions given by Chan *et al.* for comparison (see Figure 18.6). The agreement is strikingly good, except for the last two quadrature points closest to the edge, and even these are not in error by much. It is understandable that the quadrature collocation method can be somewhat confused by the presence of an edge, since it nowhere is given information about the exact location of the edge; only information at the collocation points is being used.

The theory of the IEs used was based on the assumption of Lyapunov-smooth surfaces for the reason that then $\boldsymbol{r}\cdot\hat{\boldsymbol{n}}/r^{1+\alpha} \to 0$ ($\alpha > 0$) as $r \to 0$, and this makes the double layer kernel only weakly singular. With edges or corners present, this does not hold any more. The numerical experiment here indicates that no severe difficulties need arise from the presence of edges, although the tractions will (in general) be singular. In practice concentrating more quadrature points

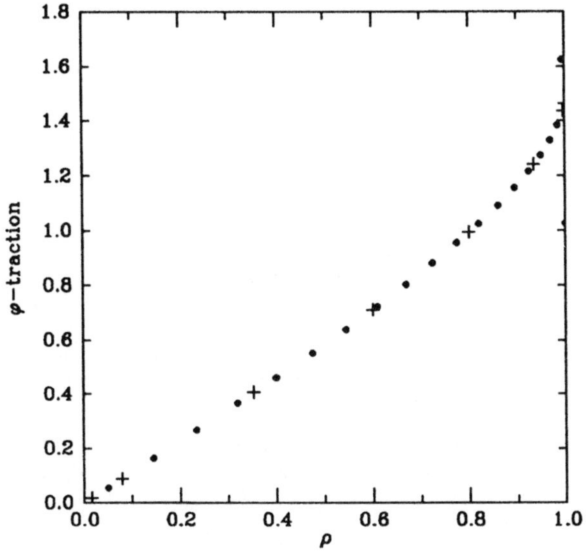

Figure 18.6: Tractions with singularity removed (factored).

where the tractions are large will help in dealing with any numerical difficulties. This is because the quadrature weights will become small and distribute large tractions to a large number of components in the discrete approximation, so it will no longer have large elements. This procedure also allows us to numerically observe the severity of the traction singularity. Similar considerations apply to multisurface problems that are close to lubrication problems, where the local tractions are almost singular. This is essentially the numerical analogue of local stretching of the surface coordinate system about the singular point(s), which could be used analytically in asymptotic analysis. With other general numerically applied formulations, not carrying out the asymptotic analysis separately, such stretching would lead to ill-conditioning caused by too close collocation points.

Finally, the inclusion monotonicity of resistance for Stokes flows noted by Hill and Power [30] — although it has been verified only for smooth surface shapes — suggests strongly that corners and edges may be rounded inwards and outwards to get corresponding bounds for the resistance. Therefore problems with corners may be avoided also by this means if it becomes necessary. Another implication of the inclusion monotonicity is that resistance problems are well-posed relative to surface shape data, meaning that small deviations in particle shape will generally produce only small deviations in resistance when one of the shapes can be bound inside and outside by the other suitably shrunk or expanded. The local surface tractions, however, are not equally well behaved, as

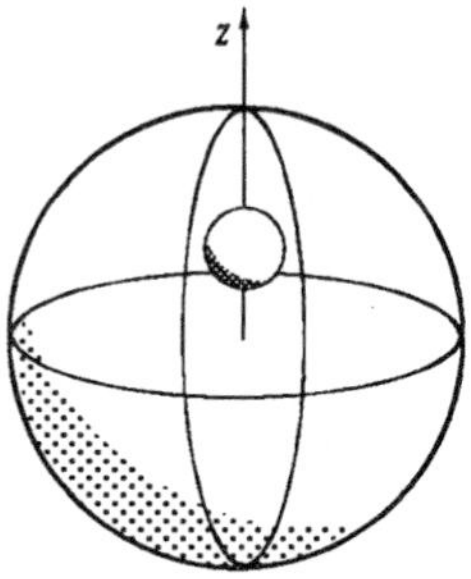

Figure 18.7: A sphere in a spherical container.

can be seen by approximating a smooth surface by a jagged one. For this reason the method introduced is superior to others, since the density of collocation points can be made high in the vicinity of areas of interest without causing catastrophic ill-conditioning. This type of behavior can only be expected of IEs of the second kind.

18.10.4 A Container Problem

A suitable model problem for testing the canonical equations with Fourier decomposition is that of a spherical particle concentrically in a spherical container. In general, the container null function is not explicitly known, and therefore it cannot be used for singularity subtraction; the outside null functions for the container can, however, be used again to get integral identities. The main difference between a container and a particle surface, in this respect, is just the direction in which the surface normal used in the double layer kernel is pointed.

Analytical results, obtained with stream functions are given in Happel and Brenner [29, p.133]. Choosing the container radius to be 1.25 times that of the particle corresponds to $\lambda = 0.8$ in the notation of the reference above (see Figure 18.7), and for this case the reference gives 138.224 as the factor by which the resistance for translation is increased relative to that in unbounded fluid. As the factor indicates, there is a very strong interaction at such a small separation. In this sense the chosen test case is fairly difficult.

The computations were carried out with equal numbers of points on the particle and the container, these being 40, 60, and 80. Using points equidistant in the axial z-coordinate gives the factors 306.4, 160.7, and 146.4. Discretization equidistant in the polar angle seems more reasonable, emphasizing more the polar areas. This gives the much better results 149.7, 143.0, and 140.8, with the same numbers of quadrature points.

This example again illustrates the importance of the choice of the parametrization. Even a bad choice will eventually give results with the desired accuracy as the mesh is refined, but a good choice can save a lot in terms of the problem

Points	equistant-z	equidistant-θ
40	306.4	149.7
60	160.7	143.0
80	146.4	140.8

Table 18.1: Drag on a sphere settling in a spherical container (exact: 138.224).

size and required CPU time. This example also verifies that the canonical equations are correctly formulated for problems with a container.

As discussed earlier, due to the symmetric shape of a sphere, the double layer operator on it is self-adjoint. From this it follows that the container null function on a sphere is actually a multiple of the surface normal, since this is the null function of the adjoint single layer problem. The same result can be seen from the fact that the container null function must remain unchanged by any rotation, since these would by symmetry also give null functions and the null space is one-dimensional. Therefore in this case using the container surface normal to remove the container null function happens to be an even better choice than it usually would be.

18.11 Possibilities for Improvement and Generalization

In the calculations with prolate spheroids, it is noted that the numerical solution becomes more difficult as the shape is more elongated. It seems that this is not so much due to the high curvature of the surface near the poles, since the method easily handles high curvature in other test cases. It is suggested that the reason for added difficulties here is just the lack of distributing the singularities inside the particle, instead of concentrating them on the centroid only. Another possible reason is the way in which the singularity subtraction is carried out, with functions some of which use the radial coordinate ρ. Distributing the singularities inside the particles corresponds to choosing a "more robust" set of basis functions to span the deficient part of the range of the double layer operator. The easiest way to implement a distribution would be to have an *a priori* fixed distribution of Stokeslets (rotlets) that gets multiplied by the total force (torque). A more sophisticated approach would automatically determine the distribution from a solution with a coarse mesh. For oblate particles, preserving axisymmetry would require the use of ring distributions of singularities. These modifications have not been tested so far.

One way to find a suitable distribution of singularities could be the method of Dabros [12]; for a more advanced adaptive variation of the method, see Han and Olson [28]. The approximate solution from this method can then be improved upon with the bordered equations. This two-stage "predictor-corrector" method can be necessary if one wants to handle difficult shapes or to inspect traction singularities or near-singularities. Current investigations in this direction have yielded promising results.

Exercises

Exercise 18.1 Computer Algebra.
The program package MACSYMA was mentioned in connection with the analytical transformations done before programming. Find out about the availability of computer algebra programs and what they are capable of.

Exercise 18.2 Nyström Method.
The Nyström method of interpolation can be used when solving second-kind IE:s with quadrature collocation. In fact, using quadrature collocation is often called *solution by Nyström's method.* The basic idea is that after discretizing with quadrature, one continuous variable remains in the (approximate) equation. By a small rearrangement this approximate equation provides a "natural" interpolation formula by which values of $x(s)$ can be found once the values $x(t_i)$ are known. With singular kernels this kind of interpolation does not sound too good any more. Check if you can figure out how singularity subtraction could help to some extent. For a good treatment of Nyström's method with continuous kernels, see the book by Atkinson [2].

Exercise 18.3 Traction Singularities.
Try to give a verbal explanation for the tractions being singular at the edges of the rotating cylinder. Similarly, try to explain the traction profile computed for the grooved sphere. Find out the exact analytical corresponding traction profile for a sphere without a groove, and draw it in the same picture for comparison.

Chapter 19

Three-Dimensional Numerical Results

There are three objectives behind the numerical examples (mostly drawn from the work of Karrila *et al.* [39] and Fuentes [20]) presented in this chapter. First, we test the CDL-BIEM concept itself, by comparing these results with those obtained from other well-established analytical, numerical, and experimental results. These tests include a selection of mobility and resistance relations for one-, two-, and three-sphere geometries and sedimentation experiments for regular polyhedra. Second, we examine the iterative algorithm to demonstrate that the solution obtained is identical numerically to that obtained by Gaussian elimination and to show that the rate of convergence is consistent with that predicted by a spectral analysis of the continuous equations. For all cases in which the spectral radius is known, we will see that the method converges according to these estimates. Finally, we test the multisphere mobility algorithm on a parallel computer (Sequent Symmetry). The results indicate that computational loads can be balanced quite evenly in a multiple-processor environment, and speedups close to the ideal limit are achieved.

The discussions in Chapters 15 through 17 emphasized the continuous equations, and the conclusions reached in those sections are independent of the choice of the discretization scheme. Actual numerical results are of course dependent on the discretization method. The results presented here are all for triangular boundary elements with characteristic functions as the basis. While relatively large numbers of boundary elements are required, our main objective here is to demonstrate the behavior of some discretized version of CDL-BIEM for a representative sample of three-dimensional structures. The number of elements can be reduced substantially by the use of more elaborate (higher order) basis elements [6, 11], but a price must be paid in the form of more elaborate pre- and post-processing steps, especially in connection with the construction of parallel algorithms and the Riesz–Lorentz construction of the surface tractions. Other approaches to discretizations are possible, for example, global collocation over the surface, as shown in the previous chapter and the original literature (Karrila and Kim [38]). Global collocation is competitive for simple shapes, such as spheres, and in axisymmetric geometries. However, for more complex structures, the simple discretization may be preferable, not only on the basis of generality, but also because of the low computational cost of vectorized operations performed as parallel tasks.

19.1 Discretization with Constant Elements

Our example centers on discretization of the following boundary integral equation for the mobility of a single particle.

$$\varphi(\boldsymbol{x}) + \mathcal{K}\varphi(\boldsymbol{x}) + \sum_{\nu=1}^{6} \varphi^{(\nu)} \langle \varphi^{(\nu)}, \varphi \rangle = \left(\boldsymbol{F} - \frac{1}{2}\boldsymbol{T} \times \nabla\right) \cdot \frac{\mathcal{G}(\boldsymbol{x} - \boldsymbol{x}_c)}{8\pi\mu} , \qquad (19.1)$$

where $\boldsymbol{F}$ and $\boldsymbol{T}$ are the hydrodynamic forces and torques exerted by the fluid on the particle, and $\varphi^{(\nu)}$, $\nu = 1, 2, \ldots, 6$ are the null solutions given explicitly by

$$\begin{aligned}
\varphi^{(1)} &= \frac{\boldsymbol{e}_1}{\sqrt{A}} , && \text{RBM translation in the } x\text{-direction} \\
\varphi^{(2)} &= \frac{\boldsymbol{e}_2}{\sqrt{A}} , && \text{RBM translation in the } y\text{-direction} \\
\varphi^{(3)} &= \frac{\boldsymbol{e}_3}{\sqrt{A}} , && \text{RBM translation in the } z\text{-direction} \\
\varphi^{(4)} &= \frac{x_2\boldsymbol{e}_3 - x_3\boldsymbol{e}_2}{\sqrt{I_1}} , && \text{RBM rotation about the } x\text{-axis} \\
\varphi^{(5)} &= \frac{x_3\boldsymbol{e}_1 - x_1\boldsymbol{e}_3}{\sqrt{I_2}} , && \text{RBM rotation about the } y\text{-axis} \\
\varphi^{(6)} &= \frac{x_1\boldsymbol{e}_2 - x_2\boldsymbol{e}_1}{\sqrt{I_3}} . && \text{RBM rotation about the } z\text{-axis}
\end{aligned}$$

Here, the origin and directions of the coordinate axes are taken as the "particle shell" center of mass and the principal directions of the "particle shell" moment of inertia tensor. With the above scaling, where A is the particle surface area and I_1, I_2, and I_3 are the moments of inertia about the principal axes, *i.e.*,

$$I_1 = \int_S (y^2 + z^2)\, dS , \quad I_2 = \int_S (z^2 + x^2)\, dS , \quad I_3 = \int_S (x^2 + y^2)\, dS ,$$

the densities are orthonormal with respect to the natural inner product, *i.e.*,

$$\left\langle \varphi^{(\mu)}, \varphi^{(\nu)} \right\rangle = \delta_{\mu\nu} .$$

For the unit sphere, $I_1 = I_2 = I_3 = 8\pi/3$.

Now suppose that the surface is divided into N boundary elements. The exact shape of these elements is not important in the subsequent discussion, although in the codes discussed later on, the elements are triangular. We identify these N elements with the integer label j, with $1 \leq j \leq N$. We approximate φ as piece-wise constant, so that on each element (say element j) the x, y, and z components of the surface density φ constitute three unknowns, for a total of

$3N$ unknowns. These we denote by $\varphi_\alpha(j)$, ($\alpha = 1, 2, 3$; and $1 \leq j \leq N$). We may adopt the convention of ordering these unknowns into a vector array as follows:

$$\begin{pmatrix} \varphi_1(1) \\ \vdots \\ \varphi_1(N) \\ \hline \varphi_2(1) \\ \vdots \\ \varphi_2(N) \\ \hline \varphi_3(1) \\ \vdots \\ \varphi_3(N) \end{pmatrix},$$

that is, we load all the x-components, then all the y-components, then all the z-components.[1]

Fortunately we also have $3N$ equations, since on each boundary element there are three equations obtained from the boundary conditions for v_x, v_y, and v_z. Following the same ordering convention as was used with the unknowns, we order the $3N$ equations as follows:

the N x-equations for elements 1 to N

the N y-equations for elements 1 to N

the N z-equations for elements 1 to N .

An example program, **CDLBIEM.FOR**, is available over the Internet network, and the rest of this section will be easier to follow with it in hand. The procedure for obtaining this from the *anonymous* user account on "Flossie" is as described in the beginning of this book.

```
% ftp flossie.che.wisc.edu
Name:    anonymous
Anonymous user OK. Enter real ident...  yourname
*  cd chapter19
*  get cdlbiem.for
*  bye
```

We consider how each term in Equation 19.1 loads into the system matrix. The first term on the LHS simply leads to the 3×3 block matrix,

$$\left(\begin{array}{c|c|c} I_N & \mathbf{0} & \mathbf{0} \\ \hline \mathbf{0} & I_N & \mathbf{0} \\ \hline \mathbf{0} & \mathbf{0} & I_N \end{array}\right),$$

[1] In some instances, *e.g.*, for parallel computer architectures, it may be more efficient to order the unknowns in triads associated with each boundary element.

with $N \times N$ identity matrix along the block diagonal.

This double layer term requires a little more work. We write this approximately as

$$\mathcal{K}\varphi = \int_{S(y)} \boldsymbol{K}(\boldsymbol{x},\boldsymbol{y}) \cdot \varphi(\boldsymbol{y})\, dS(\boldsymbol{y}) \sim \sum_{j=1}^{N} [\boldsymbol{K}(\boldsymbol{x},\boldsymbol{y}(j)) \cdot \varphi(\boldsymbol{y}(j))]\, \Delta A(j) ,$$

where $\Delta A(j)$ is the area of the j-th boundary element. The kernel $\boldsymbol{K}(\boldsymbol{x},\boldsymbol{y})$ is given by

$$\boldsymbol{K}(\boldsymbol{x},\boldsymbol{y}) = \frac{3}{2\pi} \frac{\boldsymbol{n}(\boldsymbol{y}) \cdot (\boldsymbol{x}-\boldsymbol{y})}{|\boldsymbol{x}-\boldsymbol{y}|^3} \frac{(\boldsymbol{x}-\boldsymbol{y})}{|\boldsymbol{x}-\boldsymbol{y}|} \frac{(\boldsymbol{x}-\boldsymbol{y})}{|\boldsymbol{x}-\boldsymbol{y}|} .$$

Note that in the program listing **CDLBIEM**, $\boldsymbol{y}$ (as **y()**) is also used as the "dummy variable" of integration. In what follows, we should keep in mind that we are setting the boundary condition on element i, while focusing our attention on the contribution from boundary element j. This, of course, leads to the i,j element for each of the nine submatrices in the 3×3 block matrix.

The integrand, $\boldsymbol{K}(\boldsymbol{x},\boldsymbol{y}) \cdot \varphi$, is

$$\frac{3}{2\pi} \frac{\boldsymbol{n}(\boldsymbol{y}) \cdot (\boldsymbol{x}-\boldsymbol{y})}{|\boldsymbol{x}-\boldsymbol{y}|^3} \left(\frac{(x_\beta - y_\beta)}{|\boldsymbol{x}-\boldsymbol{y}|} \cdot \varphi_\beta \right) \frac{(\boldsymbol{x}-\boldsymbol{y})}{|\boldsymbol{x}-\boldsymbol{y}|} ,$$

and if the block matrix is written as

$$\left(\begin{array}{c|c|c} A^{(11)} & A^{(12)} & A^{(13)} \\ \hline A^{(21)} & A^{(22)} & A^{(23)} \\ \hline A^{(31)} & A^{(32)} & A^{(33)} \end{array} \right) ,$$

we find that

$$A_{ij}^{(\alpha\beta)} = \frac{3}{2\pi} \frac{\boldsymbol{n}(\boldsymbol{y}(j)) \cdot (\boldsymbol{x}(i) - \boldsymbol{y}(j))}{|\boldsymbol{x}(i) - \boldsymbol{y}(j)|^3} \frac{(x_\alpha(i) - y_\alpha(j))}{|\boldsymbol{x}(i) - \boldsymbol{y}(j)|} \frac{(x_\beta(i) - y_\beta(j))}{|\boldsymbol{x}(i) - \boldsymbol{y}(j)|} .$$

Note that if $i = j$, that is, if we look at the contribution from the boundary element at which we are evaluating the boundary condition, the kernel vanishes identically, since

$$\lim_{y \to x} \boldsymbol{n}(\boldsymbol{y}) \cdot (\boldsymbol{x}-\boldsymbol{y}) = \boldsymbol{0}$$

on a Lyapunov surface. This simplification does not occur in the single layer distribution, and a special set of instructions must be included in the code to calculate the contribution from this element [72]. For CDL-BIEM with nonplanar elements and/or more sophisticated basis functions (and quadratures), this contribution to the integral also requires special logic in the codes, resulting in a corresponding degradation of the parallel algorithm (see Exercise 19.1 for an estimate of the contribution from a "polar cap" on a sphere).

We now consider the Wielandt deflation terms associated with particle translations. With $\nu = 1, 2, 3$, we write

$$\begin{aligned} \varphi^{(\nu)} \langle \varphi^{(\nu)}, \varphi \rangle &= \frac{e_\nu}{A} \int_S e_\nu \cdot \varphi(y)\, dS(y) = \frac{e_\nu}{A} \int_S \varphi_\nu(y)\, dS(y) \\ &\sim \frac{e_\nu}{A} \sum_{j=1}^{N} \varphi_\nu(j) \Delta A(j) \,. \end{aligned}$$

Thus for $\nu = 1, 2, 3$ the Wielandt deflation terms contribute the following 3×3 block diagonal matrix:

$$\frac{1}{A} \left(\begin{array}{c|c|c} diag(\Delta A(j)) & \mathbf{0} & \mathbf{0} \\ \hline \mathbf{0} & diag(\Delta A(j)) & \mathbf{0} \\ \hline \mathbf{0} & \mathbf{0} & diag(\Delta A(j)) \end{array} \right) .$$

Now consider the deflations associated with the rotational motions. For $\nu = 4$, the associated null solution is a rotation about the x-axis, and we write

$$\varphi^{(4)} \langle \varphi^{(4)}, \varphi \rangle = \frac{(x_2 e_3 - x_3 e_2)}{I_1} \int_S (y_2 \varphi_3 - y_3 \varphi_2)\, dS(y) \,.$$

Thus for $\nu = 4$, the Wielandt deflation terms contribute the following 3×3 block matrix:

$$\left(\begin{array}{c|c|c} \mathbf{0} & \mathbf{0} & \mathbf{0} \\ \hline \mathbf{0} & \frac{x_3(i) y_3(j) \Delta A(j)}{I_1} & -\frac{x_3(i) y_2(j) \Delta A(j)}{I_1} \\ \hline \mathbf{0} & -\frac{x_2(i) y_3(j) \Delta A(j)}{I_1} & \frac{x_2(i) y_2(j) \Delta A(j)}{I_1} \end{array} \right) .$$

The final result is obtained by including the contributions from the cases with $\nu = 5$ and $\nu = 6$, which can be obtained by cycling the indices $1, 2, 3$, wherever they appear in the above matrix (including the matrix element labels).

The RHS vector is obtained simply by evaluating the Stokeslet and rotlet at the i-th boundary element, as can be seen from the corresponding lines (147–181) in the listing.

The *deflation at the other end of the spectrum* is essential for rapid convergence of the iteration scheme. The eigenvalue, $\lambda = +1$, of $\mathcal{K}$ is deflated by using the eigenfunction of the adjoint, $\mathcal{K}^*$. Recall that this eigenfunction is simply the surface normal. The important thing to remember is that the system $(I + \mathcal{K})x = b$ has the same solution as

$$\left(I + \mathcal{K} - \frac{v v^t}{|v|^2} \right) x = \left(I - \frac{v v^t}{\lambda |v|^2} \right) b = \tilde{b} \,,$$

which (with $\lambda = 2$ for the operator $I + \mathcal{K}$) leads to the iterative scheme,

$$x_{n+1} = \left(-K + \frac{v v^t}{|v|^2} \right) x_n + \tilde{b} = -K x_n + \frac{v \langle v, x_n \rangle}{|v|^2} + \tilde{b} \,.$$

Number of elements	$8\pi\mu aU/F$
320	1.3594
1280	1.3370
5120	1.3329

Table 19.1: Effect of mesh refinement on sphere mobility (exact: 4/3).

In our problem, the inner product terms are simply

$$\begin{aligned} \frac{v\,\langle v, x_n\rangle}{|v|^2} &\rightarrow \frac{\boldsymbol{n}(\boldsymbol{x})}{A}\int_S \boldsymbol{n}(\boldsymbol{y})\cdot\boldsymbol{\varphi}(\boldsymbol{y})\; dS(\boldsymbol{y}) \sim \frac{\boldsymbol{n}(\boldsymbol{x})}{A}\sum_{j=1}^{N}(\boldsymbol{n}(\boldsymbol{y})\cdot\boldsymbol{\varphi}(\boldsymbol{y}))\Delta A(j) \\ \frac{v\,\langle v, b\rangle}{2|v|^2} &\sim \frac{\boldsymbol{n}(\boldsymbol{x})}{A}\sum_{j=1}^{N}(\boldsymbol{n}(\boldsymbol{y}(j))\cdot b)\Delta A(j)\,, \end{aligned}$$

and these are worked out in lines 196–205 (for $\langle v, b\rangle$) and 226–236 (for $\langle v, x_n\rangle$). Note that $\tilde{b}$ is designated as `bt()` in the code.

A final note: The listing **CDLBIEM** is written as if we are solving $Ax = b$ by iterating $x_{n+1} = x_n - Ax_n + b$. But the matrix is of the form $A = I + \mathcal{K}$, and in its construction, we explicitly added the identity (lines 137–145). The program should be streamlined by deleting lines 137–145 and the terms `xold()` in lines 247–252.

19.2 Resistance and Mobility of Spheres

In this section, we examine results for the resistance and mobility functions obtained by the CDLBIEM code discussed in the preceding section. We proceed with a convenient tessellation of the sphere surface: Start with an icosahedron and get subsequent mesh refinement by subdividing each triangular element into four triangular subelements repeatedly. As expected, discretization error diminishes with mesh refinement (see Table 19.1). The 320-element polyhedron reproduced the Stokes law drag with a relative error of about 1%. As an even more stringent test, we can compute the spectrum numerically and compare with the analytical result. The lower modes should be reproduced faithfully by the discretized system. In fact, as in many other related problems (*e.g.*, simulation of the frequencies of a stretched membrane with finite elements), the highest modes of the discretized system cannot be trusted, but fortunately the lowest modes contain the information of interest. Discretization of compact operators introduces large errors only for the higher modes that could not be resolved by the chosen discrete mesh. Our results for the lower modes agreed with the analytical results of Chapter 17 to within 1% as well (see Figure 17.4 in Chapter 17 for a typical example).

In Figures 19.1–19.3, we compare CDL-BIEM with other methods of computing the mobility functions for two spheres. Figure 19.1 shows the sphere mobilities as functions of the sphere-sphere separation for an external force on sphere 1, acting along the line of centers, with the second sphere force-free. This problem is described by the mobility relations (see Chapter 11),

$$6\pi\mu a U_1 = x_{11}^a F_1 \tag{19.2}$$
$$6\pi\mu a U_2 = x_{21}^a F_1 \,, \tag{19.3}$$

with all forces and motions along the line of centers. The solid curves are the exact results obtained by using bispherical coordinates. Figures 19.2 (translational velocities) and 19.3 (rotational velocities) show results for an external force on sphere 1, acting normal to the line of centers, with the second sphere force-free, as described by the mobility relations

$$6\pi\eta a \boldsymbol{U}_1 = y_{11}^a \boldsymbol{F}_1 \tag{19.4}$$
$$6\pi\eta a \boldsymbol{U}_2 = y_{21}^a \boldsymbol{F}_1 \tag{19.5}$$
$$4\pi\eta a^2 \boldsymbol{\omega}_1 = y_{11}^b \boldsymbol{F}_1 \times \boldsymbol{d} \tag{19.6}$$
$$4\pi\eta a^2 \boldsymbol{\omega}_2 = y_{21}^b \boldsymbol{F}_1 \times \boldsymbol{d} \,, \tag{19.7}$$

with all translation along the direction of the external force and the rotations about an axis orthogonal to both the line-of-centers and the direction of the external force. The solid curves are the boundary collocation results from [41] (see also Chapter 13 and the program listing, **mandr.for**).

The agreement between CDL-BIEM and the values in the literature is excellent, for all three cases, except for almost-touching spheres. The small (less than 1%) relative errors may be attributed to discretization errors, which may be eliminated with further mesh refinement. For almost-touching spheres, the meshes must resolve the gap geometry, and this may be accomplished by refining the mesh in the gap region.

Figure 19.4 shows results for three fixed spheres placed in an equilateral-triangle configuration. Our results for the drag are in good agreement with the multipole solution presented in Kim [42] over a wide range of sphere-sphere separations. If we extrapolate the CDL-BIEM results, we obtain an estimate for a cluster formed by three touching spheres, which is in good agreement with the experimental result obtained by Lasso and Weidman [44]. We should expect relative errors of less than 2%, so this better-than-expected agreement should be attributed to cancellation of errors.

19.3 Sedimentation of Platonic Solids

The robustness of our simple collocation scheme (planar triangular elements and characteristic functions as basis elements) is demonstrated in Figure 19.5 with drag-coefficient calculations for regular polyhedra. These shapes are quite commonly encountered in suspensions because of the crystal structure of the

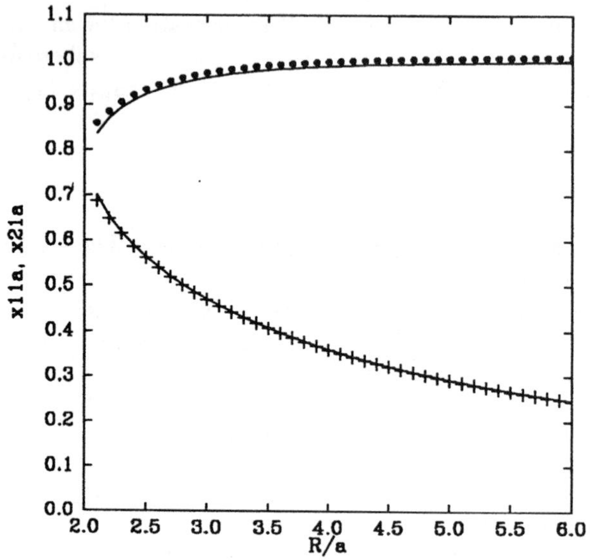

Figure 19.1: Two-sphere mobilities x^a_{11} (•) and x^a_{21} (+) *vs.* the separation (R/a); solid curves (exact solution).

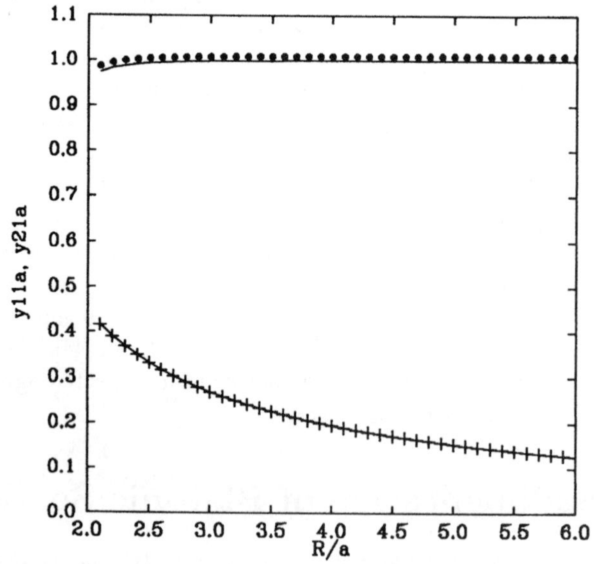

Figure 19.2: Two-sphere mobilities y^a_{11} (•) and y^a_{21} (+) *vs.* the separation (R/a); solid curves from Kim and Mifflin.

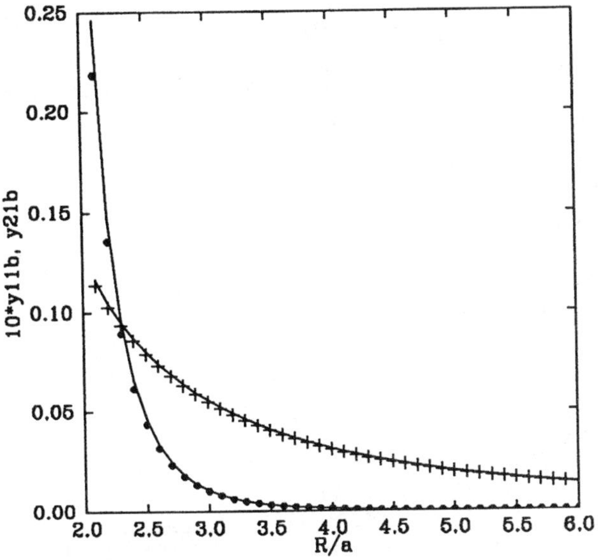

Figure 19.3: Two-sphere mobilities y_{11}^b (•) and y_{21}^b (+) *vs.* the separation (R/a); solid curves from Kim and Mifflin.

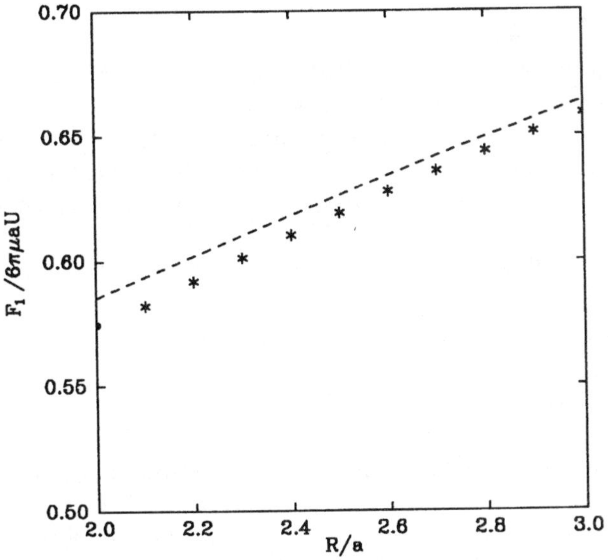

Figure 19.4: Drag on a sphere, for three spheres in an equilateral-triangle configuration: CDL-BIEM (∗); analytical (—), and experimental (•).

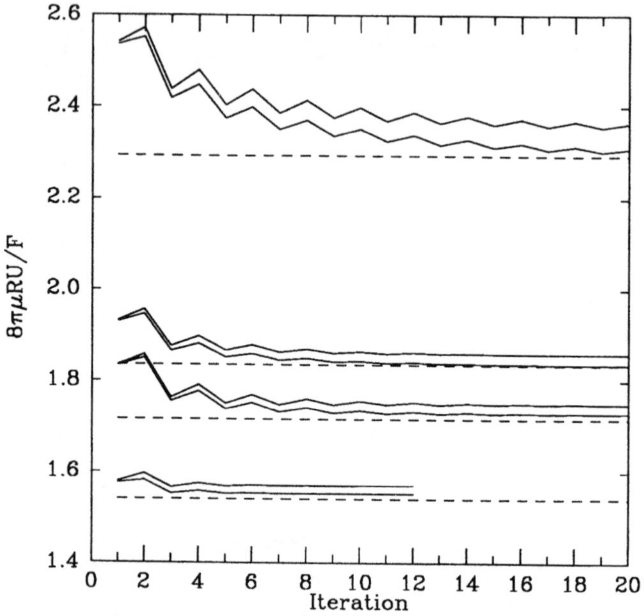

Figure 19.5: Convergence of iterations, at various discretizations. From top to bottom: tetrahedron, octahedron, cube, and icosahedron. Experiments of Pettyjohn and Christiansen (dashed lines). R is radius of circumscribed sphere.

mineral constitutents. For these shapes, the collocation equation is further simplified because, as explained earlier, *elements on the same polyhderal face* do not interact. Figure 19.5 gives the sedimentation velocities for a tetrahedron (256 and 1024 elements), octahedron (512 and 2048 elements), cube (268 and 1072 elements), and icosahedron (320 and 1280 elements). The number of iterations required to obtain convergence and the improvement obtained with mesh refinement are also shown. These results are in good agreement with the experiments of Pettyjohn and Christiansen [56], as represented in the figure (horizontal dashed lines).

As in Karrila and Kim [38] we find that CDL-BIEM works even when sharp corners (which are not allowed in the Fredholm–Riesz–Schauder theory) are present. We can explain this with heuristic reasoning based on the energy dissipation (inclusion monotonicity) theorems of Hill and Power [30]. If we round the corners by an infinitesimal amount, the drag changes only slightly, but now the Fredholm–Riesz–Schauder theory is applicable. Near the corners, we have steep velocity gradients and therefore large energy dissipation rates. These, in turn, imply eigenvalues near ± 1, and thus a slower rate of convergence for the deflated system. As expected, this effect is most pronounced for the tetrahedron and least evident for the icosahedron. However, even for the tetrahedron, the simple scheme still provides reasonable results.

Method	MicroVAX II	Astronautics ZS-1
LU Factorization	6150	360
Direct Iteration	140	8

Table 19.2: Computational times (seconds) for the 320-element mesh for the single sphere: comparison of LU factorization and iteration solutions.

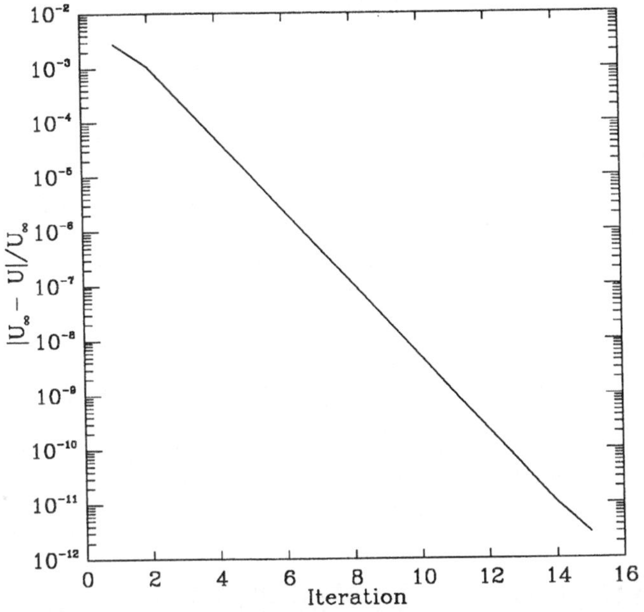

Figure 19.6: Convergence of deflated mobility problem (320 elements).

19.4 Benchmarks

For the purpose of comparison, the 320-element, single-sphere problem was solved by LU factorization (with the LINPACK routines DGECO and DGESL) and direct (Picard) iteration. Computation times are given in Table 19.2 for both the Micro VAX II and the Astronautics ZS-1. For this particular problem, the iterative scheme was roughly 45 times faster than LU factorization. The convergence of the iterative solution is shown in Figure 19.6.

The two-sphere problem converged to five significant figures after 16 iterations, with each iteration taking 6.1 seconds on the ZS-1 (in double precision). As expected, each iteration took approximately four times longer than the iterations in the single-sphere problem. Convergence rates of the two-sphere iterations are shown in Figures 19.7 and 19.8 for spheres with center-to-center

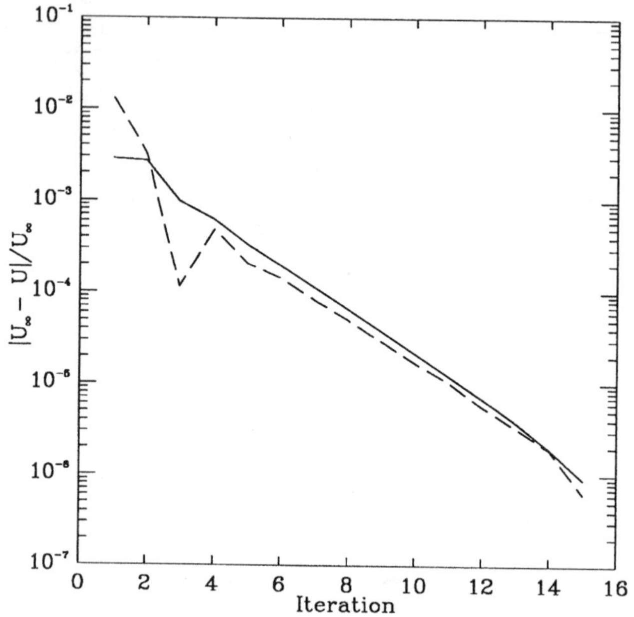

Figure 19.7: Convergence of two-sphere translational mobilities (320 elements, $R/a = 2.5$): sphere 1 (solid line), sphere 2 (dashed line).

separations $R = 2.5a$. In all cases, we used for the initial guess, $\varphi^{(0)}(\xi) = 1$. The important point here is that the rate of convergence is very fast, because the spectral radius is approximately 0.6 (see Chapter 17).

19.5 CDL-BIEM and Parallel Processing

Technological advances may increase the speed of the fastest processors by a factor of ten in the future. Ultimately, these improvements are limited by two fundamental constraints: the speed of light (which imposes an upper bound on the movement of information from one part of the computer to another) and the size of an atom (which imposes a lower bound on processor dimensions). Together, they represent a theoretical upper bound on the computational speed of a single processor.

Parallel processing offers a way of skirting these limitations, if the problem at hand can be broken into many parallel streams that can be tackled by an array of processors working independently. For problems in which this simple idea works, phenomenal gains in computational speed can be envisioned. For CDL-BIEM, the entire step, from matrix construction to Wielandt deflation to iterative solution, can be performed in parallel. Therefore, we digress here briefly to consider the current situation in parallel architectures. *The main objective of this discussion is to illustrate the point that the details of an opti-*

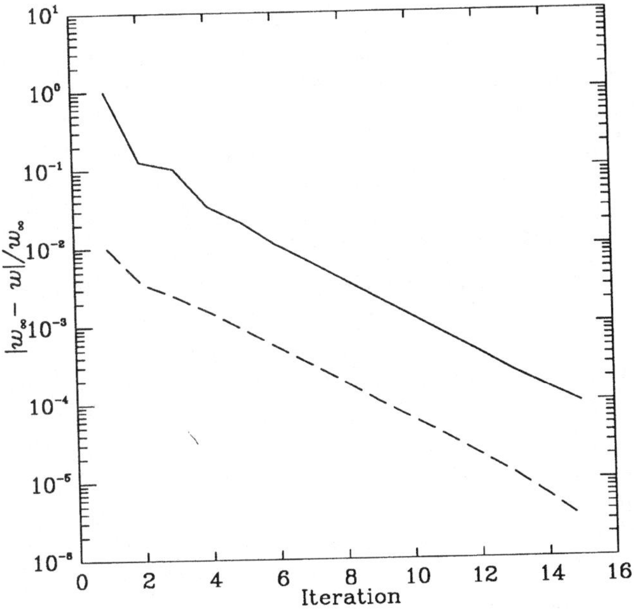

Figure 19.8: Convergence of two-sphere rotational mobilities (320 elements, $R/a = 2.5$): sphere 1 (solid line), sphere 2 (dashed line).

---

mized CDL-BIEM code will vary with computer architecture. For the forseeable future, developments in algorithms and architectures are coupled. Naturally, the scope of the discussion in this book must be restricted to general mathematical statements that are independent of architecture, but that form the basis for more detailed research for a specific computer architecture of interest. A perspective on the relation between architecture and the operations of numerical linear algebra may be gained from a number of sources: Almasi and Gottlieb [1], Bertsekas and Tsitsiklis [3], Ortega [54], and Stone [66].

Parallel computers may be classified as *coarse grained* (a small number of very powerful processors joined together) and *massively parallel* (a very large number of inexpensive, and thus less powerful, processors). Parallel computers may also be divided into categories known as *Single Instruction Multiple Data* (SIMD) and *Multiple Instruction Multiple Data* (MIMD) (as opposed to Single Instruction Single Data [SISD] conventional uniprocessor computers). These names date to Flynn's article on multiprocessor machines [19]. As the names imply, all processors on SIMD machines execute a single instruction, albeit on different data, while the processors on MIMD machines may execute differ instructions. Computational tasks that fit naturally in an SIMD environment are known as *data parallel*; matrix multiplication and image processing (FFT) are two examples. At the present time, the Connection Machines made by the Thinking Machines Corporation, with up to 2^{16} processors in parallel, are the most prominent SIMD machines.

While MIMD machines are more flexible and versatile, synchronization between processors is a major undertaking, especially as the number of processors is increased. Issues such as access to shared information by multiple processors and flow of information between processors must be resolved by the computer designer by implementing some protocol (the processors on an SIMD machine are synchronized in lockstep and the issues are quite different). As with roads, poor designs and heavy traffic can lead to flow control problems and gridlock. For an introduction to these issues, we direct the reader to [15].

Since vector dot products and matrix multiplications are data parallel, matrix construction is the major obstacle to the implementation of CDL-BIEM on SIMD machines. For simulation of a suspension of identical particles, the essential geometric information can be compressed and replicated in local memory; in fact, construction of the CDL-BIEM system matrix from the boundary element data is a data parallel operation. However, in the most general CDL-BIEM setting, the boundary element data may not fit into local memory. Another option would be to generate the matrix on another parallel computer and then perform parallel data input to the data vaults of an SIMD machine. On the other hand, with MIMD machines the time required for construction of each row is roughly the same, thus illustrating the point that details of the algorithm depend on the architecture. For yet another example, we mention that for computers with fast processors but limited memory, it may be more efficient to reconstruct the matrix at each iteration, while on most SIMD machines this tedious step should be avoided whenever possible. Finally, we can state with a high degree of confidence that with increasing problem and architecture complexity, communication between processors, that is, *flow*, rather than *processing* of information will be the bottleneck, so that reduction of information flow should be the goal in algorithm construction.

The basic premise that the simplest adaptations of CDL-BIEM are readily implemented on parallel machines has been confirmed, as was reported in Karrila *et al.* [39]. The computer used was a Sequent Symmetry, an MIMD machine consisting of 20 Intel 80386 processors, Weitek 1167 floating point accelerators and shared (common) memory. As the computational equivalent of about 60 MicroVAX IIs, the Symmetry is not the fastest computer for this problem. However, because of its low cost and rich menu of user-accessible parallel programming features, the Symmetry is a popular choice of computer scientists as a research tool for parallel processing studies.

The only essential modification required for this parallel machine was the introduction of *fork* and *barrier* statements, which split off and synchronize multiple processes. As the name implies, work on different processors stops at the barrier, thus allowing laggards to catch up with the others. This simplicity of code development is one measure of the parallel structure inherent in CDL-BIEM.

An efficiency based on actual speedup achieved divided by the number of processors used can be defined. In Table 19.3 we show the key result from

	Number of Processors		
N_s	1	8	16
1	649.68	83.89	43.55
2	1571.14	201.97	104.66
4	4146.77	530.66	273.78
8	12243.04	1559.13	797.17

Table 19.3: Benchmarks on the Sequent Symmetry: computation times (seconds) per iteration *vs.* number of spheres N_s, and number of processors.

[39]: computational times for multisphere mobilities as a function of number of (320-element) spheres and the number of processors used on the Sequent Symmetry. In all cases, efficiencies of about 96% were achieved. The iterative solutions converge so rapidly that system construction represents a significant fraction of the overall computational time, especially since matrix elements were recomputed at each iteration to conserve memory, as was the case for the results shown in Table 19.3. But since matrix construction and deflation are done in parallel, the overall algorithm remains quite efficient.

19.6 Reducing Communication Between Processors

19.6.1 Asynchronous Iterations

For small problems (say, for a cluster of several spheres), the matrix-vector multiplications of the Jacobi iterations are a straightforward operation. However when the method is directed at thousands of spheres, very large and dense systems result, say, on the level starting from 1,000,000 × 1,000,000. While it is exciting to contemplate that vector dot product computations of this magnitude can be done concurrently on parallel computers capable of executing these at the rate of 10^9 to 10^{12} floating point operations per second, we must address computational bottlenecks associated with communication limitations between processors.

Looking beyond the immediate problem at hand, we would like to address the critical problem of coordinating computational tasks (the matrix multiplication in the iterations) between hundreds and even thousands of powerful processors. As the processor components approach "supercomputer-on-a-chip" capabilities, it is the flow of information between these processors that becomes the weakest link of the *ab initio* simulation envisioned in this program.

In Figure 19.9, we show one way to minimize communications between processors. With n as the number of processors, the number of all possible connec-

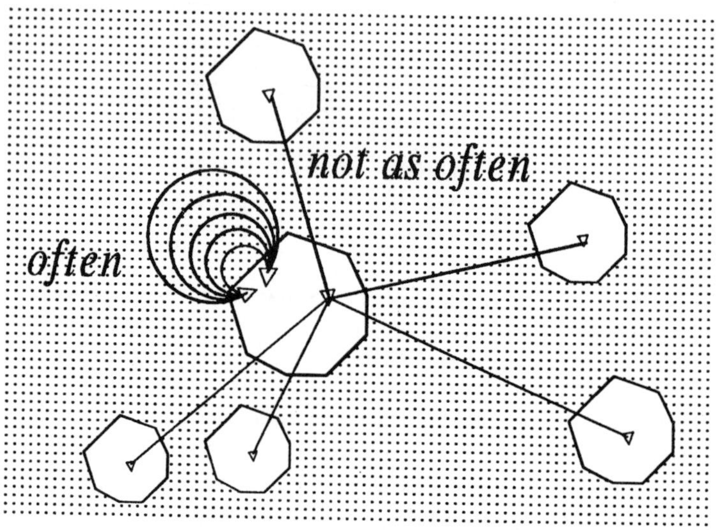

Figure 19.9: Asynchronous iterations to reduce communications and accelerate convergence.

tions grows as $n^2/2$, so that when thousands of processors are involved, communication becomes an important issue. In suspension simulation, the boundary elements on the same particle are strongly coupled, while those on different particles are less so, as we would expect on physical grounds. This implies that block Gauss–Seidel iterative strategies can be employed. For each particle, or cluster of neighboring particles, we allocate a processor with its own memory. Very rapid internal communications facilitate iterative improvements for each particle. Interactions between particles are obtained in a slower cycle, but the number of such cycles is greatly reduced. This approach raises two critical questions:

Acceleration. Do local iterations accelerate the rate of convergence? In calculating the rate of convergence, it seems natural to count only the "global iterations"; local iterations do not require communication between processors and are thus assigned zero cost.

Convergence. If information from distant parts of the suspension is not received and updated as often as local information, will the time delay cause non-convergence?

Fortunately, the answer to both of these is the favorable one: Local iterations greatly accelerate the rate of convergence; and despite the time delay from

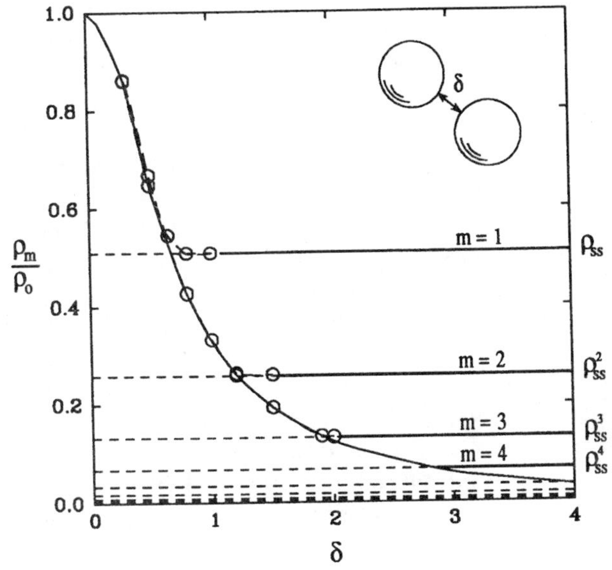

Figure 19.10: Collapsing the spectral radius with m local iterations.

distant regions, the iterations still converge to the correct solution. The key to both questions lies in the eigensystem of the double layer operator as perturbed by hydrodynamic interactions, which, for spheres, was addressed in Chapter 17. Since all combinations of multiparticle clusters involve a spectral radius less than one, the block Gauss–Seidel iterations will converge to a local fixed point. Furthermore, the action of the local iterations is now clear. Local iterations essentially map diagonal blocks to powers of the same block, and since these are the dominant factors behind the spectral radius, m local iterations should shrink the spectral radius ρ essentially as $\rho \rightarrow \rho^{m+1}$, as shown in Figure 19.10.

Indeed, the two-particle model problem suggests a fortuitous confluence of events. In the most natural parallel computational strategy envisioned for distributed processors, the time available for local iterations increases with particle-particle separation, so we can perform more local iterations. But with increasing particle-particle separations, there is also an enhanced benefit from such frequent local iterations. Indeed, when particles are greatly separated, with other particles in between, the ratio of local to global communication becomes so large that the time delay is essentially infinite; the interactions are screened. Since results obtained with and without screening can be compared, we have made a connection to one of the important concepts in effective medium theories. The ramifications for *ab initio* suspension simulation are significant.

19.6.2 Compactification of Distant Information

In the preceding discussion, we considered reducing the frequency of information exchange to address communication bottlenecks. Here, we consider reduction of the volume of information, *i.e.*, ***compactification*** of the information from distant regions, before sending them as message packets. Our idea bears strong resemblance to the multipole method employed by Greengard and Rokhlin to generate fast parallel codes for molecular dynamics simulations [26]. For suspended particles, the multipole expansions are a quite natural concept. Instead of the moment expansions of the intermolecular force laws for molecules in a cell volume, we do the same for the double layer density distribution over the surfaces of distant particles. The net effect is a great reduction in the amount of information required to characterize matrix elements linking all boundary elements on particles α with those on particle β. We simply store the coordinates of just the particle centers, and the first few multipole coefficients: force, torque, stresslet, quadrupoles, *etc.* But as always, these approximations to speed up the simulations can be introduced in a controlled fashion by performing tests with and without the approximation.

Exercises

Exercise 19.1 The Double Layer Density on a Small Polar Cap.
Consider the constant double layer density $\varphi = e_z$ distributed over the "polar cap" $0 \leq \theta \leq c \leq \pi$. Derive an analytical expression for $\mathcal{K}\varphi$ at the "north pole". Show that $\varphi + \mathcal{K}\varphi = 0$ when $c = \pi$ and that in the limit of small c, the result vanishes as $O(c^{3/2})$.

Exercise 19.2 The RBM Tractions of Platonic Solids.
Using the method illustrated in Chapter 16 involving the theorems of Lorentz and Riesz, determine numerically the RBM tractions of the platonic solids. Use mesh refinement to infer the nature of the surface tractions in the vicinity of the vertices.

Exercise 19.3 Rectangular Prisms.
Examine the numerical behavior of CDL-BIEM for rectangular prisms ranging from cubes to slender ribbons. Is it possible to draw general conclusions concerning the spectral radius of the deflated system as a function of the shape parameters of the prism?

References

[1] G. S. Almasi and A. Gottlieb. *Highly Parallel Computing.* Benjamin/Cummings, New York, 1989.

[2] K. E. Atkinson. *A Survey of Numerical Methods for the Solution of Fredholm Integral Equations of the Second Kind.* SIAM, Philadelphia, 1976.

[3] D. P. Bertsekas and J. N. Tsitsiklis. *Parallel and Distributed Computation.* Prentice-Hall, Englewood Cliffs, NJ, 1989.

[4] R. B. Bird, W. E. Stewart, and E. N. Lightfoot. *Transport Phenomena.* Wiley, New York, 1960.

[5] E. Bodewig. *Matrix Calculus.* Interscience, New York, 1956.

[6] C. A. Brebbia and J. Dominguez. *Boundary Elements An Introductory Course.* McGraw-Hill, New York, 1989.

[7] H. Brenner. The Stokes resistance of an arbitrary particle — IV. Arbitrary fields of flow. *Chem. Eng. Sci.*, 19:703–727, 1964.

[8] L. J. Briggs and A. N. Lowan. *Tables of Associated Legendre Functions.* Columbia University Press, New York, 1945.

[9] P. C. Chan, R. J. Leu, and N. H. Zargar. On the solution for the rotational motion of an axisymmetric rigid body at low Reynolds number with application to a finite length cylinder. *Chem. Eng. Commun.*, 49:145–163, 1986.

[10] T. A. Cruse. Numerical solutions in three dimensional electrostatics. *Intl. J. Solids Structures*, 5:1259–1274, 1969.

[11] T. A. Cruse. An improved boundary-integral method for three dimensional elastic stress analysis. *Computers and Structures*, 4:741–754, 1974.

[12] T. Dabros. A singularity method for calculating hydrodynamic forces and particle velocities in low-Reynolds-number flows. *J. Fluid Mech.*, 156:1–21, 1985.

[13] L. M. Delves and J. Walsh. *Numerical Solution of Integral Equations.* Clarendon Press, Oxford, 1974.

[14] J. Dieudonne. *Foundations of Modern Analysis.* Academic Press, New York, 1960.

[15] A. Dinning. A survey of synchronization methods for parallel computers. *Computers*, 22:66–76, 1989.

[16] S. R. Dungan and H. Brenner. Force and torque on a body in an unbounded Stokes flow expressed explicitly in terms of respective quadratures of the pressure and vorticity fields. *Chem. Eng. Commun.*, 82:103–110, 1989.

[17] S. Fenyo and H. W. Stolle. *Theorie und Praxis der Linearen Integralgleichungen 4 (Theory and Practice of Linear Integral Equations 4).* Birkhauser, Basel, 1984.

[18] H. E. Fettis. A new method for computing toroidal harmonics. *Mathematics of Computation*, 24:667–670, 1970.

[19] M. J. Flynn. Very high-speed computers. *Proc. of the IEEE*, 54:1901–1909, 1966.

[20] Y. O. Fuentes. *Parallel Computational Strategies for Multiparticle Interactions in Stokes Flows.* Ph. D. Dissertation, University of Wisconsin, 1990.

[21] A. Friedman. *Foundations of Modern Analysis.* Dover reprint, New York, 1982.

[22] P. Ganatos, R. Pfeffer, and S. Weinbaum. A numerical-solution technique for three-dimensional Stokes flows, with application to the motion of strongly interacting spheres in a plane. *J. Fluid Mech.*, 84:79–111, 1978.

[23] F. R Gantmacher. *The Theory of Matrices.* Chelsea, New York, 1960.

[24] W. Gautschi. Computational aspects of three-term recurrence relations. *SIAM Review*, 9:24–82, 1967.

[25] I. S. Gradshteyn and I. M. Ryzhik. *Table of Integrals, Series, and Products.* Academic Press, New York, 1980.

[26] L. Greengard and V. Rokhlin. A fast algorithm for particle simulations. *J. Comp. Physics*, 73:325–348, 1987.

[27] N. M. Gunter. *Potential Theory.* Frederick Ungar Publishing, New York, 1967.

[28] P. S. Han and M. D. Olson. An adaptive boundary element method. *Int. J. for Num. Meth. in Eng.*, 24:1187–1202, 1987.

[29] J. Happel and H. Brenner. *Low Reynolds Number Hydrodynamics.* Martinus Nijhoff, The Hague, 1983.

[30] R. Hill and G. Power. Extremum principles for slow viscous flow and the approximate calculation of drag. *Q. J. Mech. and Appl. Math*, 9:313–319, 1956.

[31] A. S. Householder. *The Theory of Matrices in Numerical Analysis.* Dover, New York, 1975.

[32] R. Hsu and P. Ganatos. The motion of a rigid body in viscous fluid bounded by a plane wall. *J. Fluid Mech.*, 207:29–72, 1989.

[33] M. A. Jaswon and G. T. Symm. *Integral Equation Methods in Potential Theory and Elasticity.* Academic Press, New York, 1977.

[34] G. B. Jeffery. On the steady rotation of a solid of revolution in a viscous fluid. *Proc. London Math. Soc.*, 14(2):327–338, 1915.

[35] D. J. Jeffrey and Y. Onishi. Calculation of the resistance and mobility functions for two unequal rigid spheres in low-Reynolds-number flow. *J. Fluid Mech.*, 139:261–290, 1984.

[36] L. V. Kantorovich and G. P. Akilov. *Functional Analysis.* Pergamon Press, New York, 1982.

[37] S. J. Karrila. *Linear Operator Theory Applied to Fast Computational Strategies for Hydrodynamic Interactions in Viscous Flows.* Ph.D. Dissertation, University of Wisconsin, Madison, WI, 1988.

[38] S. J. Karrila and S. Kim. Integral equations of the second kind for Stokes flow: direct solution for physical variables and removal of inherent accuracy limitations. *Chem. Eng. Commun.*, 82:123–161, 1989.

[39] S. J. Karrila, Y. O. Fuentes, and S. Kim. Parallel computational strategies for hydrodynamic interactions between rigid particles of arbitrary shape in a viscous fluid. *J. Rheology*, 33:913–947, 1989.

[40] S. Kim. Singularity solutions for ellipsoids in low-Reynolds-number flows: with applications to the calculation of hydrodynamic interactions in suspensions of ellipsoids. *Intl. J. Multiphase Flow*, 12:469–491, 1986.

[41] S. Kim and R. T. Mifflin. The resistance and mobility functions of two equal spheres in low-Reynolds-number flow. *Phys. Fluids*, 28:2033–2045, 1985.

[42] S. Kim. Stokes flow past three spheres: an analytic solution. *Phys. Fluids*, 30(8):2309–2314, 1987.

[43] U. A. Ladyzhenskaya. *The Mathematical Theory of Viscous Incompressible Flow.* Gordon and Breach, New York, 1963.

[44] I. A. Lasso and P. D. Weidman. Stokes drag on hollow cylinders and conglomerates. *Phys. Fluids*, 29:3921–3934, 1986.

[45] H. A. Lorentz. Ein allgemeiner Satz, die Bewegung einer reibenden Flüssigkeit betreffend, nebst einigen Anwendungen desselben (A general theorem concerning the motion of a viscous fluid and a few applications from it). *Versl. Kon. Akad. Wetensch. Amsterdam*, 5:168–174, 1896. Also in *Abhandlungen über Theoretische Physik*, 1:23–42 (1907) and *Collected Papers*, 4:7–14, Martinus Nijhoff, The Hague, 1937.

[46] J. H. C. Luke. Convergence of a multiple reflection method for calculating Stokes flow in a suspension. *SIAM J. Appl. Math.*, 49:1635–1651, 1989.

[47] W. Magnus and F. Oberhettinger. *Formulas and Theorems for the Special Functions of Mathematical Physics.* Chelsea, New York, 1949.

[48] R. Mathon and R. L. Johnston. The approximate solution of elliptic boundary-value problems by fundamental solutions. *SIAM J. Numer. Anal.*, 14:638–650, 1977.

[49] M. Mayr. On the numerical solution of axisymmetric elasticity problems using an integral equation approach. *Mech. Res. Comm.*, 3:393–398, 1976.

[50] M. Mayr, W. Drexler, and G. Kuhn. A semianalytical boundary integral approach for axisymmetric elastic bodies with arbitrary boundary conditions. *Intl. J. Solids Structures*, 16:863–871, 1980.

[51] S. G. Mikhlin. *Linear Integral Equations.* Hindustan Publishing Corp., Delhi, 1960.

[52] S. G. Mikhlin. *Mathematical Physics, an Advanced Course.* North-Holland Publishing Co., New York, 1970.

[53] F. K. G. Odqvist. Über die Randwertaufgaben der Hydrodynamik zäher Flüssigkeiten (On the boundary value problems in hydrodynamics of viscous fluids). *Math. Z.*, 32:329–375, 1930.

[54] J. M. Ortega. *Introduction to Parallel and Vector Solution of Linear Systems.* Plenum Press, New York, 1988.

[55] I. G. Petrovskii. *Lectures on the Theory of Integral Equations.* Graylock Press, Rochester, NY, 1957.

[56] E. S. Pettyjohn and E. B. Christiansen. Effect of particle shape on free-settling rates of isometric particles. *Chem. Engr. Prog.*, 44:157–172, 1948.

[57] H. Power and G. Miranda. Second kind integral equation formulation of Stokes' flows past a particle of arbitrary shape. *SIAM J. Appl. Math*, 47:689–698, 1987.

[58] C. Pozrikidis. The instability of a moving viscous drop. *J. Fluid Mech.*, 210:1–21, 1990.

[59] J. M. Rallison. Note on the Faxén relations for a particle in Stokes flow. *J. Fluid Mech.*, 88:529–533, 1978.

[60] D. Ramkrishna and N. R. Amundson. *Linear Operator Methods in Chemical Engineering.* Prentice-Hall, Englewood Cliffs, NJ, 1985.

[61] F. J. Rizzo and D. J. Shippy. A boundary integral approach to potential and elasticity problems for axisymmetric bodies with arbitrary boundary conditions. *Mech. Res. Comm.*, 6:99–103, 1979.

[62] W. Rudin. *Functional Analysis.* Tata McGraw-Hill Publishing Comp, New Delhi, 1978.

[63] R. Schmitz and B. U. Felderhof. Friction matrix for two spherical particles with hydrodynamic interaction. *Physica*, 113A:103–116, 1982.

[64] D. J. Shippy, F. J. Rizzo, and R. K. Nigam. A boundary integral equation method for axisymmetric elastostatic bodies under arbitrary surface loads. In *Innovative Numerical Analysis for the Engineering Sciences (ed. R. Shaw, et al.)* University Press of Virginia, Charlotte, VA, 1980.

[65] C. Snow. *Hypergeometric and Legendre Functions with Applications to Integral Equations of Potential Theory.* NBS Appl. Math. Series 19, Washington DC, 1952.

[66] H. S. Stone. *High-Performance Computer Architecture.* Addison-Wesley, Reading, MA, 1987.

[67] A. H. Stroud. *Approximate Calculation of Multiple Integrals.* Prentice Hall, Englewood Cliffs, NJ, 1971.

[68] Symbolics Inc. *VAX UNIX MACSYMA Reference Manual: Version 11.* Symbolics, Inc., 1985.

[69] E. T. Whittaker and G. N. Watson. *A Course of Modern Analysis.* Cambridge University Press, New York, 1963.

[70] J. Wimp. *Computation with Recurrence Relations.* Pitman Advanced Publishing Progr, Boston, 1984.

[71] B. J. Yoon and S. Kim. Note on the direct calculation of mobility functions for two equal-sized spheres in Stokes flow. *J. Fluid Mech.*, 185:437–446, 1987.

[72] G. K. Youngren and A. Acrivos. Stokes flow past a particle of arbitrary shape: a numerical method of solution. *J. Fluid Mech.*, 69:377–403, 1975.

Notation

It is necessary to mention that the notation fluctuates for the following reasons:

- We want to stick to the notations of the original sources in order to make the book a useful reference for those following current literature, having certain adopted conventions.

- Simultaneously we wish to adhere to some fixed notations, as regards the book as a whole, to make the text coherent without too much duplication of necessary definitions

- We do not wish to overload the notation with multiple definitions within one section, even though following just conventions from the literature would lead to such "context-sensitive" notation.

For this reason several symbols below have multiple meanings, all of which it would be impractical to list. The meanings given below are meant to be looked up whenever the immediate context within some section does not clarify the definition used.

Accents

$\hat{}$	per unit mass
	unit vector
	Fourier transformed
	transformed according to Equation 12.6
$\dot{}$	time derivative
$\bar{}$	closure of a set
	label
$\tilde{}$	used as a label
	indicates no summation over this repeated index
	separately defined transpose of a high-rank tensor

Mathematical Symbols

$\imath$, i	imaginary unit $\sqrt{-1}$
∇	Nabla operator

D/Dt	material derivative (Equation 1.2)
$\oint$	integration over closed contour or surface
^{c}	complex conjugation
^{o}	interior of a set
^{t}	(pre- or post-) transposing operator
O	order symbol for asymptotic rate bound
$\cdot$	dot product in 3-D space
$\times$	cross product in 3-D space
$[\cdot,\cdot]$	closed interval (set on real line)
	discrete inner product
$(\cdot,\cdot)$	open interval (set on real line)
$\langle\cdot,\cdot\rangle$	inner product
$\langle\cdot\rangle$	average
$\vert$	used to precede condition in conditional probabilities
$\vert\cdot\vert$	absolute value
	Euclidean norm
	surface area
$\rightarrow$	goes to, approaches
$\leftarrow$	substitution of "right object" for "left object"
$\gg$	much greater than
$\sim$	asymptotically equal
$\Vert$	parallel
$\perp$	orthogonal
$\triangle$	difference
Δ	function (Table 3.2, p. 55)
	area (of a boundary element)
*	adjoint
	complex conjugate
	label
	reflected field
tr	trace, sum of diagonal elements of a matrix
$\leq$	for symmetric matrices the partial ordering based on positivity of the difference matrix

Calligraphic

$\mathcal{D}^n_m$	see Equation 18.43
$\mathcal{E}^n_m$	see Equation 18.44
$\mathcal{F}$	Fourier transform operator
	linear functional
$\mathcal{F}^n_m$	see Equation 18.45
$\mathcal{G}$	Oseen tensor (Equation 2.9)
$\mathcal{G}_n$	Gegenbauer function (p. 101)
$\mathcal{H}_n$	Gegenbauer function (p. 101)

$\mathcal{H}$	Stokeson (p. 52)
	deflated double layer operator, such that $1 + \mathcal{H}$ is invertible
$\tilde{\mathcal{H}}$	$\mathcal{H}$ further deflated (Equation 17.58)
$\mathcal{I}$	function defined by an integral (p. 71)
$\mathcal{K}$	linear operator
$\mathcal{K}_n$	solution of Gegenbauer's Equation (p. 101)
$\mathcal{M}$	mirror reflection operator (Equation 12.4)
$\mathcal{P}$	pressure field of Oseen tensor (Equation 2.10), also equal to the flow field of a point source
	orthogonal projection
$\mathcal{P}_L$	"left" projection (Equation 17.52)
$\mathcal{P}_R$	"right" projection (Equation 17.53)
$\mathcal{S}$	linear functional
	symmetry parameter (p. 326)
$\mathcal{T}$	linear functional

Special Labels

Amb	ambient
b	buoyancy
	bubble
Br	Brownian
c	constant
	container
cr	center of resistance
D	disturbance
e	external
E	related to fundamental (focal) ellipse (see p. 54)
eff	effective (see p. 30)
f	fluid
g	gravity
h	homogeneous solution
H	hydrodynamic
	stress coefficient (Section 5.6)
m	mass
p	particle(s)
	particular solution
RBM	rigid-body motion
∞	asymptotic, ambient
r	physical component in direction of e_r
	rotary
s	surface
S	Smoluchowski slip velocity (p. 142)
t	total

w	wall
x	physical component in direction of e_x
y	physical component in direction of e_y
z	physical component in direction of e_z
θ	physical component in direction of e_θ
ϕ	physical component in direction of e_ϕ
$\perp$	perpendicular to axis of symmetry
$\|$	parallel to axis of symmetry

Roman Symbols

a	parameter, for ellipsoids see p. 53
A	Hamaker's constant
	area
b	parameter, for ellipsoids see p. 53
c	parameter, often concentration, for ellipsoids see p. 53
C	space of continuous functions
D	diffusion coefficient
	differential operator (Table 3.2, p. 55)
	dissipation rate in given volume
D_x^2	differential operator (p. 62)
e	eccentricity of a spheroid (p. 61)
	the fundamental charge
E	rate of energy dissipation
	transformed eccentricity of a spheroid (Table 3.5, p. 65)
	strength of electric field
E^2	streamfunction operator (Equation 4.42)
f	function
	singularity density function in Equation 3.4
F	function
g	(given) function (in an equation)
G	ellipsoidal harmonic (Table 3.2, p. 55)
H	scaled ellipsoidal harmonic, see Equation 3.7
I	function defined by an integral (p. 71)
	interval
$I_{\rho\rho}$, etc.	components of Fourier decomposed double layer kernel, see Section 18.7.1
J	Jacobian for multidimensional integration
k	Boltzmann constant
	thermal conductivity
	parameter, for Fourier analysis see Equation 18.25
K	kernel function of an integral equation
L	representative length
L_2	Hilbert space of square summable functions
l	parameter, for Fourier analysis see Equation 18.24

m	mass
	meters
	interfacial mobility (Section 9.4)
n	concentration (number per unit volume) of a chemical species
N	null space of an operator (*i.e.*, kernel, the inverse image of zero)
p	pressure
P	Peclet number (p. 127)
	probability density
p_n	solid spherical harmonic of order n
P_n^m	associated Legendre function
P_n	spherical harmonic
	Legendre polynomial
q	function (Table 3.2, p. 55)
Q_n	second-kind Legendre function
$Q_{m-1/2}^n$	toroidal function (references in Section 18.4)
r	radial coordinate of spherical polars, distance from origin
	aspect ratio of a spheroid
r_{xy}	Euclidean distance of $\boldsymbol{x}$ and $\boldsymbol{y}$
R	gas constant
	curvature radius
	separation of particle centers
	range of an operator (*i.e.*, set of images)
Re	Reynolds number
S	surface
Sl	Strouhal number (p. 147)
t	time
	parameter
T	temperature (absolute)
U	internal energy
V	volume
	representative speed
x	Cartesian coordinate
	length of vector $\boldsymbol{x}$
X^A, etc.	resistance functions, see Equations 5.25–30
y	Cartesian coordinate
Y^A, etc.	resistance functions, see Equations 5.25–30
Y_n^m	spherical harmonic (p. 94)
$\tilde{Y}_n^m$	normalized spherical harmonic (p. 94)
z	Cartesian coordinate
	valence of a chemical species
Z^M	resistance function, see Equation 5.30

Bold Roman

$\mathbf{a}$	acceleration
	tensor element in mobility matrix (p. 115)
$\mathbf{A}$	tensor in resistance matrix (p. 109)
$\mathbf{b}$	tensor element in mobility matrix (p. 115)
$\mathbf{B}$	tensor in resistance matrix (p. 109)
$\mathbf{c}$	tensor element in mobility matrix (p. 115)
$\mathbf{C}$	tensor in resistance matrix (p. 109)
$\mathbf{d}$	displacement
	unit vector along symmetry axis
$\mathbf{e}$	rate of deformation tensor
	unit vector of some coordinate system
$\mathbf{E}$	electric field
	body force field
$\mathbf{F}$	force
$\mathbf{g}$	tensor element in mobility matrix (p. 115)
$\mathbf{G}$	tensor in resistance matrix (p. 109)
$\mathbf{h}$	tensor element in mobility matrix (p. 115)
$\mathbf{H}$	tensor in resistance matrix (p. 109)
$\mathbf{k}$	argument of 3-D Fourier transformed function (Ex. 2.9)
$\mathbf{L}$	linear operator
$\mathbf{m}$	tensor element in mobility matrix (p. 115)
$\mathbf{M}$	tensor in resistance matrix (p. 109)
$\mathbf{n}$	unit surface normal
$\hat{\mathbf{n}}$	unit surface normal pointing into the interior fluid domain (*i.e.*, the region physically occupied by fluid)
$\mathbf{P}$	strength of multipole moment
$\mathbf{q}$	heat flux
$\mathbf{Q}$	electric charge
$\mathbf{r}$	position vector
$\mathbf{r}_{xy}$	vector from $\mathbf{x}$ to $\mathbf{y}$ $(= \mathbf{y} - \mathbf{x})$
$\mathbf{U}$	translation velocity
$\mathbf{S}$	stresslet
$\mathbf{T}$	torque or the corresponding second order tensor
	traction surface field
$\mathbf{v}$	velocity field
$\mathbf{V}$	single layer generated velocity field (Equation 15.4, p. 377)
$\mathbf{W}$	double layer generated velocity field (Equation 15.6, p. 377)
$\mathbf{x}$	Cartesian position vector

Greek

α	real number
α_t, α_i	scale factors dependent on surface areas (Section 17.6)
β	real number
	ratio of sphere radii
γ	Euler's constant
$\dot{\gamma}$	shear rate
δ	Kronecker delta
	Dirac delta function
ϵ	transformed eccentricity of a spheroid (Table 3.5, p. 65)
	electric permittivity (ϵ_0 for vacuum)
	gap distance
	small real number
ζ	inverse mobility
	zeta potential
	Riemann zeta function
θ	polar spherical coordinate
	angular cylindrical coordinate (Section 9.2)
κ	bulk viscosity (footnote on p. 8)
	inverse Debye length (p. 137)
	viscosity ratio (Chapter 6)
	negative of the ratio of interior and exterior dissipation rates (Section 17.2.1)
λ	eigenvalue
	viscosity ratio
	ellipsoidal coordinate (Table 3.2, p. 55)
	ratio of length scales (Chapter 6)
Λ	transformed viscosity ratio (p. 250)
μ	viscosity
ν	kinematic viscosity (μ/ρ)
ξ	parameter describing separation of two spheres (p. 272)
π	pi, 3.14159...
π_n	solid spherical harmonic (p. 87)
Π	pressure field generated by a double layer (Equation 15.7, p. 377)
ρ	density
	radial cylindrical coordinate
ρ_e	charge density
Υ	differential operator (p. 228)
ϕ	angular spherical or cylindrical coordinate
	electric potential
ϕ_i	basis functions in some expansion
Φ	spherical harmonic
Φ_v	dissipation rate per unit volume (p. 14)

χ	polar spherical coordinate (Section 9.2)
χ_n	solid spherical harmonic of order n
ψ	Stokes stream function
ψ_0	surface potential (electric)
ω_i	quadrature weight on approximate integration
Ω	set
	pressure field generated by a single layer (Equation 15.5, p. 377)

Bold Greek

$\boldsymbol{\delta}$	identity tensor (footnote on p. 8)
$\boldsymbol{\epsilon}$	alternating tensor (footnote on p. 28)
$\boldsymbol{\eta}$	position vector of a surface point
$\boldsymbol{\xi}$	position vector of a surface point
$\boldsymbol{\sigma}$	stress tensor
$\boldsymbol{\Sigma}$	stress field of the Oseen–Burgers tensor (Section 2.4.1)
$\boldsymbol{\varphi}$	double layer density
$\boldsymbol{\psi}$	single layer density
$\boldsymbol{\omega}$	particle angular velocity
$\boldsymbol{\Omega}$	fluid angular velocity

Index

Engineering

DE RE METALLICA, Georgius Agricola. The famous Hoover translation of greatest treatise on technological chemistry, engineering, geology, mining of early modern times (1556). All 289 original woodcuts. 638pp. 6¾ x 11. 60006-8

FUNDAMENTALS OF ASTRODYNAMICS, Roger Bate et al. Modern approach developed by U.S. Air Force Academy. Designed as a first course. Problems, exercises. Numerous illustrations. 455pp. 5⅜ x 8½. 60061-0

DYNAMICS OF FLUIDS IN POROUS MEDIA, Jacob Bear. For advanced students of ground water hydrology, soil mechanics and physics, drainage and irrigation engineering, and more. 335 illustrations. Exercises, with answers. 784pp. 6⅛ x 9¼. 65675-6

THEORY OF VISCOELASTICITY (Second Edition), Richard M. Christensen. Complete, consistent description of the linear theory of the viscoelastic behavior of materials. Problem-solving techniques discussed. 1982 edition. 29 figures. xiv+364pp. 6⅛ x 9¼. 42880-X

MECHANICS, J. P. Den Hartog. A classic introductory text or refresher. Hundreds of applications and design problems illuminate fundamentals of trusses, loaded beams and cables, etc. 334 answered problems. 462pp. 5⅜ x 8½. 60754-2

MECHANICAL VIBRATIONS, J. P. Den Hartog. Classic textbook offers lucid explanations and illustrative models, applying theories of vibrations to a variety of practical industrial engineering problems. Numerous figures. 233 problems, solutions. Appendix. Index. Preface. 436pp. 5⅜ x 8½. 64785-4

STRENGTH OF MATERIALS, J. P. Den Hartog. Full, clear treatment of basic material (tension, torsion, bending, etc.) plus advanced material on engineering methods, applications. 350 answered problems. 323pp. 5⅜ x 8½. 60755-0

A HISTORY OF MECHANICS, René Dugas. Monumental study of mechanical principles from antiquity to quantum mechanics. Contributions of ancient Greeks, Galileo, Leonardo, Kepler, Lagrange, many others. 671pp. 5⅜ x 8½. 65632-2

STABILITY THEORY AND ITS APPLICATIONS TO STRUCTURAL MECHANICS, Clive L. Dym. Self-contained text focuses on Koiter postbuckling analyses, with mathematical notions of stability of motion. Basing minimum energy principles for static stability upon dynamic concepts of stability of motion, it develops asymptotic buckling and postbuckling analyses from potential energy considerations, with applications to columns, plates, and arches. 1974 ed. 208pp. 5⅜ x 8½. 42541-X

METAL FATIGUE, N. E. Frost, K. J. Marsh, and L. P. Pook. Definitive, clearly written, and well-illustrated volume addresses all aspects of the subject, from the historical development of understanding metal fatigue to vital concepts of the cyclic stress that causes a crack to grow. Includes 7 appendixes. 544pp. 5⅜ x 8½. 40927-9

Mathematics

FUNCTIONAL ANALYSIS (Second Corrected Edition), George Bachman and Lawrence Narici. Excellent treatment of subject geared toward students with background in linear algebra, advanced calculus, physics, and engineering. Text covers introduction to inner-product spaces, normed, metric spaces, and topological spaces; complete orthonormal sets, the Hahn-Banach Theorem and its consequences, and many other related subjects. 1966 ed. 544pp. 6⅛ x 9¼. 40251-7

ASYMPTOTIC EXPANSIONS OF INTEGRALS, Norman Bleistein & Richard A. Handelsman. Best introduction to important field with applications in a variety of scientific disciplines. New preface. Problems. Diagrams. Tables. Bibliography. Index. 448pp. 5⅜ x 8½. 65082-0

VECTOR AND TENSOR ANALYSIS WITH APPLICATIONS, A. I. Borisenko and I. E. Tarapov. Concise introduction. Worked-out problems, solutions, exercises. 257pp. 5⅝ x 8¼. 63833-2

THE ABSOLUTE DIFFERENTIAL CALCULUS (CALCULUS OF TENSORS), Tullio Levi-Civita. Great 20th-century mathematician's classic work on material necessary for mathematical grasp of theory of relativity. 452pp. 5⅝ x 8¼. 63401-9

AN INTRODUCTION TO ORDINARY DIFFERENTIAL EQUATIONS, Earl A. Coddington. A thorough and systematic first course in elementary differential equations for undergraduates in mathematics and science, with many exercises and problems (with answers). Index. 304pp. 5⅜ x 8½. 65942-9

FOURIER SERIES AND ORTHOGONAL FUNCTIONS, Harry F. Davis. An incisive text combining theory and practical example to introduce Fourier series, orthogonal functions and applications of the Fourier method to boundary-value problems. 570 exercises. Answers and notes. 416pp. 5⅜ x 8½. 65973-9

COMPUTABILITY AND UNSOLVABILITY, Martin Davis. Classic graduate-level introduction to theory of computability, usually referred to as theory of recurrent functions. New preface and appendix. 288pp. 5⅜ x 8½. 61471-9

ASYMPTOTIC METHODS IN ANALYSIS, N. G. de Bruijn. An inexpensive, comprehensive guide to asymptotic methods–the pioneering work that teaches by explaining worked examples in detail. Index. 224pp. 5⅜ x 8½ 64221-6

APPLIED COMPLEX VARIABLES, John W. Dettman. Step-by-step coverage of fundamentals of analytic function theory–plus lucid exposition of five important applications: Potential Theory; Ordinary Differential Equations; Fourier Transforms; Laplace Transforms; Asymptotic Expansions. 66 figures. Exercises at chapter ends. 512pp. 5⅜ x 8½. 64670-X

INTRODUCTION TO LINEAR ALGEBRA AND DIFFERENTIAL EQUATIONS, John W. Dettman. Excellent text covers complex numbers, determinants, orthonormal bases, Laplace transforms, much more. Exercises with solutions. Undergraduate level. 416pp. 5⅜ x 8½. 65191-6